AF710059

IET MATERIALS AND DEVICES 13

The Handbook of Electrical Resistivity

Other volumes in this series:

Volume 8	**Physics and technology of heterojunction devices** D.V. Morgan and R.H. Williams (Editors)
Volume 9	**Electrical degradation and breakdown in polymers** L.A. Dissado and J.C. Fothergill
Volume 12	**Handbook of microlithography, micromachining and microfabrication, 2 volumes** P. Rai-Choudhury (Editor)

The Handbook of Electrical Resistivity

New materials and pressure effects

Edited by
G. Dyos

The Institution of Engineering and Technology

Published by The Institution of Engineering and Technology, London, United Kingdom

The Institution of Engineering and Technology is registered as a Charity in England & Wales (no. 211014) and Scotland (no. SC038698).

© 2012 The Institution of Engineering and Technology

First published 2012

This publication is copyright under the Berne Convention and the Universal Copyright Convention. All rights reserved. Apart from any fair dealing for the purposes of research or private study, or criticism or review, as permitted under the Copyright, Designs and Patents Act 1988, this publication may be reproduced, stored or transmitted, in any form or by any means, only with the prior permission in writing of the publishers, or in the case of reprographic reproduction in accordance with the terms of licences issued by the Copyright Licensing Agency. Enquiries concerning reproduction outside those terms should be sent to the publisher at the undermentioned address:

The Institution of Engineering and Technology
Michael Faraday House
Six Hills Way, Stevenage
Herts, SG1 2AY, United Kingdom

www.theiet.org

While the authors and publisher believe that the information and guidance given in this work are correct, all parties must rely upon their own skill and judgement when making use of them. Neither the authors nor publisher assumes any liability to anyone for any loss or damage caused by any error or omission in the work, whether such an error or omission is the result of negligence or any other cause. Any and all such liability is disclaimed.

The moral rights of the authors to be identified as authors of this work have been asserted by them in accordance with the Copyright, Designs and Patents Act 1988.

British Library Cataloguing in Publication Data
A catalogue record for this product is available from the British Library

ISBN 978-1-84919-149-4 (hardback)
ISBN 978-1-84919-117-3 (PDF)

Typeset in India by MPS Limited
Printed in the UK by CPI Group (UK) Ltd, Croydon, CR0 4YY

Contents

Preface

Since the publication of the *Electrical Resistivity Handbook* by the IET, edited by G.T. Dyos and T. Farrell in 1992, interest in high-temperature superconductivity has grown. I was approached and commissioned by the Institution of Engineering and Technology to produce a further volume covering the electrical resistivity of new materials produced in the last two decades and the effects of pressure on resistivity.

I would be pleased to receive any copies of original referenced papers describing the resistivity of materials which might be of interest for inclusion in a further volume at a later date.

I would like to thank my wife Muriel who has supported me during this project, often only seeing me when I emerged from the study at mealtimes. Also I would like to thank Professor J. Lawton for his support and helpful discussions and IET staff for sorting out problems in the INSPEC searches.

Gordon T. Dyos
GT Innovations
Ysgubor Newydd
Llandrillo, Corwen, UK LL21 0SY

Chapter 1
Introduction

G. Dyos and T. Farrell

1.1 Continuation of the *Electrical Resistivity Handbook*

This volume incorporates two sections dealing with the resistivity of elements, compounds and alloys as a function of temperature and pressure and a third section dealing mainly with the resistance of various materials.

The compilation of this volume was made easier by the search facilities of the IET's INSPEC database containing some 13 million papers. From which the leading authors of approximately 3000 papers were contacted by email and postal services. It was very disappointing that only 35% bothered to reply and some 30 authors wanted a free copy of the book! It was very much appreciated by the editor for the plaudits received from university professors for undertaking this work.

1.2 Explanation of the graphical format

The vast majority of the data presented here has been gathered from technical papers published in the last two decades. Some graphical data, which were included in the *Electrical Resistivity Handbook*, have been included in this volume, where they cover a larger temperature range.

The data has been obtained from copies of the pertinent paper, either in graphical or in tabular format. In the case of the graphical format, the graphs have been scanned into the computer and the axes calibrated and the data coordinates used to produce a graph based on a standard graphical format. The resistivity data is expressed in units of ohm metre (ohm m) $\times 10^{-8}$, temperature in kelvin (K), pressure in gigapascals (GPa) and resistance in ohms.

1.3 Material classification

The catalogue system used in this volume is the same as that used in the *Electrical Resistivity Handbook*. First, the materials are listed in alphabetical order. Where a compound has several elements they are listed in order of decreasing atomic percentage. Should a compound have elements all having equal atomic percentage, then they are listed in alphabetical order.

Data which has been presented in the original paper in graphical form and then extracted by scanning and replotting is indicated by an asterisk (*) against the reference number for that graph. It should be noted that only data obtained from bulk and crystalline materials has been evaluated. No data from thin films has been used.

It should be noted that the majority of authors of papers do not indicate whether an allowance for thermal expansion has been applied to the resistivity of the compound.

All the compounds in this handbook are expressed in terms of atomic percentages, that is, the number of atoms of an element present in the compound expressed as a percentage of the total number of atoms present in one molecule of the compound. Where compounds have been listed in the original reference as weight percentages, these have been converted into atomic percentages.

1.4 Electronic conduction in solids

To aid the understanding of the electrical resistivity data on the materials presented in this handbook and to assist with its estimation for materials whose exact composition is not listed, some fundamental concepts of electronic conduction in solids are required. These concepts are presented at the descriptive level; the reader who requires a more detailed quantitative approach should consult the standard text books [1–5] or the cited references.

Electronic conduction is concerned with the motion of electrons within the solid under the influence of an applied electric field. Of necessity, therefore, the electrons must be free to move; experience shows that metals are good conductors of electricity and thus some of the electrons in the metal can be regarded as ‘free’. The earliest approach to quantifying the electrical conduction in metals, recognising that electrons were responsible for the transport of charge, was that of Drude [6], who treated the metal as an electron gas. When an electric field, E, is applied to the metal, the electrons of charge $-e$ are accelerated by the electrostatic force eE. After they have travelled for a certain average time, τ, the electrons suffer a collision with the atoms of the solid and are effectively arrested, transferring their kinetic energy to the solid. From simple mechanics, the mean drift velocity of the electrons, v_D, is given by

$$v_D = \frac{Ee\tau}{m}$$

where m is the mass of the electron. If the volume density of the electrons is n, then the electrical current per unit area, J, is given by

$$J = nev_D = \frac{ne^2\tau E}{m} \tag{1.1}$$

This is a statement of Ohm's law, where the electrical conductivity ρ is defined as $J = \sigma E$.

The electrical resistivity $\rho = 1/\sigma$ is therefore

$$\rho = \frac{m}{ne^2\tau} \tag{1.2}$$

To obtain an estimate of ρ from (1.2), the values of n and τ are required. Taking the 'relaxation time' τ to be given by λ/v_0, where λ is the electron mean free path and v_0 the thermal velocity of the electrons, then from the equipartition law $mv_0^2/2 = 3k_BT/2$ and

$$\rho = \frac{(3k_BmT)^{1/2}}{n\lambda e^2} \tag{1.3}$$

If it is assumed that each atom contributes one electron to the conduction process, n is the number density of atoms in the solid, and if λ is taken to be the interatomic spacing, then (1.3) gives values of ρ in the 10^{-6} to 10^{-7} ohm m range, which covers the room temperature experimentally observed resistivities for most metals.

Although this simple classical Drude theory gives Ohm's law and taking the power dissipation to be approximately mv_D^2/τ is consistent with Joule heating (it can also be used to derive the relationship between the electrical and thermal conductivities of metals – the Wiedemann–Franz law), it cannot readily explain the differences in the electrical resistivity upon alloying, nor does it predict the correct variation of resistivity with temperature. The Drude theory is restricted to metals; it cannot explain the electrical resistivity of semiconductors. However, the important point to emerge from this approach is the concept of collisions or scattering of electrons and the significance of τ, the 'relaxation time'.

The quantum mechanics approach pioneered by Sommerfeld [7] assumes the free electrons of the metal to be moving in a constant internal potential and that the electrons in this potential well obey the laws of quantum mechanics rather than the laws of classical mechanics. Essentially, this well consists of a quasi-continuum of energy levels and when the valence electrons are 'poured' into the well they progressively fill these levels up to a certain energy, E_F, known as the Fermi level. This value of E_F varies from metal to metal but is typically a few electron volts ($1\ \text{eV} = 1.6 \times 10^{-19}$ J). In this quantum Fermi gas model, the number of electrons which take part in the conduction is restricted to those lying within the energy range k_BT of the Fermi level, where k_B is Boltzmann's constant. Thus, at room temperature, for a monovalent metal with $E_F \approx 5$ eV, the number of electrons is only 5×10^{-3} times the total number of valence electrons. The average drift velocity of the classical mechanics is replaced by the Fermi velocity, $v_F \approx (2E_F/m)^{1/2}$, which for $E_F \approx 5$ eV is about 1.3×10^6 m/s. If these mean values are substituted into (1.3), and the mean free path, λ, is again taken to be the interatomic spacing, then the Sommerfeld model predicts a resistivity which is several orders of magnitude larger than the experimental values. However, the Fermi gas model derives the Wiedemann–Franz law with remarkable agreement with experiment and successfully accounts for electronic specific heat. Thus, the difficulty it encounters in describing conductivity must lie in the definition of the mean free path. In order to reconcile the prediction of the Fermi gas model with experiment, the electron mean free path must be larger than interatomic spacings.

The Fermi gas model, in its simplicity, treated the electrons as being in a constant potential well. Solids are comprised of crystals, the sizes of which are large compared with atomic dimensions (with the exception of amorphous metals). Each ion within the crystal lattice has its own potential well, and as the ions are arranged periodically, there is a periodic arrangement of potential wells.

If one atomic potential well is considered in isolation, then there are discrete electron energy levels such as those that are encountered in the hydrogen atom. If two such wells are brought into close proximity, then from the Pauli exclusion principle which forbids any two electrons from occupying the same energy level (neglecting electron spin), the isolated discrete energy levels split into two levels. When a large number, N, of such wells are brought into close proximity, which is the case in a crystal lattice, then the original discrete energy level is broadened into a quasi-continuous energy band (actually comprising N energy levels). The gaps between the original isolated discrete energy levels may be preserved, but are smaller; in some cases the gaps no longer exist and the energy bands overlap. This 'Band Theory of Solids' provides the framework for understanding the classification of solids into metals, semiconductors and insulators.

If the electron is considered to be a wave travelling perpendicular to a set of lattice planes, then Bragg reflection will occur at certain wavelengths. For these wavelengths, λ_F, the wave is not propagated through the crystal and therefore they are forbidden to the conduction electrons of the crystal. The wave vectors, k, corresponding to the wavelengths ($k = 2\pi/\lambda_F$) occur at the boundaries of the Brillouin zone (the unit cell of the reciprocal lattice) and constitute energy gaps.

The Brillouin zone model relates the electron energy levels to the crystal structure. In regions close to the zone boundaries the variation of electron energy with its wave vector deviates from that where the periodic potential can be treated as constant, as a direct result of energy gaps. One consequence is that the mass of the electron is no longer that of the free electron; if the Fermi level occurs in such a region, the mass of the electron should be replaced by an effective mass, m^*, and this can be quite different from m.

The quantum mechanical approach provides the basis for the understanding of solids in terms of their conduction behaviour. The band theory and Brillouin zone model are primarily concerned with electronic structure, whereas electron conduction is concerned with the motion of the electrons under the influence of an externally applied electric field within that structure. The solution to the vexing problem of the mean free path being considerably larger than atomic dimensions was provided by Bloch [8], who considered, via a Fourier analysis, the propagation of electron waves in the lattice structure. If the structure is perfectly periodic, no scattering occurs, whereas deviations from periodicity give rise to scattering and hence to electrical resistance.

If the potential wells at the lattice sites are identical and the lattice sites themselves are regularly spaced, then the electron wave propagates almost undisturbed; in other words, very little scattering occurs, which is equivalent to a mean free path which is much longer than interatomic distances. In fact, in a perfectly periodic structure, the electron wave is not disturbed at all and in such a structure the resistivity would be zero (this is not to be confused with superconductivity, where the resistivity is zero in non-periodic structures). A perfectly periodic structure is never realised in practice although it is approached in very pure single crystal solids at temperatures close to absolute zero.

Thus, in normal solid electronic conductors, the electrical resistivity arises from departures from periodicity in the lattice potential wells and there are many ways in which this can occur. These include

(i) displacement of the atoms due to their thermal motion;
(ii) imperfections in the crystal lattice, such as dislocations, vacancies and grain boundaries in polycrystalline solids;
(iii) the substitution of foreign atoms at the lattice sites.

The departures from periodicity arising from (i) and (ii) are characteristic of pure metals whilst all three sources contribute to the resistivity of metallic alloys.

1.5 Pure metals

In a well-annealed pure metal at low temperature, the density of the imperfections is low and their contribution to the electrical resistivity is small. Dislocations, which arise for example by cold working, give rise to a temperature-independent resistivity which is usually negligible in comparison with the resistivity at high temperatures where phonon scattering dominates (see below) but may contribute significantly to the resistivity at cryogenic temperatures. For this reason, when studying pure metals it is preferable to use well-annealed specimens. On the other hand, vacancies are thermally activated; their number density increases with temperature. At room and cryogenic temperatures, the vacancy contribution to the resistivity is negligibly small and only starts to become significant at temperatures approaching the melting point. However, this is somewhat academic in the sense that the vacancy contribution to the resistivity is inherent to the metal and temperature.

At finite temperatures, the atoms vibrate about their mean positions. Effectively, the interatomic spacing is no longer constant; as the temperature increases, so does the amplitude of the vibrations, giving rise to greater changes in the interatomic spacing and hence larger departures from periodicity. The result is a temperature variation of the mean free path and hence of the electrical resistivity.

The vibration of the atoms can be regarded as waves propagating through the crystal lattice. By treating the lattice vibrations as being quantised (the quantum of energy is known as a phonon), it can be shown that there is a cut-off phonon frequency; phonons of frequency larger than the cut-off frequency are not allowed. The temperature corresponding to this frequency is known as the Debye temperature (usually denoted by θ_D).

Now the scattering of the electrons is essentially the interaction of the electron waves (which are defined by the periodicity of the lattice) and the displacements of the lattice sites which can be described by lattice waves. The characteristic electron energy E_F (the Fermi energy) is equivalent to many thousands of degrees kelvin, whilst the Debye temperature is typically a few hundred degrees kelvin. Thus, the effect of temperature on the electronic structure of a solid is small compared with its effect on the lattice and, consequently, for scattering purposes, the effect of temperature on the lattice only need be considered.

At temperatures above θ_D, all lattice modes are excited and contribute to the scattering of the electrons. The phonon occupancy of the excited lattice modes increases linearly with temperature and consequently, as the mean free path is inversely proportional to the number of scattering centres, the resistivity increases linearly with temperature. At temperatures much lower than θ_D not all lattice modes are excited; only those with frequency v less than $k_B T/h$ (where h is Planck's constant) will be excited, and at very low temperatures the number of excited modes falls off as $(T/\theta_D)^3$. Thus, for temperatures lower than θ_D, the resistivity varies more rapidly than linear, and it was shown by Bloch [9] that at these very low temperatures the resistivity varies as the fifth power of temperature. Gruneisen [10] later developed the Bloch analysis to give an expression for the resistivity which covered the entire temperature range.

Some pure metals can exist with different crystal structures known as allotropes. Each allotrope has its own Brillouin zone and thus its own characteristic resistivity, which may exhibit a different temperature dependence. Such phase changes are more often encountered in metallic alloys and these are discussed later.

Whilst electron scattering by phonons is the dominant process in determining the temperature variation of the resistivity of most metals, other temperature-dependent scattering processes, such as electron–electron scattering, which gives a T^2 dependence for the resistivity at low temperatures, occur, and these are often seen in the magnetic and transition metals. Magnetic phase transformations occur in the ferromagnetic metals at the Curie temperature, T_c. These transitions are associated with a continuous decay in long-range magnetic order, which vanishes at T_c. As the temperature is increased and T_c is approached, the resistivity rises at an increased rate reflecting the increase in disorder. At T_c, where magnetic disorder is complete, there is a change in slope of the resistivity/temperature curve.

The abrupt increase in the electrical resistivity of many metals at the melting point in passing from the solid to the liquid state, and its steady increase with temperature in the liquid state, is associated with the absence of long-range crystal structure and the increase in disorder in the liquid state.

1.6 Metallic alloys

1.6.1 Dilute alloys

When a 'foreign' atom is substituted for an atom of the host metal, the potential well at the substitution site is changed. The contributions to the change in the potential well due to the substitution arise from the change in valency and change in atomic volume. The valence electrons from the foreign atom merge with the conduction electron number of the host with the result that the potential well at the substitution site has a different charge from that which exists at the host sites; in the mathematical treatment, the potential well has a different depth. The change in atomic volume causes the spatial dimensions of the impurity potential well to differ from that of the host.

When foreign atoms are randomly substituted into the host lattice and the crystal structure of the host is preserved, then provided that the foreign atoms do not interact with each other, the electrical resistivity increases in proportion to the number of foreign atoms added. Such a situation is realised in practice in dilute alloys where the alloy addition is typically less than 5%. Moreover, with such dilute alloys, the foreign atoms vibrate as though they were host atoms and consequently the additional contribution to the resistivity is temperature independent. This was first recognised experimentally in the 1860s by Matthiessen and Vogt and a simple rule, known as Matthiessen's rule, which applies to many dilute alloys, can be stated:

$$\rho_a(T) = \rho_p(T) + \rho(c) \tag{1.5}$$

where $\rho_a(T)$ and $\rho_p(T)$ are the temperature variations of the resistivities of the dilute alloy and the host and $\rho(c)$ is the concentration dependent resistivity of the foreign atoms, which is almost linear in concentration. There are deviations from this rule even for very dilute alloys, but with the exception of transition metal alloys involving the magnetic metals, the deviations are very small and for practical purposes can often be regarded as negligible.

1.6.2 Concentrated alloys

When the number of substitutional foreign atoms becomes large, for example when their concentration is greater than 5%, the assumption that they do not interact with each other is no longer valid and Matthiessen's rule breaks down. The distinction between host and foreign atom becomes obscure (with a 50 atomic percent binary alloy there is no distinction) and thus the $\rho_p(T)$ term of (1.4) loses its original significance. That is not to say the resistivity is temperature independent but rather the temperature dependence is not necessarily that of the major constituent of a concentrated alloy. As the concentration of foreign atoms increases, then so does $\rho(c)$ of (1.4) although it is no longer linear with concentration, and the result is that the contribution from the foreign atoms dominates.

To complicate matters further, according to the equilibrium phase diagram the alloy may exist as a single or multiphase, ordered or disordered solid solution depending on concentration and temperature. The data is presented in this handbook in terms of resistivity against temperature. Whilst this is the preferred means of presentation, it is sometimes easier to discuss the resistivity at a given temperature as a function of concentration.

The simplest situation in concentrated alloys is that of a binary system which exhibits complete solid solubility series at a given temperature. For such a disordered alloy, Nordheim [11] showed that the resistivity was of the form $\rho \propto c(l - c)$, which can be written more generally as $\rho = A + c(1 - Bc)$, where A and B are constants which depend on the alloy series and c is the concentration of one of the components of the alloy. The Nordheim rule specifically relates to the impurity resistivity of concentrated alloys. With such alloys, impurity resistivity normally dominates and the rule can be applied to the total resistivity, especially for room temperature and below.

The above account relates to substitutional impurities which are randomly distributed in the host lattice. There are some alloy systems where for certain compositions the impurity atoms are located at distinct lattice positions, creating a long-range 'ordered' state with a degree of periodicity which is absent in the random state. At these compositions, where such a 'superlattice' exists, the resistivity of the ordered state is somewhat less than that of the random state. Generally, the random state is preferred at high temperatures and the long-range ordered state only occurs below a critical temperature. Thus, care should be exercised when using Nordheim's rule to estimate the resistivity of an alloy.

Phase-equilibrium diagrams of alloy systems indicate that the crystal structure of the alloy often changes with temperature. As mentioned above, these allotropes have different electronic structures which give rise to differences in the electrical resistivity, and this is manifest as a discontinuity in the resistivity/temperature plot, the discontinuity occurring at the boundary of the phase field. In a particular phase field, Nordheim's rule can in principle be applied, but the constants A and B will differ from phase to phase.

Alloy systems do not necessarily exist in a single phase. Indeed the mechanical properties of many engineering alloys depend on the coexistence of two or more phases. Each phase has its own characteristic resistivity which may exhibit a different temperature variation. For the binary two-phase systems, the resistivity may be regarded as lying between two limits corresponding to the two phases being elongated and aligned parallel to the current flux and where the two phases are platelets perpendicular to the current flow. If the volume fractions of the phases are known together with the characteristic resistivities of the individual single-phase alloys which bound the two phase fields, then the limits can be evaluated by regarding them as parallel and series networks. There is no general rule which governs whether the series or parallel configuration will exhibit the greater resistivity and in any case it is highly unlikely that these two somewhat idealised situations will be encountered in practice; likewise, there is no general rule to indicate whether or not the resistivity of the two-phase alloy will be greater or less than that of one of the individual bounding single-phase alloys; however, it might be anticipated that the resistivity would be less than that of the single-phase disordered alloy of the same composition and this is often observed.

In alloy systems where magnetic transformations occur, the change in long-range order is reflected in the resistivity and is often manifest as a change in slope of the resistivity/temperature curve at the critical or transition temperature. In general, the magnitude of the effect of long-range order on the resistivity is difficult to estimate. Ferromagnetic alloys usually contain one of the ferromagnetic pure metals whilst anti-ferromagnetic long-range ordering can be found in many of the alloy systems involving transition and rare-earth metals.

1.7 Semiconductors

As mentioned above, the band theory of solids provides a framework for classifying solids into metals, semiconductors and insulators, the last of which are not relevant to this handbook. In metals, there are continuous energy states above the Fermi energy due either to the bands not being completely filled or to the bands overlapping in energy, whereas in semiconductors there exists an energy gap, E_g, between the valence and conduction bands with the Fermi energy lying in the gap. In the pure elemental semiconductors such as germanium and silicon, electrons must be thermally excited into the conduction band (and as a result an equal number of holes created in the valence band) for these materials to behave as intrinsic conductors. At temperatures of around room temperature and below, the thermal energy, $k_B T$ (<0.03 eV), is small compared with E_g (≈ 0.7 eV for germanium, ≈ 1.1 eV for silicon), and as the number of carriers varies according to $\exp(-E_g/2k_B T)$, these materials are poor conductors; at absolute zero no thermal excitation can occur and they behave as insulators. Increasing the temperature causes the number of carriers to increase exponentially and hence the resistivity to fall. Small amounts of impurities can give rise to donor and acceptor levels within the energy gap which are very close, in energy terms, to the conduction band and valence band, respectively; the energy separation, E_i, may be comparable to the room temperature thermal energy ($k_B T = 0.026$ eV). As the number of carriers now varies as $\exp(-E_i/2k_B T)$, the result is that many more electrons are excited from the donor level into the conduction band (or correspondingly many more holes created in the valence band) at lower temperatures, leading to n-type (electron) and p-type (hole) conductivity. This conductivity, which is proportional to the concentration of impurities, can be several orders of magnitude larger than the intrinsic conductivity. On the other hand, the addition of some impurities to an already impure semiconductor can reduce the total number of carriers via compensation. Thus, the influence of impurities on the resistivity is in marked contrast to the case of metals. Because impurity concentration is usually very low and these materials are normally used in single crystal form where the defect concentration is low, phonon scattering is dominant, but the variation of resistivity with temperature is most strongly influenced by the variation in the number of carriers. For these reasons only the high temperature resistivity of the elemental semiconductors is given in the handbook.

1.8 Summary

To summarise these fundamental concepts, the resistivity of the pure metals is well understood. Reliable experimental data exists over a wide range of temperature and this in general conforms to theoretical expectations. For dilute alloys (less than 5 atomic percent), provided the increase in residual resistivity per atomic percentage impurity addition is known, the resistivity can be estimated with reasonable confidence from Matthiessen's rule. This rule holds remarkably well for many alloy systems although there are exceptions, notably when the host metal is one of the ferromagnets.

In concentrated alloys the situation is much more complicated. While Nordheim's rule can be used to obtain a fair estimate of the resistivity of some concentrated binary alloys, it has limited applicability. This is due to the many factors which influence the resistivity; these include atomic and magnetic long-range order, changes in crystal structure (or phase) as the alloy composition and temperature vary and coexistence of two or more phases. All these factors together make it almost impossible to make even a fair estimate of the resistivity of a given alloy, let alone its temperature dependence. If the resistivity of an alloy of near composition and similar heat treatment is known, then by using the principles outlined above, a reasonable estimate of the resistivity of the alloy in question may be obtained.

The purpose of this handbook is to provide a comprehensive compilation of experimental data in graphical form of resistivity versus temperature which we hope will be of assistance in this respect. Failing this, one must resort to the experimental determination of the resistivity; chapter 2 deals with some of the experimental methods of measuring resistivity.

1.9 Abbreviations

The graphical data presented in this book is based on either single crystals, bulk characteristics or alloy format. Some materials incorporate unusual resistivity characteristics when pretreated and these have been annotated with the abbreviations listed below.

AP	As prepared
MSG	Monoclinic space group
n/a	Data not available
OA	Oxygen annealed
OSG	Orthoclinic space group
Q	Quench temperature, K
δ	Thermal correction factor

Chapter 2
Measurement techniques

T. Farrell

2.1 Introduction

There are many techniques available for the determination of electrical resistivity. The simplest and most widely used is the so-called four probe method in which the resistance of a portion of the sample is measured when a direct current passes through the sample. This technique involves two current contacts and two voltage contacts attached to the sample. The four probe technique is applicable to the vast majority of metals and alloys and a variation of the method, the Van der Pauw technique, is widely used for the determination of the resistivity of semiconductor wafers. The technique is applicable over a wide range of temperature. However, difficulties may be experienced with contacts if, for example, the material is brittle or chemically reactive (as it may be in the molten state). For such materials, methods which do not rely on making physical contact to the sample are preferable; the most widely used non-contacting techniques are those based on eddy currents induced in the sample when it is placed in a magnetic field which varies with time. In the following paragraphs, the four probe technique and some of the eddy current techniques are briefly described.

The theoretical background to the four probe and eddy current methods can be demonstrated by the application of two of Maxwell's equations:

$$\operatorname{curl} \boldsymbol{E} = -\frac{\partial \boldsymbol{B}}{\partial t} = -\mu\mu_0 \frac{\partial \boldsymbol{H}}{\partial t} \tag{2.1}$$

$$\operatorname{curl} \boldsymbol{H} = \boldsymbol{J} + \frac{\partial \boldsymbol{D}}{\partial t} = \frac{\boldsymbol{E}}{\rho} + \varepsilon\varepsilon_0 \frac{\partial \boldsymbol{E}}{\partial t} \tag{2.2}$$

In these equations, $\boldsymbol{E}$ is the electric field, $\boldsymbol{D}$ is the electric displacement, $\boldsymbol{H}$ is the magnetic field, $\boldsymbol{B}$ is the magnetic induction, ε_0 and μ_0 are the permittivity and permeability of free space, ε and μ are the relative permittivity and permeability of the sample and $\boldsymbol{J}$ is the current in unit cross-sectional area of the sample of resistivity ρ.

For a conducting sample, the displacement current $\partial \boldsymbol{D}/\partial t$ is negligible in comparison with $\boldsymbol{J}$ and thus (2.2) becomes

$$\operatorname{curl} \boldsymbol{H} = \boldsymbol{J} = \frac{\boldsymbol{E}}{\rho} \tag{2.2a}$$

Equation 2.2*a* is the basis for the four probe method. The expression $\boldsymbol{J} = \boldsymbol{E}/\rho$ is simply a statement of Ohm's law. Thus, writing $\boldsymbol{J} = I/A_S$ where I is the current in the sample of uniform cross-section A_S and V the voltage measured by two impedance probes separated by a distance L, the resistance R is

$$R = \rho \frac{L}{A_S} \tag{2.3}$$

If the sample is cylindrical of radius a, the resistance is

$$R = \left(\frac{\rho}{a^2}\right)\frac{L}{\pi} \tag{2.3a}$$

The basis for the eddy current methods is seen by substituting (2.1) into (2.2*a*):

$$\operatorname{curl}\operatorname{curl}\boldsymbol{H} = \nabla^2\boldsymbol{H} = \frac{1}{\rho}\operatorname{curl}\boldsymbol{E} = -\frac{\mu\mu_0}{\rho}\frac{\partial H}{\partial t} \tag{2.4}$$

For a sample having isotropic magnetic and electrical properties, which is true of most polycrystalline materials, and where the magnetic field is aligned along the principal sample dimension (for example, along the axis of a cylinder, say the z direction), the vector equation (2.4) becomes scalar, and since $B = \mu\mu_0 H$ and the magnetic flux, ϕ, in the sample is $\phi = A_S B$, we have

$$\frac{\partial\phi}{\partial t} = -\frac{\rho}{\mu\mu_0}\frac{\partial^2\phi}{\partial z^2} \tag{2.4a}$$

Equation (2.4*a*) is the standard diffusion equation where the flux diffusivity (which has the dimensions of $(\text{length})^2/\text{time}$) is $\rho/\mu\mu_0$. It governs how the magnetic flux in the sample changes with a time-dependent magnetic field. Solutions to the diffusion equation for various geometrical arrangements and different boundary conditions (for example, the time dependence of the magnetic field) are given in standard texts [12]. By way of illustration, consider a cylindrical sample of radius a which experiences a uniform magnetic flux, ϕ_0, which is suddenly removed at time $t = 0$; the average flux at time t is given by

$$\phi = \frac{4\phi_0}{a^2}\sum_{n=1}^{\infty}\frac{1}{\beta_n^2}\exp\left(-\beta_n^2\left(\frac{\rho}{\mu\mu_0 a^2}\right)t\right) \tag{2.5}$$

where the coefficients β_v are the positive roots of $\boldsymbol{J}_0(\beta) = 0$, $\boldsymbol{J}_0$ being the zero-order Bessel function. Thus, by observing the time dependence of the flux decay, the resistivity can be deduced. Details of this method along with other diffusion eddy current based methods involving periodic magnetic fields are discussed below. It is seen that in both (2.3) and (2.5) a term involving ρ/a^2 appears. The same term appears in other eddy current methods, which being diffusion based must have a term involving $(\text{length})^2$ – or cross-sectional area – and this is important when considering the accuracy of the determination of ρ.

2.2 Four probe methods

These methods rely on the measurement of the resistance of the portion of the sample between the two voltage probes. For (2.3) to be applicable, the cross-sectional area of the sample should be uniform and the voltage probes should be connected to the sample in such a way that the length, L, can be accurately determined.

The preferred method of measuring the resistance is to use a Wheatstone bridge, which compares the unknown resistance of the sample with that of three standard resistors, one of which is a variable standard. As the bridge is a null method, calibration of the detector is not necessary. The accuracy with which the null point can be determined depends not only on the sensitivity of the detector but also on the relative magnitudes of the resistances of the arms of the bridge. A good general rule for optimum bridge operation is that the resistances in the four arms should be of the same order. With a well-designed bridge and sensitive detector, the bridge sensitivity can be very high and errors in the determination of resistivity arising from the resistance measurement can be regarded as being negligible.

Alternatively, the sample resistance can be measured using a constant current and measuring the probe voltage difference using a potentiometer or digital voltmeter; the latter has the advantage that it can be calibrated to give a direct reading of the resistance. To avoid zero errors it is usual practice to incorporate current reversal and take the average value of the two readings. Zero errors can arise not only from an incorrectly adjusted voltmeter but also from thermoelectric effects if a temperature gradient exists along the sample and the voltage leads, usually copper, have a different thermoelectric power from that of the sample being measured.

The accuracy with which the resistivity is determined from the resistance measurements can be dominated by the errors involved in the measurements of the sample cross-section and the separation of the voltage probes. This depends on the magnitude of A_S and L, which in turn is influenced by the magnitude of the resistivity through the need to have an accurate measurement of the resistance. For a pure metal with a resistivity of 5×10^{-8} ohm m, through which a current of 1 A passes and where the voltage is determined to an accuracy of 10 nV, then for a resistance measurement to an accuracy of 0.01%, the voltage between the probes should be 100 μV, giving a resistance of 10^{-4} ohm. From (2.3), L/A_S should be 2×10^3/m and thus can be satisfied realistically with a cylindrical sample of diameter 8 mm and with $L = 100$ mm. The power dissipated in the sample is 0.1 mW and this will not cause any significant heating.

Thus, for the majority of metals and alloys, the resistivity at temperatures above room temperature, cylindrical samples measuring about 10 mm in diameter and with L of about 50 to 100 mm are ideal shapes. Clearly, if this demands too large a quantity of material, the sample dimensions should be scaled down accordingly, but this will have the effect of increasing the errors in the measurement of L and A_S. At lower temperatures, particularly with annealed pure metals, the resistivity can be as low as 2×10^{-11} ohm m and this usually requires the sample to be in the form of a thin wire.

The measurement of L usually presents no problems. If mechanical clamps embodying 'knife-edge' contacts are used as the voltage probes, then the separation of appropriate datum marks on the clamps can be determined to plus or minus 0.02 mm with a travelling microscope; thus, the error introduced in the measurement of L is less than 0.1%.

The determination of A_S is more demanding. First, A_S should be uniform, and this can be checked by measuring the sample diameter (for the cylindrical sample) at various positions along L. It is preferable to use a non-contacting method for this measurement (if a micrometer is used, the contact pressure can lead to an erroneously low value for the diameter). Optical projection techniques and photographic enlargements can be used to get an accuracy of better than 0.1% in A_S, provided, of course, the sample is prepared to this accuracy.

Thus, provided care is taken in the measurement of L and A_S, the resistivity can readily be determined with an absolute accuracy of around 0.1% at and above room temperature. At cryogenic temperatures, in particular at liquid helium temperatures and with pure metals, the error in the determination of the resistance is maintained at negligible proportions by employing voltage detection equipment which involves superconducting elements and has much higher sensitivity.

2.3 Temperature variation of resistivity

With the four probe technique, the temperature variation of the resistivity can be determined quite simply by thermally attaching the sample to a heat sink, the temperature of which is controlled. The sample temperature can be measured using a suitable thermocouple attached to one of the voltage probes and any temperature gradients can be measured using a differential thermocouple with its two junctions thermally anchored to the two voltage probes; in this way, the mean sample temperature can be determined. The design of the furnace is not considered here, but for high-temperature work, a vacuum furnace is usually used to reduce the thermal load and prevent oxidation.

To overcome problems with thermoelectrically generated voltages, the voltage leads should be continuous to the wall of the vacuum vessel; likewise, for accurate temperature measurements it is preferable to use continuous thermocouple leads.

When varying the temperature, the change in sample dimensions due to thermal expansion should be taken into account, especially since L and A_S are usually carefully determined at room temperature. To illustrate the magnitude of the error which may be introduced by the neglect of thermal expansion, consider a cylindrical sample to which (2.3*a*) applies. At any temperature T, the resistance is

$$R_T = \rho_T \frac{L_0}{\pi a_0^2} \frac{(1+\alpha\Delta T)}{(1+\alpha\Delta T)^2} = \rho_T = \frac{L_0}{\pi a_0^2}(1+\alpha\Delta T)^{-1}$$

where a is the thermal expansion coefficient, R_T is the measured resistance and ρ_T is the resistivity. Thus, the percentage error introduced is $\alpha\Delta T \times 100$. Taking a representative value of $\alpha = 15 \times 10^{-6}$ for most metals and alloys, it is readily appreciated that the error can be comparable to the accuracy in the determination of the resistivity for a ΔT of 100 K.

2.4 Eddy current methods

Several techniques which rely on induced eddy currents for the determination of resistivity have evolved, the simplest of which correspond to the situations where the applied magnetic field is periodic, for example varying sinusoidally with time, and where the field is a step function being suddenly switched from zero to a preset constant value or conversely switched from a steady value to zero. Whilst these elegant techniques overcome the need for electrical contact to the sample, they do have characteristic features which can give misleading results.

First, it is important to recognise that the eddy currents are not uniformly distributed throughout the sample when a periodic magnetic field is applied. They decay exponentially from the sample surface with a measure of the decay being the 'skin depth'. The skin depth, δ, is given by

$$\delta = \left(\frac{\pi f \mu \mu_0}{\rho}\right)^{-1/2}$$

and is the depth at which the eddy currents have fallen to $1/e$ of their surface value. If the whole of the sample is to be examined, then, in the case of the periodically varying magnetic field, the frequency, f, should be chosen so that δ is not significantly less than the sample dimension, which, in the case of a cylinder with the magnetic field parallel to the cylinder axis, is the cylinder radius. If δ is significantly less than the sample dimensions, then surface features such as roughness and contaminant films can affect the resistivity measurements. It must be emphasised that the only situation where uniform current density occurs is in the direct current measurements on uniform cross-section samples as discussed in the four probe method.

Second, the magnetic field can directly influence the motion of the electrons within the sample (in addition to 'indirect' influence via induction which gives rise to the eddy currents), leading to additional resistance over that caused by scattering. This 'magnetoresistance' is dependent on the magnitude of the applied magnetic field and its occurrence can be checked by varying the field strength.

These techniques are well suited to the determination of the resistivity of weakly magnetic materials where the relative permeability, μ, can safely be regarded as unity. However, some caution should be exercised when dealing with ferromagnetic and strongly paramagnetic materials or materials which exhibit magnetic transformations where μ is neither constant nor unity.

The most convenient means of applying a magnetic field is to place the conducting sample in a coil. With the step function methods, the resistivity is determined by observing the change of flux in the sample as a function of time after the field has suddenly been applied or removed; this is described for the case of field removal by (2.5). For periodically varying fields, the difference between the impedance of the loaded and empty coil is measured.

2.5 Step function methods

The basis of this method has already been mentioned. A convenient arrangement is to insert the sample into a coil, apply a direct current to the coil for sufficient time to ensure complete flux penetration into the sample and then de-energise the coil by opening a switch. A simple circuit is shown in Fig. 2.1. The flux decay is detected by a secondary coil wound over the central region of the energising coil. The time dependence of the average flux, ϕ, in a cylindrical sample after removing the magnetic field is given by (2.5). If there are N turns on the second coil then the voltage induced in that coil is

$$-N\frac{\partial\phi}{\partial t}$$

Using the relationship between ϕ and magnetic field H

$$\phi = \pi a^2 \mu\mu_0 H$$

and differentiating (2.5) with respect to time and multiplying by N, the secondary coil voltage is obtained:

$$V = 4\pi N\rho H_0 \sum_{n=1}^{\infty} \exp\left(-\beta_n^2\left(\frac{\rho}{\mu\mu_0 a^2}\right)t\right)$$

The voltage decays as a sum of exponential factors. The first three coefficients β_n^2 have the values 5.784, 30.470 and 74.892; thus, for longer times, the decay is a simple exponential governed by β_1^2. To illustrate this, consider the exponential term involving β_2 to contribute a maximum of 0.1% of the contribution from β_1. For this condition to apply:

$$\exp\left(-\left(24\cdot 676\left(\frac{\rho}{\mu\mu_0 a^2}\right)t\right)\right) \leq 10^{-3}$$

Thus, for a material with $\rho = 5\times 10^{-8}$ ohm m, $a = 5\times 10^{-3}$ m and $\mu = l$, the time should be greater than 1.75×10^{-4} s in which case the sum of the exponential

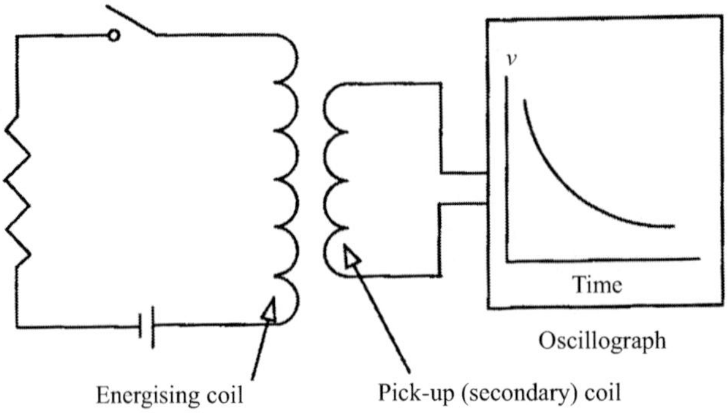

Figure 2.1 Simple circuit for the step function method

factors is approximately 0.2. The measured voltage at this time is $1.25 \times 10^{-5} H_0$ V and after $t = 5 \times 10^{-4}$ s the voltage has decayed to $6.25 \times 10^{-7} H_0$ V. The voltage/time variation can be measured and stored using a suitable oscilloscope.

2.6 Periodically varying fields

The impedance, Z_L, of a magnetic field coil into which a conducting sample is placed (a loaded coil) comprises resistive and reactive components:

$$Z_L = R_L + iX_L$$

The resistance R_L and reactance X_L of the loaded coil are given by

$$R_L = R_0 + R_1$$
$$X_L = X_0 + X_1$$

where R_0 and X_0 refer to the empty solenoid and R_1 and X_1 are the resistive and reactive contributions from the eddy currents induced in the sample. In fact, R_1 and X_1 measure the out-of-phase and in-phase flux in the sample.

In general

$$R_1 = A' X_0 F_R(y)$$
$$X_1 = A' X_0 F_X(y)$$

or $$\frac{X_1}{R_1} = \frac{F_X(y)}{F_R(y)} = F_{XR}(y)$$

where the constant A' is a measure of the ratio of the cross-sectional area of the sample to that of the coil and y is a parameter involving the skin depth and the dimension of the sample. Analytical expressions of the coil functions $F_X(y)$, $F_R(y)$ and hence $F_{XR}(y)$ can be derived for the simpler geometrical shapes such as 'infinite' cylinders, in which case y is given by

$$y = \frac{\sqrt{2}a}{\delta} = (2\pi\mu\mu_0)^{1/2}\left(\frac{fa^2}{\rho}\right)^{1/2}$$

where a is the radius, as before, and the coil 'filling factor', A', is simply A_S/A_C, where A_C is the cross-sectional area of the coil. The coil functions for cylindrical geometry are shown in Figs. 2.2 and 2.3 and for such a geometry these figures can be used directly to obtain ρ once X_0, X_1 and R_1 have been measured.

Samples of irregular geometry, for which analytic expressions cannot be derived, can be used but the coil/sample system should be calibrated over a range of frequencies using a 'standard' specimen of known resistivity, ρ_S, having identical shape to that to be determined. The experimentally determined R_S/X_0, X_S/X_0 and X_S/R_S give $A'F_R(y)$, $A'F_X(y)$ and $F_{XR}(y)$, respectively (X_S and R_S refer to the calibration with the standard resistivity specimen). There are advantages in using $F_{XR}(y)$ as this eliminates the need to know the coil filling factor which, for non-simple geometries, is difficult to calculate.

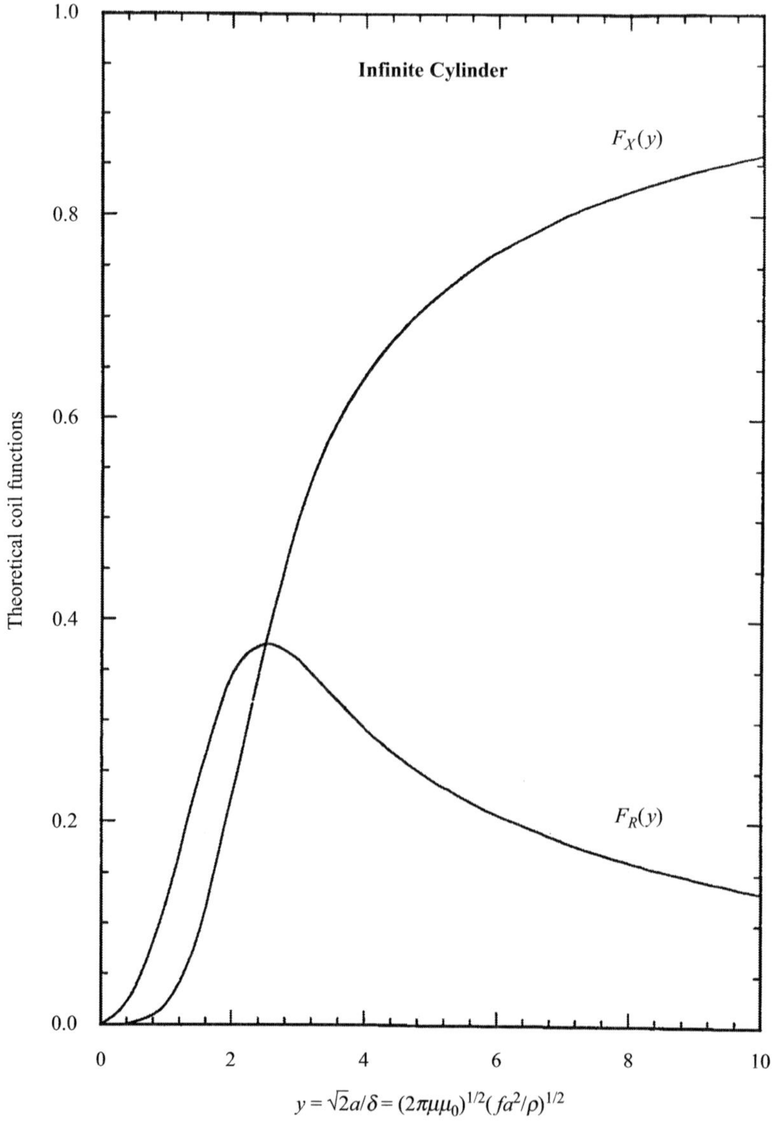

Figure 2.2 Coil functions $F_R(y)$ and $F_X(y)$ for an infinite cylinder

Having determined the empty coil parameter and obtained the coil functions, a single frequency measurement is sufficient to determine the resistivity of the unknown sample. The choice of frequency is important for two reasons. First, as mentioned beforehand, it is desirable that $\delta > a$, and second, since the resistivity determination involves the subtraction of the unloaded coil measurements from the

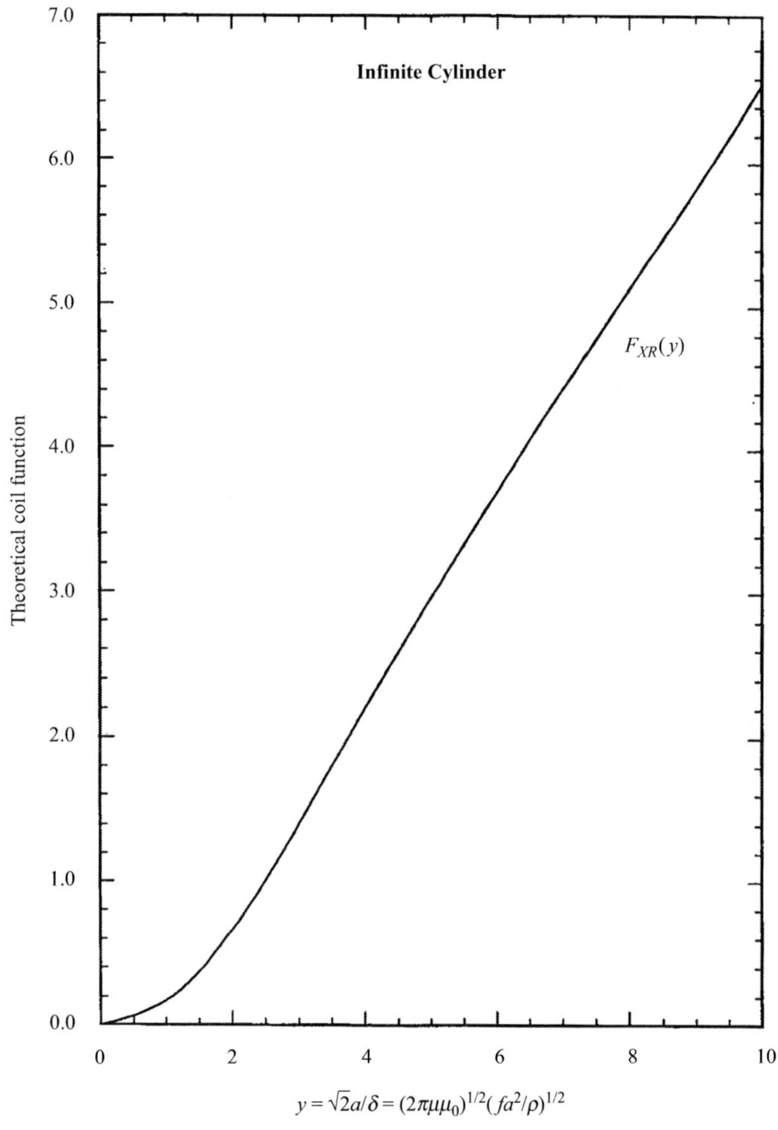

Figure 2.3 Coil function $F_{XR}(y)$ for an infinite cylinder

loaded coil measurements, it is desirable to have a large difference between the two sets of measurements. This can be illustrated by reference to Fig. 2.2 and again taking a cylindrical sample of radius 5 mm and resistivity of 5×10^{-8} ohm m. $F_X(y)$ and $F_R(y)$ are both greater than 0.2 in the range from $y = 2$ to 5.5 and this corresponds to $f = 1$ to 7.5 kHz. At $f = 1.0$ kHz, the skin depth in the above material is 3.6 mm, whilst at $f = 7.5$ kHz, the skin depth is 1.3 mm. If the criterion of $\delta = a = 5$ mm is used, the frequency is 500 Hz giving $y = 1.4$; $F_X(y)$ and $F_R(y)$ are reduced to 0.07 and 0.22, respectively. This implies that the maximum value of X_L/X_0 is 1.07 (taking $A' = 1$), which imposes a more stringent demand on the experimental determination of X_L and X_0. At higher frequencies (7.5 kHz), the experimental demands are relaxed but skin depth considerations may affect the result.

2.7 Inductive techniques

2.7.1 *Self-inductance*

The coil should be longer than the sample to avoid end effects, thus ensuring uniformity of the magnetic flux in the sample. Coil design details are beyond the scope of this account, but taking into consideration such factors as low thermal dissipation, maximum coil filling factor, small magnetic fields, etc., then for a cylindrical sample of diameter about 10 mm, R_0 and $2\pi f X_0$ ($f = 10^3$ Hz) can each be a few ohms, giving a coil impedance of around 10 ohms.

The loaded and unloaded coil impedances are measured using an appropriate variable frequency bridge. Two self-inductance bridge networks are shown in Fig. 2.4. The Anderson bridge, Fig. 2.4*b*, has the advantage that it avoids the use of variable standard inductors or capacitors. Taking the coil current to be in the milliampere range (for low thermal dissipation and low magnetic fields) and following the general rule for optimum bridge operation, the driving voltage of the audio-frequency oscillator is in the 0.1 to 1.0 V range. Using a tuned detector, the sensitivity of the bridge can be higher than 0.005%, and with an appropriately chosen frequency and good coil filling factor, R_1 and $2\pi f X_1$, can be around $0.15 \times 2\pi f X_0$ implying that R_1 and X_1 can be measured to an accuracy of 0.05%. The error in y is that associated with the accuracy to which a can be measured and this has been discussed in connection with the four probe method.

The uncertainty in ρ arises from the uncertainties in y^2 (and hence a^2) and if $F_X(y)$ and $F_R(y)$ are used the uncertainties in A', which involve both a^2 and d^2 (d is the coil diameter); if $F_{XR}(y)$ is used the uncertainties are reduced as A' is not required. If a non-simple shape is used which necessitates experimental determination of $F_X(y)$ and $F_R(y)$, the errors are compounded by the dimensions of both the standard and unknown samples. Taking all these factors into consideration the accuracy in ρ when determined using the self-inductance technique is likely to be in the region of 0.1% with the lowest uncertainties being achieved with a uniform cross-section cylindrical sample.

2.7.2 *Mutual inductance*

The use of a secondary coil wound over the central region of the primary coil has the advantage of eliminating end effects and offers the prospect of examining regions of the sample along its length. The analysis of the coils for mutual inductance is quite complex. As an approximation, the reactance and resistance of a suitable unloaded

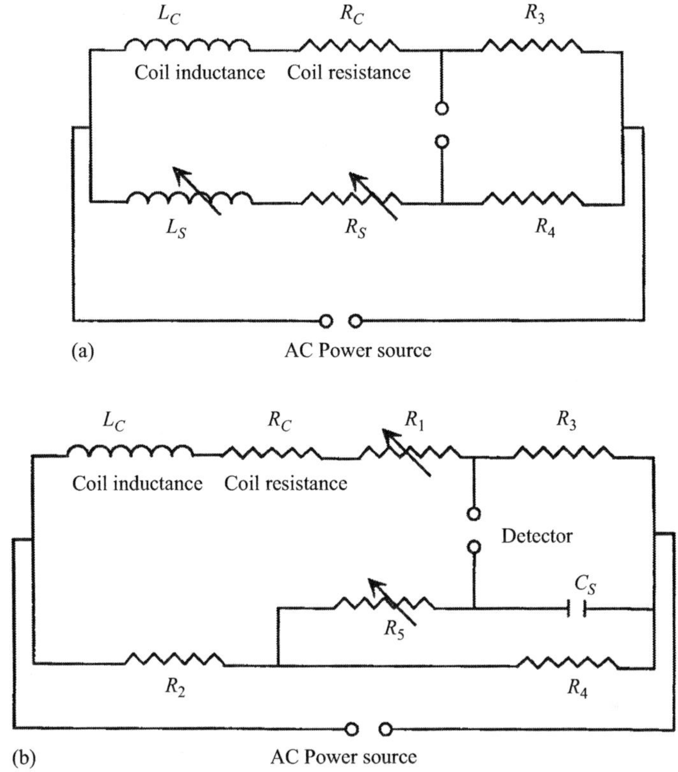

Figure 2.4 Typical bridges for the determination of self-inductance. (a) Typical bridge using standard inductors; (b) Anderson bridge

coil system is scaled down from the self-inductance and resistance of the primary coil by the turns ratio. Thus, a typical system will have a mutual inductance in the region of 100 μH (as opposed to a few millihenries for the single-coil self-inductance).

The determination of mutual inductance is most readily achieved by making a direct comparison with a variable standard mutual inductance, connecting the primary windings of both standard and unknown in series to an audio-frequency oscillator and connecting the secondary windings in series opposition so that the induced voltages oppose each other. In principle, the detector which is in series with the secondaries shows a null reading when the mutual inductances and unknown inductors are equal. In practice, because a resistance element exists, it is almost impossible to obtain a balance. The so-called Hartshorn bridge, shown in Fig. 2.5, overcomes this difficulty. The technique requires a balance to be obtained in the unloaded system by adjusting M_S and R. The sample is then inserted and the balance is restored by varying M_S and R by ΔM_S and ΔR from the unloaded condition. Recognising that R_1^+ and X_1^+ are measures of the out-of-phase and in-phase flux associated with the eddy currents, R_1^+ is given by ΔM_S and X_1^+ by $\Delta R/2\pi f$. As with the self-inductance technique,

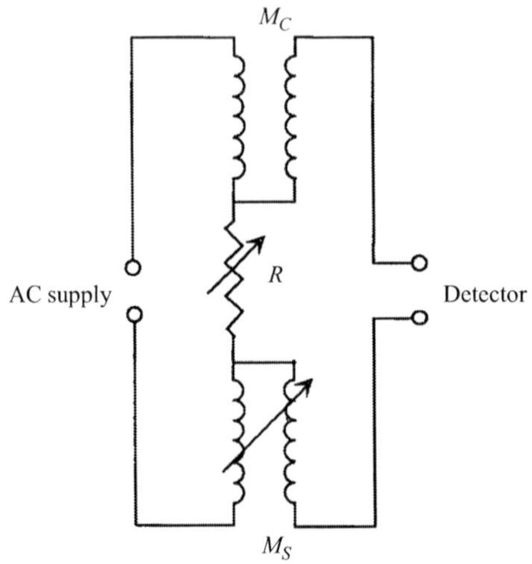

Figure 2.5 The Hartshorn mutual inductance bridge. M_C = mutual inductance of the coil, M_S = variable standard mutual inductance

the value of y is then obtained from the calibrated coil function or, if cylindrical samples are used, from the theoretical coil functions shown in Figs. 2.2 and 2.3.

The mutual inductance technique relies on exactly the same parameters as the self-inductance method; the uncertainties involved are again those associated with the measurement of sample and coil dimensions and hence the achievable accuracy in the determination of the resistivity is similar.

2.7.3 *Inductive techniques and temperature variations*

The primary purpose of using eddy current techniques is to avoid physical contact with the sample. When determining the resistivity as a function of temperature, the sample should be in an isothermal enclosure with the temperature of the enclosure monitored with an appropriate thermometric device such as a thermocouple, resistance thermometer or pyrometer. The use of an isothermal enclosure virtually eliminates thermoelectrically generated voltages.

When the highest accuracy is sought, the effects of thermal expansion must be taken into account and these effects enter through expansion of the sample and the coils (in the four probe method only the expansion of the sample needs to be considered).

The reactance of the unloaded coil is related to its dimension via $X_0 = \text{constant} \times d_c^2/l_c$, where d_c and l_c are the coil diameter and length, respectively. The effect of an increase in temperature, ΔT, is to increase X_0 by $\alpha\Delta T \times 100\%$ where α may be taken to be the coefficient of expansion of the coil windings. The magnitude of the change is similar to that discussed in connection with the variation of resistance with temperature in the four probe method.

Neglecting the lead resistances, R_0 is proportional to l_w/d_w^2, where l_w and d_w are the wire length and diameter. Assuming the coil is tightly wound then $d_w \approx B_c l_c$. and $l_c \approx C_c l_c$ where B_c and C_c are constants involving the number of turns on the coil and the number of layers of windings. R_0 is therefore proportional to d_c/l_c^2 and the effect of a temperature increase of ΔT is to decrease R_0 by $\alpha\Delta T \times 100\%$. Again, the magnitude of the effect can easily exceed the accuracy with which the resistivity can be determined and corrections for thermal expansion should be made.

The sample dimensions enter through the parameter y which is proportional to $(r/a^2)^{1/2}$; thus, if the thermal expansion behaviour of the sample is known, a simple correction can be applied for y.

However, it is not just a simple matter of correcting the impedance measurements for thermal expansion as is the case with the resistance measurements in the four probe method. Since R_0 and L_0 change with temperature, the bridge balance conditions also change. Thus, for the most reliable results, the implication is that the equipment should have provision for withdrawing the sample from the coil to enable R_0 and L_0 as well as R_L and X_L to be measured at each temperature.

The coil filling factor is also affected by thermal expansion. If F_X, F_R are used to determine the resistivity, then a simple correction for thermal expansion (sample/coil) can be applied; this is not necessary if F_{XR} is used.

This account of experimental techniques concludes with a few remarks about an interesting method of resistivity measurement which involves induced eddy currents and as such avoids the use of electrical contact. Unlike the eddy current techniques discussed above, it does not rely on electrical measurements. The technique involves a rotating uniform magnetic field such as that produced by the stators of a polyphase induction motor. When a cylindrical sample is placed in the field, the induced eddy currents interact with the flux causing the sample to experience a torque, Γ, given by

$$\Gamma = \frac{\pi\omega_r\mu^2\mu_0^2 l a^4 H^2}{4\rho}$$

where ω_r is the relative angular velocity of H and the sample. If the sample's motion is restrained by a torsion wire, it will rotate to a fixed position where Γ is balanced by the reaction in the torsion wire, enabling the above expression to be used with ω_f, the angular velocity of the field, replacing ω_r. Measurement of the deflection by the incorporation of a mirror into the torsion wire suspension and the use of optical levers gives an accurate measurement of Γ. The resistivity, ρ, is then determined using the above expression. This technique is more sensitive to the sample dimensions than the electrical techniques and is thus inherently less accurate. Furthermore, when using this method to determine the temperature variation of the resistivity, it is important that corrections for thermal expansion be applied.

Chapter 3
Explanation of graphs

G. Dyos

3.1 Data analysis

The majority of the data presented in this handbook has been gleaned from material published over the last two decades. The data has been extracted from the original published paper or book. The interpretation of the data is that of the original author except where an obvious error has occurred in the publishing of tabular material. Otherwise, every effort has been made to ensure the accuracy of the transcripted original data.

3.2 Catalogue system for materials

The catalogue system used in this book is based on two levels. First, the materials are listed in alphabetical order; where a material has several elements, they are listed in order of decreasing atomic percentage. Should a material have elements all having equal atomic percentage then they are listed in alphabetical order.

Data which has been presented in the original paper in graphical form and extracted is indicated by an asterisk (*) against the reference number in question. Data presented, in the original form, in tabular format is indicated by the absence of an asterisk against the reference number. Data from various sources which has been surveyed by an author to determine a best fit of that data is indicated by an 'R' after the reference. Data included in that survey is not represented in this handbook.

If the material is a disordered crystalline compound, the letter 'D' is placed after the reference number and similarly the letter 'A' is used if the material is amorphous. It should be noted that the vast majority of authors do not indicate whether an allowance for thermal expansion has been applied to the resistivity of the material. Where the purity of an element has not been stated in the original paper, but the information greatly extends the range of resistivity, this has been denoted by '%na'.

3.3 Material composition

The purity of the elements in this book is expressed in terms of the weight percentage of the elements present.

All the compounds listed in this handbook are expressed in terms of atomic percentages, that is, the number of atoms of an element expressed as a percentage of the total number of atoms present in one molecule of the material.

Where compounds have been listed in the original reference as weight percentages, these have been converted into atomic percentages.

Chapter 4

Text references

[1] Pollock D.D. *Electrical Conduction in Solids: An Introduction* (Ohio, American Society for Metals, 1985)

[2] Rossiter P.L. *The Electrical Resistivity of Metals and Alloys* (Cambridge, Cambridge University Press, 1987)

[3] Schroder K. *Handbook of Electrical Resistivities of Binary Metallic Alloys* (Florida, CRC Press, 1983)

[4] Meaden G.T. *The Electrical Resistance of Metals* (London, Heywood Books, 1963)

[5] Kittel C. *Introduction to Solid State Physics* (New York, John Wiley & Sons, 1976)

[6] Drude P. 'Zur elektronentheorie der metalle'. *Ann. Phys.* 1900, vol. 1, p. 566

[7] Sommerfeld A. 'Zur elektronentheorie der metalle auf grund der Fermischen statistik'. *Z. Phys.* 1928, vol. 47, p. 1

[8] Bloch F. 'Uber die quantenmechanik der elektronen in kristallgittern'. *Z. Phys.* 1928, vol. 52, p. 555

[9] Bloch F. 'Zum elektrischen widerstandsgesetz bie tiefen temperaturen'. *Z. Phys.* 1930, vol. 59, p. 208

[10] Gruneisen E. 'Die abhangigkeit das elektrischen widerstandes reiner metalle van der temperatur'. *Ann. Phys.* 1933, vol. 16, p. 530

[11] Nordheim L. 'Zur elektronentheorie der metalle'. *Ann. Phys.* 1931, vol. 9, no. 5, pp. 607–78

[12] Carslaw H.S., Jaeger J.S. *Conduction of Heat in Solids* (Oxford, Oxford University Press, 1959)

Chapter 5

Index of materials

Resistivity – pressure effects

Normalised resistance, pressure and resistance effects

Chapter 6
Material resistivity graphs

6.1 Electrical resistivity as a function of temperature

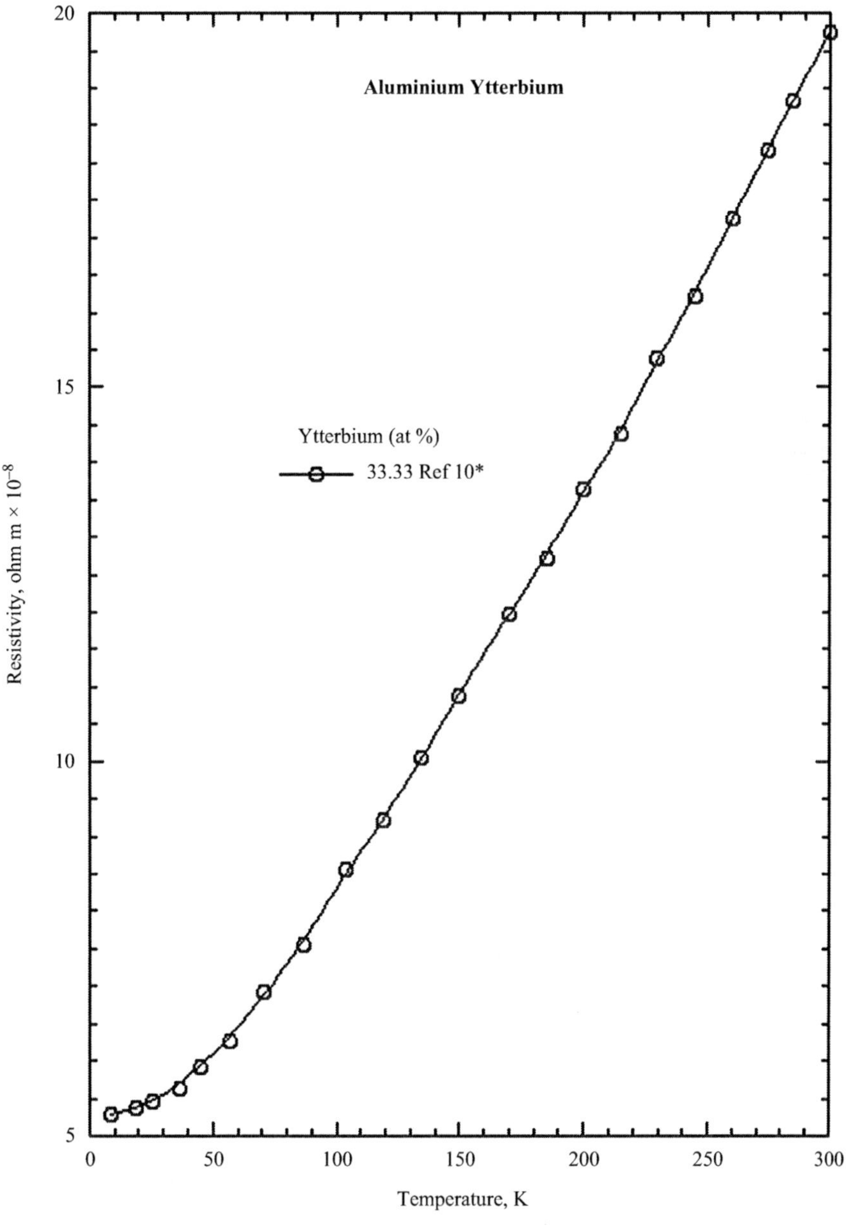

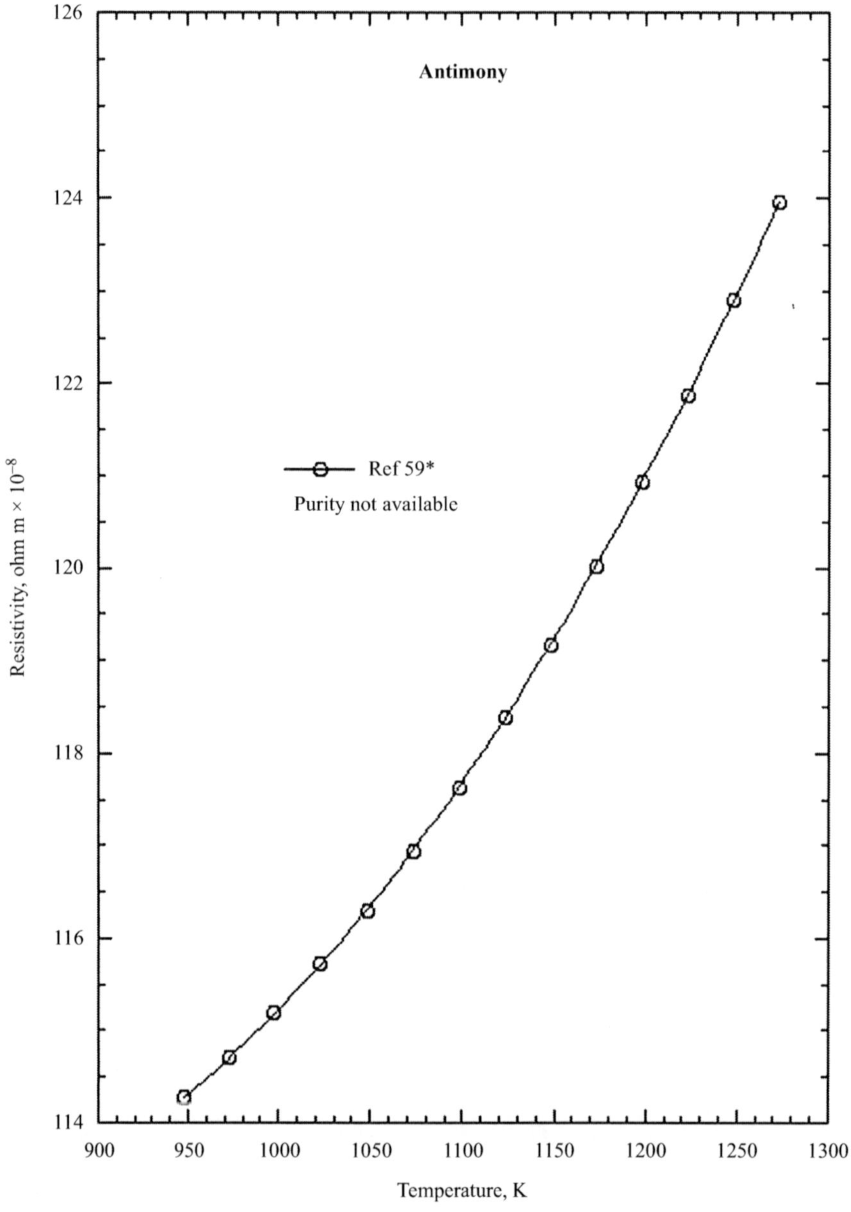
Antimony
Ref 59*
Purity not available
Resistivity, ohm m × 10⁻⁸
126
124
122
120
118
116
114
900
950
1000
1050
1100
1150
1200
1250
1300
Temperature, K

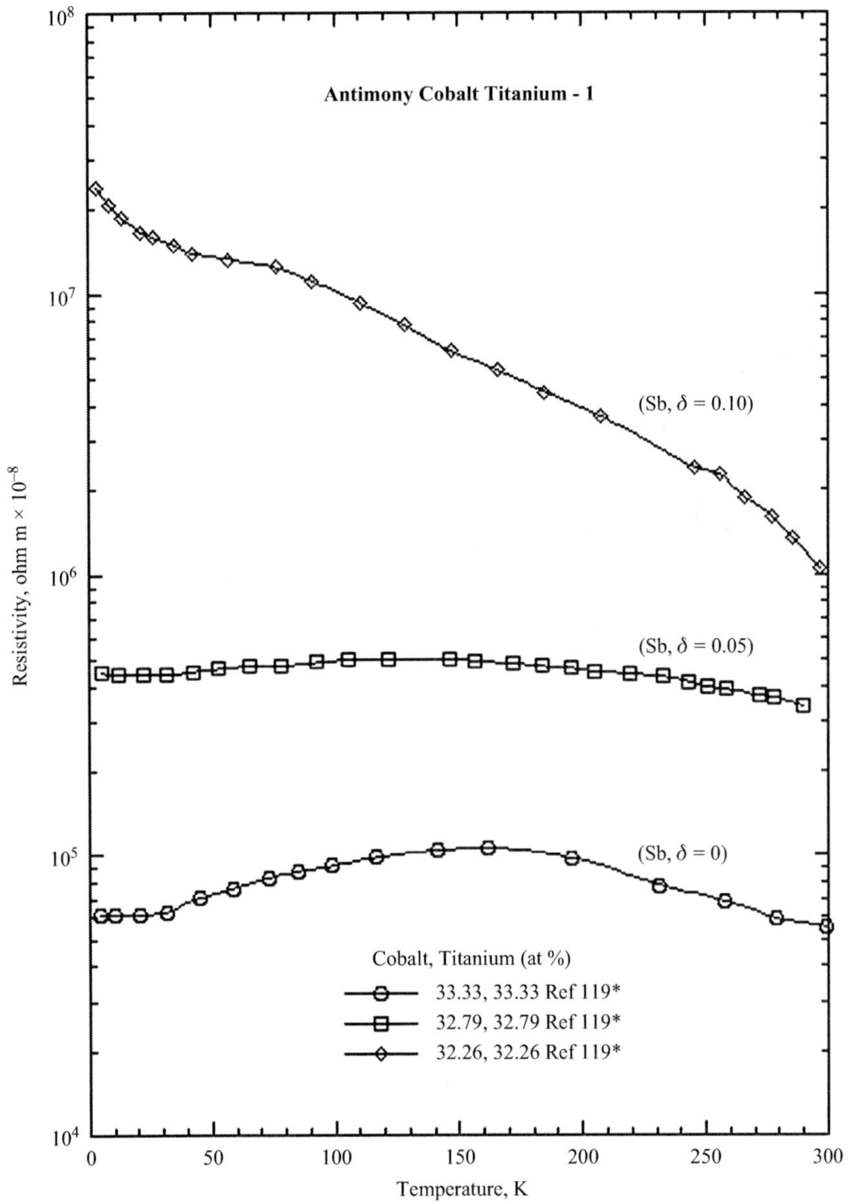
Antimony Cobalt Titanium - 1
Resistivity, ohm m × 10⁻⁸
Temperature, K
(Sb, δ = 0.10)
(Sb, δ = 0.05)
(Sb, δ = 0)
Cobalt, Titanium (at %)
33.33, 33.33 Ref 119*
32.79, 32.79 Ref 119*
32.26, 32.26 Ref 119*

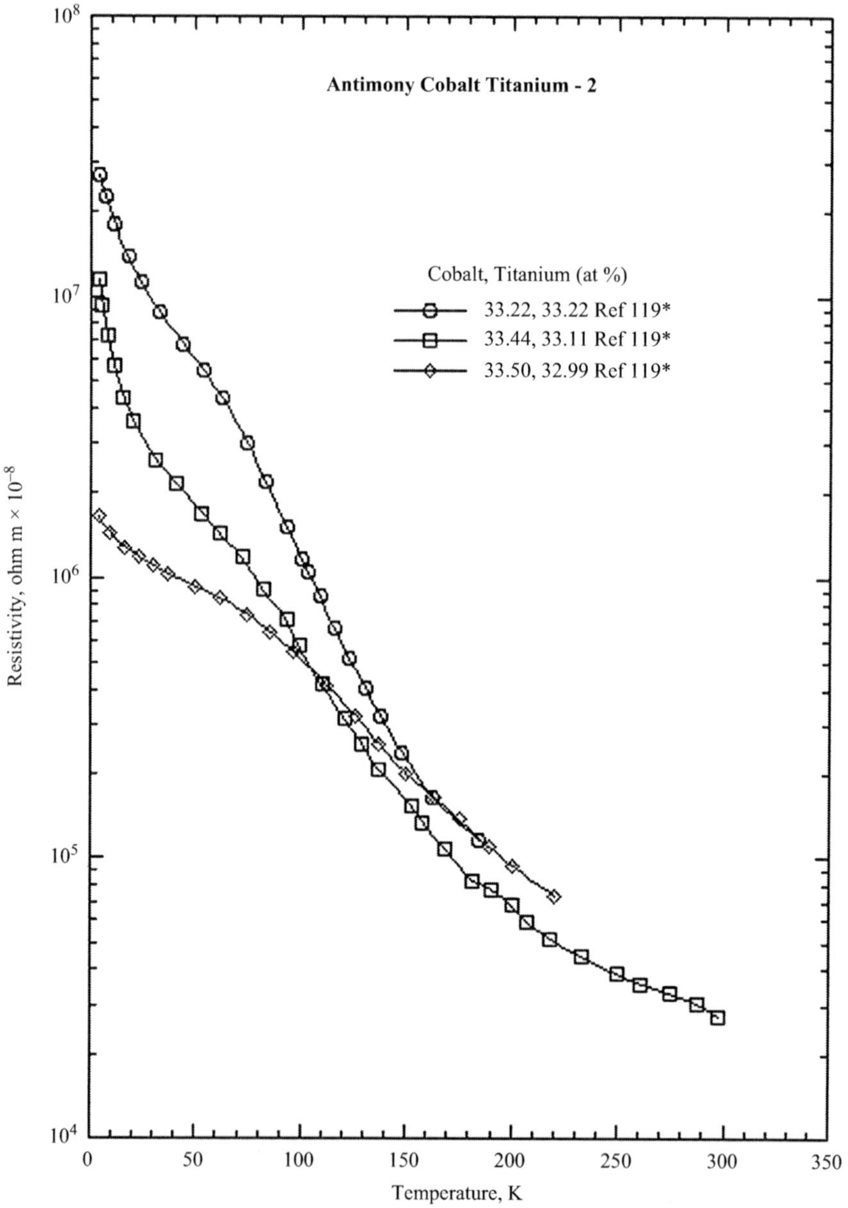
Antimony Cobalt Titanium - 2
Cobalt, Titanium (at %)
33.22, 33.22 Ref 119*
33.44, 33.11 Ref 119*
33.50, 32.99 Ref 119*
Resistivity, ohm m × 10⁻⁸
Temperature, K
10^8
10^7
10^6
10^5
10^4
0
50
100
150
200
250
300
350

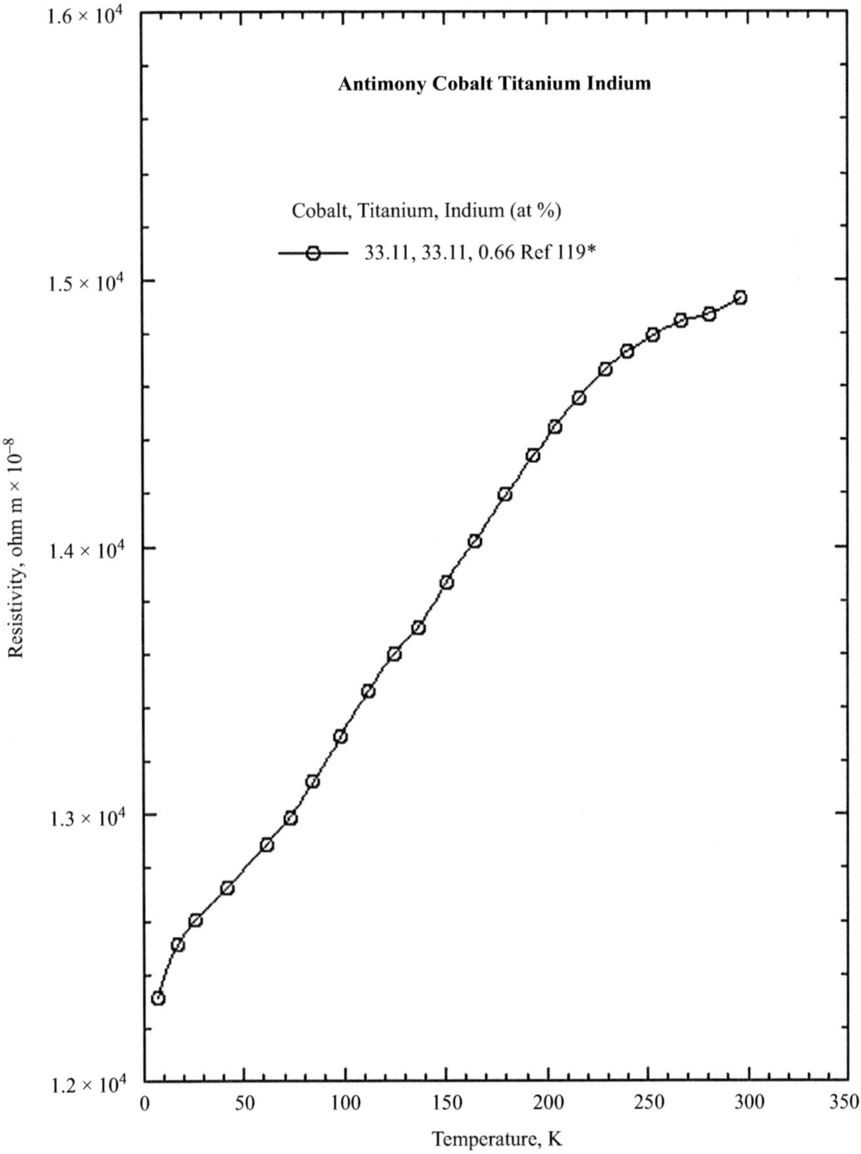
Antimony Cobalt Titanium Indium
Cobalt, Titanium, Indium (at %)
33.11, 33.11, 0.66 Ref 119*
Resistivity, ohm m × 10⁻⁸
1.6 × 10⁴
1.5 × 10⁴
1.4 × 10⁴
1.3 × 10⁴
1.2 × 10⁴
0
50
100
150
200
250
300
350
Temperature, K

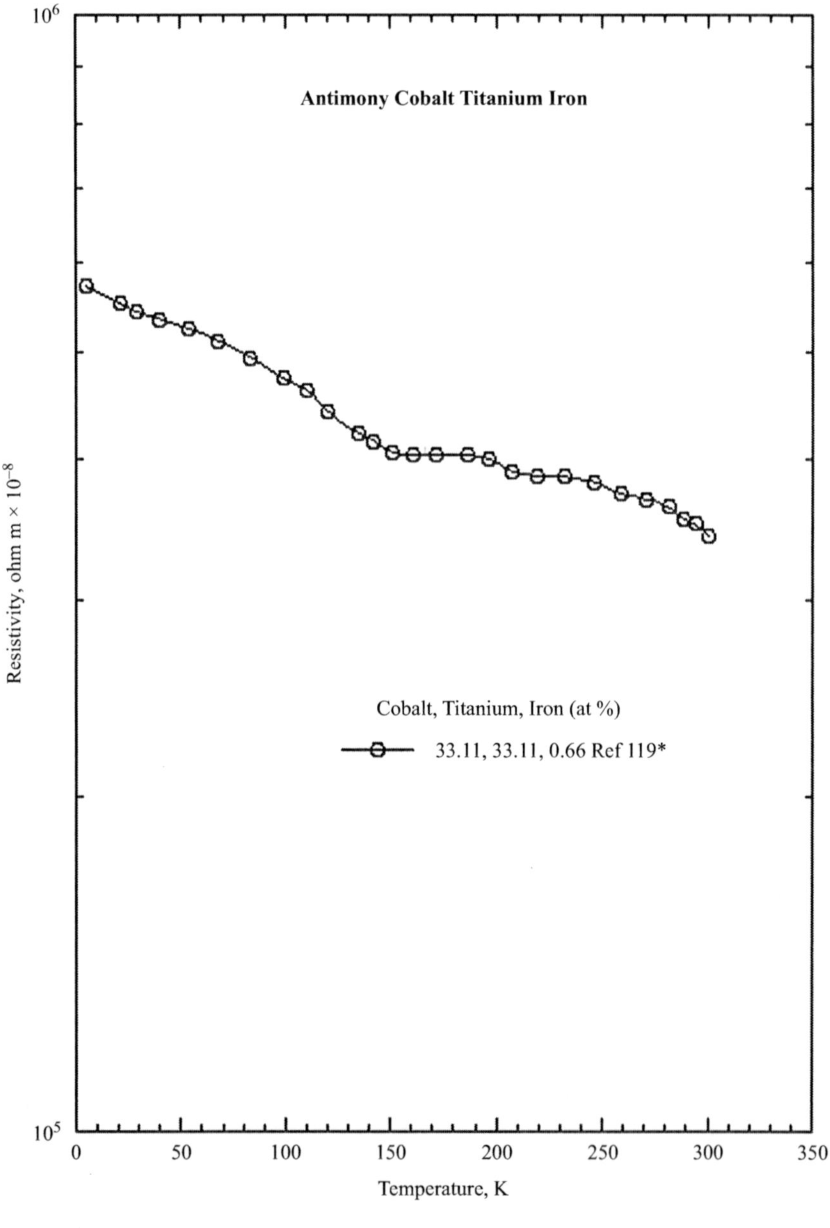

Antimony Cobalt Titanium Iron
Cobalt, Titanium, Iron (at %)
33.11, 33.11, 0.66 Ref 119*
Resistivity, ohm m × 10^-8
Temperature, K
10^6
10^5
0
50
100
150
200
250
300
350

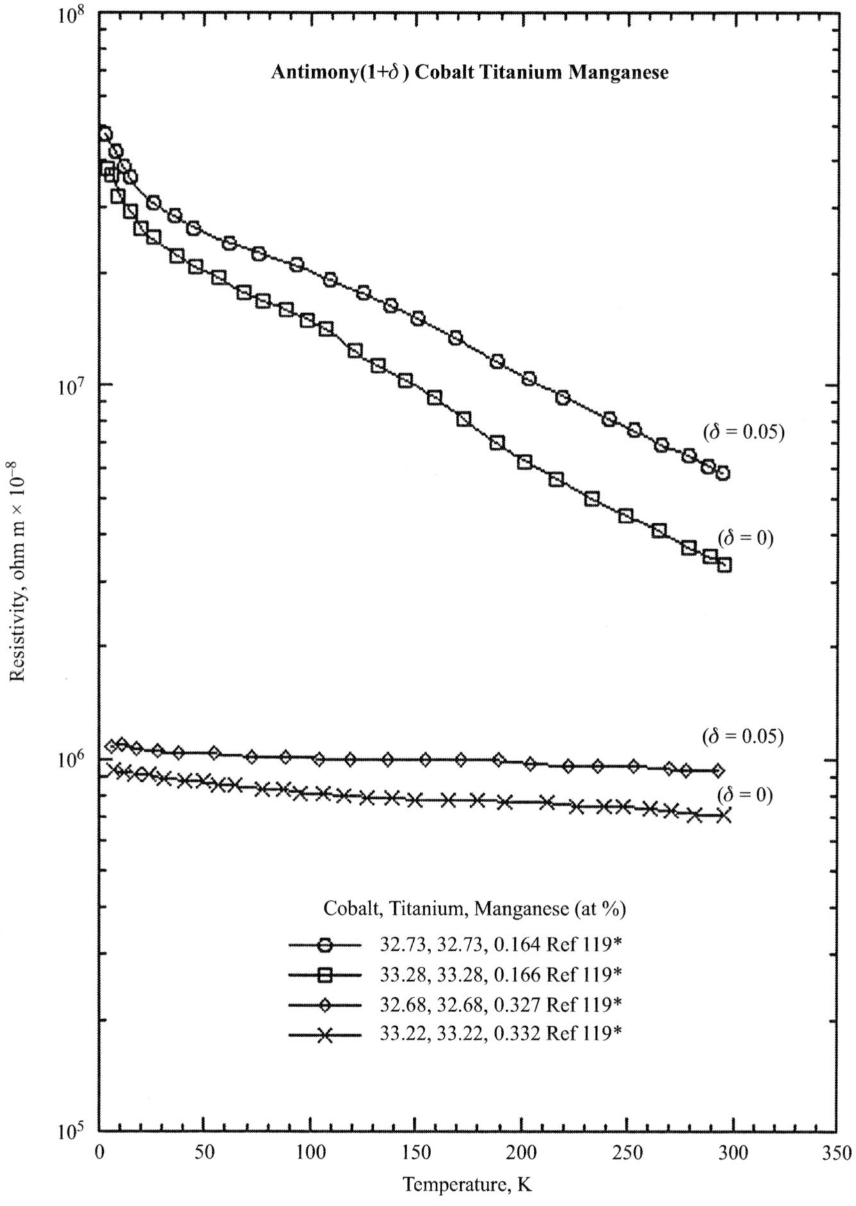
Antimony(1+δ) Cobalt Titanium Manganese
Resistivity, ohm m × 10⁻⁸
Temperature, K
10^8
10^7
10^6
10^5
0
50
100
150
200
250
300
350
(δ = 0.05)
(δ = 0)
(δ = 0.05)
(δ = 0)
Cobalt, Titanium, Manganese (at %)
32.73, 32.73, 0.164 Ref 119*
33.28, 33.28, 0.166 Ref 119*
32.68, 32.68, 0.327 Ref 119*
33.22, 33.22, 0.332 Ref 119*

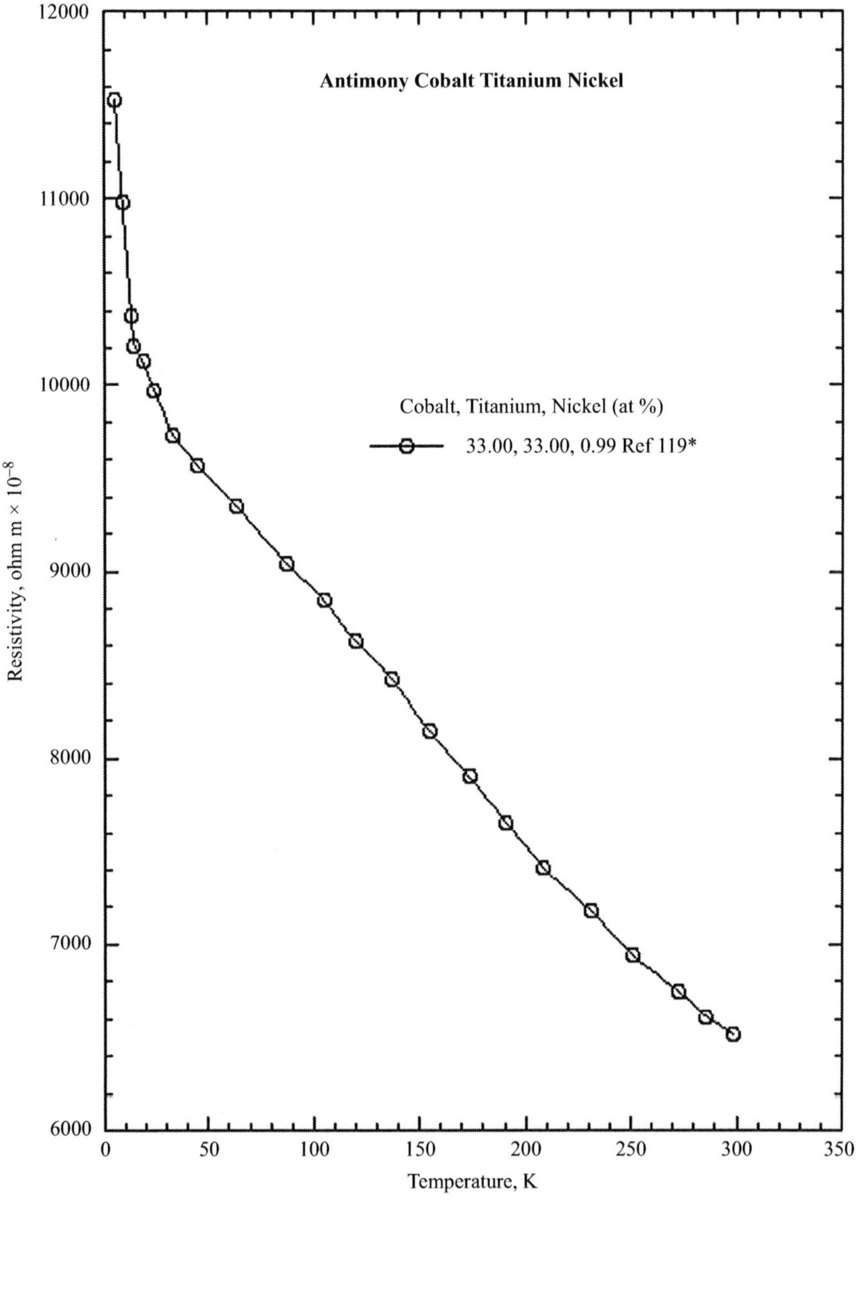
Antimony Cobalt Titanium Nickel
Cobalt, Titanium, Nickel (at %)
33.00, 33.00, 0.99 Ref 119*
Resistivity, ohm m × 10^−8
Temperature, K
12000
11000
10000
9000
8000
7000
6000
0
50
100
150
200
250
300
350

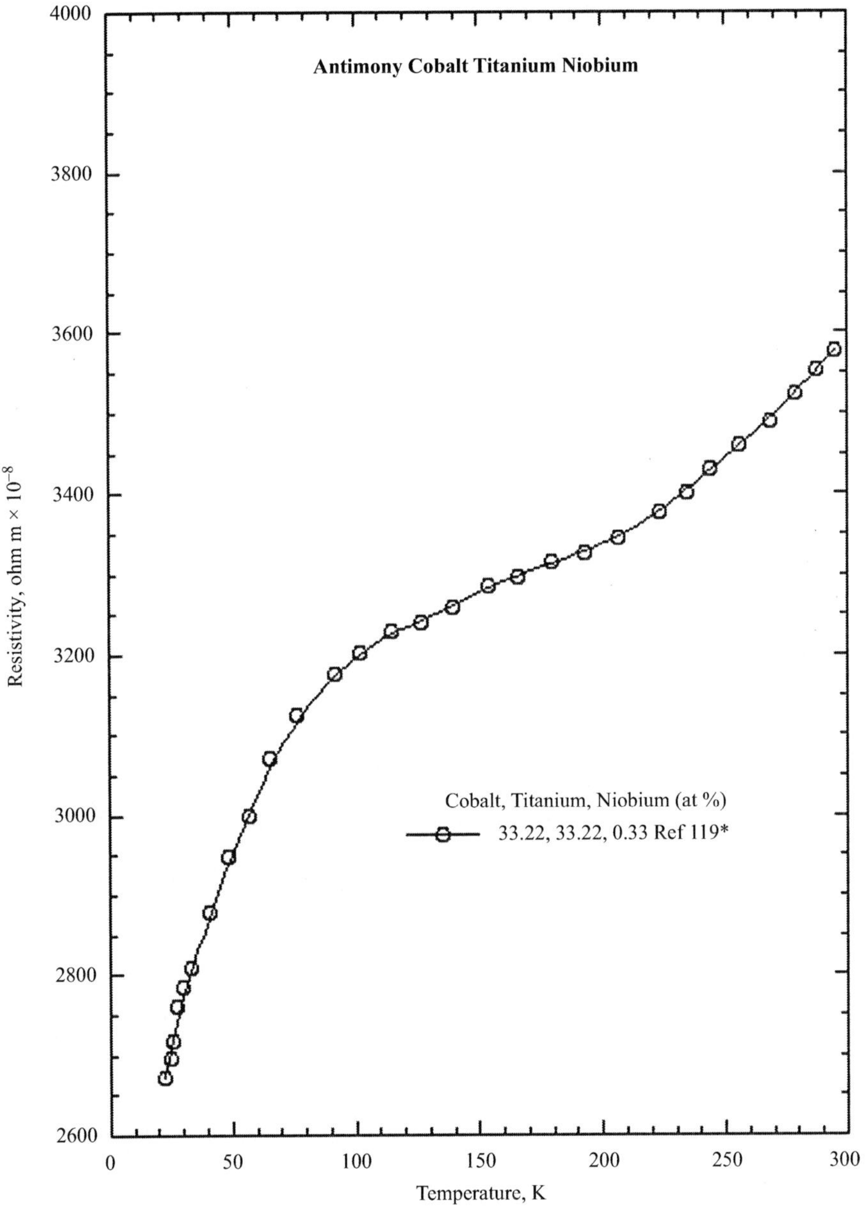
Antimony Cobalt Titanium Niobium
Cobalt, Titanium, Niobium (at %)
33.22, 33.22, 0.33 Ref 119*
Resistivity, ohm m × 10^{-8}
Temperature, K
4000
3800
3600
3400
3200
3000
2800
2600
0
50
100
150
200
250
300

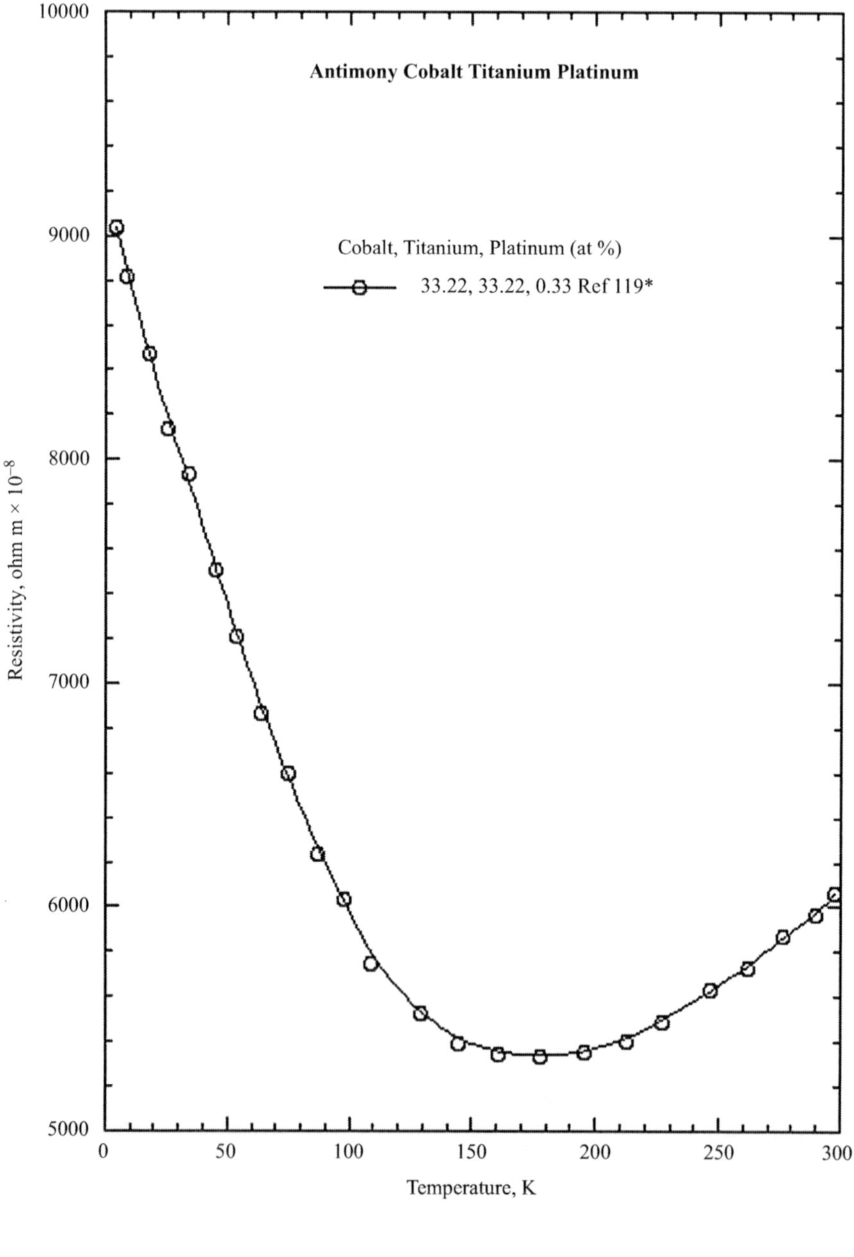
Antimony Cobalt Titanium Platinum
Cobalt, Titanium, Platinum (at %)
33.22, 33.22, 0.33 Ref 119*
Resistivity, ohm m × 10⁻⁸
10000
9000
8000
7000
6000
5000
0
50
100
150
200
250
300
Temperature, K

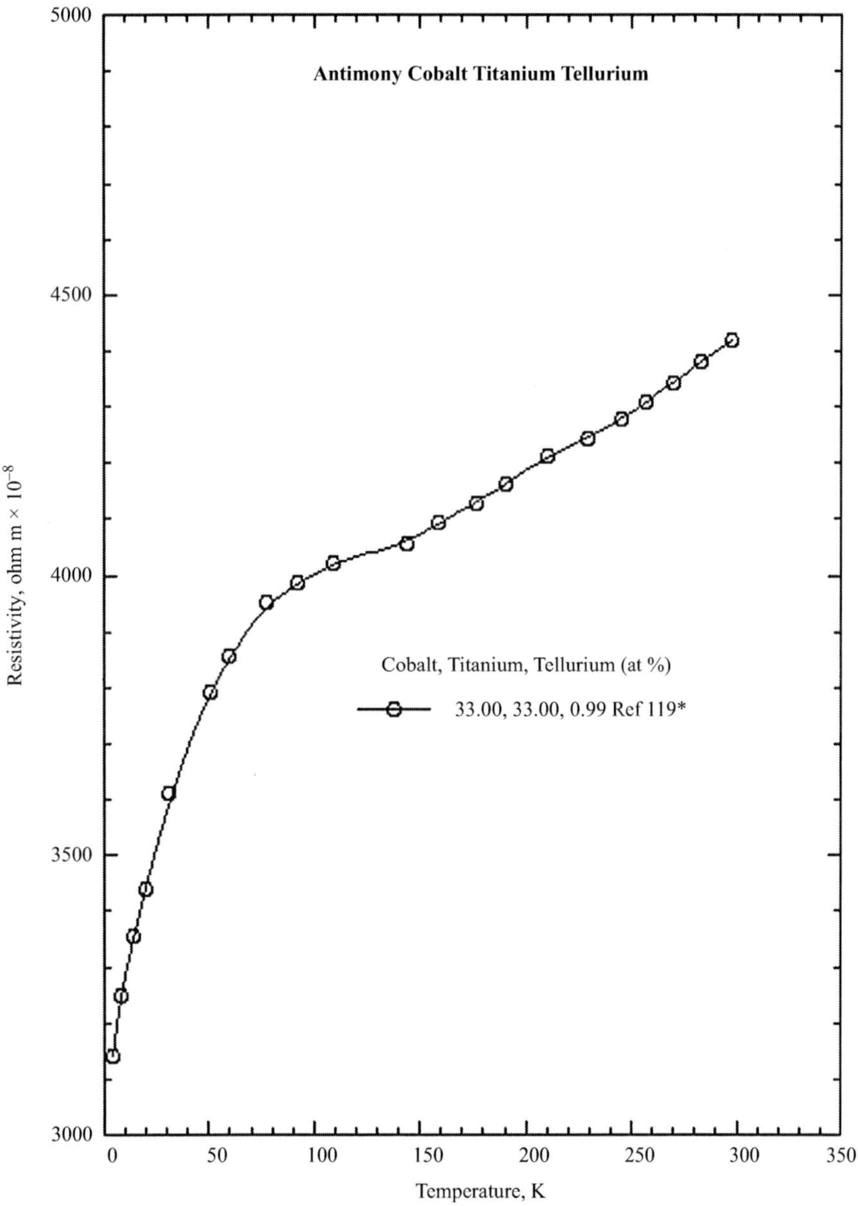
Antimony Cobalt Titanium Tellurium
Cobalt, Titanium, Tellurium (at %)
33.00, 33.00, 0.99 Ref 119*
Resistivity, ohm m × 10⁻⁸
Temperature, K
5000
4500
4000
3500
3000
0
50
100
150
200
250
300
350

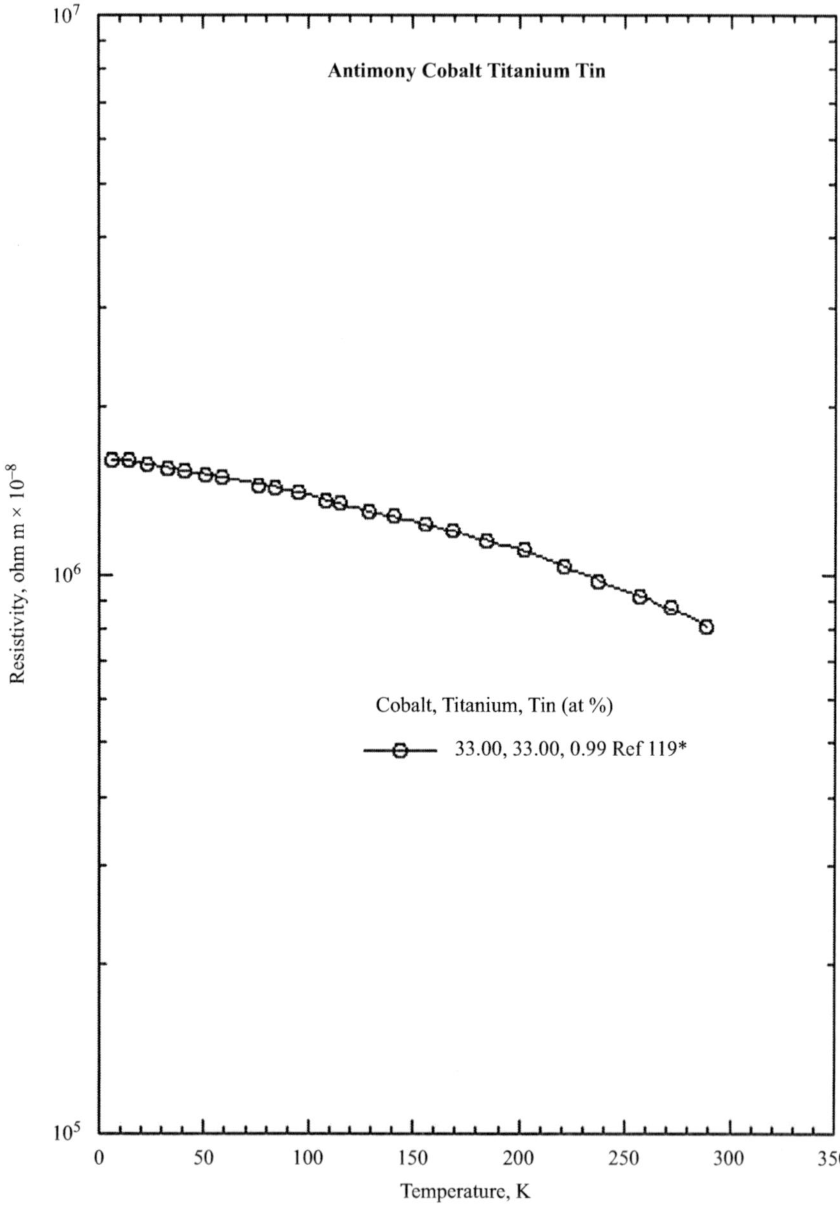

Antimony Cobalt Titanium Tin
Cobalt, Titanium, Tin (at %)
33.00, 33.00, 0.99 Ref 119*
Resistivity, ohm m × 10⁻⁸
10⁷
10⁶
10⁵
0
50
100
150
200
250
300
350
Temperature, K

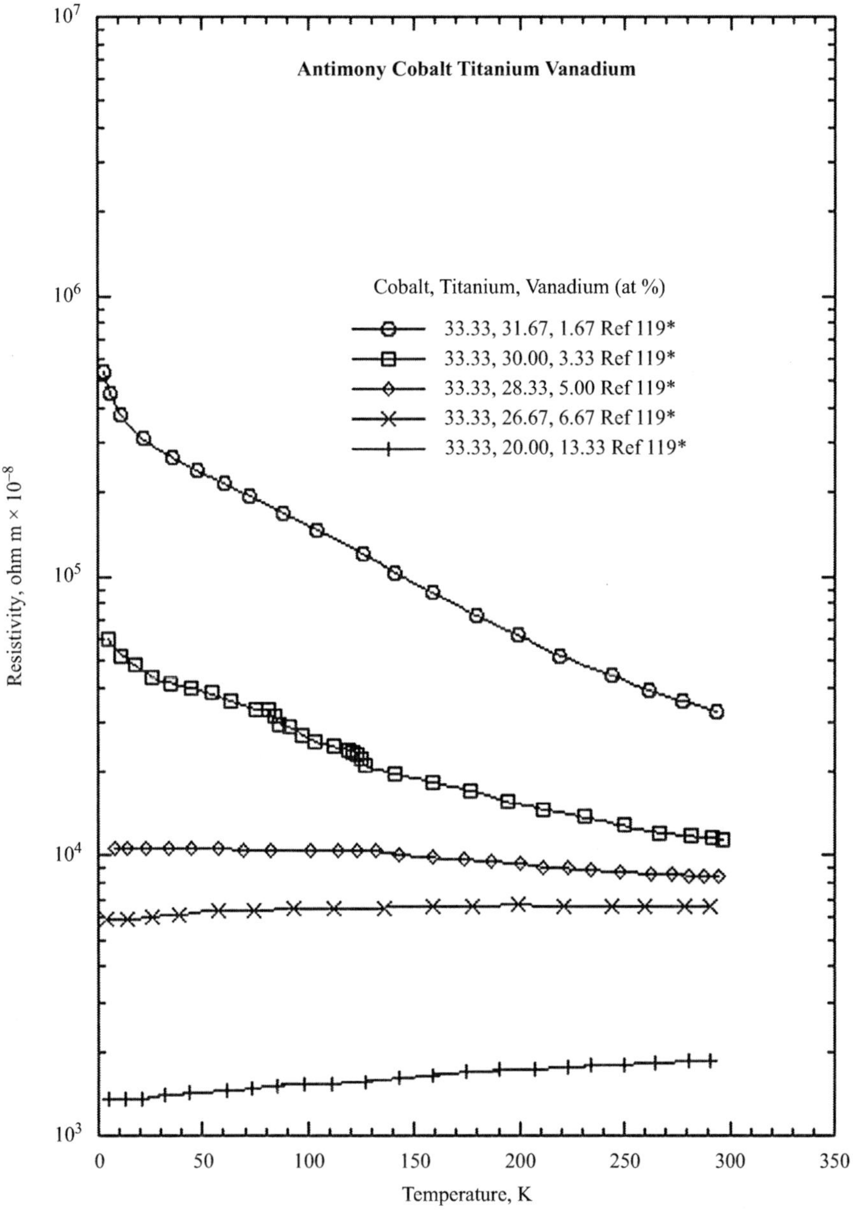

Antimony Cobalt Titanium Vanadium
Cobalt, Titanium, Vanadium (at %)
33.33, 31.67, 1.67 Ref 119*
33.33, 30.00, 3.33 Ref 119*
33.33, 28.33, 5.00 Ref 119*
33.33, 26.67, 6.67 Ref 119*
33.33, 20.00, 13.33 Ref 119*
Resistivity, ohm m × 10^-8
Temperature, K

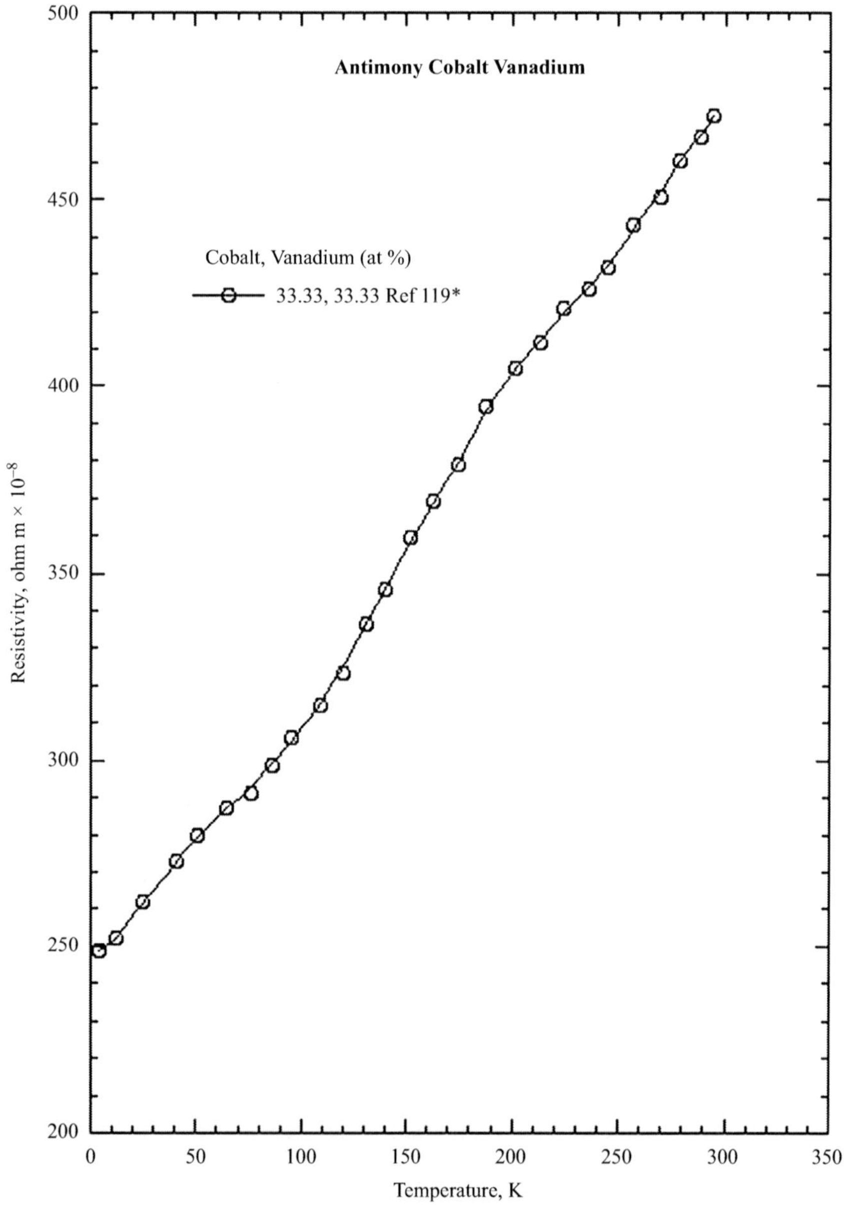
Antimony Cobalt Vanadium
Cobalt, Vanadium (at %)
33.33, 33.33 Ref 119*
Resistivity, ohm m × 10⁻⁸
Temperature, K
500
450
400
350
300
250
200
0
50
100
150
200
250
300
350

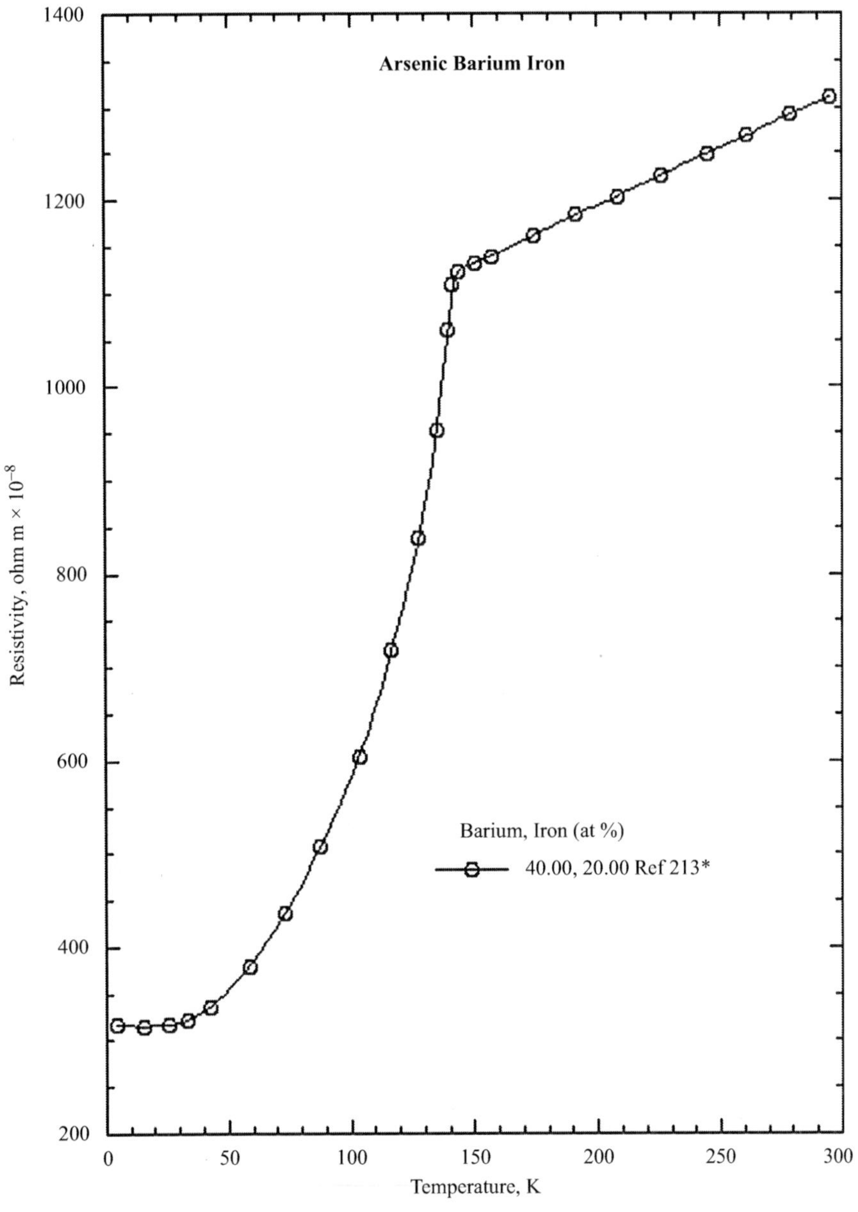
Arsenic Barium Iron
Resistivity, ohm m × 10⁻⁸
Temperature, K
Barium, Iron (at %)
40.00, 20.00 Ref 213*
1400
1200
1000
800
600
400
200
0
50
100
150
200
250
300

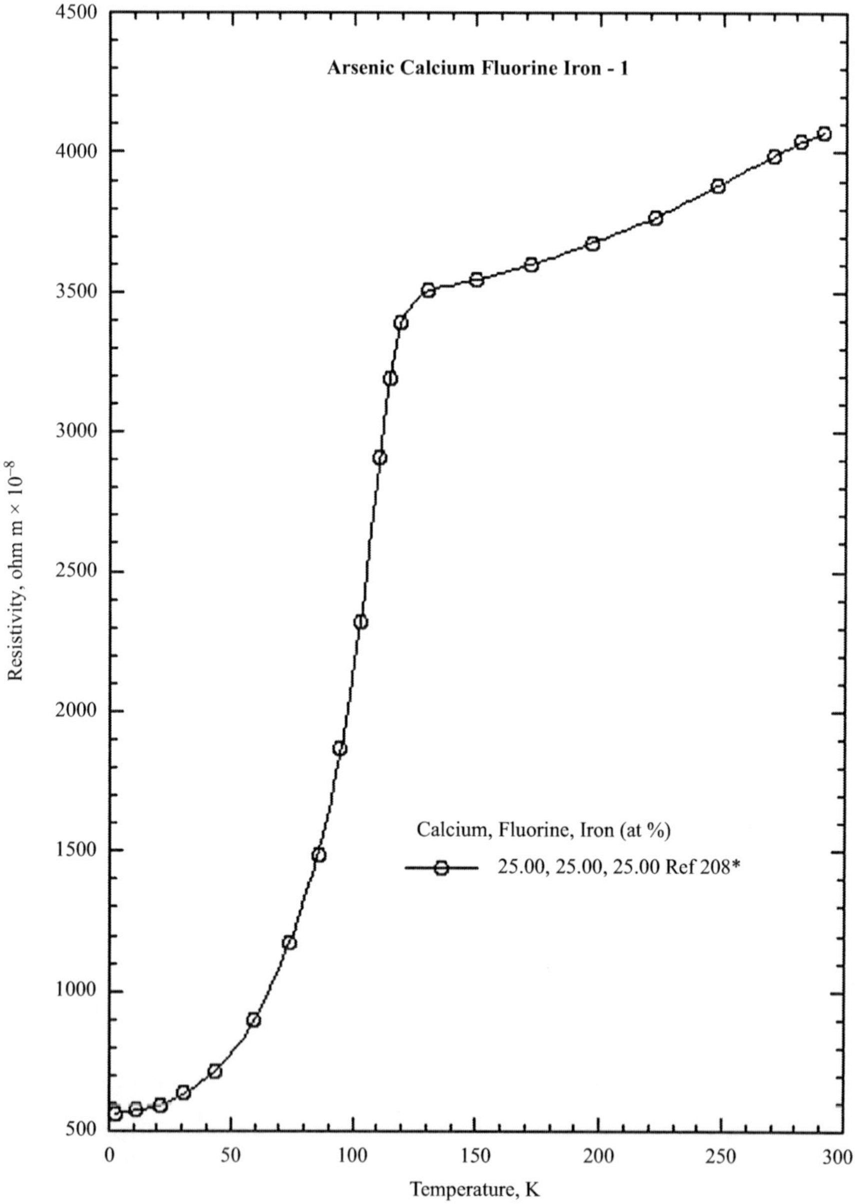
Arsenic Calcium Fluorine Iron - 1
Calcium, Fluorine, Iron (at %)
25.00, 25.00, 25.00 Ref 208*
Resistivity, ohm m × 10⁻⁸
Temperature, K
4500
4000
3500
3000
2500
2000
1500
1000
500
0
50
100
150
200
250
300

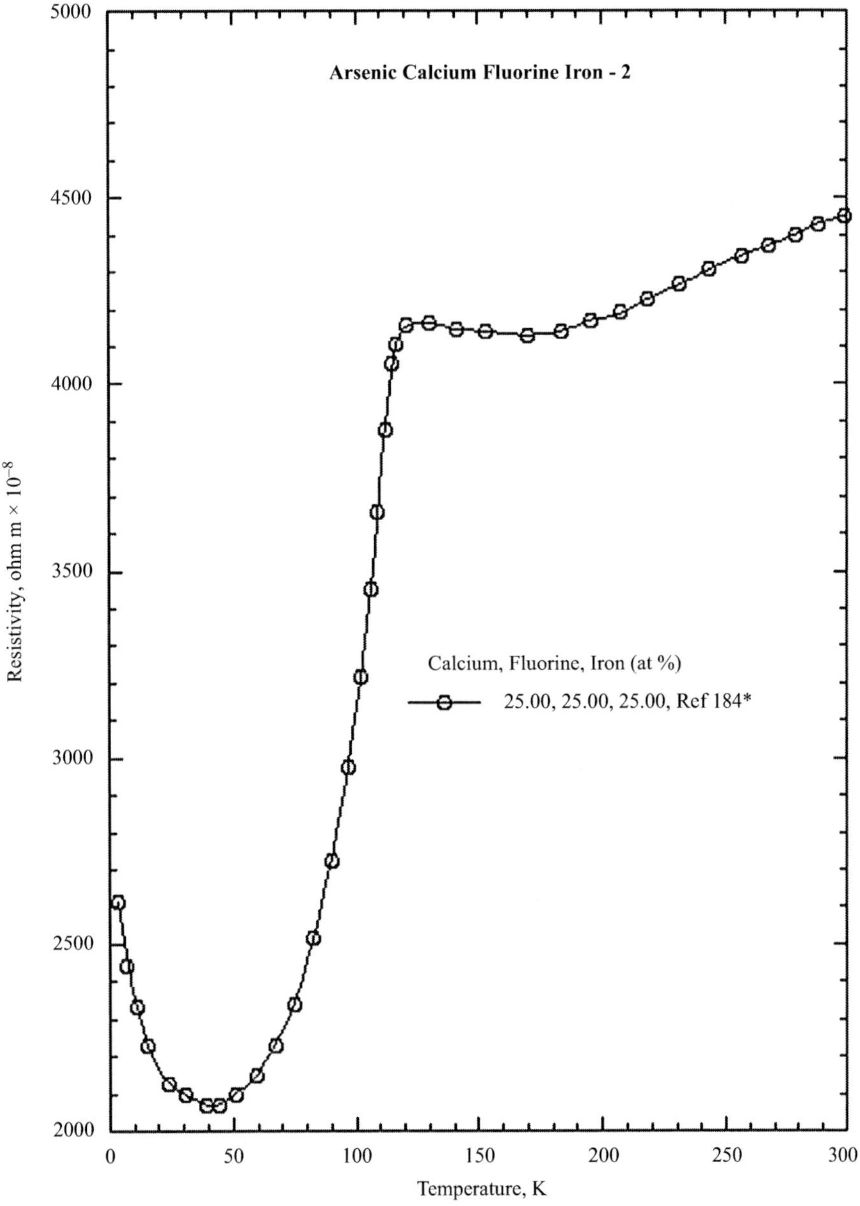
Arsenic Calcium Fluorine Iron - 2
Calcium, Fluorine, Iron (at %)
25.00, 25.00, 25.00, Ref 184*
Resistivity, ohm m × 10⁻⁸
Temperature, K
5000
4500
4000
3500
3000
2500
2000
0
50
100
150
200
250
300

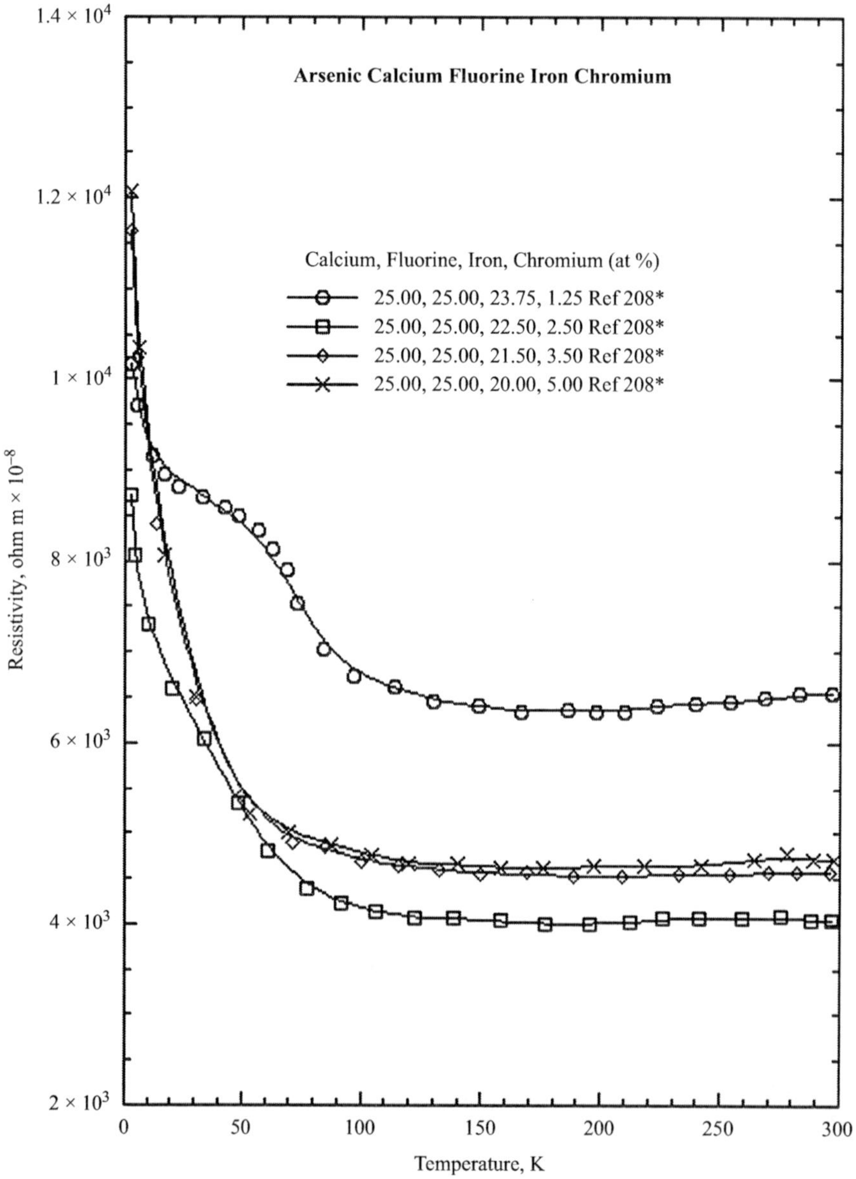
Arsenic Calcium Fluorine Iron Chromium
Calcium, Fluorine, Iron, Chromium (at %)
25.00, 25.00, 23.75, 1.25 Ref 208*
25.00, 25.00, 22.50, 2.50 Ref 208*
25.00, 25.00, 21.50, 3.50 Ref 208*
25.00, 25.00, 20.00, 5.00 Ref 208*
1.4 × 10⁴
1.2 × 10⁴
1 × 10⁴
8 × 10³
6 × 10³
4 × 10³
2 × 10³
Resistivity, ohm m × 10⁻⁸
0
50
100
150
200
250
300
Temperature, K

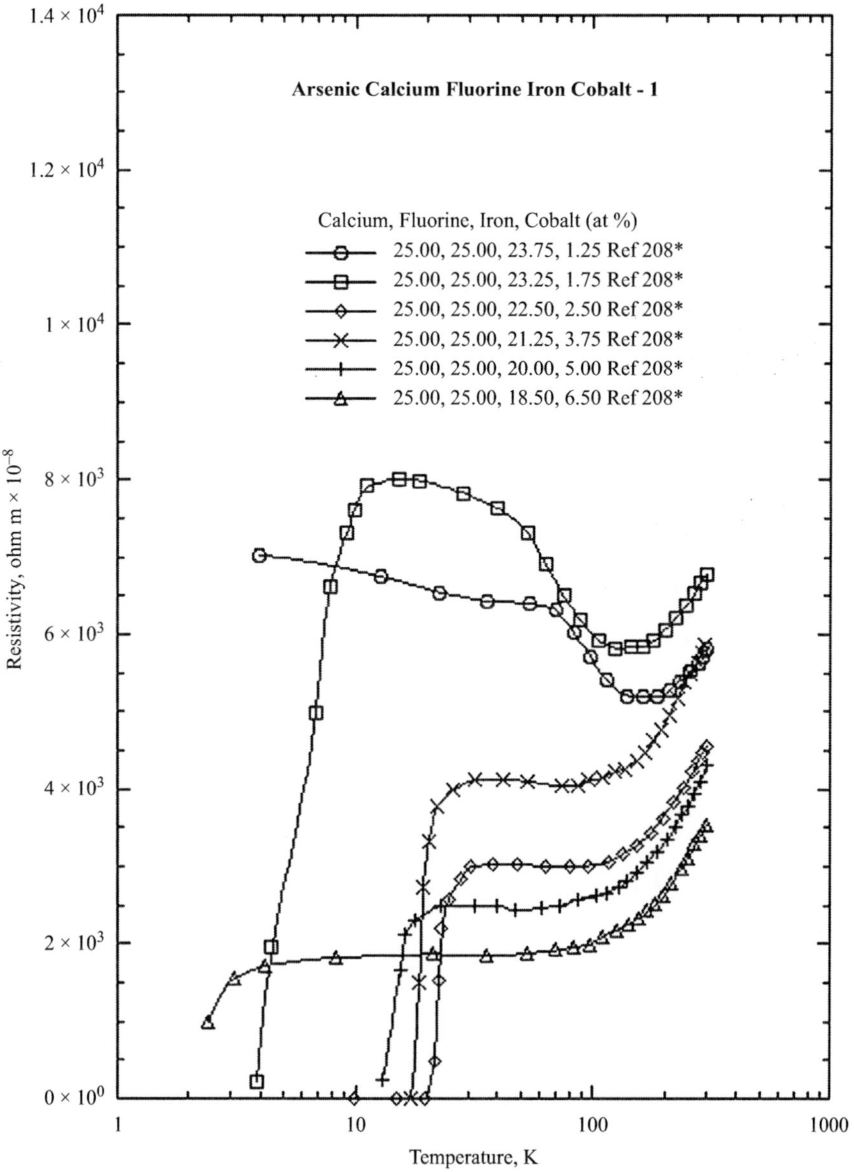
Arsenic Calcium Fluorine Iron Cobalt - 1
Calcium, Fluorine, Iron, Cobalt (at %)
25.00, 25.00, 23.75, 1.25 Ref 208*
25.00, 25.00, 23.25, 1.75 Ref 208*
25.00, 25.00, 22.50, 2.50 Ref 208*
25.00, 25.00, 21.25, 3.75 Ref 208*
25.00, 25.00, 20.00, 5.00 Ref 208*
25.00, 25.00, 18.50, 6.50 Ref 208*
Resistivity, ohm m × 10⁻⁸
Temperature, K
1.4 × 10⁴
1.2 × 10⁴
1 × 10⁴
8 × 10³
6 × 10³
4 × 10³
2 × 10³
0 × 10⁰
1
10
100
1000

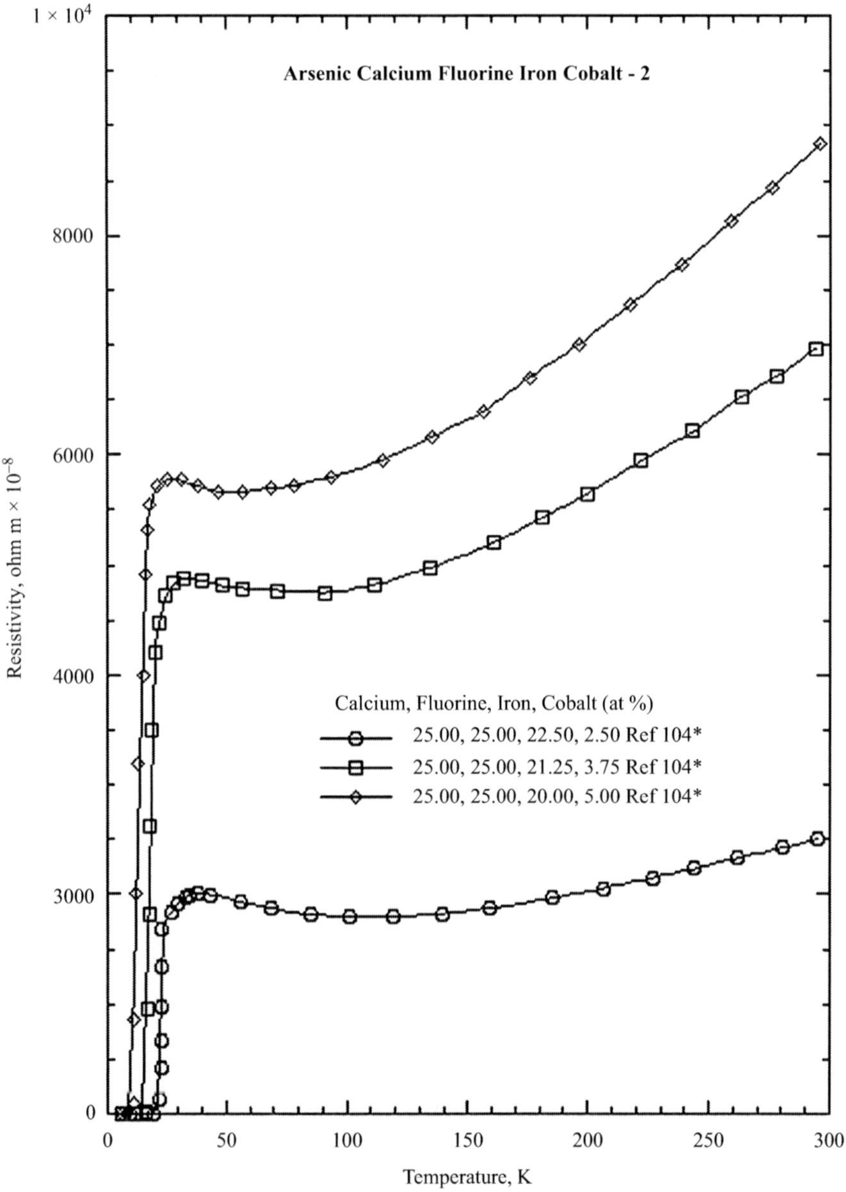
Arsenic Calcium Fluorine Iron Cobalt - 2
Calcium, Fluorine, Iron, Cobalt (at %)
25.00, 25.00, 22.50, 2.50 Ref 104*
25.00, 25.00, 21.25, 3.75 Ref 104*
25.00, 25.00, 20.00, 5.00 Ref 104*
Resistivity, ohm m × 10⁻⁸
Temperature, K
1 × 10⁴
8000
6000
4000
3000
0
0
50
100
150
200
250
300

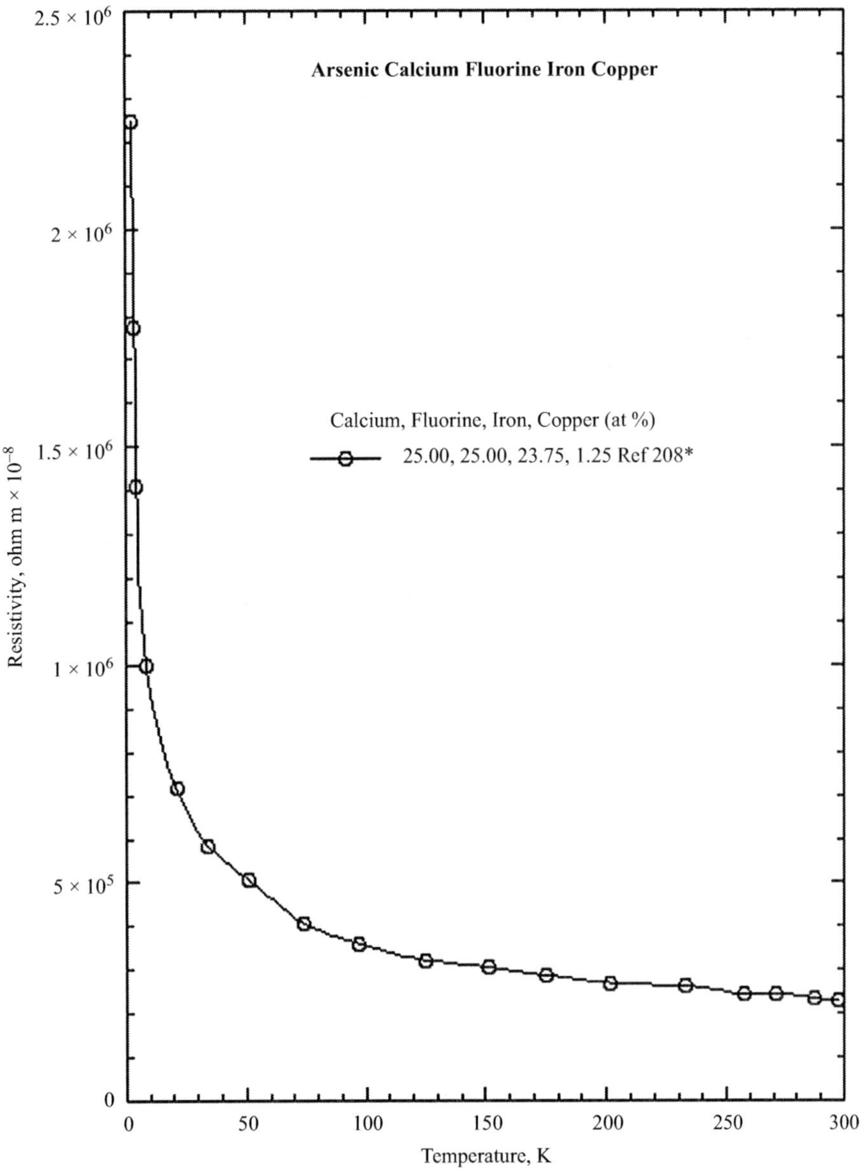

Arsenic Calcium Fluorine Iron Copper
Calcium, Fluorine, Iron, Copper (at %)
25.00, 25.00, 23.75, 1.25 Ref 208*
2.5 × 10^6
2 × 10^6
1.5 × 10^6
1 × 10^6
5 × 10^5
0
Resistivity, ohm m × 10^−8
0
50
100
150
200
250
300
Temperature, K

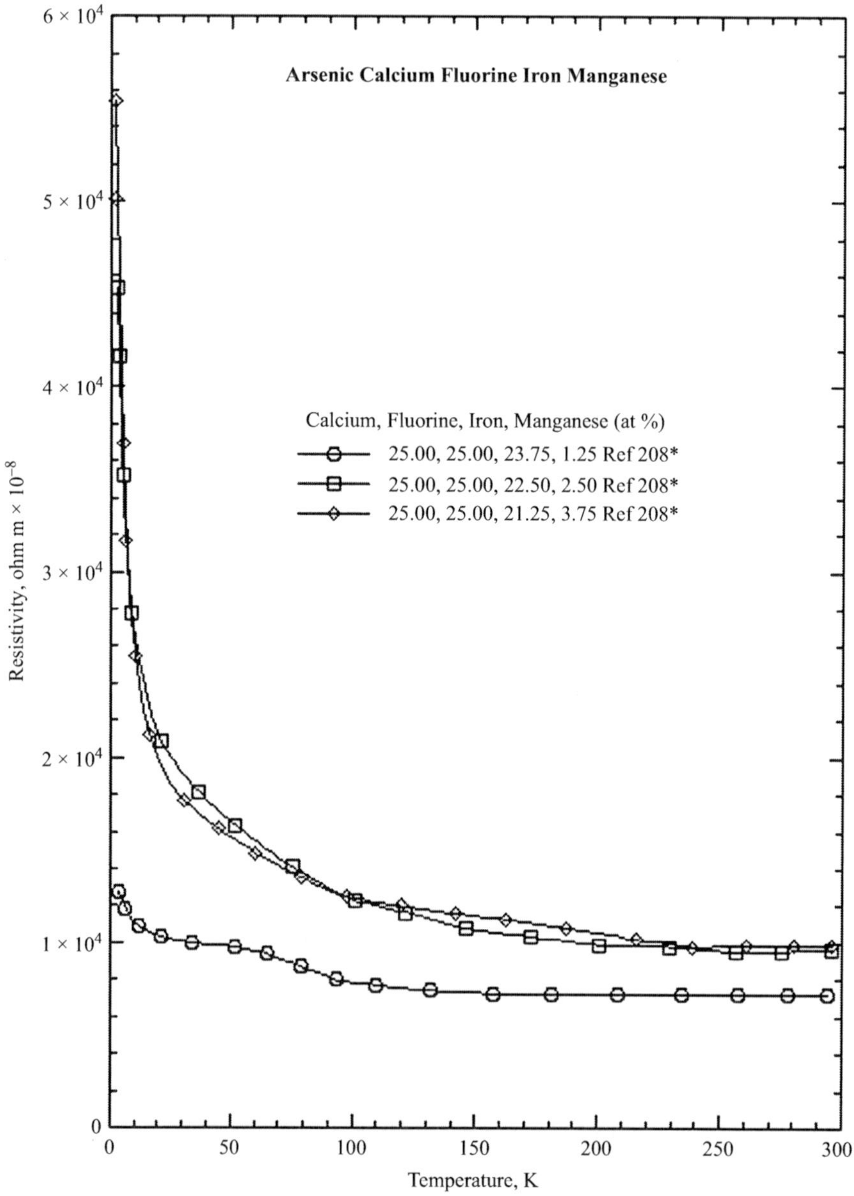
Arsenic Calcium Fluorine Iron Manganese
Calcium, Fluorine, Iron, Manganese (at %)
25.00, 25.00, 23.75, 1.25 Ref 208*
25.00, 25.00, 22.50, 2.50 Ref 208*
25.00, 25.00, 21.25, 3.75 Ref 208*
Resistivity, ohm m × 10⁻⁸
Temperature, K
6 × 10⁴
5 × 10⁴
4 × 10⁴
3 × 10⁴
2 × 10⁴
1 × 10⁴
0
50
100
150
200
250
300

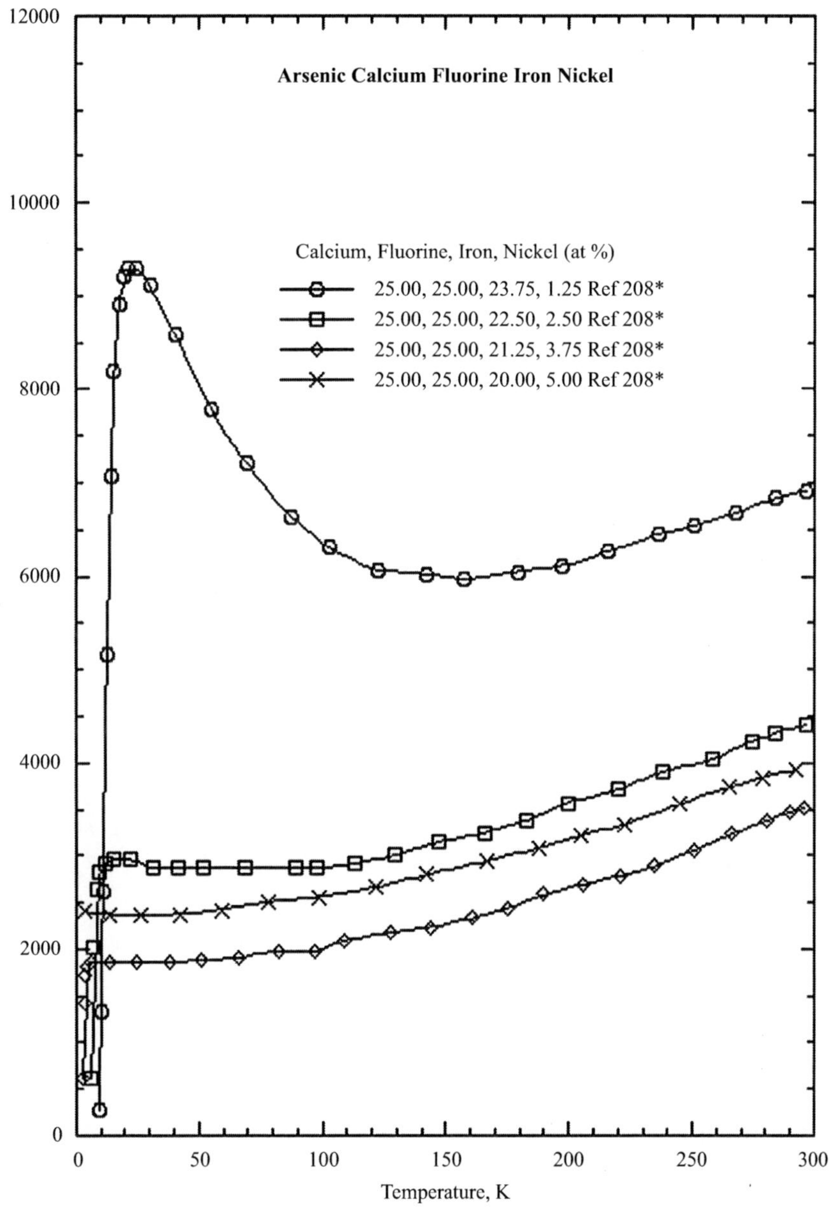
Arsenic Calcium Fluorine Iron Nickel
Calcium, Fluorine, Iron, Nickel (at %)
25.00, 25.00, 23.75, 1.25 Ref 208*
25.00, 25.00, 22.50, 2.50 Ref 208*
25.00, 25.00, 21.25, 3.75 Ref 208*
25.00, 25.00, 20.00, 5.00 Ref 208*
Resistivity, ohm m × 10−8
Temperature, K
0
2000
4000
6000
8000
10000
12000
0
50
100
150
200
250
300

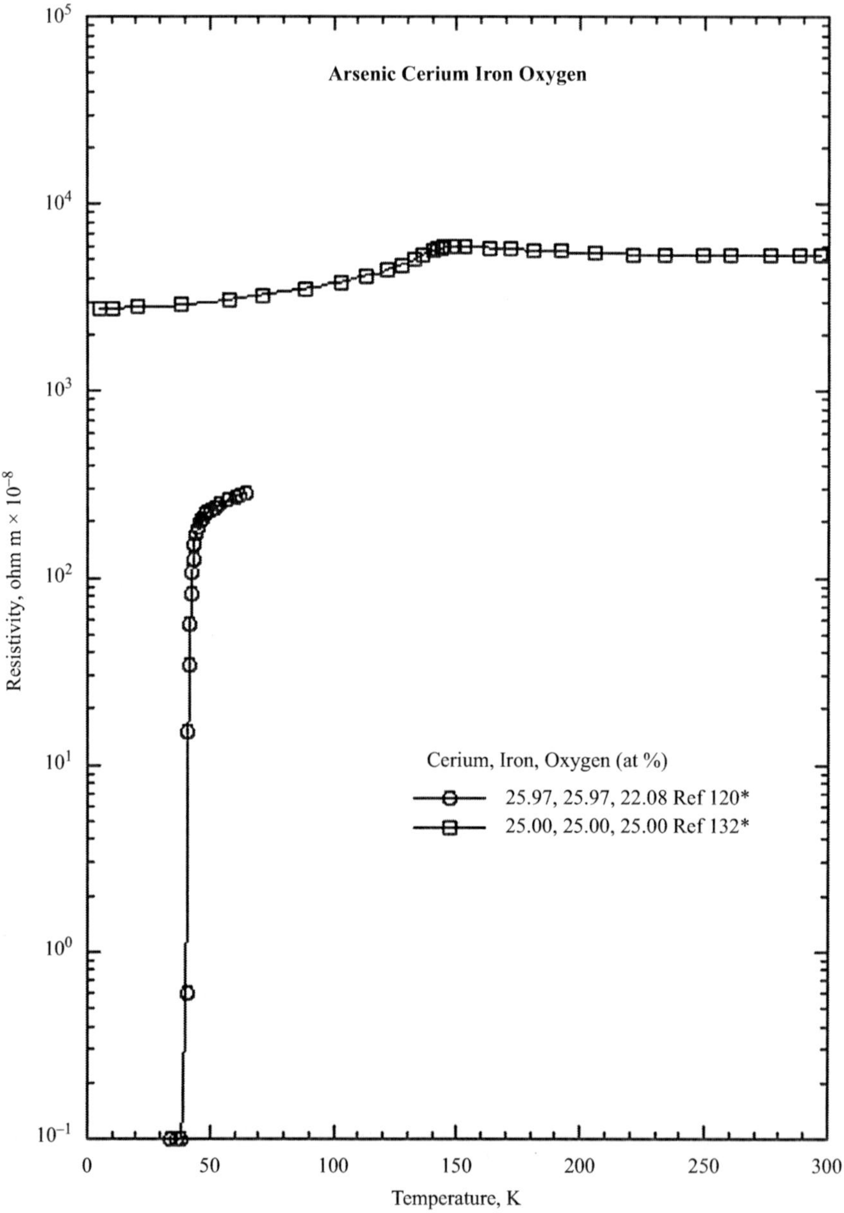
Arsenic Cerium Iron Oxygen
Cerium, Iron, Oxygen (at %)
25.97, 25.97, 22.08 Ref 120*
25.00, 25.00, 25.00 Ref 132*
Resistivity, ohm m × 10⁻⁸
Temperature, K
10⁵
10⁴
10³
10²
10¹
10⁰
10⁻¹
0
50
100
150
200
250
300

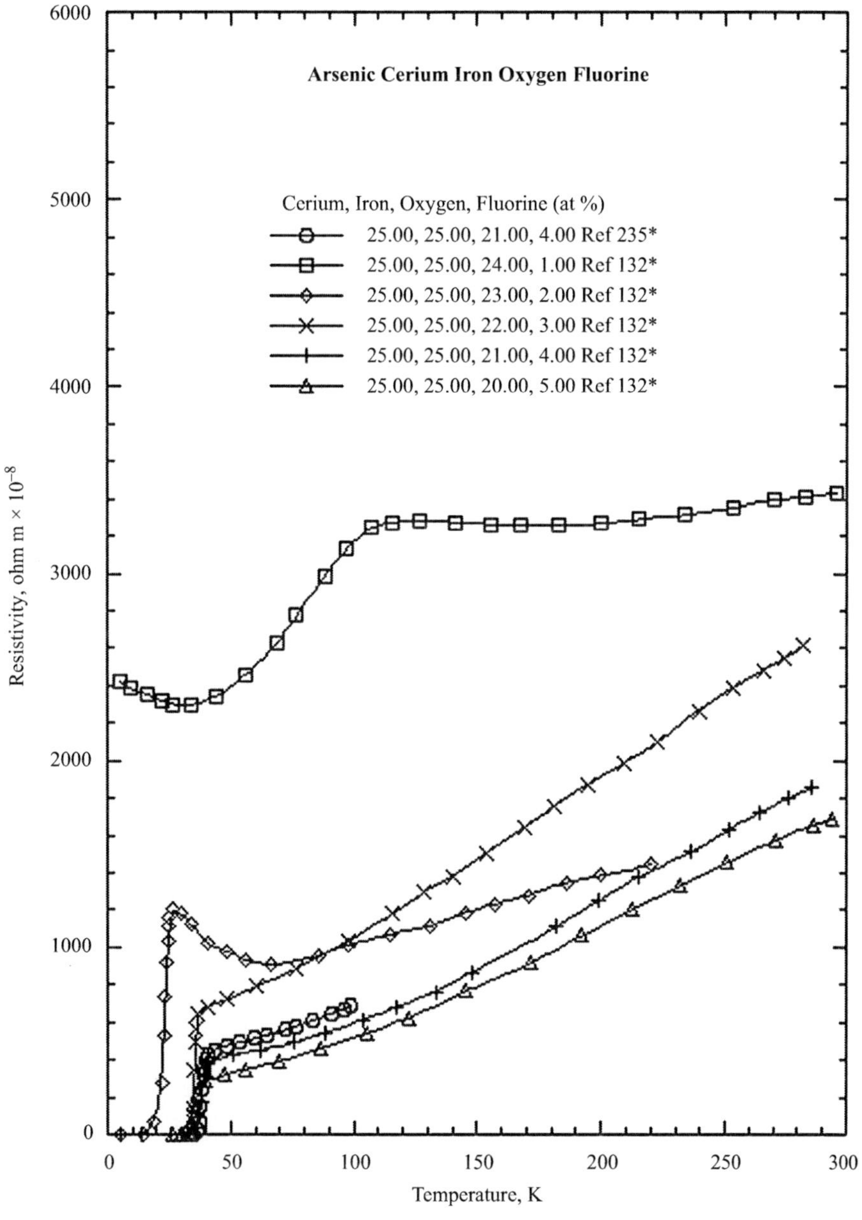
Arsenic Cerium Iron Oxygen Fluorine
Cerium, Iron, Oxygen, Fluorine (at %)
25.00, 25.00, 21.00, 4.00 Ref 235*
25.00, 25.00, 24.00, 1.00 Ref 132*
25.00, 25.00, 23.00, 2.00 Ref 132*
25.00, 25.00, 22.00, 3.00 Ref 132*
25.00, 25.00, 21.00, 4.00 Ref 132*
25.00, 25.00, 20.00, 5.00 Ref 132*
Resistivity, ohm m × 10⁻⁸
Temperature, K
0
1000
2000
3000
4000
5000
6000
0
50
100
150
200
250
300

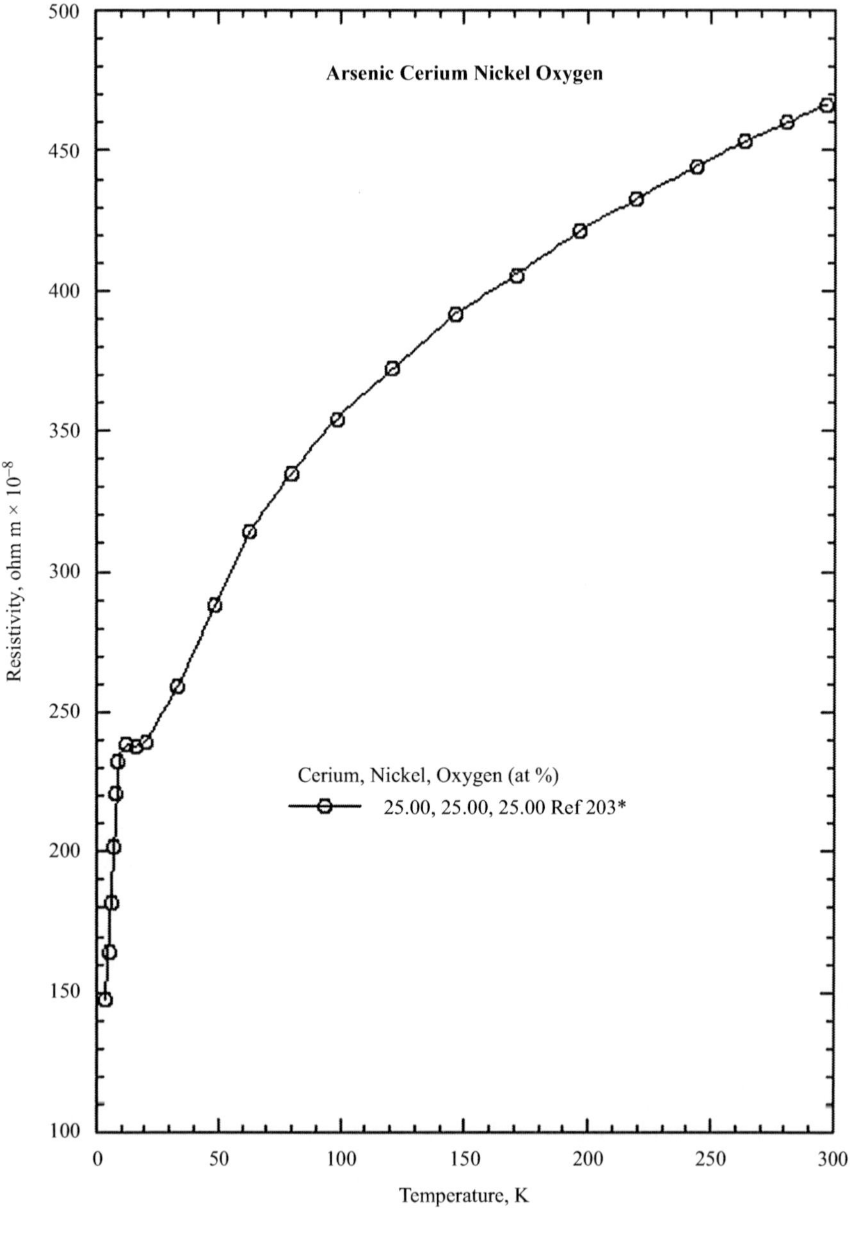
Arsenic Cerium Nickel Oxygen
Cerium, Nickel, Oxygen (at %)
25.00, 25.00, 25.00 Ref 203*
Resistivity, ohm m × 10^−8
Temperature, K
500
450
400
350
300
250
200
150
100
0
50
100
150
200
250
300

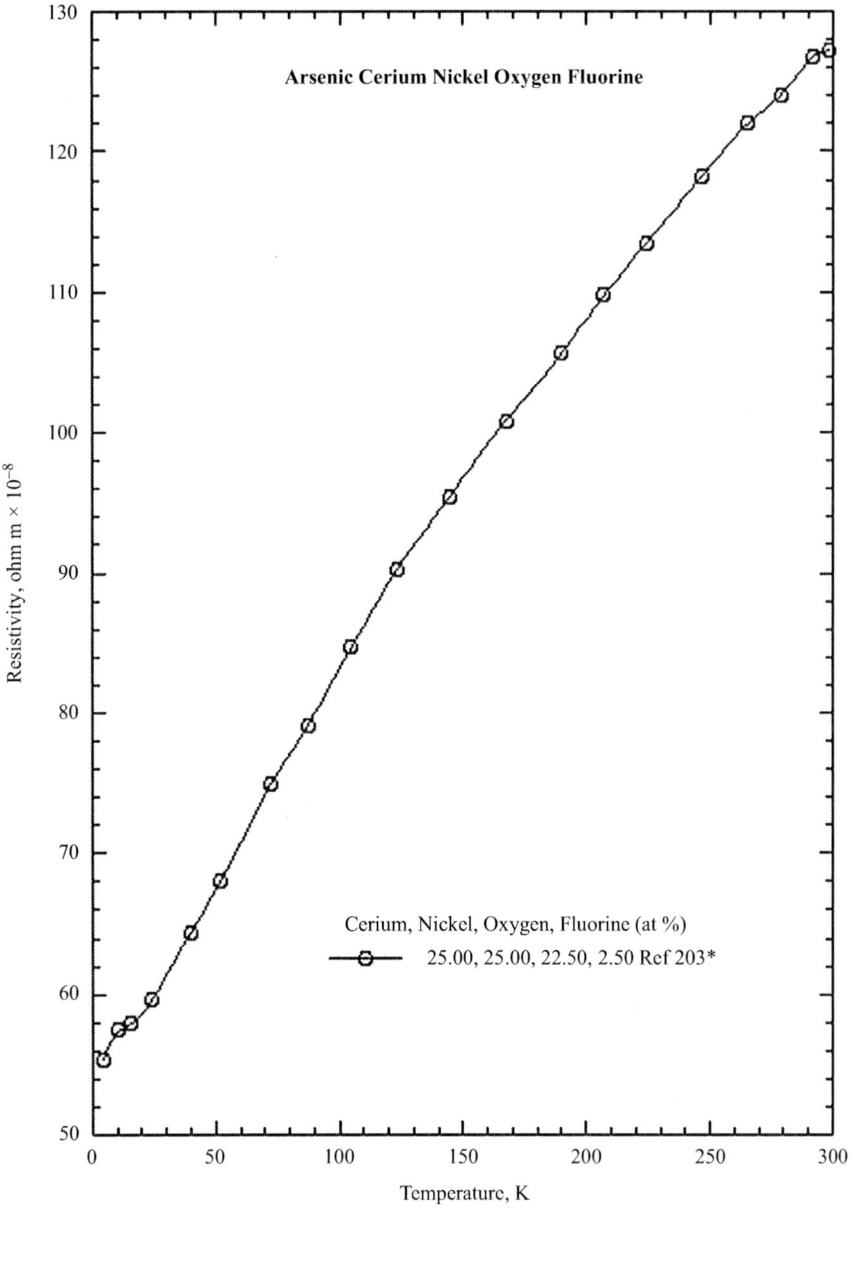

Arsenic Cerium Nickel Oxygen Fluorine
Cerium, Nickel, Oxygen, Fluorine (at %)
25.00, 25.00, 22.50, 2.50 Ref 203*
Resistivity, ohm m × 10⁻⁸
Temperature, K
130
120
110
100
90
80
70
60
50
0
50
100
150
200
250
300

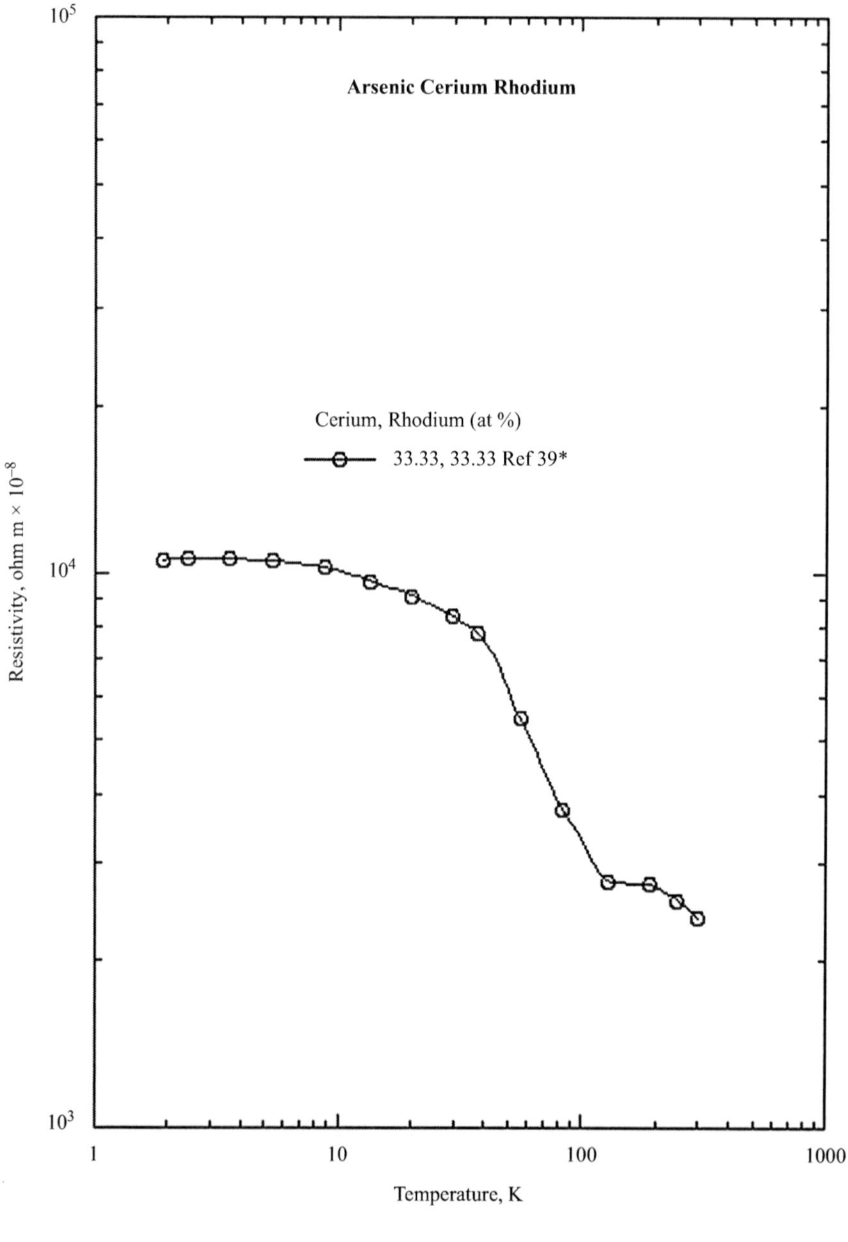
Arsenic Cerium Rhodium
Cerium, Rhodium (at %)
33.33, 33.33 Ref 39*
Resistivity, ohm m × 10⁻⁸
10⁵
10⁴
10³
1
10
100
1000
Temperature, K

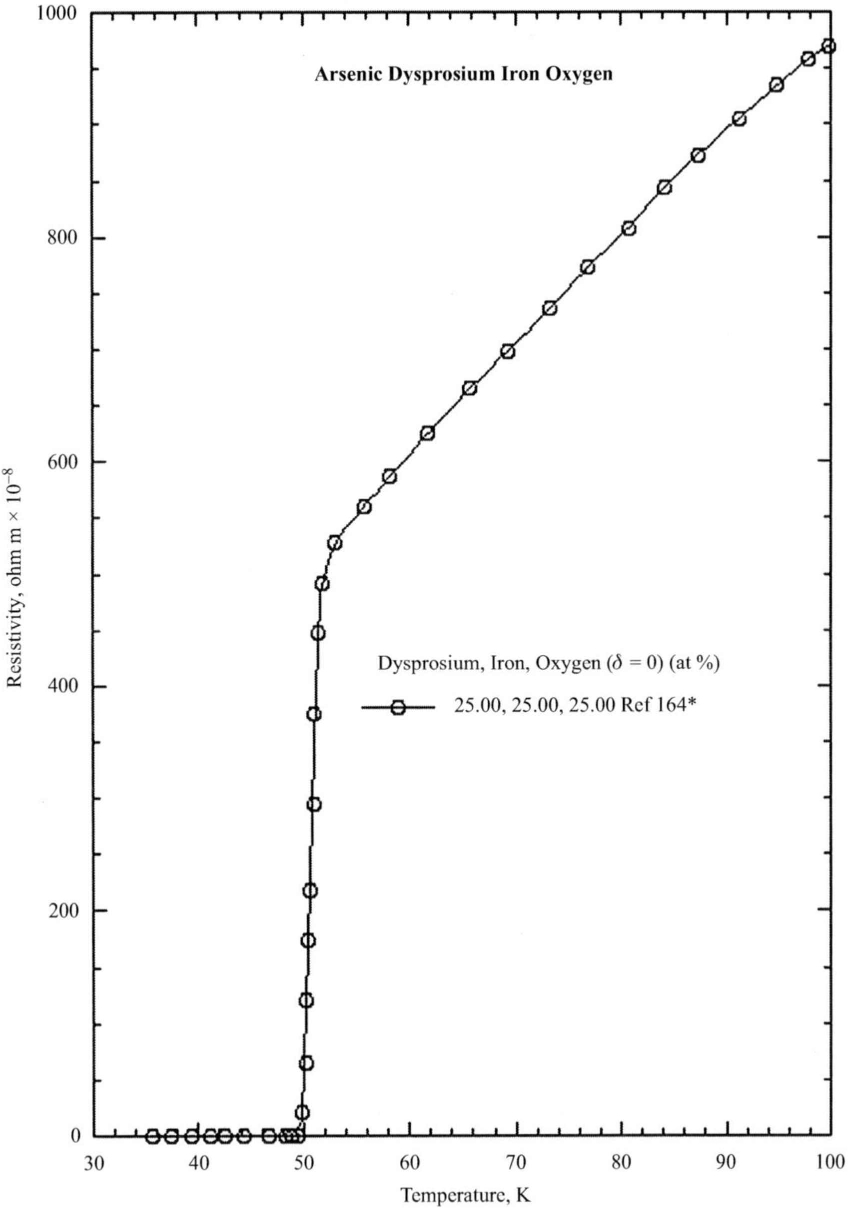
Arsenic Dysprosium Iron Oxygen
Dysprosium, Iron, Oxygen (δ = 0) (at %)
25.00, 25.00, 25.00 Ref 164*
Resistivity, ohm m × 10⁻⁸
Temperature, K
0
200
400
600
800
1000
30
40
50
60
70
80
90
100

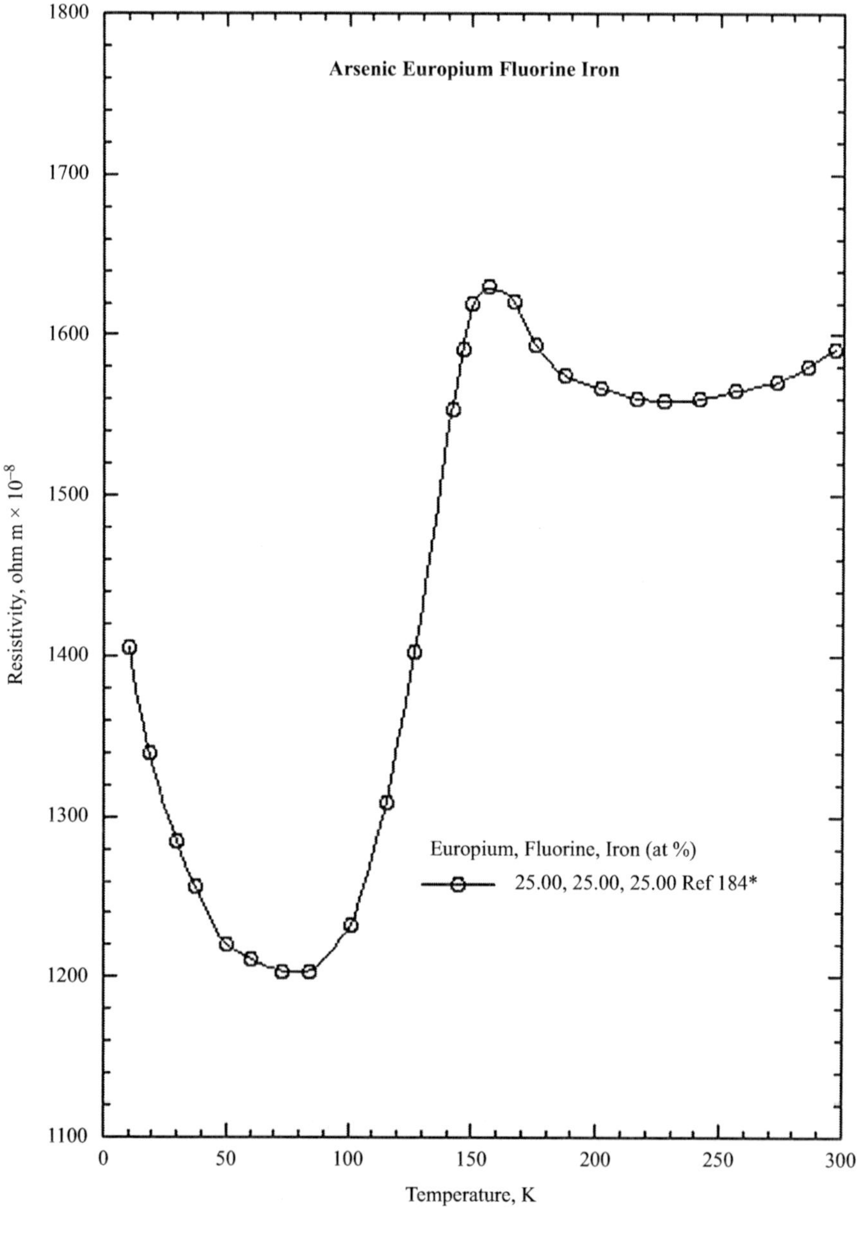

Arsenic Europium Fluorine Iron
Europium, Fluorine, Iron (at %)
25.00, 25.00, 25.00 Ref 184*
Resistivity, ohm m × 10⁻⁸
1800
1700
1600
1500
1400
1300
1200
1100
0
50
100
150
200
250
300
Temperature, K

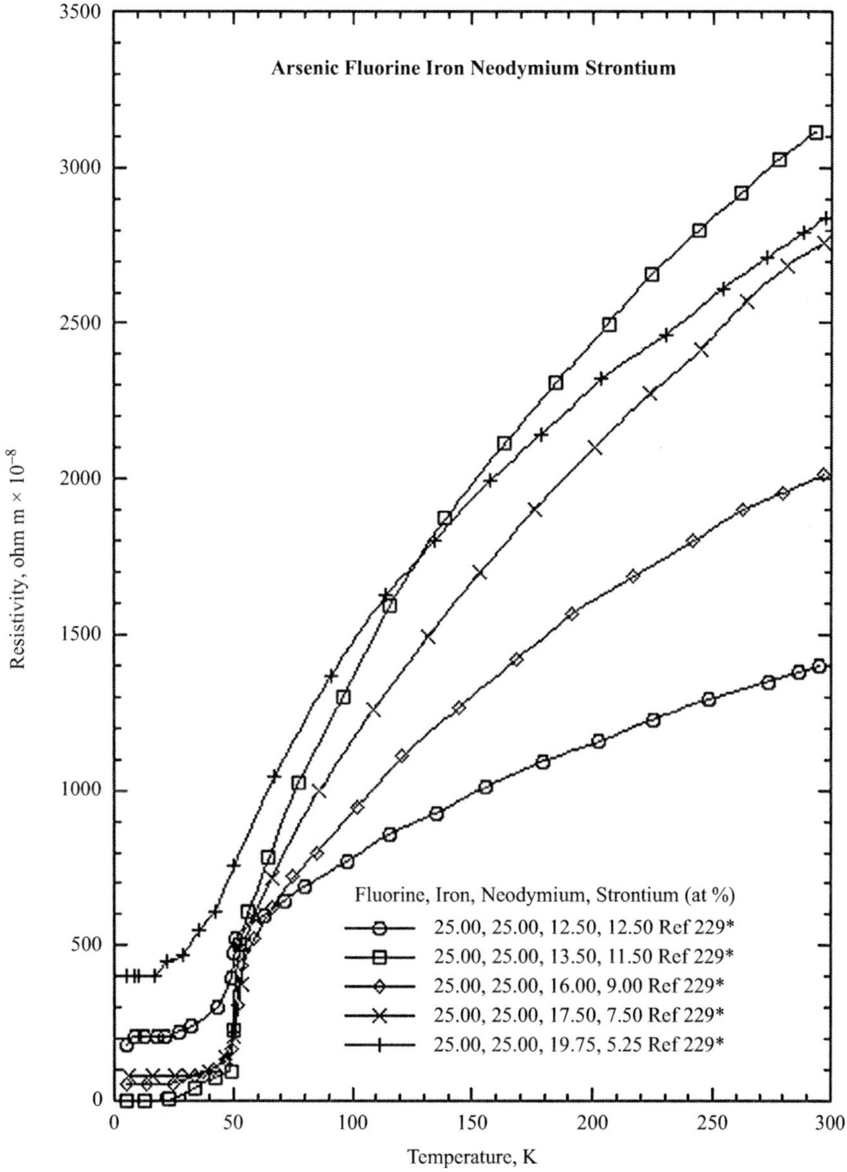
Arsenic Fluorine Iron Neodymium Strontium
Resistivity, ohm m × 10⁻⁸
Temperature, K
Fluorine, Iron, Neodymium, Strontium (at %)
25.00, 25.00, 12.50, 12.50 Ref 229*
25.00, 25.00, 13.50, 11.50 Ref 229*
25.00, 25.00, 16.00, 9.00 Ref 229*
25.00, 25.00, 17.50, 7.50 Ref 229*
25.00, 25.00, 19.75, 5.25 Ref 229*
0
500
1000
1500
2000
2500
3000
3500
0
50
100
150
200
250
300

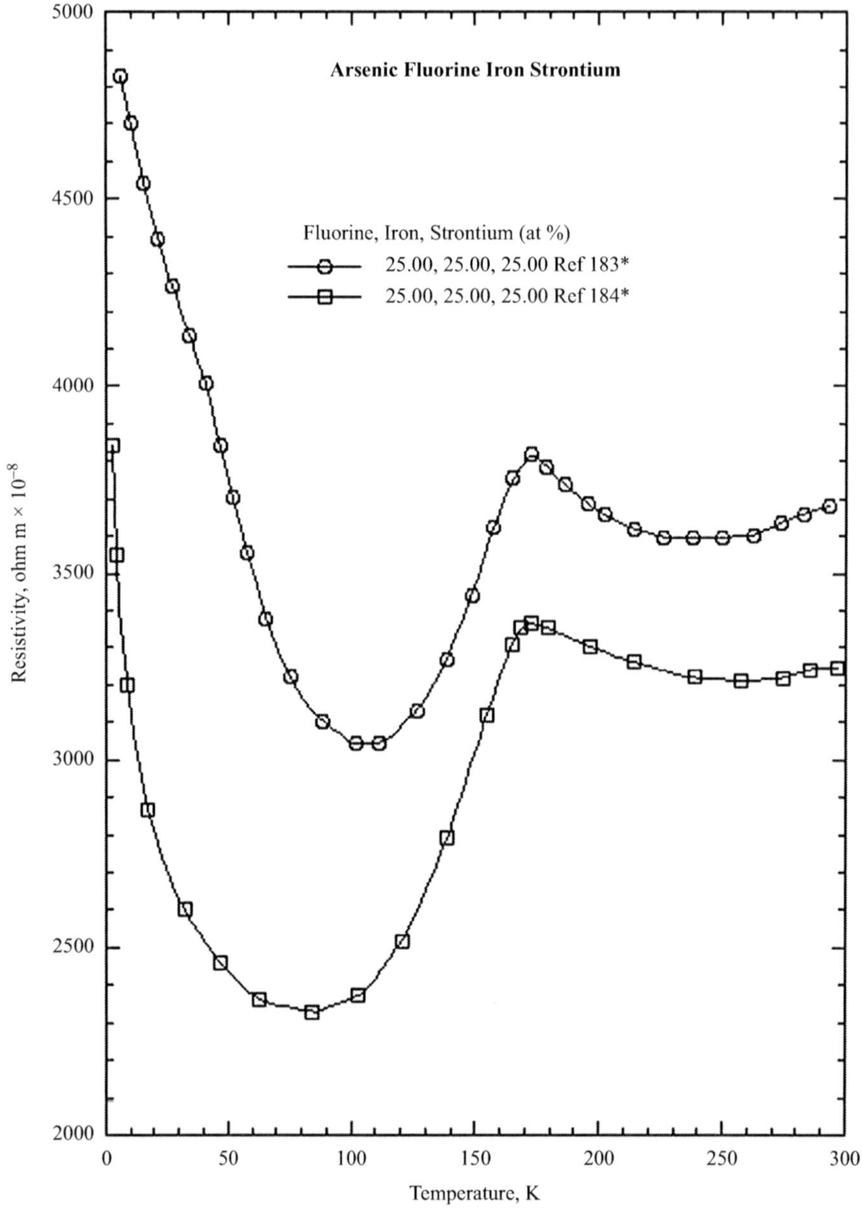
Arsenic Fluorine Iron Strontium
Fluorine, Iron, Strontium (at %)
25.00, 25.00, 25.00 Ref 183*
25.00, 25.00, 25.00 Ref 184*
Resistivity, ohm m × 10⁻⁸
Temperature, K
5000
4500
4000
3500
3000
2500
2000
0
50
100
150
200
250
300

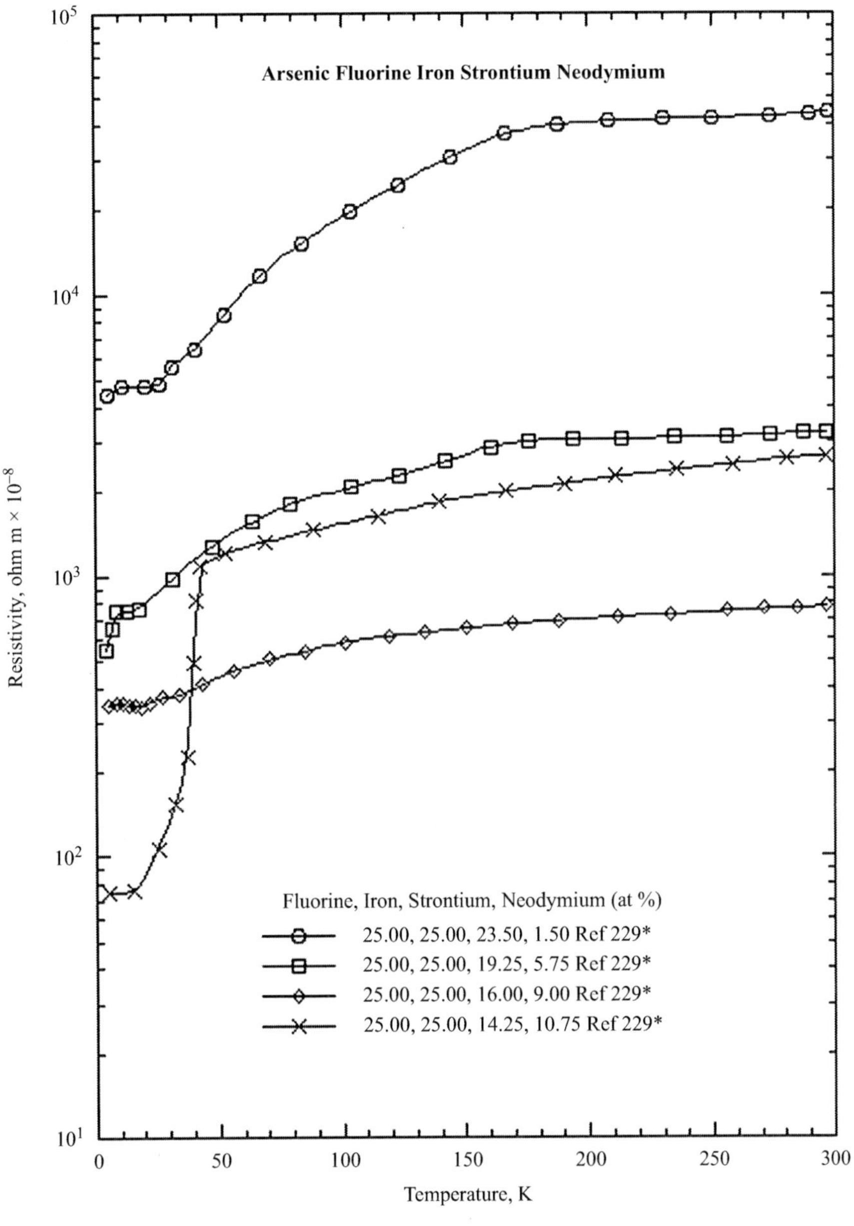
Arsenic Fluorine Iron Strontium Neodymium
Resistivity, ohm m × 10⁻⁸
Temperature, K
10⁵
10⁴
10³
10²
10¹
0
50
100
150
200
250
300
Fluorine, Iron, Strontium, Neodymium (at %)
25.00, 25.00, 23.50, 1.50 Ref 229*
25.00, 25.00, 19.25, 5.75 Ref 229*
25.00, 25.00, 16.00, 9.00 Ref 229*
25.00, 25.00, 14.25, 10.75 Ref 229*

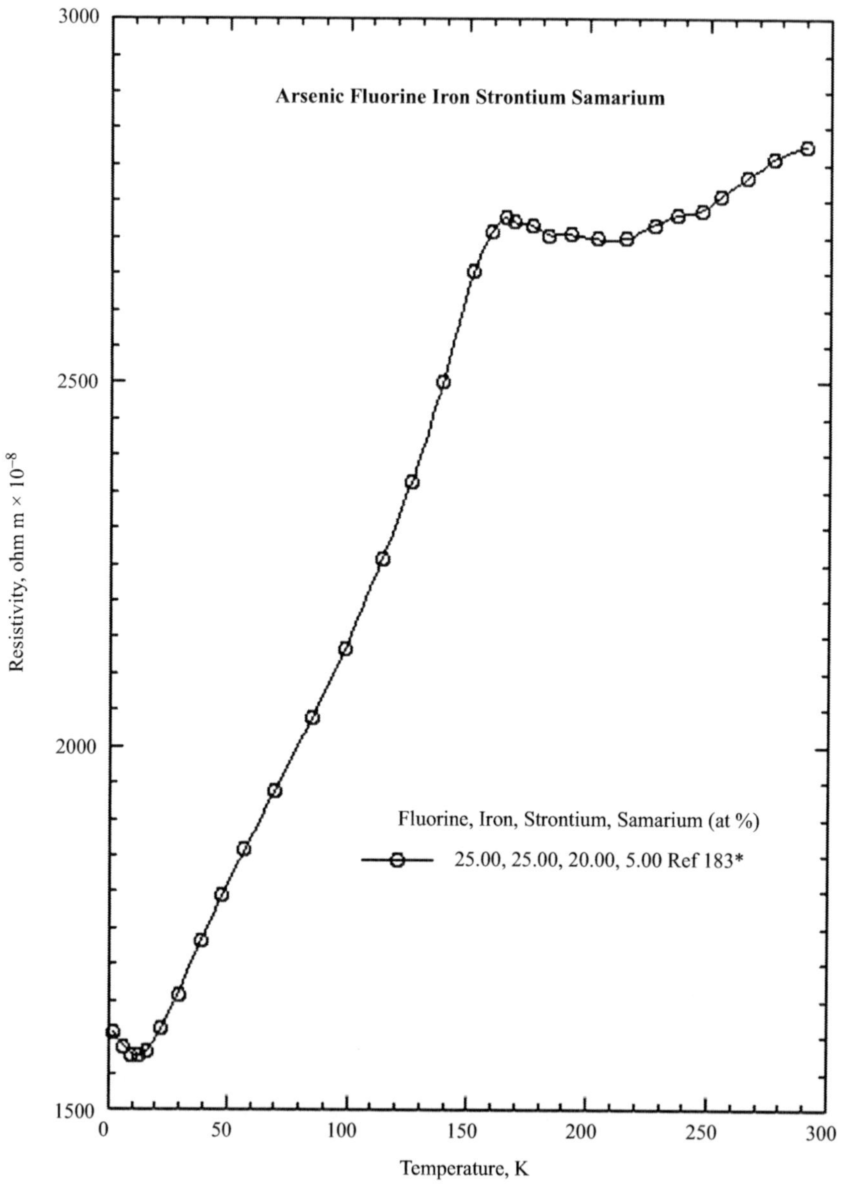
Arsenic Fluorine Iron Strontium Samarium
Fluorine, Iron, Strontium, Samarium (at %)
25.00, 25.00, 20.00, 5.00 Ref 183*
Resistivity, ohm m × 10⁻⁸
3000
2500
2000
1500
0
50
100
150
200
250
300
Temperature, K

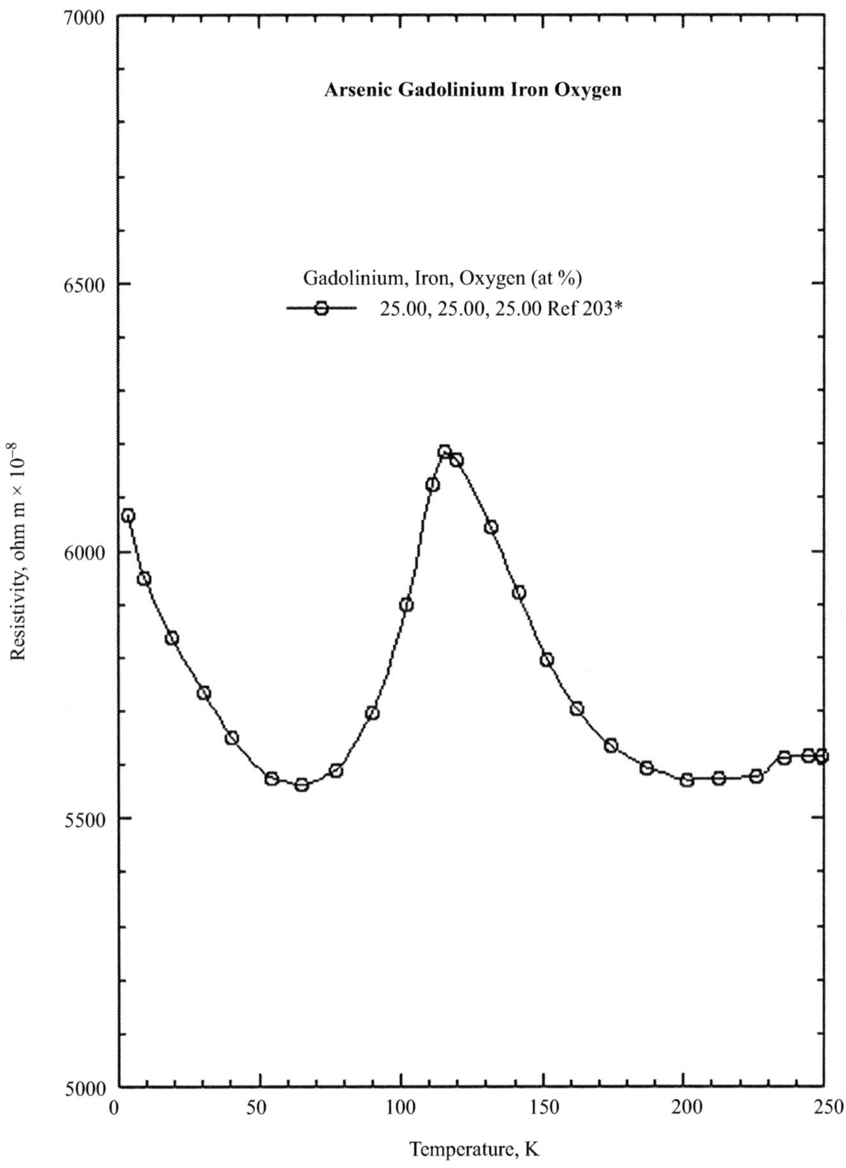
Arsenic Gadolinium Iron Oxygen
Gadolinium, Iron, Oxygen (at %)
25.00, 25.00, 25.00 Ref 203*
Resistivity, ohm m × 10⁻⁸
Temperature, K
7000
6500
6000
5500
5000
0
50
100
150
200
250

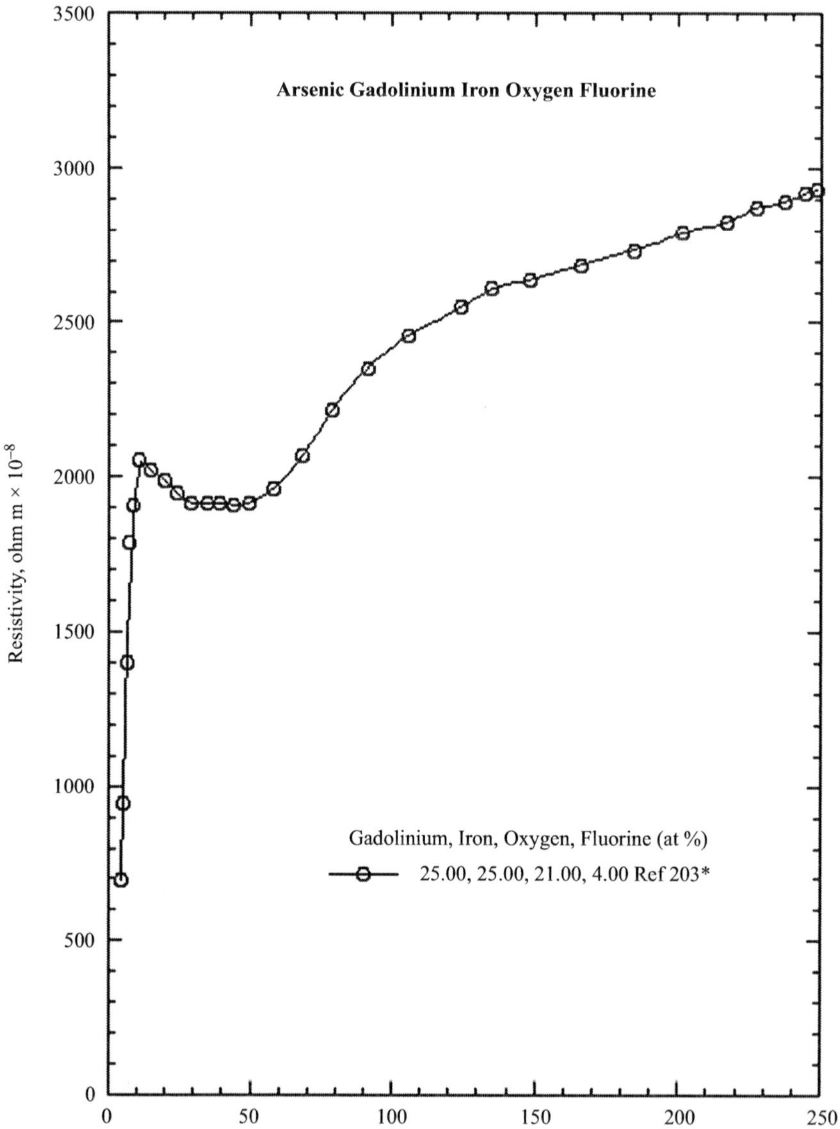
Arsenic Gadolinium Iron Oxygen Fluorine
Resistivity, ohm m × 10⁻⁸
3500
3000
2500
2000
1500
1000
500
0
0
50
100
150
200
250
Temperature, K
Gadolinium, Iron, Oxygen, Fluorine (at %)
25.00, 25.00, 21.00, 4.00 Ref 203*

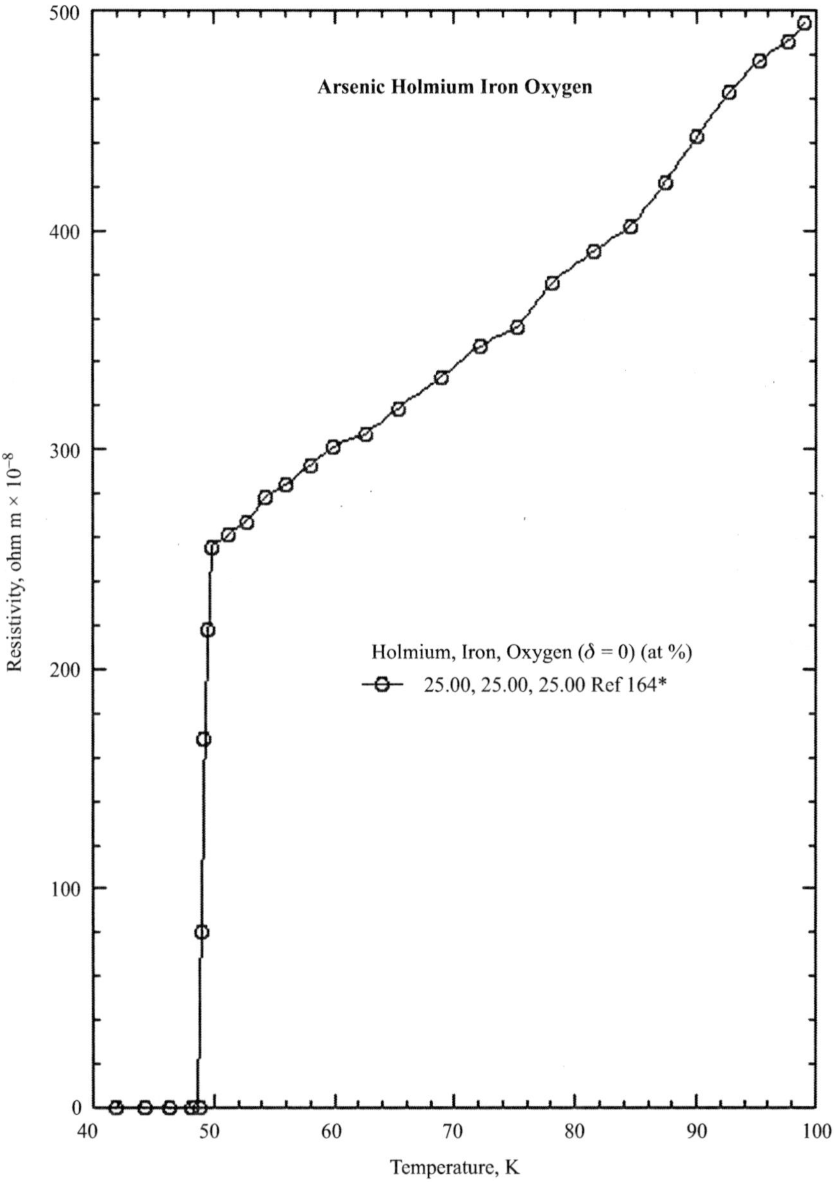
Arsenic Holmium Iron Oxygen
Holmium, Iron, Oxygen (δ = 0) (at %)
25.00, 25.00, 25.00 Ref 164*
Resistivity, ohm m × 10⁻⁸
Temperature, K
0
100
200
300
400
500
40
50
60
70
80
90
100

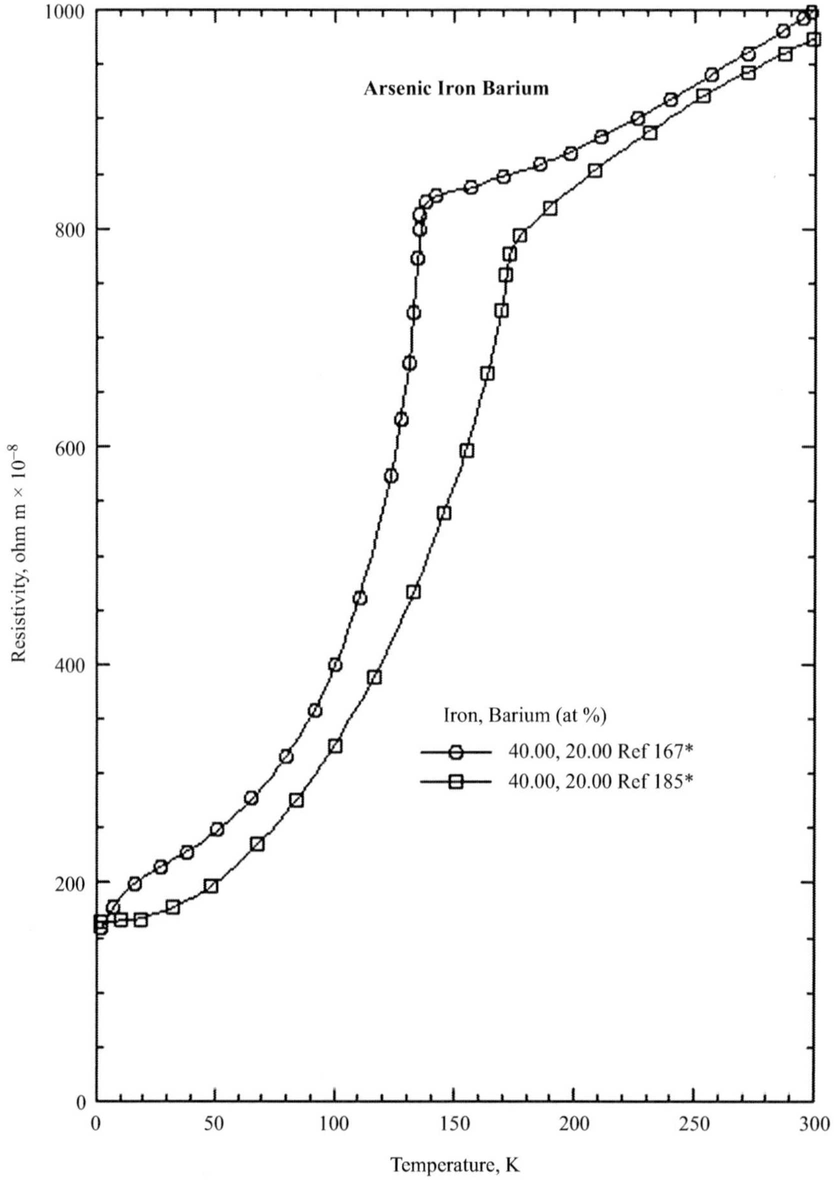
Arsenic Iron Barium
Iron, Barium (at %)
40.00, 20.00 Ref 167*
40.00, 20.00 Ref 185*
Resistivity, ohm m × 10⁻⁸
Temperature, K
0
200
400
600
800
1000
0
50
100
150
200
250
300

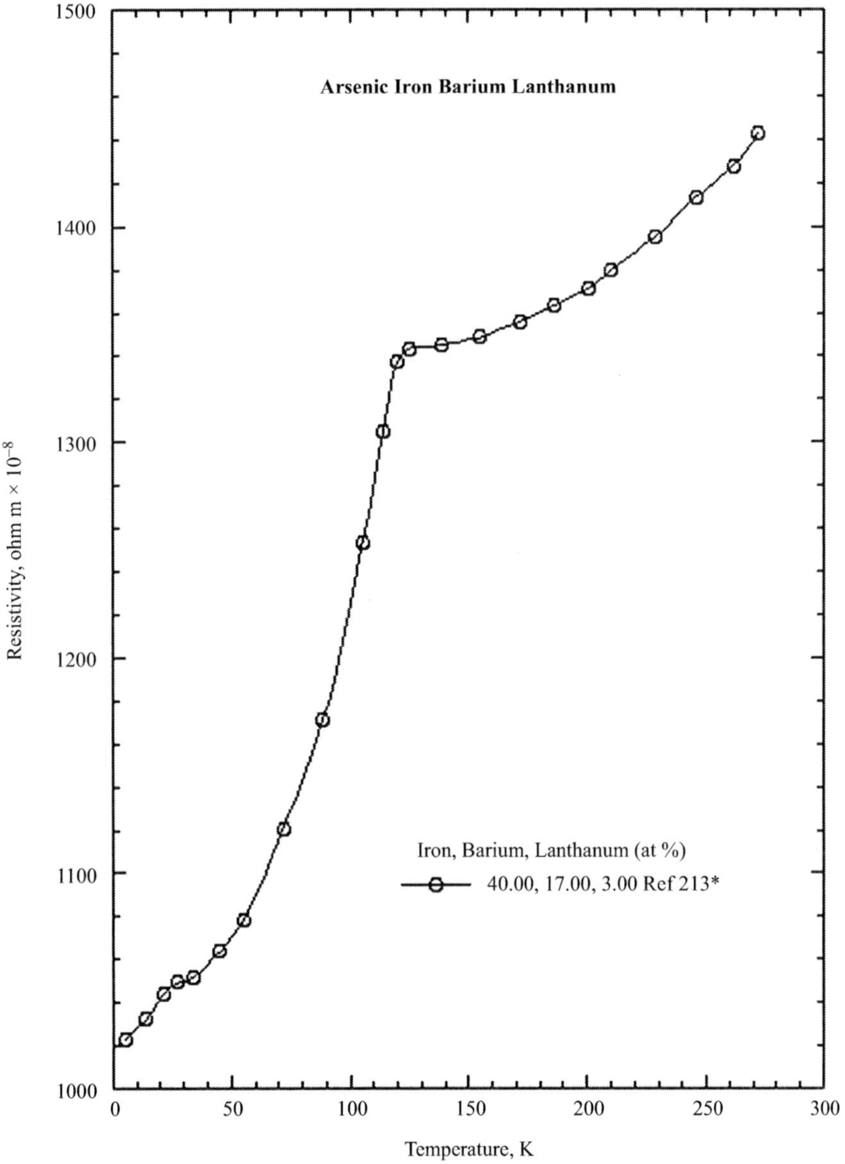
Arsenic Iron Barium Lanthanum
Iron, Barium, Lanthanum (at %)
40.00, 17.00, 3.00 Ref 213*
Resistivity, ohm m × 10^{-8}
Temperature, K
1000
1100
1200
1300
1400
1500
0
50
100
150
200
250
300

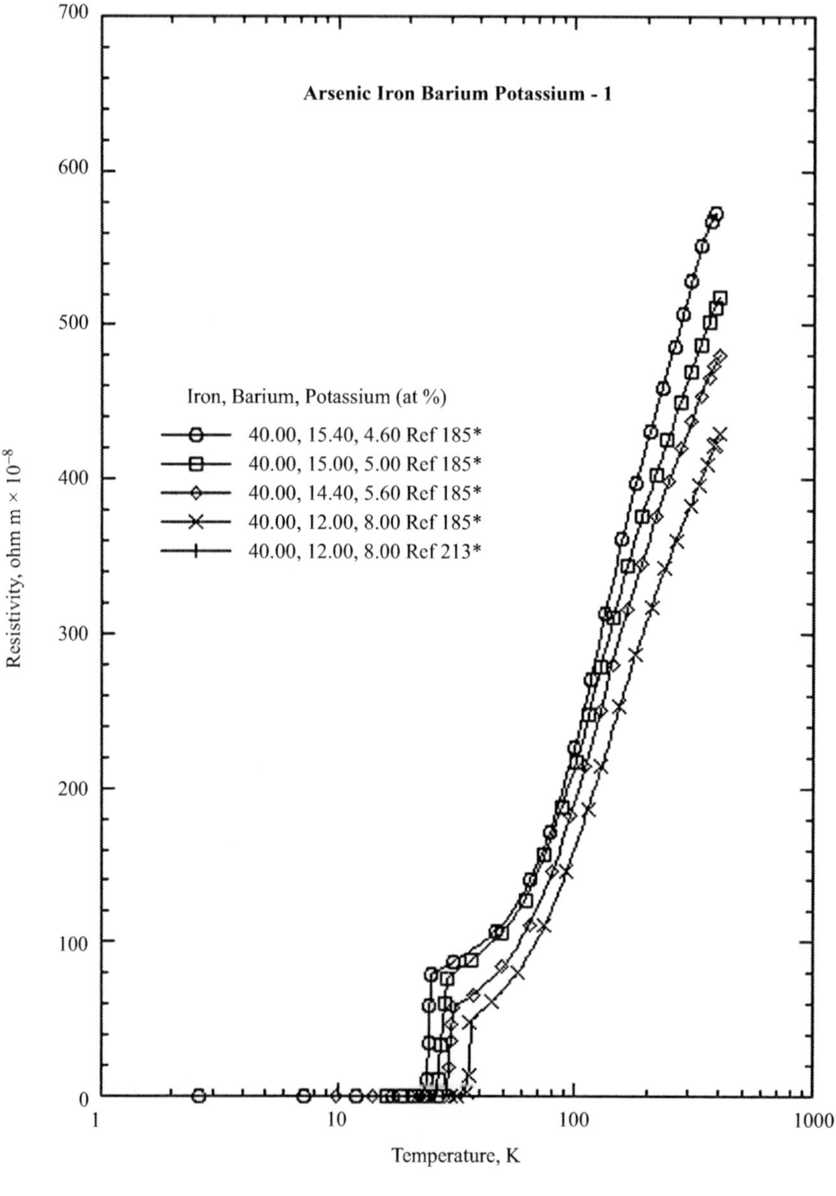
Arsenic Iron Barium Potassium - 1
Iron, Barium, Potassium (at %)
40.00, 15.40, 4.60 Ref 185*
40.00, 15.00, 5.00 Ref 185*
40.00, 14.40, 5.60 Ref 185*
40.00, 12.00, 8.00 Ref 185*
40.00, 12.00, 8.00 Ref 213*
Resistivity, ohm m × 10⁻⁸
Temperature, K
0
100
200
300
400
500
600
700
1
10
100
1000

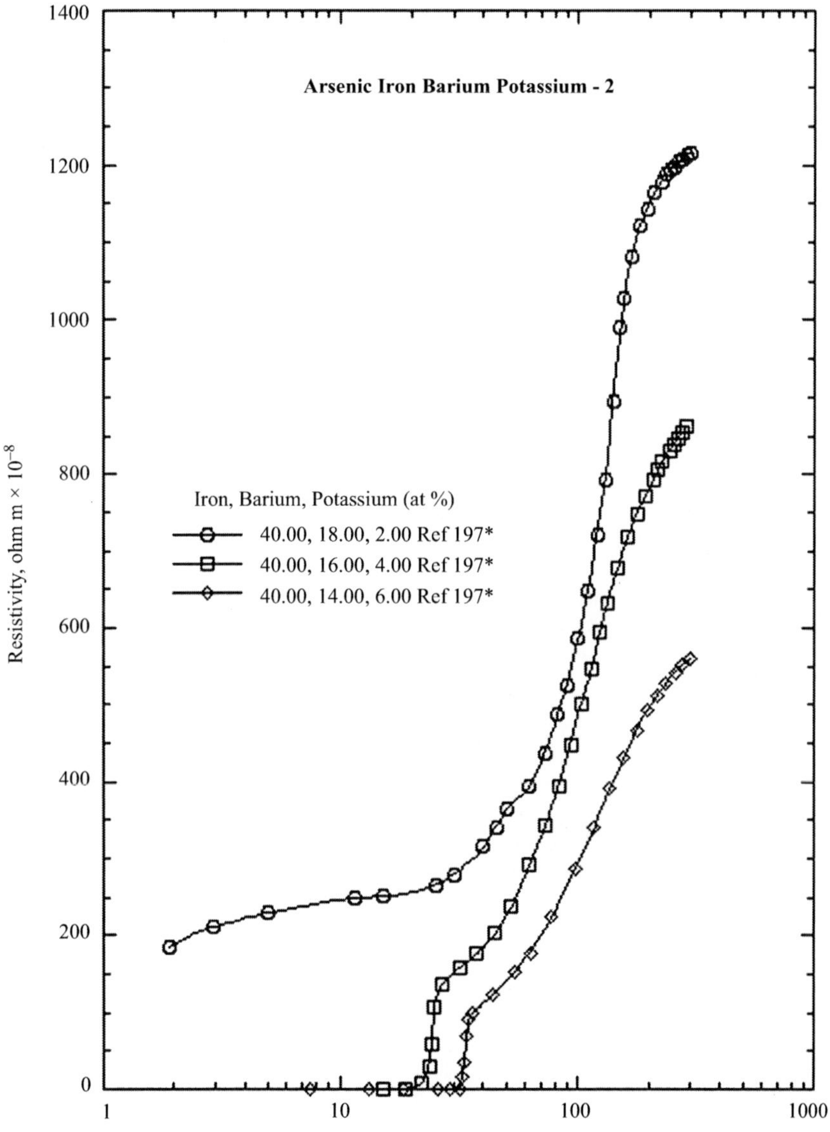
Arsenic Iron Barium Potassium - 2
Iron, Barium, Potassium (at %)
40.00, 18.00, 2.00 Ref 197*
40.00, 16.00, 4.00 Ref 197*
40.00, 14.00, 6.00 Ref 197*
Resistivity, ohm m × 10⁻⁸
Temperature, K
0
200
400
600
800
1000
1200
1400
1
10
100
1000

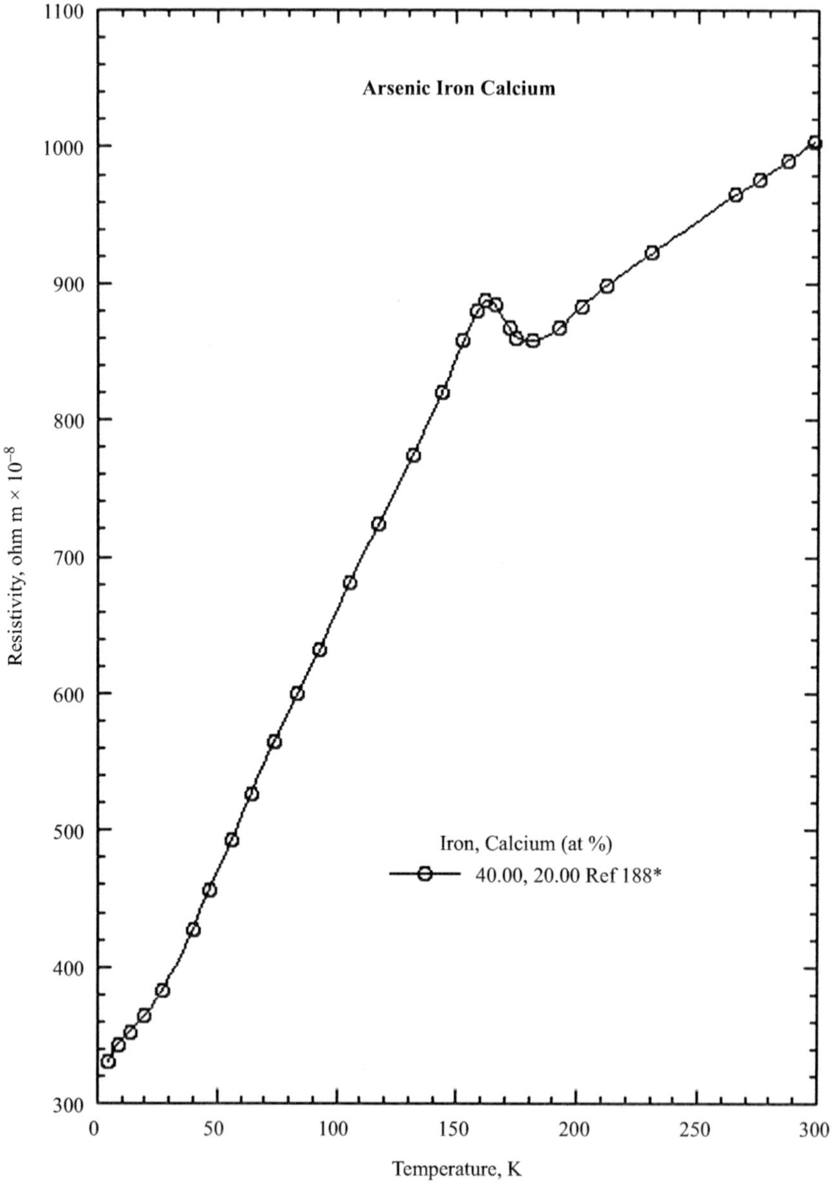
Arsenic Iron Calcium
Resistivity, ohm m × 10⁻⁸
1100
1000
900
800
700
600
500
400
300
0
50
100
150
200
250
300
Temperature, K
Iron, Calcium (at %)
40.00, 20.00 Ref 188*

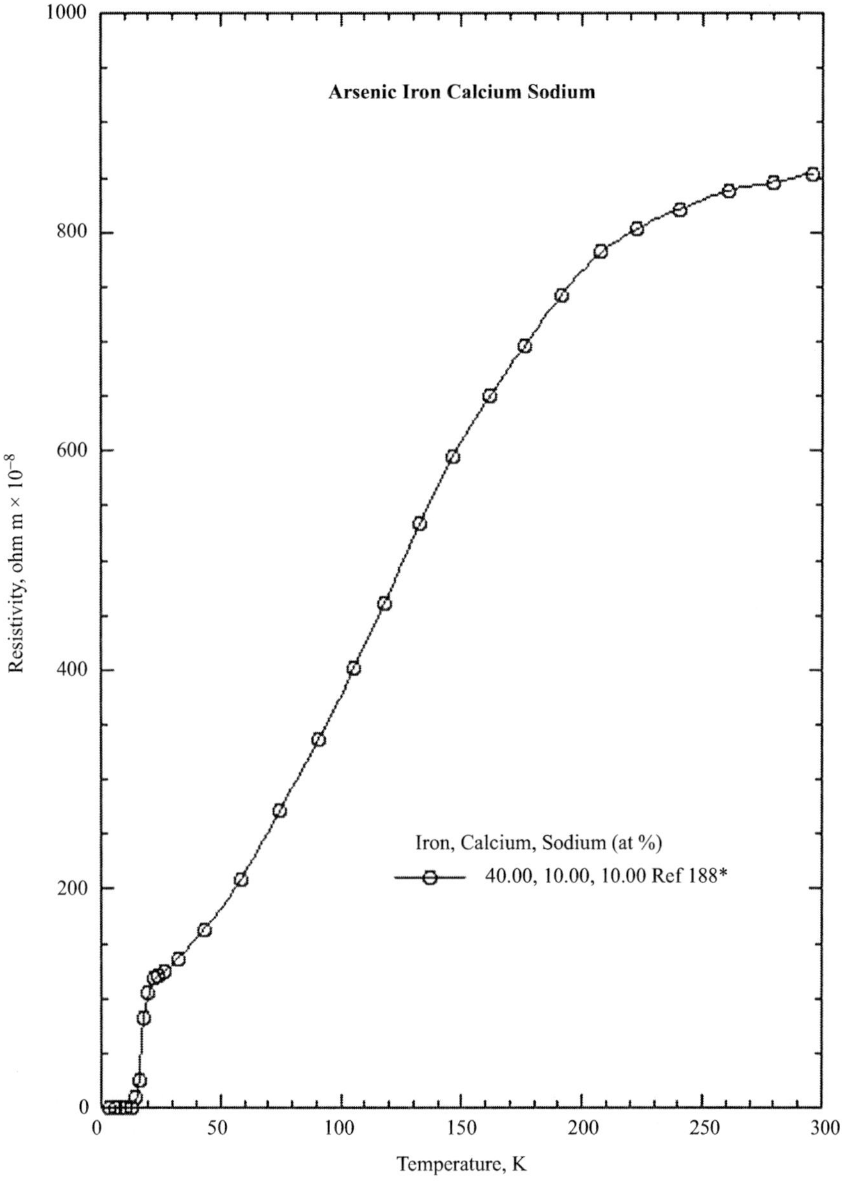

Arsenic Iron Calcium Sodium
Iron, Calcium, Sodium (at %)
40.00, 10.00, 10.00 Ref 188*
Resistivity, ohm m × 10^{-8}
Temperature, K
0
200
400
600
800
1000
0
50
100
150
200
250
300

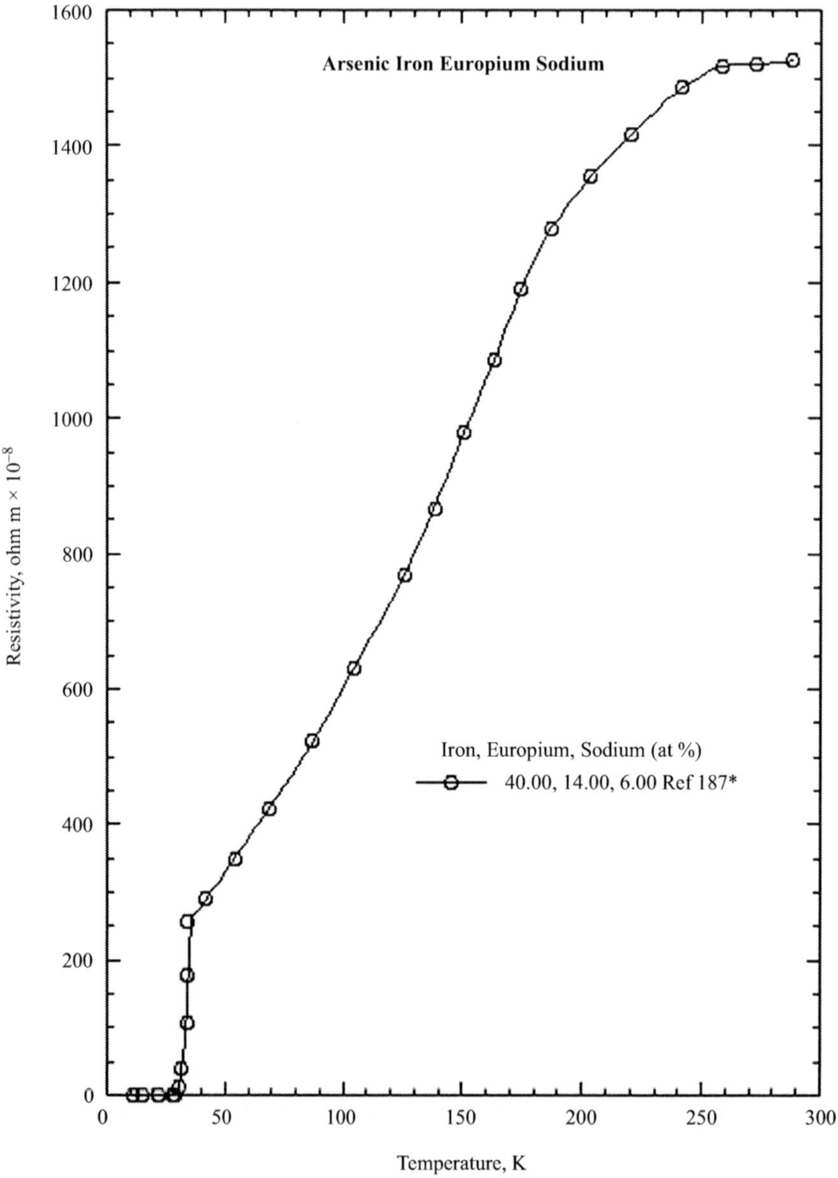
Arsenic Iron Europium Sodium
Iron, Europium, Sodium (at %)
40.00, 14.00, 6.00 Ref 187*
Resistivity, ohm m × 10⁻⁸
Temperature, K
0
200
400
600
800
1000
1200
1400
1600
0
50
100
150
200
250
300

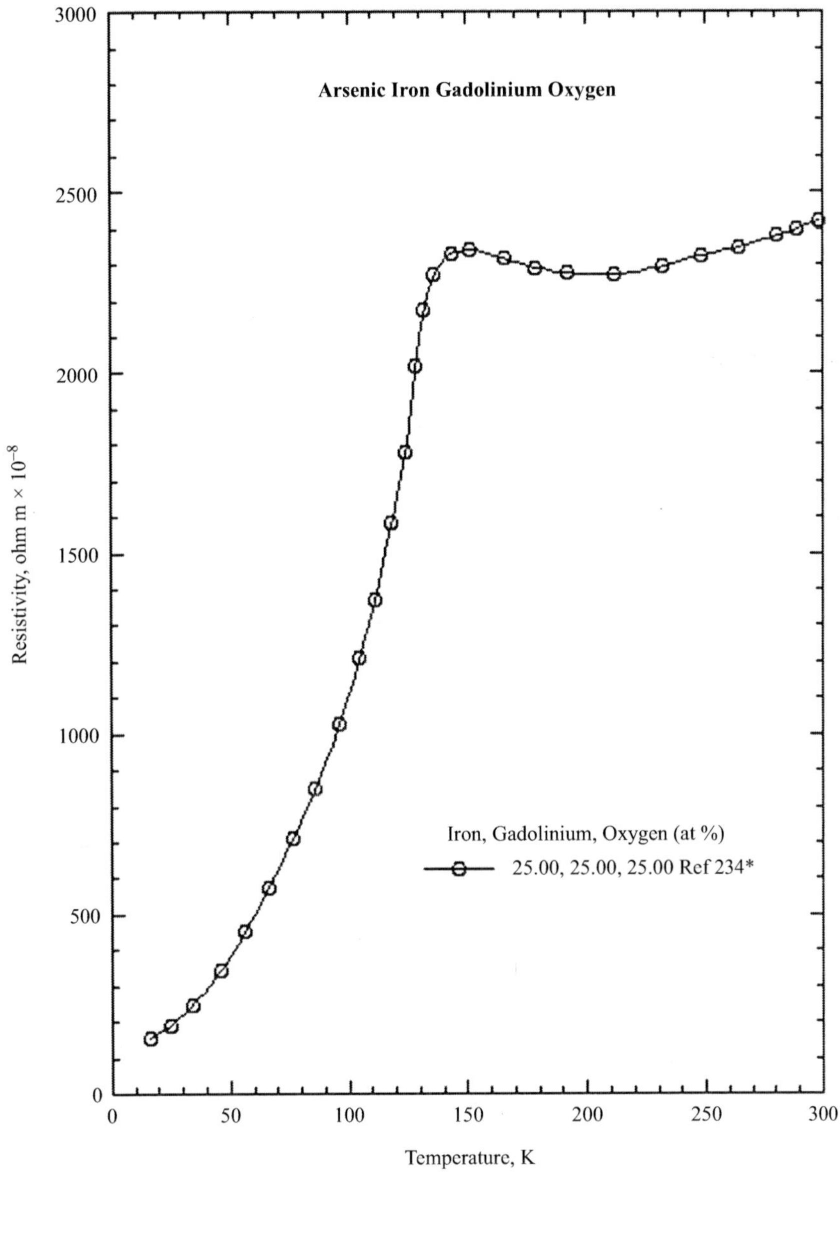
Arsenic Iron Gadolinium Oxygen
Iron, Gadolinium, Oxygen (at %)
25.00, 25.00, 25.00 Ref 234*
Resistivity, ohm m × 10⁻⁸
Temperature, K
0
500
1000
1500
2000
2500
3000
0
50
100
150
200
250
300

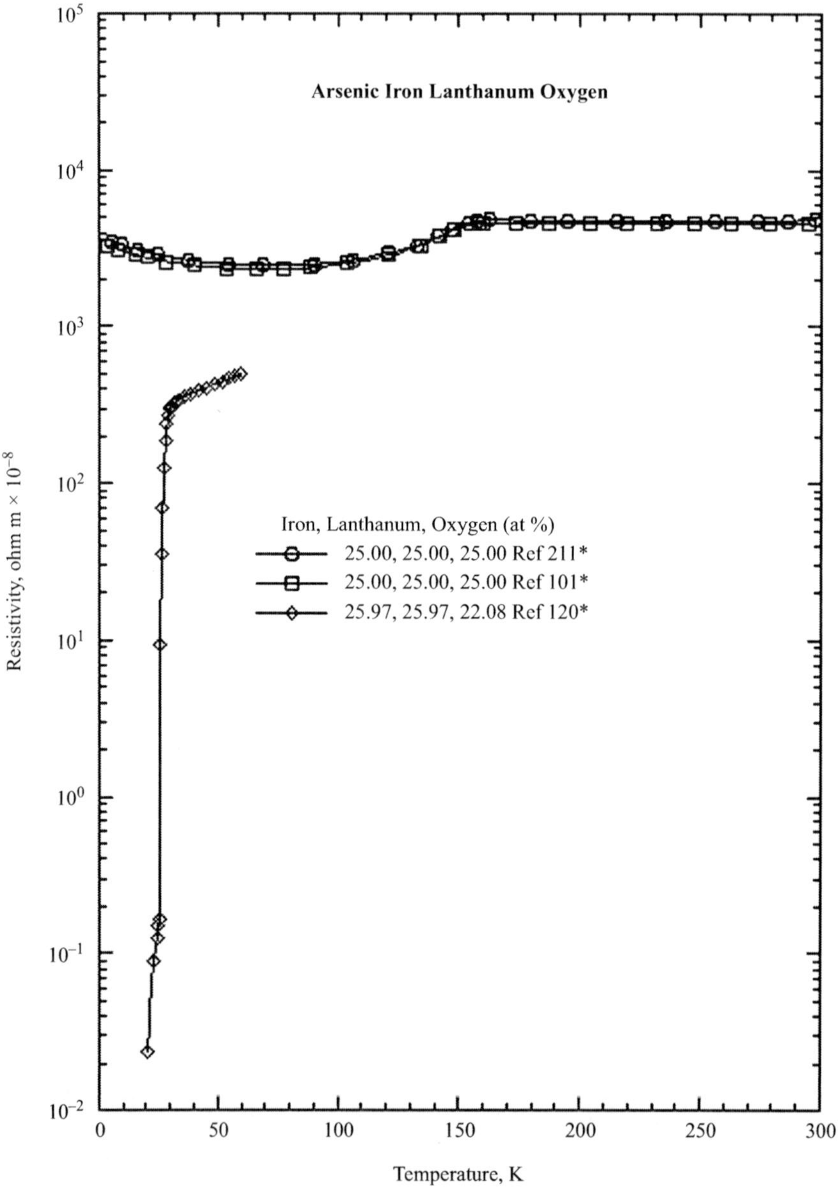

Arsenic Iron Lanthanum Oxygen
Iron, Lanthanum, Oxygen (at %)
25.00, 25.00, 25.00 Ref 211*
25.00, 25.00, 25.00 Ref 101*
25.97, 25.97, 22.08 Ref 120*
Resistivity, ohm m × 10−8
Temperature, K
10^5
10^4
10^3
10^2
10^1
10^0
10^−1
10^−2
0
50
100
150
200
250
300

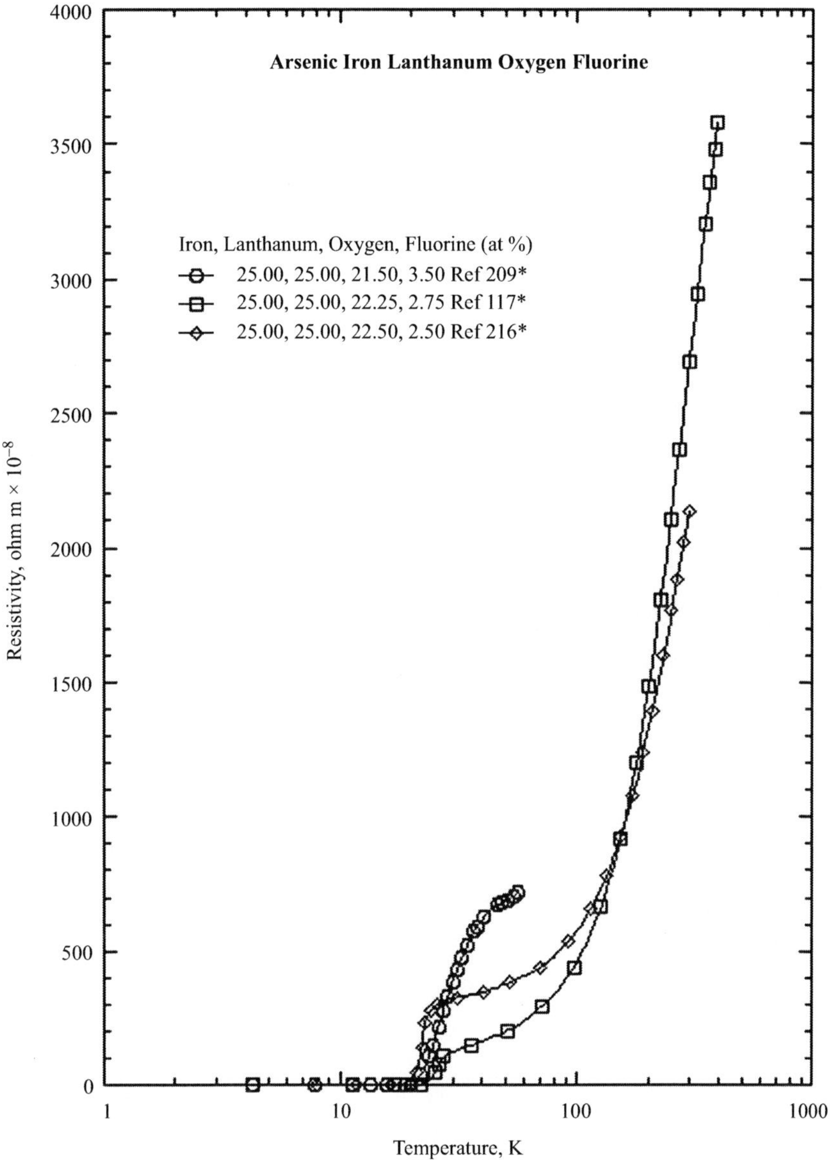
Arsenic Iron Lanthanum Oxygen Fluorine
Iron, Lanthanum, Oxygen, Fluorine (at %)
25.00, 25.00, 21.50, 3.50 Ref 209*
25.00, 25.00, 22.25, 2.75 Ref 117*
25.00, 25.00, 22.50, 2.50 Ref 216*
Resistivity, ohm m × 10⁻⁸
4000
3500
3000
2500
2000
1500
1000
500
0
1
10
100
1000
Temperature, K

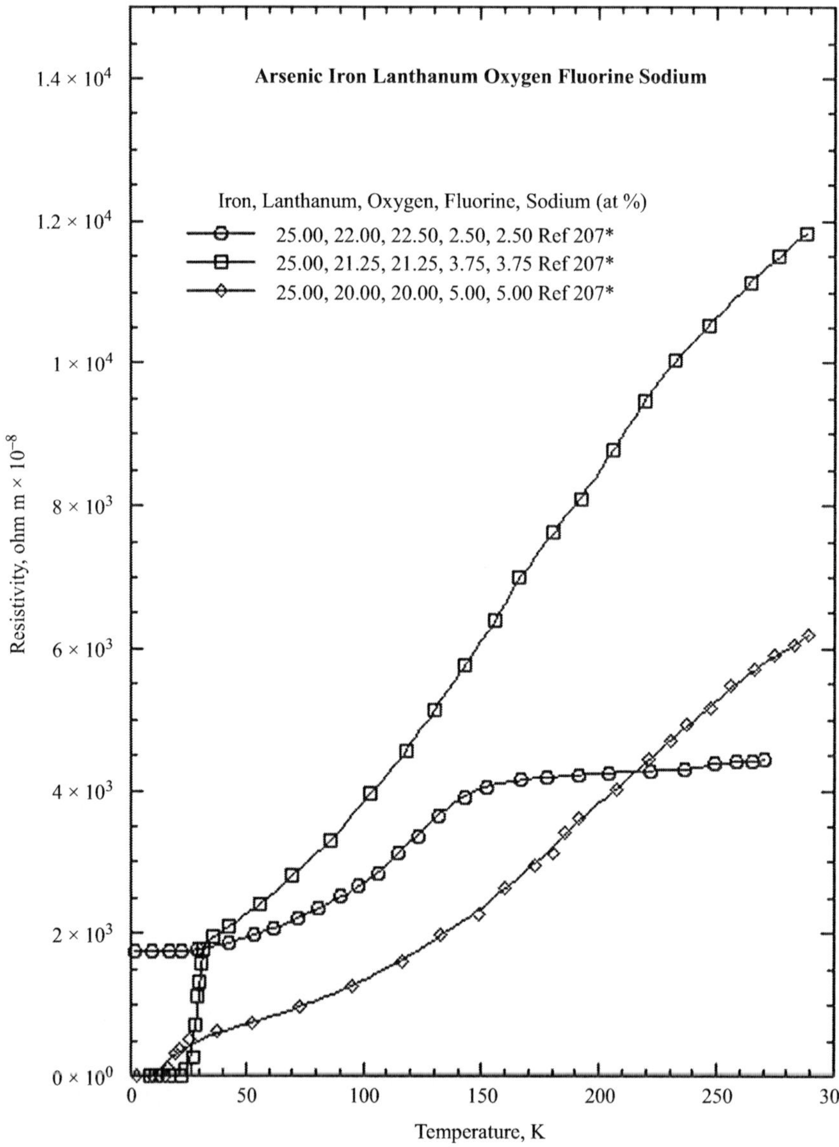
Arsenic Iron Lanthanum Oxygen Fluorine Sodium
Iron, Lanthanum, Oxygen, Fluorine, Sodium (at %)
25.00, 22.00, 22.50, 2.50, 2.50 Ref 207*
25.00, 21.25, 21.25, 3.75, 3.75 Ref 207*
25.00, 20.00, 20.00, 5.00, 5.00 Ref 207*
Resistivity, ohm m × 10⁻⁸
Temperature, K
1.4 × 10⁴
1.2 × 10⁴
1 × 10⁴
8 × 10³
6 × 10³
4 × 10³
2 × 10³
0 × 10⁰
0
50
100
150
200
250
300

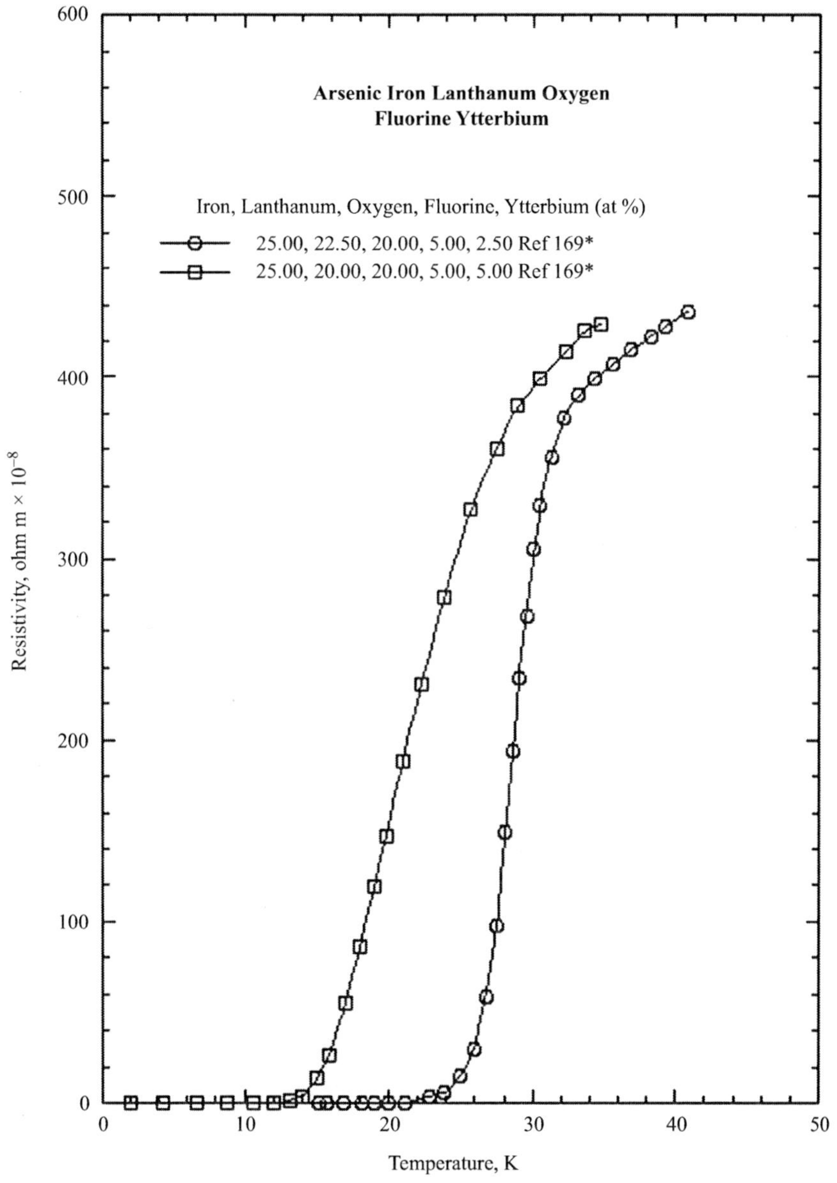

Arsenic Iron Lanthanum Oxygen Fluorine Ytterbium
Iron, Lanthanum, Oxygen, Fluorine, Ytterbium (at %)
25.00, 22.50, 20.00, 5.00, 2.50 Ref 169*
25.00, 20.00, 20.00, 5.00, 5.00 Ref 169*
Resistivity, ohm m × 10⁻⁸
Temperature, K
0
100
200
300
400
500
600
0
10
20
30
40
50

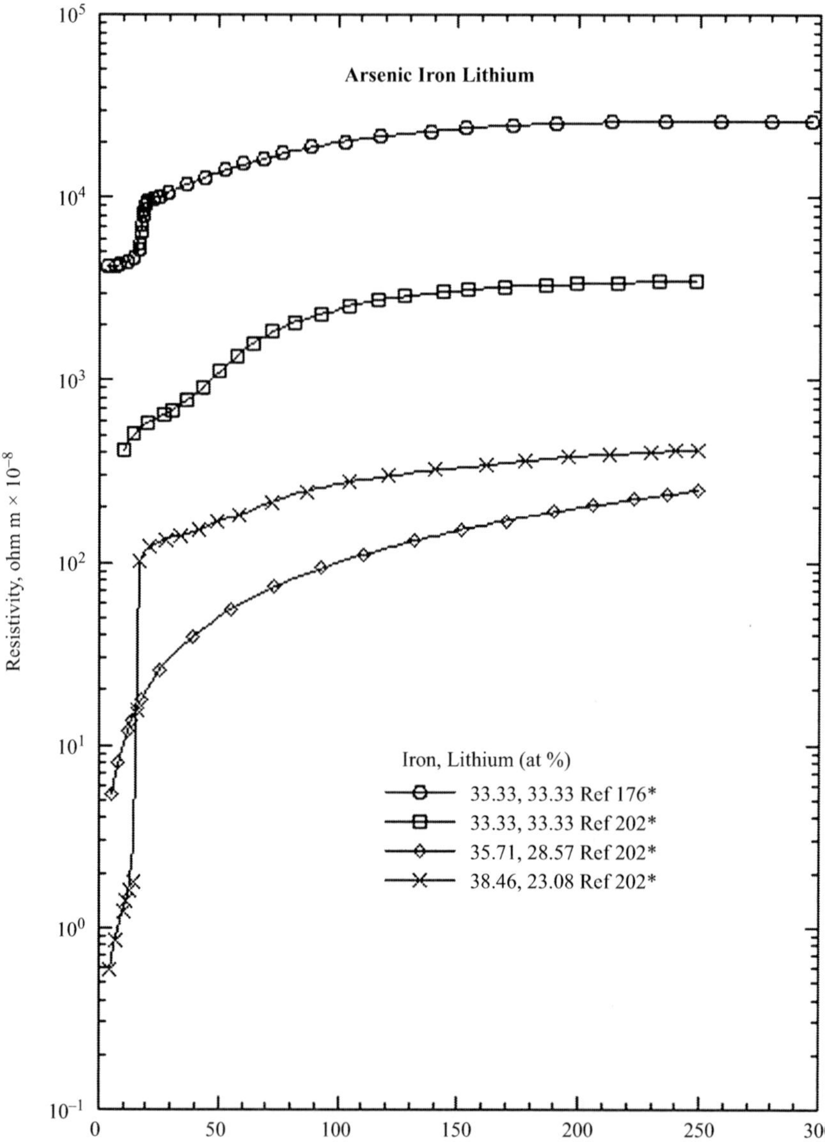
Arsenic Iron Lithium
Resistivity, ohm m × 10⁻⁸
Temperature, K
Iron, Lithium (at %)
33.33, 33.33 Ref 176*
33.33, 33.33 Ref 202*
35.71, 28.57 Ref 202*
38.46, 23.08 Ref 202*
0
50
100
150
200
250
300

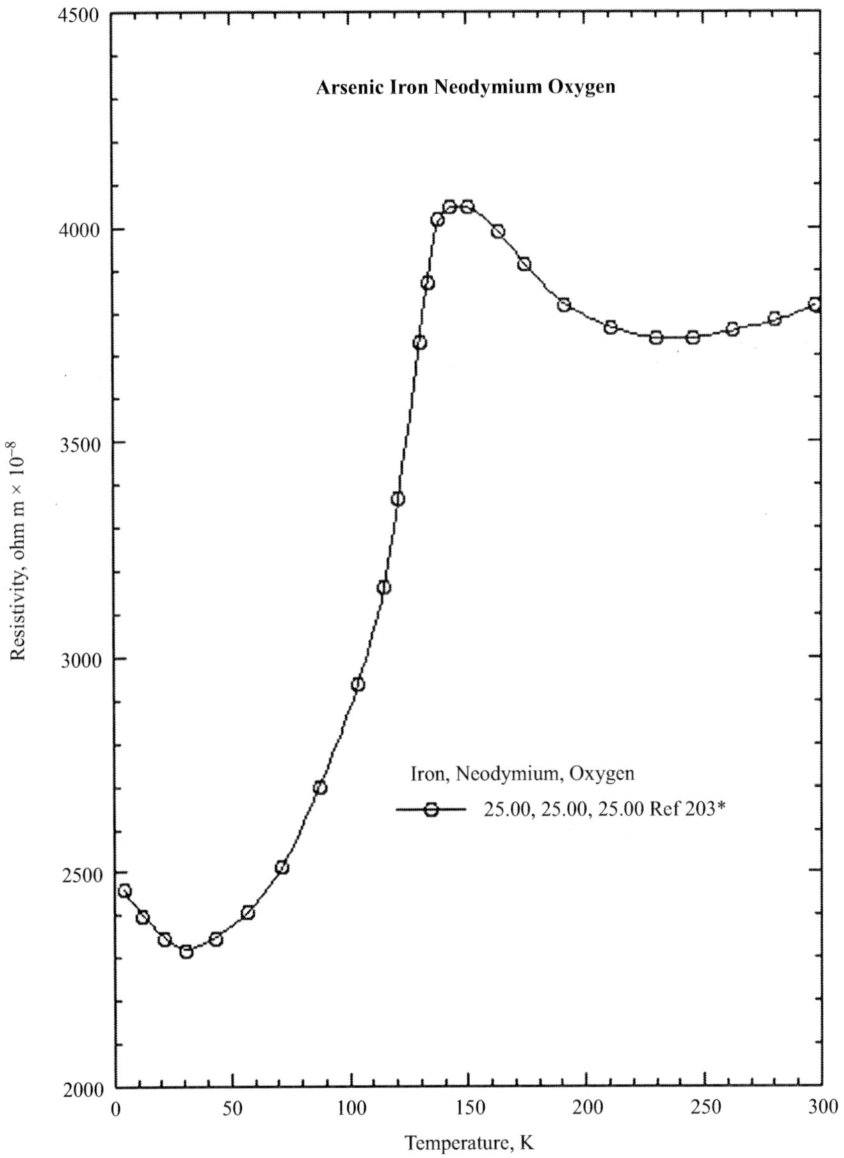
Arsenic Iron Neodymium Oxygen
Iron, Neodymium, Oxygen
25.00, 25.00, 25.00 Ref 203*
Resistivity, ohm m × 10⁻⁸
Temperature, K
4500
4000
3500
3000
2500
2000
0
50
100
150
200
250
300

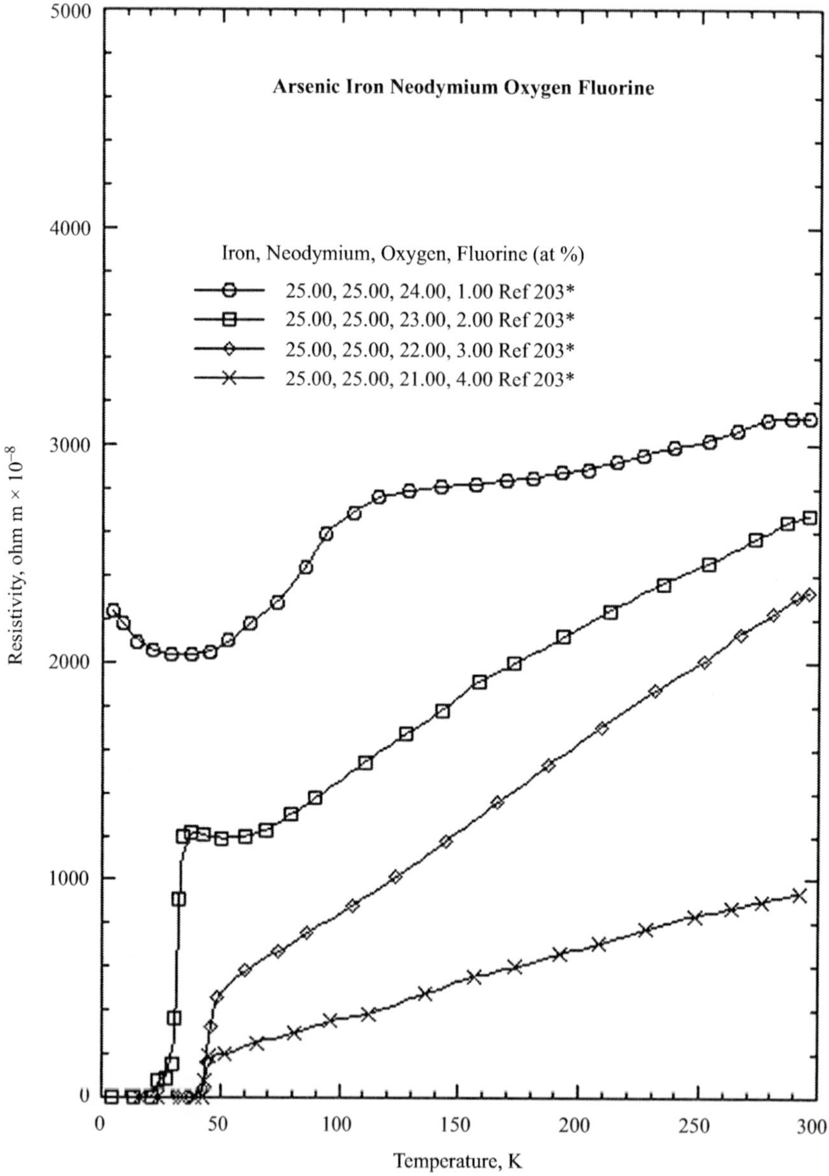

Arsenic Iron Neodymium Oxygen Fluorine
Iron, Neodymium, Oxygen, Fluorine (at %)
25.00, 25.00, 24.00, 1.00 Ref 203*
25.00, 25.00, 23.00, 2.00 Ref 203*
25.00, 25.00, 22.00, 3.00 Ref 203*
25.00, 25.00, 21.00, 4.00 Ref 203*
Resistivity, ohm m × 10⁻⁸
Temperature, K
0
1000
2000
3000
4000
5000
0
50
100
150
200
250
300

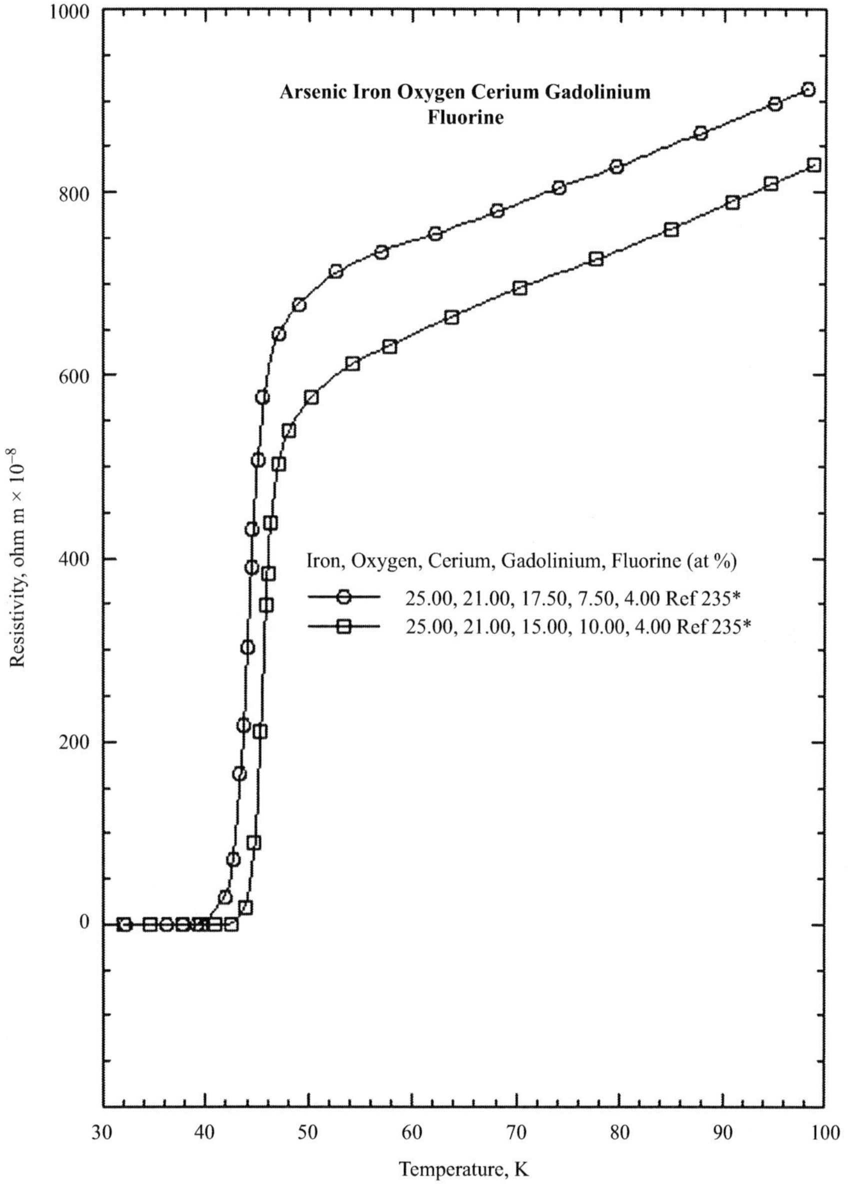
Arsenic Iron Oxygen Cerium Gadolinium Fluorine
Iron, Oxygen, Cerium, Gadolinium, Fluorine (at %)
25.00, 21.00, 17.50, 7.50, 4.00 Ref 235*
25.00, 21.00, 15.00, 10.00, 4.00 Ref 235*
Resistivity, ohm m × 10⁻⁸
1000
800
600
400
200
0
30
40
50
60
70
80
90
100
Temperature, K

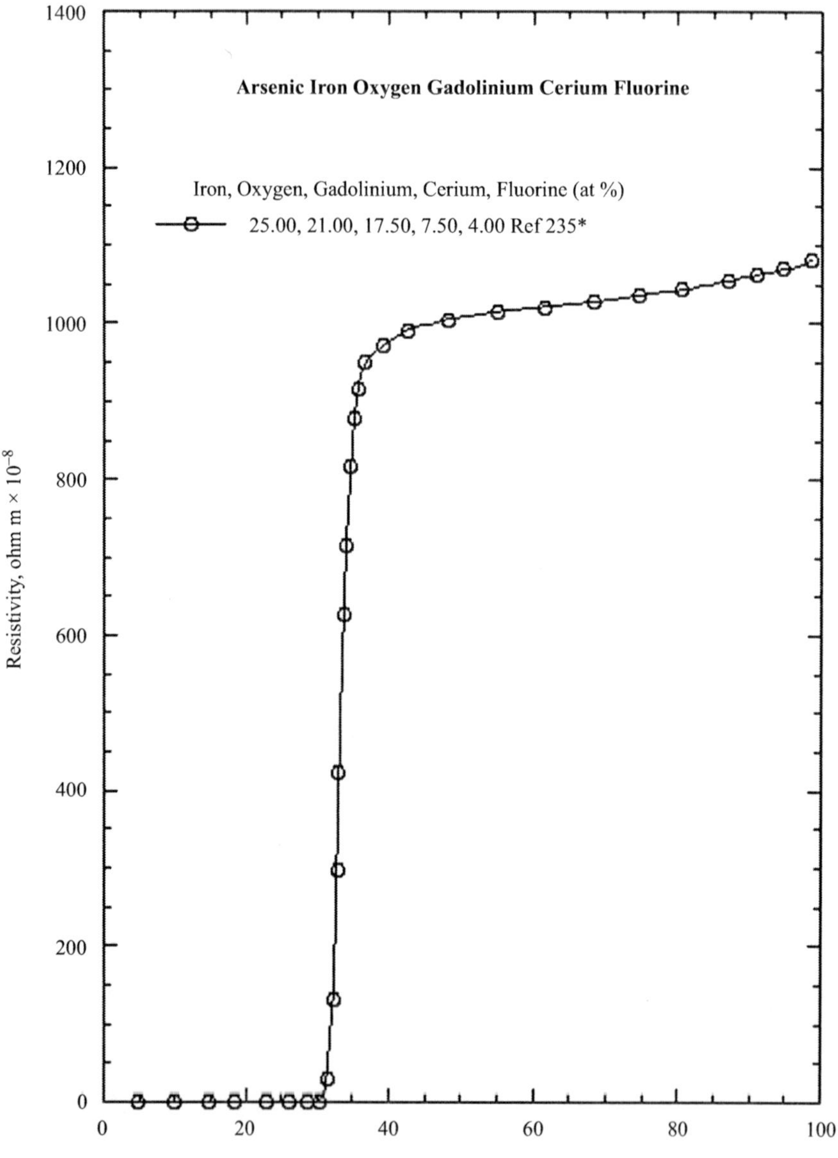
Arsenic Iron Oxygen Gadolinium Cerium Fluorine
Iron, Oxygen, Gadolinium, Cerium, Fluorine (at %)
25.00, 21.00, 17.50, 7.50, 4.00 Ref 235*
Resistivity, ohm m × 10⁻⁸
0
200
400
600
800
1000
1200
1400
0
20
40
60
80
100
Temperature, K

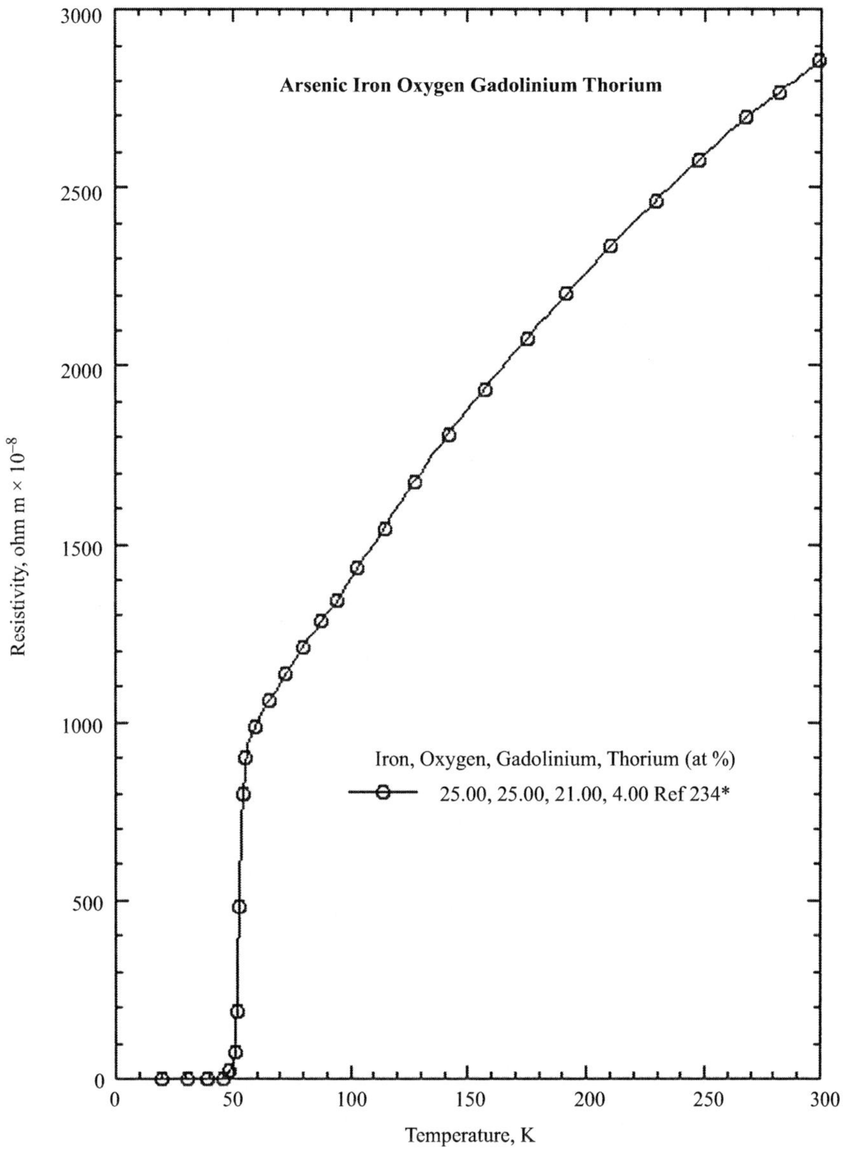

Arsenic Iron Oxygen Gadolinium Thorium
Iron, Oxygen, Gadolinium, Thorium (at %)
25.00, 25.00, 21.00, 4.00 Ref 234*
Resistivity, ohm m × 10⁻⁸
Temperature, K
0
500
1000
1500
2000
2500
3000
0
50
100
150
200
250
300

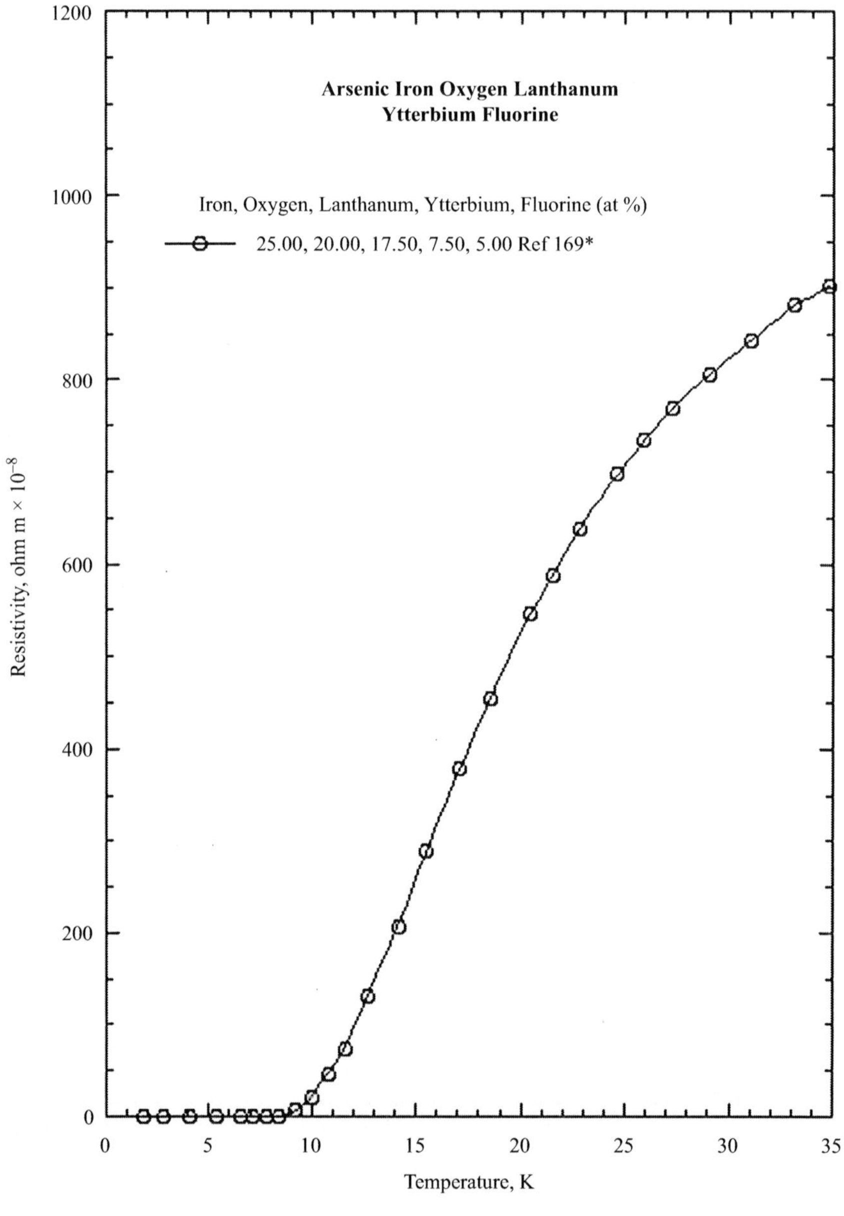
Arsenic Iron Oxygen Lanthanum
Ytterbium Fluorine
Iron, Oxygen, Lanthanum, Ytterbium, Fluorine (at %)
25.00, 20.00, 17.50, 7.50, 5.00 Ref 169*
Resistivity, ohm m × 10−8
1200
1000
800
600
400
200
0
0
5
10
15
20
25
30
35
Temperature, K

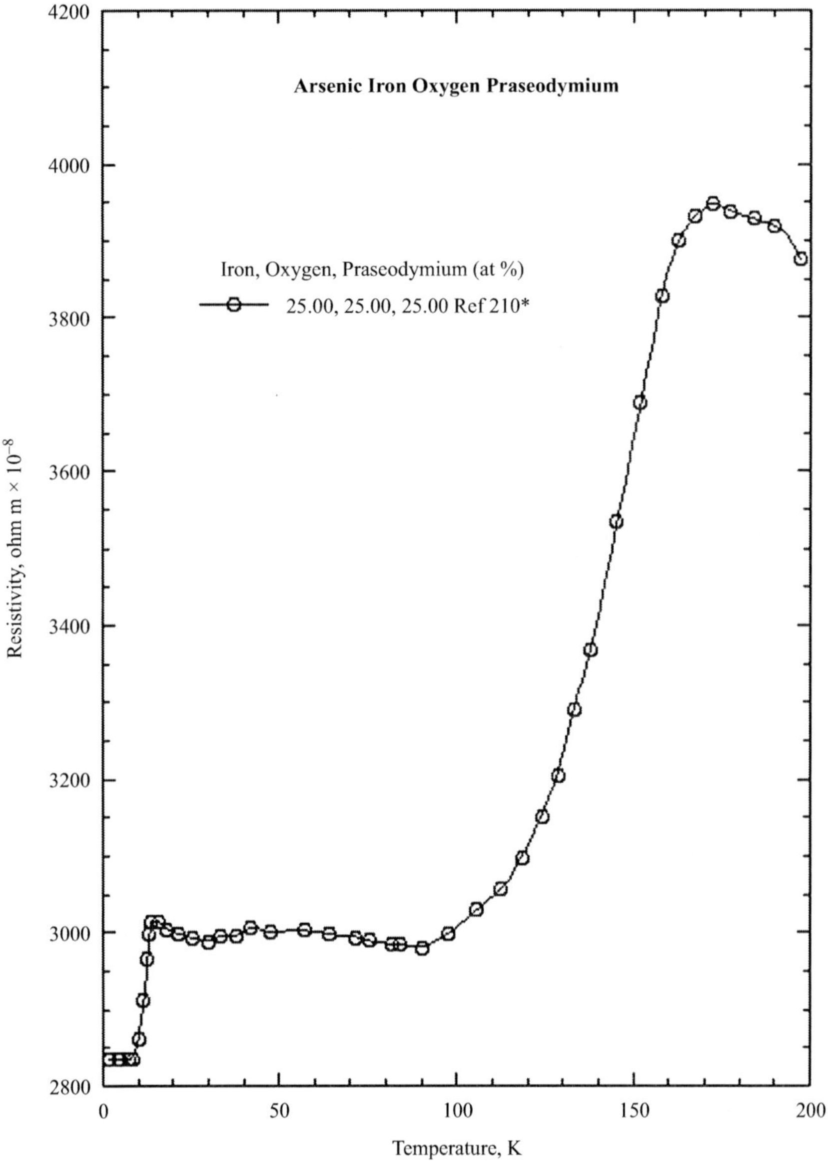
Arsenic Iron Oxygen Praseodymium
Iron, Oxygen, Praseodymium (at %)
25.00, 25.00, 25.00 Ref 210*
Resistivity, ohm m × 10⁻⁸
Temperature, K
4200
4000
3800
3600
3400
3200
3000
2800
0
50
100
150
200

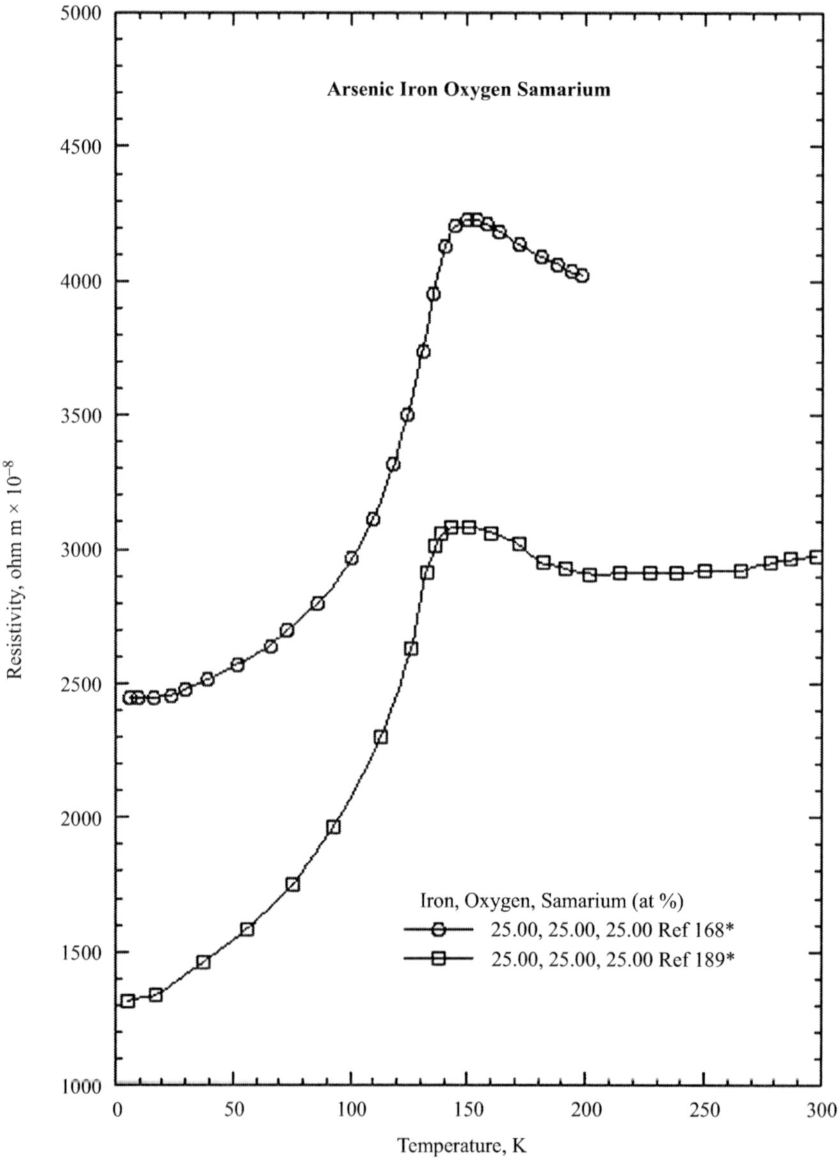
Arsenic Iron Oxygen Samarium
Resistivity, ohm m × 10^{-8}
Temperature, K
Iron, Oxygen, Samarium (at %)
25.00, 25.00, 25.00 Ref 168*
25.00, 25.00, 25.00 Ref 189*
1000
1500
2000
2500
3000
3500
4000
4500
5000
0
50
100
150
200
250
300

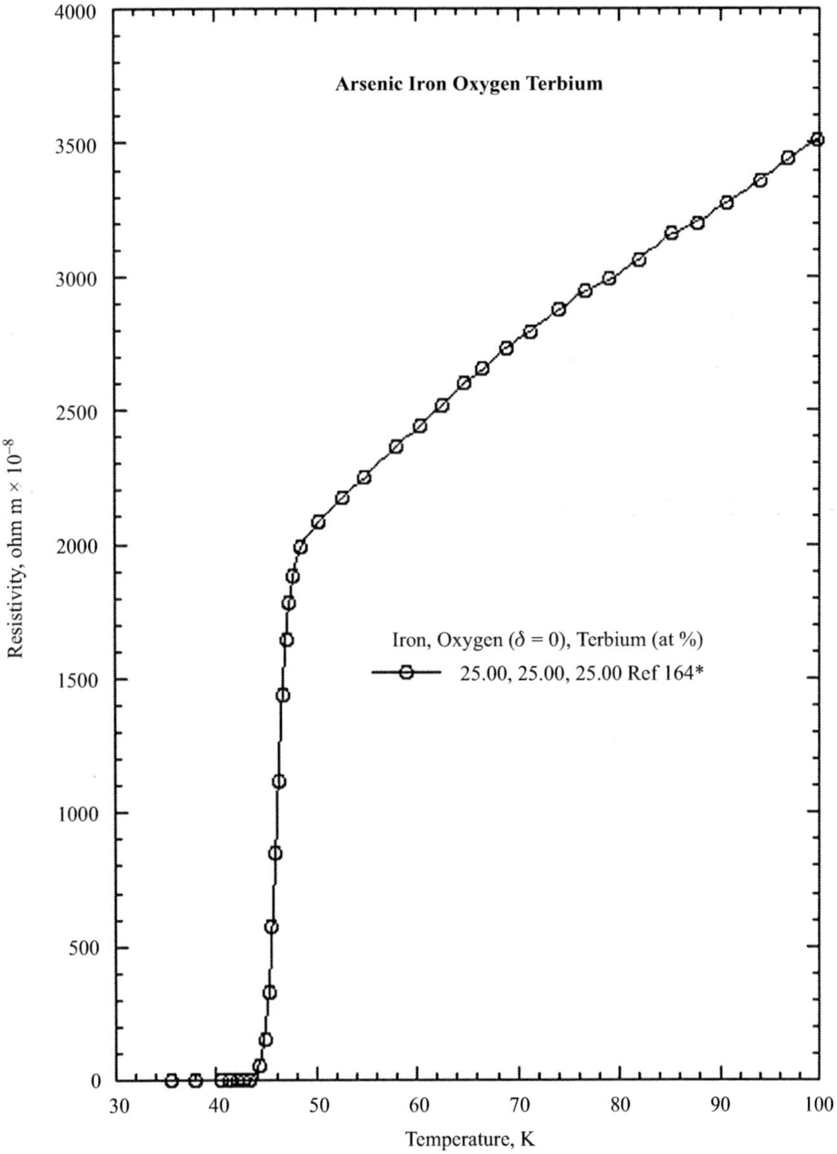
Arsenic Iron Oxygen Terbium
Iron, Oxygen (δ = 0), Terbium (at %)
25.00, 25.00, 25.00 Ref 164*
Resistivity, ohm m × 10⁻⁸
4000
3500
3000
2500
2000
1500
1000
500
0
30
40
50
60
70
80
90
100
Temperature, K

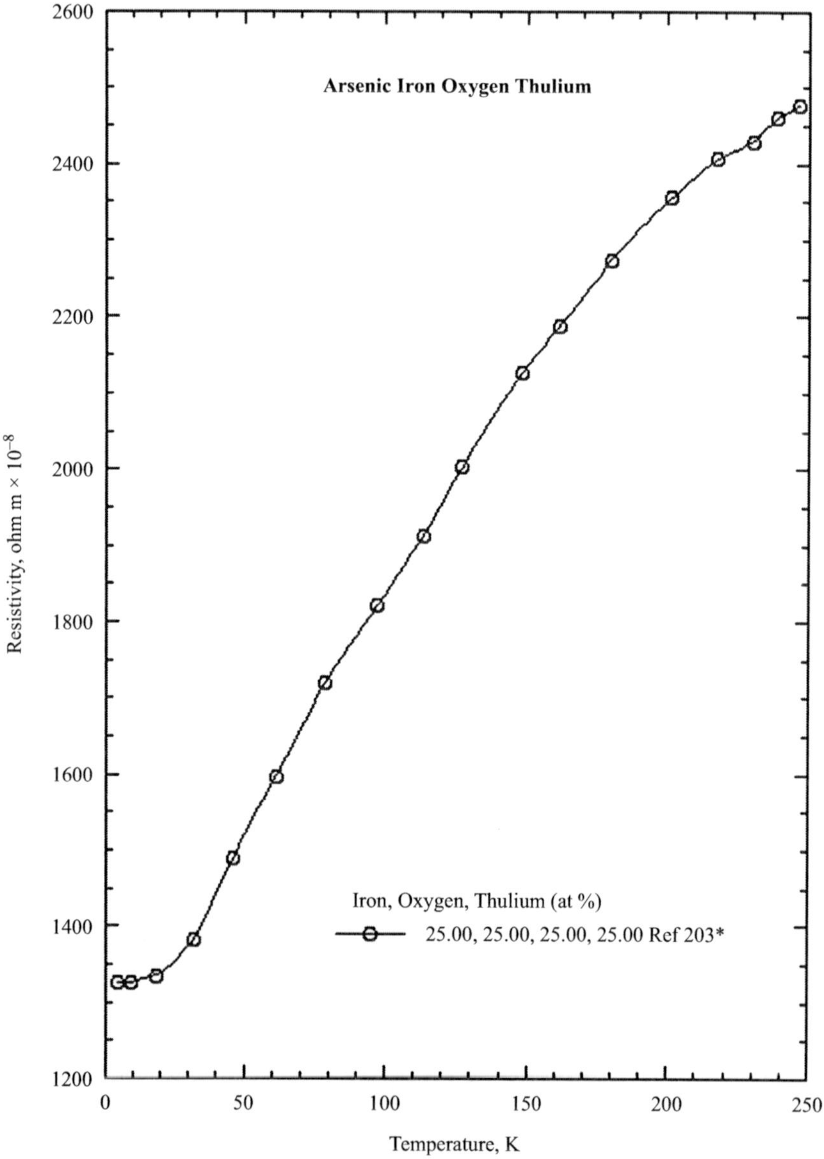
Arsenic Iron Oxygen Thulium
Iron, Oxygen, Thulium (at %)
25.00, 25.00, 25.00, 25.00 Ref 203*
Resistivity, ohm m × 10^-8
Temperature, K
2600
2400
2200
2000
1800
1600
1400
1200
0
50
100
150
200
250

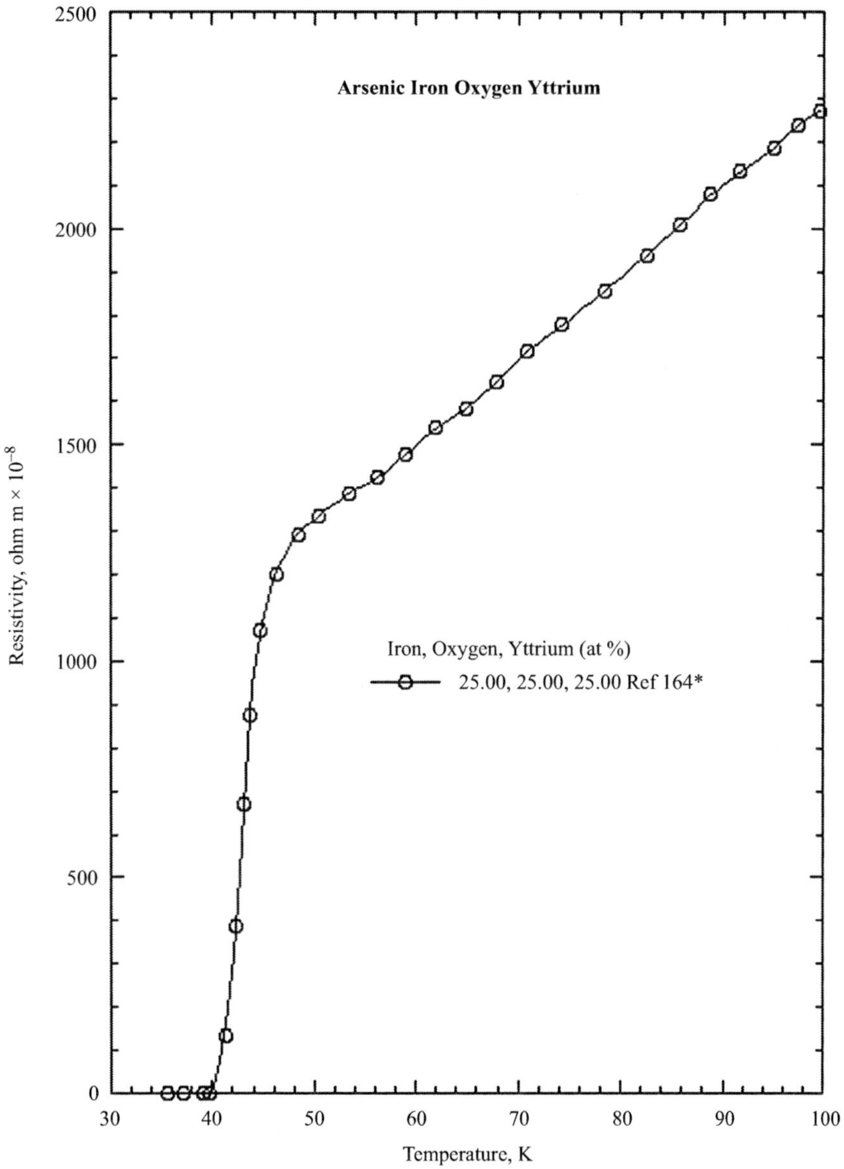
Arsenic Iron Oxygen Yttrium
Iron, Oxygen, Yttrium (at %)
25.00, 25.00, 25.00 Ref 164*
Resistivity, ohm m × 10⁻⁸
Temperature, K
0
500
1000
1500
2000
2500
30
40
50
60
70
80
90
100

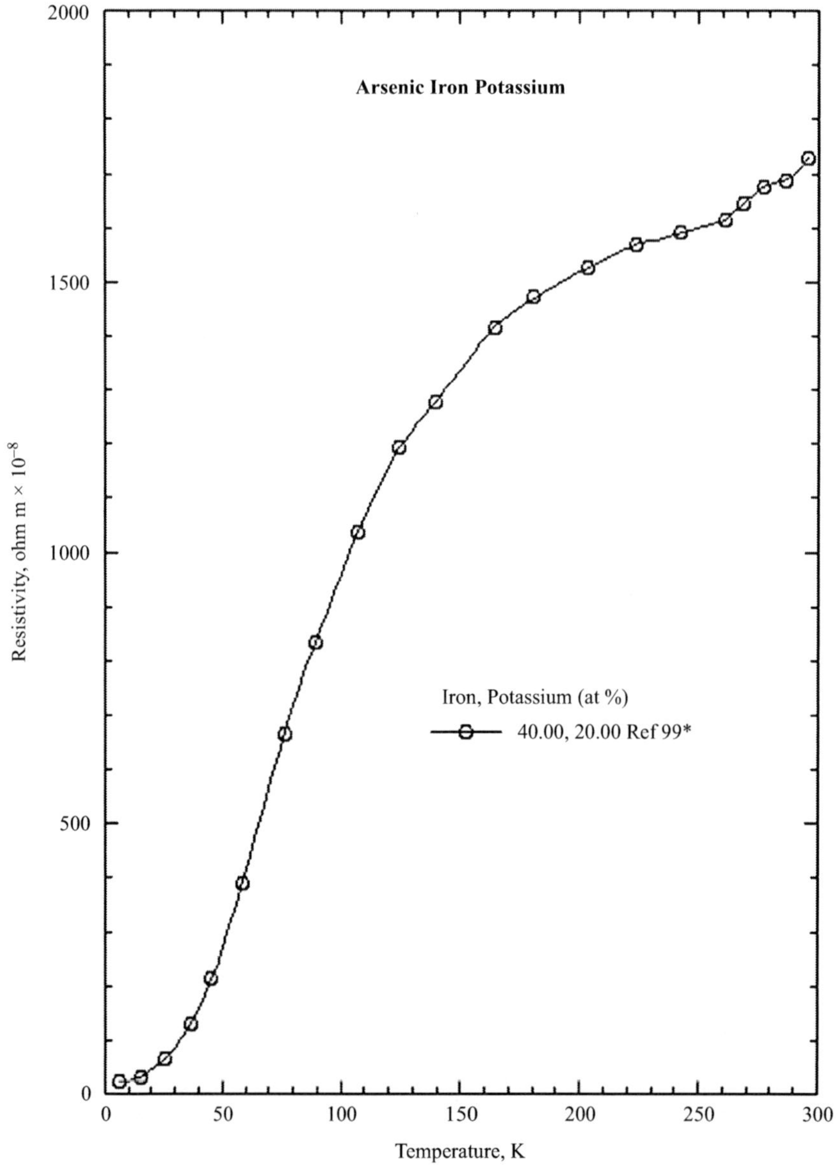

Arsenic Iron Potassium
Iron, Potassium (at %)
40.00, 20.00 Ref 99*
Resistivity, ohm m × 10–8
Temperature, K
0
500
1000
1500
2000
0
50
100
150
200
250
300

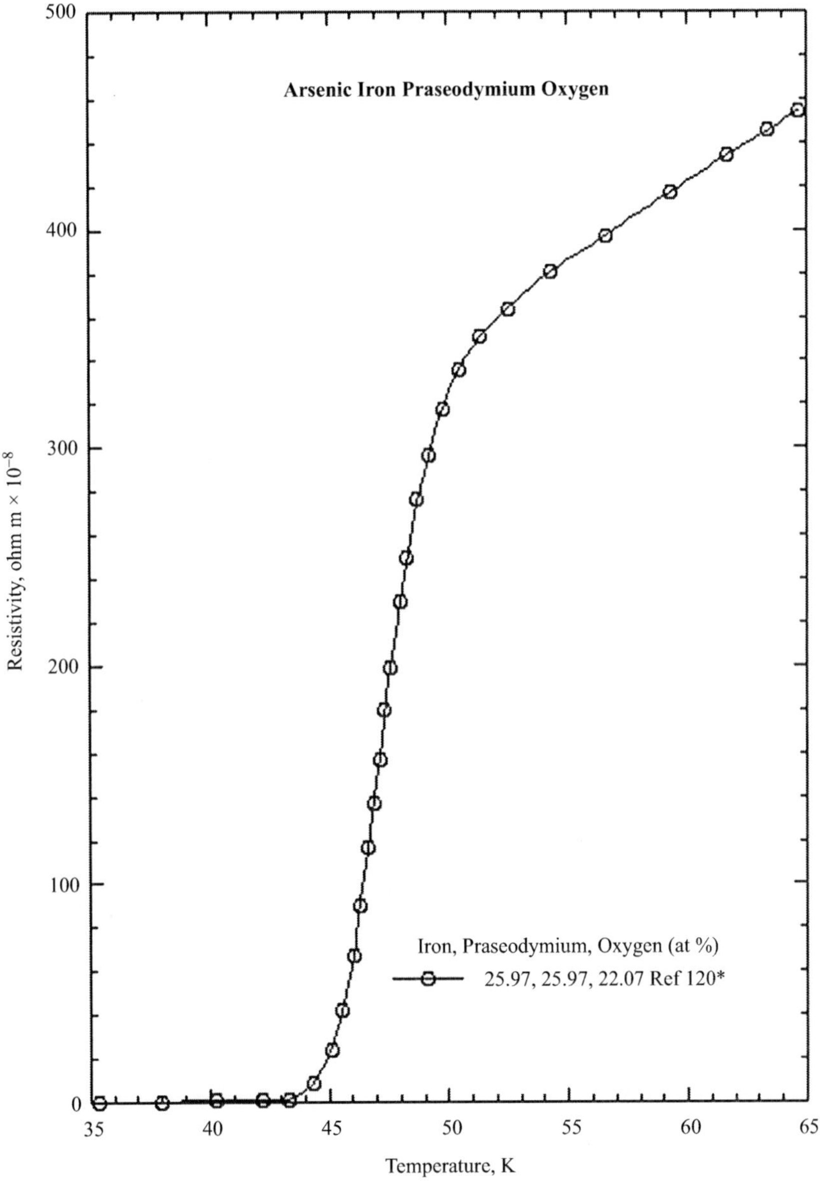
Arsenic Iron Praseodymium Oxygen
Iron, Praseodymium, Oxygen (at %)
25.97, 25.97, 22.07 Ref 120*
Resistivity, ohm m × 10⁻⁸
Temperature, K
0
100
200
300
400
500
35
40
45
50
55
60
65

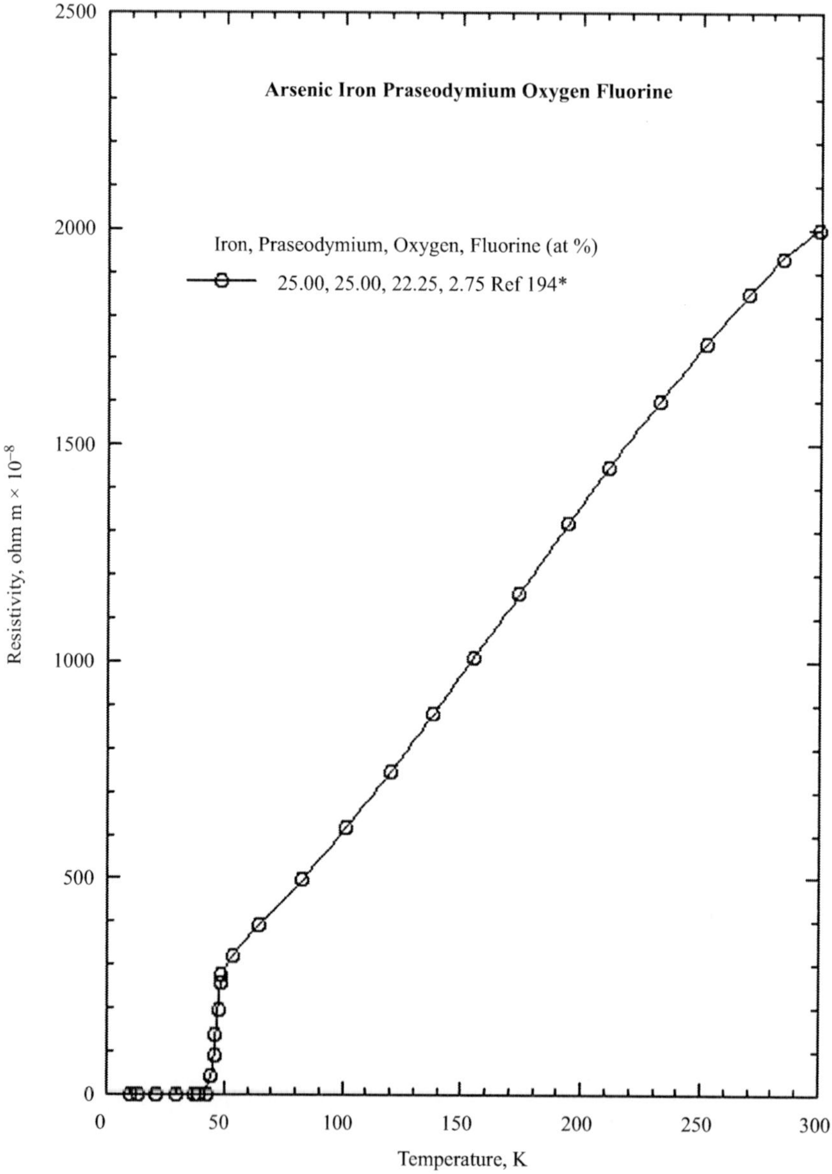
Arsenic Iron Praseodymium Oxygen Fluorine
Iron, Praseodymium, Oxygen, Fluorine (at %)
25.00, 25.00, 22.25, 2.75 Ref 194*
Resistivity, ohm m × 10⁻⁸
Temperature, K
0
500
1000
1500
2000
2500
0
50
100
150
200
250
300

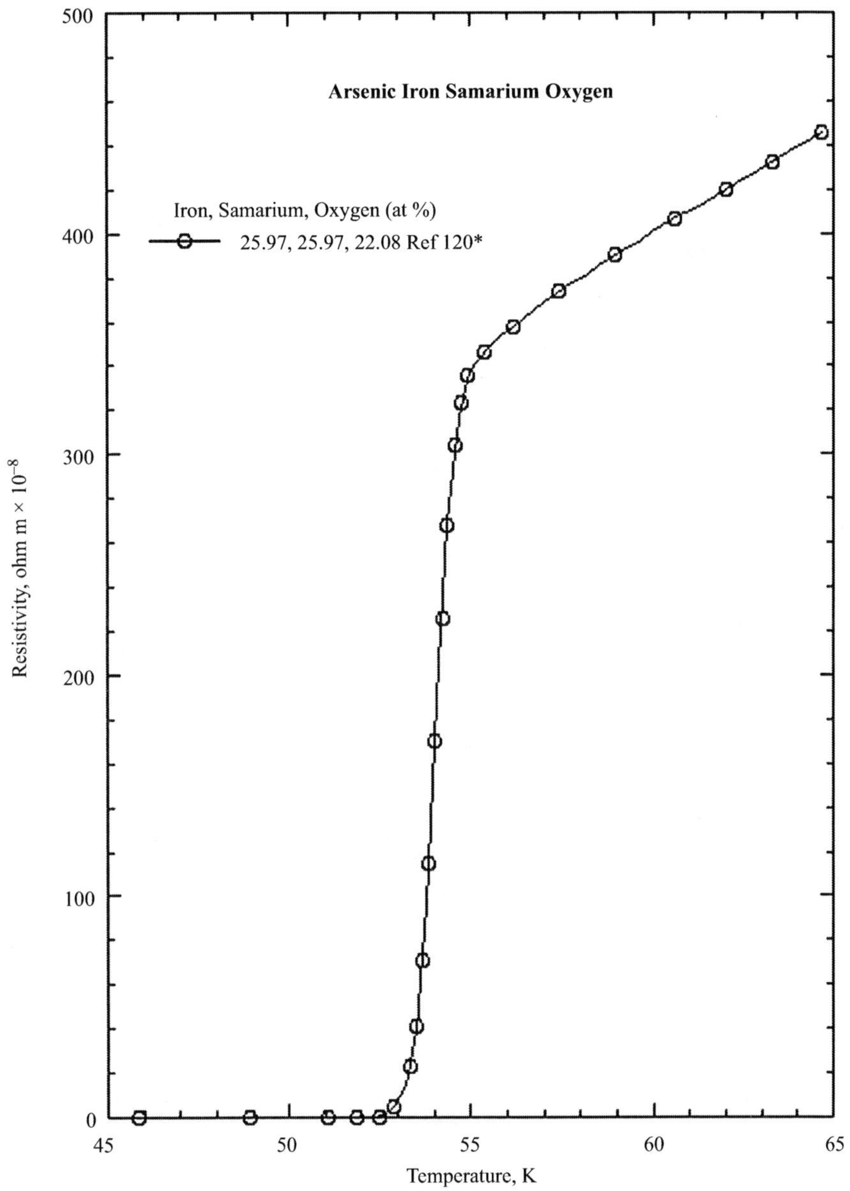
Arsenic Iron Samarium Oxygen
Iron, Samarium, Oxygen (at %)
25.97, 25.97, 22.08 Ref 120*
Resistivity, ohm m × 10^{-8}
Temperature, K
0
100
200
300
400
500
45
50
55
60
65

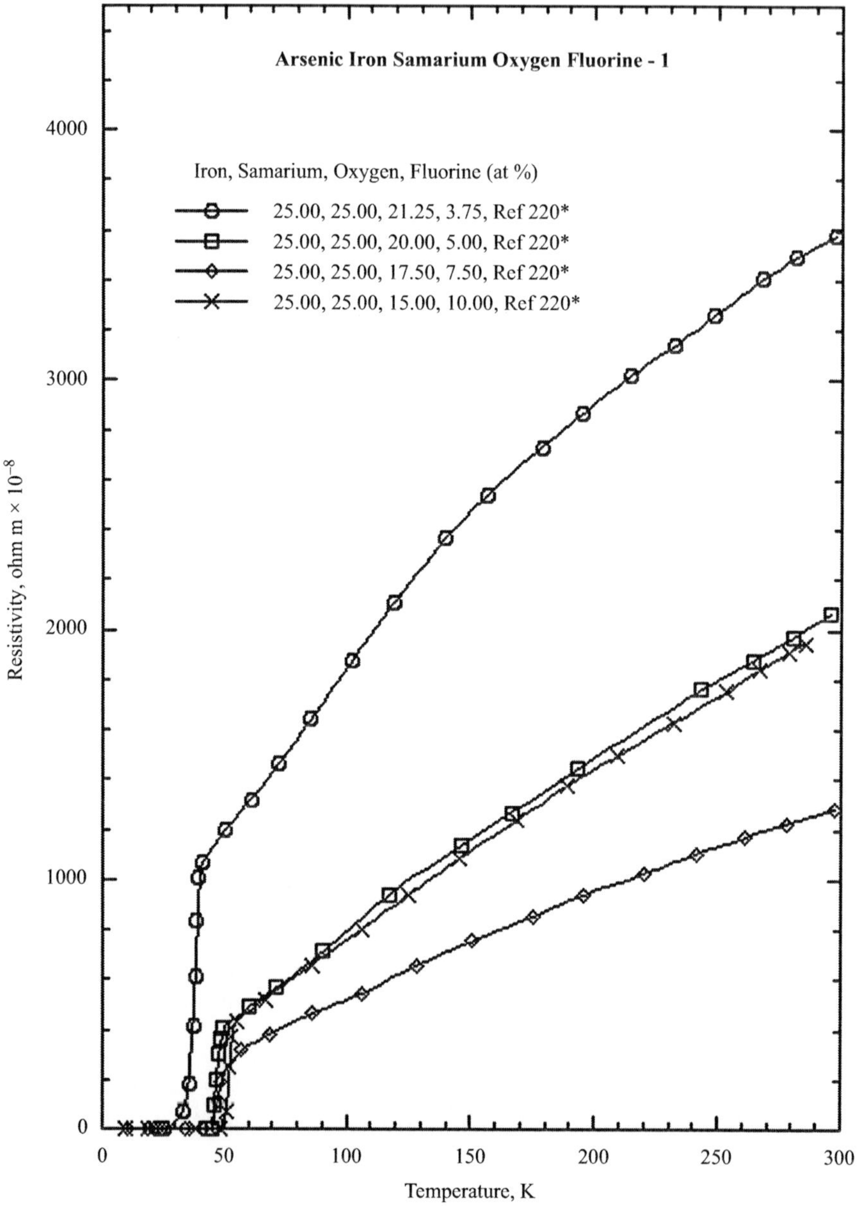

Arsenic Iron Samarium Oxygen Fluorine - 1
Iron, Samarium, Oxygen, Fluorine (at %)
25.00, 25.00, 21.25, 3.75, Ref 220*
25.00, 25.00, 20.00, 5.00, Ref 220*
25.00, 25.00, 17.50, 7.50, Ref 220*
25.00, 25.00, 15.00, 10.00, Ref 220*
Resistivity, ohm m × 10^{-8}
0
1000
2000
3000
4000
0
50
100
150
200
250
300
Temperature, K

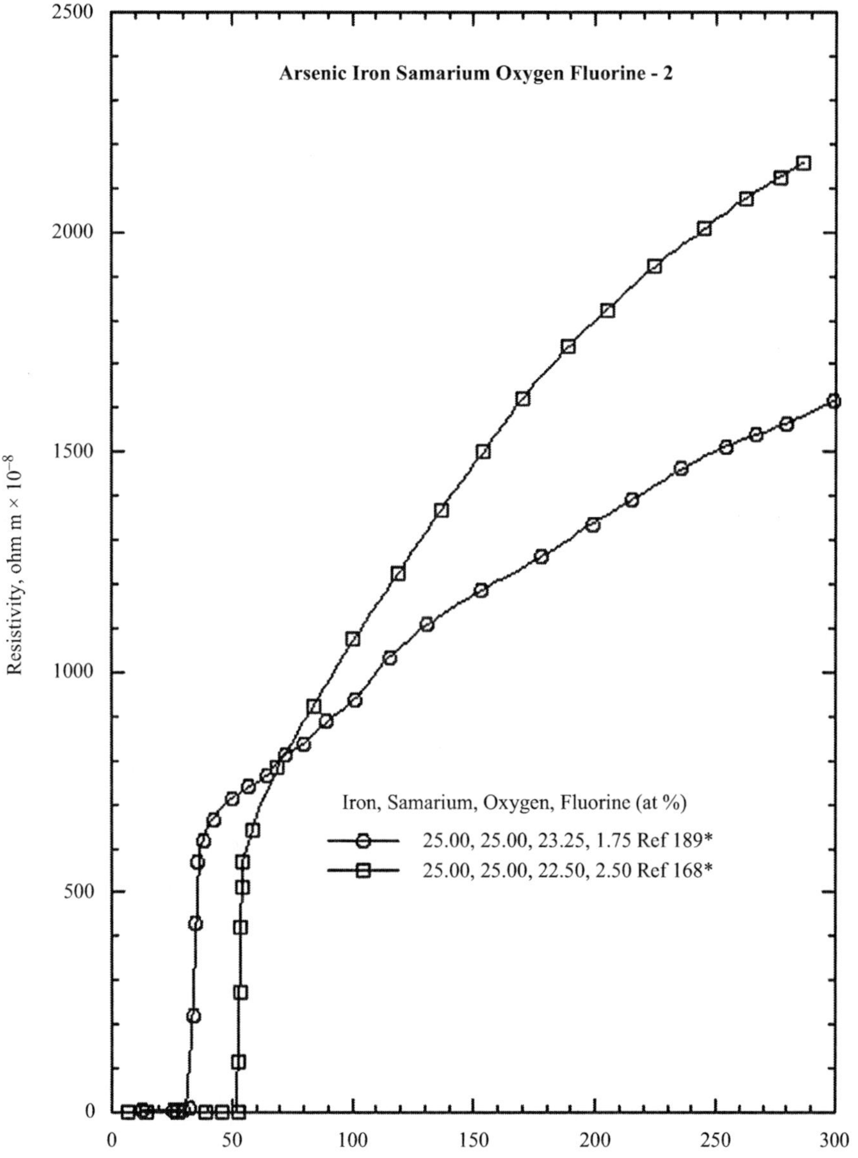

Arsenic Iron Samarium Oxygen Fluorine - 2
Iron, Samarium, Oxygen, Fluorine (at %)
25.00, 25.00, 23.25, 1.75 Ref 189*
25.00, 25.00, 22.50, 2.50 Ref 168*
Resistivity, ohm m × 10⁻⁸
Temperature, K
0
500
1000
1500
2000
2500
0
50
100
150
200
250
300

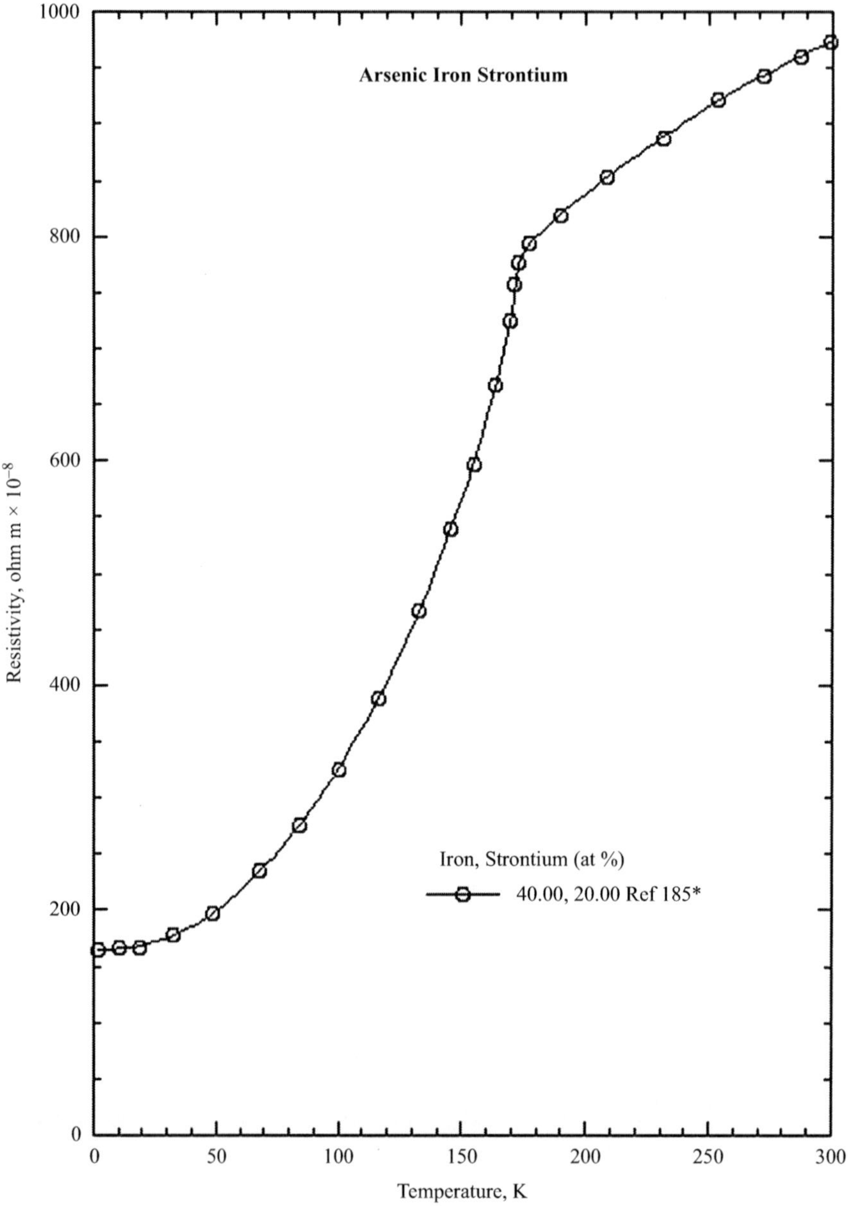
Arsenic Iron Strontium
Iron, Strontium (at %)
40.00, 20.00 Ref 185*
Resistivity, ohm m × 10⁻⁸
Temperature, K
0
200
400
600
800
1000
0
50
100
150
200
250
300

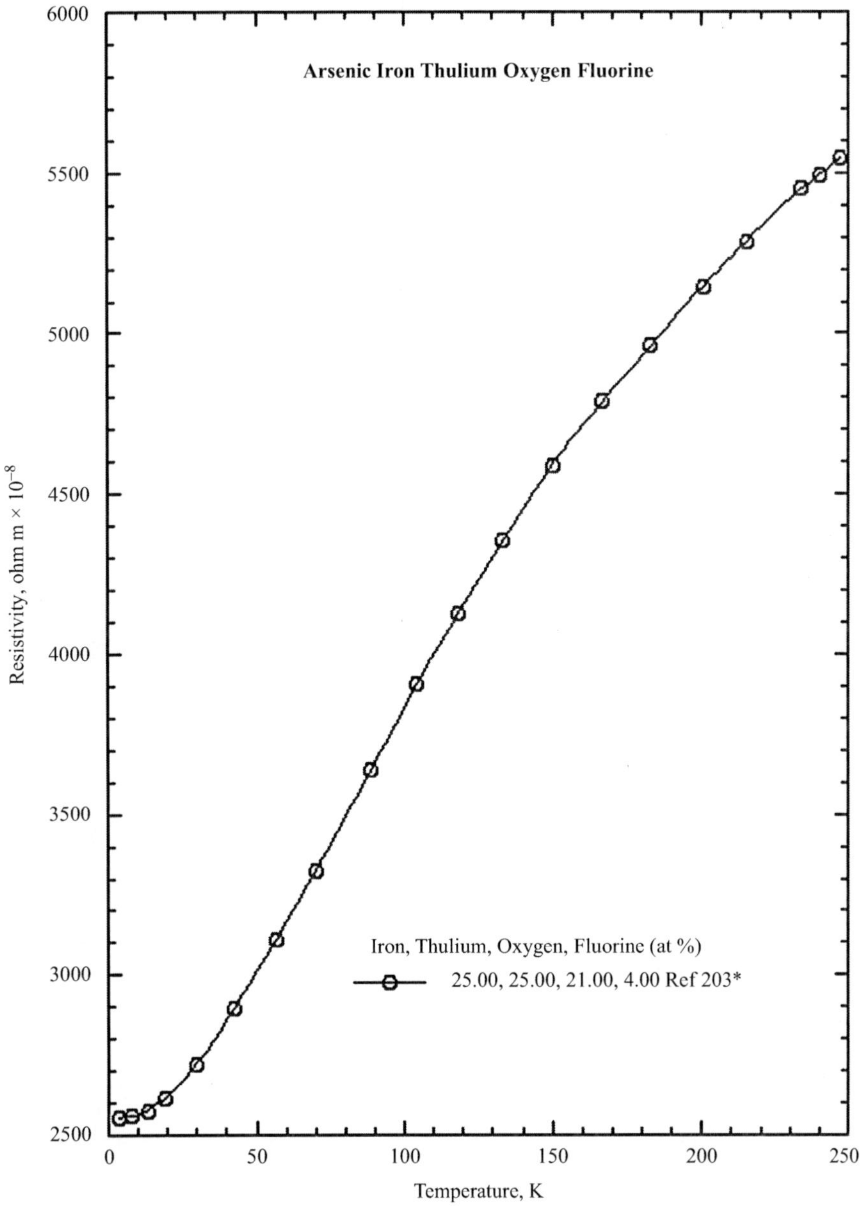

Arsenic Iron Thulium Oxygen Fluorine
Iron, Thulium, Oxygen, Fluorine (at %)
25.00, 25.00, 21.00, 4.00 Ref 203*
Resistivity, ohm m × 10−8
Temperature, K
6000
5500
5000
4500
4000
3500
3000
2500
0
50
100
150
200
250

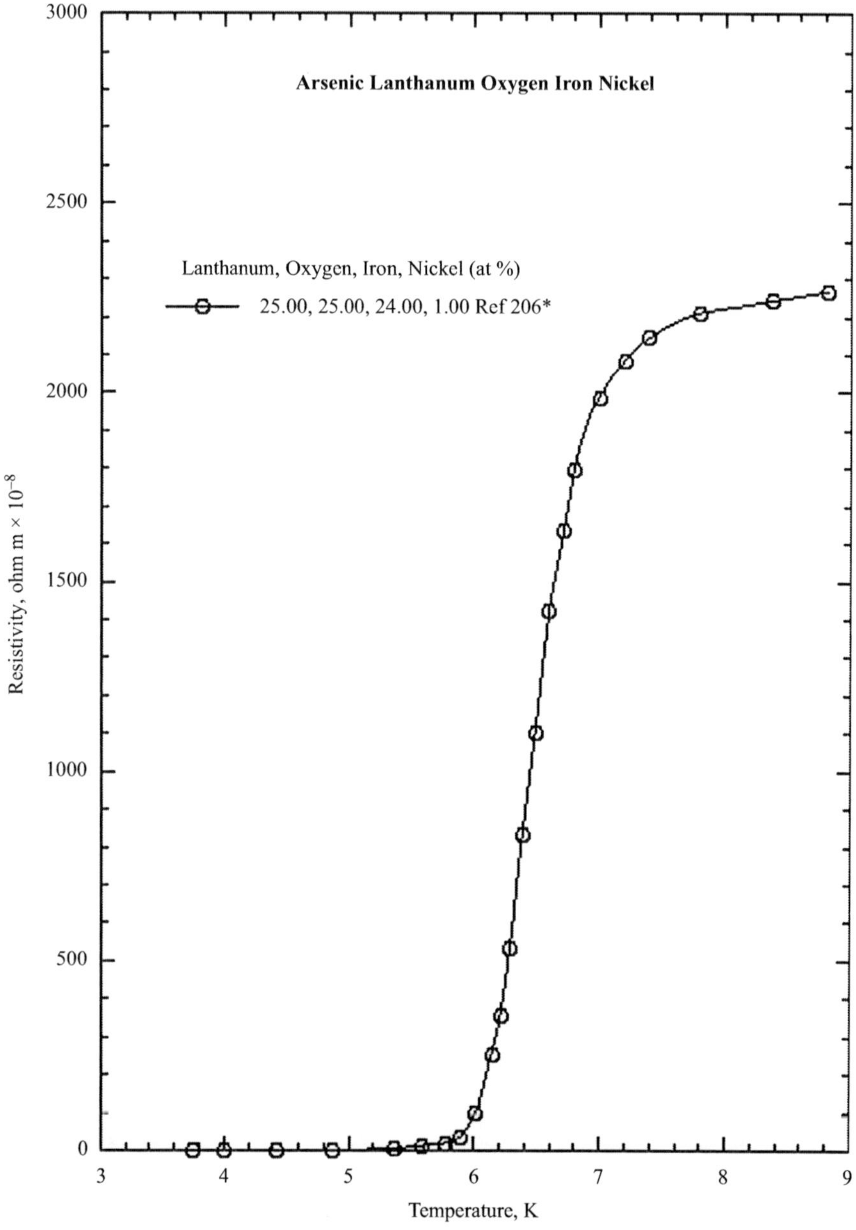
Arsenic Lanthanum Oxygen Iron Nickel
Lanthanum, Oxygen, Iron, Nickel (at %)
25.00, 25.00, 24.00, 1.00 Ref 206*
Resistivity, ohm m × 10⁻⁸
Temperature, K
3000
2500
2000
1500
1000
500
0
3
4
5
6
7
8
9

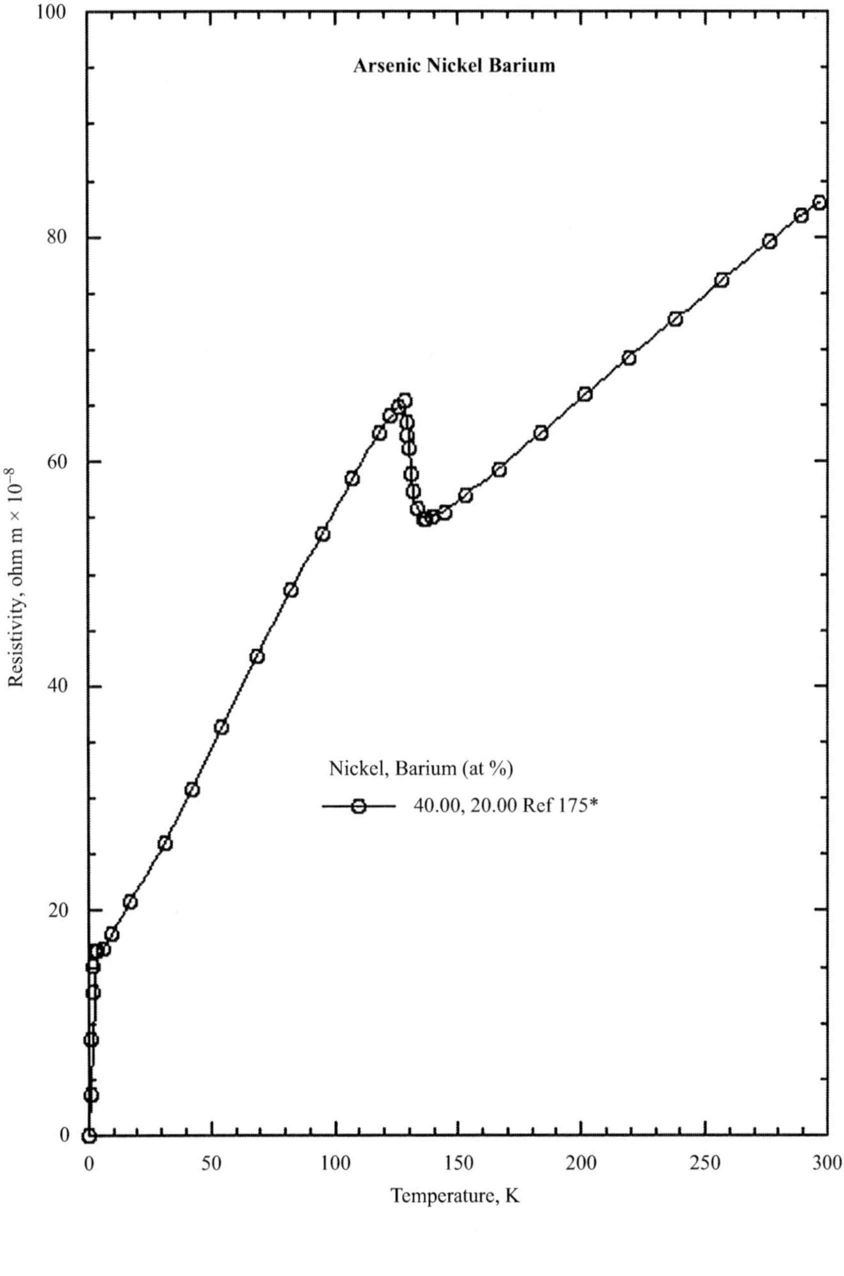
Arsenic Nickel Barium
Resistivity, ohm m × 10−8
Temperature, K
Nickel, Barium (at %)
40.00, 20.00 Ref 175*
0
20
40
60
80
100
0
50
100
150
200
250
300

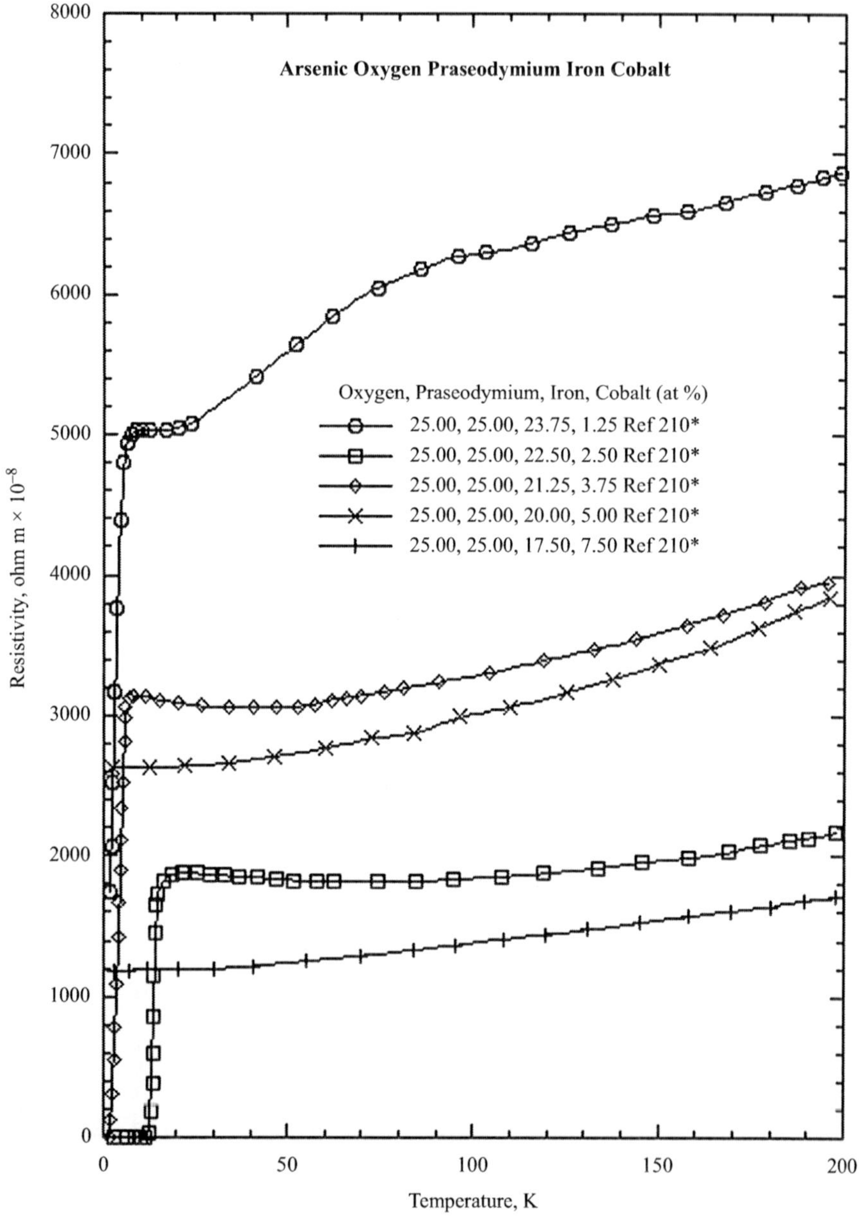
Arsenic Oxygen Praseodymium Iron Cobalt
Oxygen, Praseodymium, Iron, Cobalt (at %)
25.00, 25.00, 23.75, 1.25 Ref 210*
25.00, 25.00, 22.50, 2.50 Ref 210*
25.00, 25.00, 21.25, 3.75 Ref 210*
25.00, 25.00, 20.00, 5.00 Ref 210*
25.00, 25.00, 17.50, 7.50 Ref 210*
Resistivity, ohm m × 10−8
Temperature, K
0
1000
2000
3000
4000
5000
6000
7000
8000
0
50
100
150
200

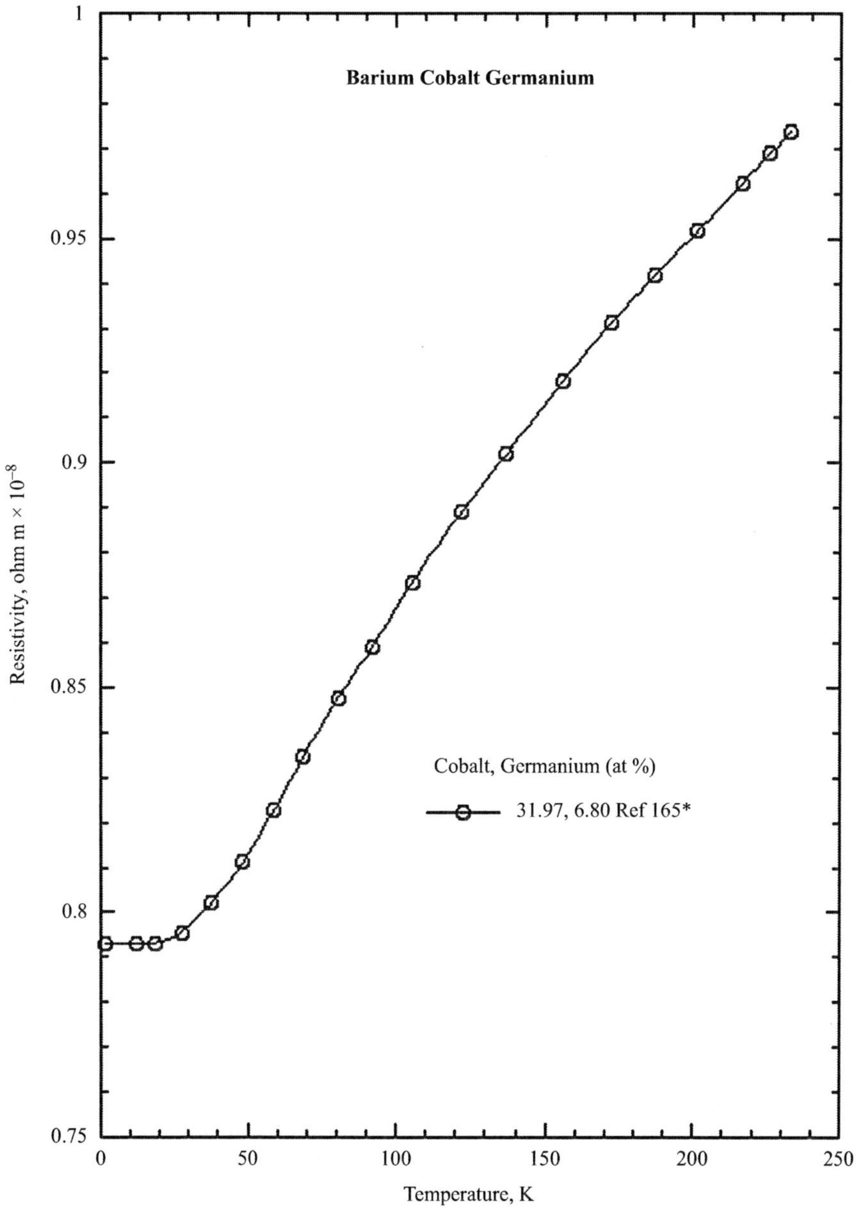

Barium Cobalt Germanium
Cobalt, Germanium (at %)
31.97, 6.80 Ref 165*
Resistivity, ohm m × 10⁻⁸
Temperature, K
1
0.95
0.9
0.85
0.8
0.75
0
50
100
150
200
250

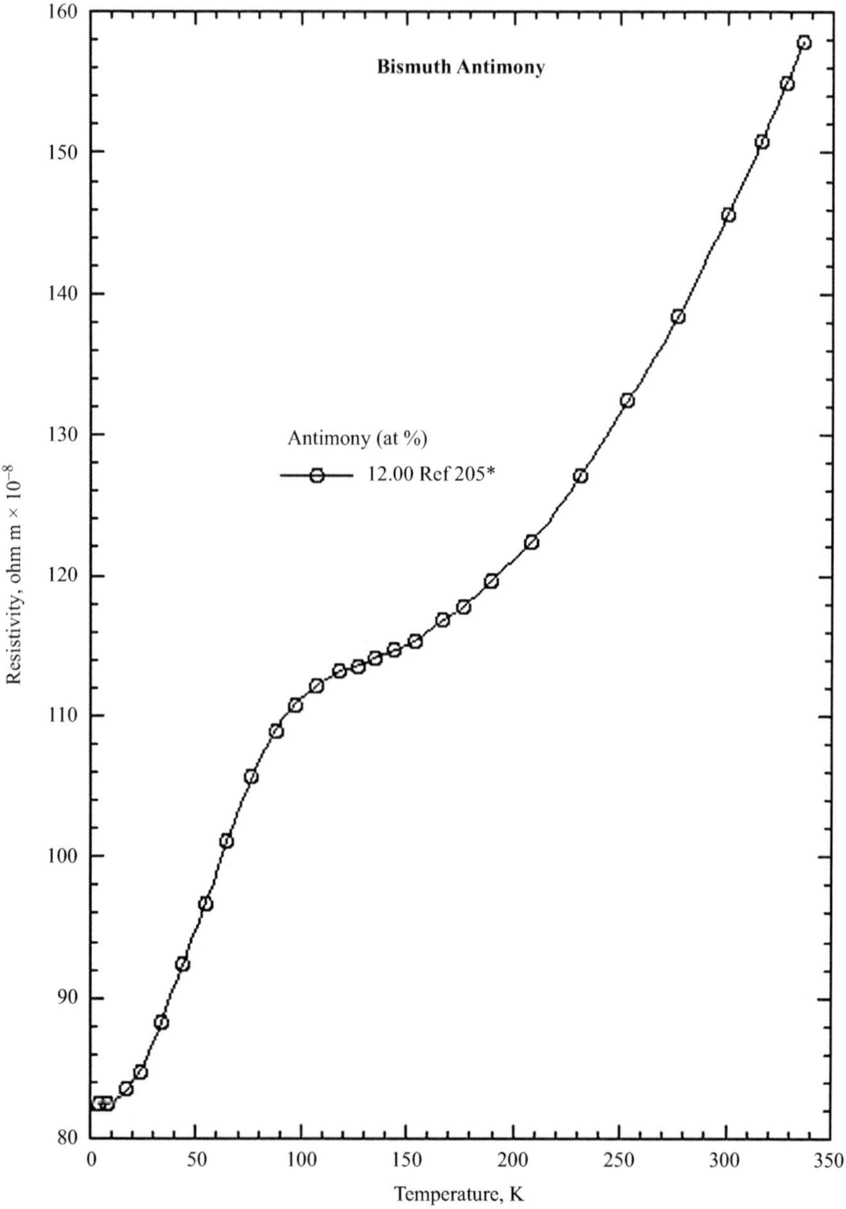
Bismuth Antimony
Antimony (at %)
12.00 Ref 205*
Resistivity, ohm m × 10⁻⁸
Temperature, K
160
150
140
130
120
110
100
90
80
0
50
100
150
200
250
300
350

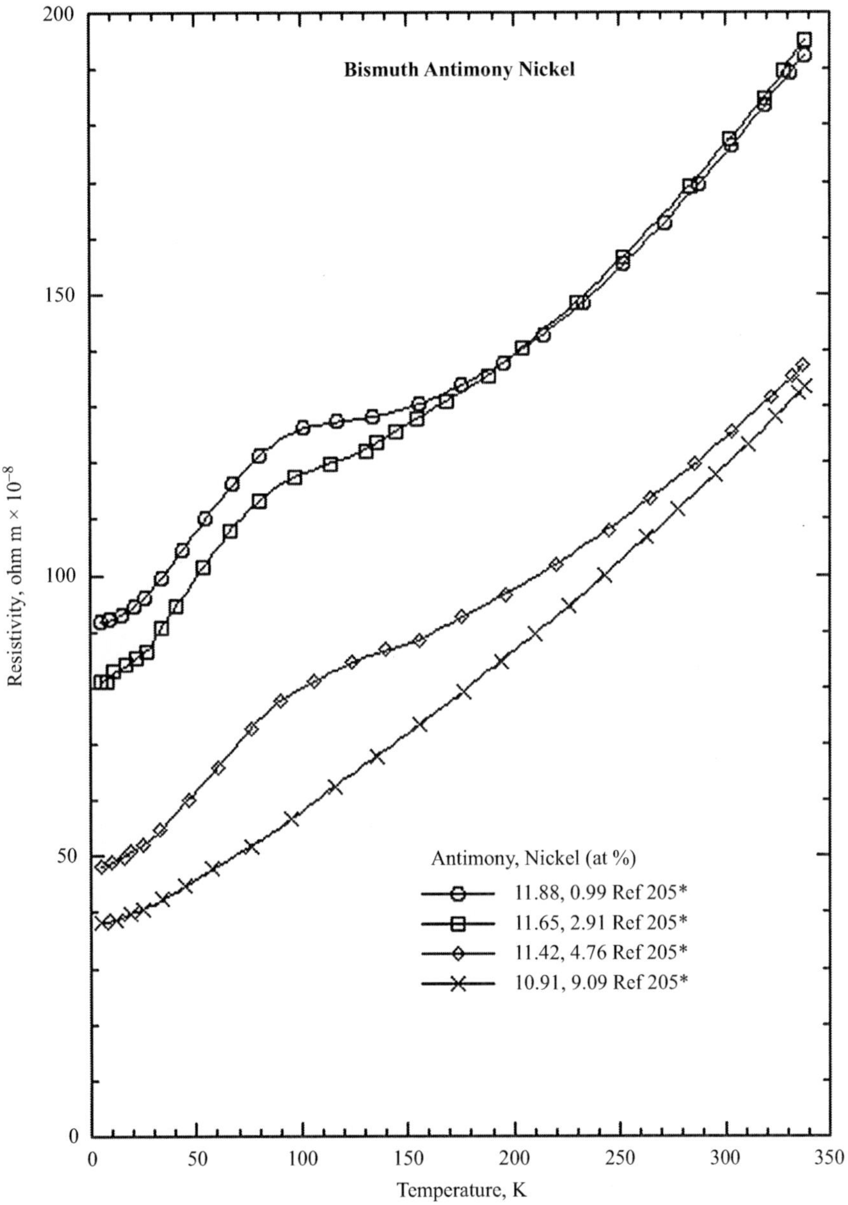
Bismuth Antimony Nickel
Antimony, Nickel (at %)
11.88, 0.99 Ref 205*
11.65, 2.91 Ref 205*
11.42, 4.76 Ref 205*
10.91, 9.09 Ref 205*
Resistivity, ohm m × 10^{-8}
Temperature, K
0
50
100
150
200
250
300
350

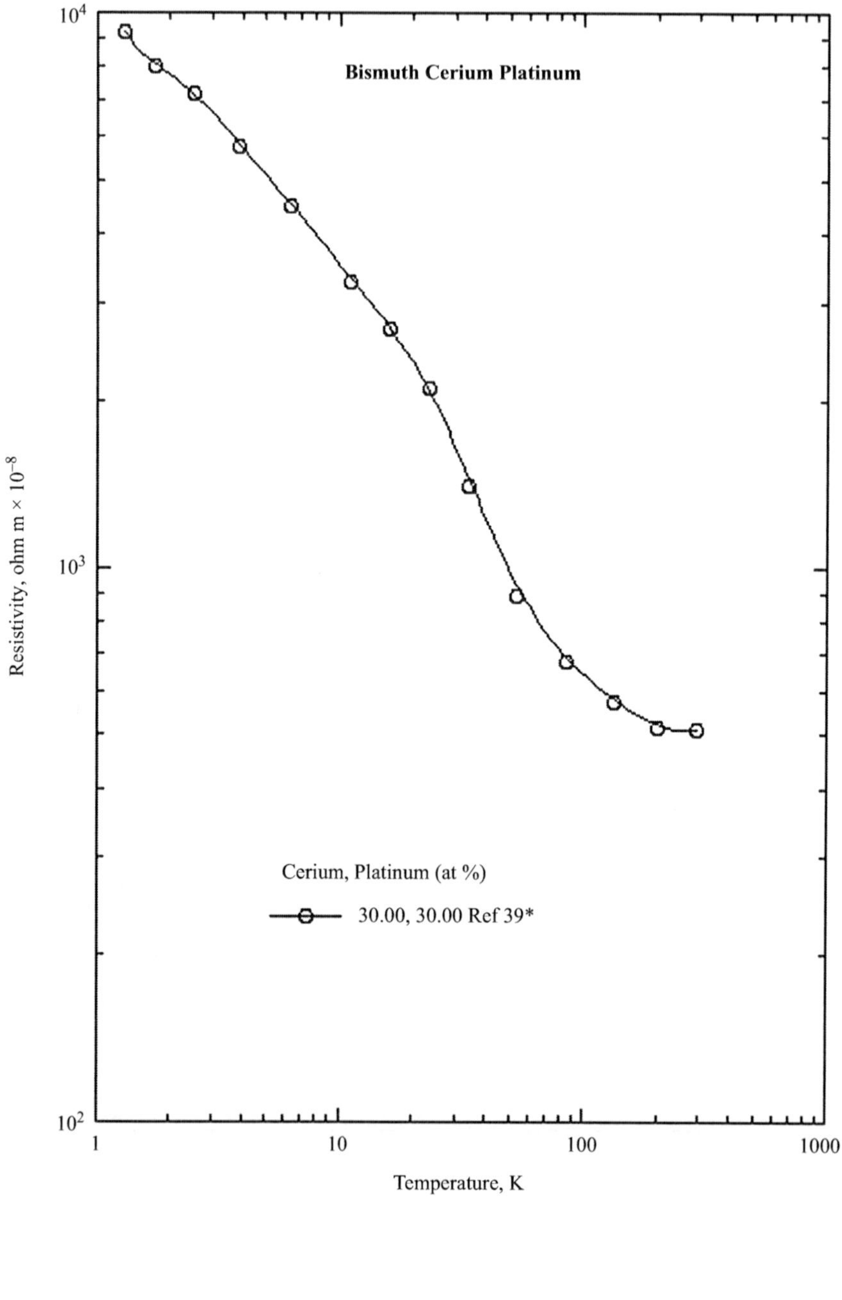
Bismuth Cerium Platinum
Cerium, Platinum (at %)
30.00, 30.00 Ref 39*
Resistivity, ohm m × 10⁻⁸
Temperature, K
10⁴
10³
10²
1
10
100
1000

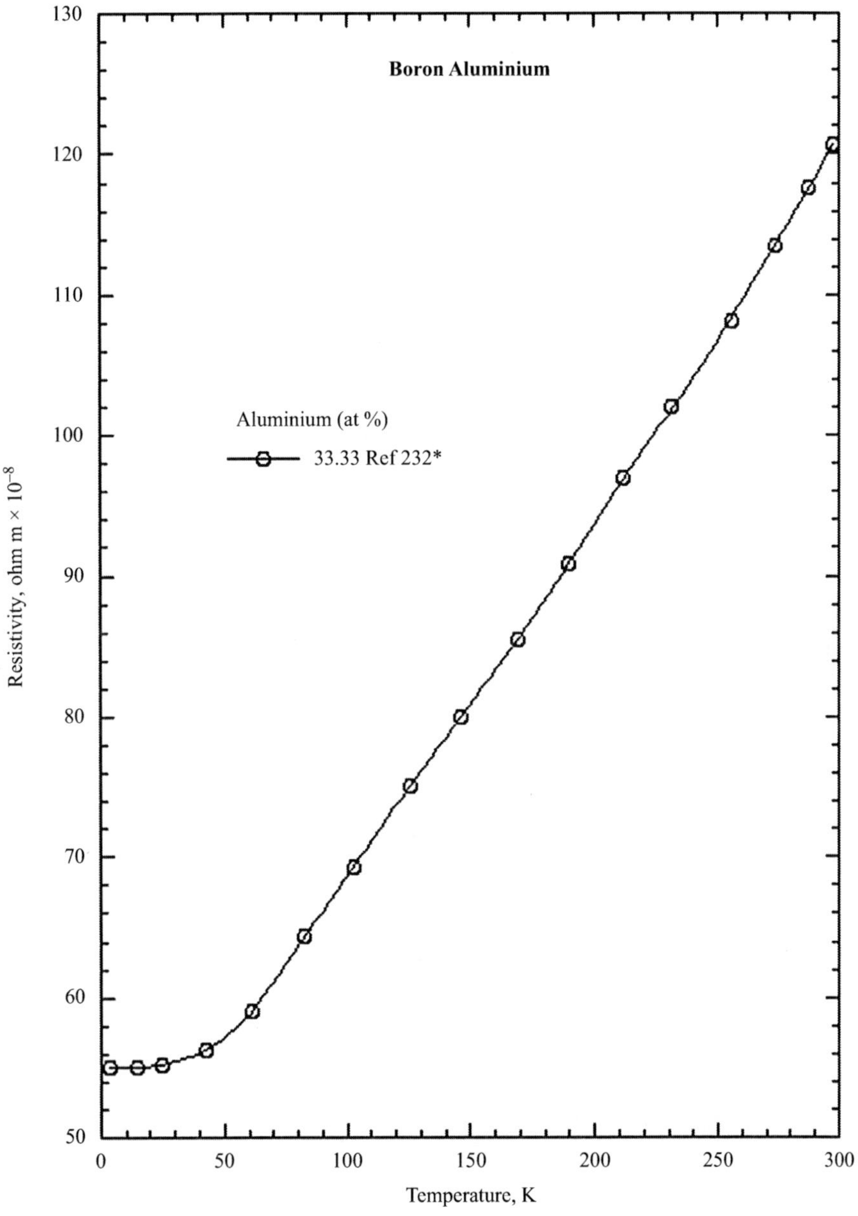

Boron Aluminium
Aluminium (at %)
33.33 Ref 232*
Resistivity, ohm m × 10⁻⁸
Temperature, K
130
120
110
100
90
80
70
60
50
0
50
100
150
200
250
300

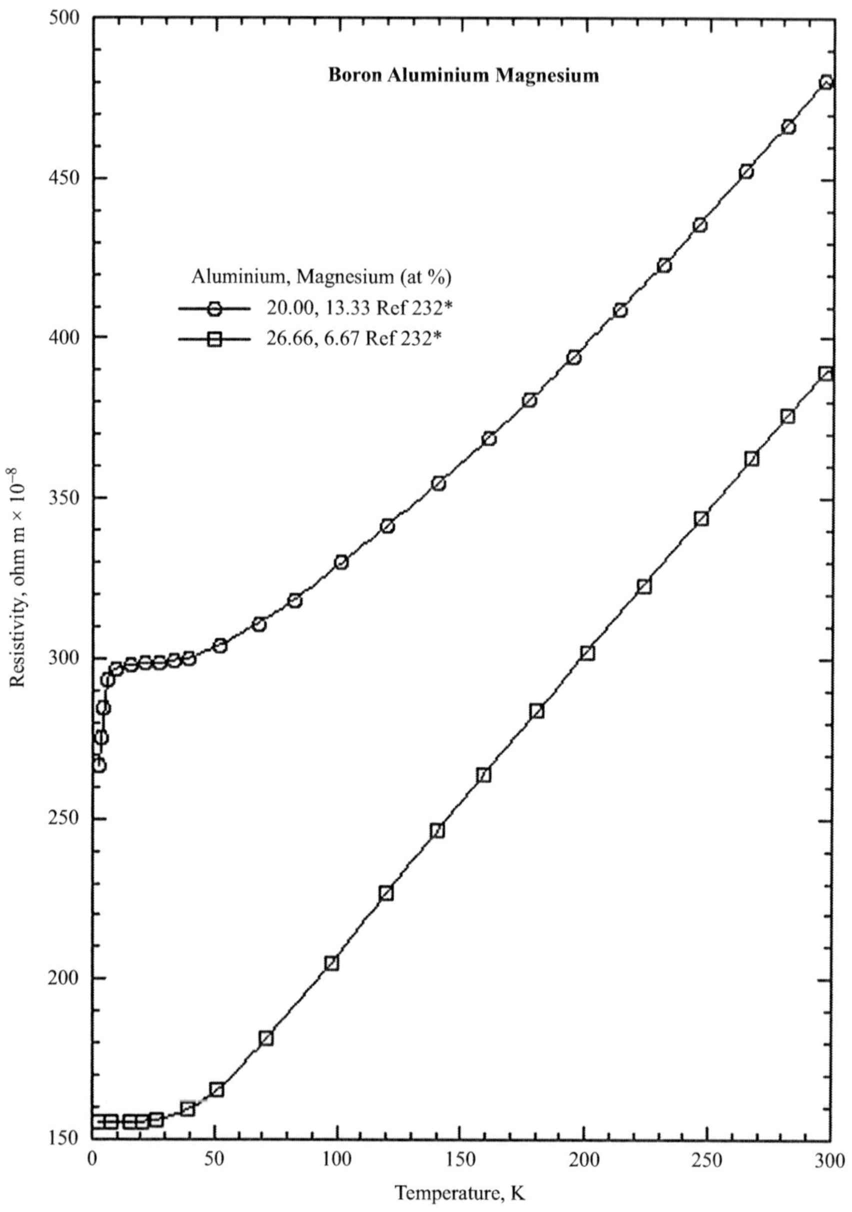
Boron Aluminium Magnesium
Aluminium, Magnesium (at %)
20.00, 13.33 Ref 232*
26.66, 6.67 Ref 232*
Resistivity, ohm m × 10⁻⁸
Temperature, K
500
450
400
350
300
250
200
150
0
50
100
150
200
250
300

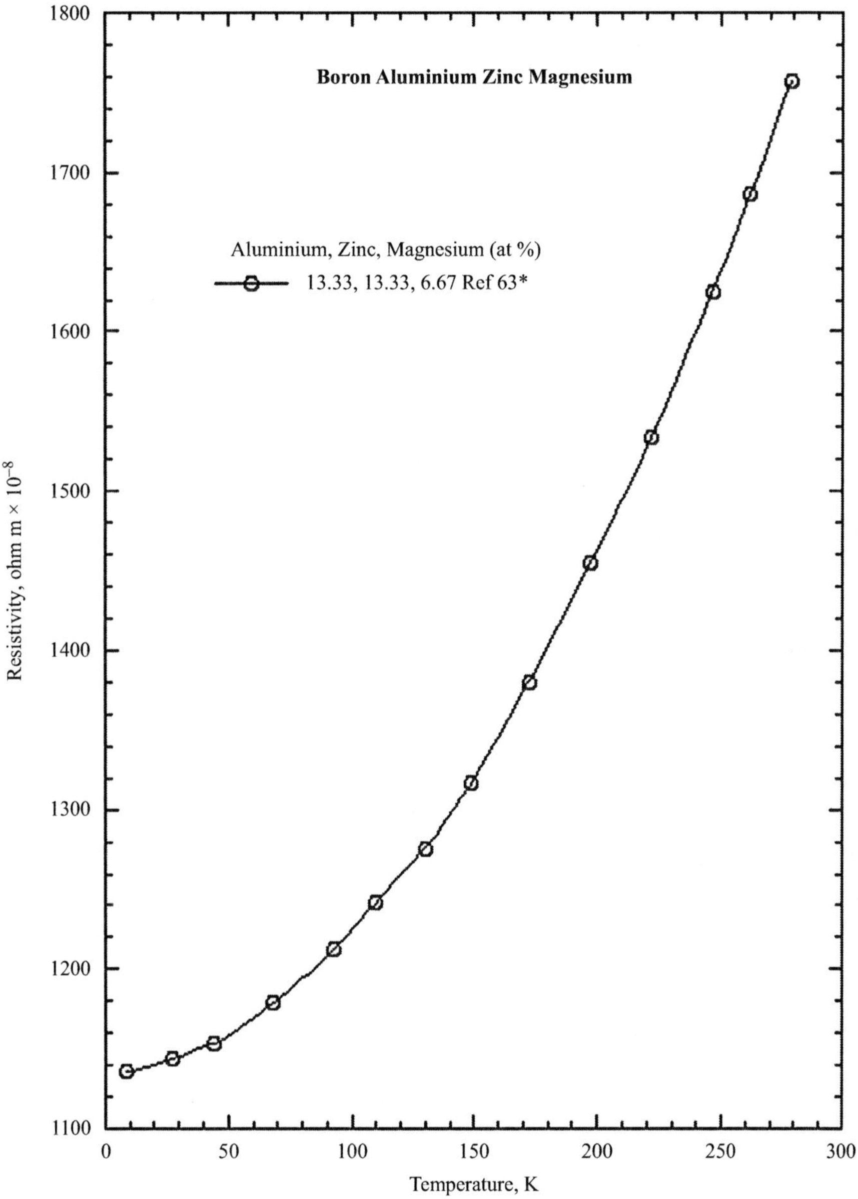
Boron Aluminium Zinc Magnesium
Aluminium, Zinc, Magnesium (at %)
13.33, 13.33, 6.67 Ref 63*
Resistivity, ohm m × 10⁻⁸
Temperature, K
1800
1700
1600
1500
1400
1300
1200
1100
0
50
100
150
200
250
300

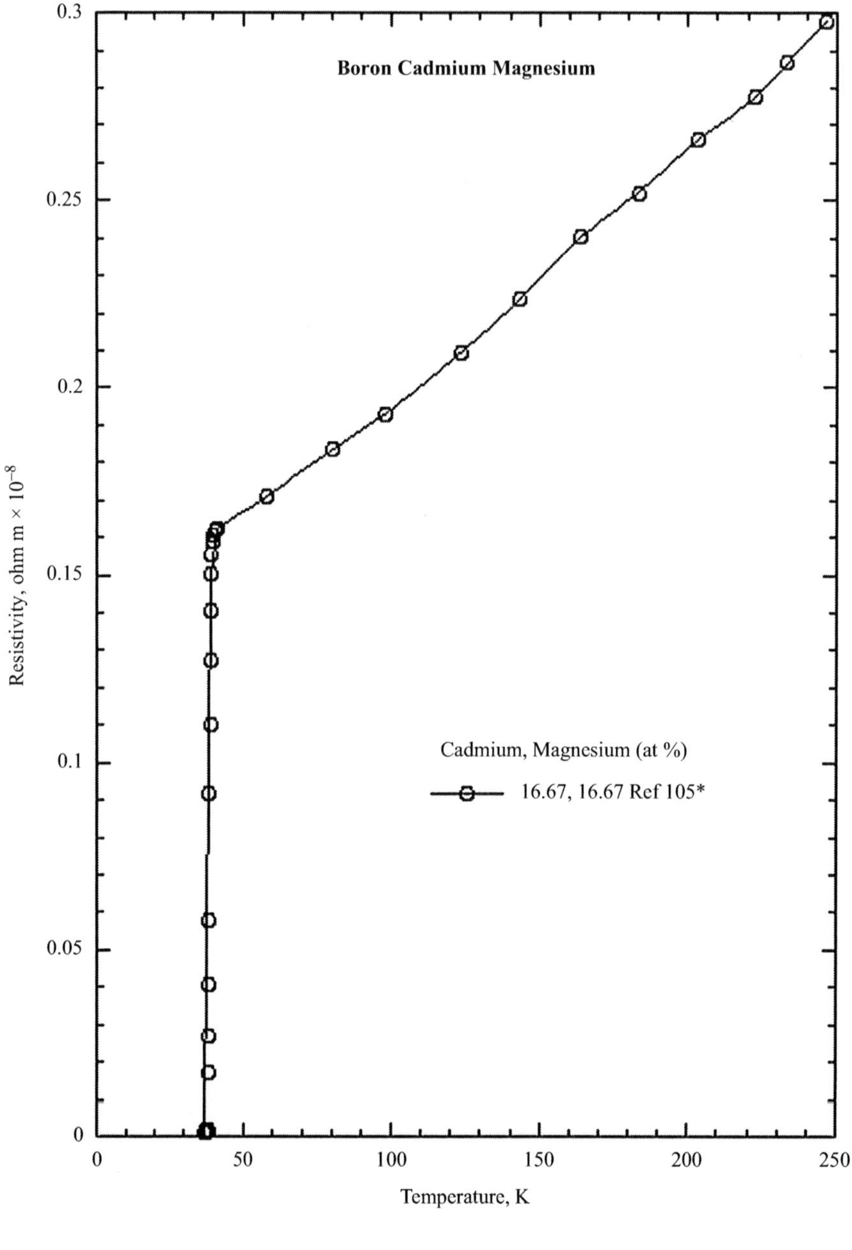
Boron Cadmium Magnesium
Cadmium, Magnesium (at %)
16.67, 16.67 Ref 105*
Resistivity, ohm m × 10⁻⁸
Temperature, K
0
0.05
0.1
0.15
0.2
0.25
0.3
0
50
100
150
200
250

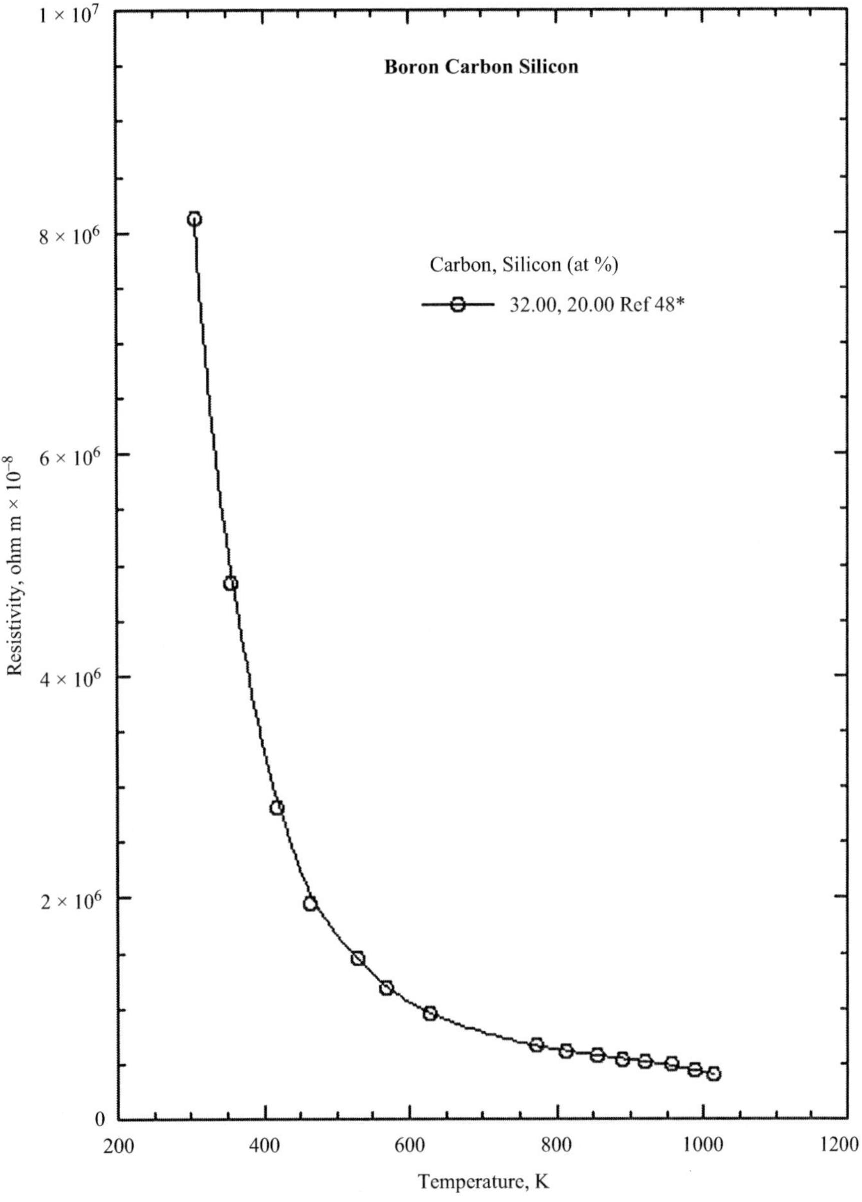
Boron Carbon Silicon
Carbon, Silicon (at %)
32.00, 20.00 Ref 48*
Resistivity, ohm m × 10⁻⁸
1 × 10⁷
8 × 10⁶
6 × 10⁶
4 × 10⁶
2 × 10⁶
0
200
400
600
800
1000
1200
Temperature, K

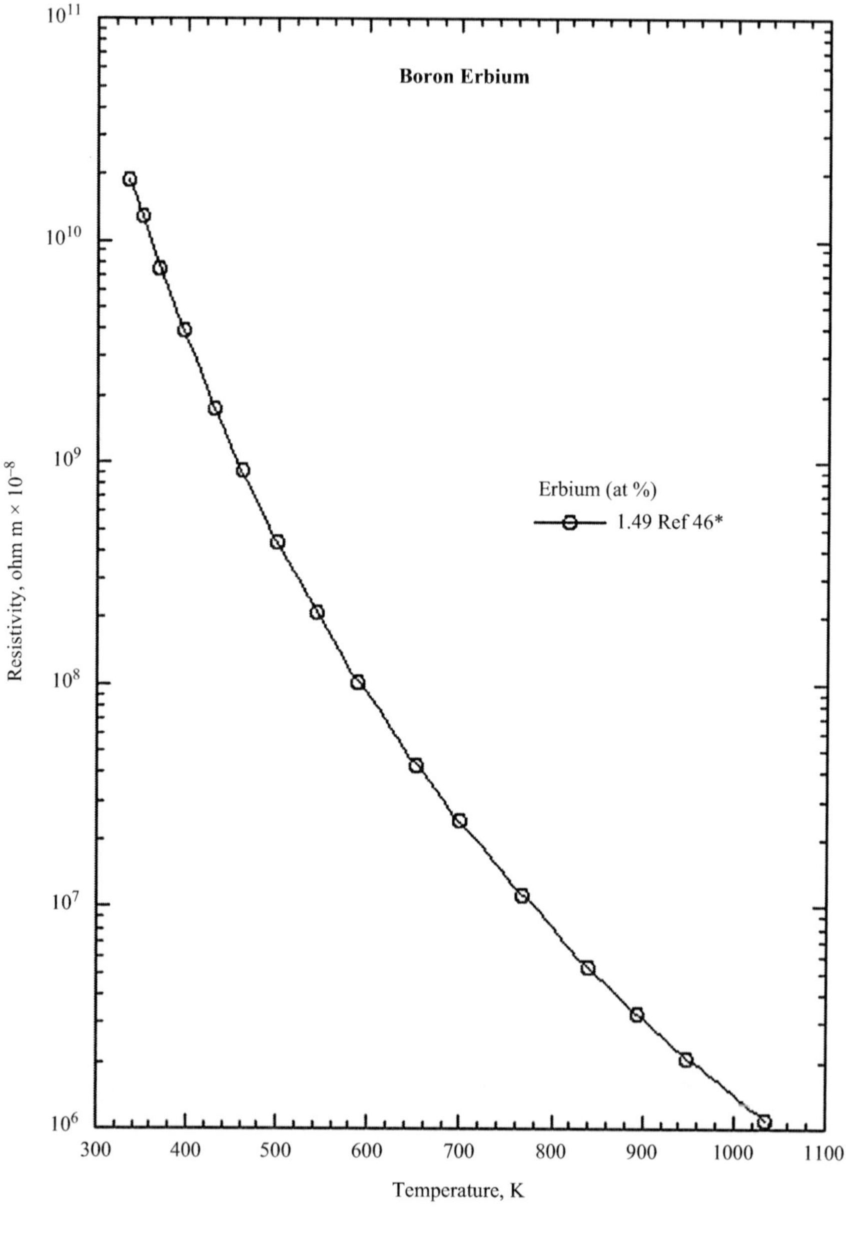
Boron Erbium
Erbium (at %)
1.49 Ref 46*
Resistivity, ohm m × 10⁻⁸
Temperature, K
300
400
500
600
700
800
900
1000
1100

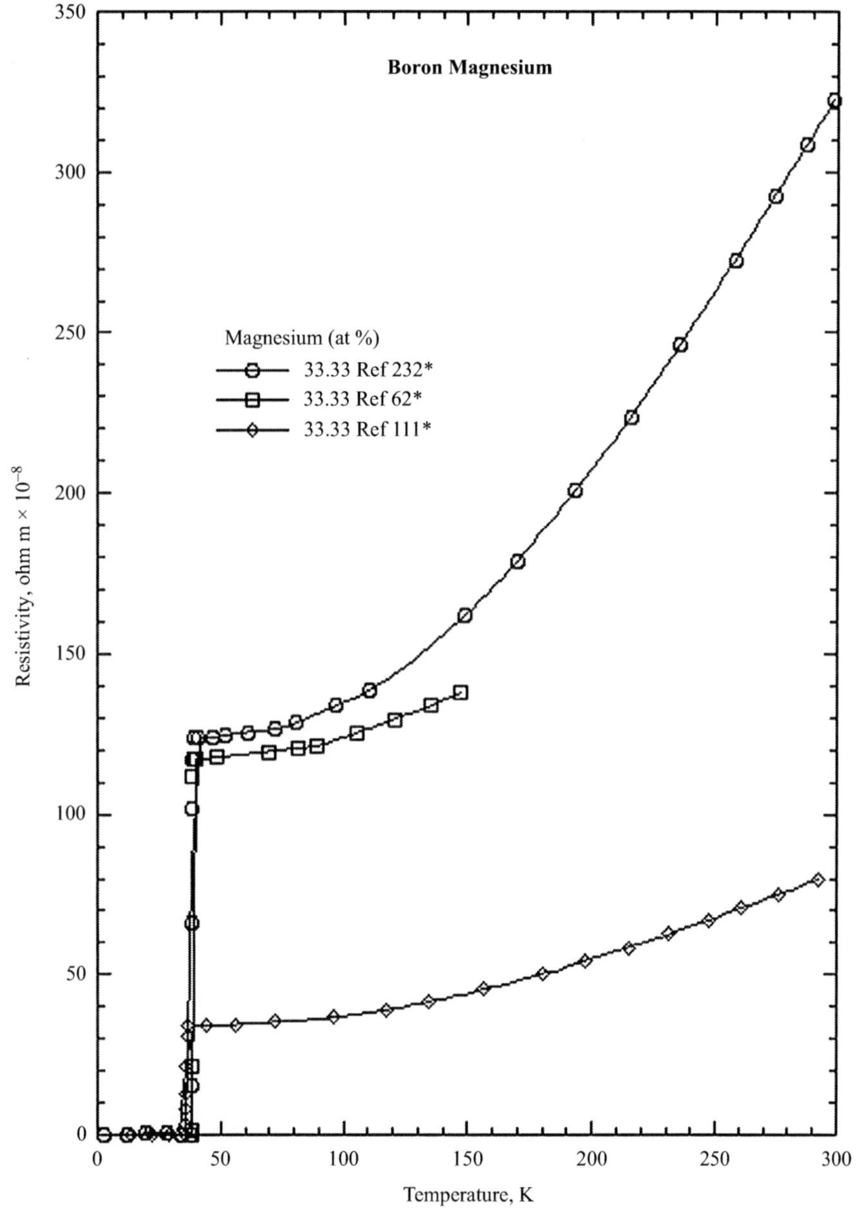
Boron Magnesium
Magnesium (at %)
33.33 Ref 232*
33.33 Ref 62*
33.33 Ref 111*
Resistivity, ohm m × 10⁻⁸
Temperature, K
0
50
100
150
200
250
300
350

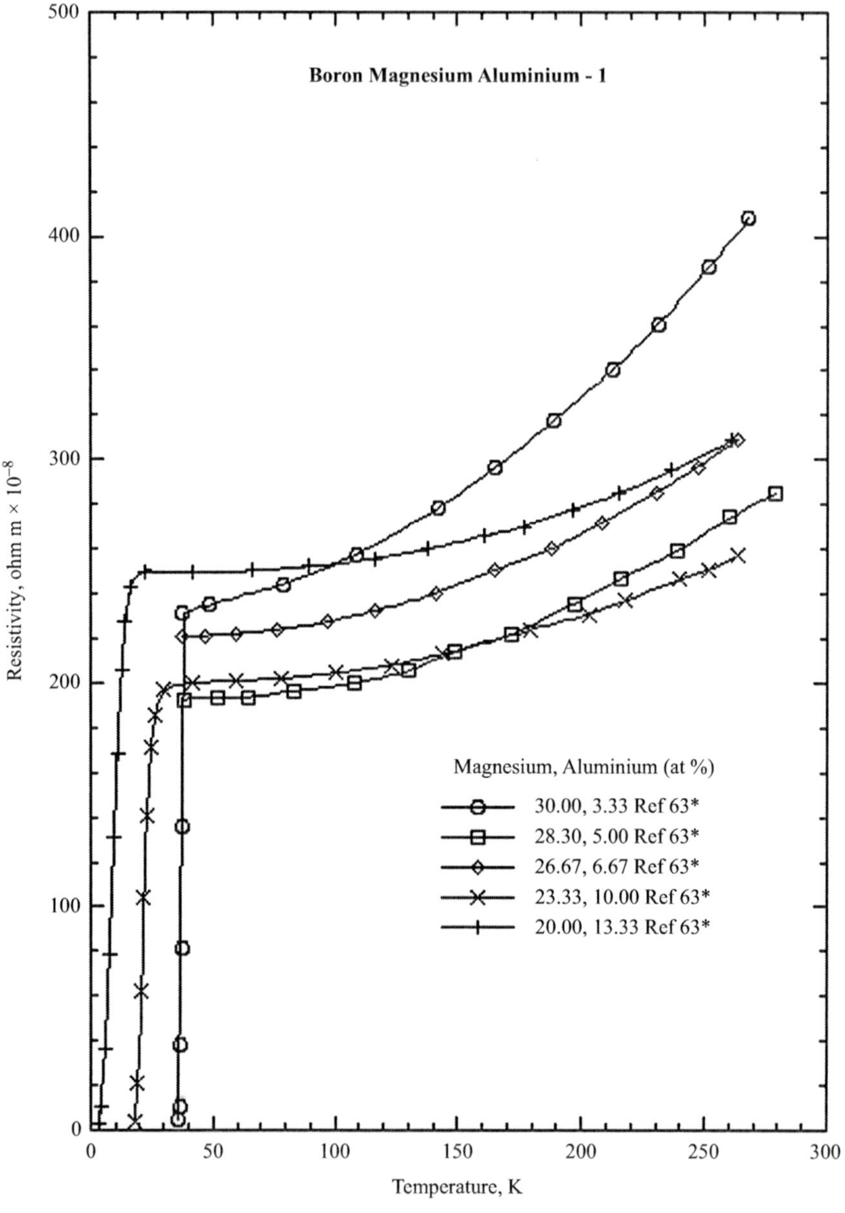
Boron Magnesium Aluminium - 1
Resistivity, ohm m × 10⁻⁸
Temperature, K
0
100
200
300
400
500
50
150
250
Magnesium, Aluminium (at %)
30.00, 3.33 Ref 63*
28.30, 5.00 Ref 63*
26.67, 6.67 Ref 63*
23.33, 10.00 Ref 63*
20.00, 13.33 Ref 63*

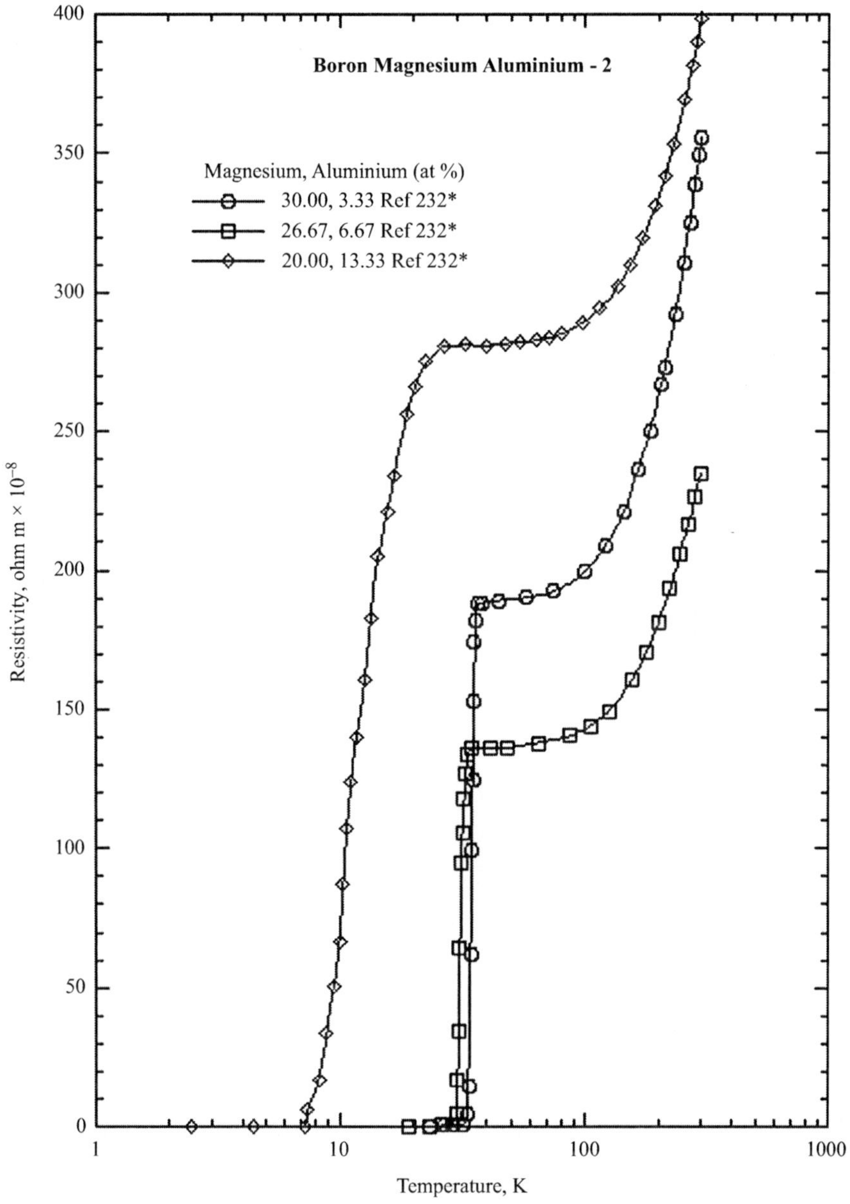
Boron Magnesium Aluminium - 2
Magnesium, Aluminium (at %)
30.00, 3.33 Ref 232*
26.67, 6.67 Ref 232*
20.00, 13.33 Ref 232*
Resistivity, ohm m × 10⁻⁸
Temperature, K
0
50
100
150
200
250
300
350
400
1
10
100
1000

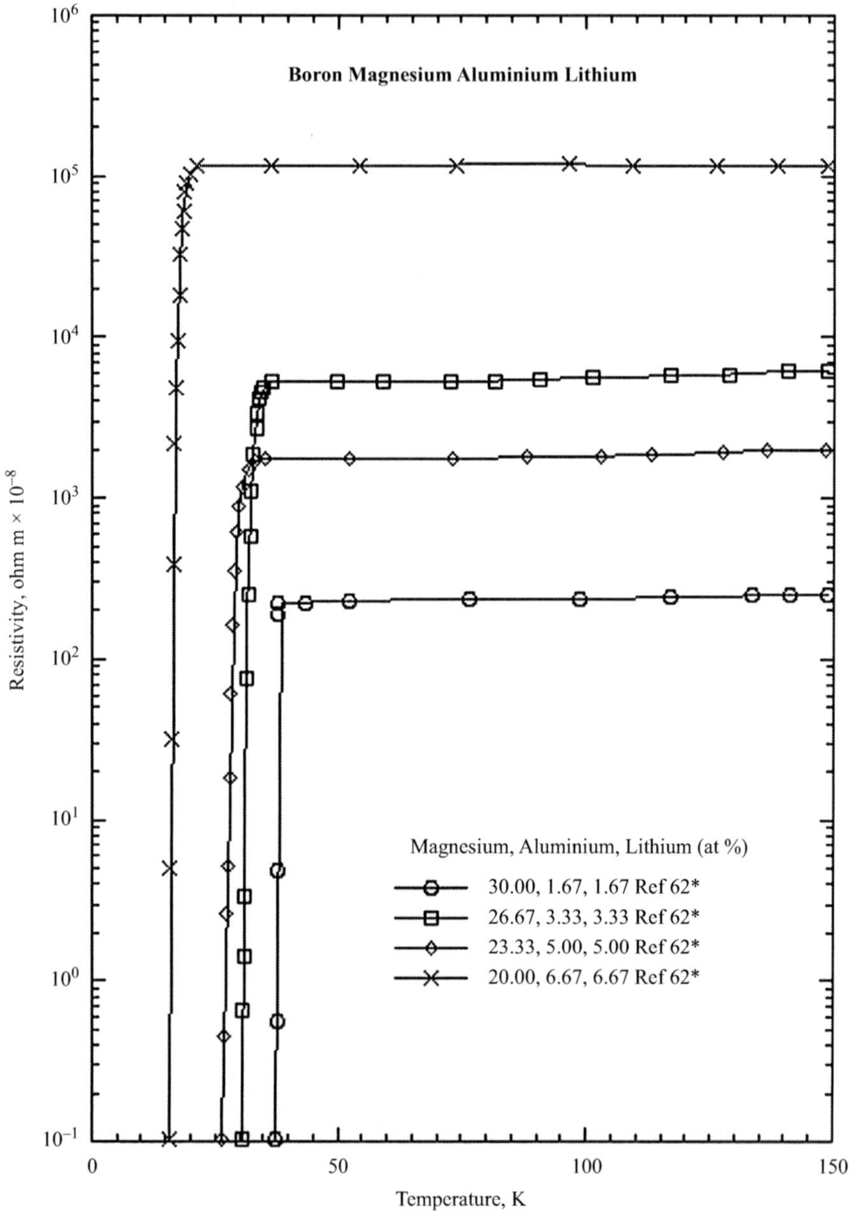
Boron Magnesium Aluminium Lithium
Resistivity, ohm m × 10⁻⁸
Temperature, K
Magnesium, Aluminium, Lithium (at %)
30.00, 1.67, 1.67 Ref 62*
26.67, 3.33, 3.33 Ref 62*
23.33, 5.00, 5.00 Ref 62*
20.00, 6.67, 6.67 Ref 62*
0
50
100
150

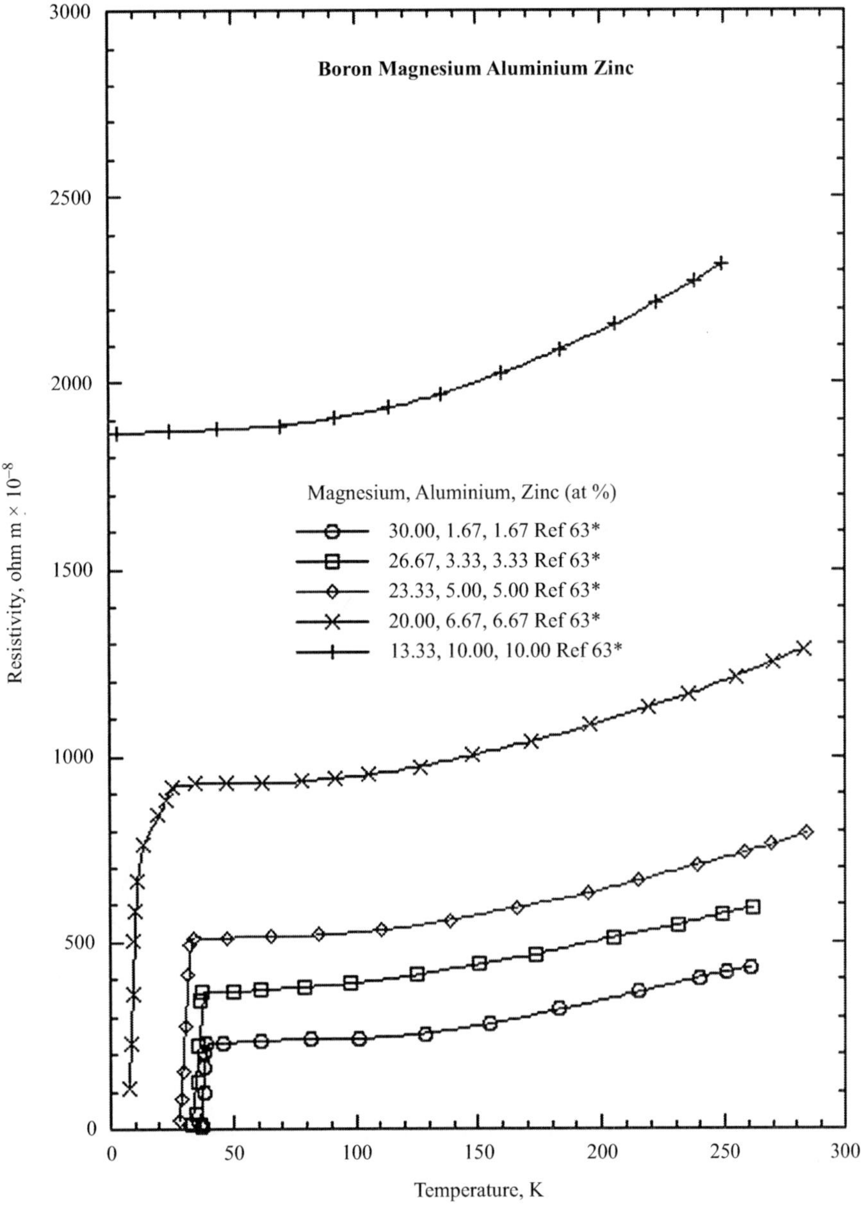

Boron Magnesium Aluminium Zinc
Magnesium, Aluminium, Zinc (at %)
30.00, 1.67, 1.67 Ref 63*
26.67, 3.33, 3.33 Ref 63*
23.33, 5.00, 5.00 Ref 63*
20.00, 6.67, 6.67 Ref 63*
13.33, 10.00, 10.00 Ref 63*
Resistivity, ohm m × 10^-8
Temperature, K
0
500
1000
1500
2000
2500
3000
0
50
100
150
200
250
300

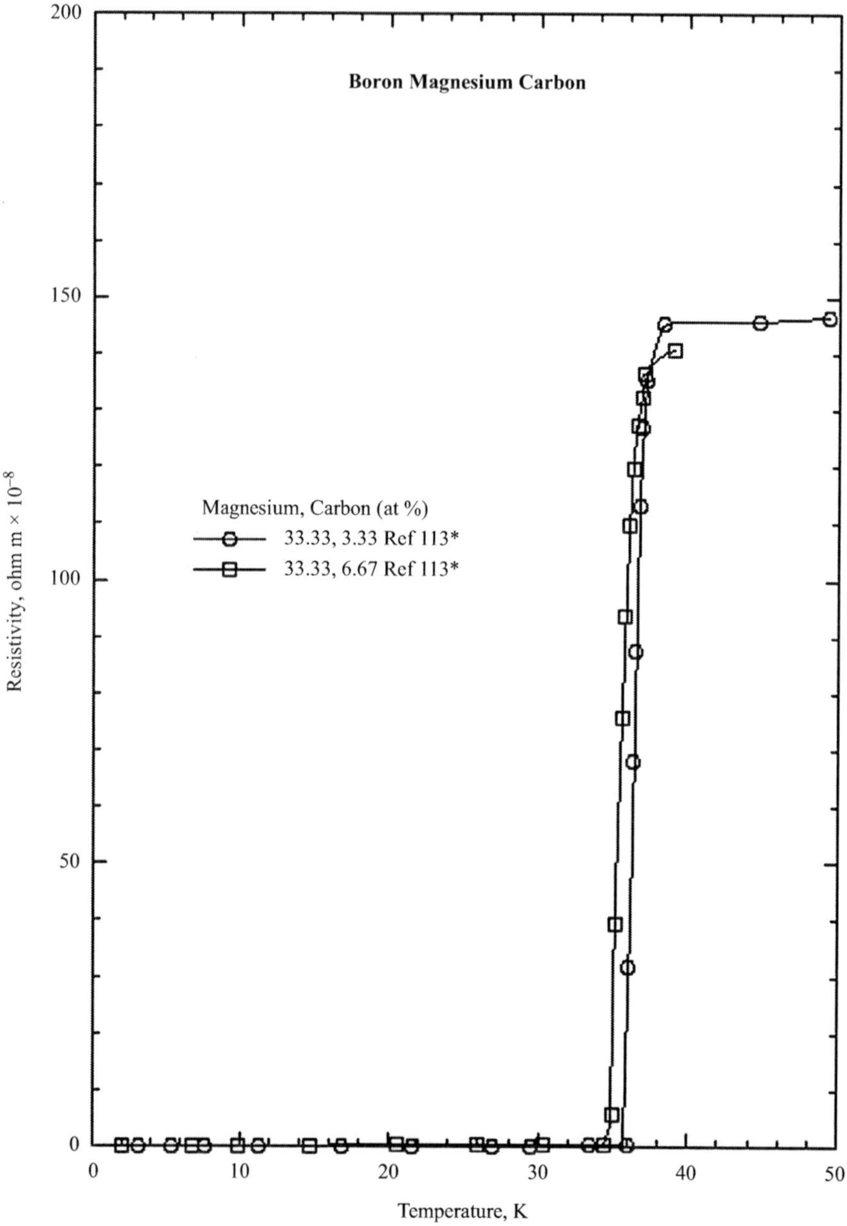
Boron Magnesium Carbon
Magnesium, Carbon (at %)
33.33, 3.33 Ref 113*
33.33, 6.67 Ref 113*
Resistivity, ohm m × 10⁻⁸
Temperature, K
0
10
20
30
40
50
0
50
100
150
200

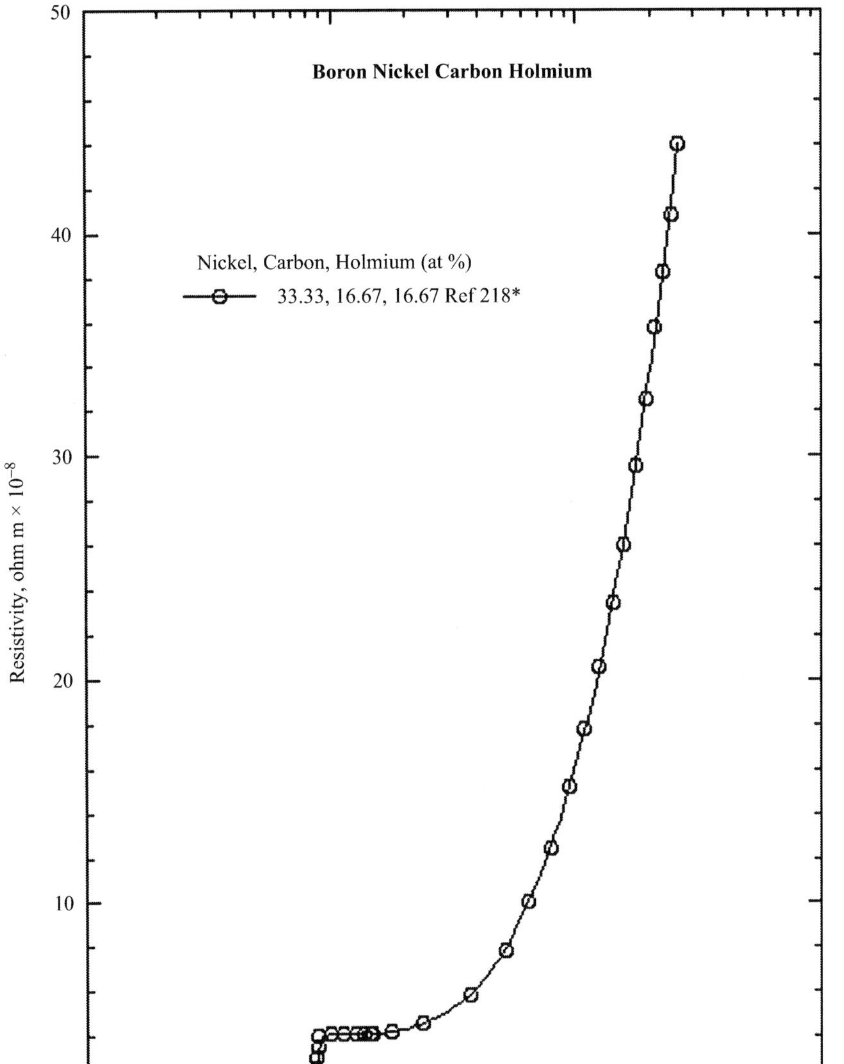
Boron Nickel Carbon Holmium
Nickel, Carbon, Holmium (at %)
33.33, 16.67, 16.67 Ref 218*
Resistivity, ohm m × 10⁻⁸
Temperature, K
0
10
20
30
40
50
1
10
100
1000

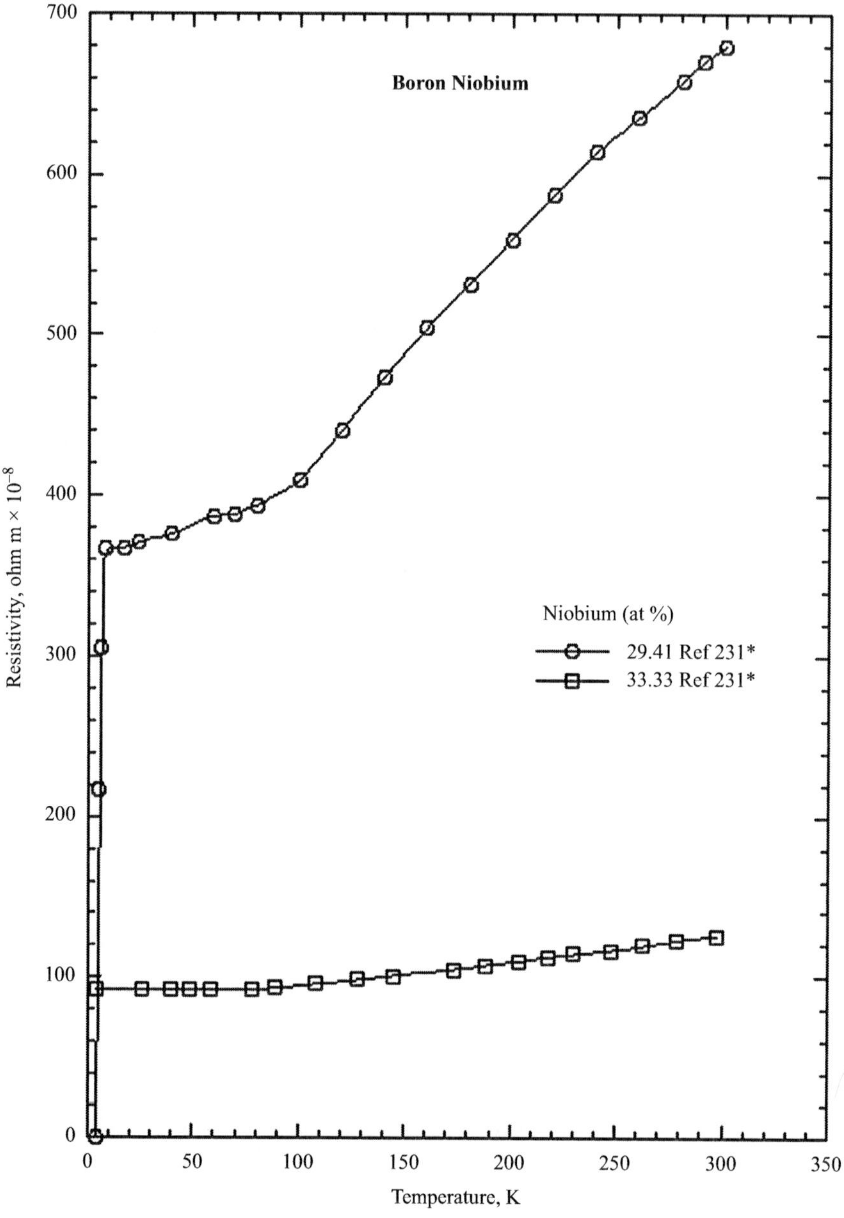
Boron Niobium
Niobium (at %)
29.41 Ref 231*
33.33 Ref 231*
Resistivity, ohm m × 10−8
Temperature, K
0
100
200
300
400
500
600
700
0
50
100
150
200
250
300
350

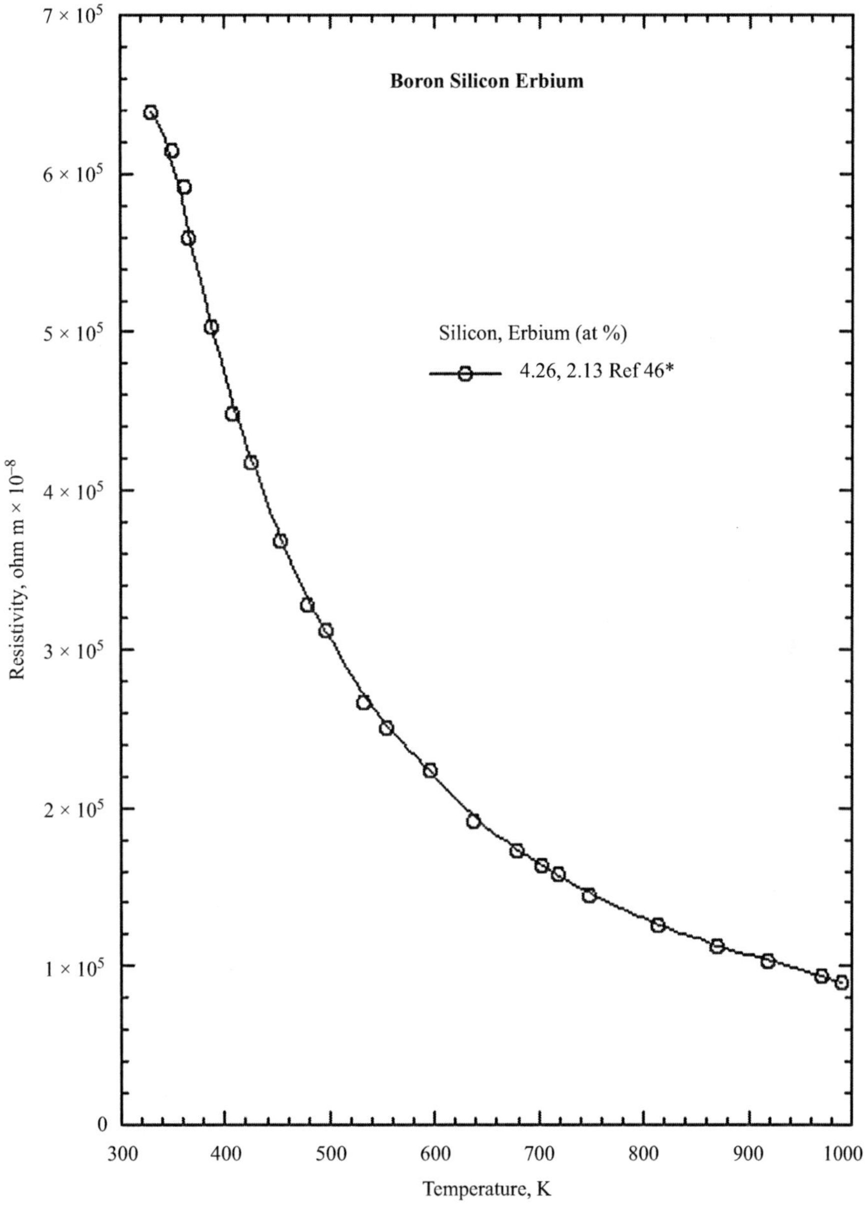

Boron Silicon Erbium
Silicon, Erbium (at %)
4.26, 2.13 Ref 46*
Resistivity, ohm m × 10^−8
7 × 10^5
6 × 10^5
5 × 10^5
4 × 10^5
3 × 10^5
2 × 10^5
1 × 10^5
0
300
400
500
600
700
800
900
1000
Temperature, K

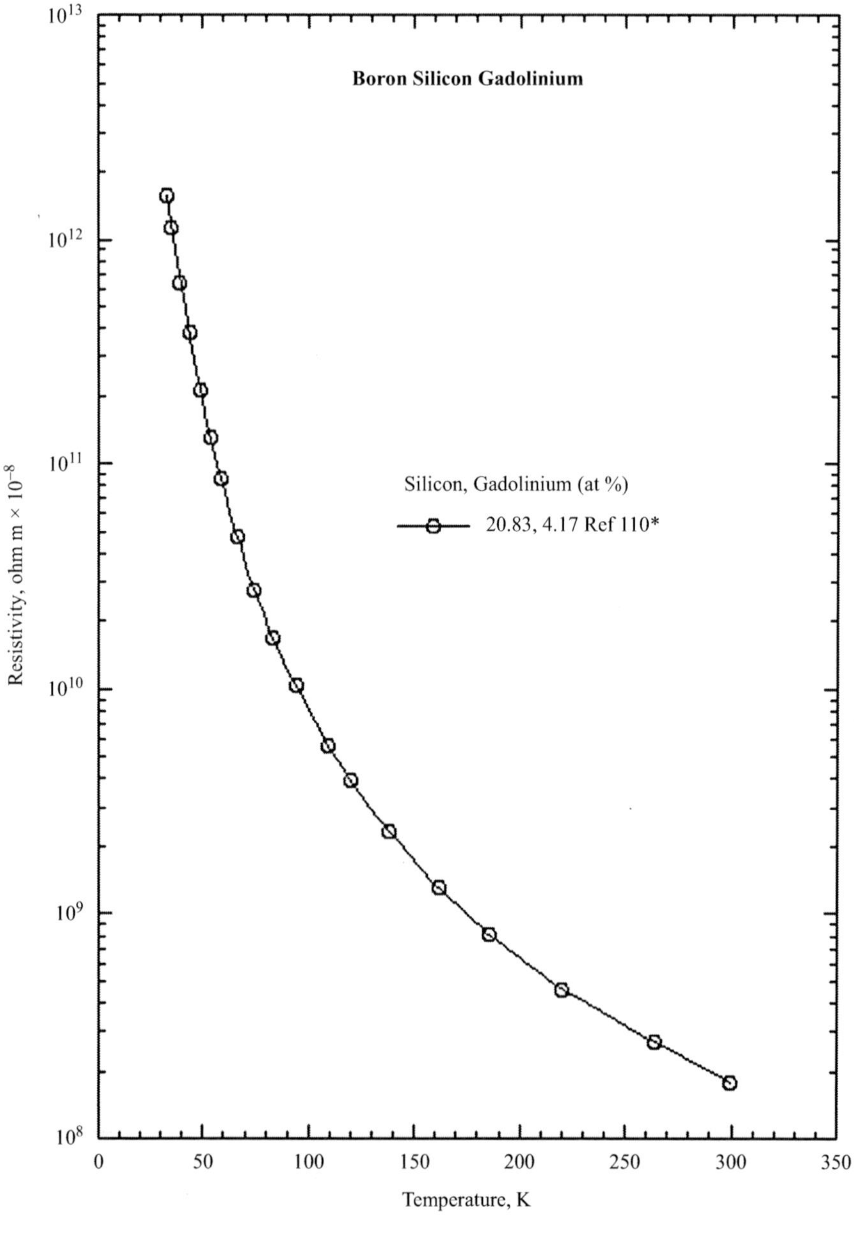
Boron Silicon Gadolinium
Silicon, Gadolinium (at %)
20.83, 4.17 Ref 110*
Resistivity, ohm m × 10⁻⁸
Temperature, K
10¹³
10¹²
10¹¹
10¹⁰
10⁹
10⁸
0
50
100
150
200
250
300
350

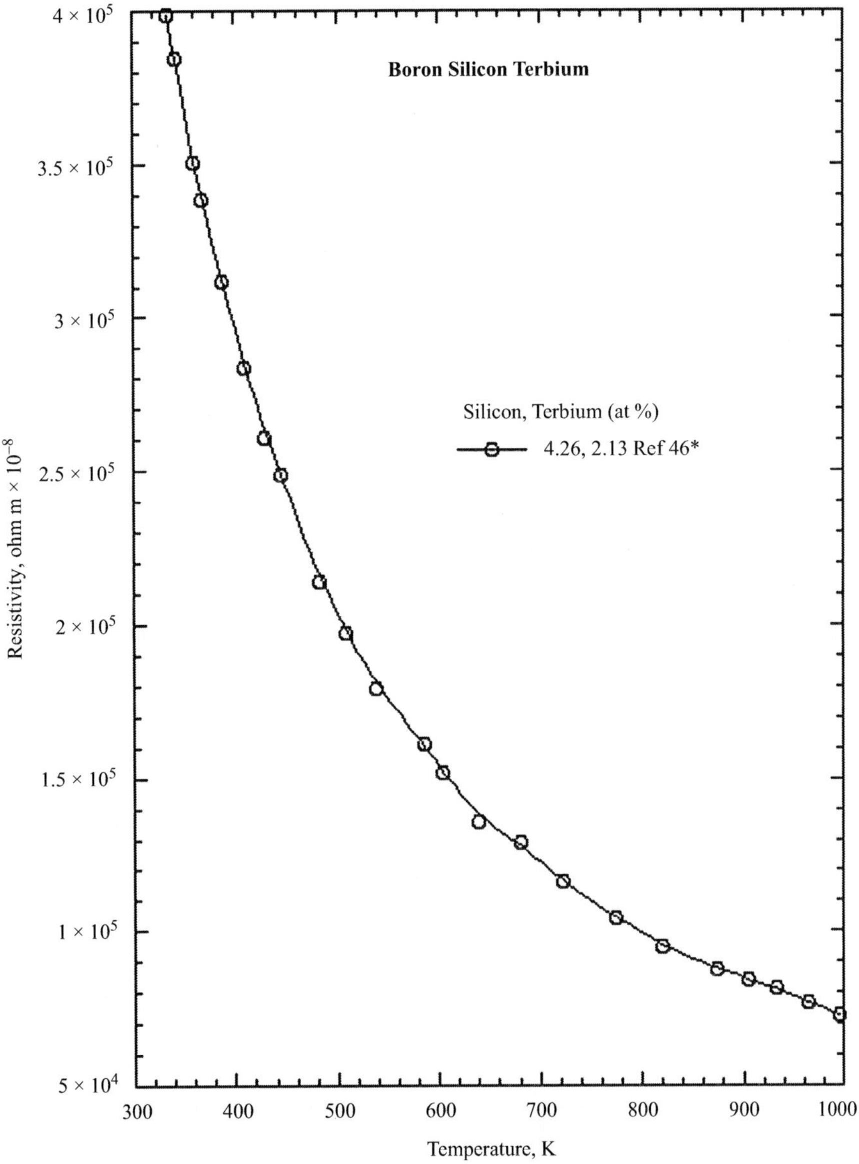

Boron Silicon Terbium
Silicon, Terbium (at %)
4.26, 2.13 Ref 46*
Resistivity, ohm m × 10⁻⁸
4 × 10⁵
3.5 × 10⁵
3 × 10⁵
2.5 × 10⁵
2 × 10⁵
1.5 × 10⁵
1 × 10⁵
5 × 10⁴
300
400
500
600
700
800
900
1000
Temperature, K

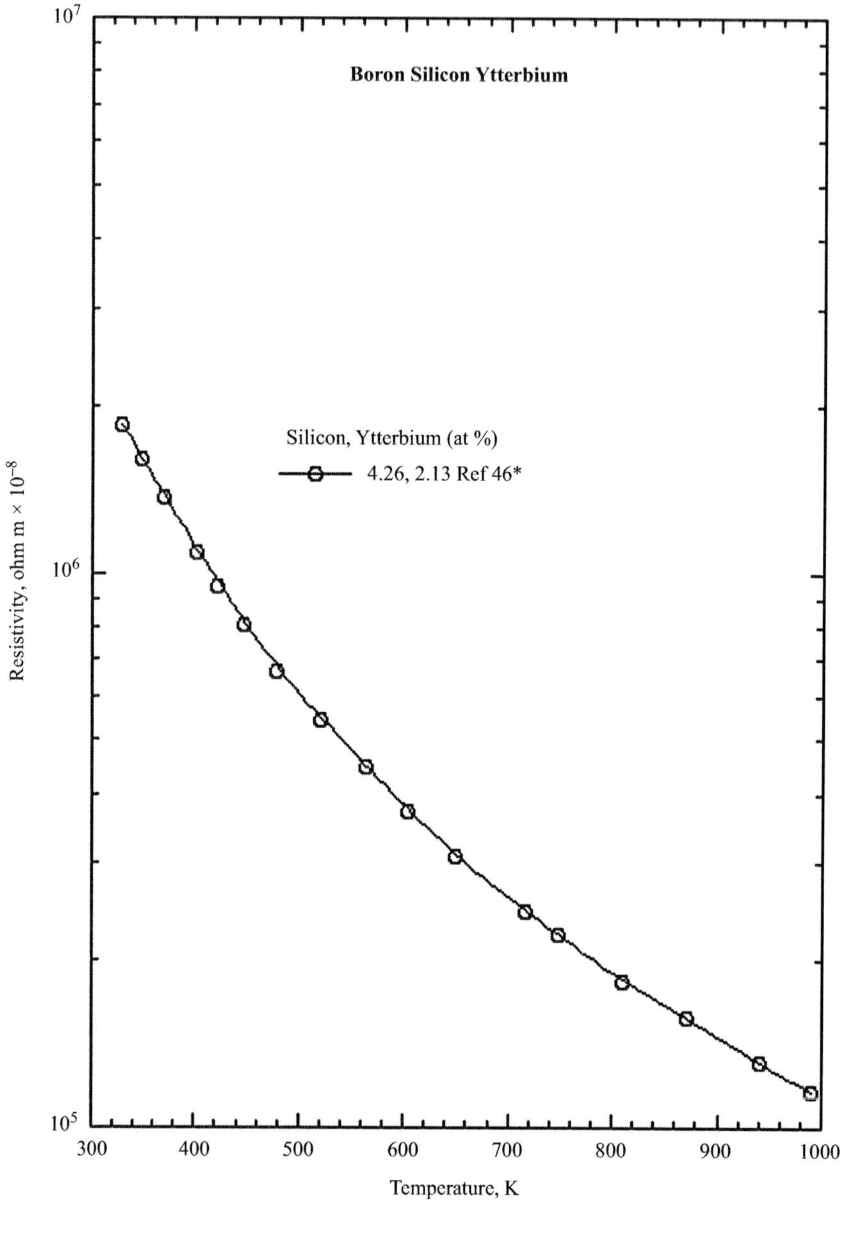
Boron Silicon Ytterbium
Silicon, Ytterbium (at %)
4.26, 2.13 Ref 46*
Resistivity, ohm m × 10⁻⁸
10^7
10^6
10^5
300
400
500
600
700
800
900
1000
Temperature, K

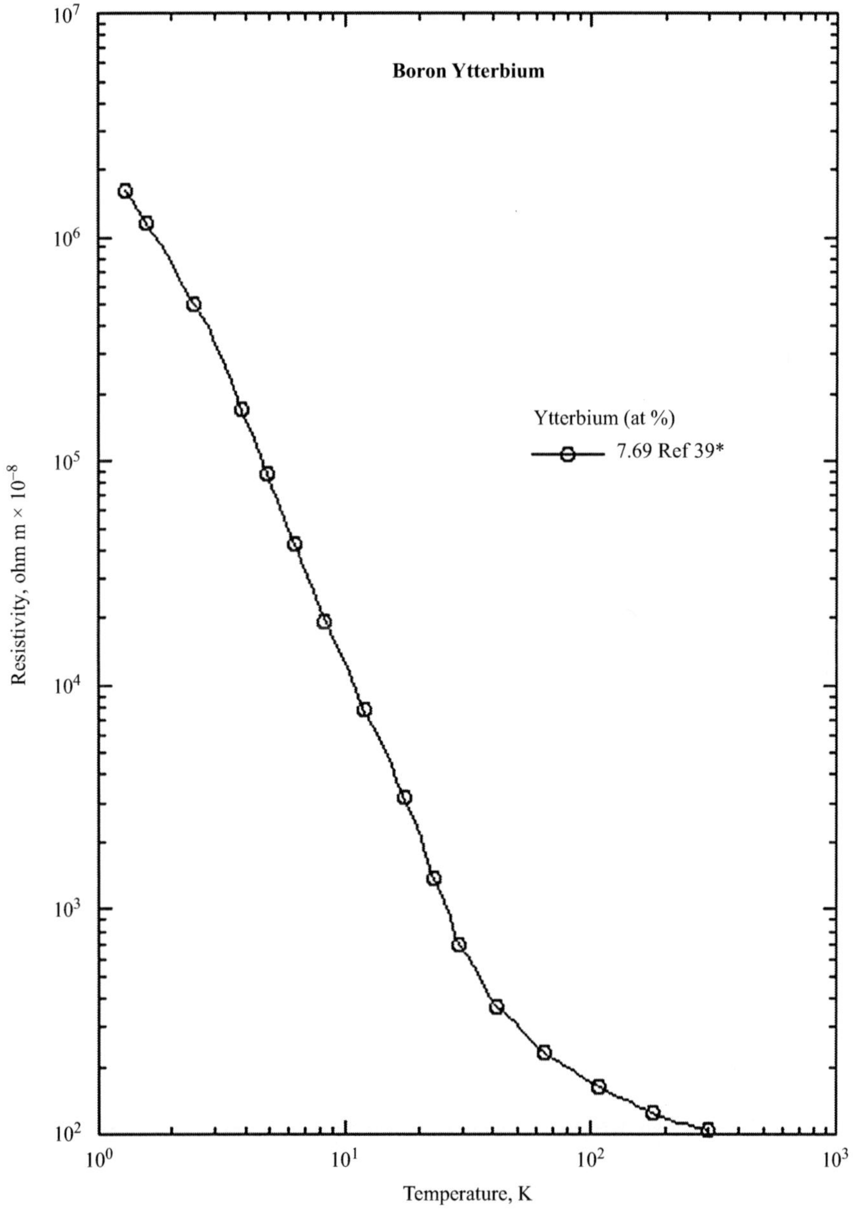

Boron Ytterbium
Ytterbium (at %)
7.69 Ref 39*
Resistivity, ohm m × 10⁻⁸
Temperature, K
10^7
10^6
10^5
10^4
10^3
10^2
10^0
10^1
10^2
10^3

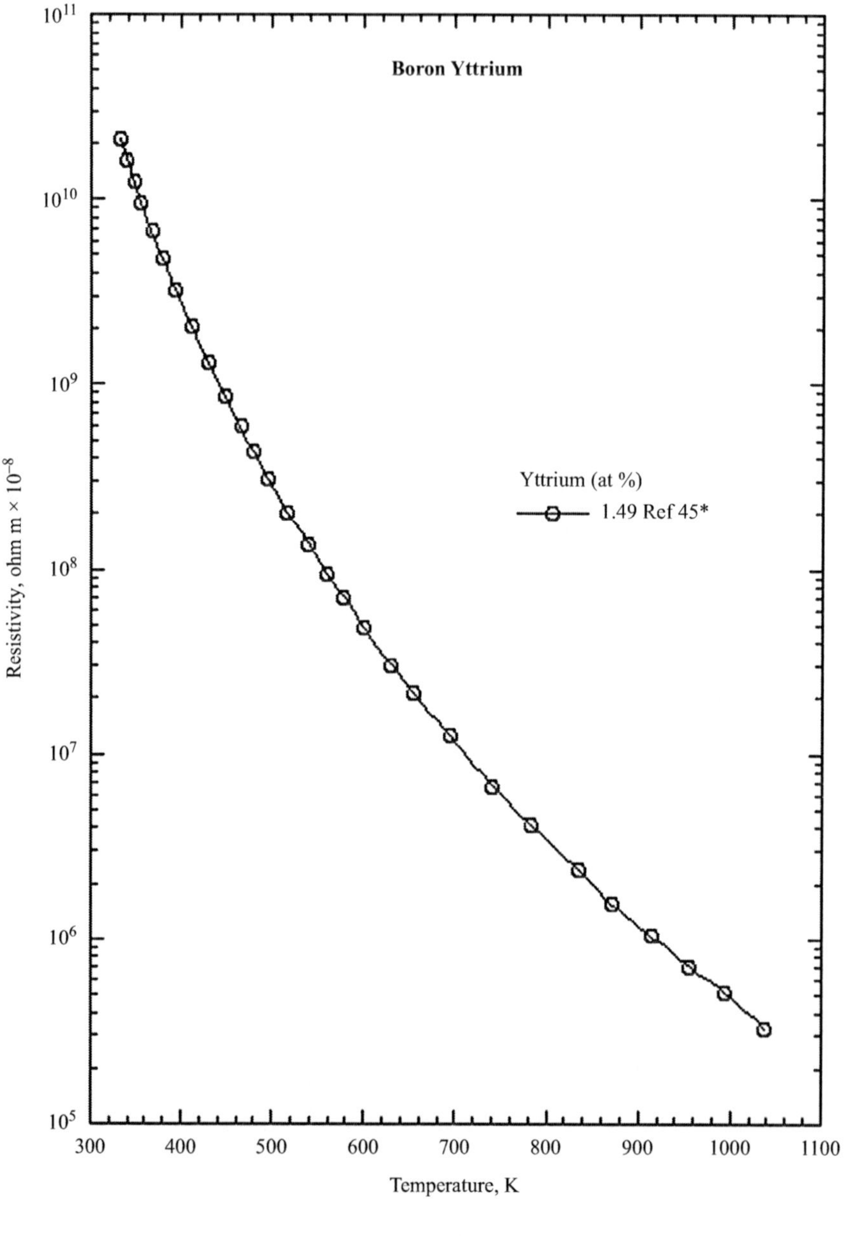
Boron Yttrium
Yttrium (at %)
1.49 Ref 45*
Resistivity, ohm m × 10⁻⁸
Temperature, K
10^11
10^10
10^9
10^8
10^7
10^6
10^5
300
400
500
600
700
800
900
1000
1100

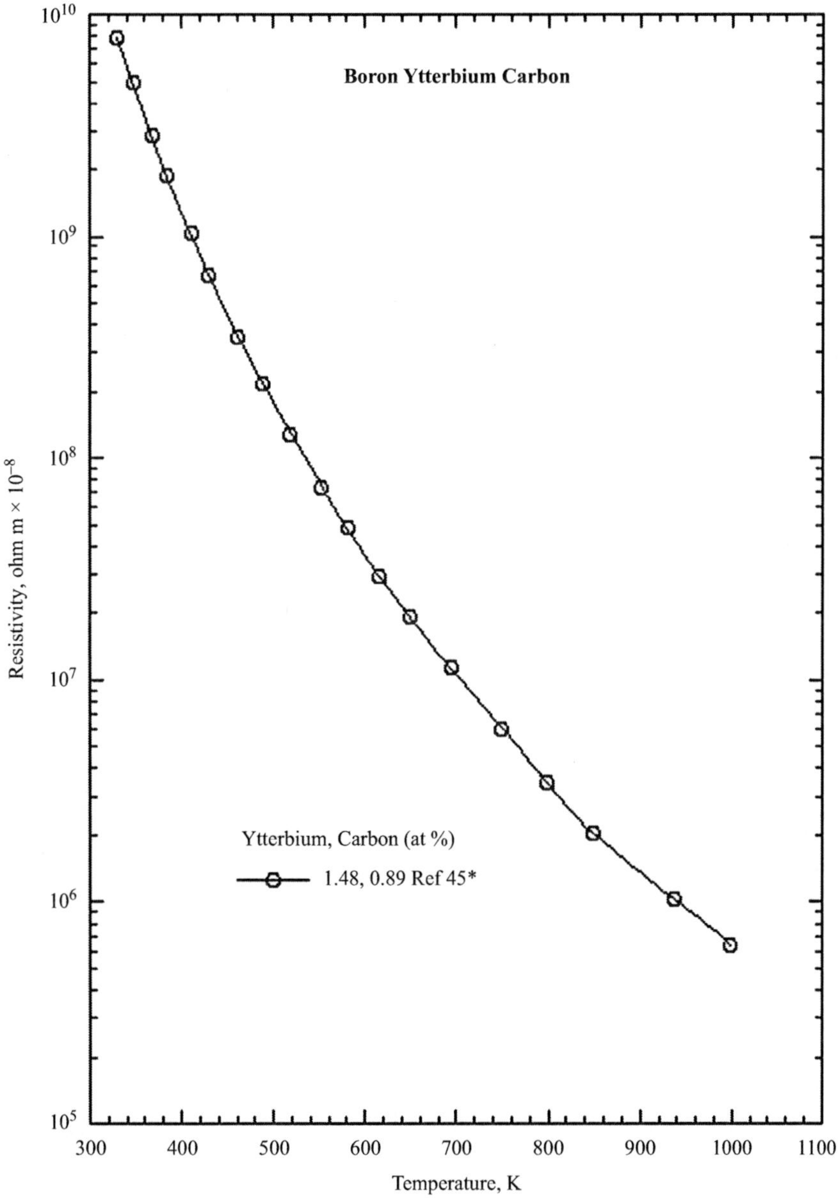
Boron Ytterbium Carbon
Ytterbium, Carbon (at %)
1.48, 0.89 Ref 45*
Resistivity, ohm m × 10⁻⁸
Temperature, K
10^10
10^9
10^8
10^7
10^6
10^5
300
400
500
600
700
800
900
1000
1100

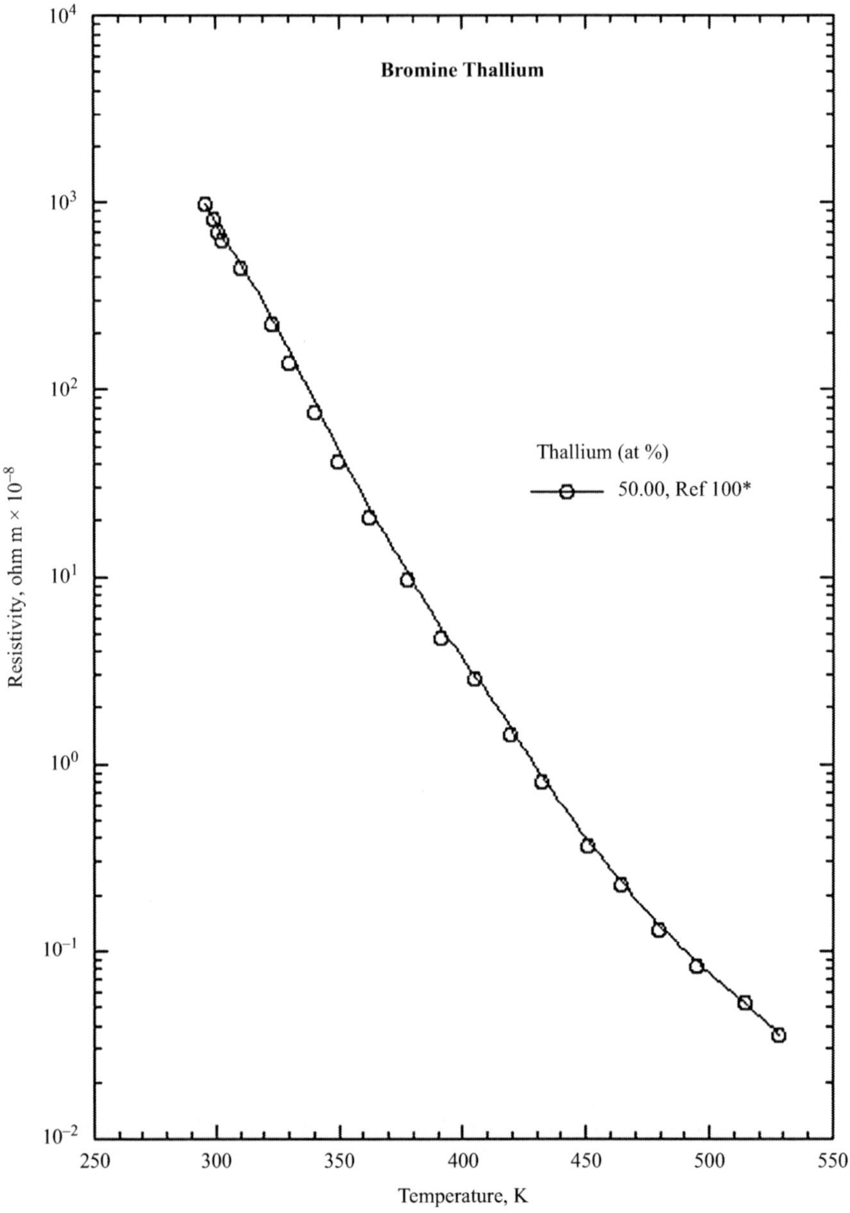
Bromine Thallium
Thallium (at %)
50.00, Ref 100*
Resistivity, ohm m × 10⁻⁸
Temperature, K
10⁴
10³
10²
10¹
10⁰
10⁻¹
10⁻²
250
300
350
400
450
500
550

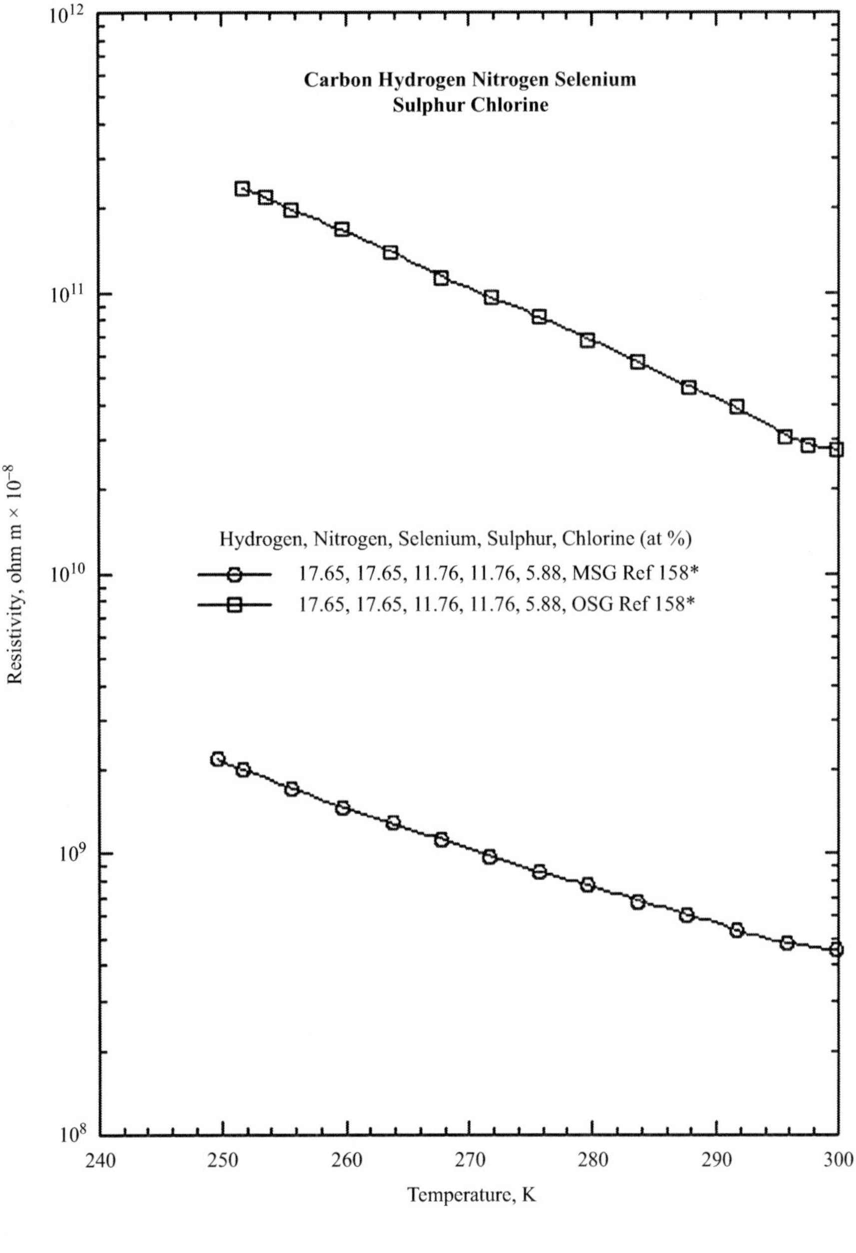

Carbon Hydrogen Nitrogen Selenium
Sulphur Chlorine
Hydrogen, Nitrogen, Selenium, Sulphur, Chlorine (at %)
17.65, 17.65, 11.76, 11.76, 5.88, MSG Ref 158*
17.65, 17.65, 11.76, 11.76, 5.88, OSG Ref 158*
Resistivity, ohm m × 10⁻⁸
10¹²
10¹¹
10¹⁰
10⁹
10⁸
240
250
260
270
280
290
300
Temperature, K

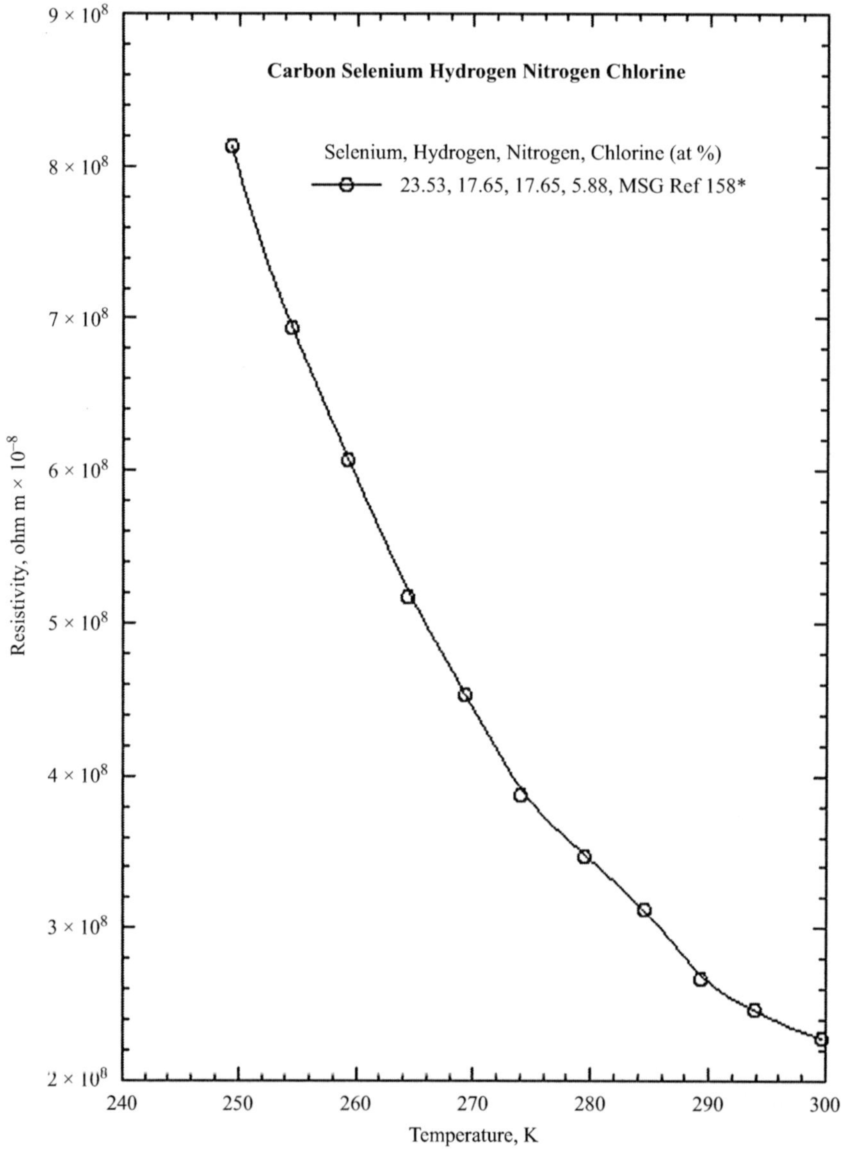
Carbon Selenium Hydrogen Nitrogen Chlorine
Selenium, Hydrogen, Nitrogen, Chlorine (at %)
23.53, 17.65, 17.65, 5.88, MSG Ref 158*
Resistivity, ohm m × 10⁻⁸
Temperature, K
9 × 10^8
8 × 10^8
7 × 10^8
6 × 10^8
5 × 10^8
4 × 10^8
3 × 10^8
2 × 10^8
240
250
260
270
280
290
300

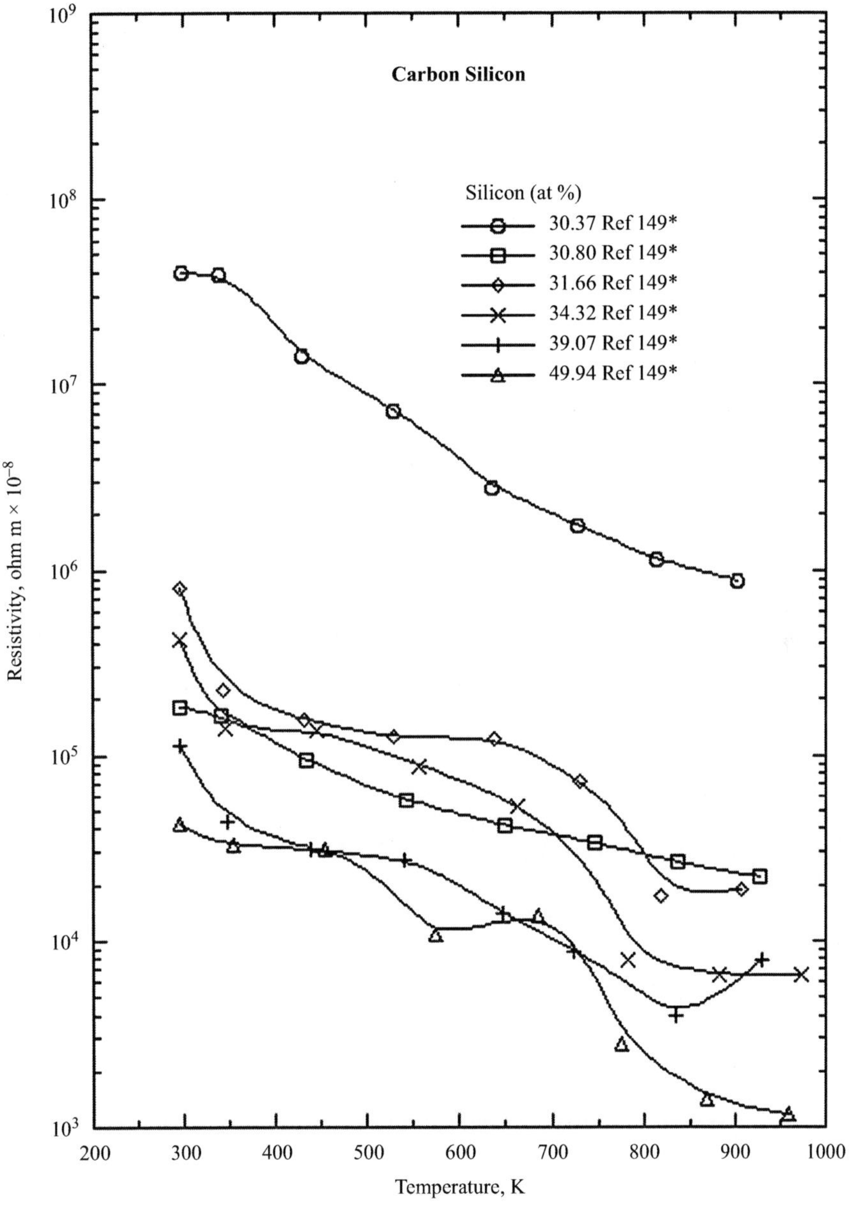

Carbon Silicon
Silicon (at %)
30.37 Ref 149*
30.80 Ref 149*
31.66 Ref 149*
34.32 Ref 149*
39.07 Ref 149*
49.94 Ref 149*
Resistivity, ohm m × 10^-8
Temperature, K
10^3
10^4
10^5
10^6
10^7
10^8
10^9
200
300
400
500
600
700
800
900
1000

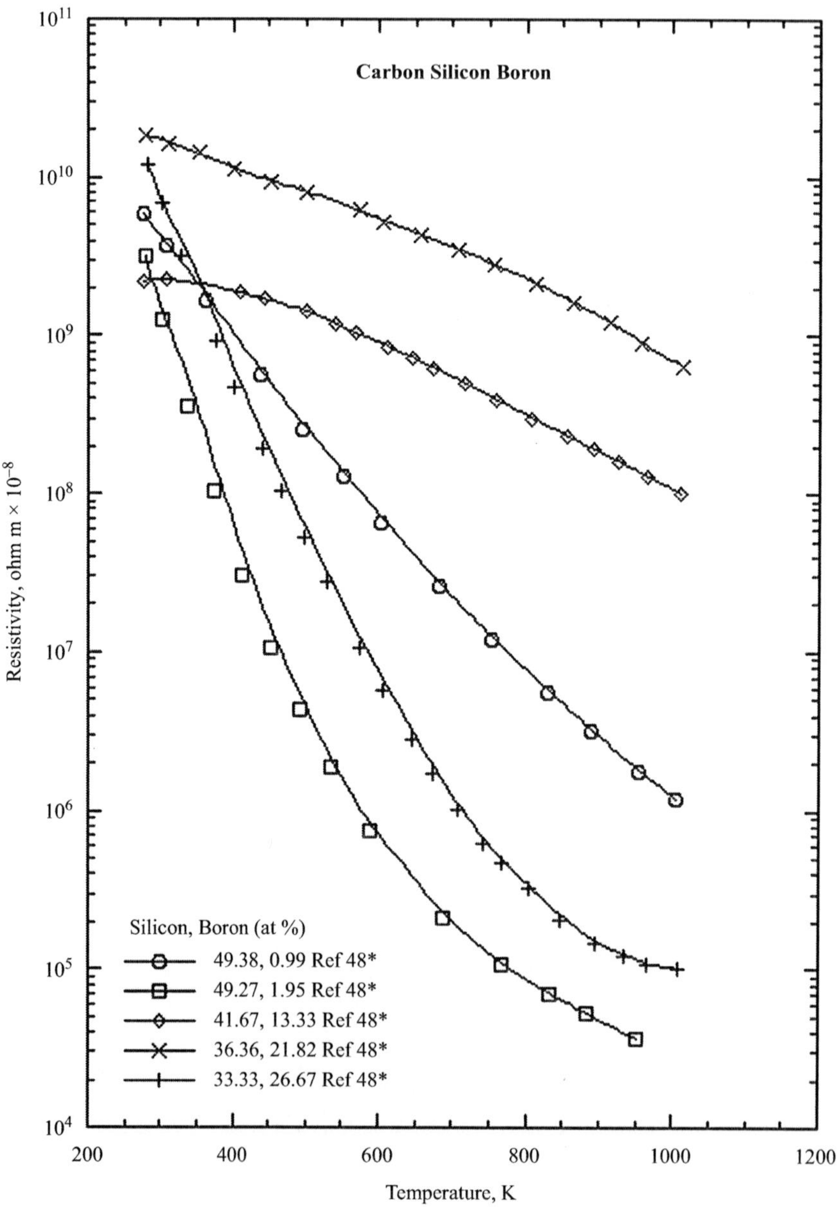
Carbon Silicon Boron
Silicon, Boron (at %)
49.38, 0.99 Ref 48*
49.27, 1.95 Ref 48*
41.67, 13.33 Ref 48*
36.36, 21.82 Ref 48*
33.33, 26.67 Ref 48*
Resistivity, ohm m × 10⁻⁸
Temperature, K
10¹¹
10¹⁰
10⁹
10⁸
10⁷
10⁶
10⁵
10⁴
200
400
600
800
1000
1200

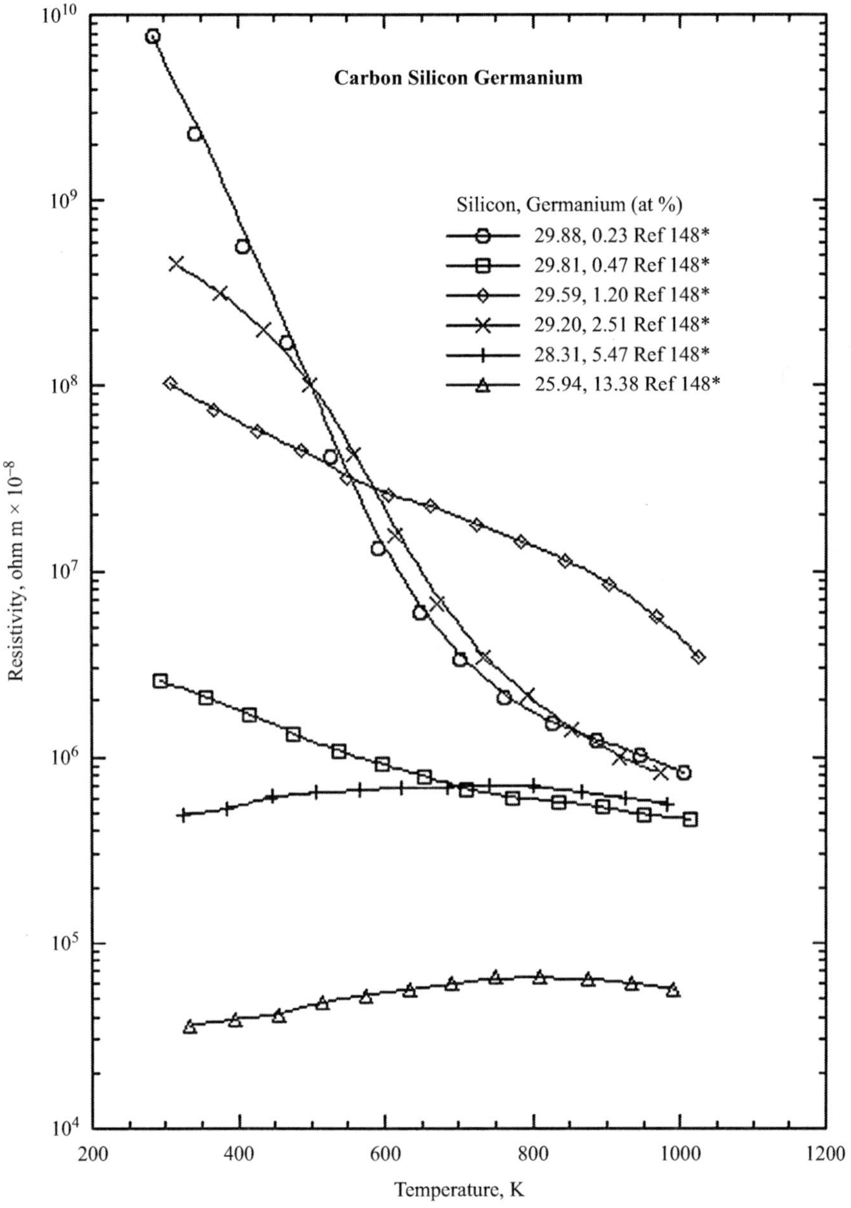

Carbon Silicon Germanium
Silicon, Germanium (at %)
29.88, 0.23 Ref 148*
29.81, 0.47 Ref 148*
29.59, 1.20 Ref 148*
29.20, 2.51 Ref 148*
28.31, 5.47 Ref 148*
25.94, 13.38 Ref 148*
Resistivity, ohm m × 10⁻⁸
Temperature, K

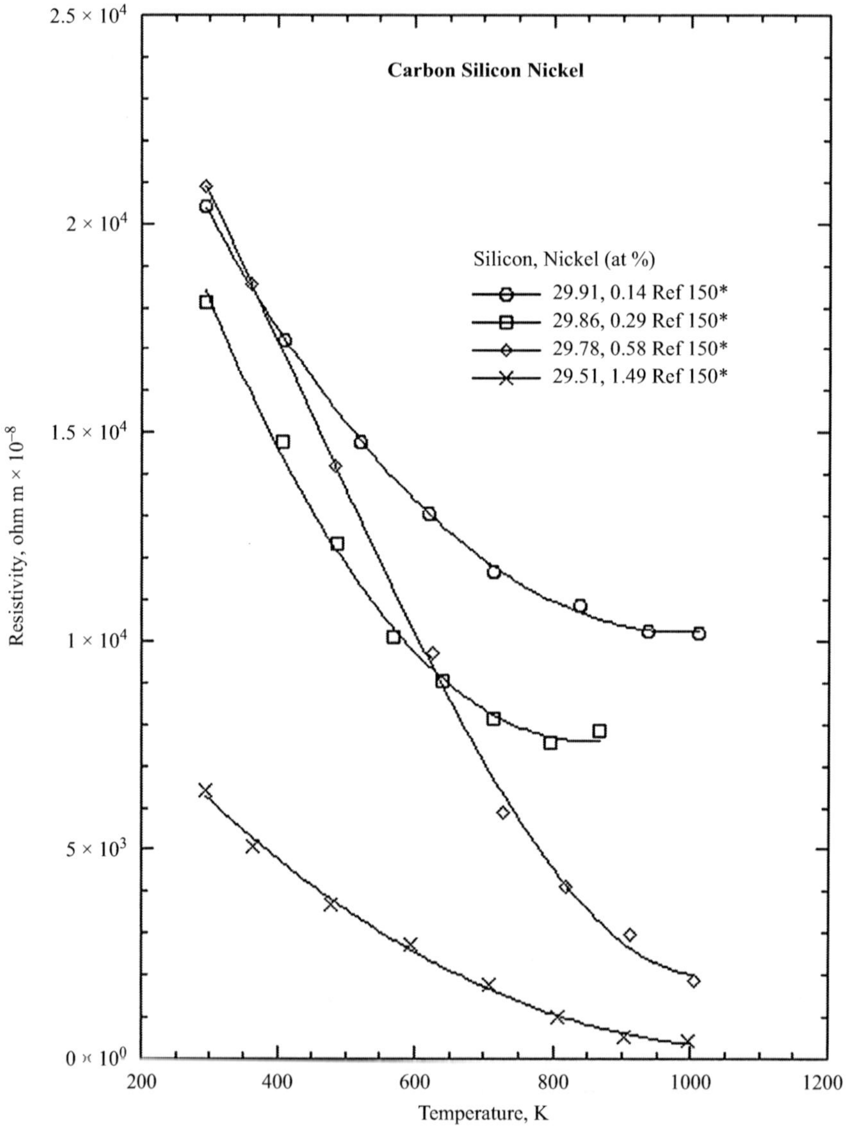
Carbon Silicon Nickel
Silicon, Nickel (at %)
29.91, 0.14 Ref 150*
29.86, 0.29 Ref 150*
29.78, 0.58 Ref 150*
29.51, 1.49 Ref 150*
Resistivity, ohm m × 10⁻⁸
Temperature, K
2.5 × 10⁴
2 × 10⁴
1.5 × 10⁴
1 × 10⁴
5 × 10³
0 × 10⁰
200
400
600
800
1000
1200

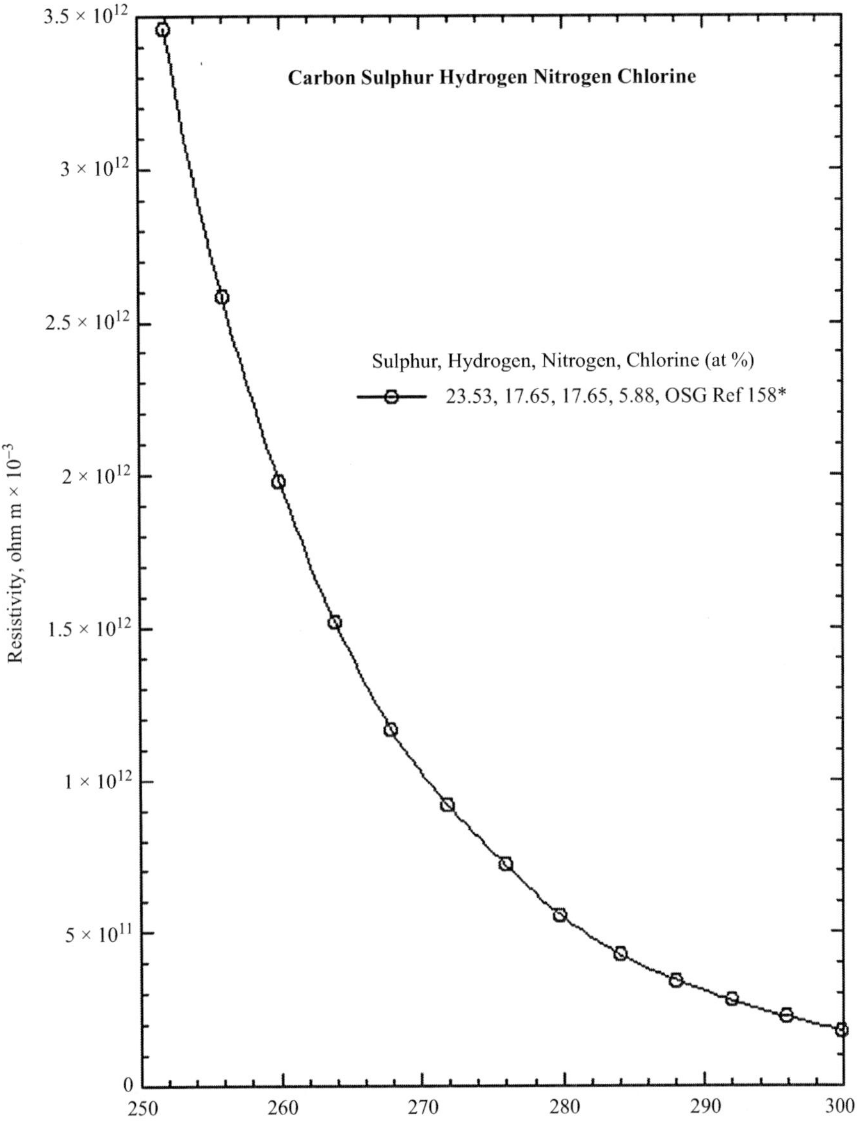
Carbon Sulphur Hydrogen Nitrogen Chlorine
Sulphur, Hydrogen, Nitrogen, Chlorine (at %)
23.53, 17.65, 17.65, 5.88, OSG Ref 158*
Resistivity, ohm m × 10⁻³
3.5 × 10¹²
3 × 10¹²
2.5 × 10¹²
2 × 10¹²
1.5 × 10¹²
1 × 10¹²
5 × 10¹¹
0
250
260
270
280
290
300
Temperature, K

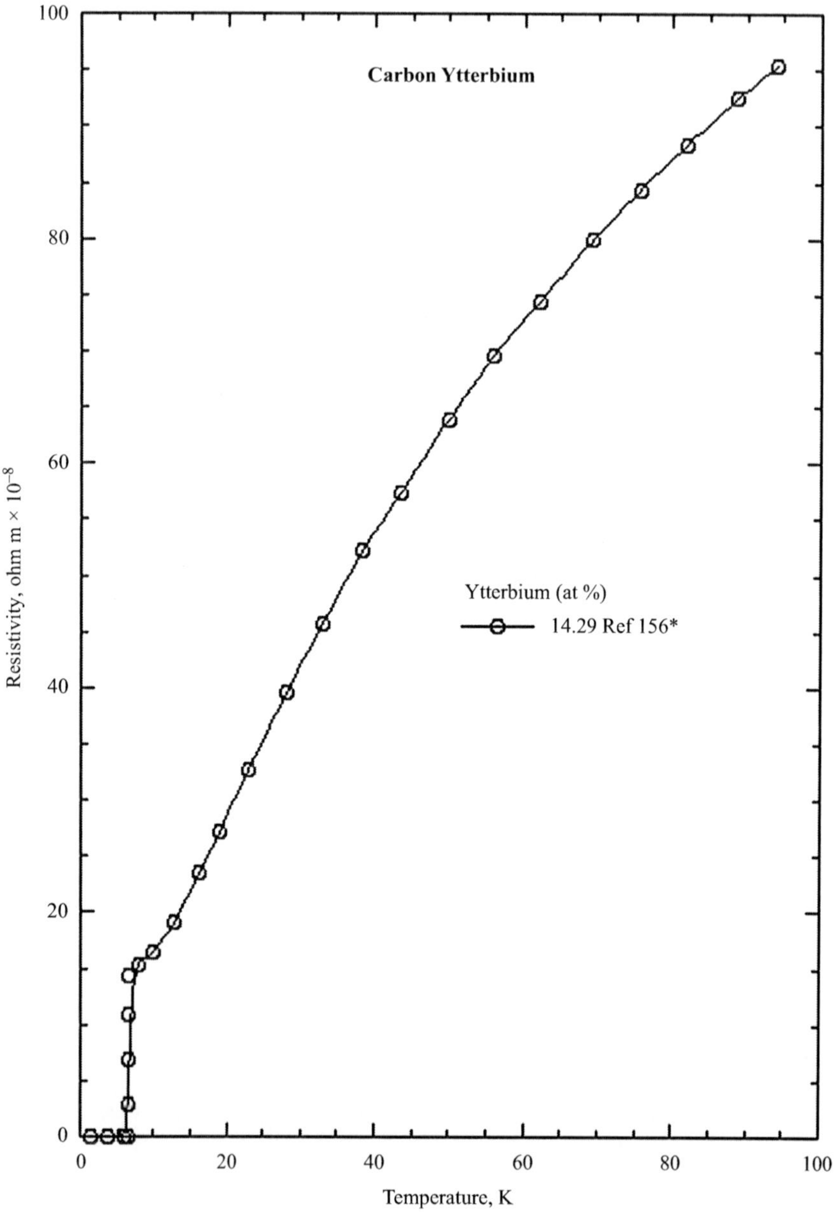
Carbon Ytterbium
Ytterbium (at %)
14.29 Ref 156*
Resistivity, ohm m × 10^{-8}
Temperature, K
0
20
40
60
80
100

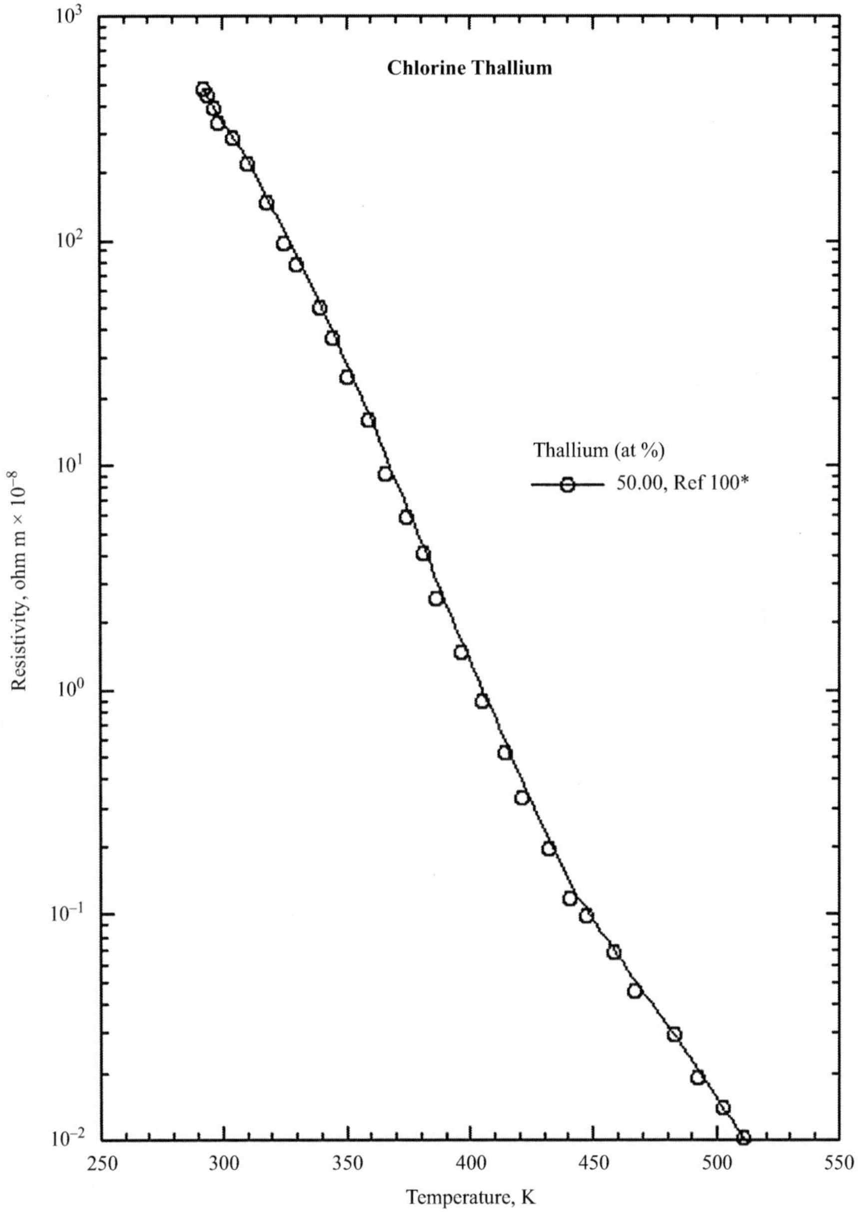

Chlorine Thallium
Thallium (at %)
50.00, Ref 100*
Resistivity, ohm m × 10⁻⁸
10³
10²
10¹
10⁰
10⁻¹
10⁻²
250
300
350
400
450
500
550
Temperature, K

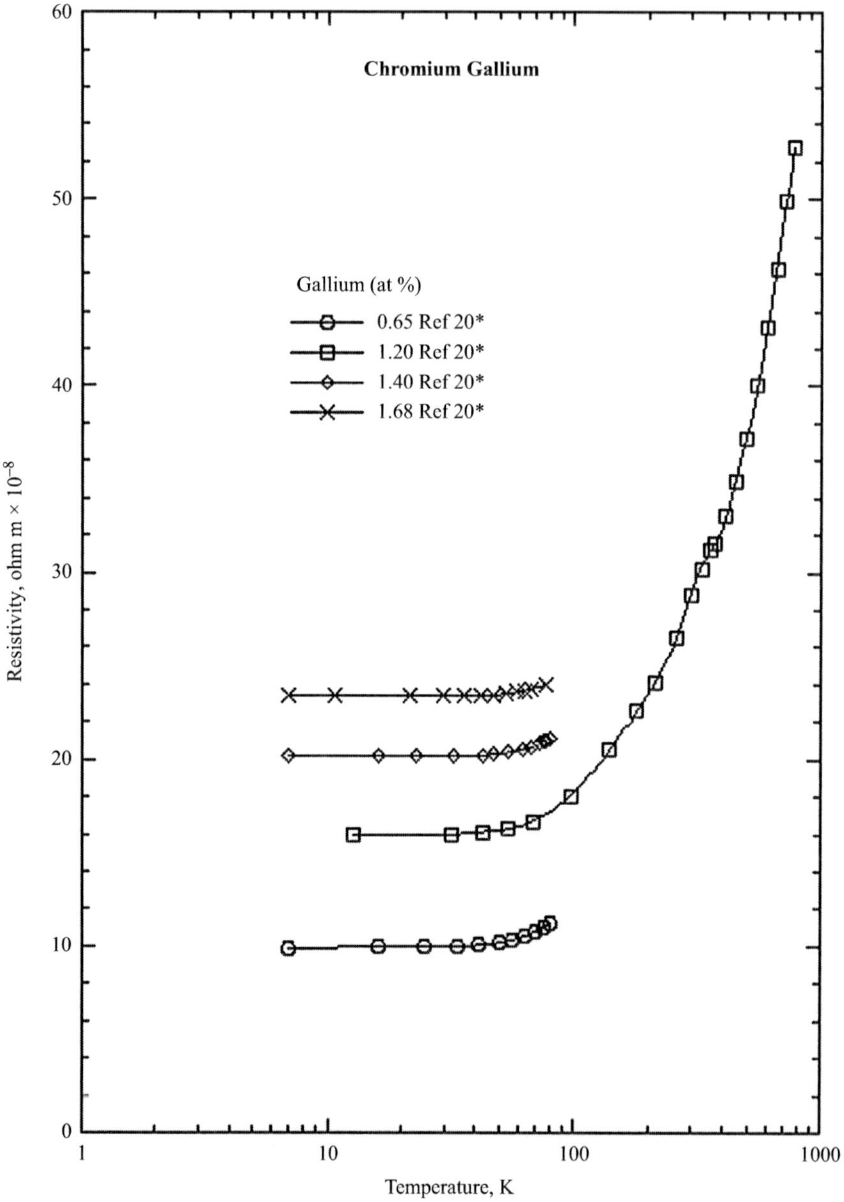
Chromium Gallium
Gallium (at %)
0.65 Ref 20*
1.20 Ref 20*
1.40 Ref 20*
1.68 Ref 20*
Resistivity, ohm m × 10⁻⁸
Temperature, K
0
10
20
30
40
50
60
1
10
100
1000

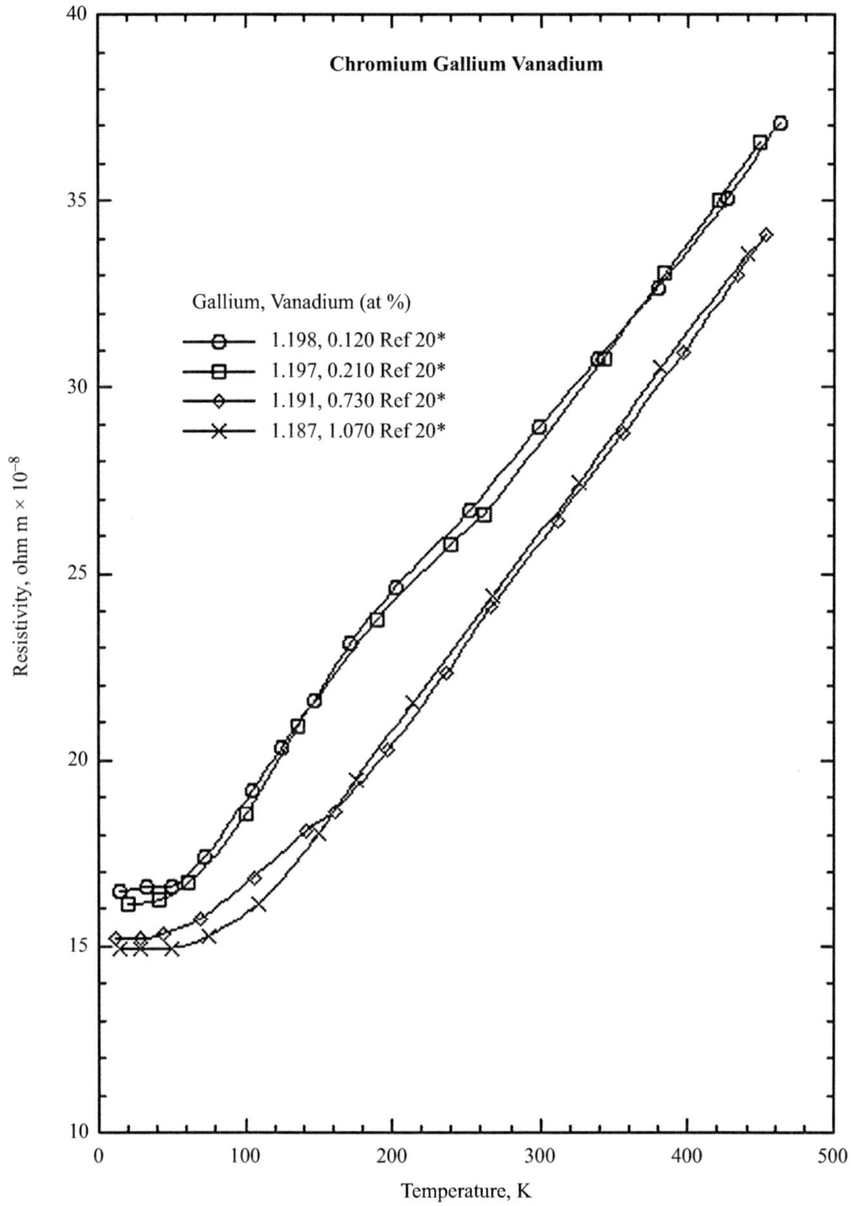

Chromium Gallium Vanadium
Gallium, Vanadium (at %)
1.198, 0.120 Ref 20*
1.197, 0.210 Ref 20*
1.191, 0.730 Ref 20*
1.187, 1.070 Ref 20*
Resistivity, ohm m × 10⁻⁸
Temperature, K
40
35
30
25
20
15
10
0
100
200
300
400
500

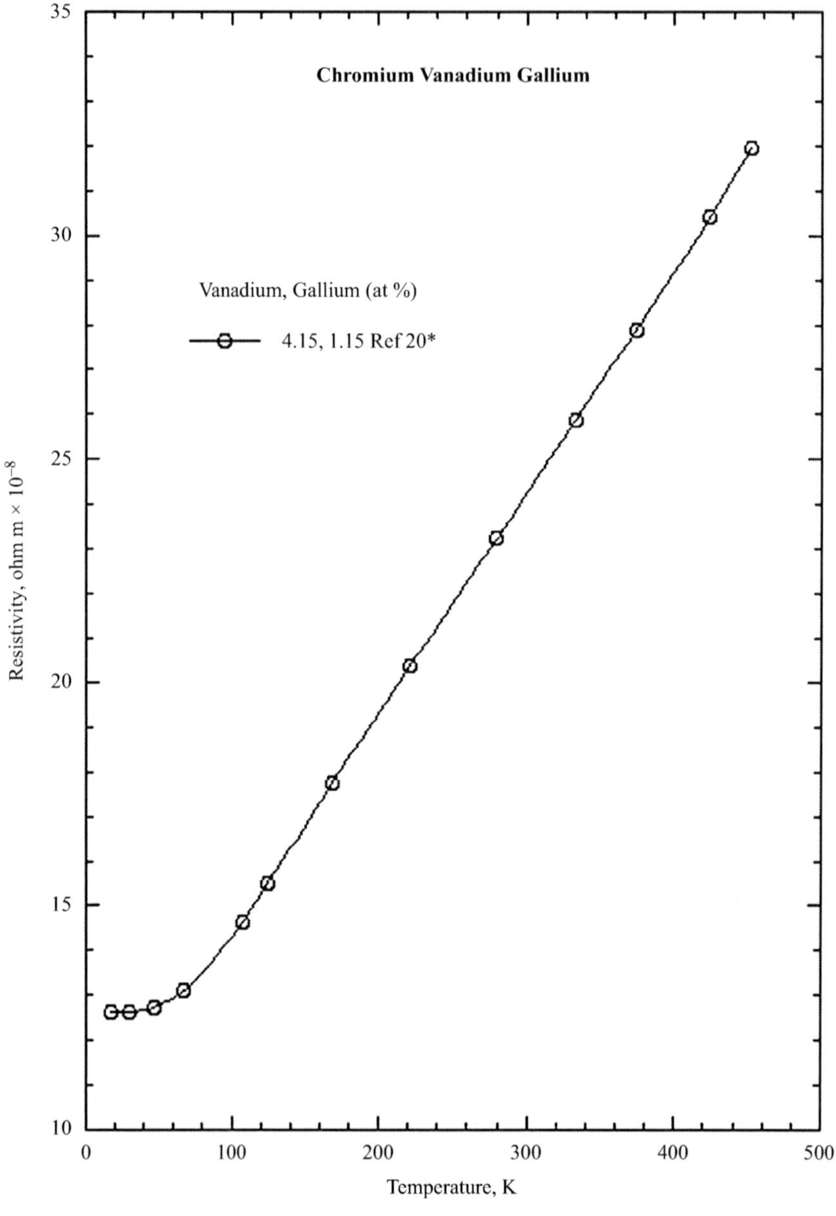
Chromium Vanadium Gallium
Vanadium, Gallium (at %)
4.15, 1.15 Ref 20*
Resistivity, ohm m × 10⁻⁸
Temperature, K
35
30
25
20
15
10
0
100
200
300
400
500

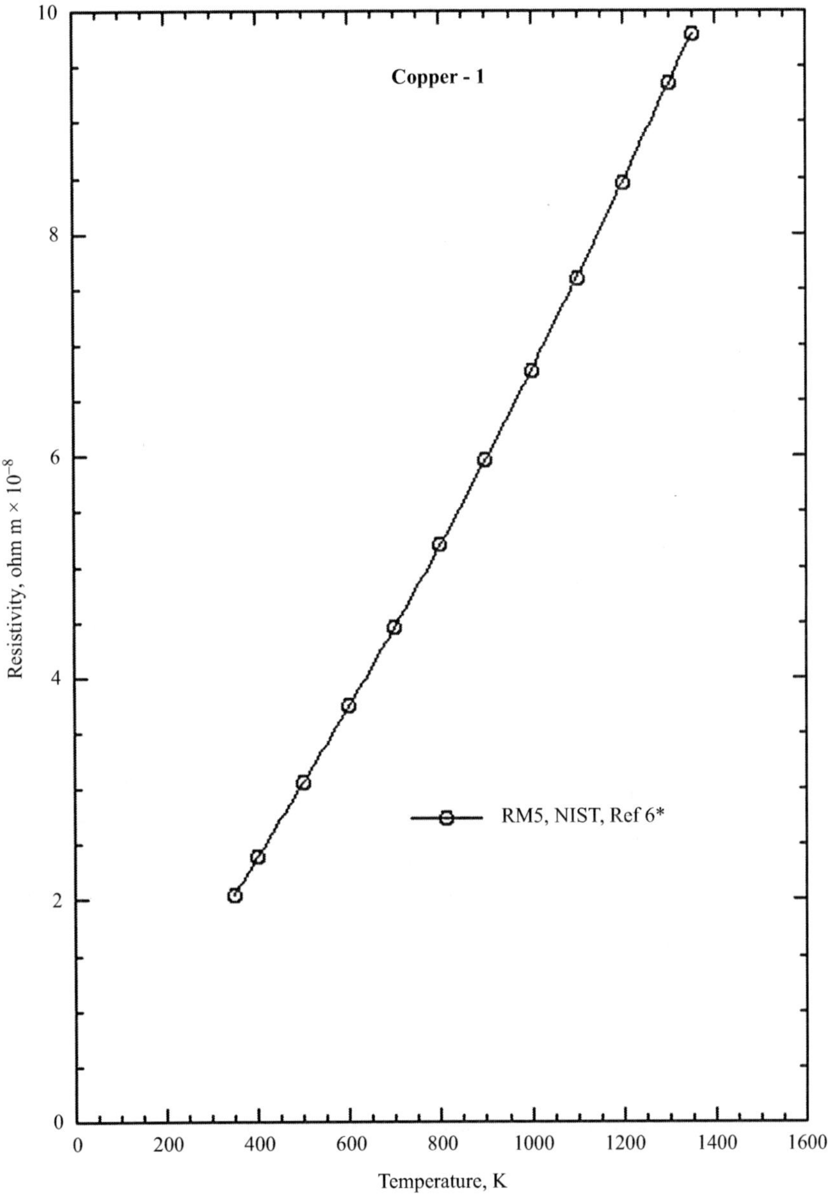
Copper - 1
RM5, NIST, Ref 6*
Resistivity, ohm m × 10⁻⁸
Temperature, K
0
2
4
6
8
10
0
200
400
600
800
1000
1200
1400
1600

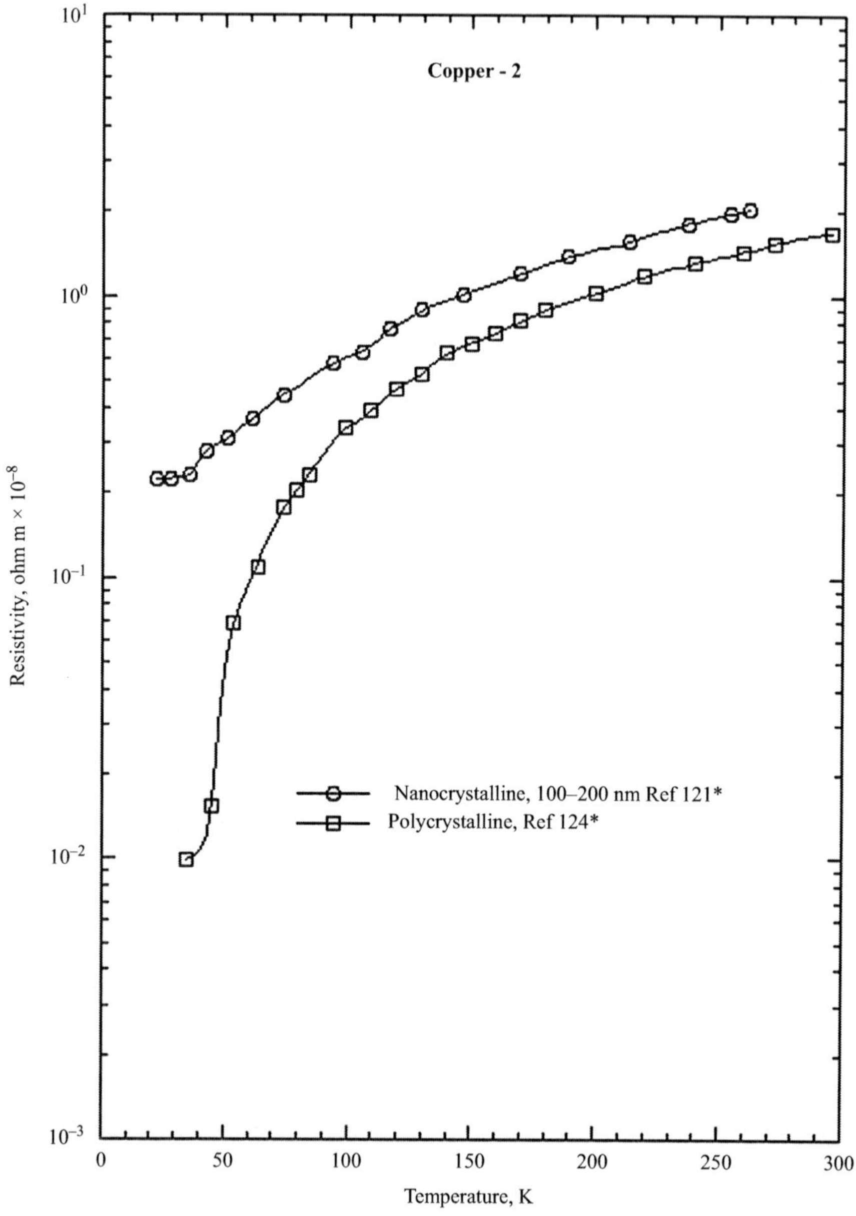
Copper - 2
Nanocrystalline, 100–200 nm Ref 121*
Polycrystalline, Ref 124*
Resistivity, ohm m × 10⁻⁸
Temperature, K
10¹
10⁰
10⁻¹
10⁻²
10⁻³
0
50
100
150
200
250
300

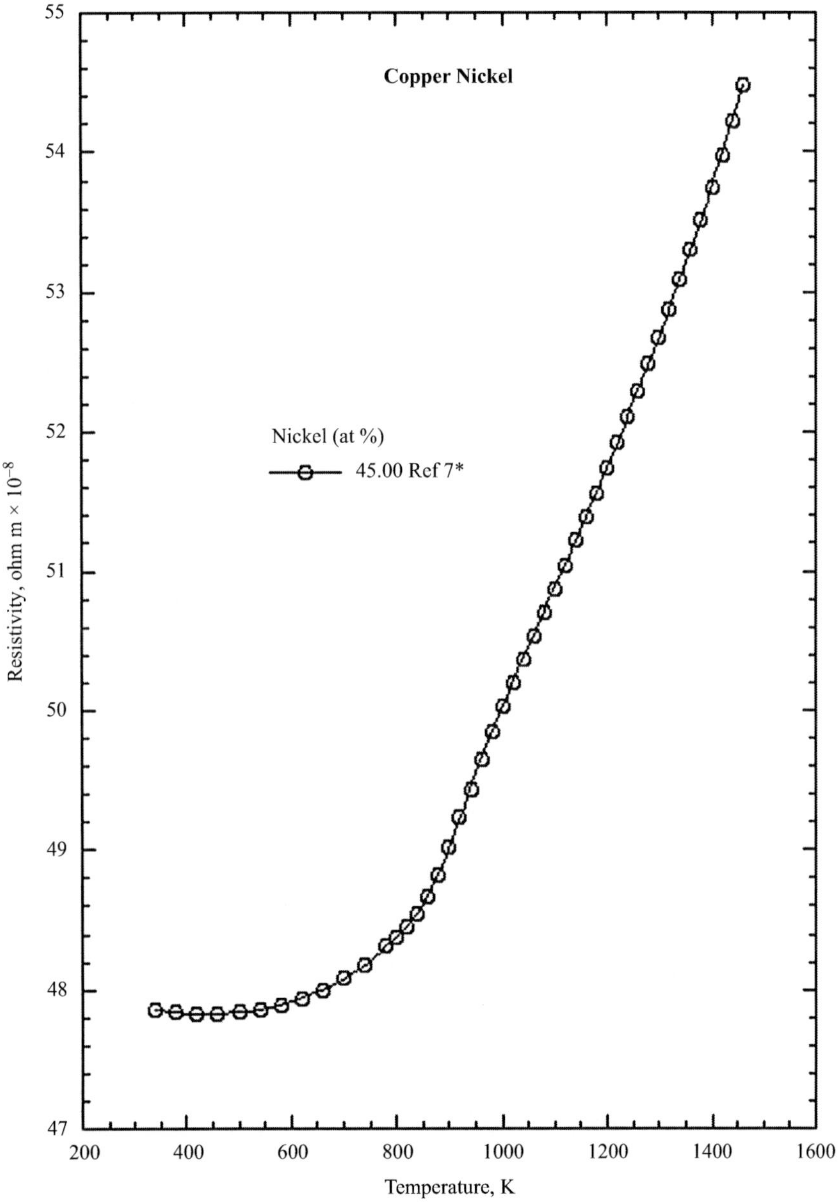
Copper Nickel
Nickel (at %)
45.00 Ref 7*
Resistivity, ohm m × 10⁻⁸
Temperature, K
55
54
53
52
51
50
49
48
47
200
400
600
800
1000
1200
1400
1600

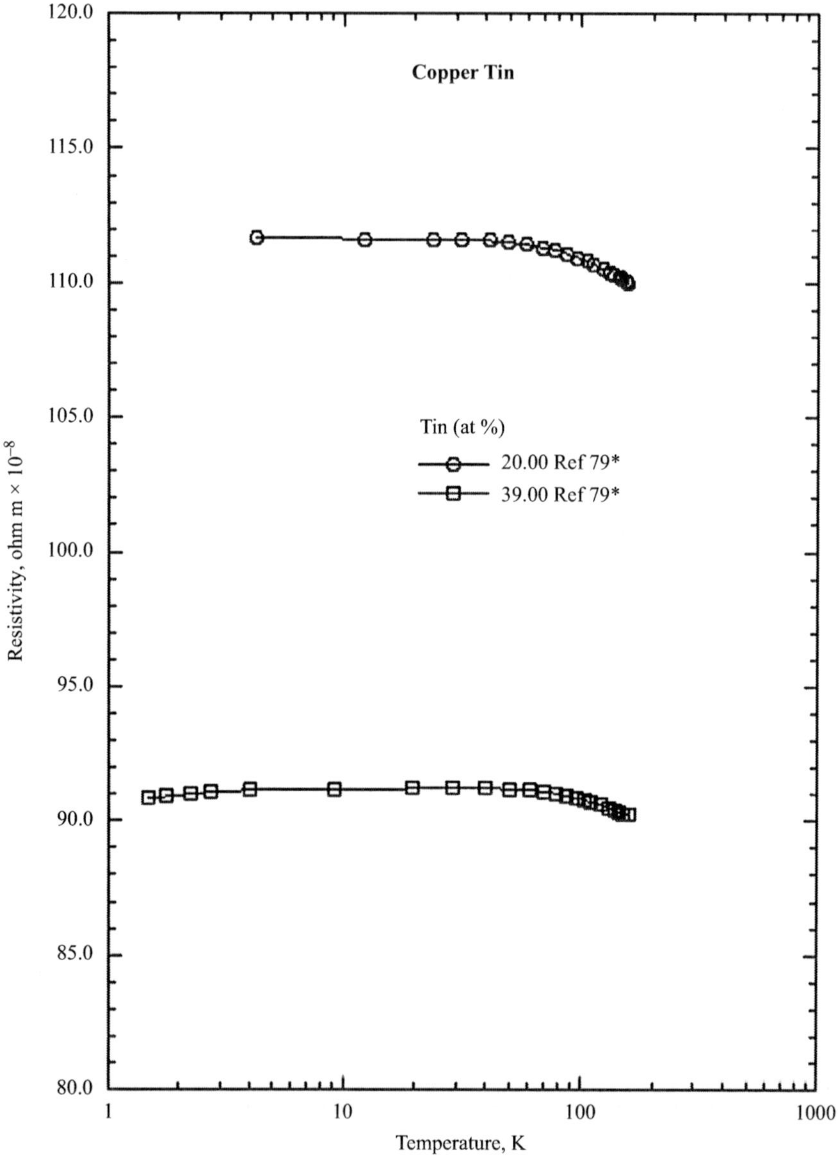
Copper Tin
Tin (at %)
20.00 Ref 79*
39.00 Ref 79*
Resistivity, ohm m × 10−8
120.0
115.0
110.0
105.0
100.0
95.0
90.0
85.0
80.0
1
10
100
1000
Temperature, K

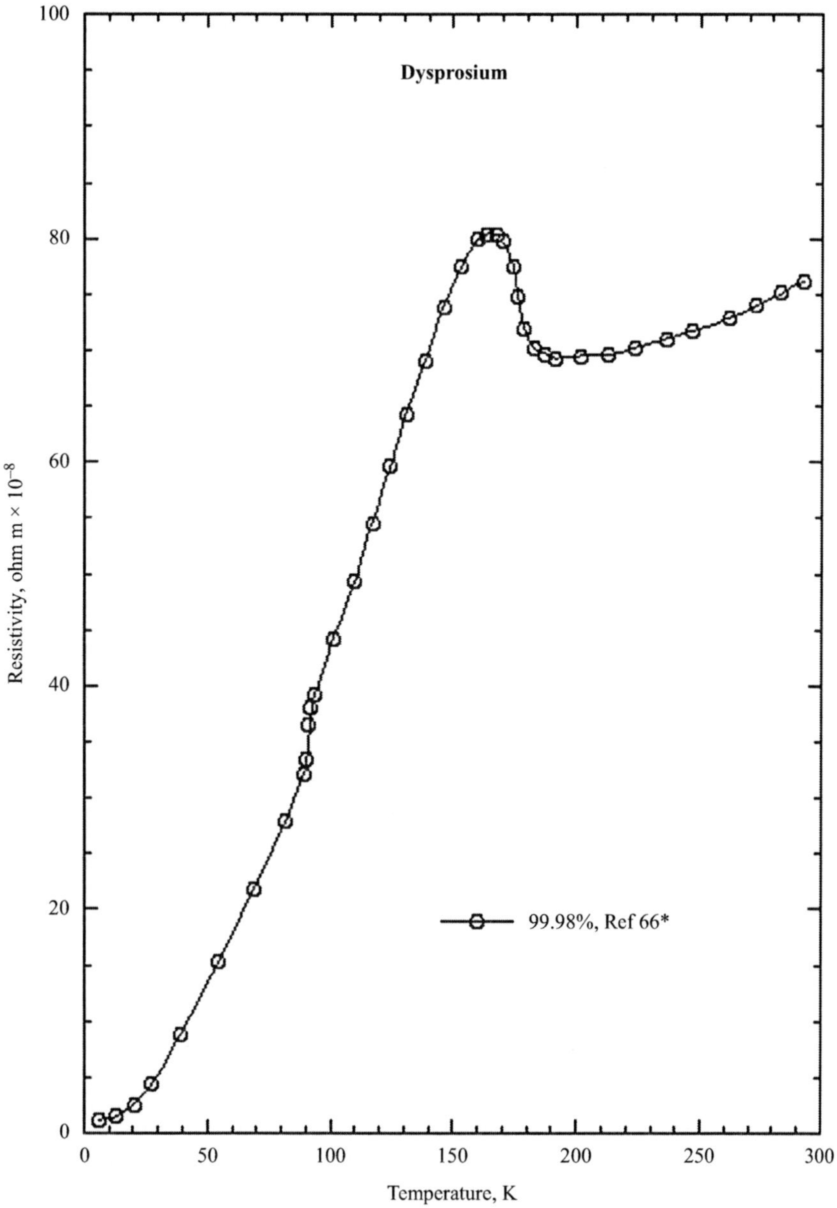
Dysprosium
99.98%, Ref 66*
Resistivity, ohm m × 10⁻⁸
Temperature, K
0
20
40
60
80
100
0
50
100
150
200
250
300

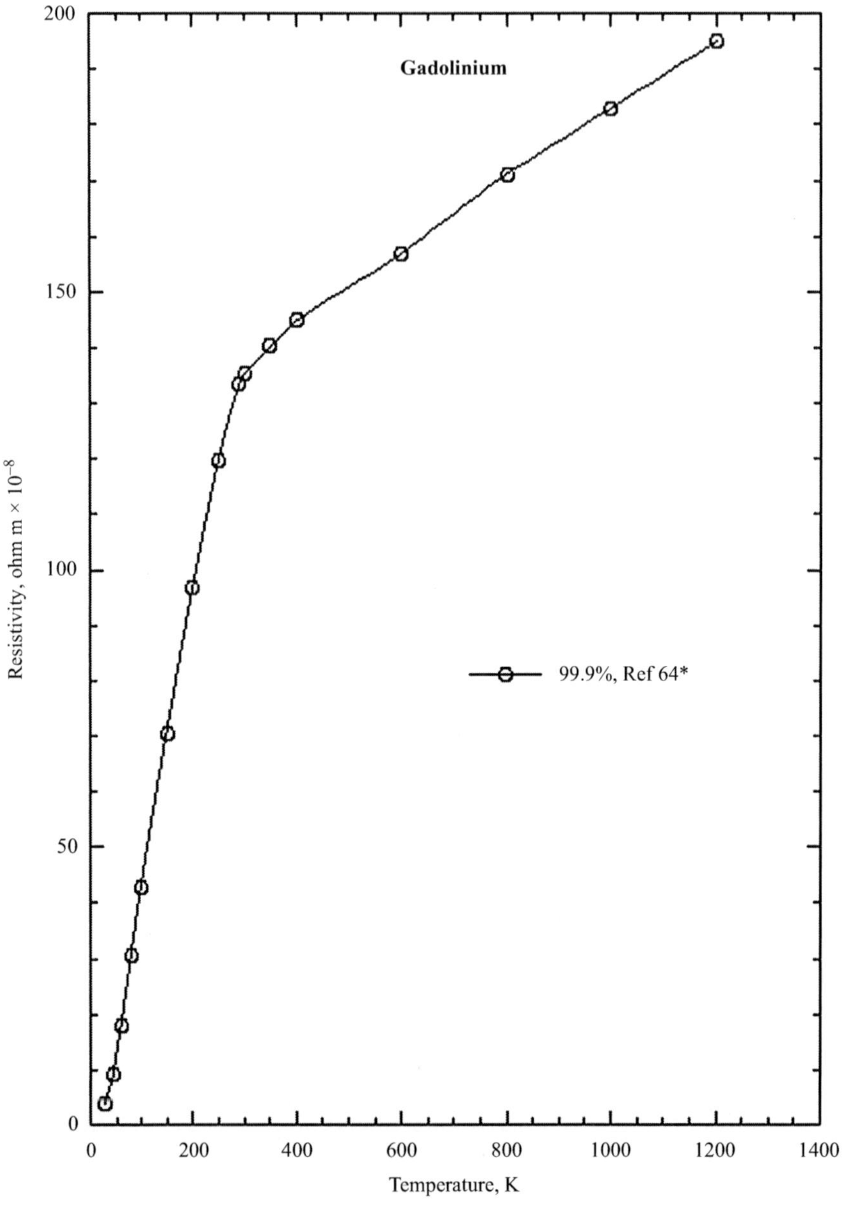
Gadolinium
99.9%, Ref 64*
Resistivity, ohm m × 10^{-8}
Temperature, K
0
50
100
150
200
0
200
400
600
800
1000
1200
1400

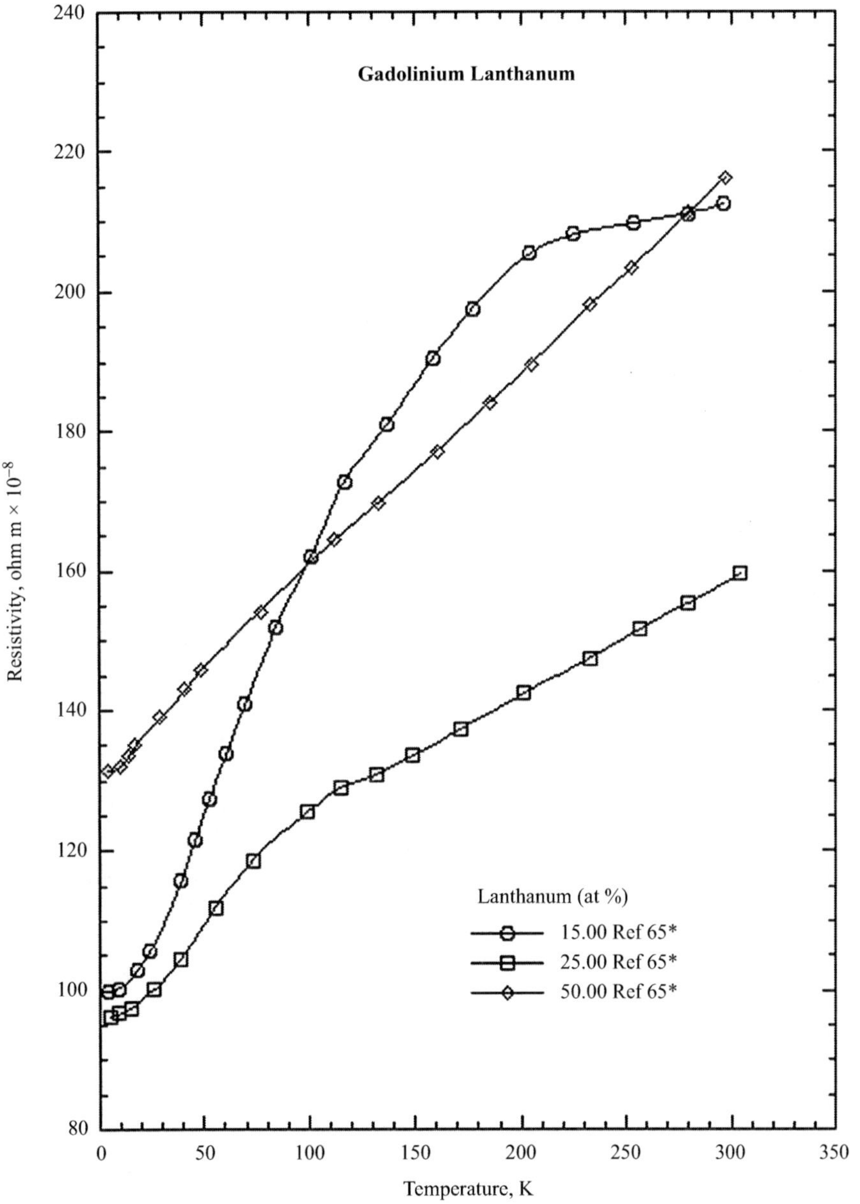
Gadolinium Lanthanum
Resistivity, ohm m × 10⁻⁸
Temperature, K
Lanthanum (at %)
15.00 Ref 65*
25.00 Ref 65*
50.00 Ref 65*
240
220
200
180
160
140
120
100
80
0
50
100
150
200
250
300
350

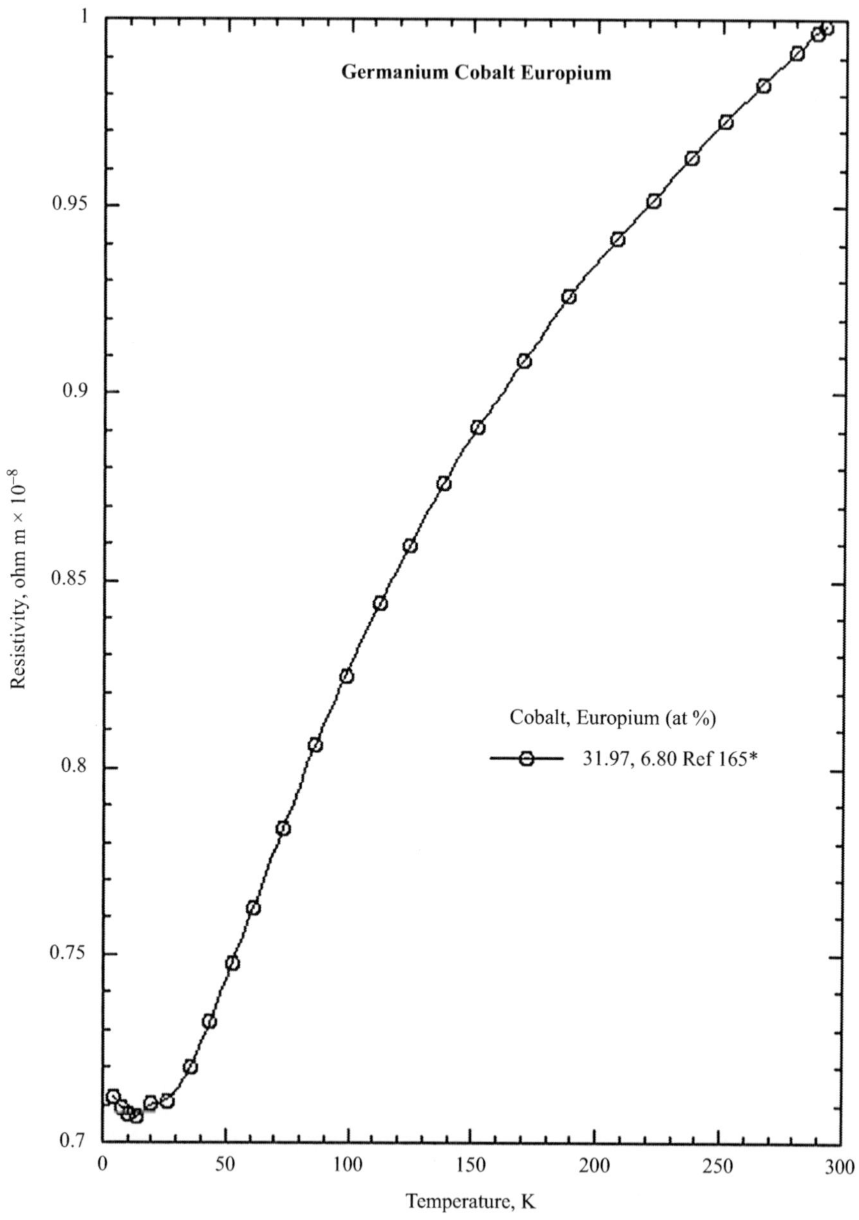
Germanium Cobalt Europium
Resistivity, ohm m × 10⁻⁸
Temperature, K
Cobalt, Europium (at %)
31.97, 6.80 Ref 165*
1
0.95
0.9
0.85
0.8
0.75
0.7
0
50
100
150
200
250
300

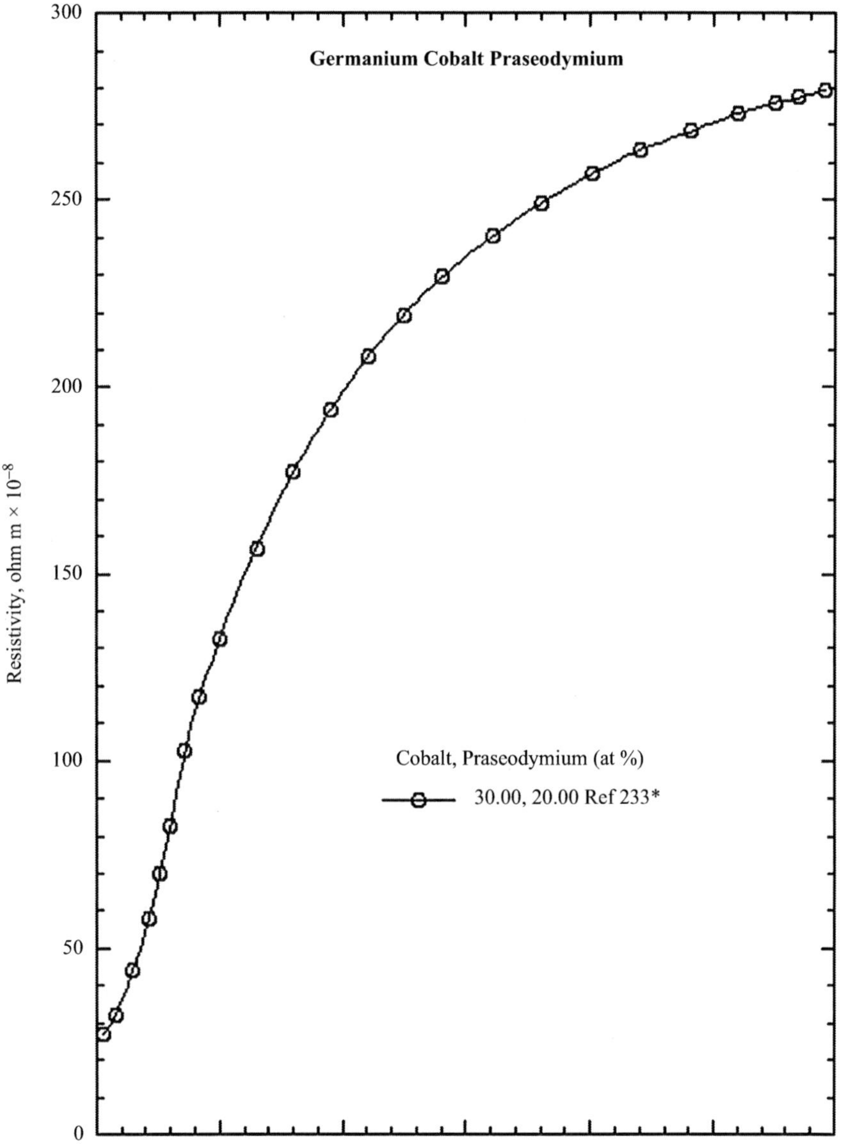
Germanium Cobalt Praseodymium
Cobalt, Praseodymium (at %)
30.00, 20.00 Ref 233*
Resistivity, ohm m × 10^−8
Temperature, K
0
50
100
150
200
250
300

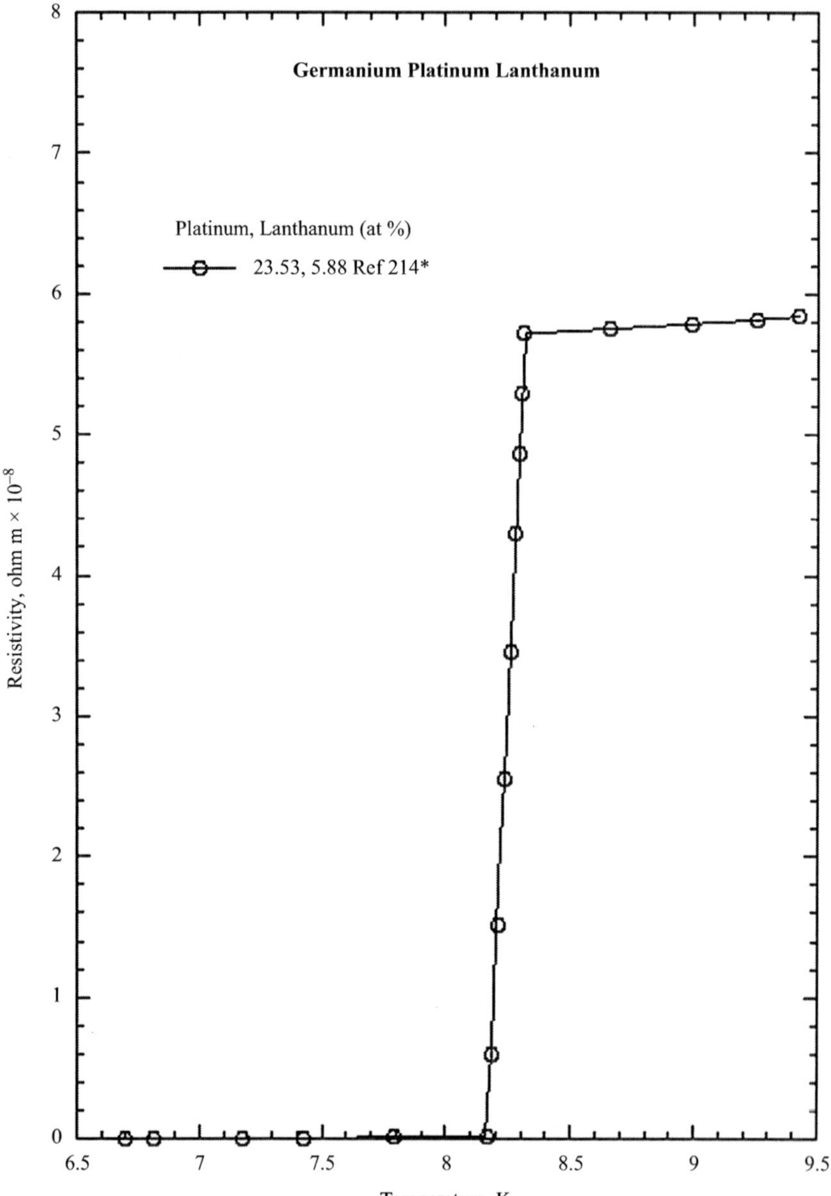
Germanium Platinum Lanthanum
Platinum, Lanthanum (at %)
23.53, 5.88 Ref 214*
Resistivity, ohm m × 10⁻⁸
Temperature, K
0
1
2
3
4
5
6
7
8
6.5
7
7.5
8
8.5
9
9.5

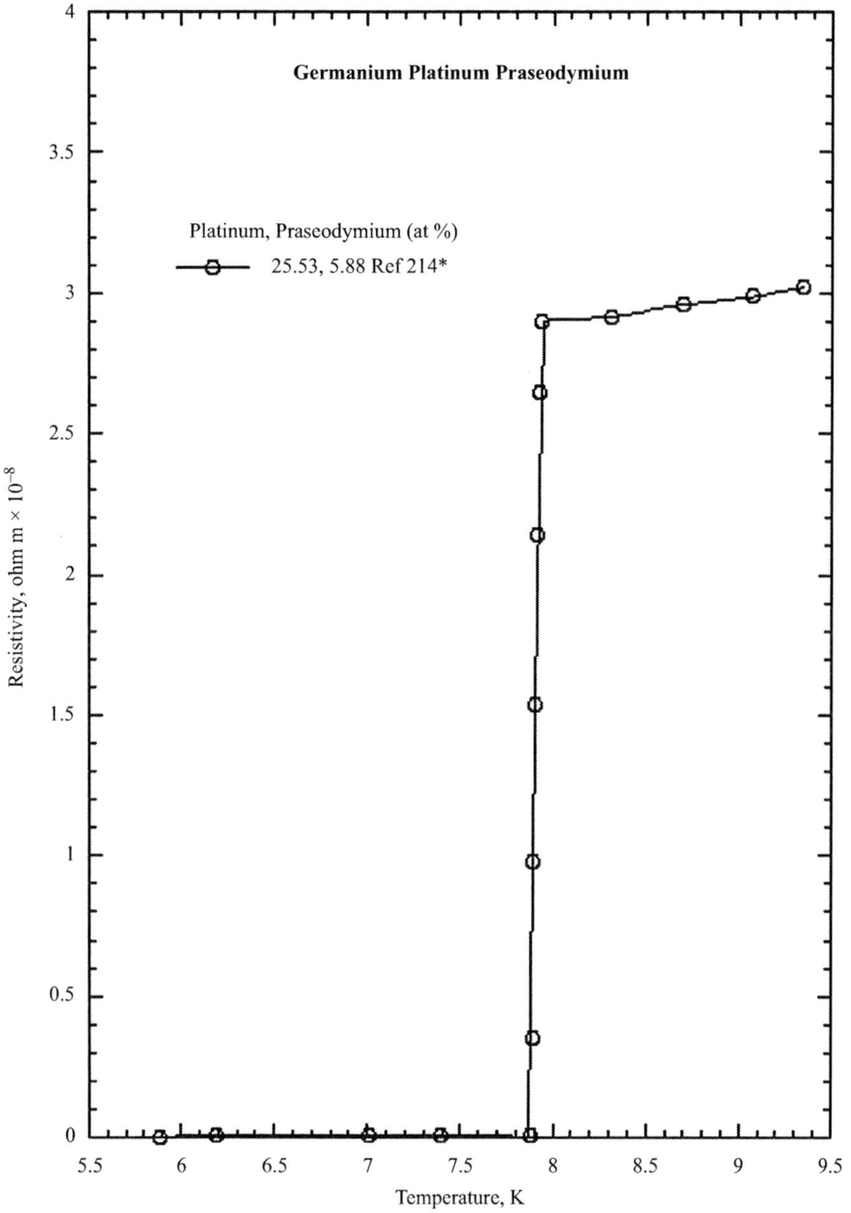
Germanium Platinum Praseodymium
Platinum, Praseodymium (at %)
25.53, 5.88 Ref 214*
Resistivity, ohm m × 10⁻⁸
Temperature, K
0
0.5
1
1.5
2
2.5
3
3.5
4
5.5
6
6.5
7
7.5
8
8.5
9
9.5

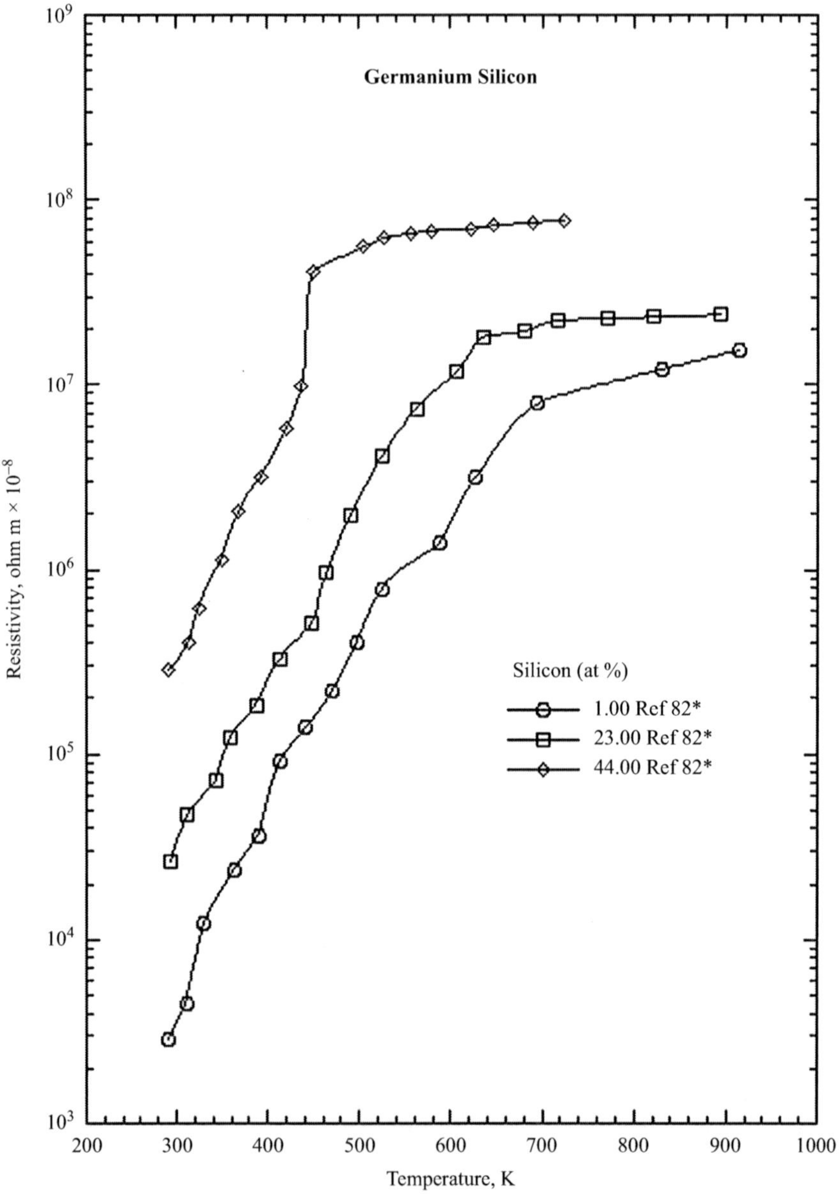
Germanium Silicon
Resistivity, ohm m × 10⁻⁸
Temperature, K
Silicon (at %)
1.00 Ref 82*
23.00 Ref 82*
44.00 Ref 82*
10³
10⁴
10⁵
10⁶
10⁷
10⁸
10⁹
200
300
400
500
600
700
800
900
1000

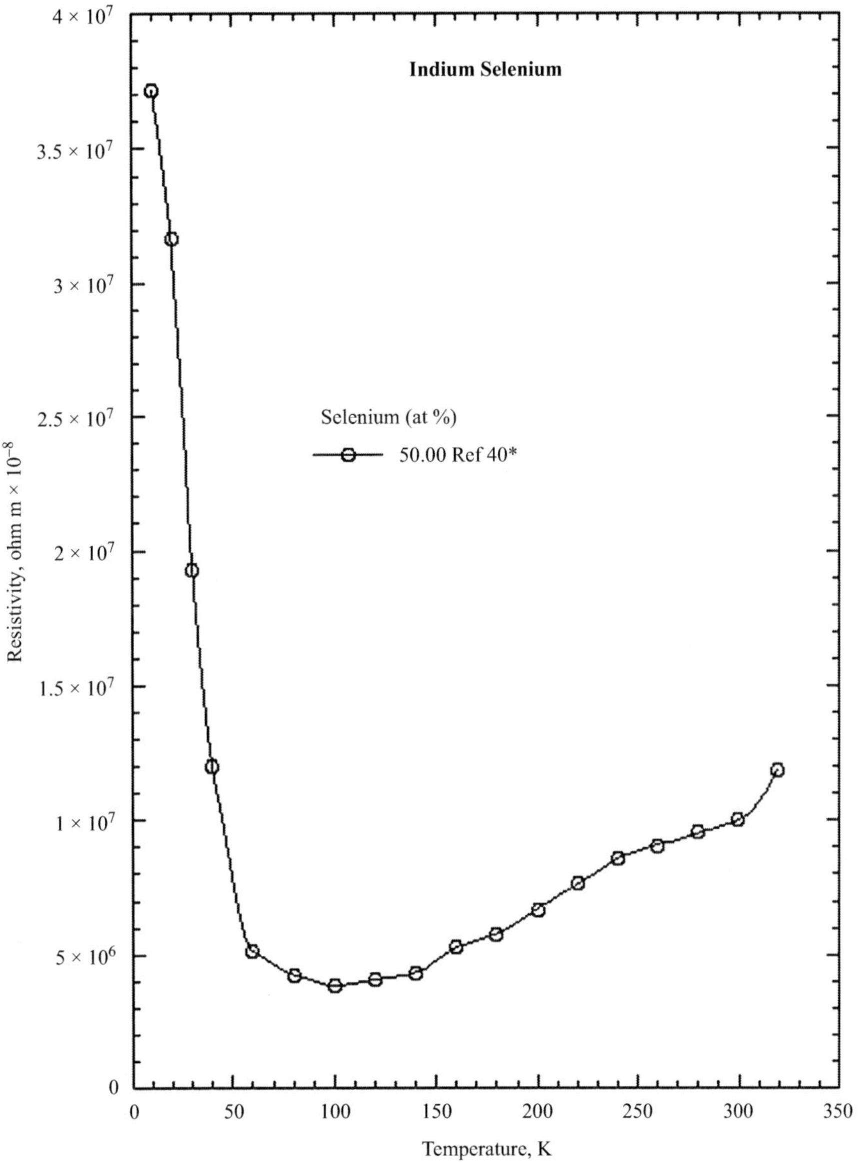
Indium Selenium
Selenium (at %)
50.00 Ref 40*
Resistivity, ohm m × 10^-8
Temperature, K
4 × 10^7
3.5 × 10^7
3 × 10^7
2.5 × 10^7
2 × 10^7
1.5 × 10^7
1 × 10^7
5 × 10^6
0
0
50
100
150
200
250
300
350

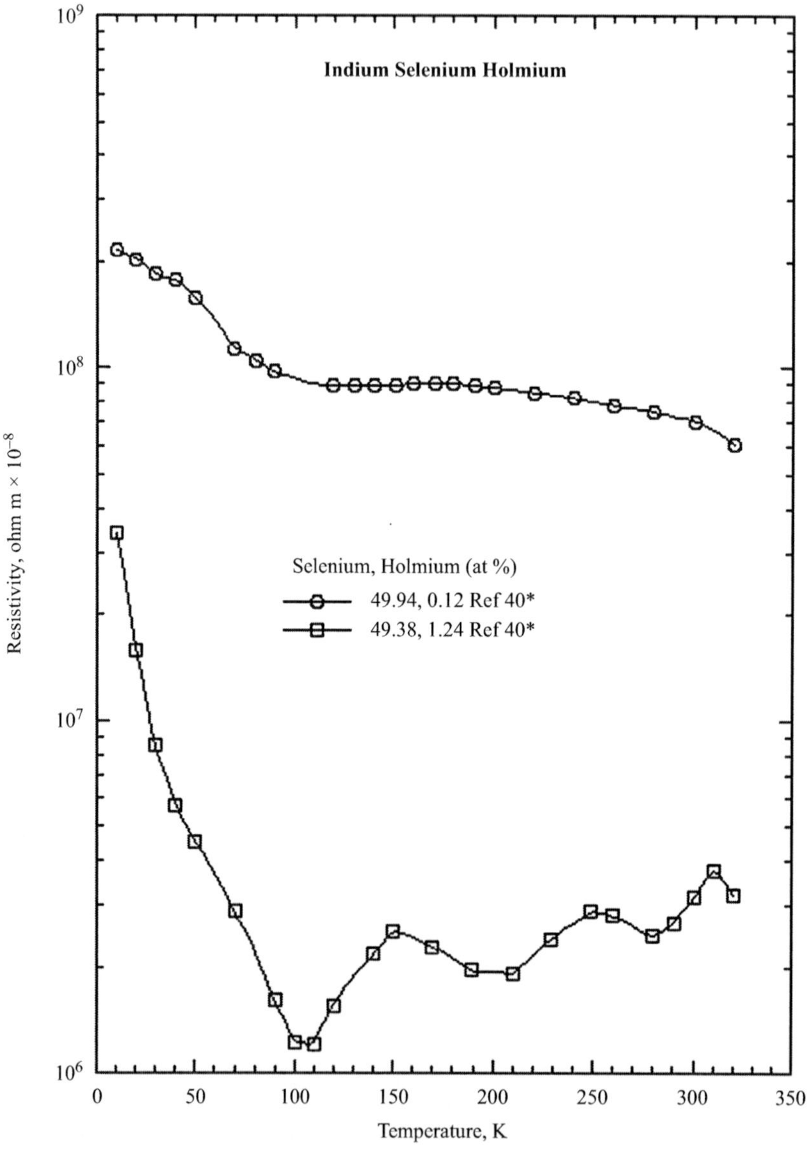
Indium Selenium Holmium
Selenium, Holmium (at %)
49.94, 0.12 Ref 40*
49.38, 1.24 Ref 40*
Resistivity, ohm m × 10⁻⁸
Temperature, K
10^9
10^8
10^7
10^6
0
50
100
150
200
250
300
350

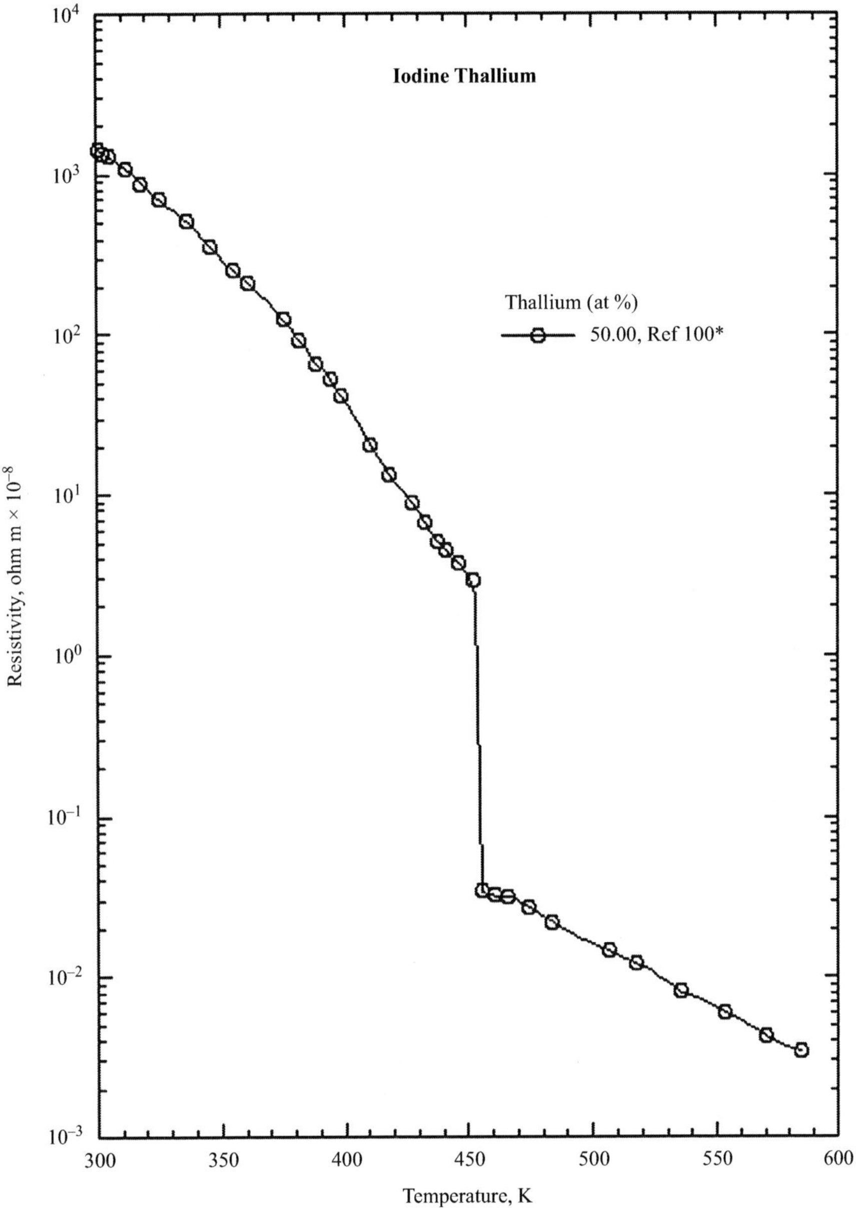
Iodine Thallium
Thallium (at %)
50.00, Ref 100*
Resistivity, ohm m × 10⁻⁸
Temperature, K
10⁴
10³
10²
10¹
10⁰
10⁻¹
10⁻²
10⁻³
300
350
400
450
500
550
600

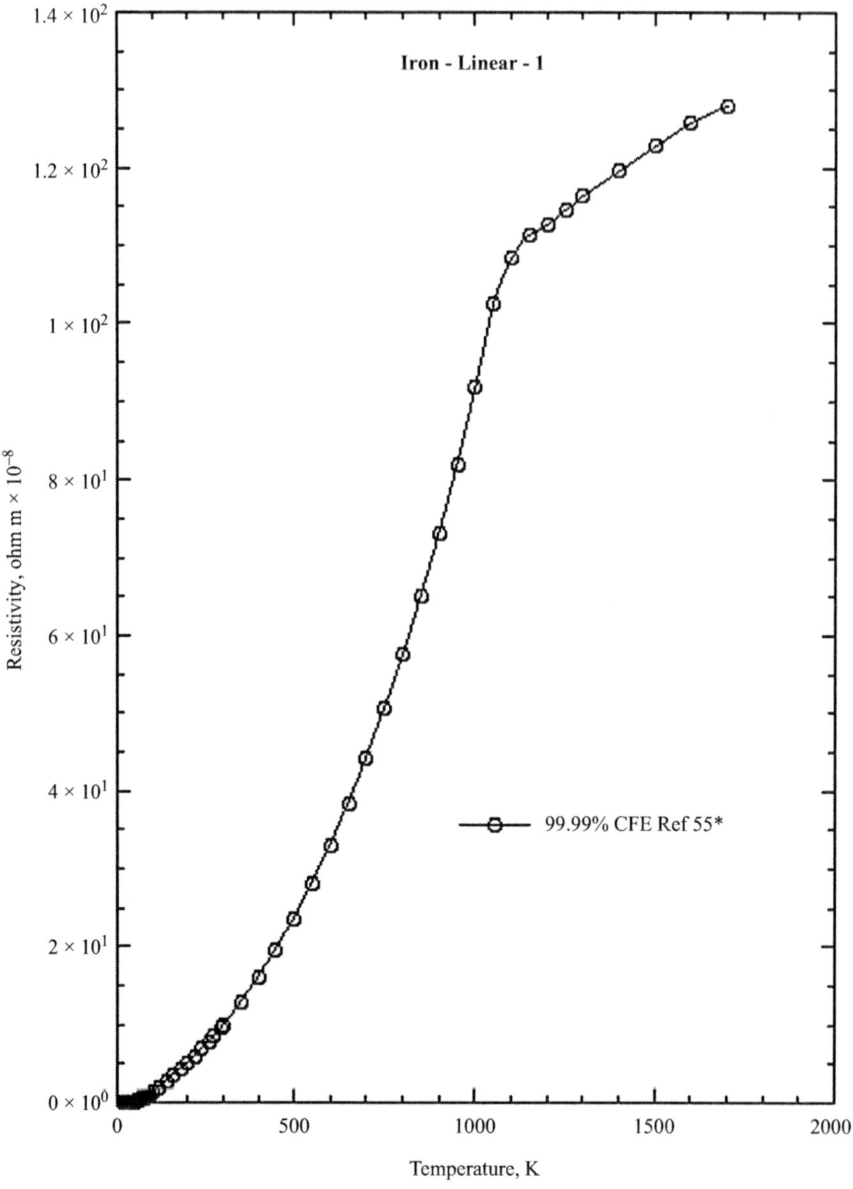
Iron - Linear - 1
99.99% CFE Ref 55*
Resistivity, ohm m × 10⁻⁸
1.4 × 10²
1.2 × 10²
1 × 10²
8 × 10¹
6 × 10¹
4 × 10¹
2 × 10¹
0 × 10⁰
0
500
1000
1500
2000
Temperature, K

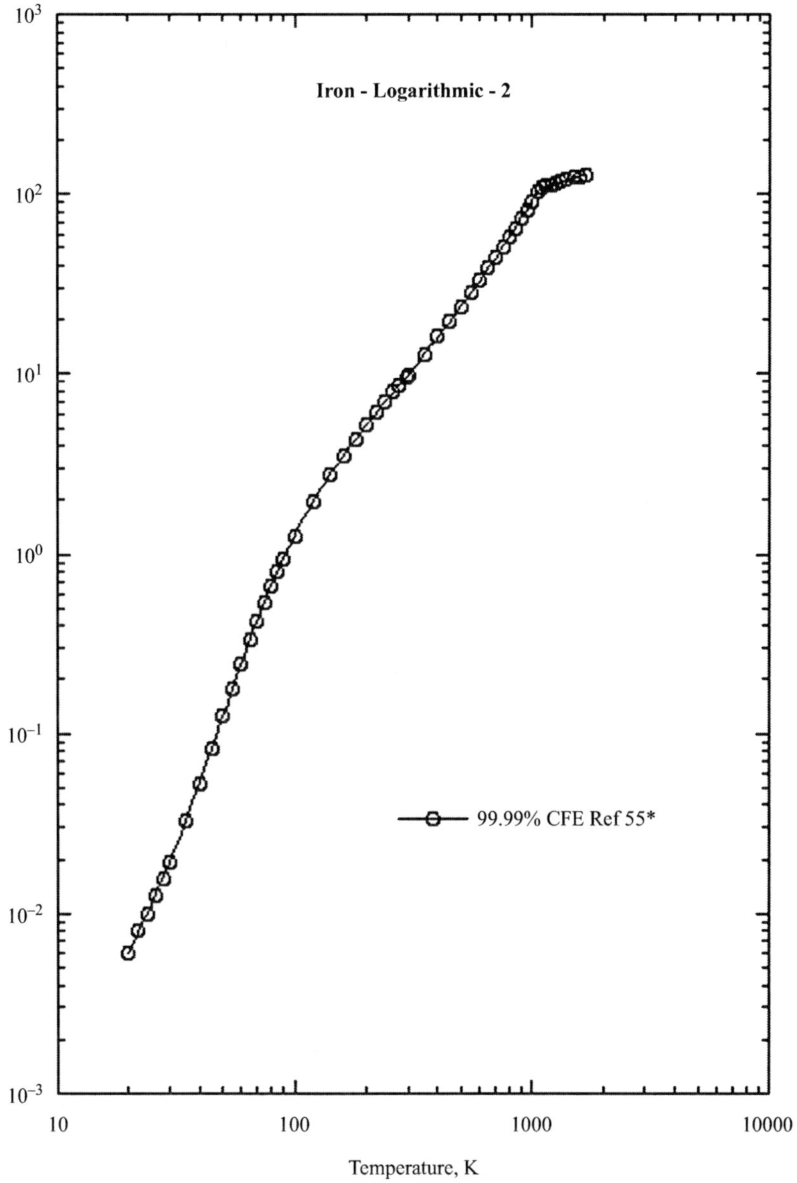
Iron - Logarithmic - 2
99.99% CFE Ref 55*
Resistivity, ohm m × 10^−8
Temperature, K
10^3
10^2
10^1
10^0
10^−1
10^−2
10^−3
10
100
1000
10000

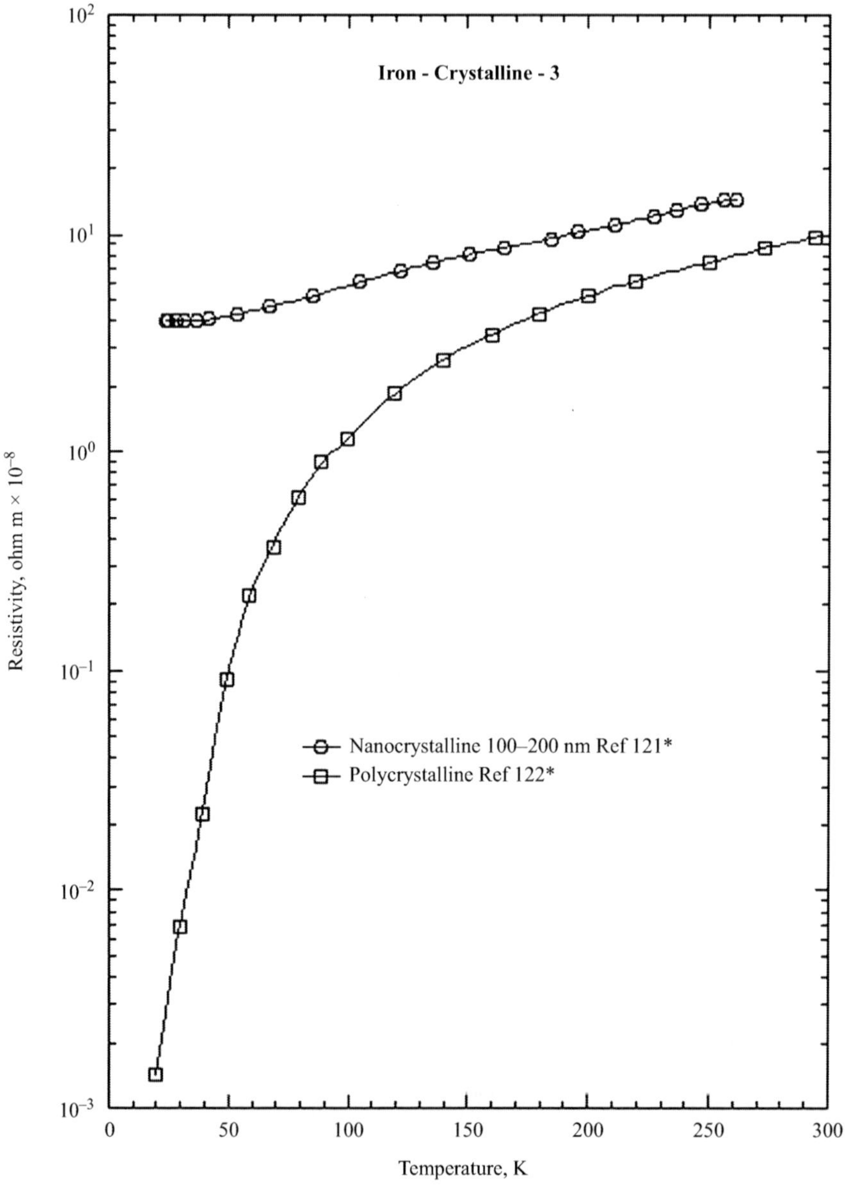
Iron - Crystalline - 3
Nanocrystalline 100–200 nm Ref 121*
Polycrystalline Ref 122*
Resistivity, ohm m × 10⁻⁸
Temperature, K
10²
10¹
10⁰
10⁻¹
10⁻²
10⁻³
0
50
100
150
200
250
300

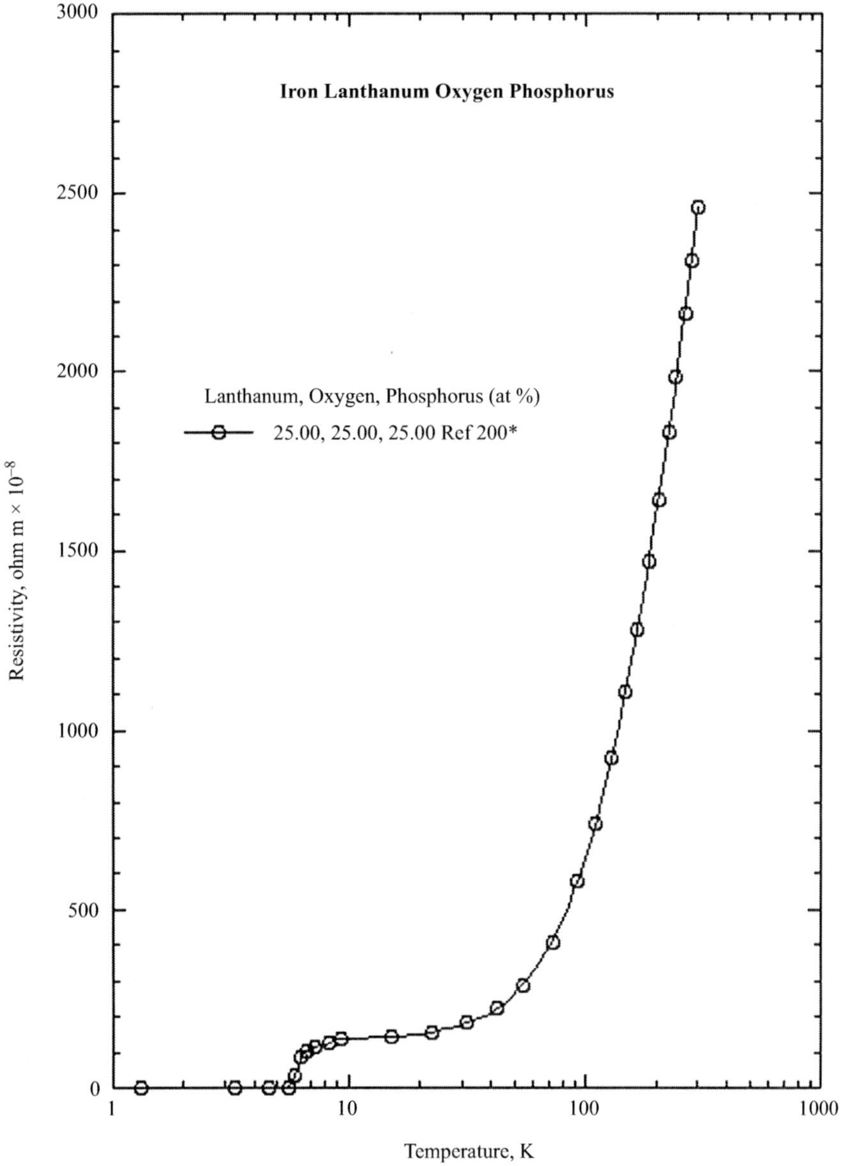

Iron Lanthanum Oxygen Phosphorus
Lanthanum, Oxygen, Phosphorus (at %)
25.00, 25.00, 25.00 Ref 200*
Resistivity, ohm m × 10−8
Temperature, K
0
500
1000
1500
2000
2500
3000
1
10
100
1000

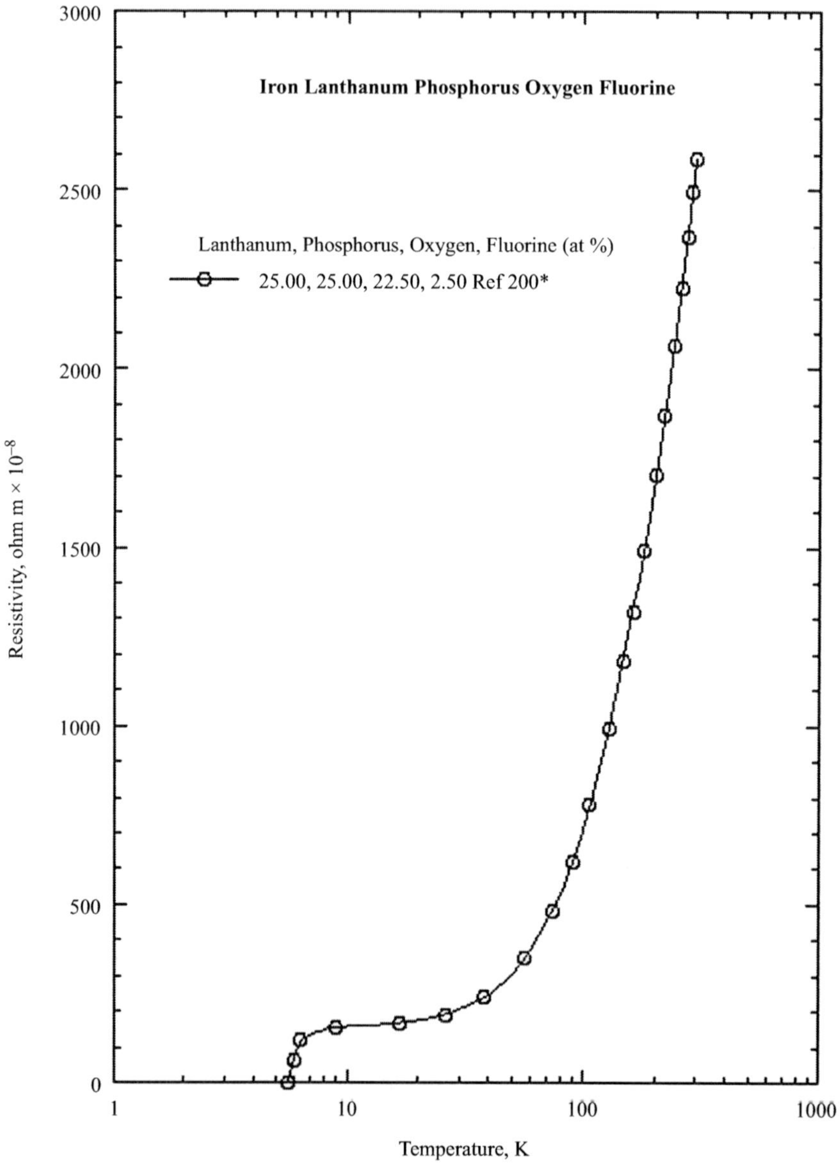
Iron Lanthanum Phosphorus Oxygen Fluorine
Lanthanum, Phosphorus, Oxygen, Fluorine (at %)
25.00, 25.00, 22.50, 2.50 Ref 200*
Resistivity, ohm m × 10^{-8}
Temperature, K
0
500
1000
1500
2000
2500
3000
1
10
100
1000

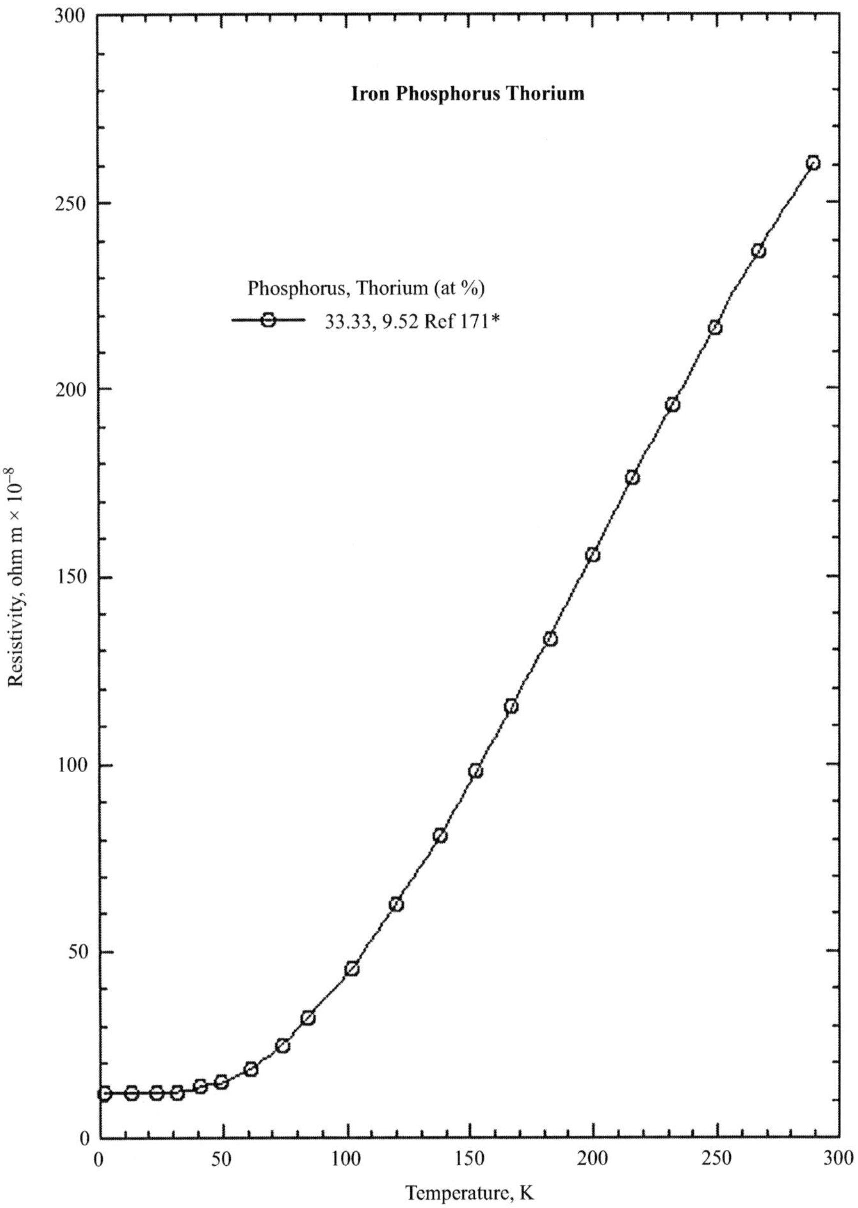
Iron Phosphorus Thorium
Phosphorus, Thorium (at %)
33.33, 9.52 Ref 171*
Resistivity, ohm m × 10⁻⁸
Temperature, K
0
50
100
150
200
250
300

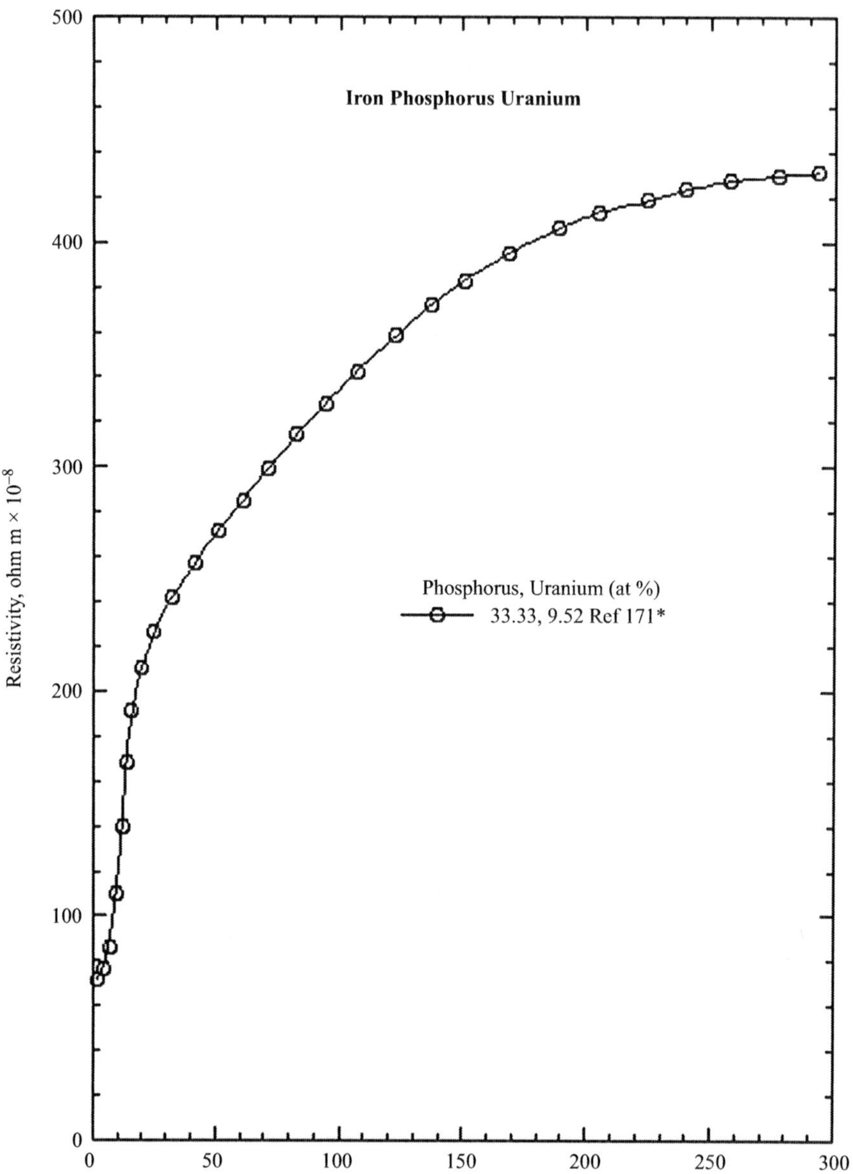
Iron Phosphorus Uranium
Phosphorus, Uranium (at %)
33.33, 9.52 Ref 171*
Resistivity, ohm m × 10−8
Temperature, K
0
100
200
300
400
500
0
50
100
150
200
250
300

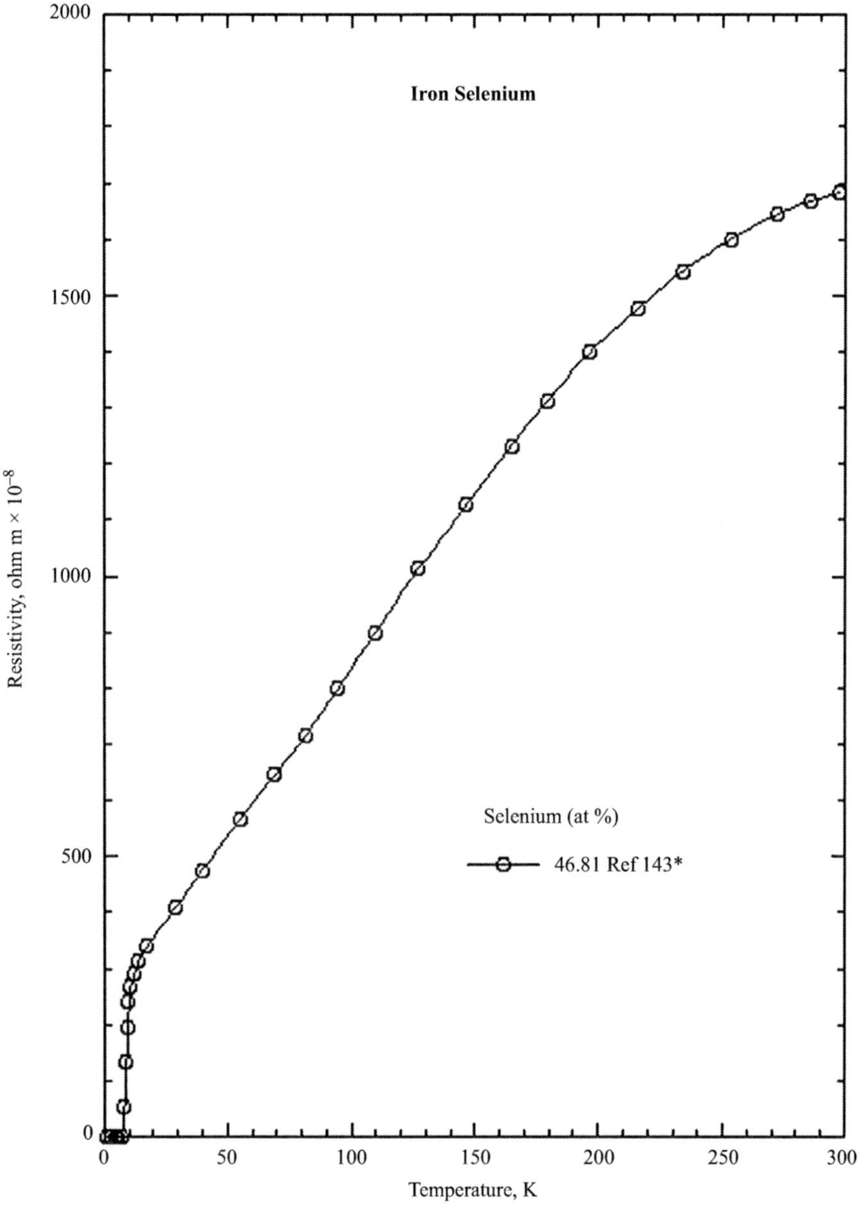
Iron Selenium
Selenium (at %)
46.81 Ref 143*
Resistivity, ohm m × 10^{-8}
Temperature, K
0
500
1000
1500
2000
0
50
100
150
200
250
300

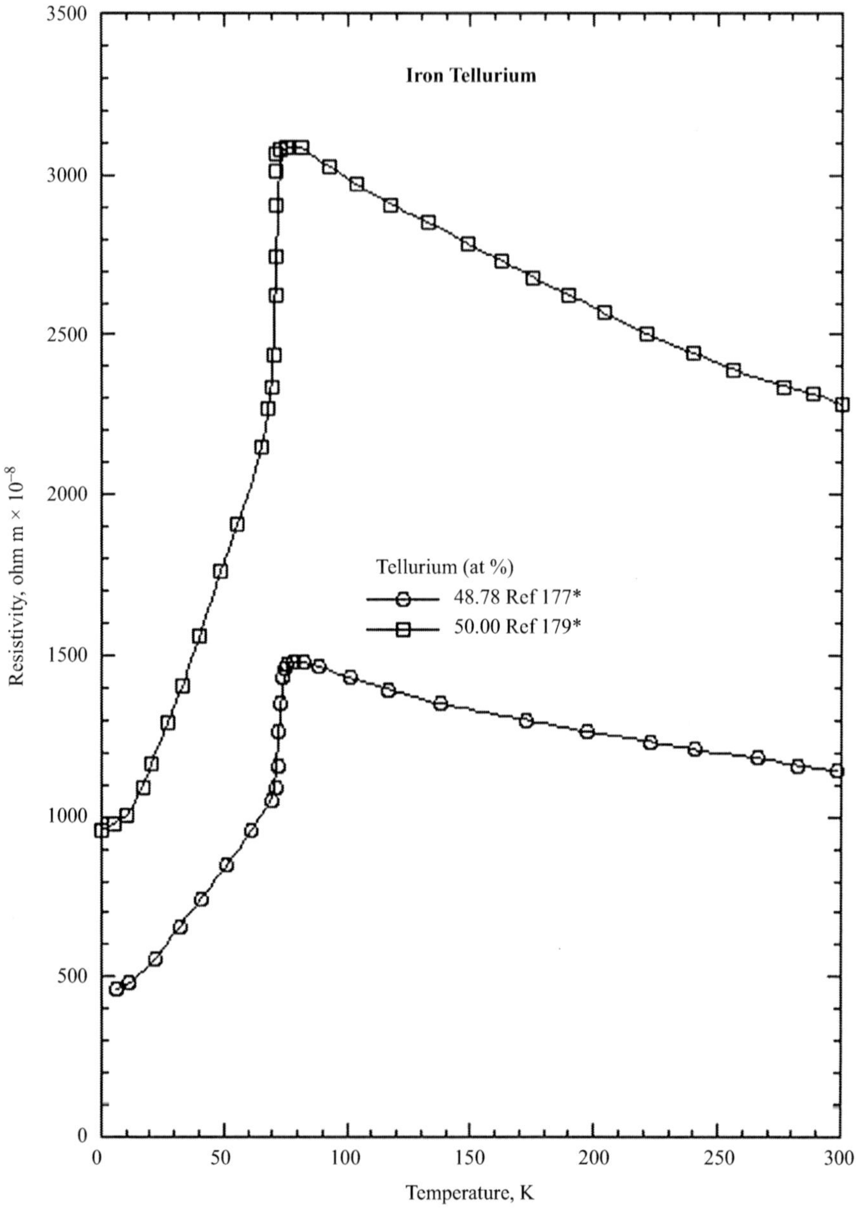

Iron Tellurium
Tellurium (at %)
48.78 Ref 177*
50.00 Ref 179*
Resistivity, ohm m × 10^-8
Temperature, K
0
500
1000
1500
2000
2500
3000
3500
0
50
100
150
200
250
300

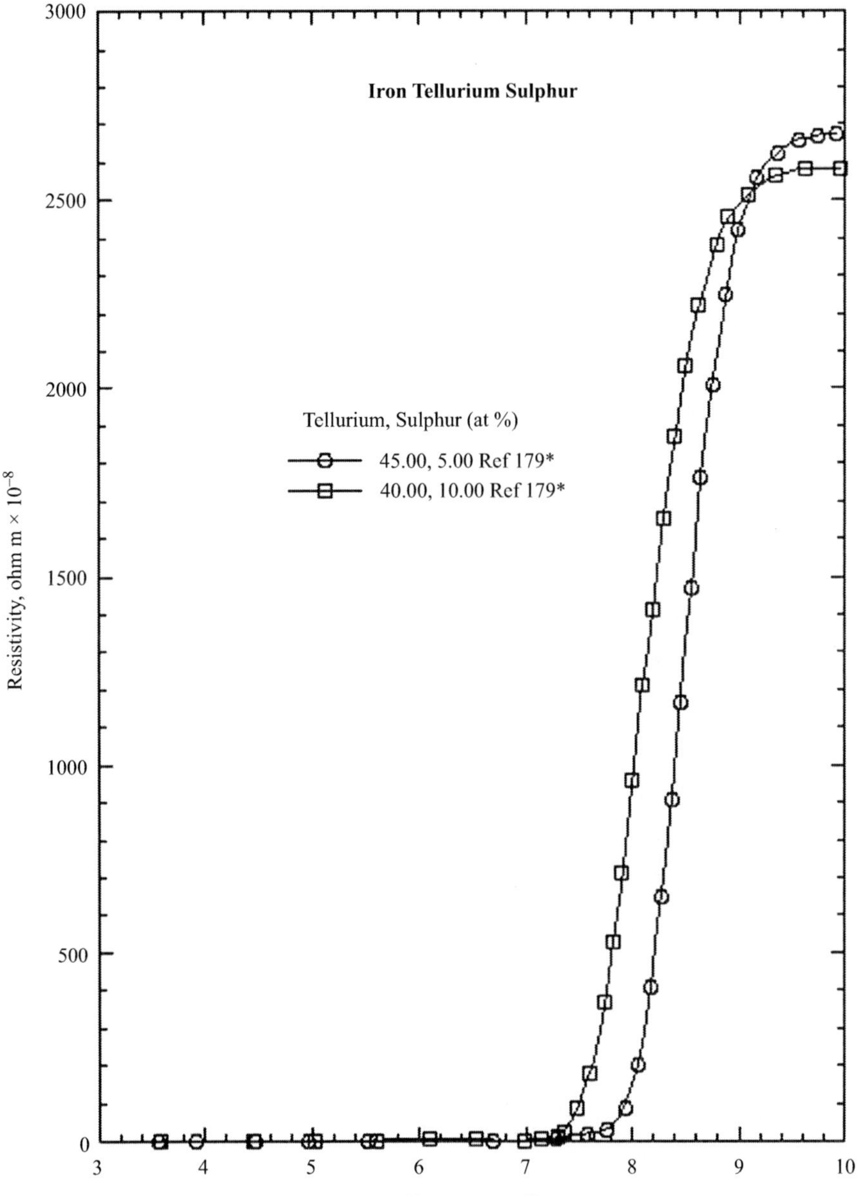
Iron Tellurium Sulphur
Tellurium, Sulphur (at %)
45.00, 5.00 Ref 179*
40.00, 10.00 Ref 179*
Resistivity, ohm m × 10^{-8}
Temperature, K
3000
2500
2000
1500
1000
500
0
3
4
5
6
7
8
9
10

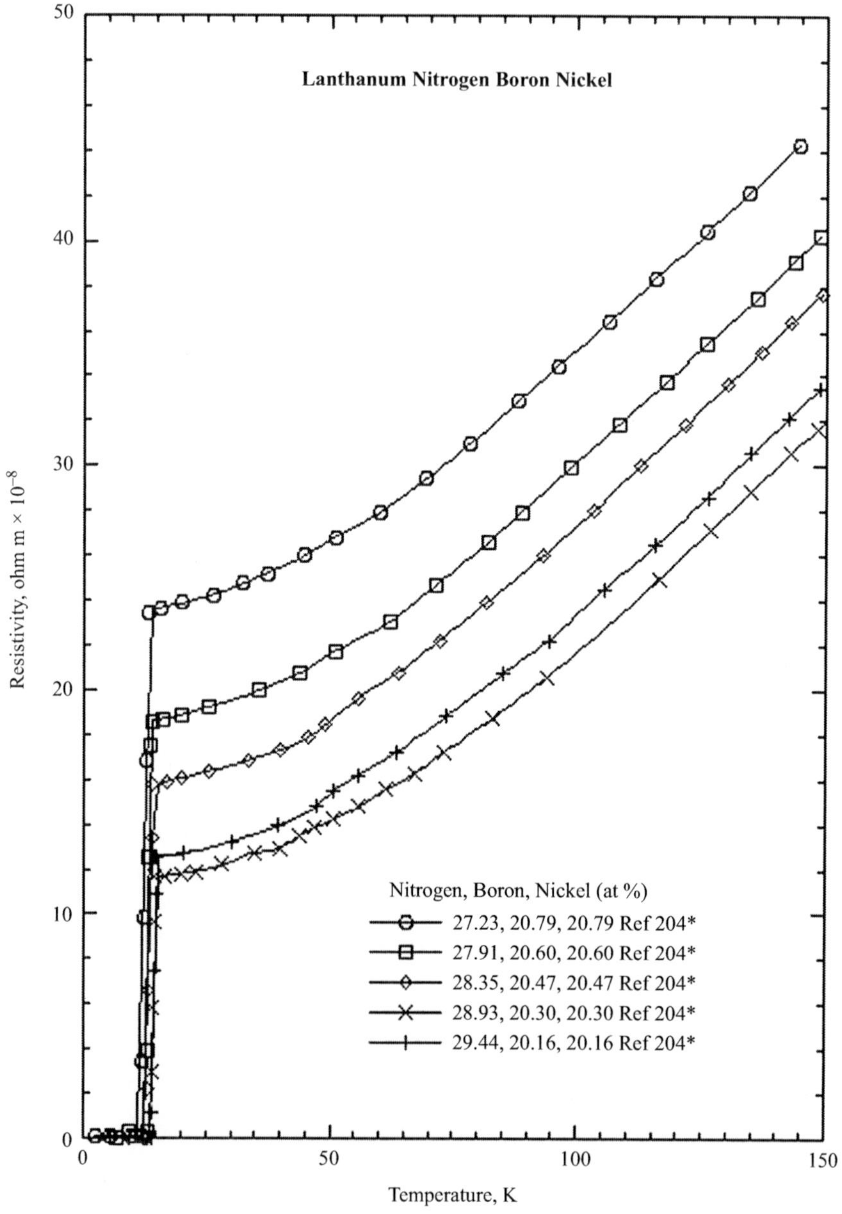
Lanthanum Nitrogen Boron Nickel
Resistivity, ohm m × 10^{-8}
Temperature, K
0
10
20
30
40
50
0
50
100
150
Nitrogen, Boron, Nickel (at %)
27.23, 20.79, 20.79 Ref 204*
27.91, 20.60, 20.60 Ref 204*
28.35, 20.47, 20.47 Ref 204*
28.93, 20.30, 20.30 Ref 204*
29.44, 20.16, 20.16 Ref 204*

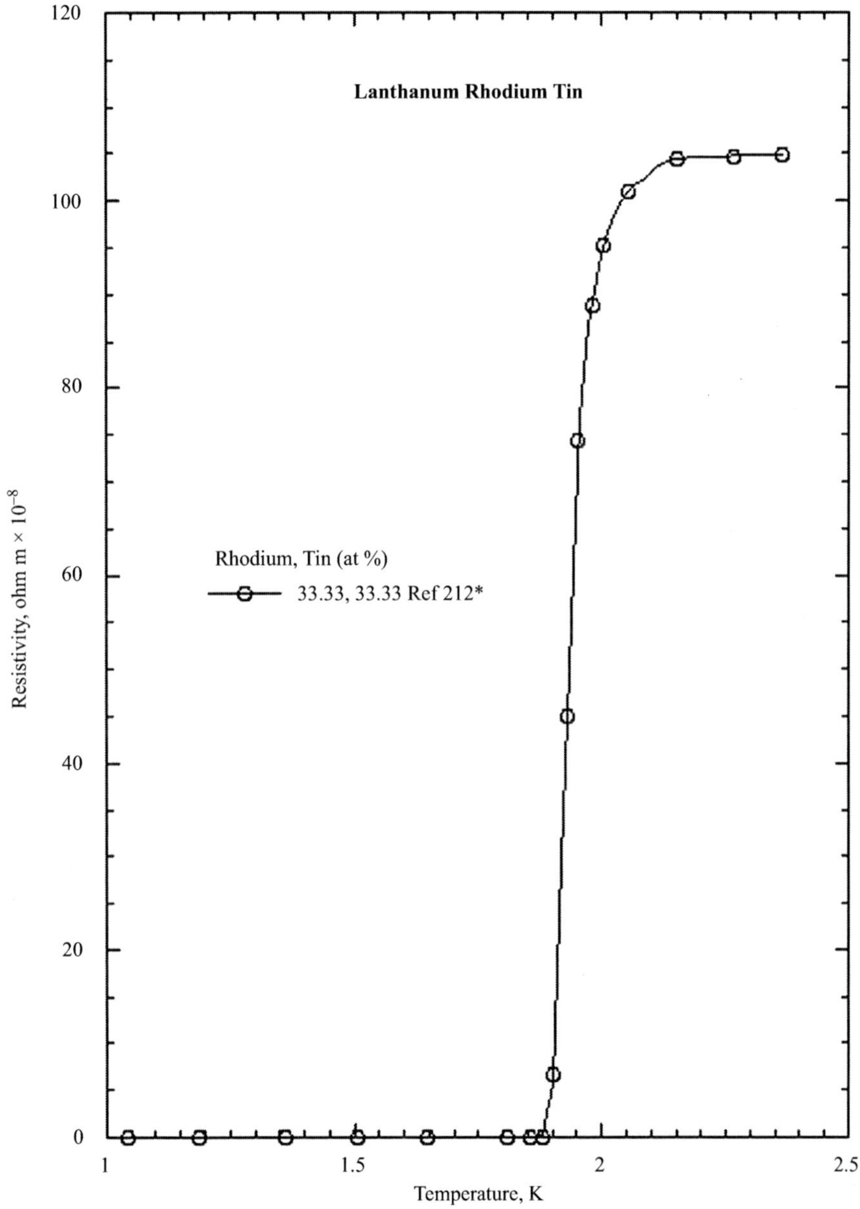
Lanthanum Rhodium Tin
Rhodium, Tin (at %)
33.33, 33.33 Ref 212*
Resistivity, ohm m × 10−8
Temperature, K
0
20
40
60
80
100
120
1
1.5
2
2.5

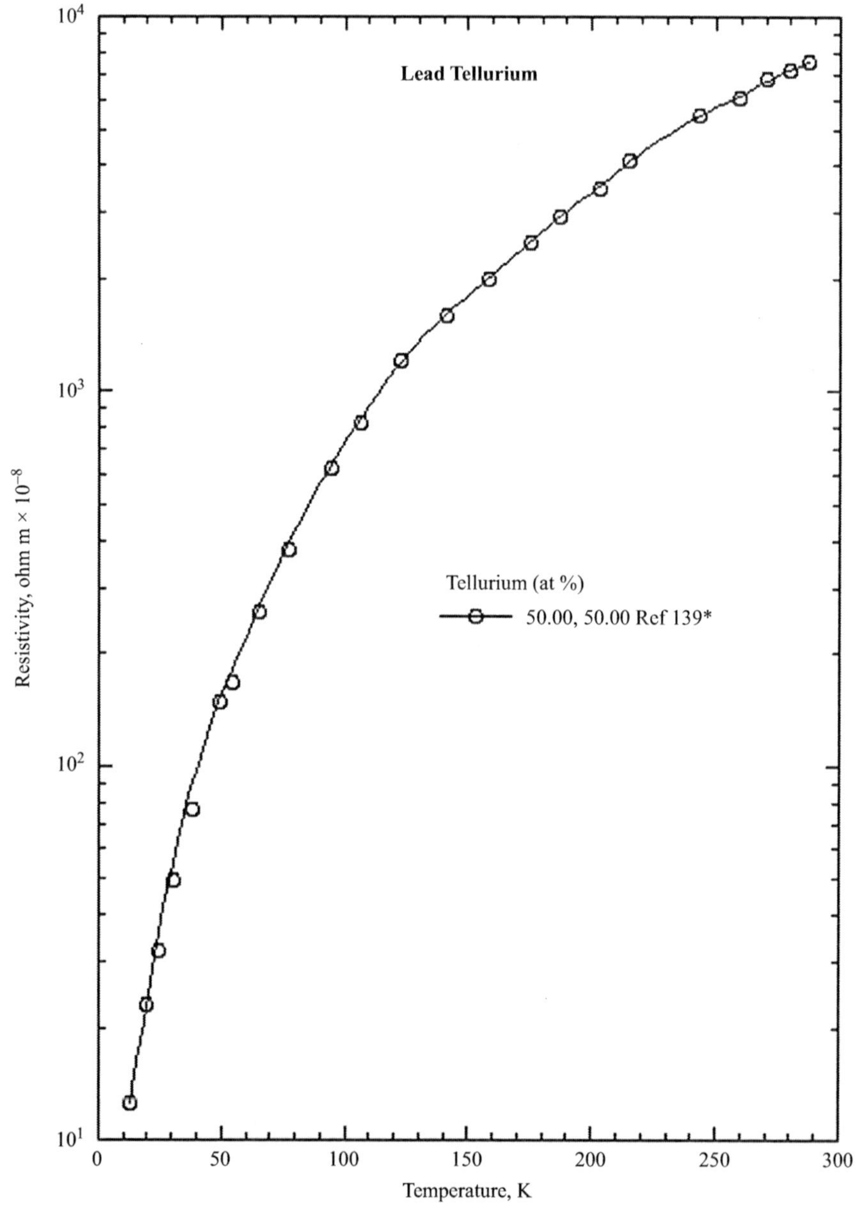
Lead Tellurium
Tellurium (at %)
50.00, 50.00 Ref 139*
Resistivity, ohm m × 10⁻⁸
Temperature, K
10¹
10²
10³
10⁴
0
50
100
150
200
250
300

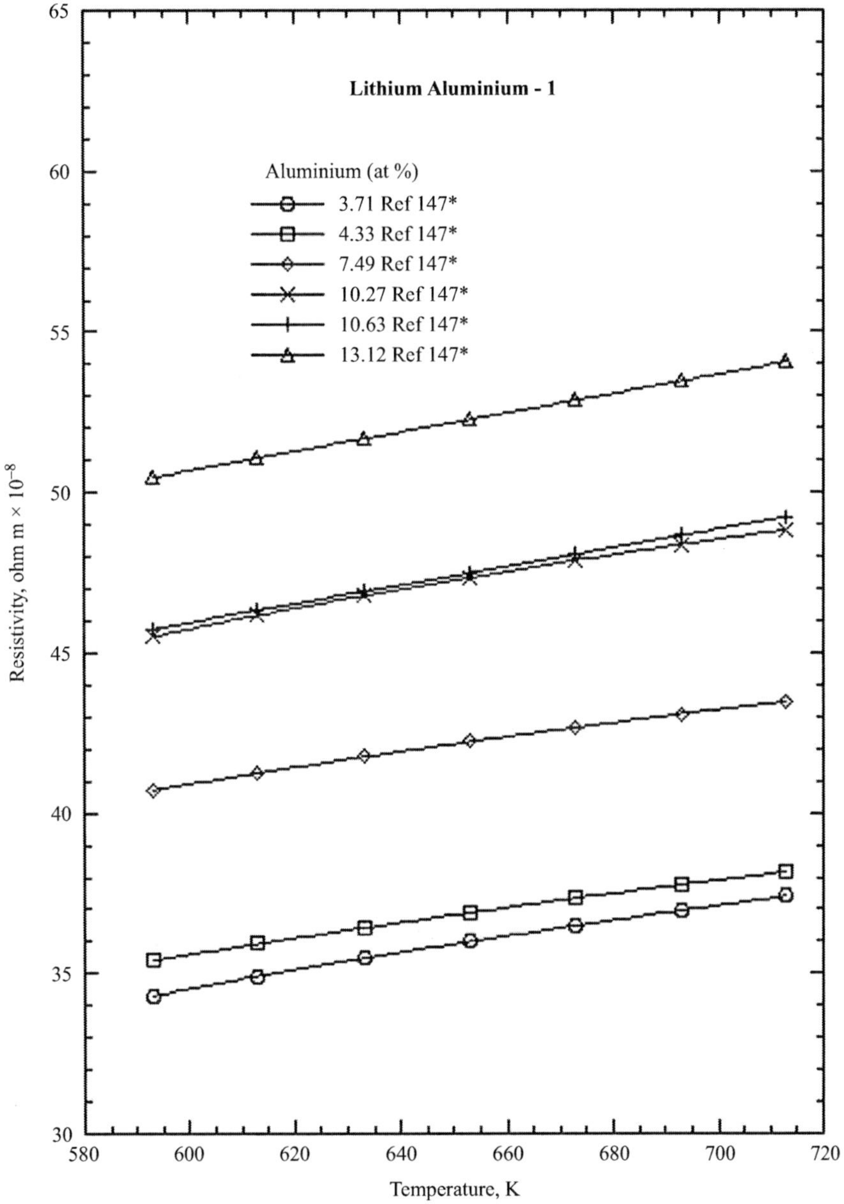
Lithium Aluminium - 1
Aluminium (at %)
3.71 Ref 147*
4.33 Ref 147*
7.49 Ref 147*
10.27 Ref 147*
10.63 Ref 147*
13.12 Ref 147*
Resistivity, ohm m × 10^{-8}
Temperature, K
65
60
55
50
45
40
35
30
580
600
620
640
660
680
700
720

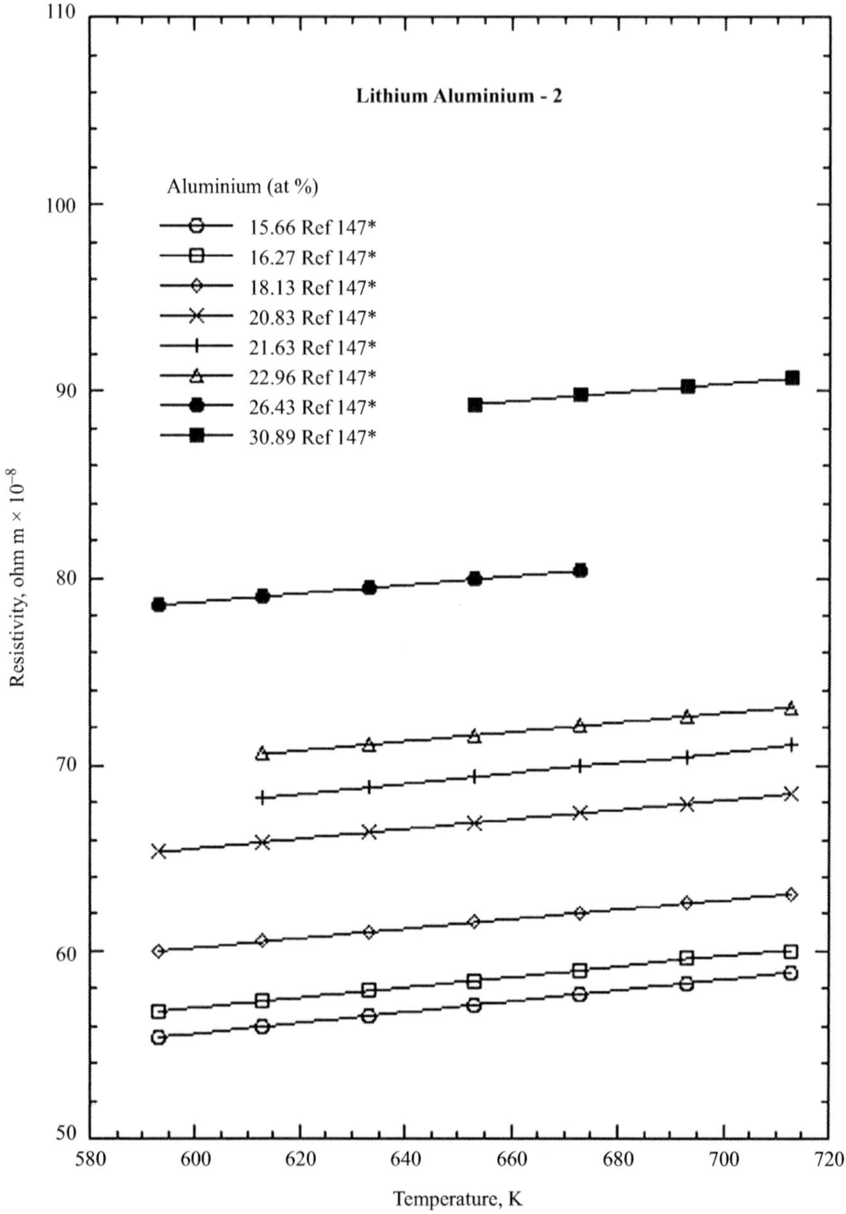
Lithium Aluminium - 2
Aluminium (at %)
15.66 Ref 147*
16.27 Ref 147*
18.13 Ref 147*
20.83 Ref 147*
21.63 Ref 147*
22.96 Ref 147*
26.43 Ref 147*
30.89 Ref 147*
Resistivity, ohm m × 10⁻⁸
Temperature, K
110
100
90
80
70
60
50
580
600
620
640
660
680
700
720

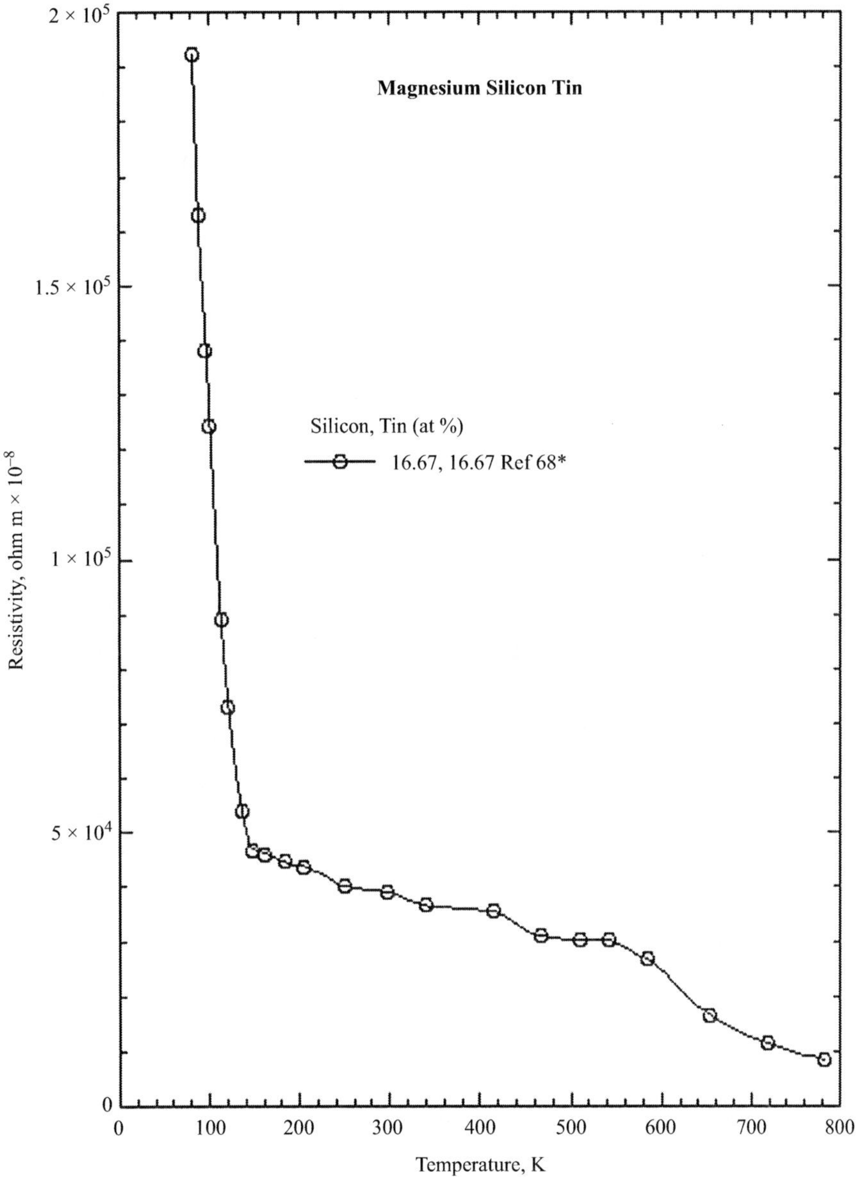
Magnesium Silicon Tin
Silicon, Tin (at %)
16.67, 16.67 Ref 68*
Resistivity, ohm m × 10⁻⁸
Temperature, K
2 × 10⁵
1.5 × 10⁵
1 × 10⁵
5 × 10⁴
0
0
100
200
300
400
500
600
700
800

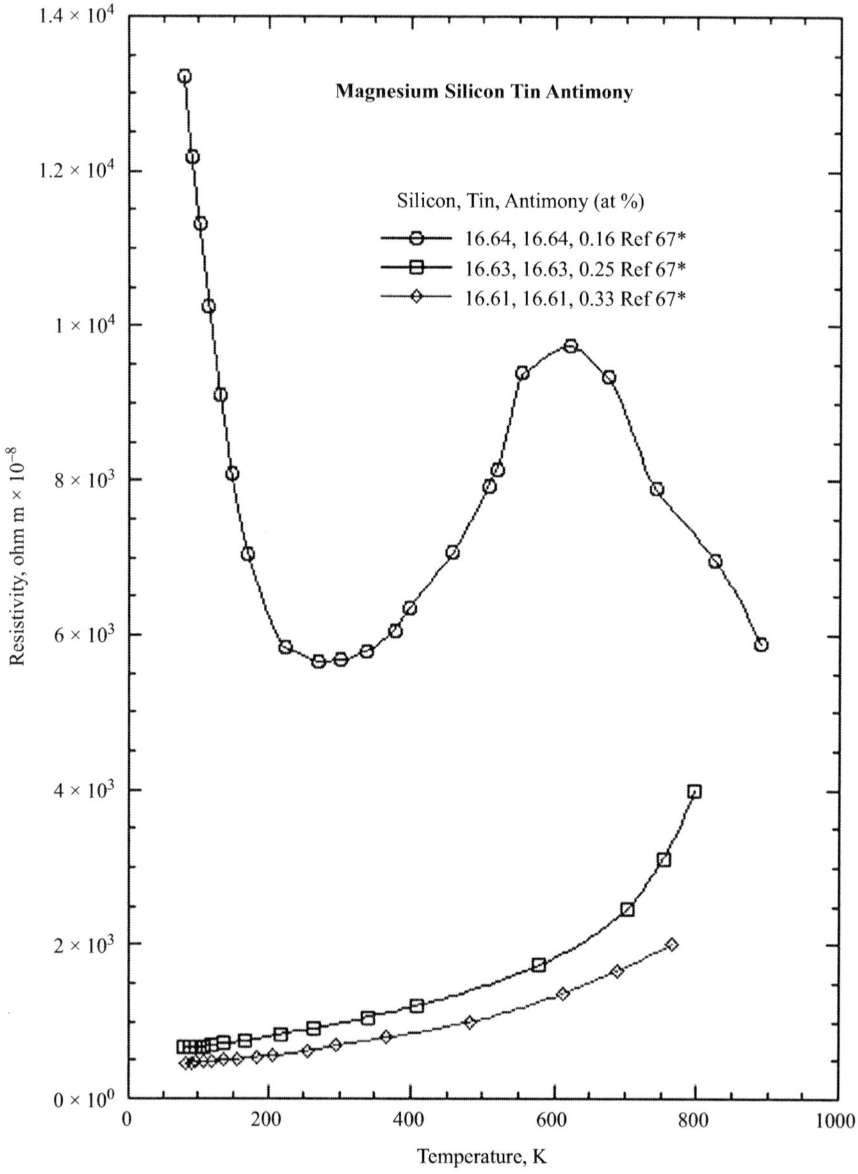
Magnesium Silicon Tin Antimony
Silicon, Tin, Antimony (at %)
16.64, 16.64, 0.16 Ref 67*
16.63, 16.63, 0.25 Ref 67*
16.61, 16.61, 0.33 Ref 67*
Resistivity, ohm m × 10⁻⁸
Temperature, K
1.4 × 10^4
1.2 × 10^4
1 × 10^4
8 × 10^3
6 × 10^3
4 × 10^3
2 × 10^3
0 × 10^0
0
200
400
600
800
1000

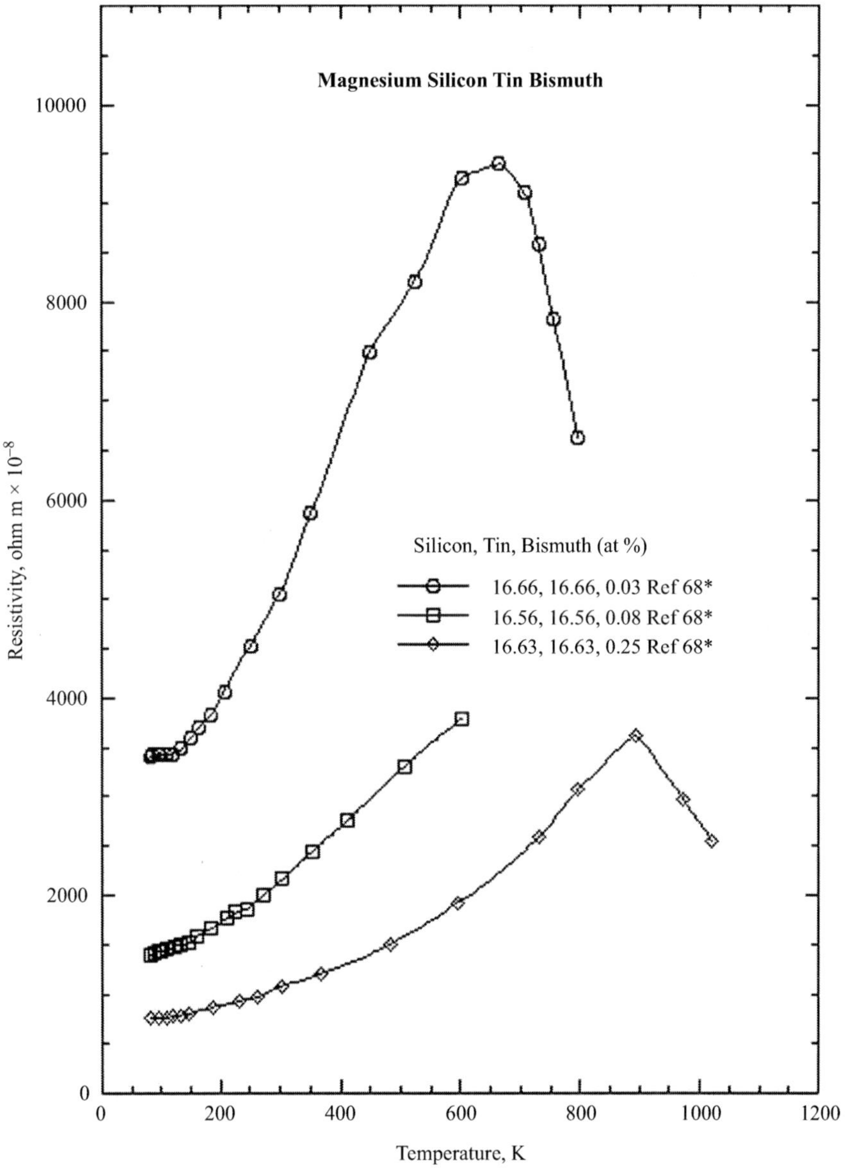

Magnesium Silicon Tin Bismuth
Resistivity, ohm m × 10^{-8}
Temperature, K
Silicon, Tin, Bismuth (at %)
16.66, 16.66, 0.03 Ref 68*
16.56, 16.56, 0.08 Ref 68*
16.63, 16.63, 0.25 Ref 68*
0
2000
4000
6000
8000
10000
0
200
400
600
800
1000
1200

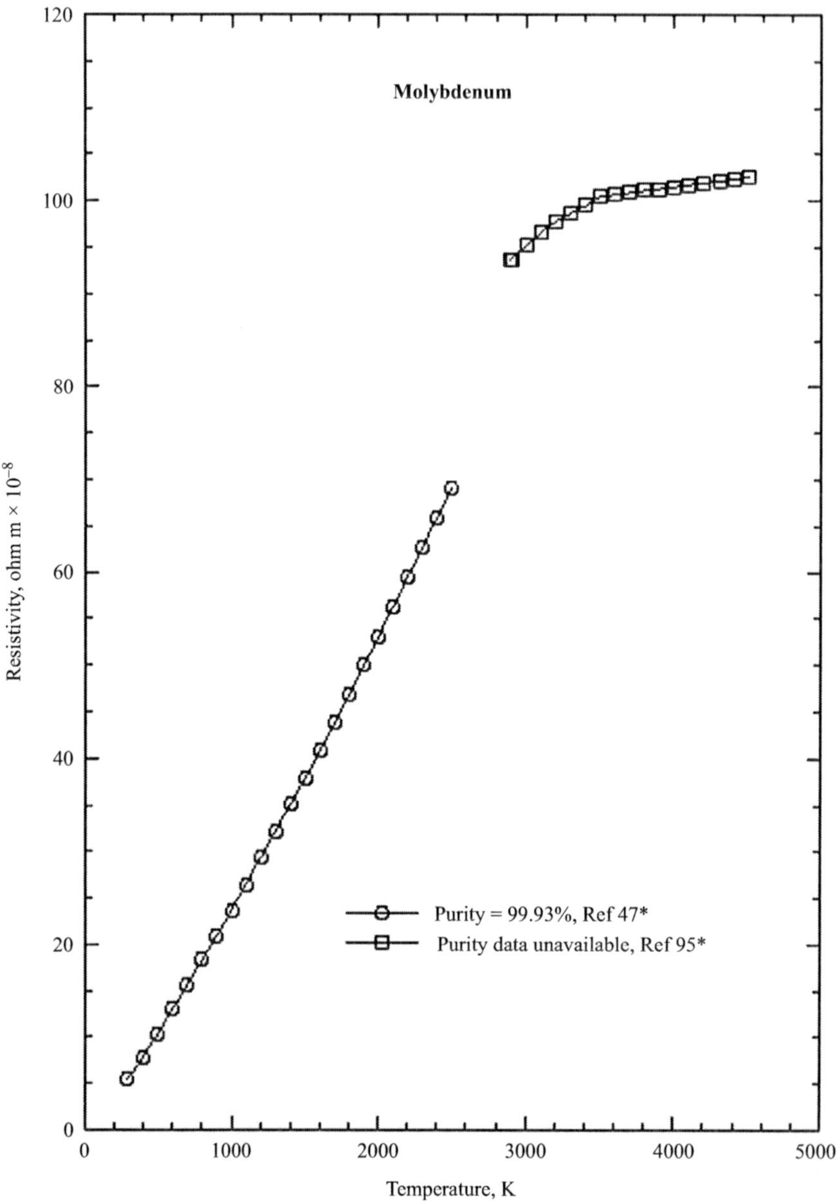
Molybdenum
Purity = 99.93%, Ref 47*
Purity data unavailable, Ref 95*
Resistivity, ohm m × 10^-8
Temperature, K
0
20
40
60
80
100
120
0
1000
2000
3000
4000
5000

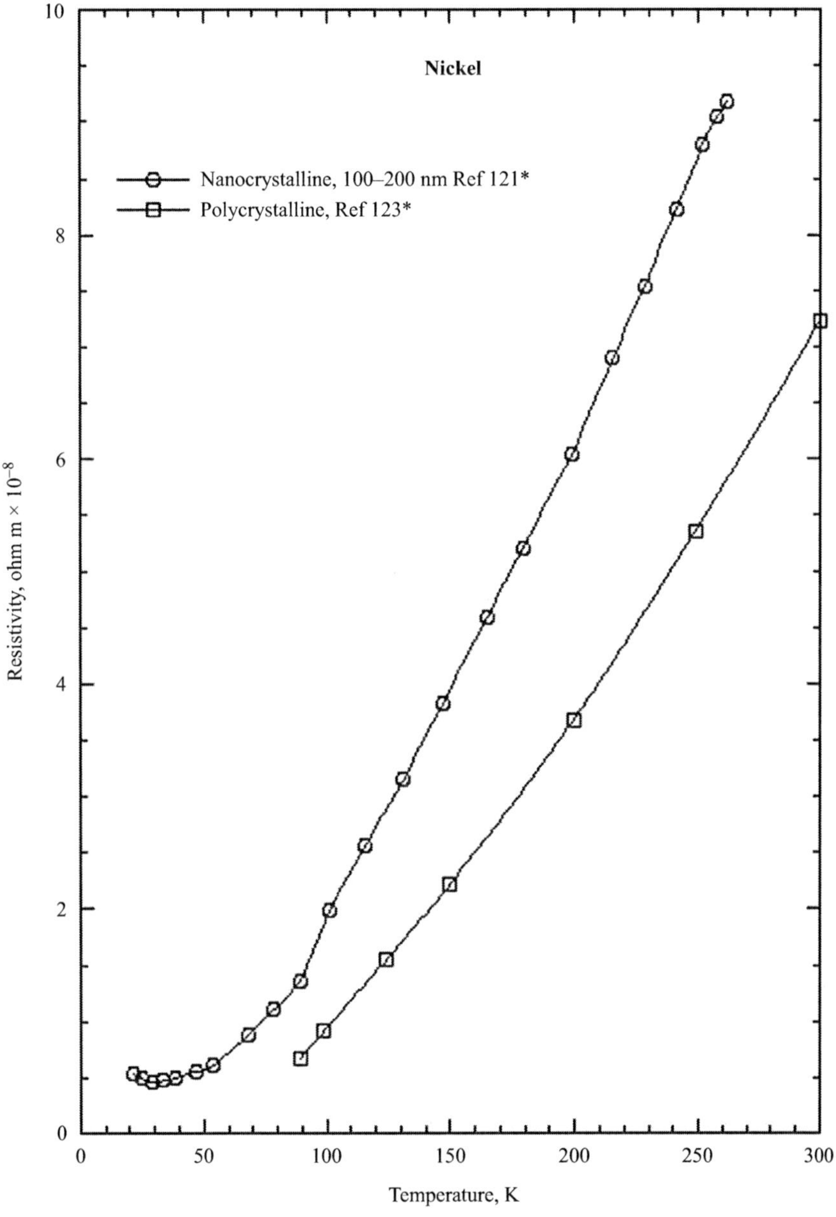
Nickel
Nanocrystalline, 100–200 nm Ref 121*
Polycrystalline, Ref 123*
Resistivity, ohm m × 10⁻⁸
Temperature, K
0
2
4
6
8
10
0
50
100
150
200
250
300

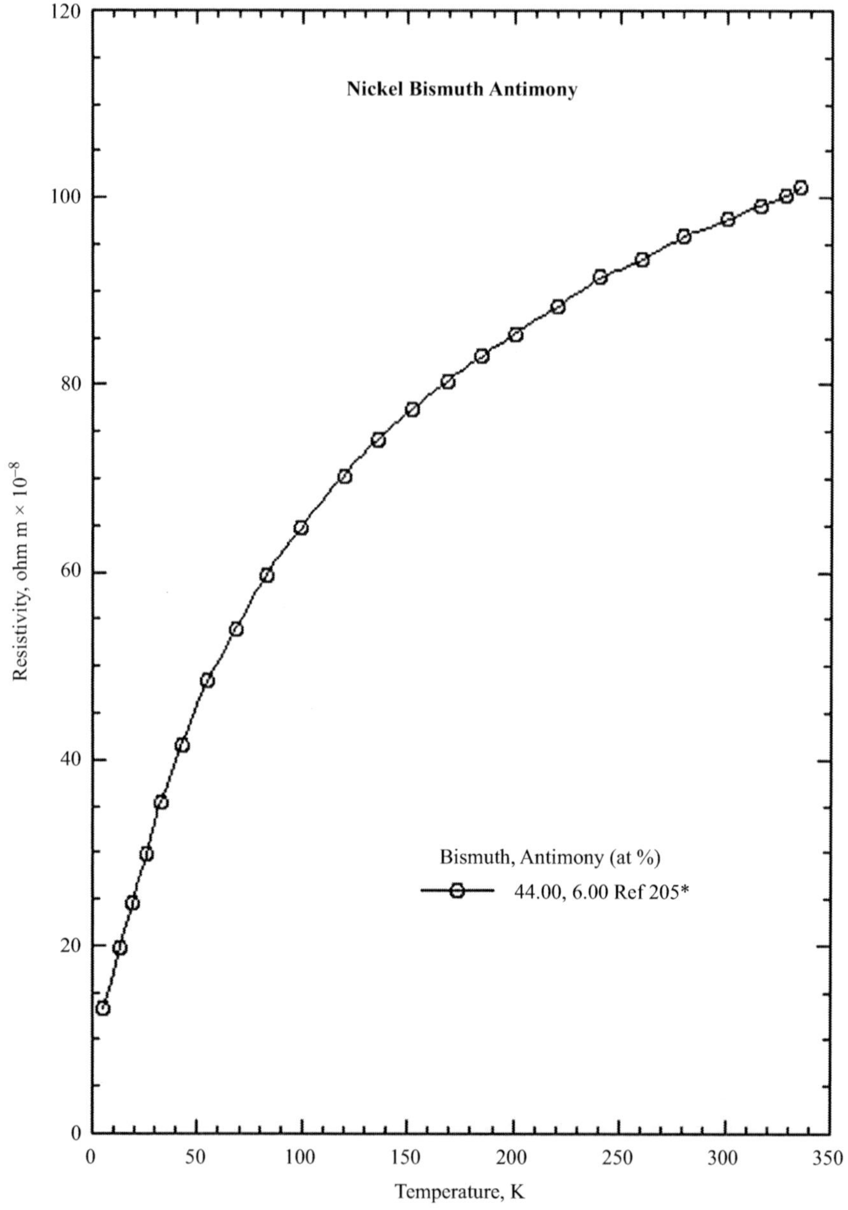
Nickel Bismuth Antimony
Bismuth, Antimony (at %)
44.00, 6.00 Ref 205*
Resistivity, ohm m × 10⁻⁸
Temperature, K
0
20
40
60
80
100
120
0
50
100
150
200
250
300
350

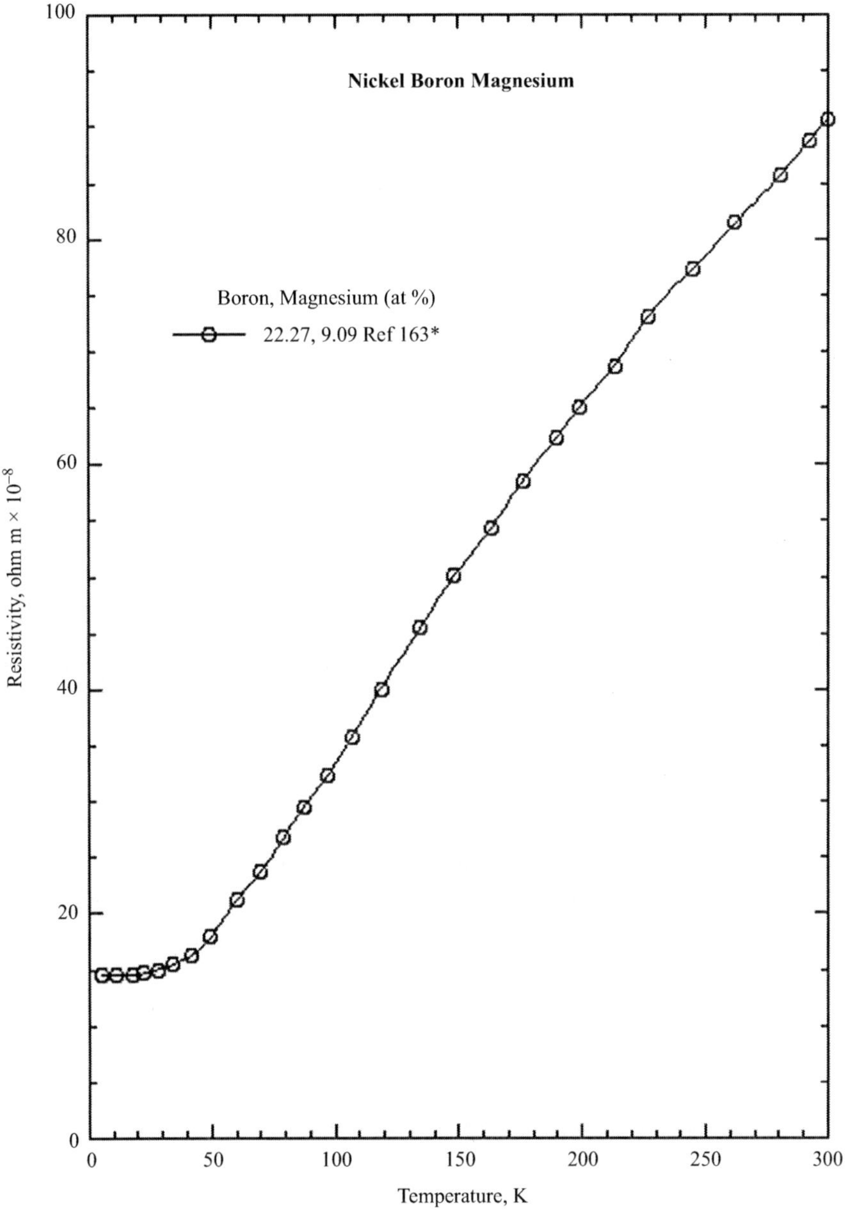
Nickel Boron Magnesium
Boron, Magnesium (at %)
22.27, 9.09 Ref 163*
Resistivity, ohm m × 10−8
Temperature, K
0
20
40
60
80
100
0
50
100
150
200
250
300

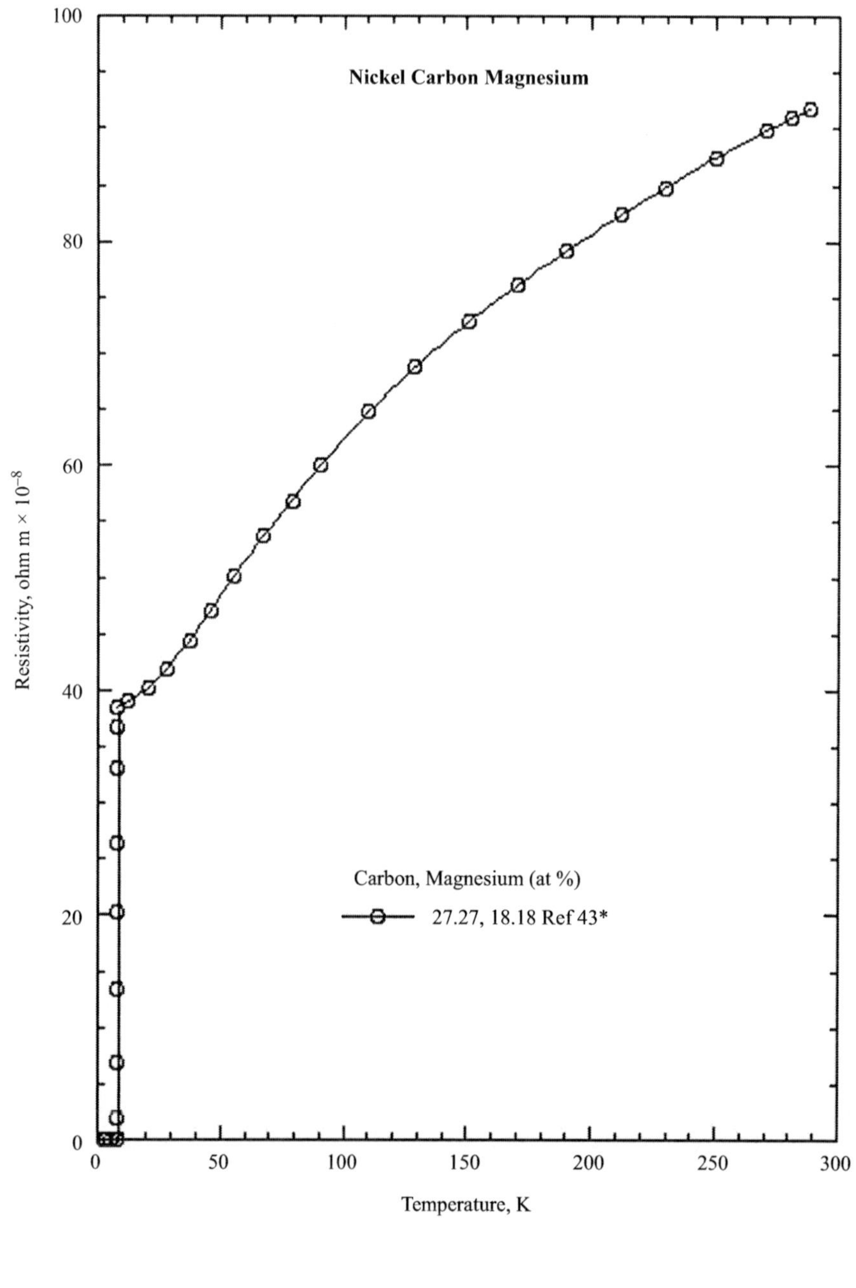
Nickel Carbon Magnesium
Resistivity, ohm m × 10⁻⁸
100
80
60
40
20
0
0
50
100
150
200
250
300
Temperature, K
Carbon, Magnesium (at %)
27.27, 18.18 Ref 43*

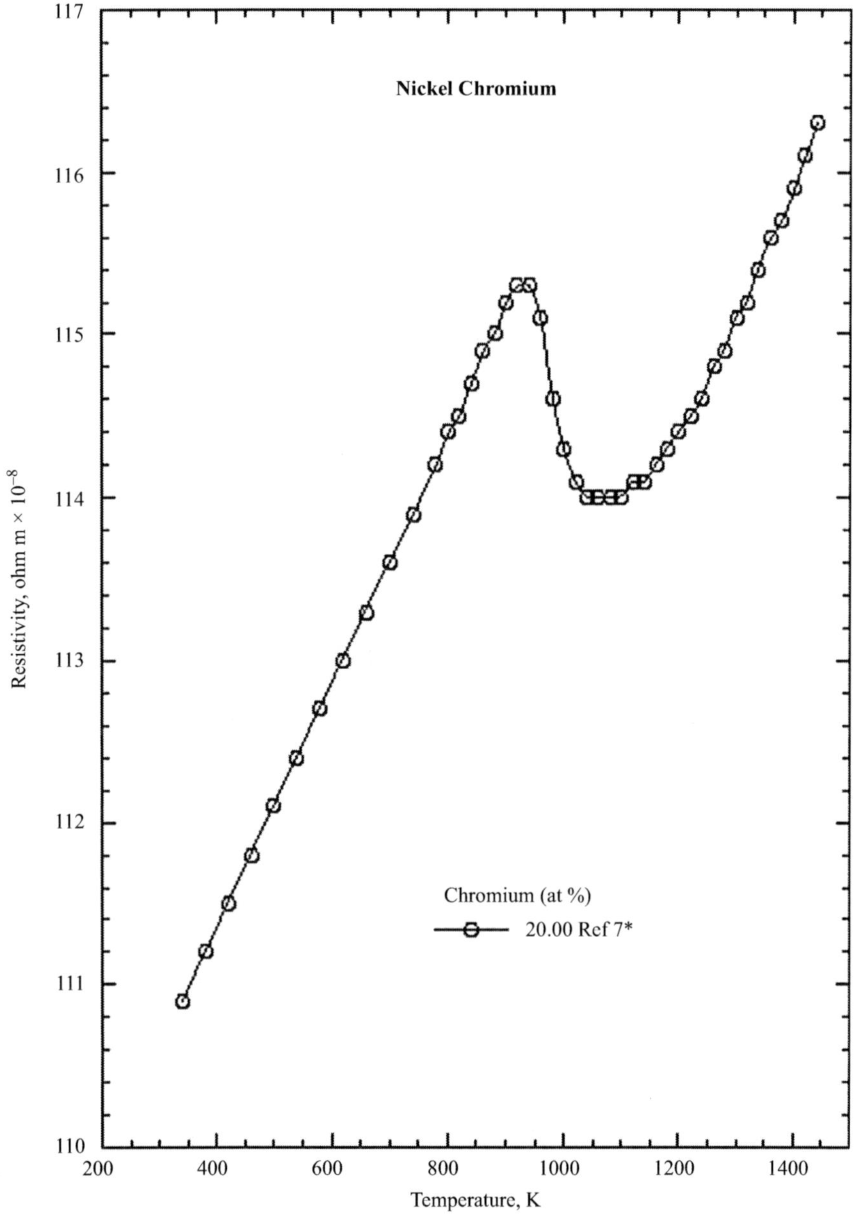
Nickel Chromium
Resistivity, ohm m × 10−8
Temperature, K
Chromium (at %)
20.00 Ref 7*
110
111
112
113
114
115
116
117
200
400
600
800
1000
1200
1400

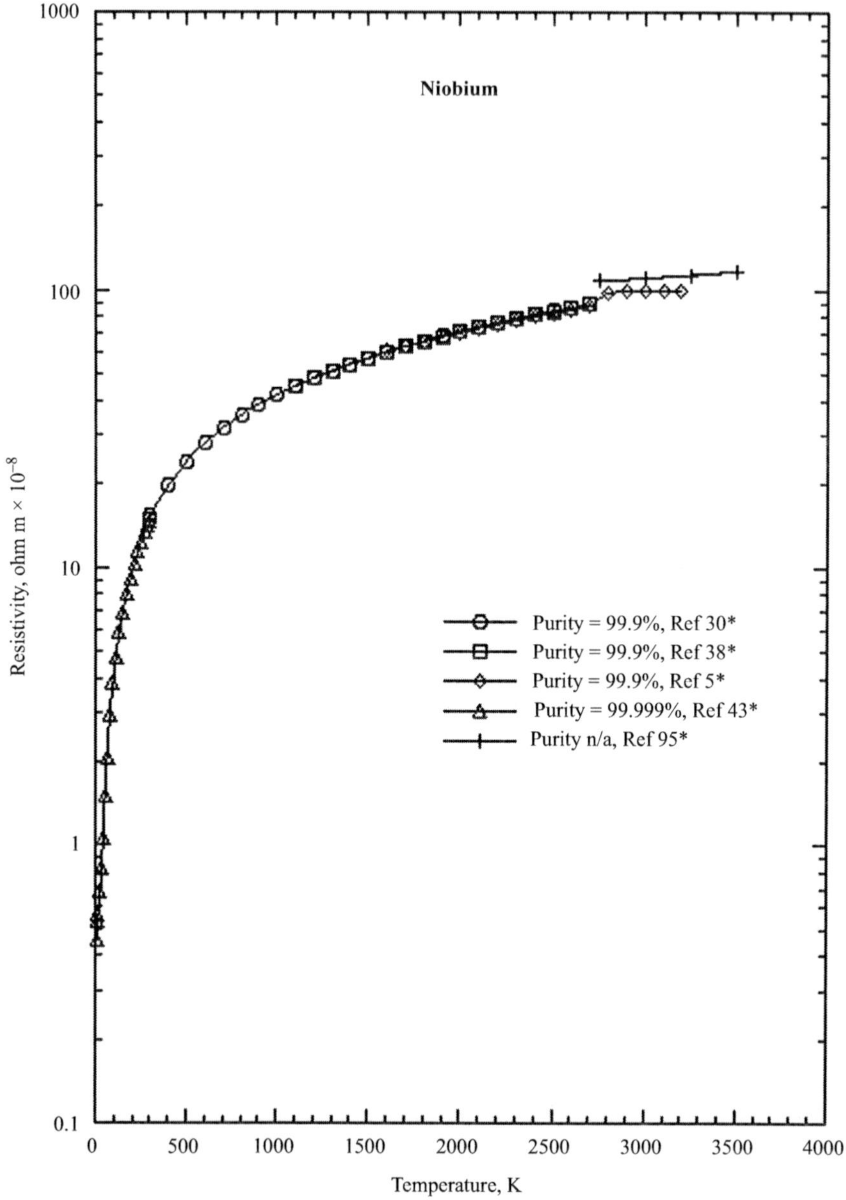
Niobium
Purity = 99.9%, Ref 30*
Purity = 99.9%, Ref 38*
Purity = 99.9%, Ref 5*
Purity = 99.999%, Ref 43*
Purity n/a, Ref 95*
Resistivity, ohm m × 10⁻⁸
Temperature, K
1000
100
10
1
0.1
0
500
1000
1500
2000
2500
3000
3500
4000

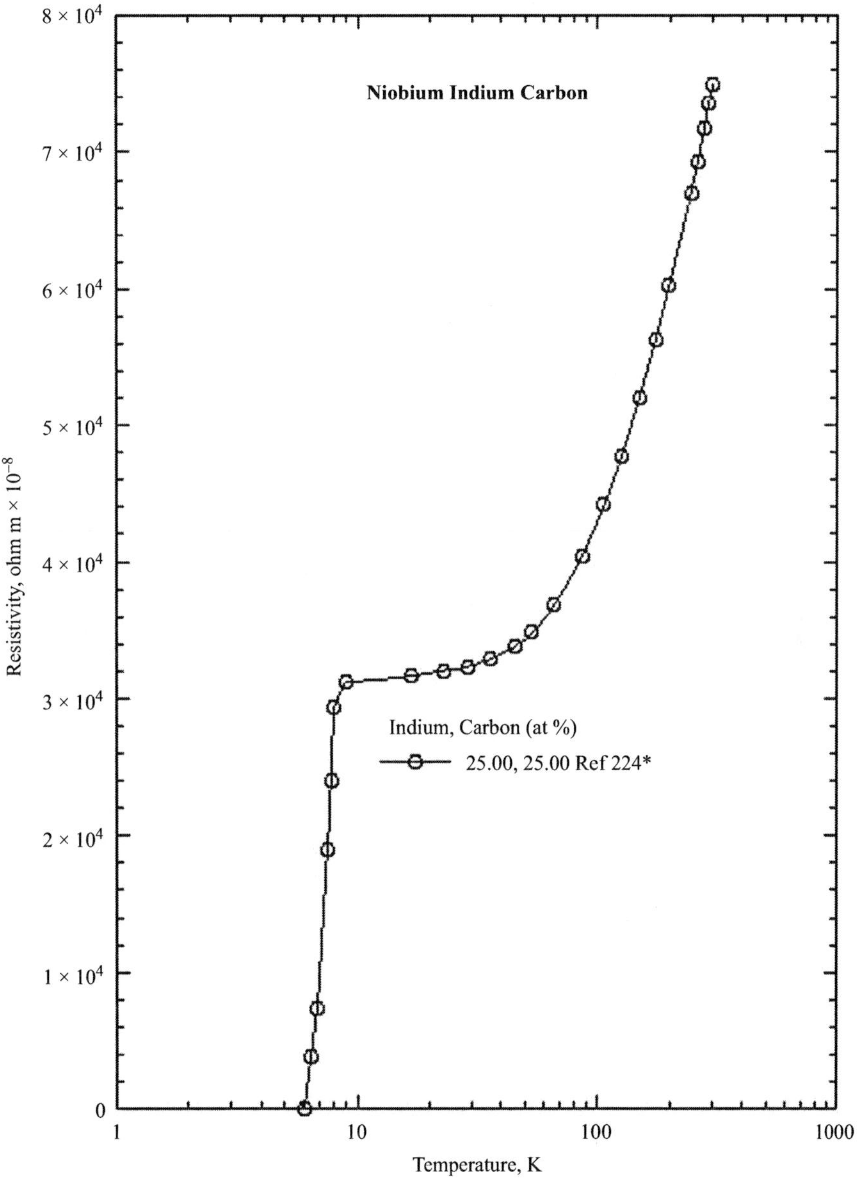

Niobium Indium Carbon
Indium, Carbon (at %)
25.00, 25.00 Ref 224*
Resistivity, ohm m × 10⁻⁸
Temperature, K
0
1 × 10⁴
2 × 10⁴
3 × 10⁴
4 × 10⁴
5 × 10⁴
6 × 10⁴
7 × 10⁴
8 × 10⁴
1
10
100
1000

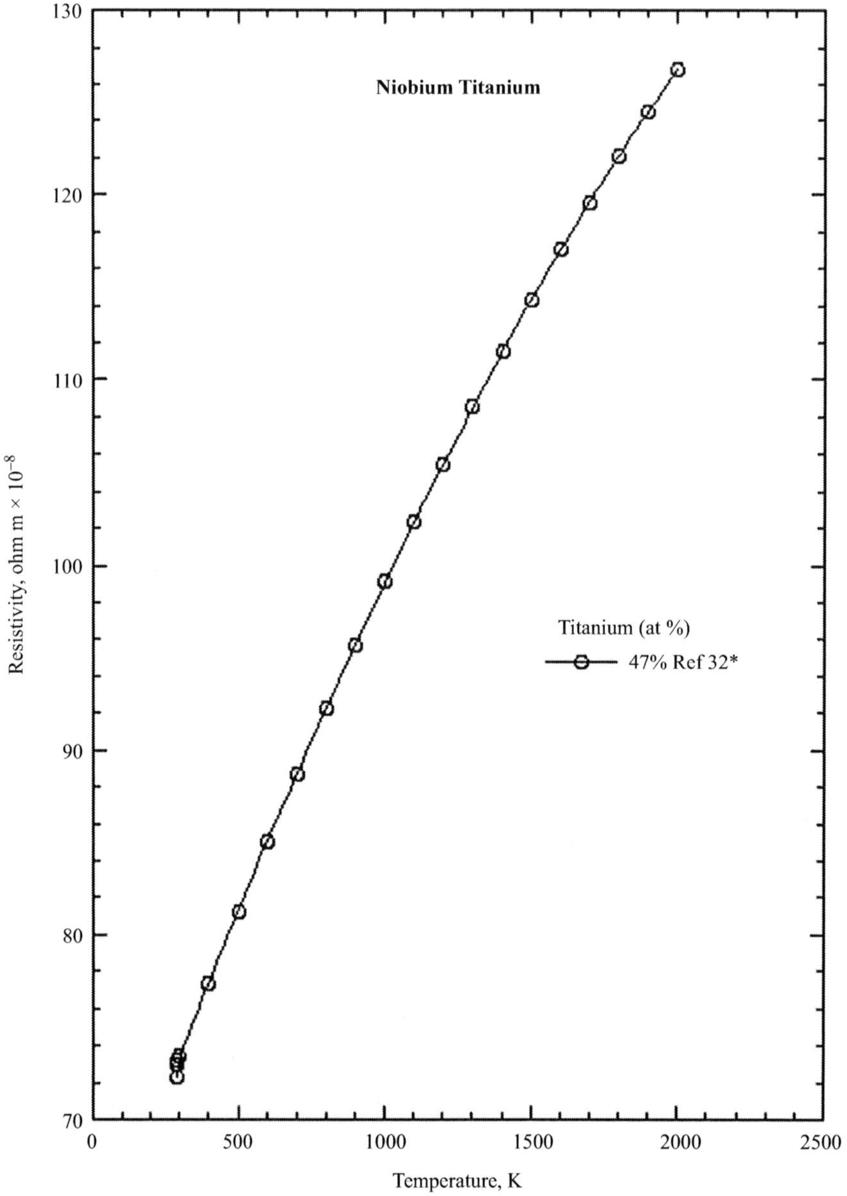
Niobium Titanium
Titanium (at %)
47% Ref 32*
Resistivity, ohm m × 10⁻⁸
Temperature, K
130
120
110
100
90
80
70
0
500
1000
1500
2000
2500

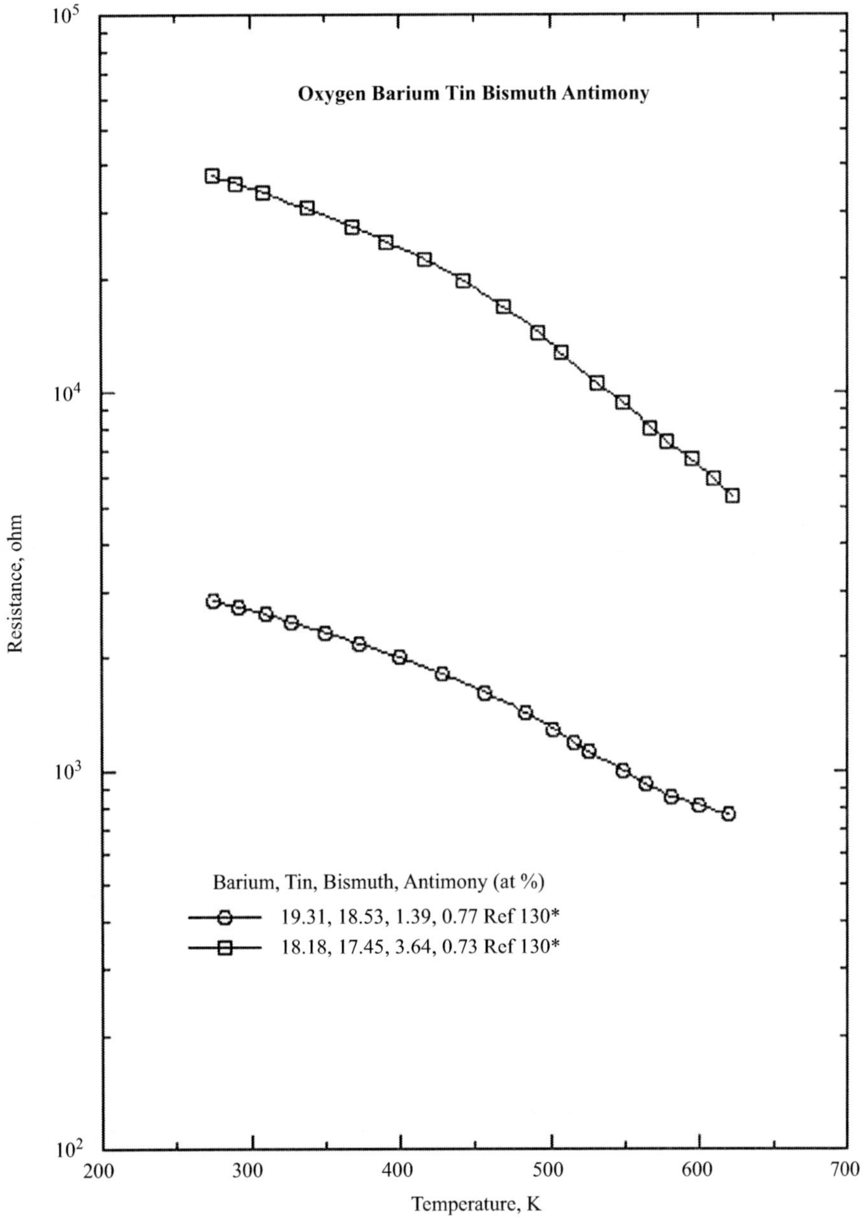
Oxygen Barium Tin Bismuth Antimony
Resistance, ohm
10^5
10^4
10^3
10^2
200
300
400
500
600
700
Temperature, K
Barium, Tin, Bismuth, Antimony (at %)
19.31, 18.53, 1.39, 0.77 Ref 130*
18.18, 17.45, 3.64, 0.73 Ref 130*

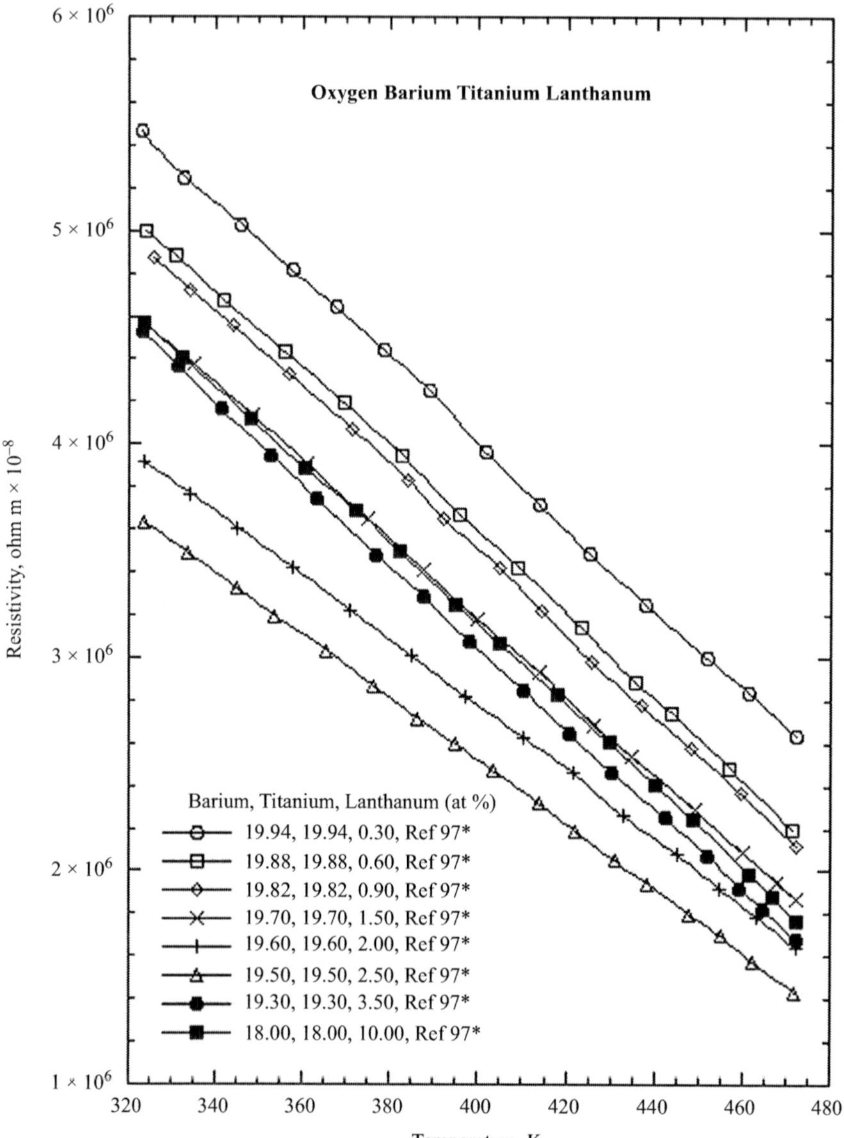

Oxygen Barium Titanium Lanthanum
Resistivity, ohm m × 10⁻⁸
Temperature, K
6 × 10⁶
5 × 10⁶
4 × 10⁶
3 × 10⁶
2 × 10⁶
1 × 10⁶
320
340
360
380
400
420
440
460
480
Barium, Titanium, Lanthanum (at %)
19.94, 19.94, 0.30, Ref 97*
19.88, 19.88, 0.60, Ref 97*
19.82, 19.82, 0.90, Ref 97*
19.70, 19.70, 1.50, Ref 97*
19.60, 19.60, 2.00, Ref 97*
19.50, 19.50, 2.50, Ref 97*
19.30, 19.30, 3.50, Ref 97*
18.00, 18.00, 10.00, Ref 97*

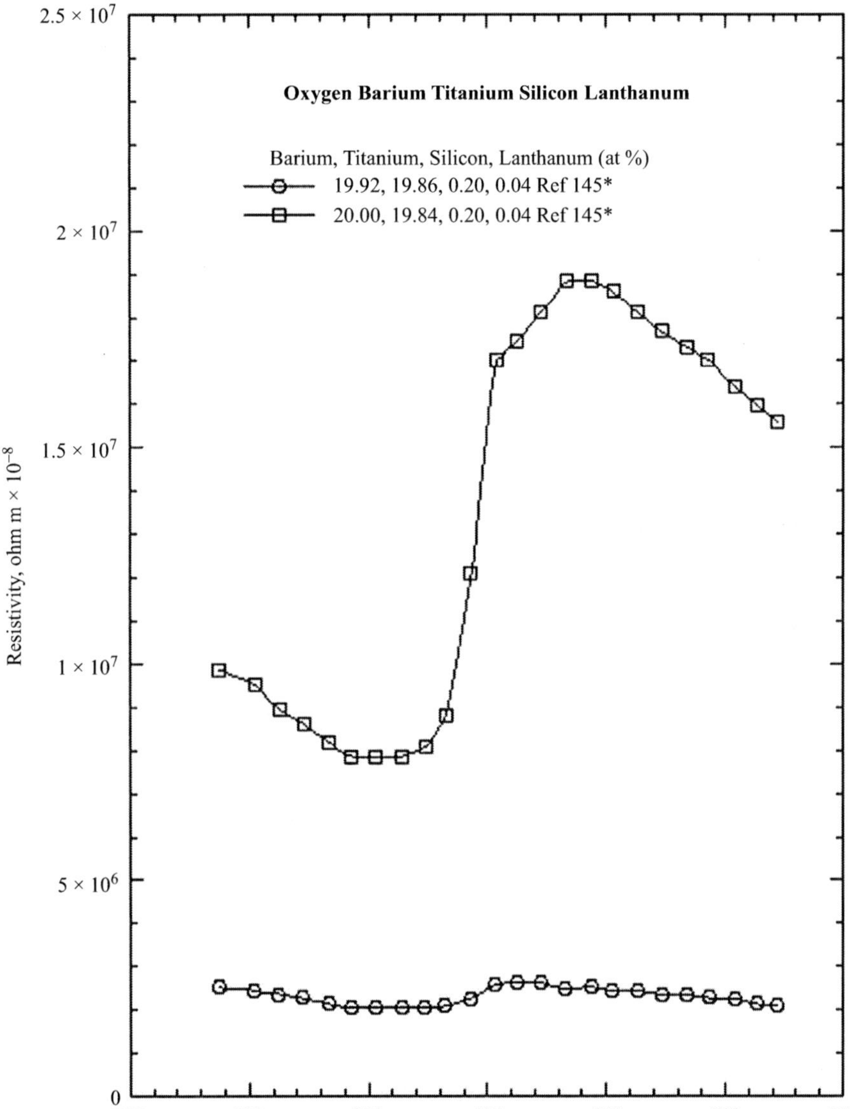
Oxygen Barium Titanium Silicon Lanthanum
Barium, Titanium, Silicon, Lanthanum (at %)
19.92, 19.86, 0.20, 0.04 Ref 145*
20.00, 19.84, 0.20, 0.04 Ref 145*
Resistivity, ohm m × 10⁻⁸
2.5 × 10⁷
2 × 10⁷
1.5 × 10⁷
1 × 10⁷
5 × 10⁶
0
250
300
350
400
450
500
550
Temperature, K

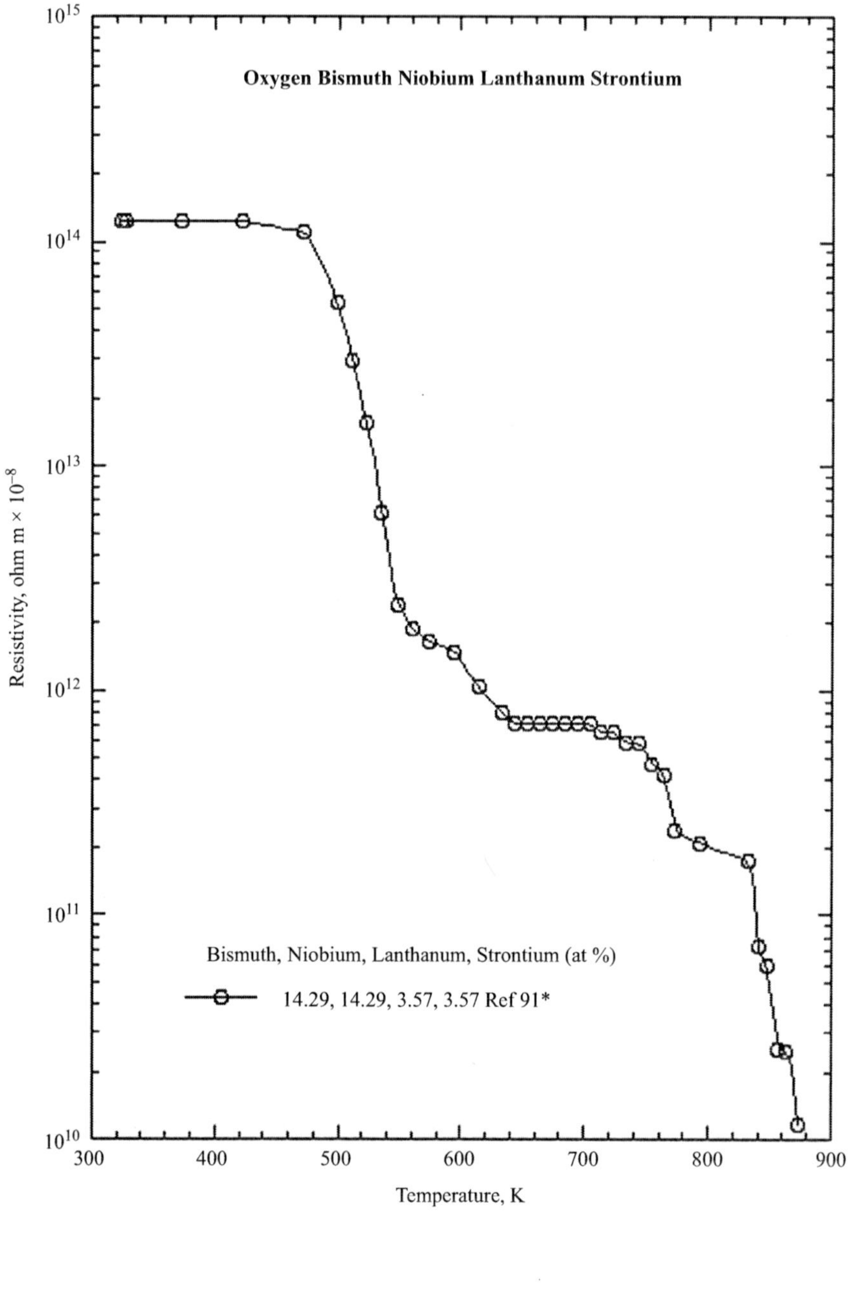
Oxygen Bismuth Niobium Lanthanum Strontium
Bismuth, Niobium, Lanthanum, Strontium (at %)
14.29, 14.29, 3.57, 3.57 Ref 91*
Resistivity, ohm m × 10⁻⁸
Temperature, K
10^15
10^14
10^13
10^12
10^11
10^10
300
400
500
600
700
800
900

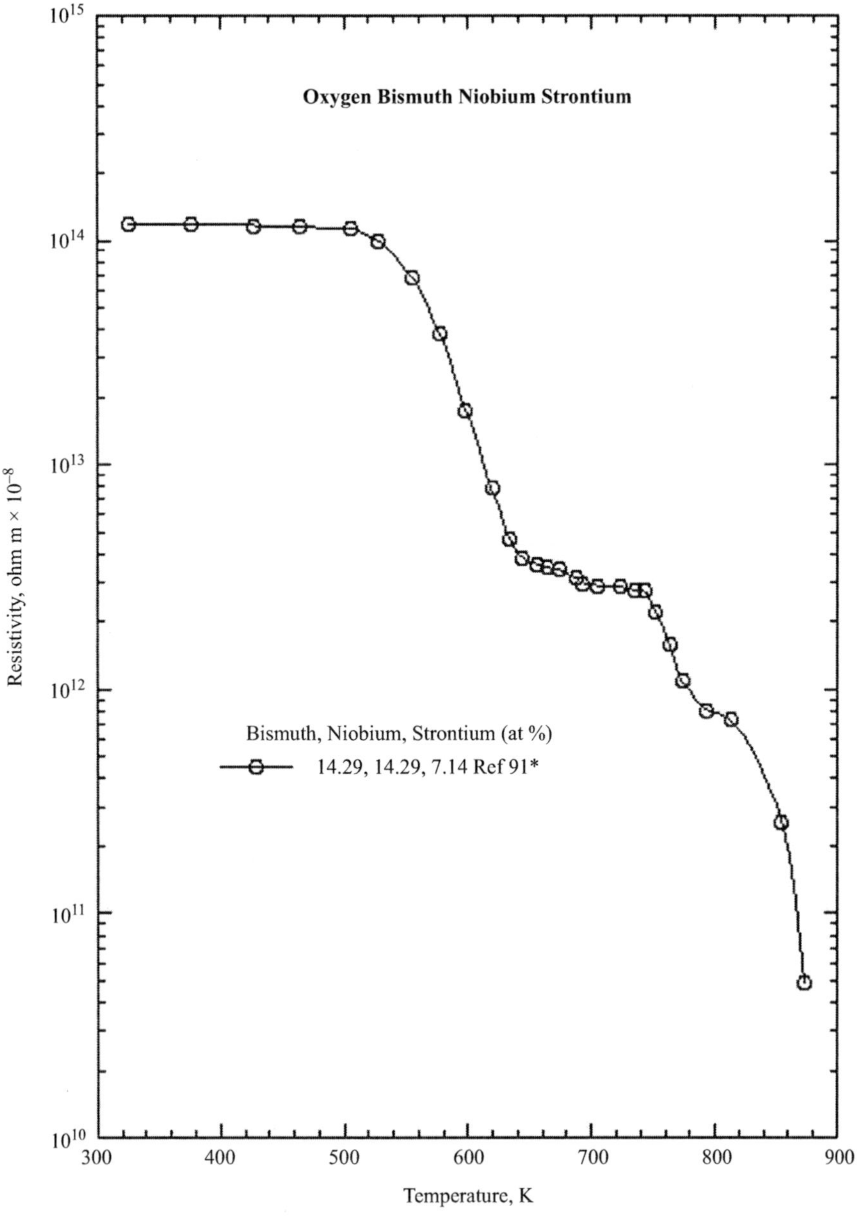
Oxygen Bismuth Niobium Strontium
Bismuth, Niobium, Strontium (at %)
14.29, 14.29, 7.14 Ref 91*
Resistivity, ohm m × 10^{-8}
Temperature, K
10^{10}
10^{11}
10^{12}
10^{13}
10^{14}
10^{15}
300
400
500
600
700
800
900

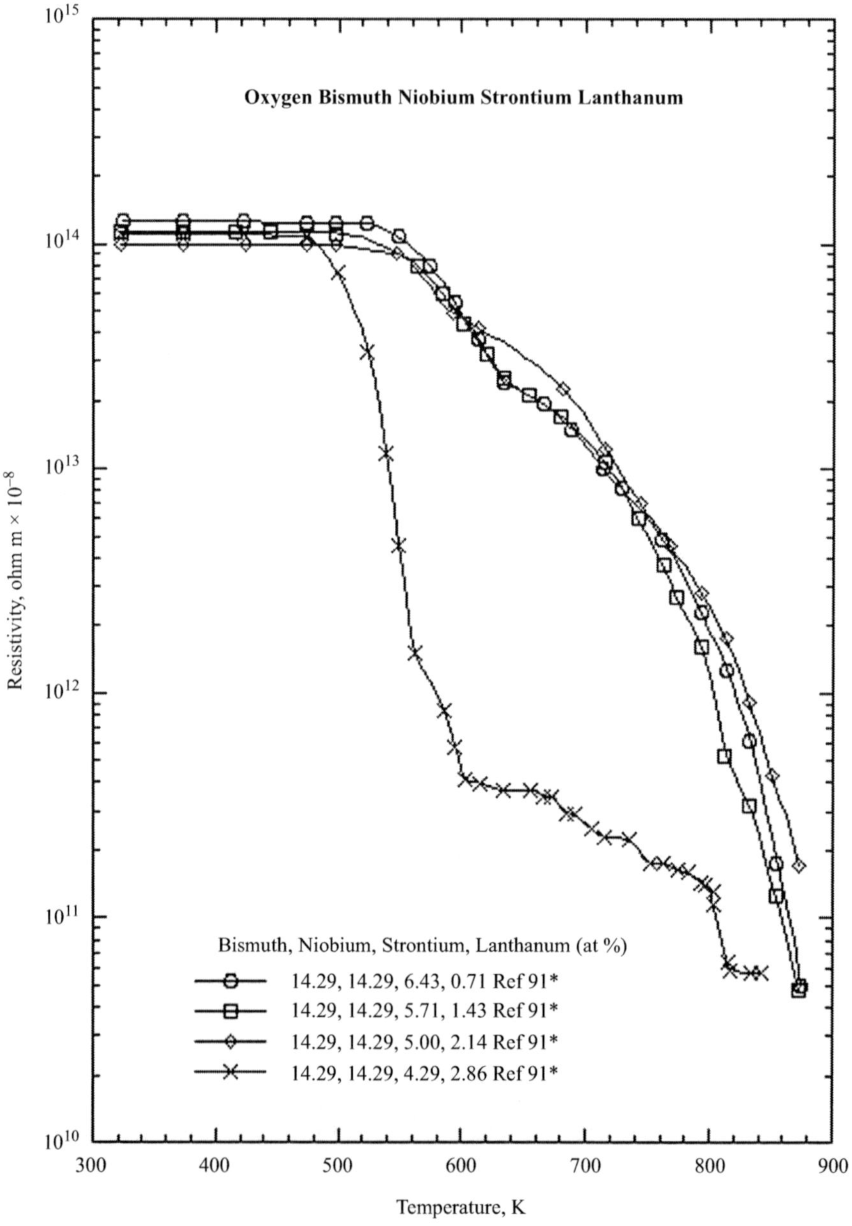
Oxygen Bismuth Niobium Strontium Lanthanum
Resistivity, ohm m × 10⁻⁸
Temperature, K
10^15
10^14
10^13
10^12
10^11
10^10
300
400
500
600
700
800
900
Bismuth, Niobium, Strontium, Lanthanum (at %)
14.29, 14.29, 6.43, 0.71 Ref 91*
14.29, 14.29, 5.71, 1.43 Ref 91*
14.29, 14.29, 5.00, 2.14 Ref 91*
14.29, 14.29, 4.29, 2.86 Ref 91*

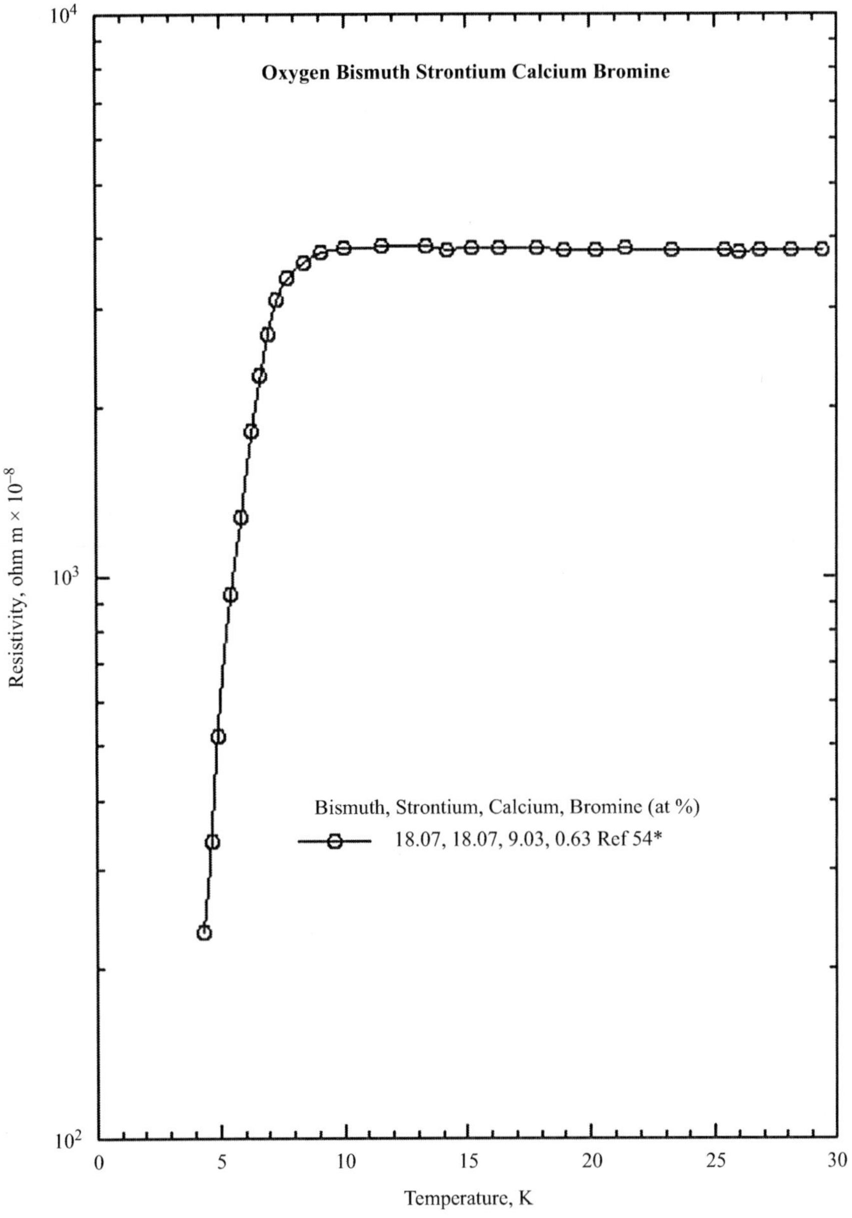
Oxygen Bismuth Strontium Calcium Bromine
Bismuth, Strontium, Calcium, Bromine (at %)
18.07, 18.07, 9.03, 0.63 Ref 54*
Resistivity, ohm m × 10⁻⁸
10⁴
10³
10²
0
5
10
15
20
25
30
Temperature, K

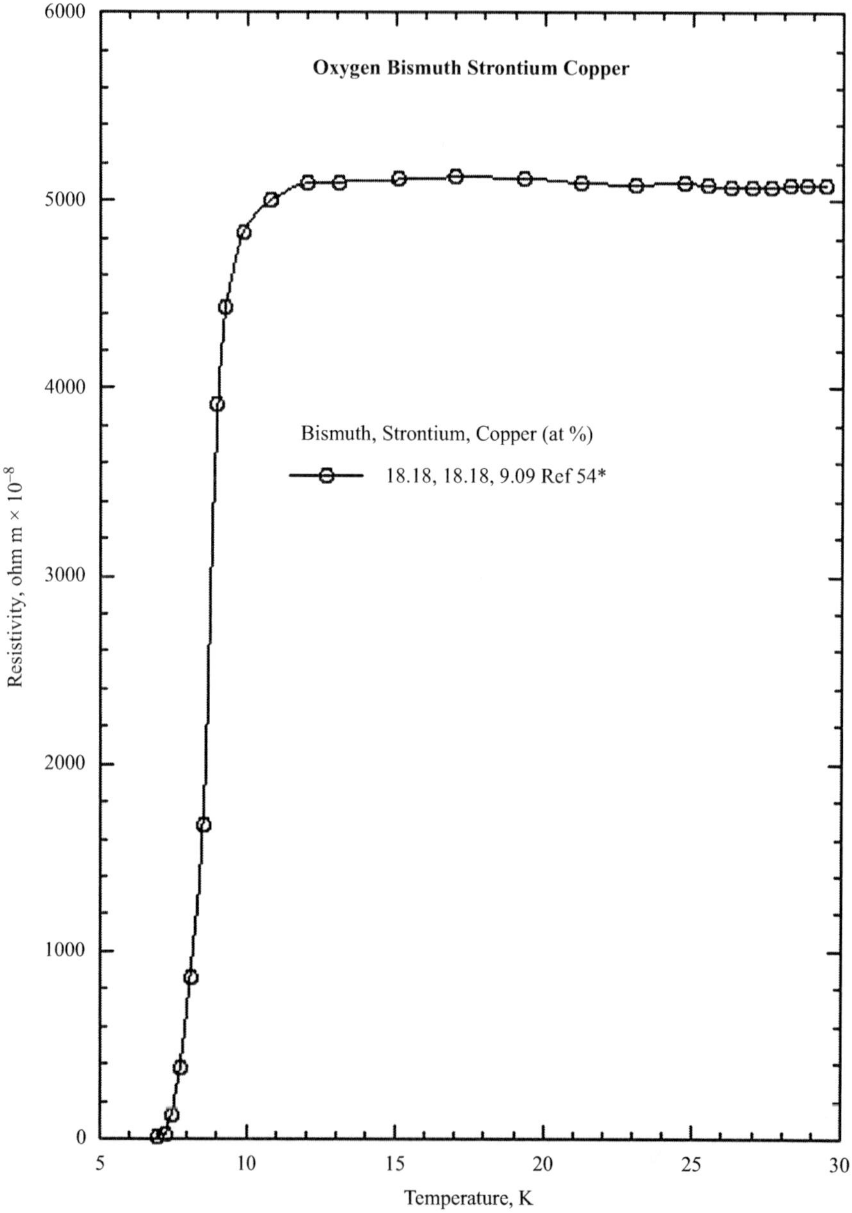
Oxygen Bismuth Strontium Copper
Bismuth, Strontium, Copper (at %)
18.18, 18.18, 9.09 Ref 54*
Resistivity, ohm m × 10^-8
Temperature, K
0
1000
2000
3000
4000
5000
6000
5
10
15
20
25
30

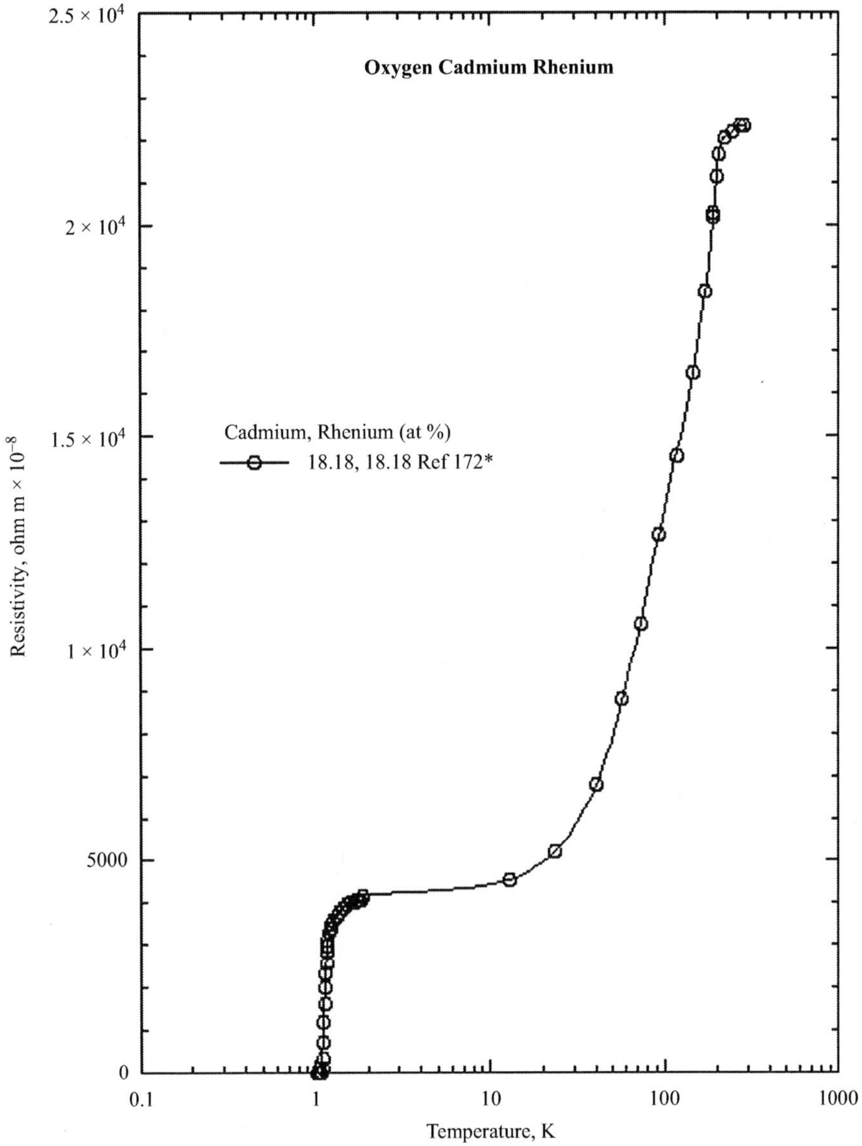

Oxygen Cadmium Rhenium
Cadmium, Rhenium (at %)
18.18, 18.18 Ref 172*
Resistivity, ohm m × 10⁻⁸
2.5 × 10⁴
2 × 10⁴
1.5 × 10⁴
1 × 10⁴
5000
0
0.1
1
10
100
1000
Temperature, K

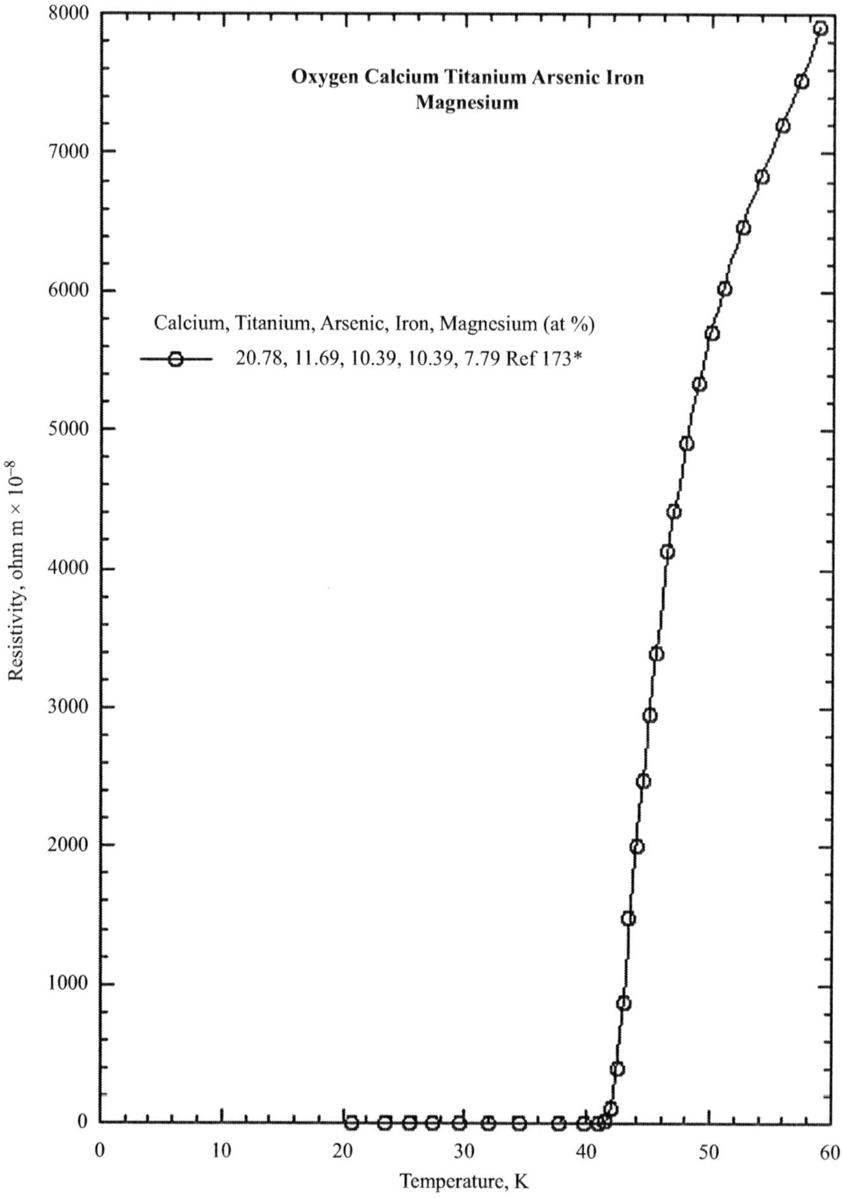
Oxygen Calcium Titanium Arsenic Iron
Magnesium
Calcium, Titanium, Arsenic, Iron, Magnesium (at %)
20.78, 11.69, 10.39, 10.39, 7.79 Ref 173*
Resistivity, ohm m × 10⁻⁸
Temperature, K
8000
7000
6000
5000
4000
3000
2000
1000
0
0
10
20
30
40
50
60

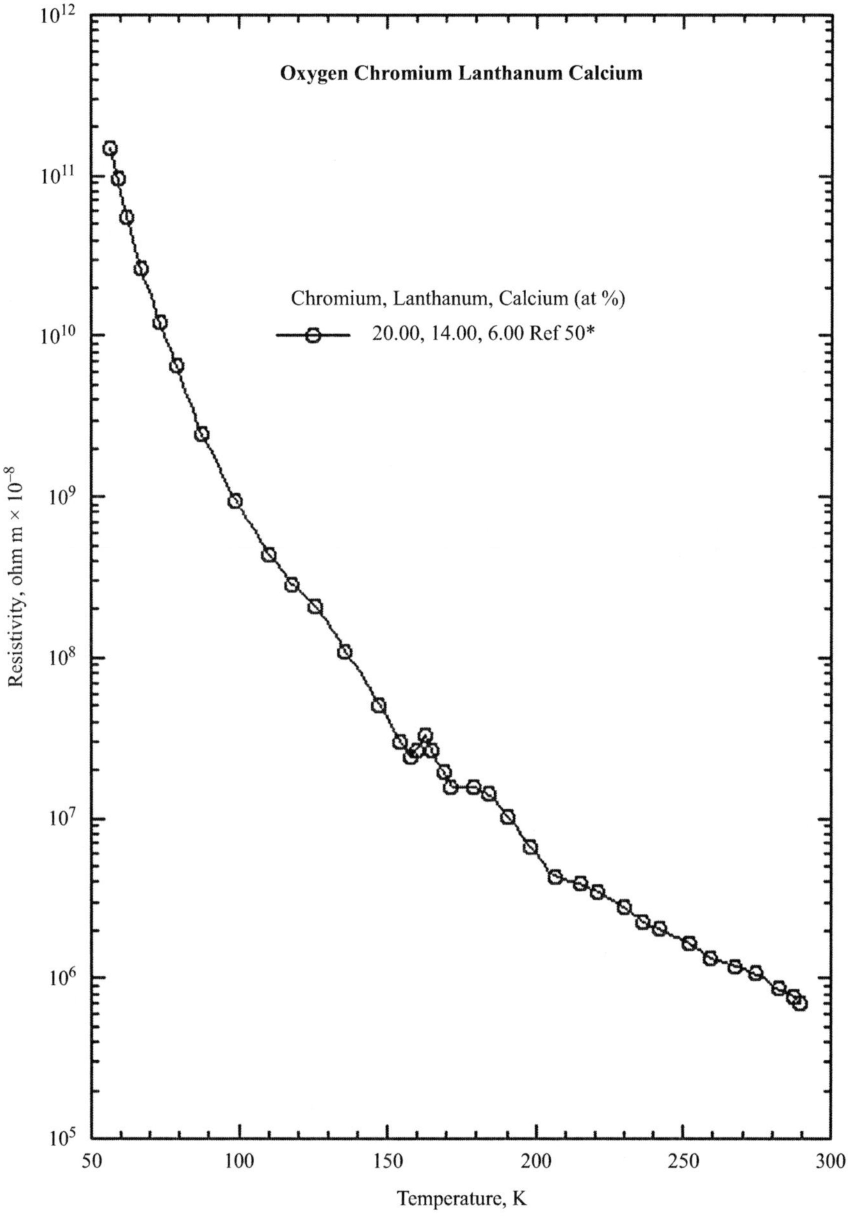
Oxygen Chromium Lanthanum Calcium
Chromium, Lanthanum, Calcium (at %)
20.00, 14.00, 6.00 Ref 50*
Resistivity, ohm m × 10⁻⁸
Temperature, K
10^12
10^11
10^10
10^9
10^8
10^7
10^6
10^5
50
100
150
200
250
300

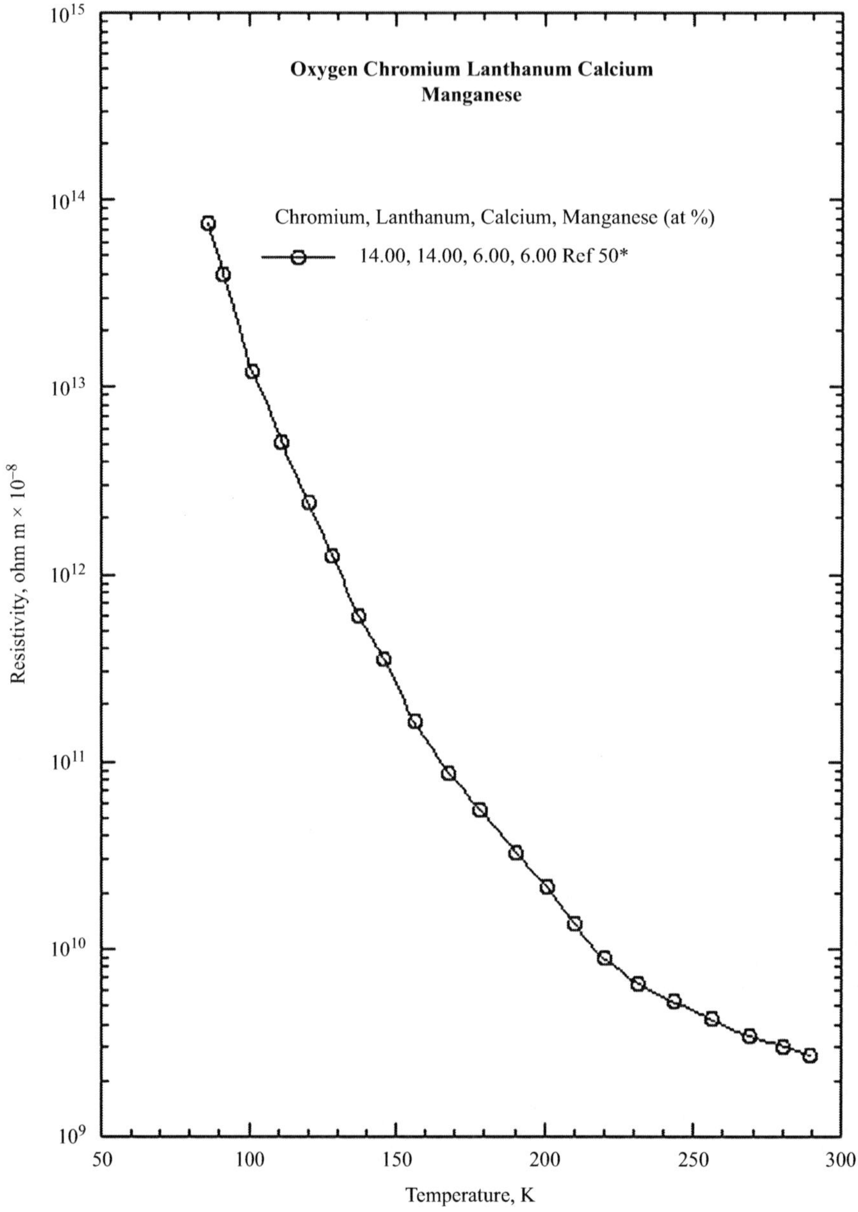
Oxygen Chromium Lanthanum Calcium
Manganese
Chromium, Lanthanum, Calcium, Manganese (at %)
14.00, 14.00, 6.00, 6.00 Ref 50*
Resistivity, ohm m × 10⁻⁸
10^15
10^14
10^13
10^12
10^11
10^10
10^9
50
100
150
200
250
300
Temperature, K

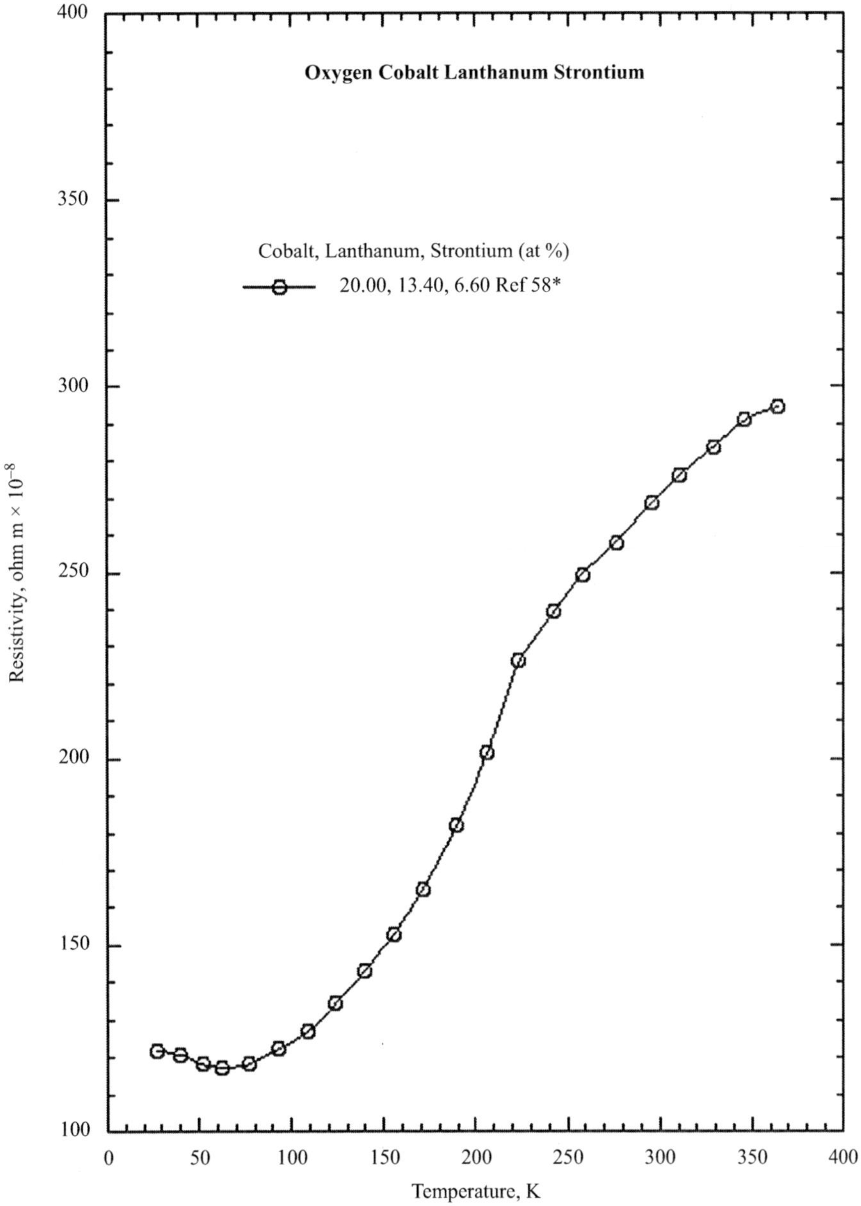
Oxygen Cobalt Lanthanum Strontium
Cobalt, Lanthanum, Strontium (at %)
20.00, 13.40, 6.60 Ref 58*
Resistivity, ohm m × 10⁻⁸
Temperature, K
400
350
300
250
200
150
100
0
50
100
150
200
250
300
350
400

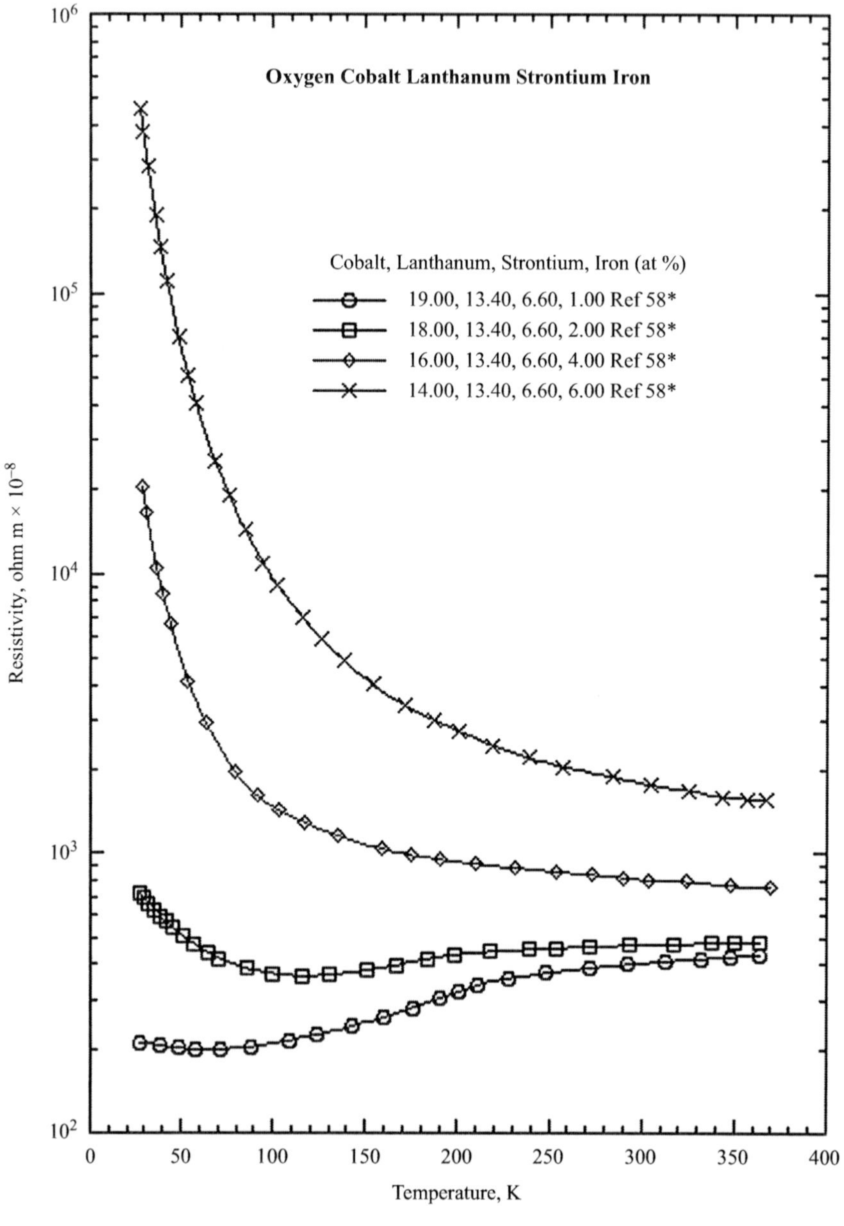
Oxygen Cobalt Lanthanum Strontium Iron
Cobalt, Lanthanum, Strontium, Iron (at %)
19.00, 13.40, 6.60, 1.00 Ref 58*
18.00, 13.40, 6.60, 2.00 Ref 58*
16.00, 13.40, 6.60, 4.00 Ref 58*
14.00, 13.40, 6.60, 6.00 Ref 58*
Resistivity, ohm m × 10⁻⁸
Temperature, K
10²
10³
10⁴
10⁵
10⁶
0
50
100
150
200
250
300
350
400

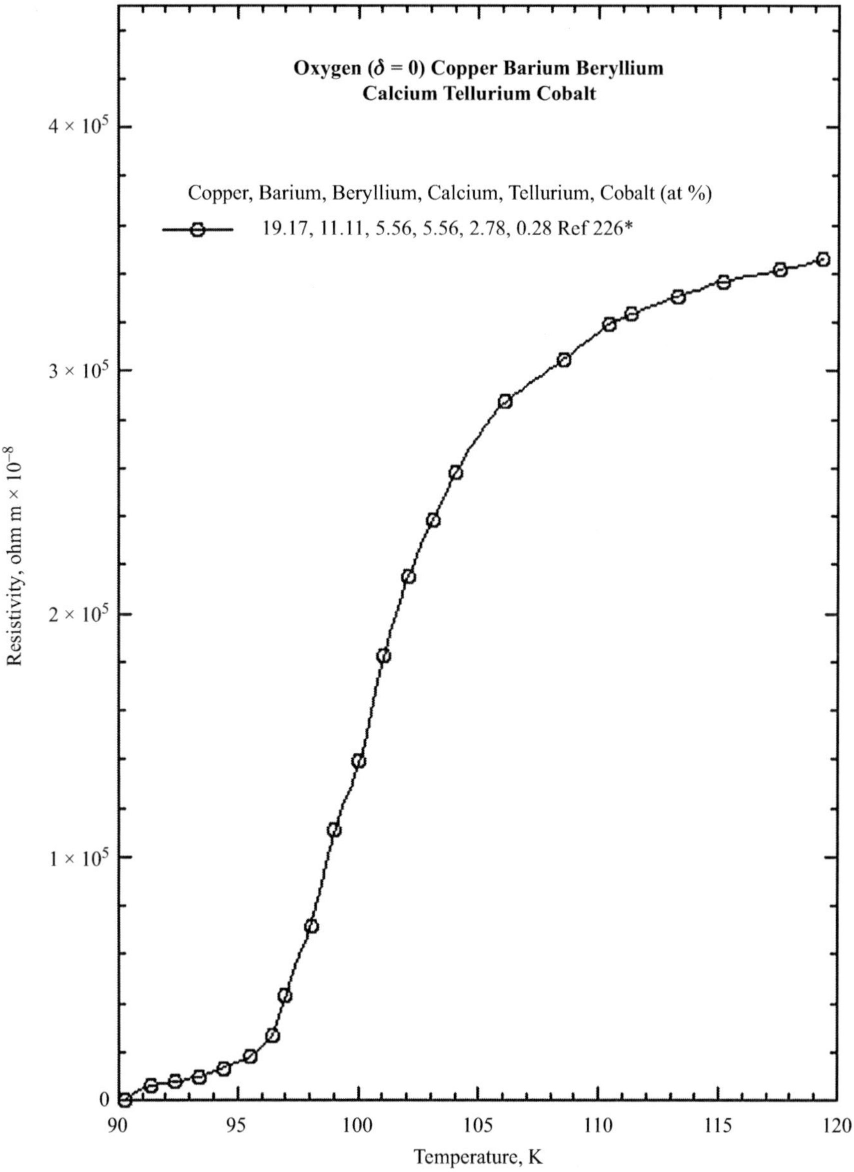
Oxygen (δ = 0) Copper Barium Beryllium
Calcium Tellurium Cobalt
Copper, Barium, Beryllium, Calcium, Tellurium, Cobalt (at %)
19.17, 11.11, 5.56, 5.56, 2.78, 0.28 Ref 226*
4 × 10^5
3 × 10^5
2 × 10^5
1 × 10^5
0
Resistivity, ohm m × 10^−8
90
95
100
105
110
115
120
Temperature, K

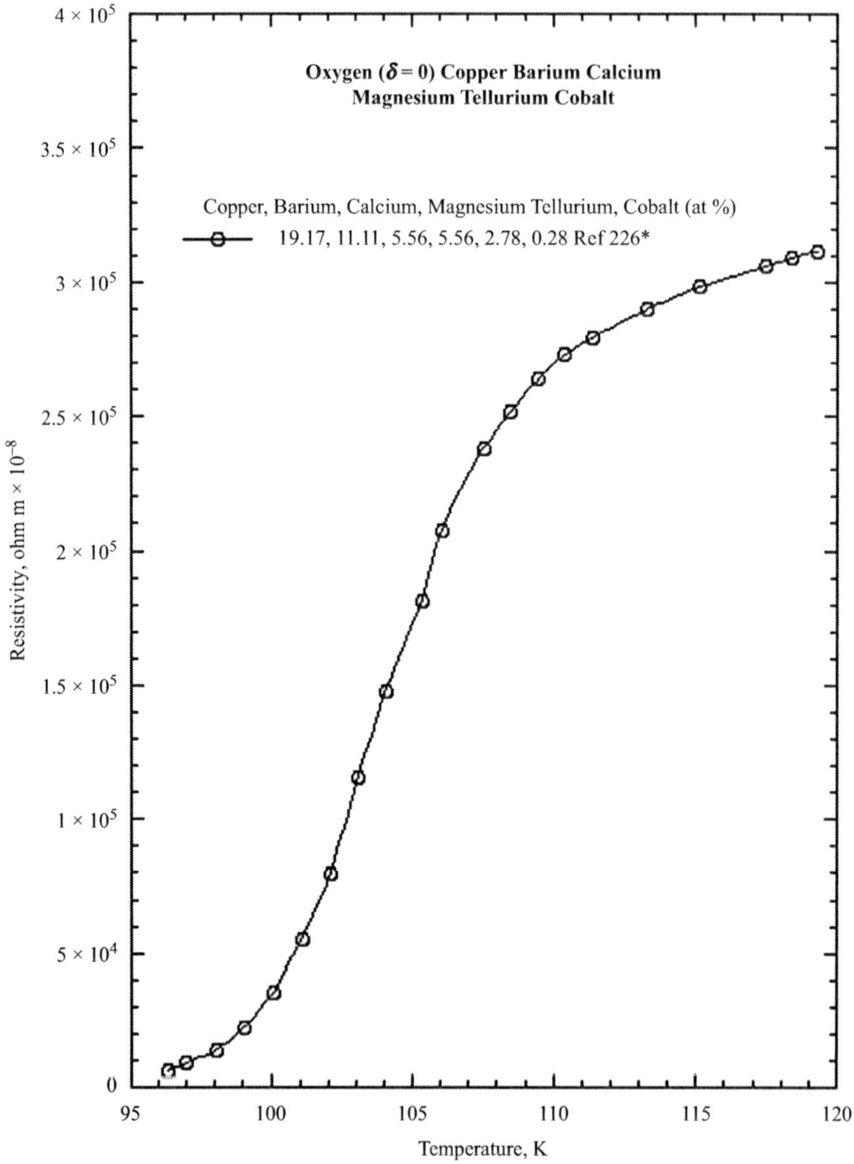
Oxygen (δ = 0) Copper Barium Calcium
Magnesium Tellurium Cobalt
Copper, Barium, Calcium, Magnesium Tellurium, Cobalt (at %)
19.17, 11.11, 5.56, 5.56, 2.78, 0.28 Ref 226*
Resistivity, ohm m × 10⁻⁸
Temperature, K
4 × 10⁵
3.5 × 10⁵
3 × 10⁵
2.5 × 10⁵
2 × 10⁵
1.5 × 10⁵
1 × 10⁵
5 × 10⁴
0
95
100
105
110
115
120

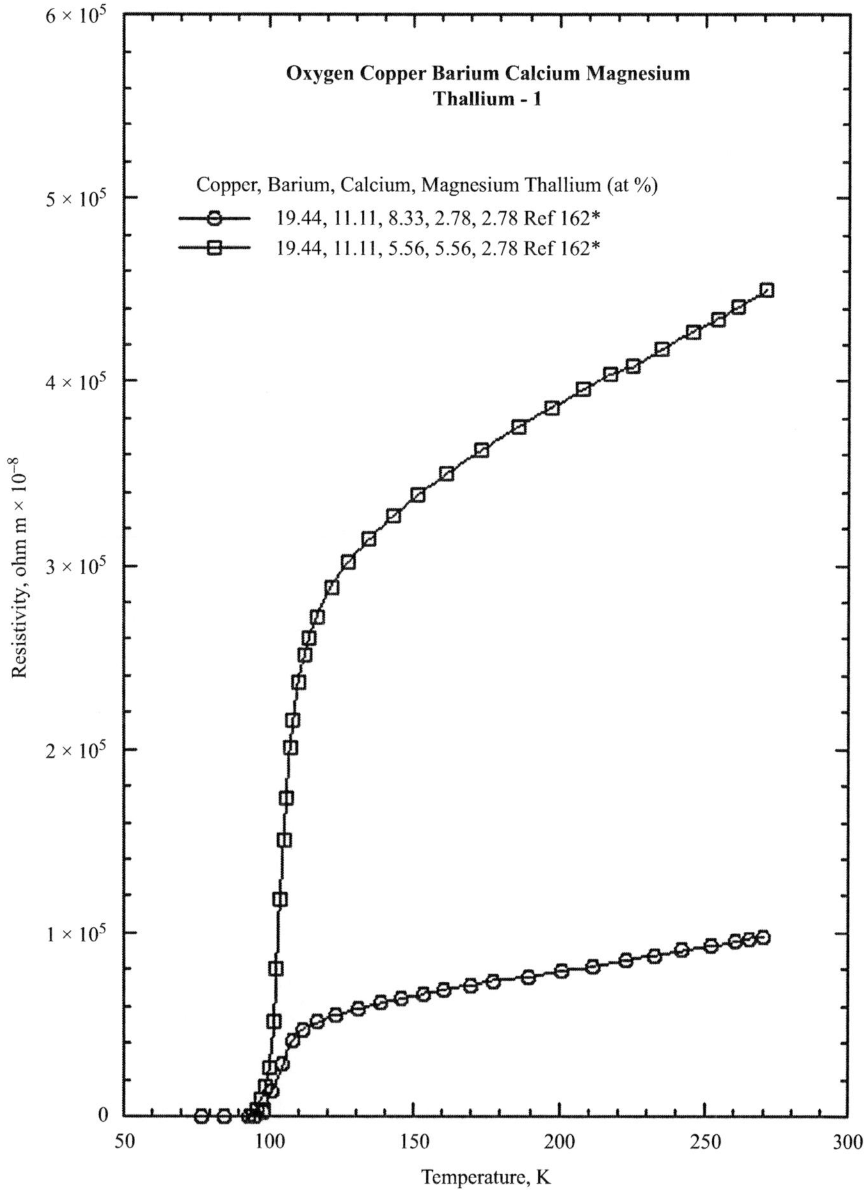

Oxygen Copper Barium Calcium Magnesium Thallium - 1
Copper, Barium, Calcium, Magnesium Thallium (at %)
19.44, 11.11, 8.33, 2.78, 2.78 Ref 162*
19.44, 11.11, 5.56, 5.56, 2.78 Ref 162*
Resistivity, ohm m × 10⁻⁸
Temperature, K
6 × 10^5
5 × 10^5
4 × 10^5
3 × 10^5
2 × 10^5
1 × 10^5
0
50
100
150
200
250
300

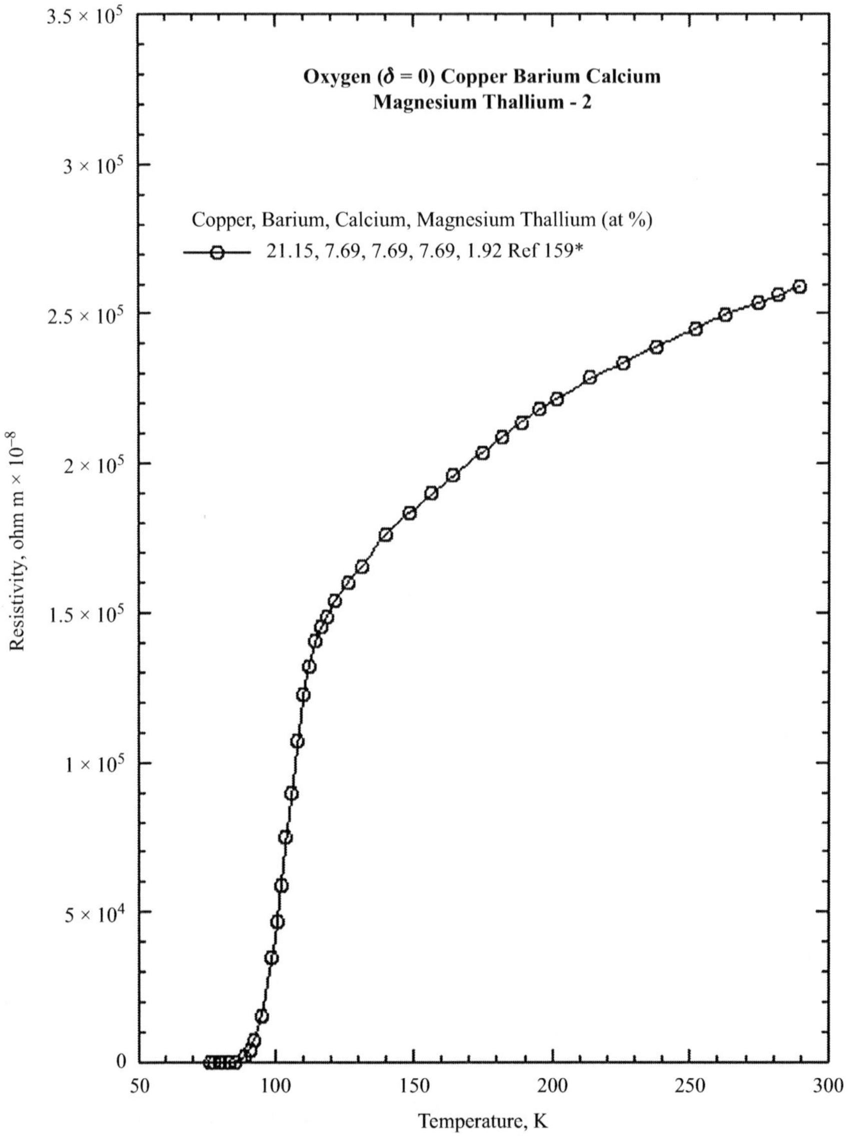
Oxygen (δ = 0) Copper Barium Calcium Magnesium Thallium - 2
Copper, Barium, Calcium, Magnesium Thallium (at %)
21.15, 7.69, 7.69, 7.69, 1.92 Ref 159*
Resistivity, ohm m × 10⁻⁸
3.5 × 10⁵
3 × 10⁵
2.5 × 10⁵
2 × 10⁵
1.5 × 10⁵
1 × 10⁵
5 × 10⁴
0
50
100
150
200
250
300
Temperature, K

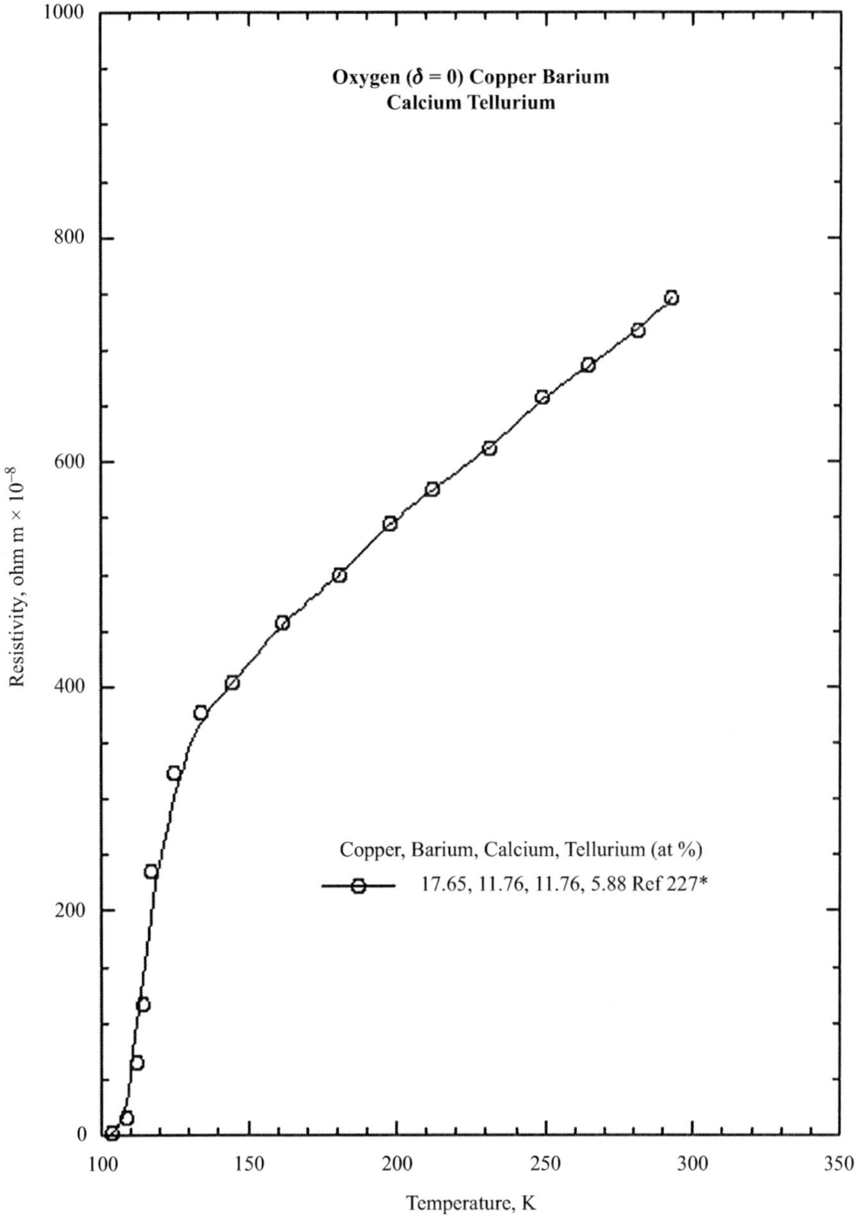
Oxygen (δ = 0) Copper Barium
Calcium Tellurium
Copper, Barium, Calcium, Tellurium (at %)
17.65, 11.76, 11.76, 5.88 Ref 227*
Resistivity, ohm m × 10⁻⁸
Temperature, K
0
200
400
600
800
1000
100
150
200
250
300
350

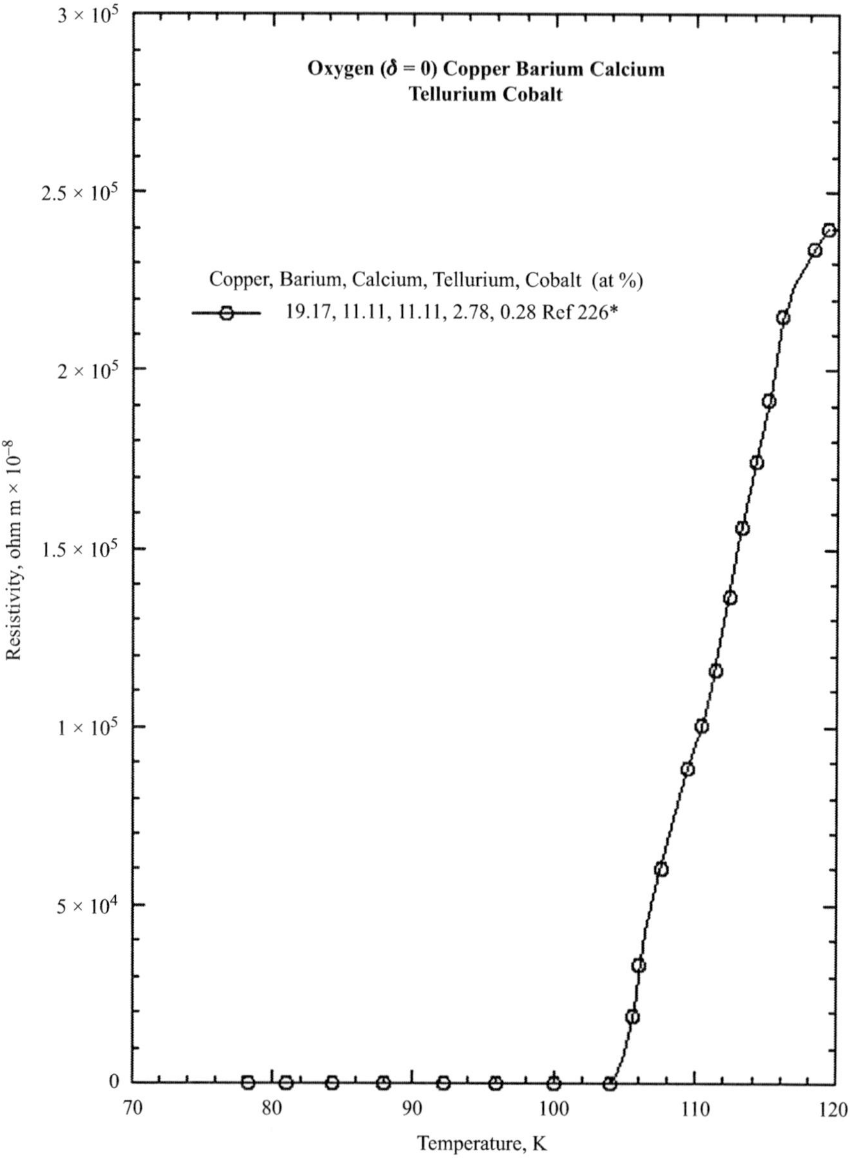
Oxygen (δ = 0) Copper Barium Calcium Tellurium Cobalt
Copper, Barium, Calcium, Tellurium, Cobalt (at %)
19.17, 11.11, 11.11, 2.78, 0.28 Ref 226*
3 × 10^5
2.5 × 10^5
2 × 10^5
1.5 × 10^5
1 × 10^5
5 × 10^4
0
Resistivity, ohm m × 10^-8
70
80
90
100
110
120
Temperature, K

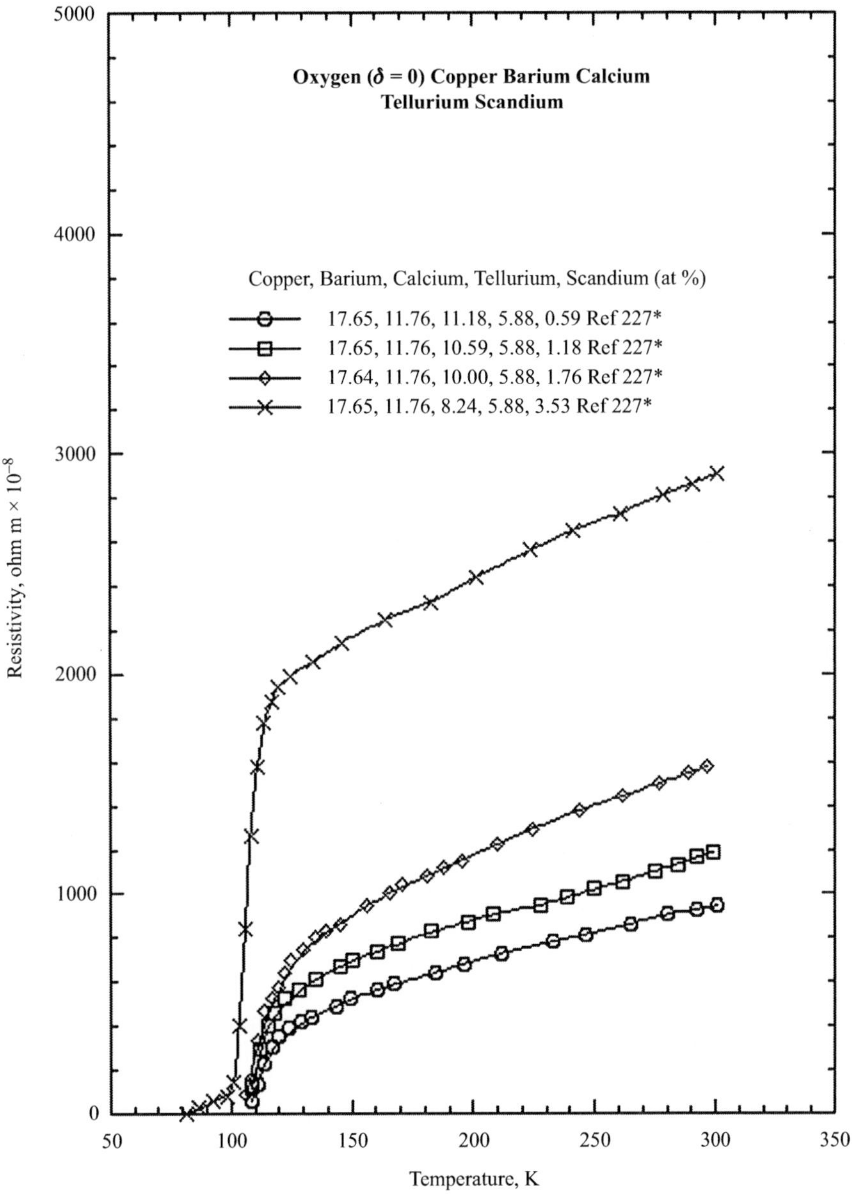

Oxygen (δ = 0) Copper Barium Calcium Tellurium Scandium
Copper, Barium, Calcium, Tellurium, Scandium (at %)
17.65, 11.76, 11.18, 5.88, 0.59 Ref 227*
17.65, 11.76, 10.59, 5.88, 1.18 Ref 227*
17.64, 11.76, 10.00, 5.88, 1.76 Ref 227*
17.65, 11.76, 8.24, 5.88, 3.53 Ref 227*
Resistivity, ohm m × 10^{-8}
Temperature, K
0
1000
2000
3000
4000
5000
50
100
150
200
250
300
350

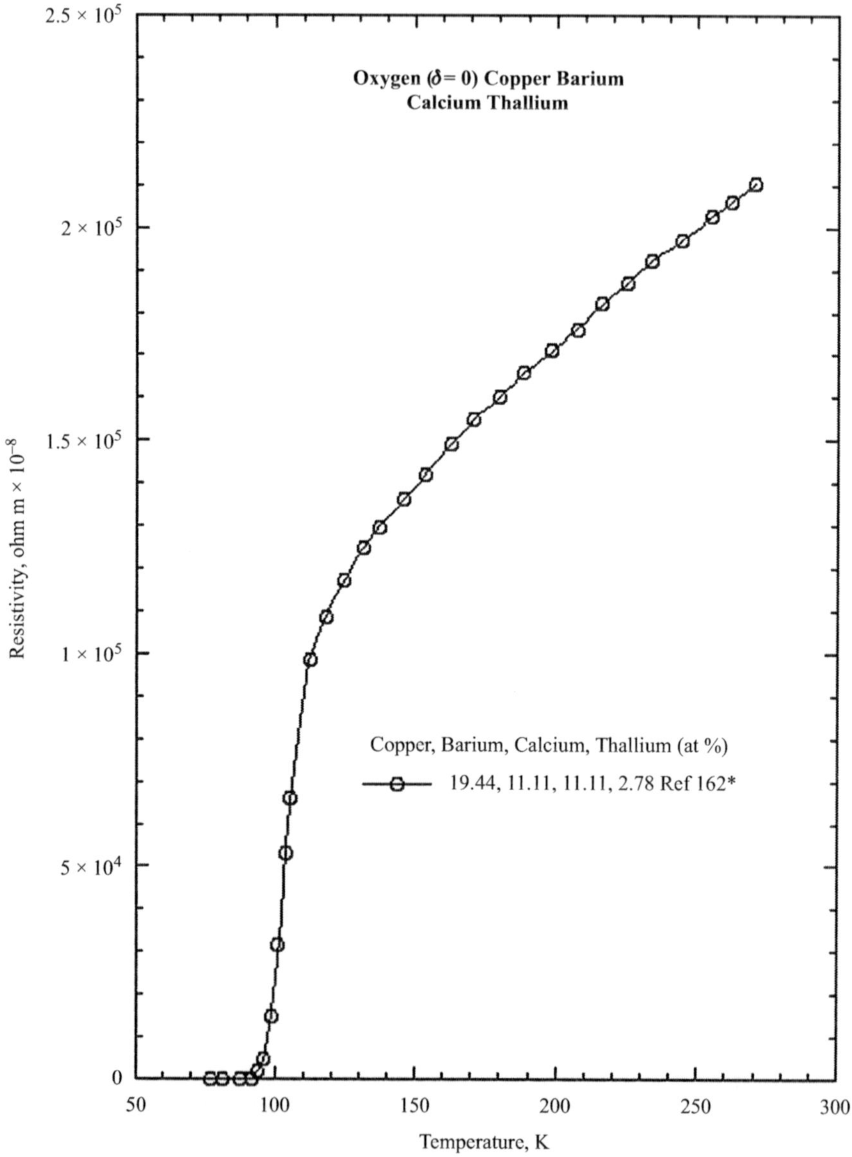
Oxygen (δ= 0) Copper Barium
Calcium Thallium
Copper, Barium, Calcium, Thallium (at %)
19.44, 11.11, 11.11, 2.78 Ref 162*
Resistivity, ohm m × 10⁻⁸
Temperature, K
2.5 × 10⁵
2 × 10⁵
1.5 × 10⁵
1 × 10⁵
5 × 10⁴
0
50
100
150
200
250
300

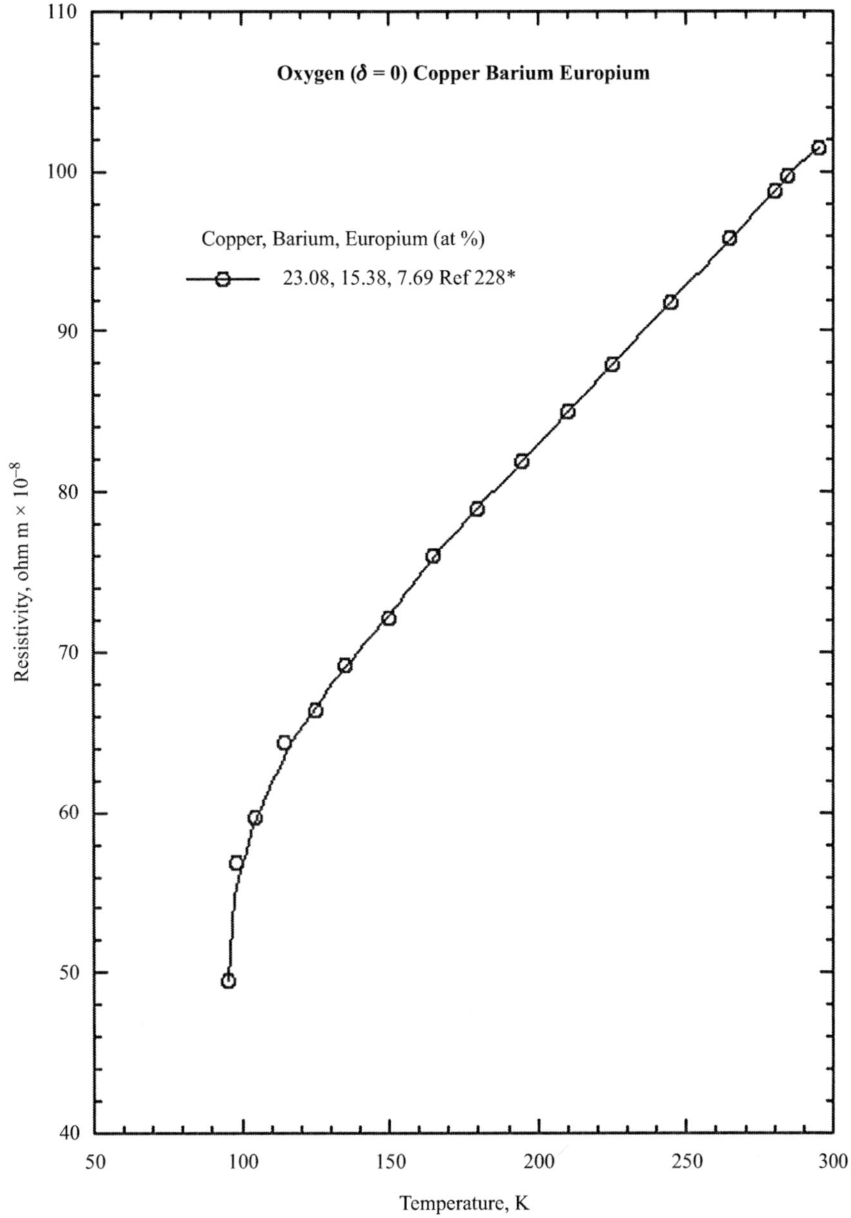

Oxygen (δ = 0) Copper Barium Europium
Copper, Barium, Europium (at %)
23.08, 15.38, 7.69 Ref 228*
Resistivity, ohm m × 10⁻⁸
110
100
90
80
70
60
50
40
50
100
150
200
250
300
Temperature, K

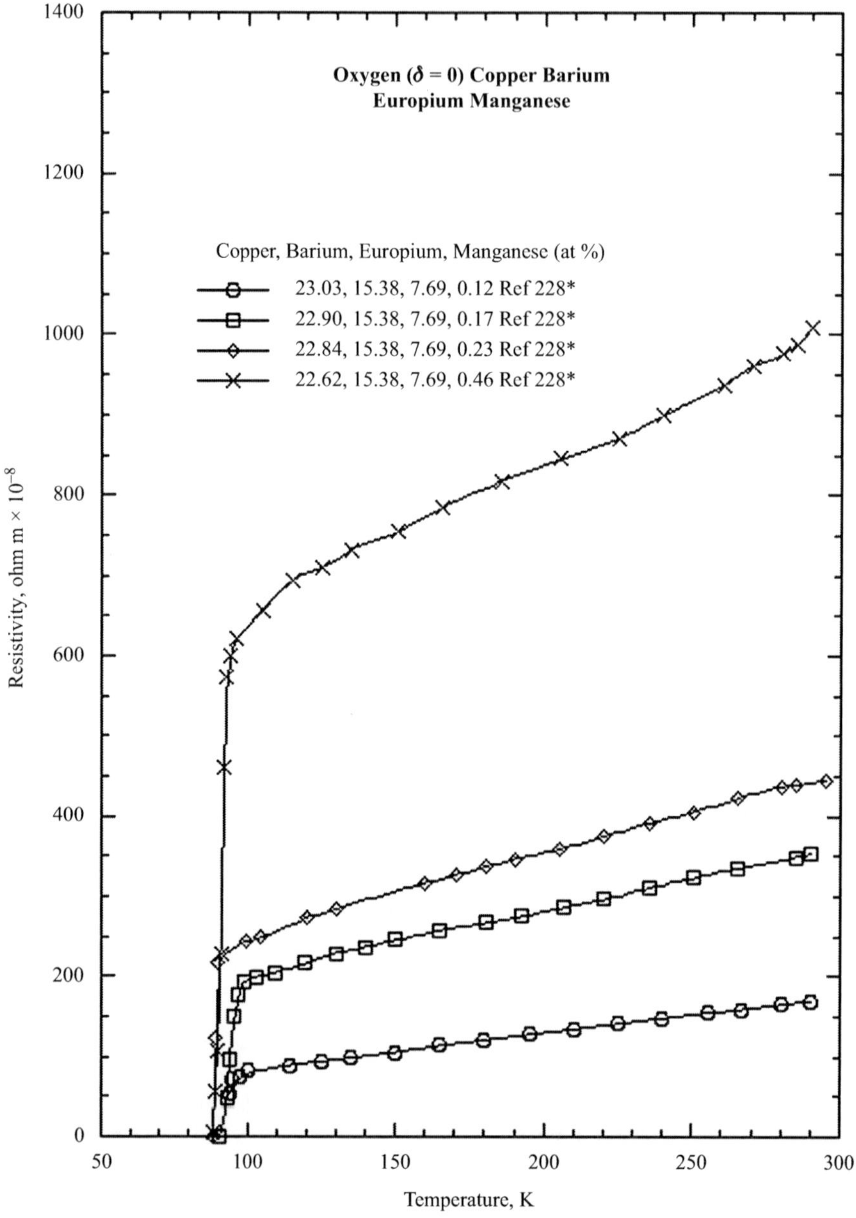
Oxygen (δ = 0) Copper Barium
Europium Manganese
Copper, Barium, Europium, Manganese (at %)
23.03, 15.38, 7.69, 0.12 Ref 228*
22.90, 15.38, 7.69, 0.17 Ref 228*
22.84, 15.38, 7.69, 0.23 Ref 228*
22.62, 15.38, 7.69, 0.46 Ref 228*
Resistivity, ohm m × 10⁻⁸
Temperature, K
0
200
400
600
800
1000
1200
1400
50
100
150
200
250
300

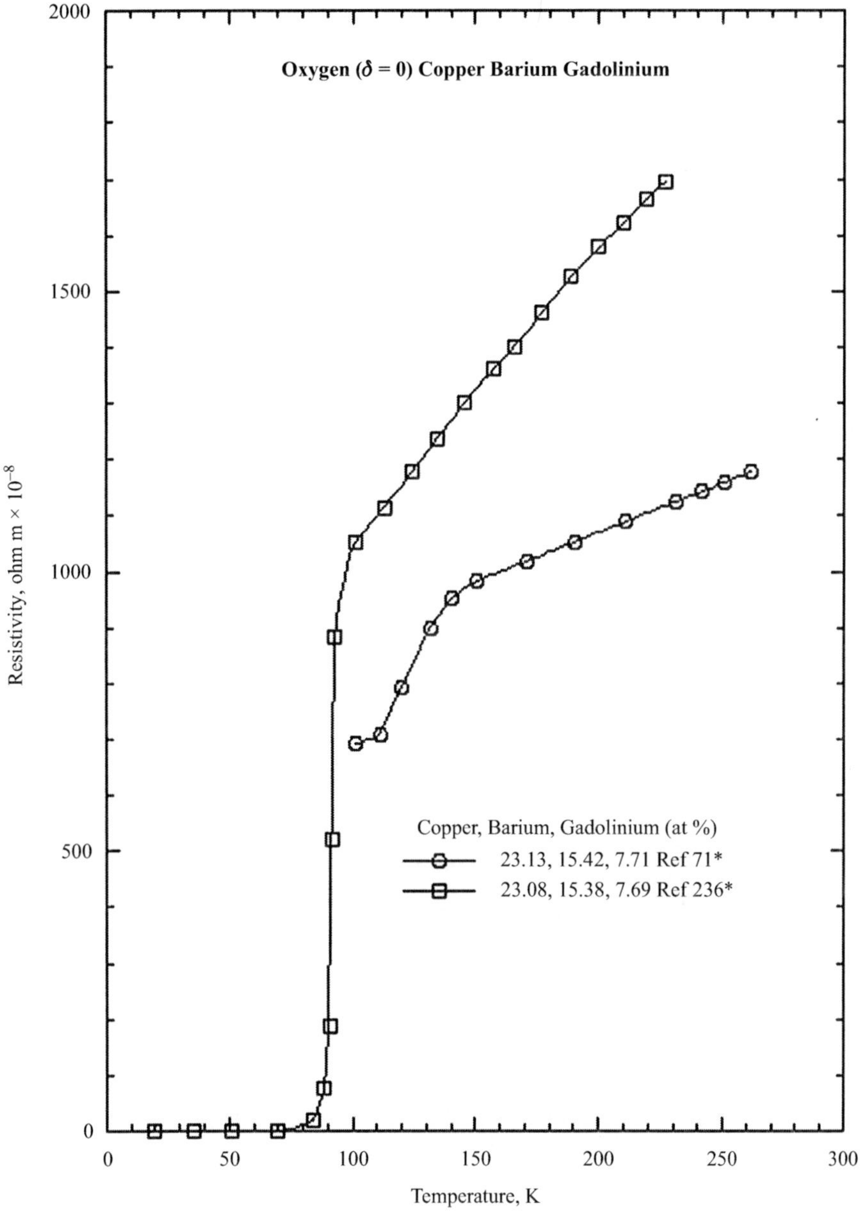
Oxygen (δ = 0) Copper Barium Gadolinium
Copper, Barium, Gadolinium (at %)
23.13, 15.42, 7.71 Ref 71*
23.08, 15.38, 7.69 Ref 236*
Resistivity, ohm m × 10^{-8}
Temperature, K
2000
1500
1000
500
0
0
50
100
150
200
250
300

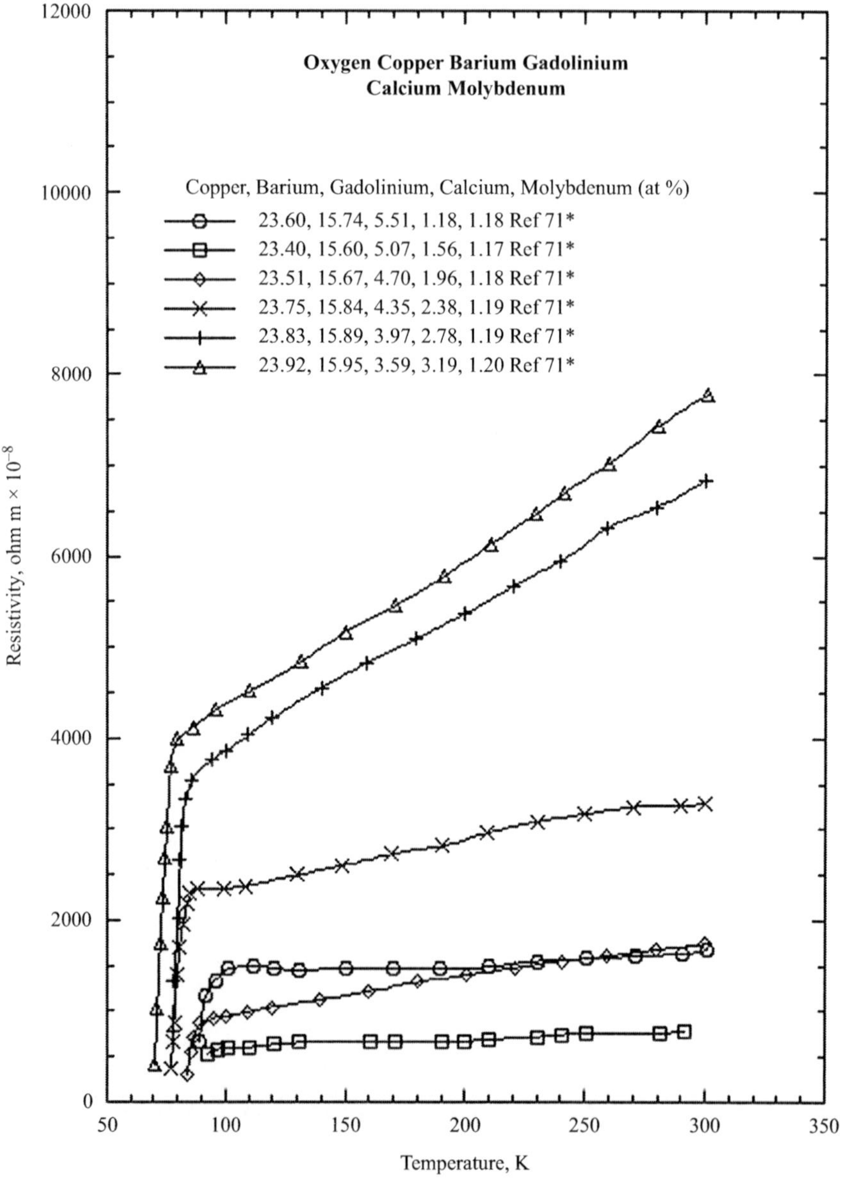
Oxygen Copper Barium Gadolinium
Calcium Molybdenum
Copper, Barium, Gadolinium, Calcium, Molybdenum (at %)
23.60, 15.74, 5.51, 1.18, 1.18 Ref 71*
23.40, 15.60, 5.07, 1.56, 1.17 Ref 71*
23.51, 15.67, 4.70, 1.96, 1.18 Ref 71*
23.75, 15.84, 4.35, 2.38, 1.19 Ref 71*
23.83, 15.89, 3.97, 2.78, 1.19 Ref 71*
23.92, 15.95, 3.59, 3.19, 1.20 Ref 71*
Resistivity, ohm m × 10^−8
Temperature, K
0
2000
4000
6000
8000
10000
12000
50
100
150
200
250
300
350

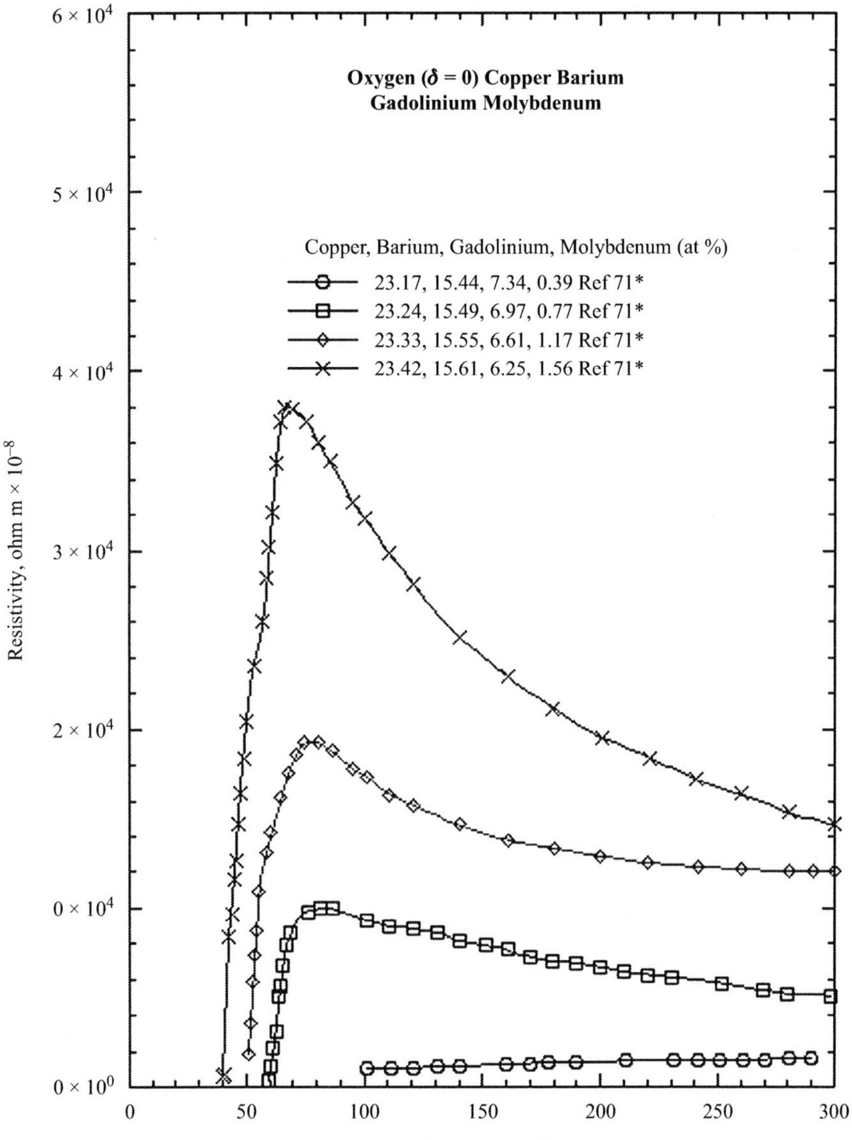

Oxygen (δ = 0) Copper Barium Gadolinium Molybdenum
Copper, Barium, Gadolinium, Molybdenum (at %)
23.17, 15.44, 7.34, 0.39 Ref 71*
23.24, 15.49, 6.97, 0.77 Ref 71*
23.33, 15.55, 6.61, 1.17 Ref 71*
23.42, 15.61, 6.25, 1.56 Ref 71*
Resistivity, ohm m × 10⁻⁸
6 × 10⁴
5 × 10⁴
4 × 10⁴
3 × 10⁴
2 × 10⁴
0 × 10⁴
0 × 10⁰
0
50
100
150
200
250
300
Temperature, K

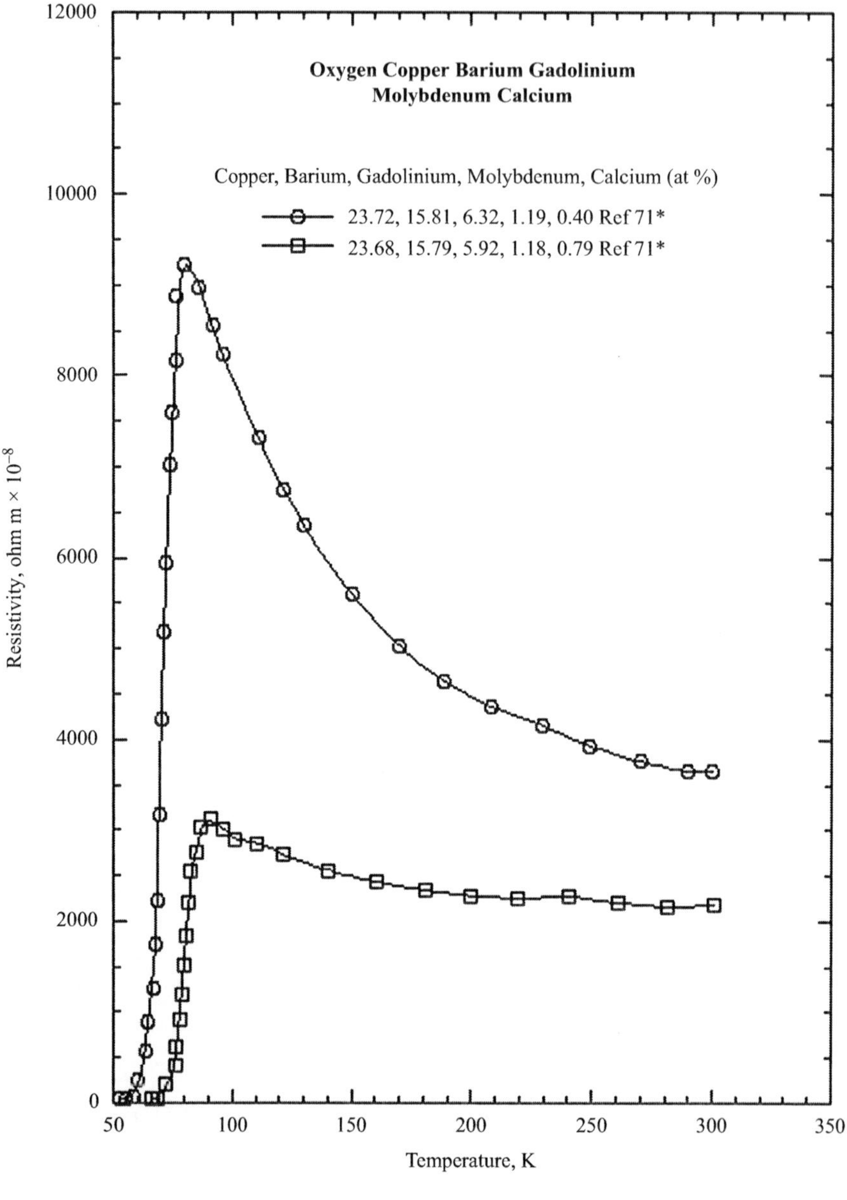
Oxygen Copper Barium Gadolinium
Molybdenum Calcium
Copper, Barium, Gadolinium, Molybdenum, Calcium (at %)
23.72, 15.81, 6.32, 1.19, 0.40 Ref 71*
23.68, 15.79, 5.92, 1.18, 0.79 Ref 71*
Resistivity, ohm m × 10⁻⁸
Temperature, K
0
2000
4000
6000
8000
10000
12000
50
100
150
200
250
300
350

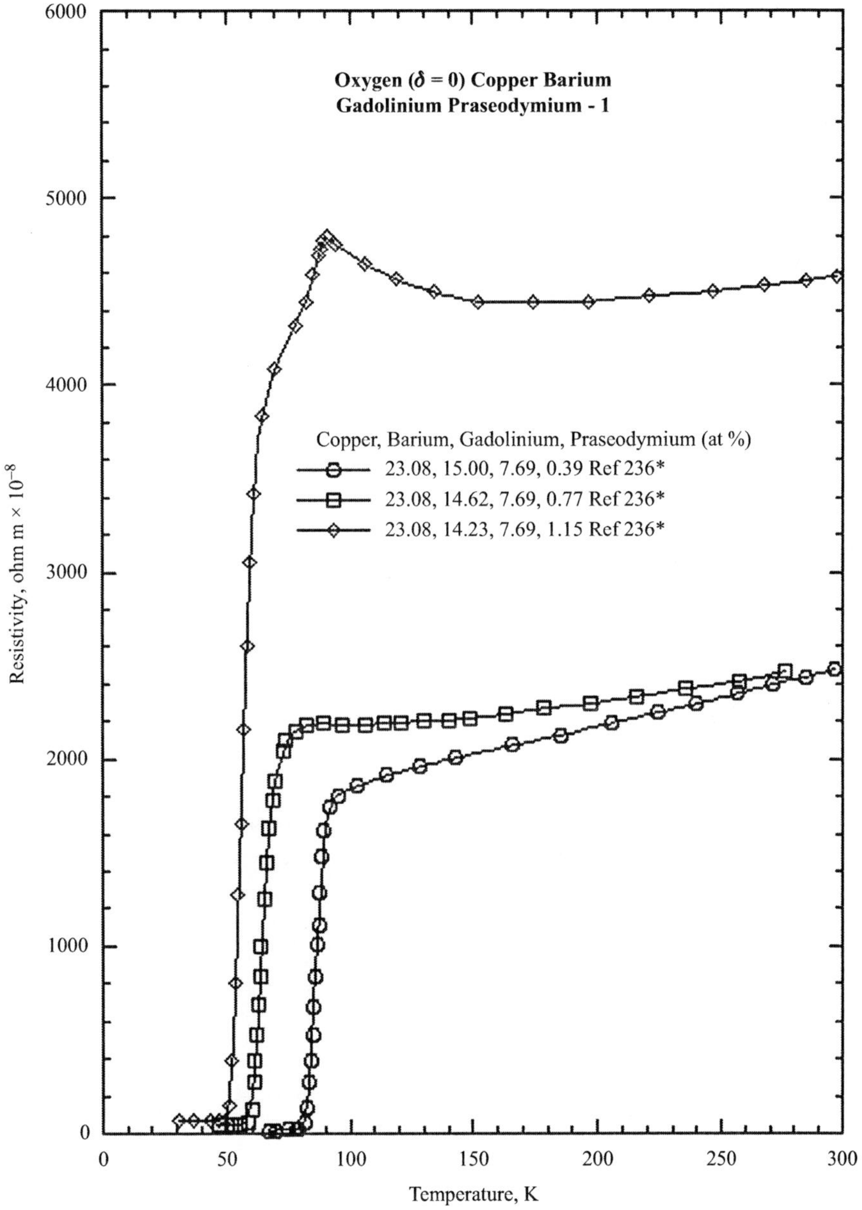

Oxygen (δ = 0) Copper Barium
Gadolinium Praseodymium - 1
Copper, Barium, Gadolinium, Praseodymium (at %)
23.08, 15.00, 7.69, 0.39 Ref 236*
23.08, 14.62, 7.69, 0.77 Ref 236*
23.08, 14.23, 7.69, 1.15 Ref 236*
Resistivity, ohm m × 10^{-8}
Temperature, K
0
1000
2000
3000
4000
5000
6000
0
50
100
150
200
250
300

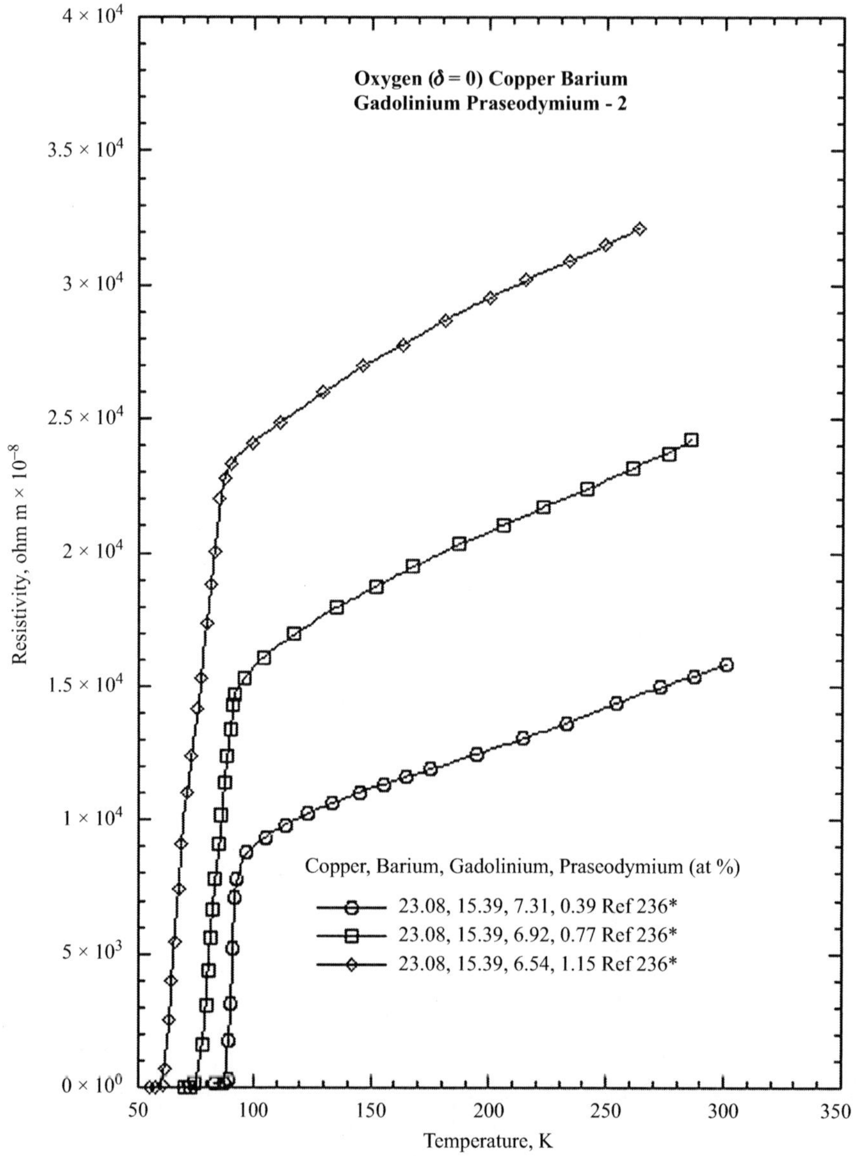
Oxygen (δ = 0) Copper Barium
Gadolinium Praseodymium - 2
Resistivity, ohm m × 10−8
Temperature, K
Copper, Barium, Gadolinium, Praseodymium (at %)
23.08, 15.39, 7.31, 0.39 Ref 236*
23.08, 15.39, 6.92, 0.77 Ref 236*
23.08, 15.39, 6.54, 1.15 Ref 236*
4 × 10^4
3.5 × 10^4
3 × 10^4
2.5 × 10^4
2 × 10^4
1.5 × 10^4
1 × 10^4
5 × 10^3
0 × 10^0
50
100
150
200
250
300
350

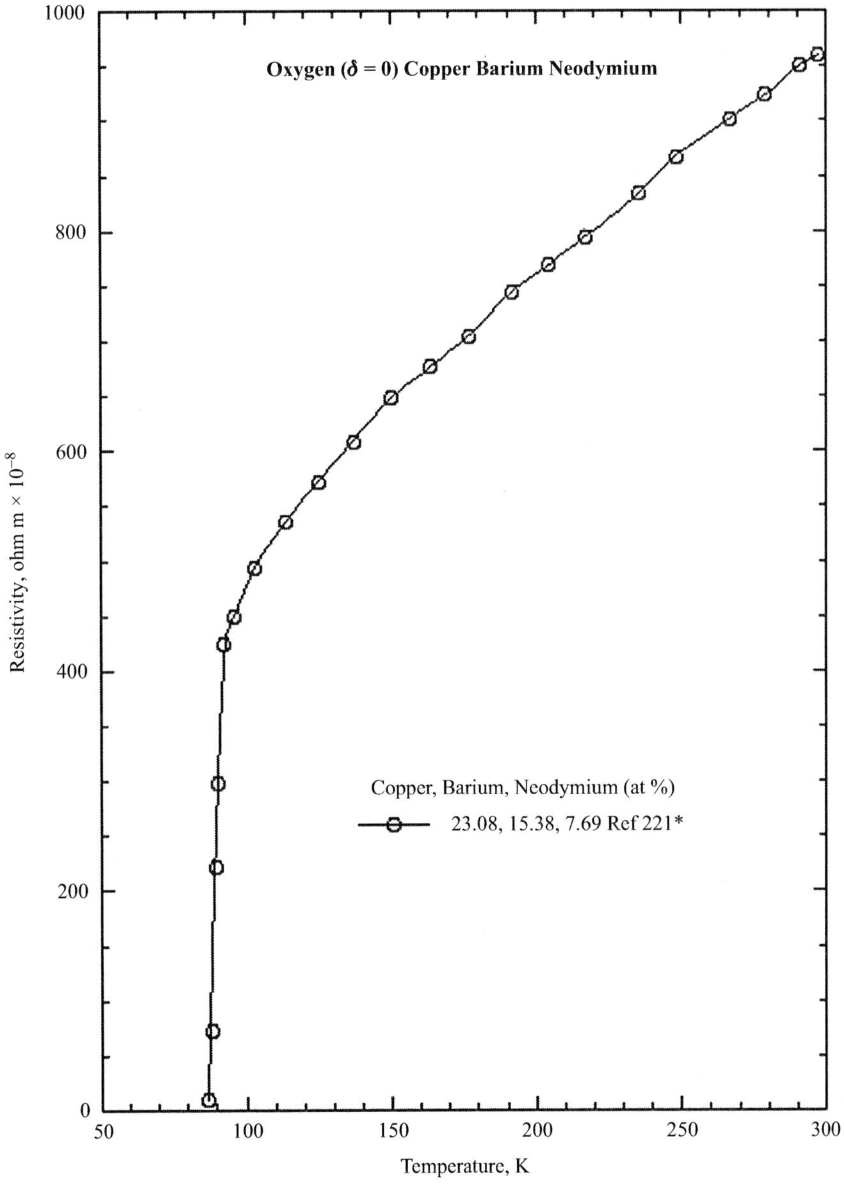
Oxygen (δ = 0) Copper Barium Neodymium
Copper, Barium, Neodymium (at %)
23.08, 15.38, 7.69 Ref 221*
Resistivity, ohm m × 10⁻⁸
Temperature, K
0
200
400
600
800
1000
50
100
150
200
250
300

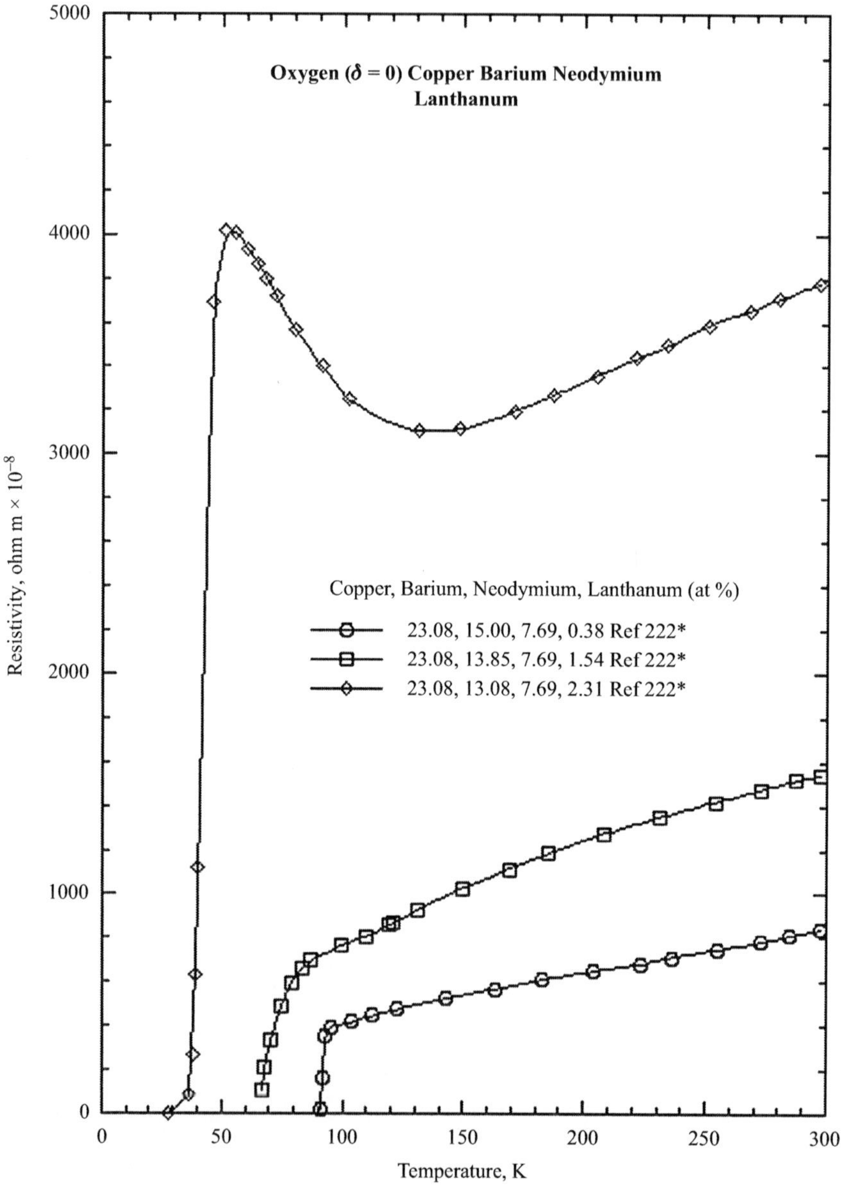
Oxygen (δ = 0) Copper Barium Neodymium Lanthanum
Copper, Barium, Neodymium, Lanthanum (at %)
23.08, 15.00, 7.69, 0.38 Ref 222*
23.08, 13.85, 7.69, 1.54 Ref 222*
23.08, 13.08, 7.69, 2.31 Ref 222*
Resistivity, ohm m × 10⁻⁸
Temperature, K
5000
4000
3000
2000
1000
0
0
50
100
150
200
250
300

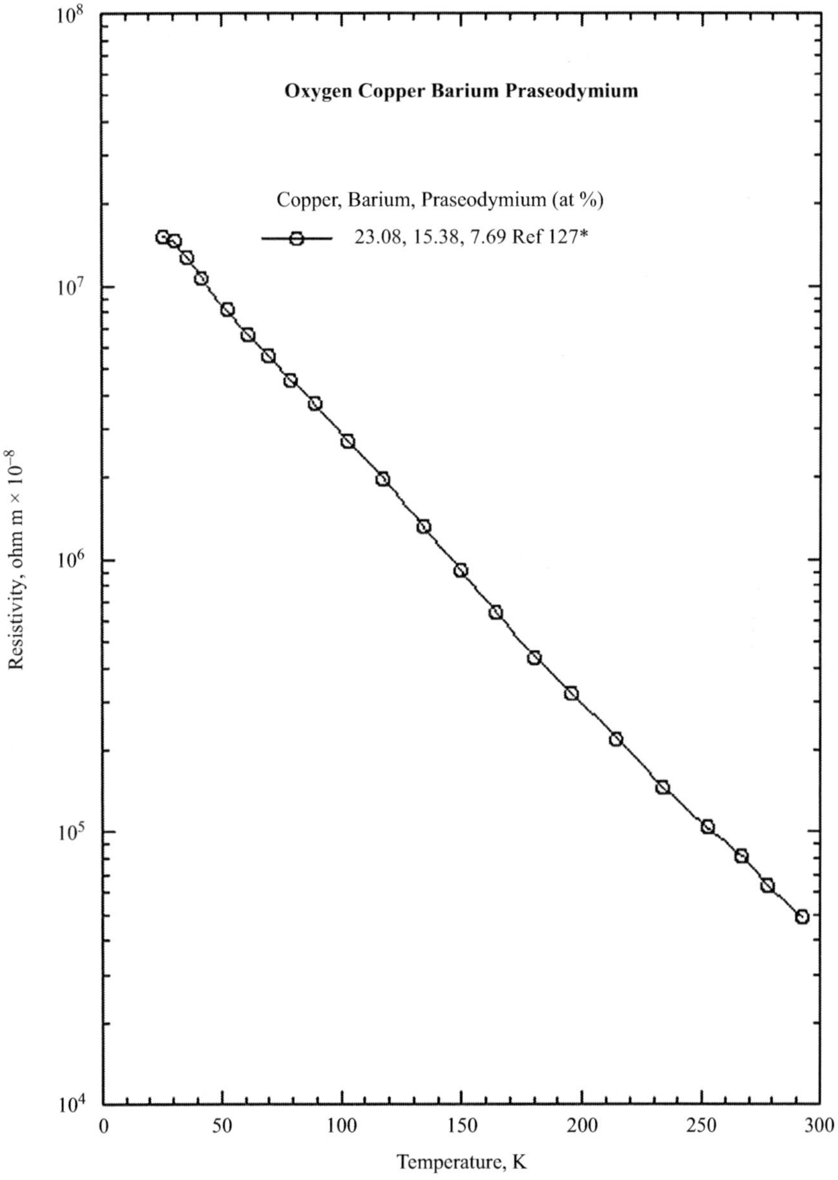
Oxygen Copper Barium Praseodymium
Copper, Barium, Praseodymium (at %)
23.08, 15.38, 7.69 Ref 127*
Resistivity, ohm m × 10^−8
Temperature, K
10^8
10^7
10^6
10^5
10^4
0
50
100
150
200
250
300

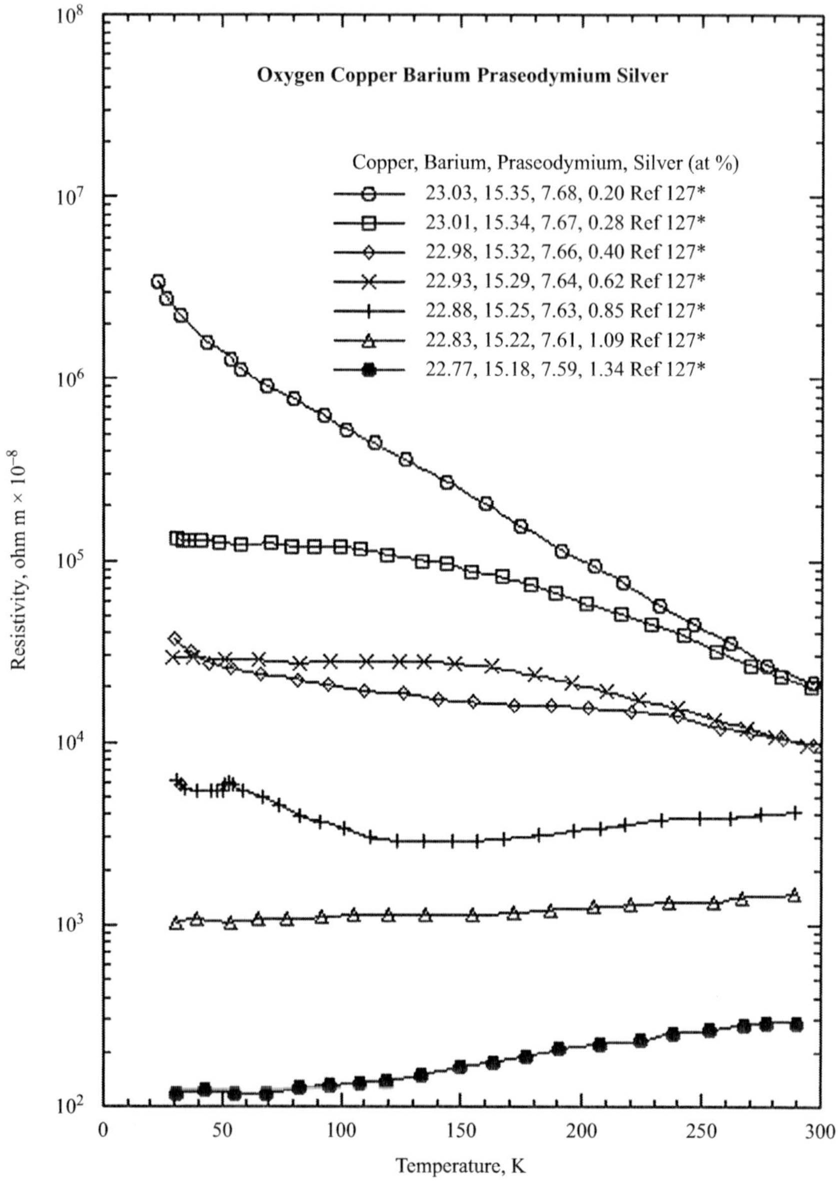
Oxygen Copper Barium Praseodymium Silver
Copper, Barium, Praseodymium, Silver (at %)
23.03, 15.35, 7.68, 0.20 Ref 127*
23.01, 15.34, 7.67, 0.28 Ref 127*
22.98, 15.32, 7.66, 0.40 Ref 127*
22.93, 15.29, 7.64, 0.62 Ref 127*
22.88, 15.25, 7.63, 0.85 Ref 127*
22.83, 15.22, 7.61, 1.09 Ref 127*
22.77, 15.18, 7.59, 1.34 Ref 127*
Resistivity, ohm m × 10⁻⁸
Temperature, K
10⁸
10⁷
10⁶
10⁵
10⁴
10³
10²
0
50
100
150
200
250
300

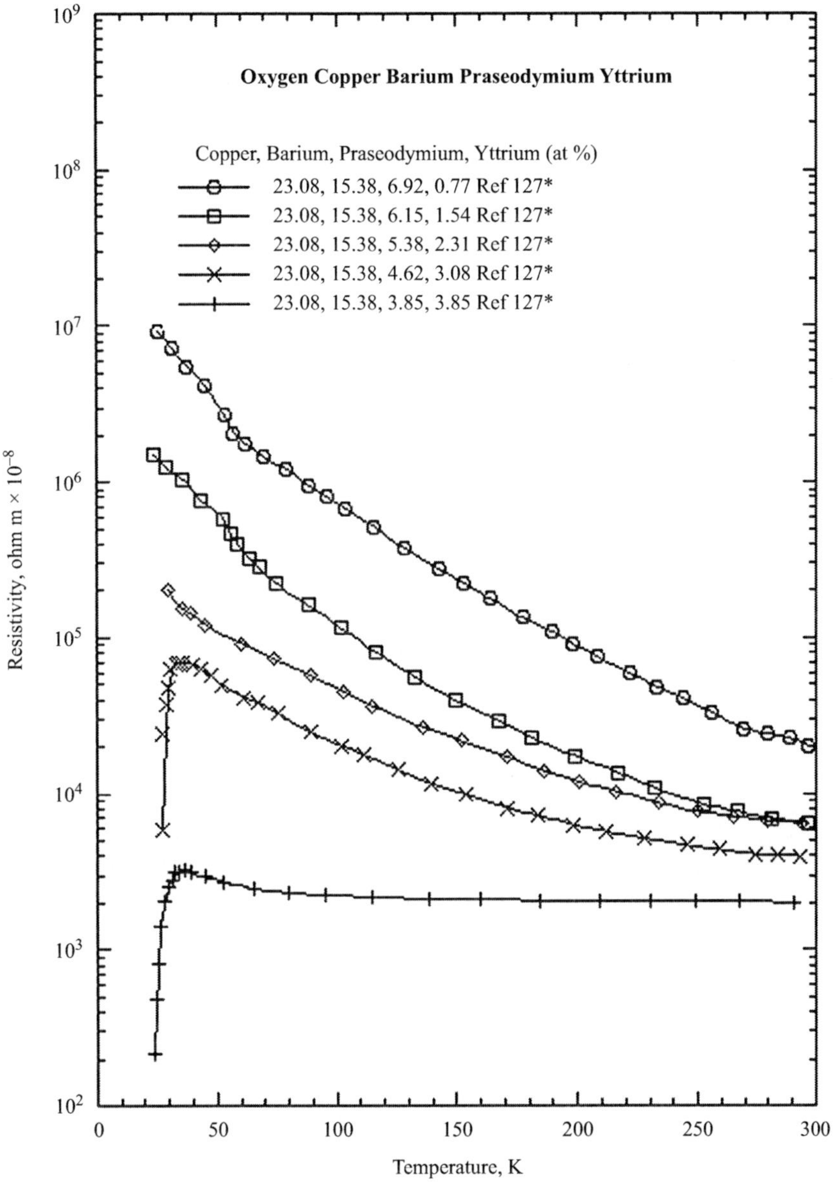
Oxygen Copper Barium Praseodymium Yttrium
Copper, Barium, Praseodymium, Yttrium (at %)
23.08, 15.38, 6.92, 0.77 Ref 127*
23.08, 15.38, 6.15, 1.54 Ref 127*
23.08, 15.38, 5.38, 2.31 Ref 127*
23.08, 15.38, 4.62, 3.08 Ref 127*
23.08, 15.38, 3.85, 3.85 Ref 127*
Resistivity, ohm m × 10⁻⁸
Temperature, K
10⁹
10⁸
10⁷
10⁶
10⁵
10⁴
10³
10²
0
50
100
150
200
250
300

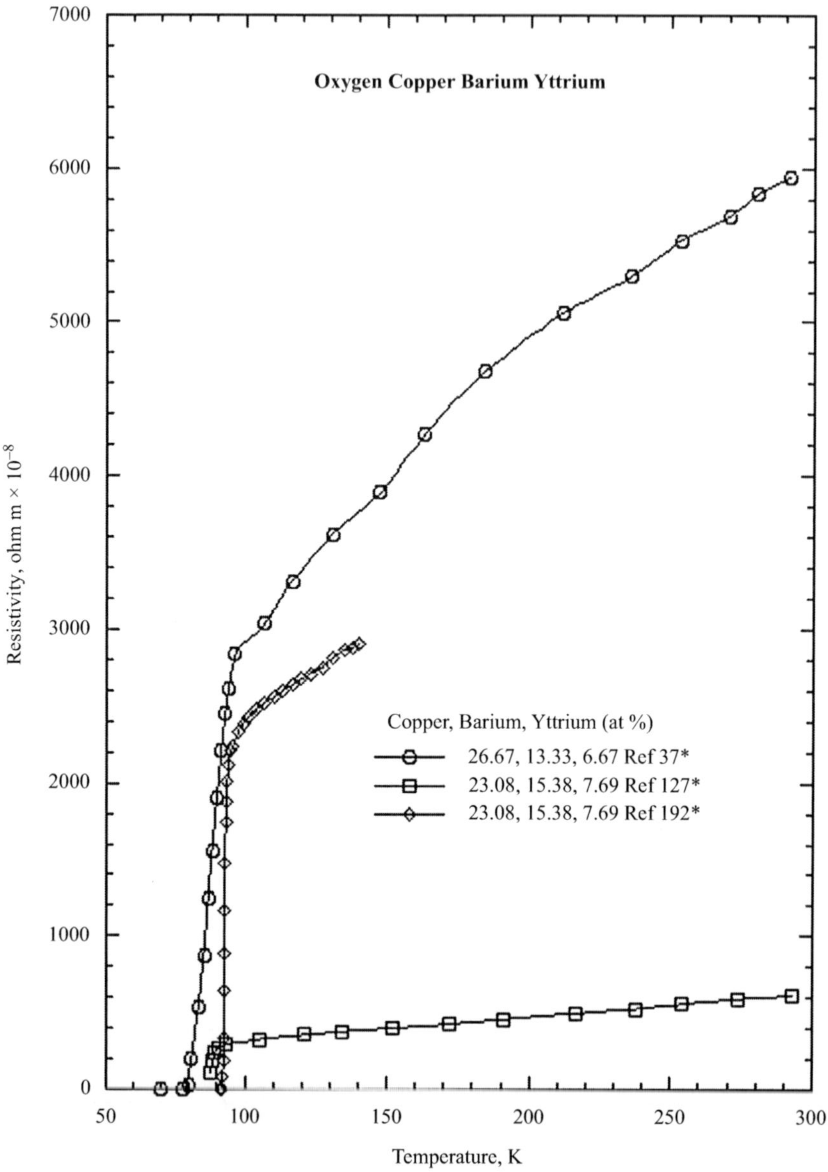
Oxygen Copper Barium Yttrium
Copper, Barium, Yttrium (at %)
26.67, 13.33, 6.67 Ref 37*
23.08, 15.38, 7.69 Ref 127*
23.08, 15.38, 7.69 Ref 192*
Resistivity, ohm m × 10⁻⁸
Temperature, K
7000
6000
5000
4000
3000
2000
1000
0
50
100
150
200
250
300

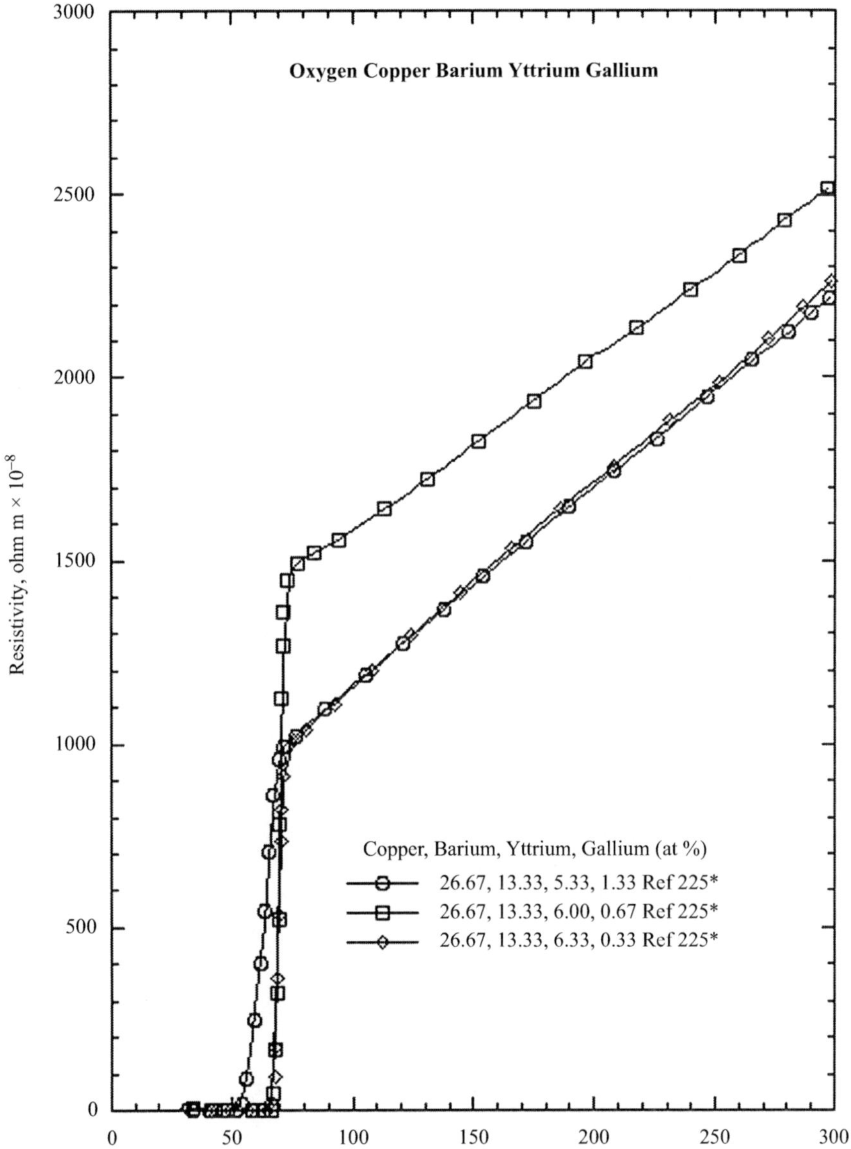
Oxygen Copper Barium Yttrium Gallium
Copper, Barium, Yttrium, Gallium (at %)
26.67, 13.33, 5.33, 1.33 Ref 225*
26.67, 13.33, 6.00, 0.67 Ref 225*
26.67, 13.33, 6.33, 0.33 Ref 225*
Resistivity, ohm m × 10⁻⁸
Temperature, K
0
500
1000
1500
2000
2500
3000
0
50
100
150
200
250
300

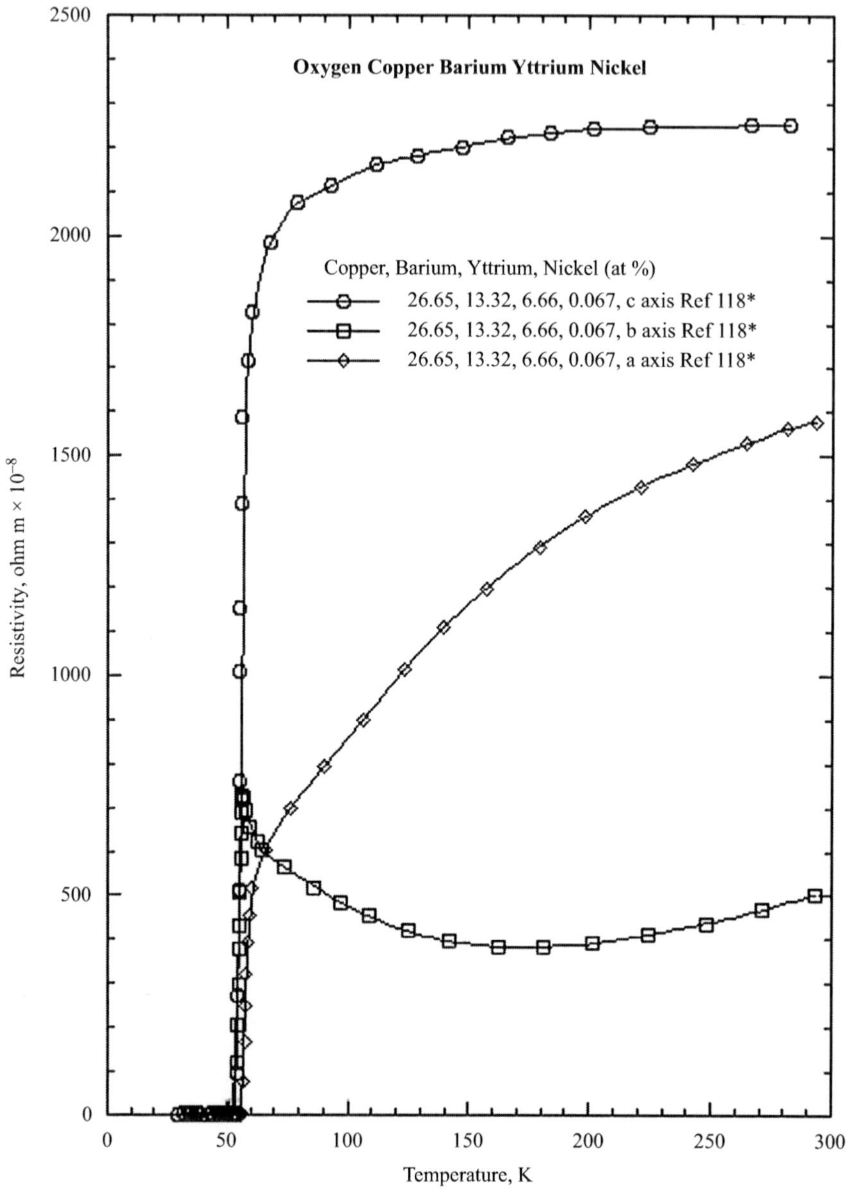
Oxygen Copper Barium Yttrium Nickel
Copper, Barium, Yttrium, Nickel (at %)
26.65, 13.32, 6.66, 0.067, c axis Ref 118*
26.65, 13.32, 6.66, 0.067, b axis Ref 118*
26.65, 13.32, 6.66, 0.067, a axis Ref 118*
Resistivity, ohm m × 10⁻⁸
Temperature, K
0
500
1000
1500
2000
2500
0
50
100
150
200
250
300

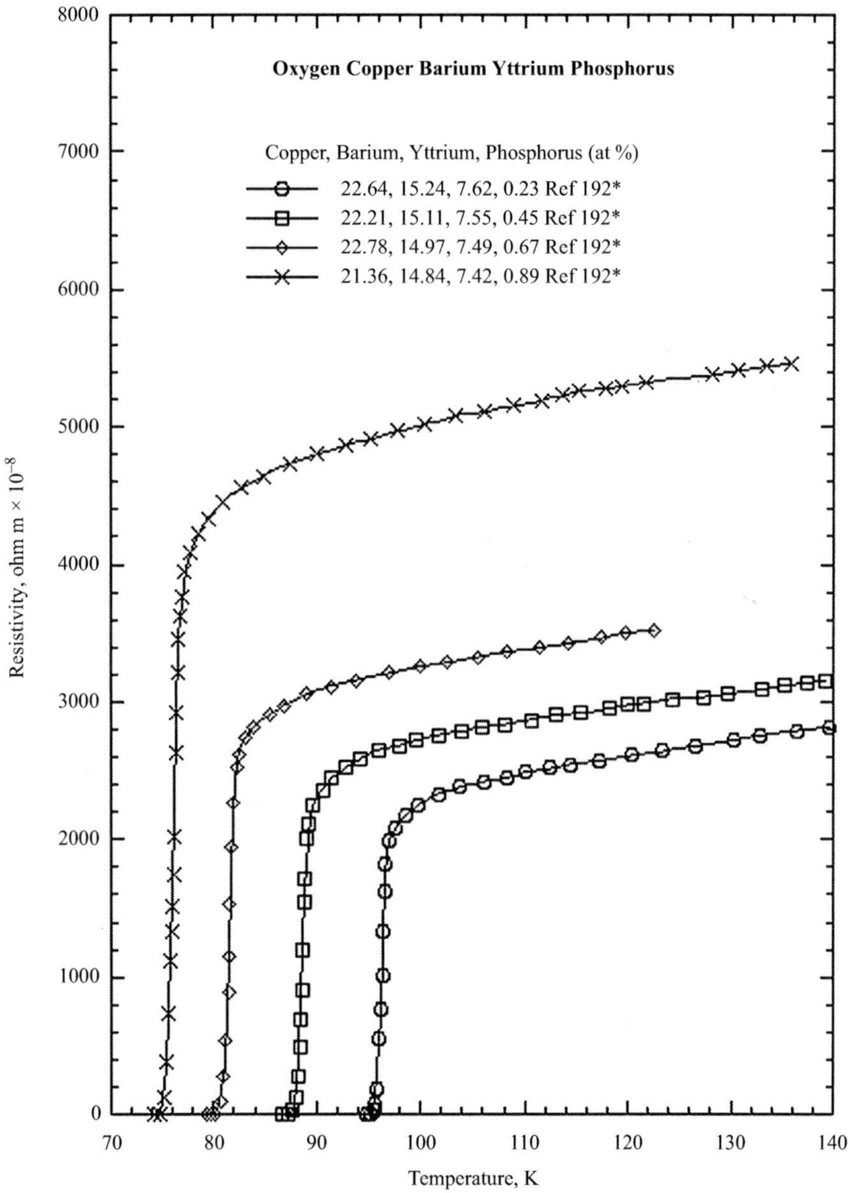
Oxygen Copper Barium Yttrium Phosphorus
Copper, Barium, Yttrium, Phosphorus (at %)
22.64, 15.24, 7.62, 0.23 Ref 192*
22.21, 15.11, 7.55, 0.45 Ref 192*
22.78, 14.97, 7.49, 0.67 Ref 192*
21.36, 14.84, 7.42, 0.89 Ref 192*
Resistivity, ohm m × 10⁻⁸
Temperature, K
0
1000
2000
3000
4000
5000
6000
7000
8000
70
80
90
100
110
120
130
140

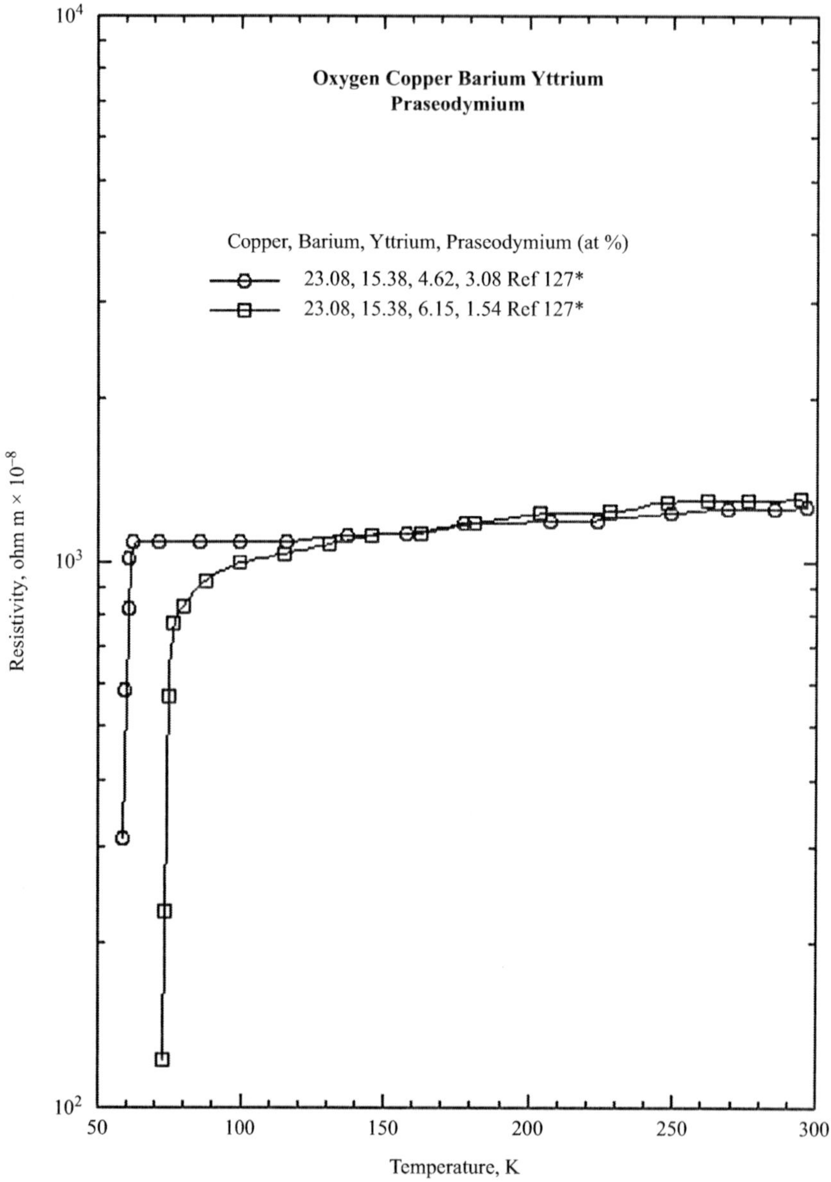
Oxygen Copper Barium Yttrium
Praseodymium
Copper, Barium, Yttrium, Praseodymium (at %)
23.08, 15.38, 4.62, 3.08 Ref 127*
23.08, 15.38, 6.15, 1.54 Ref 127*
10^4
10^3
10^2
Resistivity, ohm m × 10^{-8}
50
100
150
200
250
300
Temperature, K

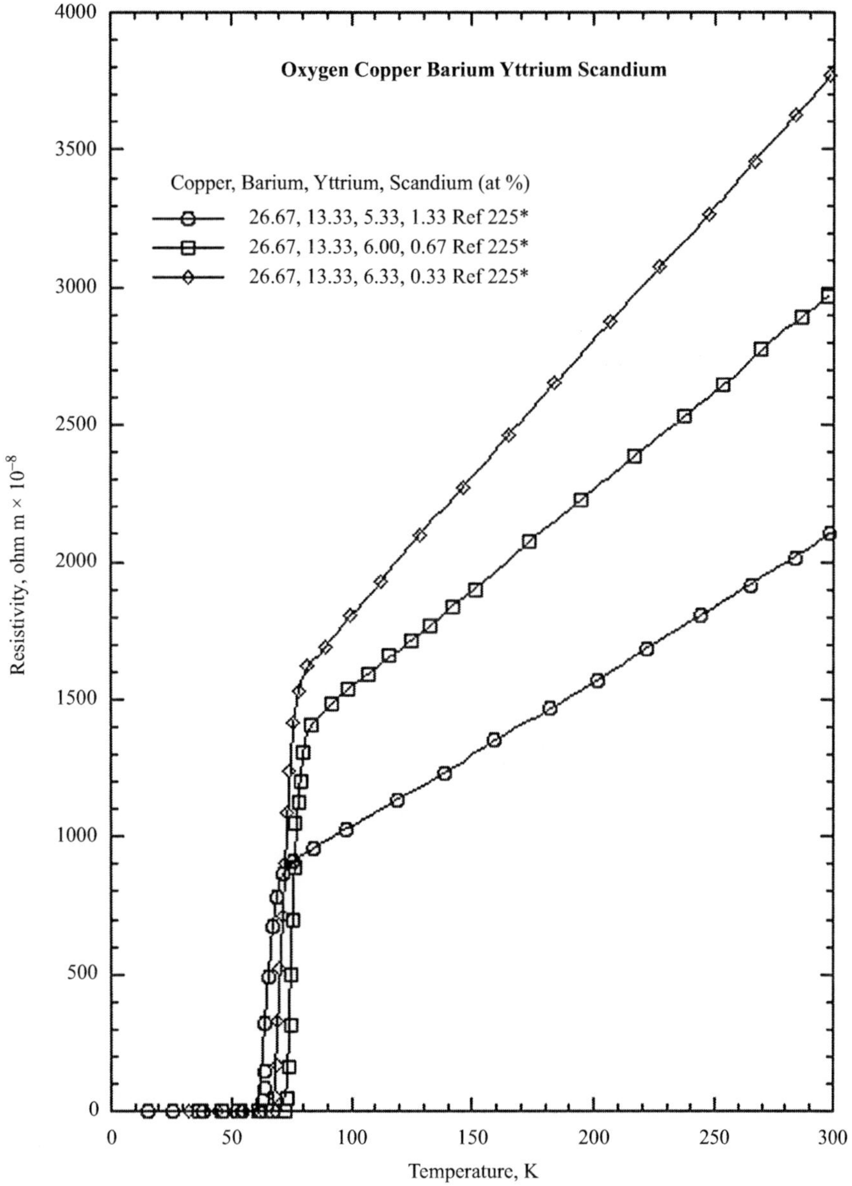
Oxygen Copper Barium Yttrium Scandium
Copper, Barium, Yttrium, Scandium (at %)
26.67, 13.33, 5.33, 1.33 Ref 225*
26.67, 13.33, 6.00, 0.67 Ref 225*
26.67, 13.33, 6.33, 0.33 Ref 225*
Resistivity, ohm m × 10⁻⁸
Temperature, K
0
500
1000
1500
2000
2500
3000
3500
4000
0
50
100
150
200
250
300

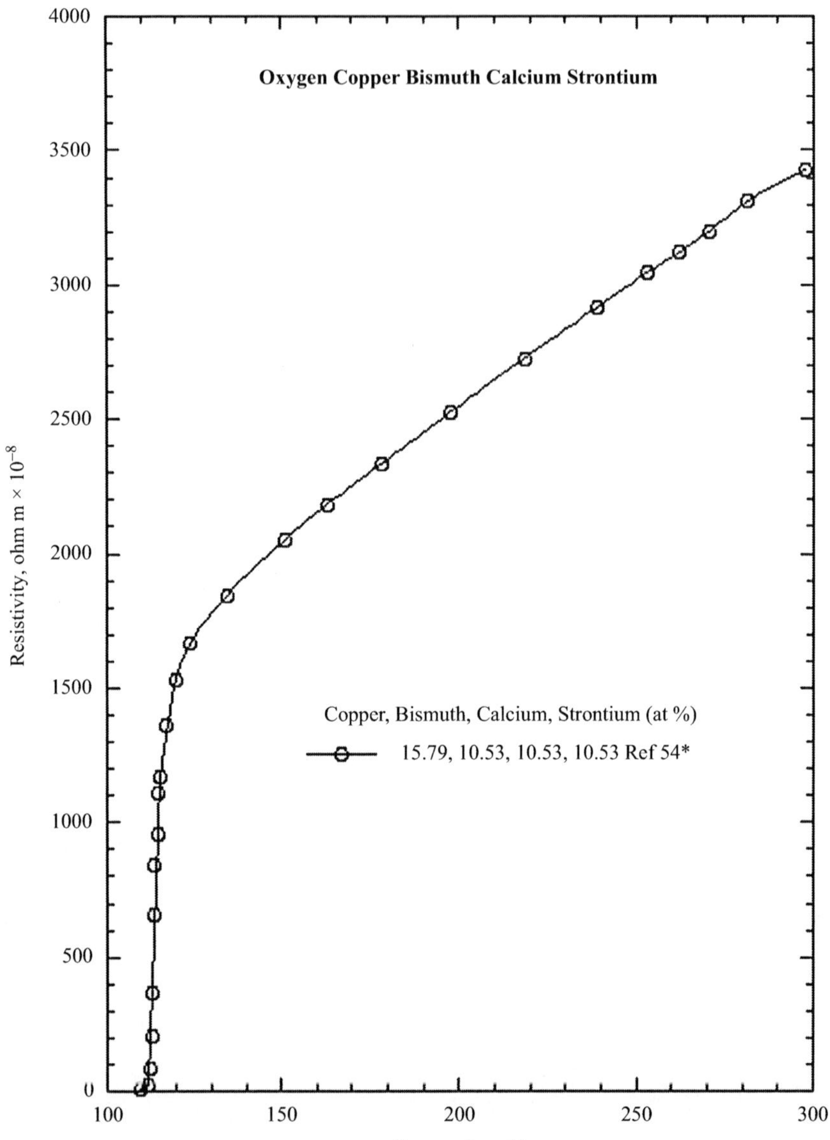
Oxygen Copper Bismuth Calcium Strontium
Resistivity, ohm m × 10^-8
Temperature, K
Copper, Bismuth, Calcium, Strontium (at %)
15.79, 10.53, 10.53, 10.53 Ref 54*
4000
3500
3000
2500
2000
1500
1000
500
0
100
150
200
250
300

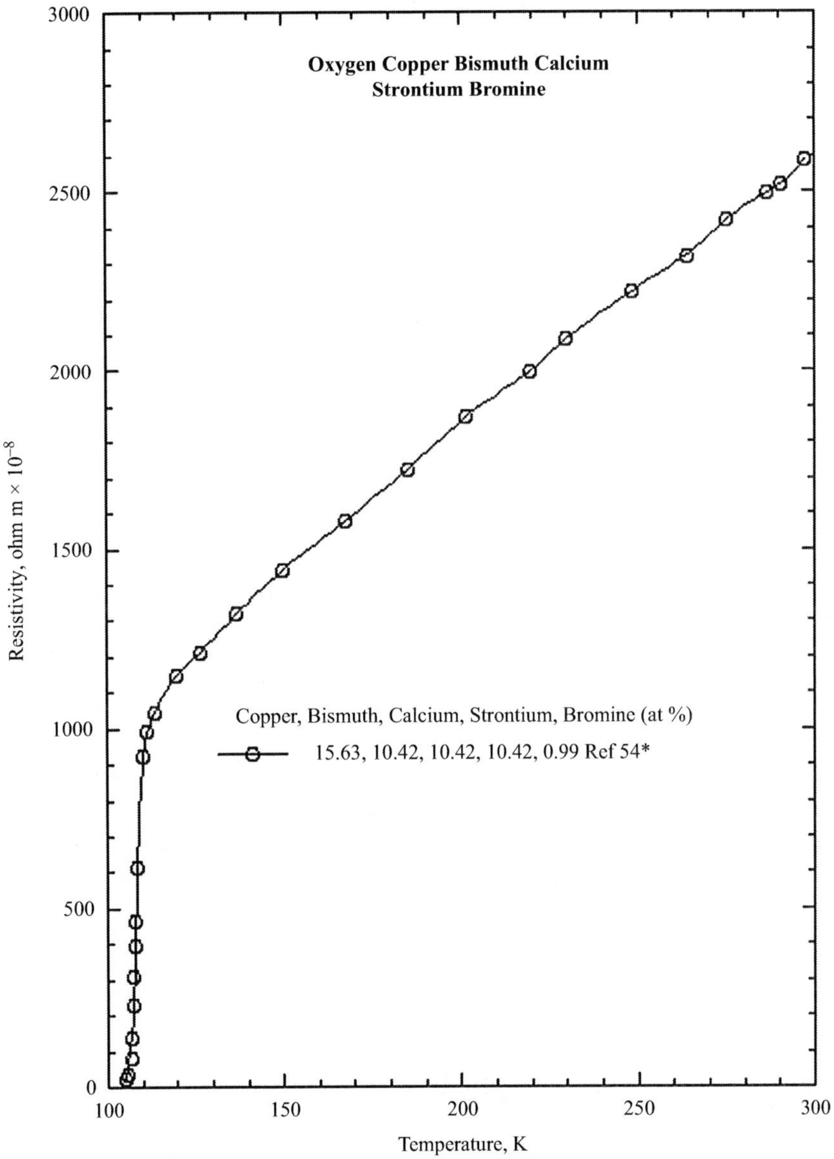
Oxygen Copper Bismuth Calcium
Strontium Bromine
Copper, Bismuth, Calcium, Strontium, Bromine (at %)
15.63, 10.42, 10.42, 10.42, 0.99 Ref 54*
Resistivity, ohm m × 10⁻⁸
Temperature, K
3000
2500
2000
1500
1000
500
0
100
150
200
250
300

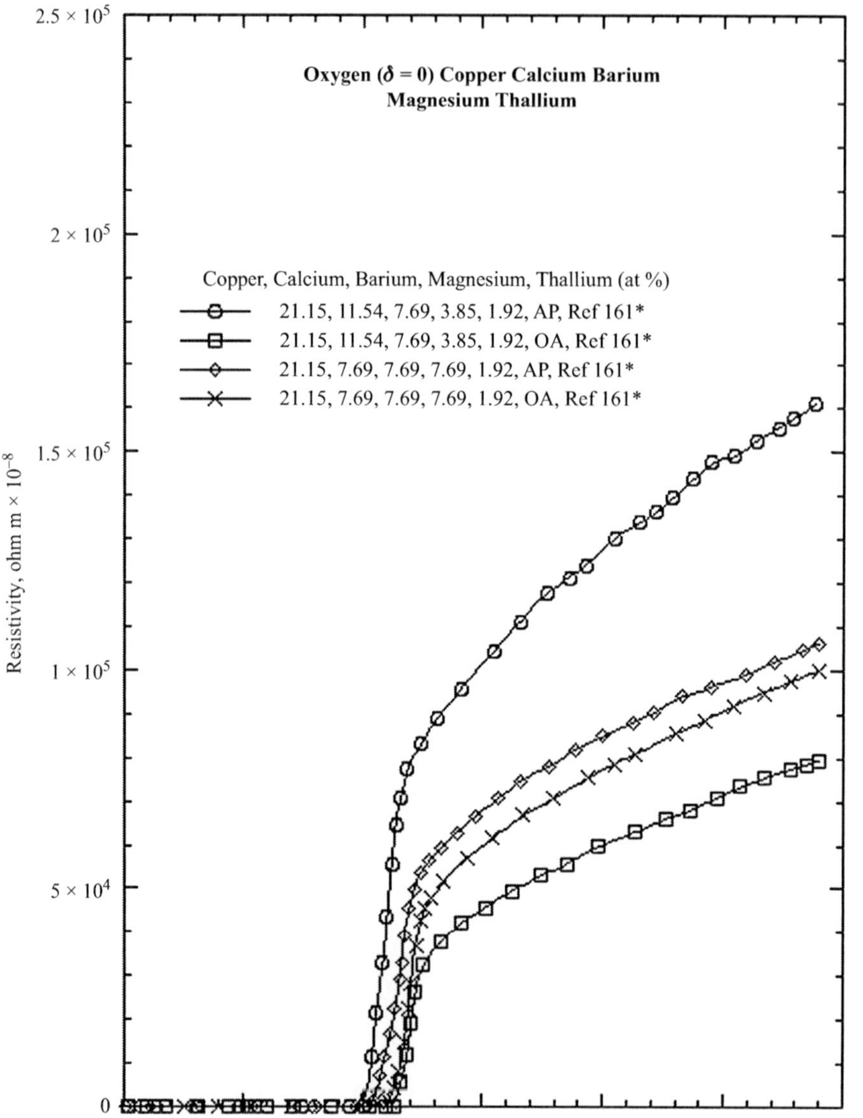

Oxygen (δ = 0) Copper Calcium Barium Magnesium Thallium
Copper, Calcium, Barium, Magnesium, Thallium (at %)
21.15, 11.54, 7.69, 3.85, 1.92, AP, Ref 161*
21.15, 11.54, 7.69, 3.85, 1.92, OA, Ref 161*
21.15, 7.69, 7.69, 7.69, 1.92, AP, Ref 161*
21.15, 7.69, 7.69, 7.69, 1.92, OA, Ref 161*
Resistivity, ohm m × 10⁻⁸
2.5 × 10⁵
2 × 10⁵
1.5 × 10⁵
1 × 10⁵
5 × 10⁴
0
0
50
100
150
200
250
300
Temperature, K

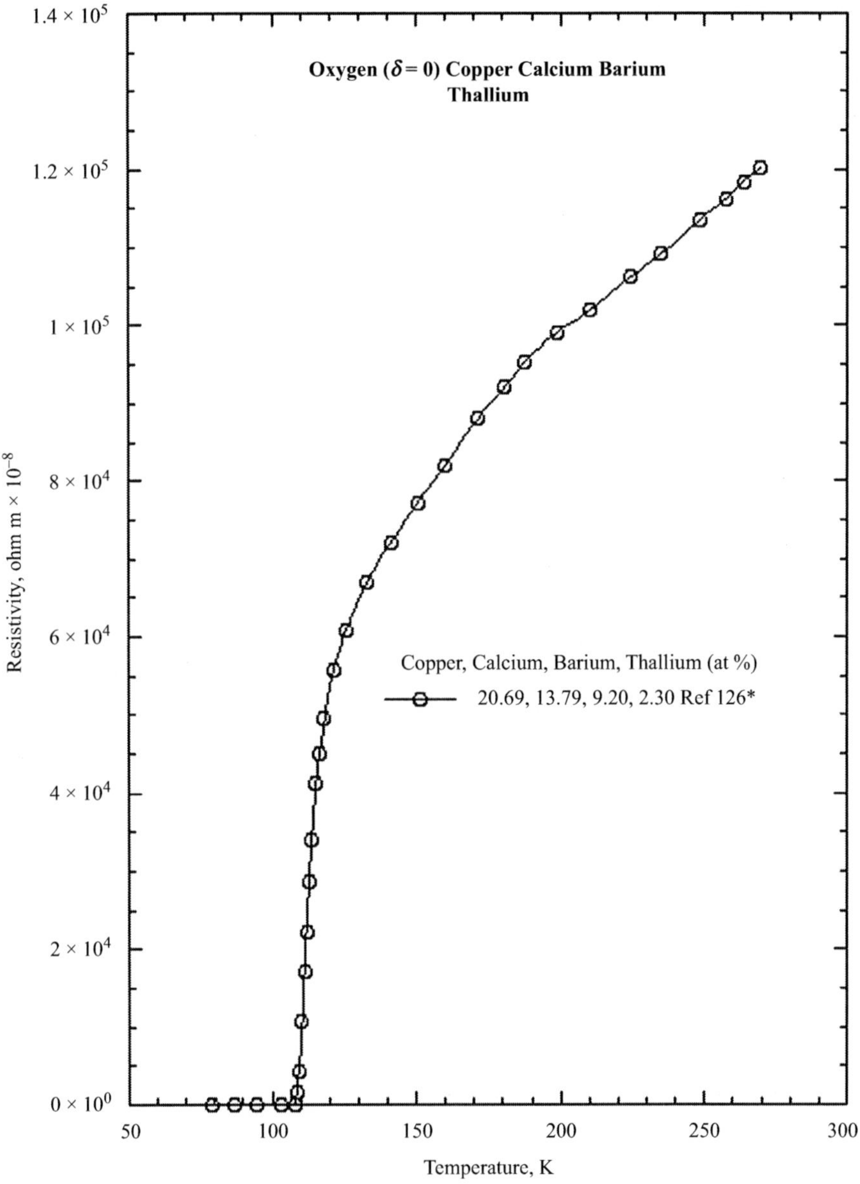
Oxygen (δ = 0) Copper Calcium Barium Thallium
Copper, Calcium, Barium, Thallium (at %)
20.69, 13.79, 9.20, 2.30 Ref 126*
Resistivity, ohm m × 10^{-8}
1.4×10^5
1.2×10^5
1×10^5
8×10^4
6×10^4
4×10^4
2×10^4
0×10^0
50
100
150
200
250
300
Temperature, K

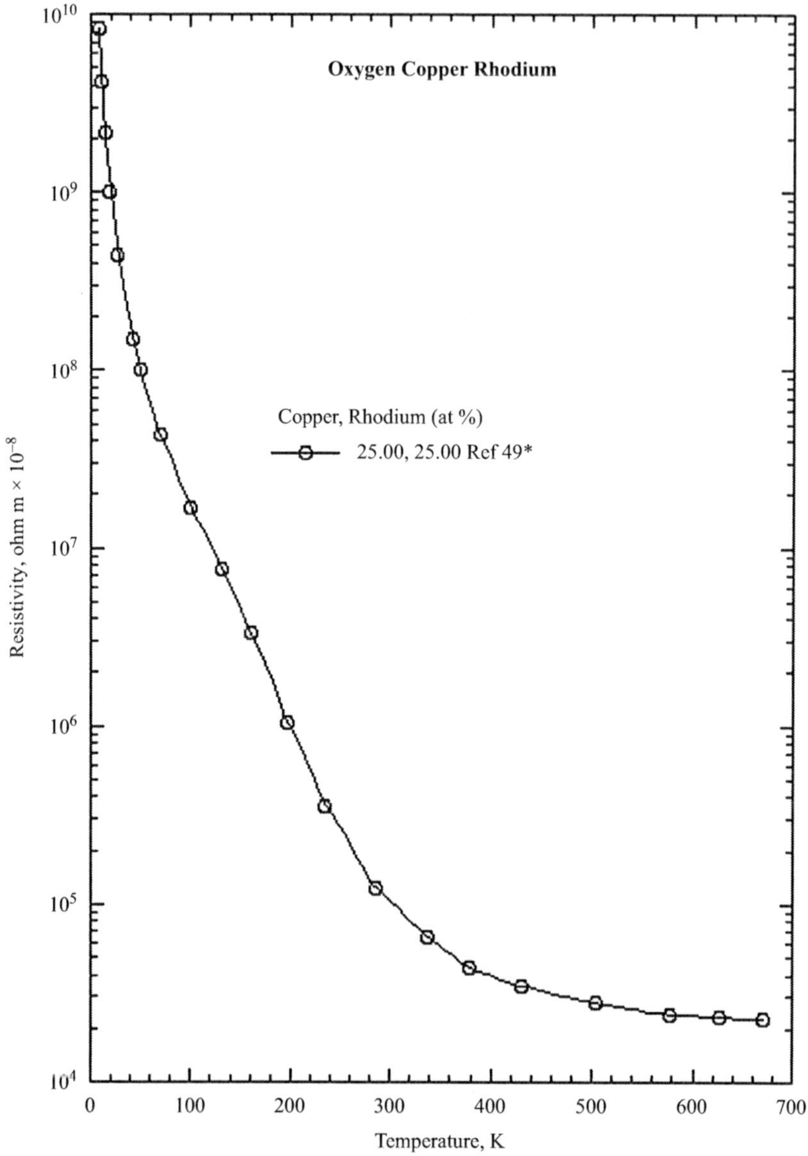
Oxygen Copper Rhodium
Copper, Rhodium (at %)
25.00, 25.00 Ref 49*
Resistivity, ohm m × 10⁻⁸
Temperature, K
10^10
10^9
10^8
10^7
10^6
10^5
10^4
0
100
200
300
400
500
600
700

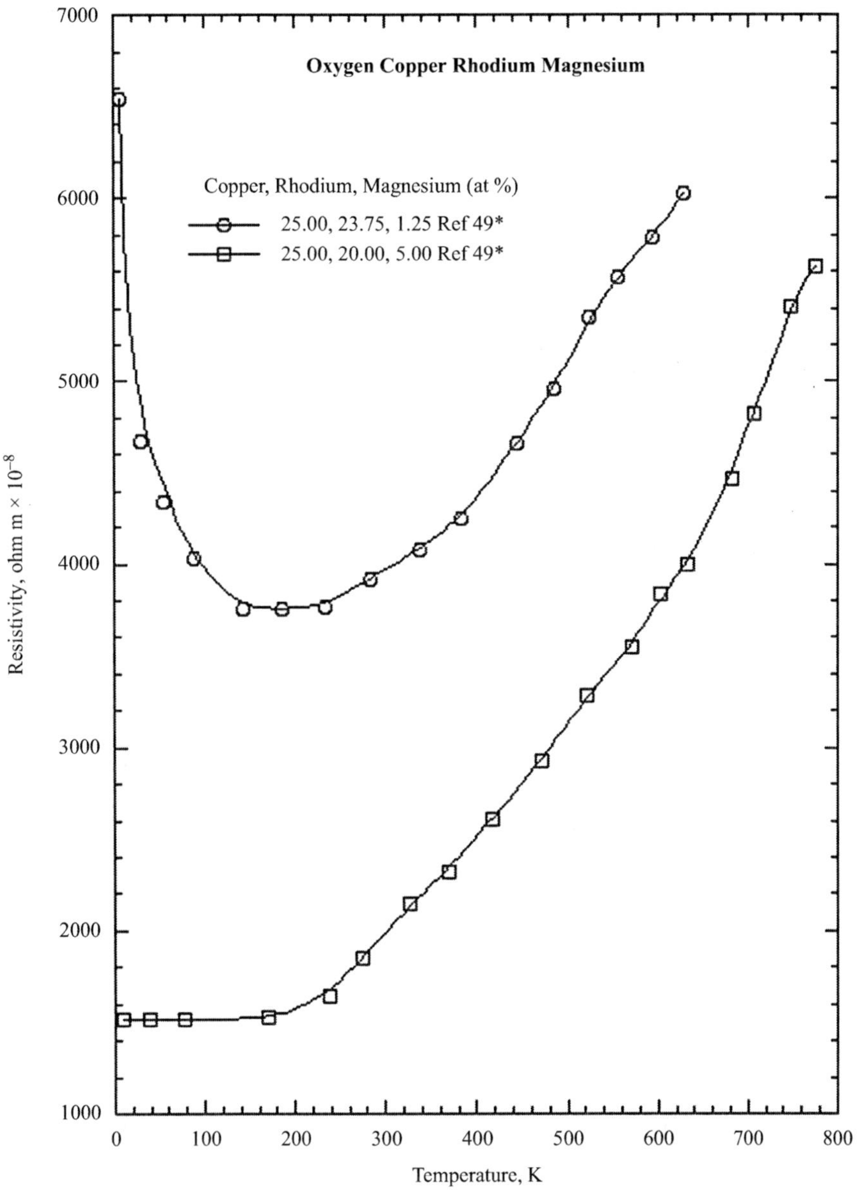
Oxygen Copper Rhodium Magnesium
Copper, Rhodium, Magnesium (at %)
25.00, 23.75, 1.25 Ref 49*
25.00, 20.00, 5.00 Ref 49*
Resistivity, ohm m × 10⁻⁸
Temperature, K
7000
6000
5000
4000
3000
2000
1000
0
100
200
300
400
500
600
700
800

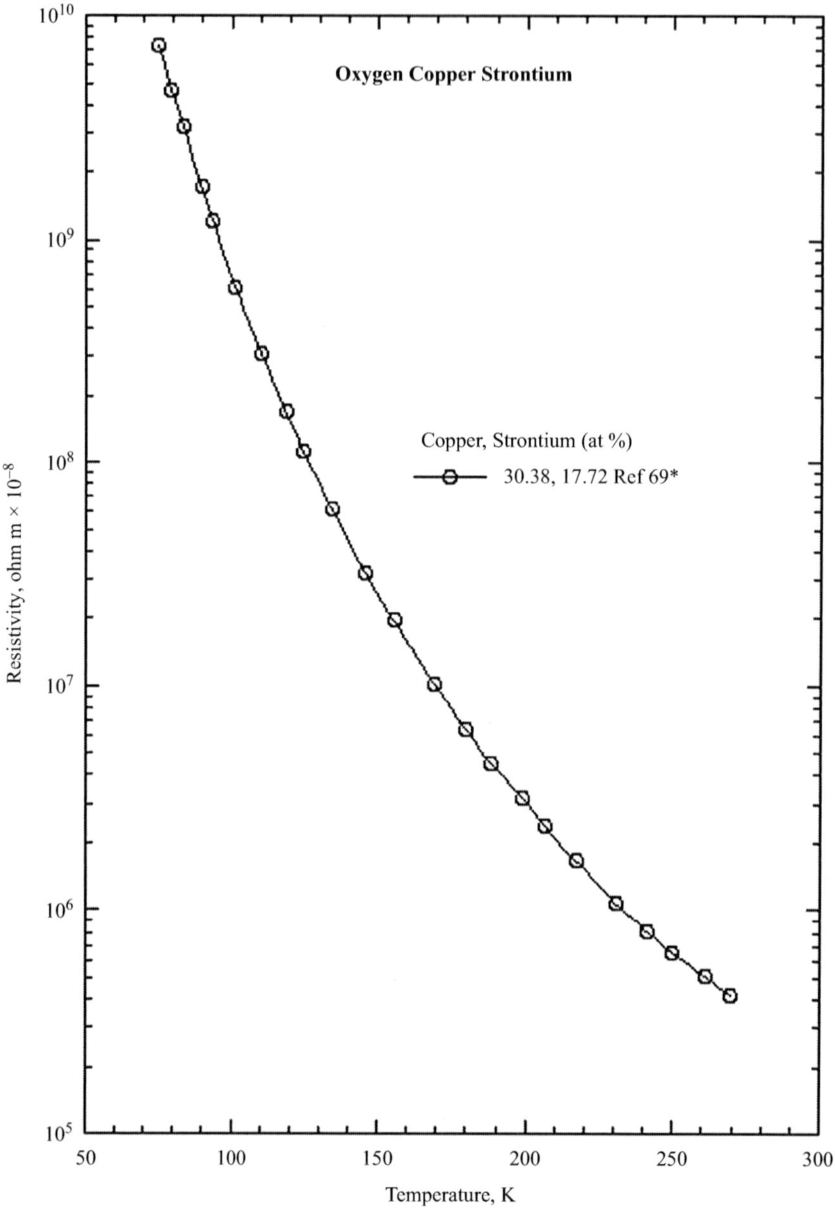

Oxygen Copper Strontium
Copper, Strontium (at %)
30.38, 17.72 Ref 69*
Resistivity, ohm m × 10⁻⁸
Temperature, K
10¹⁰
10⁹
10⁸
10⁷
10⁶
10⁵
50
100
150
200
250
300

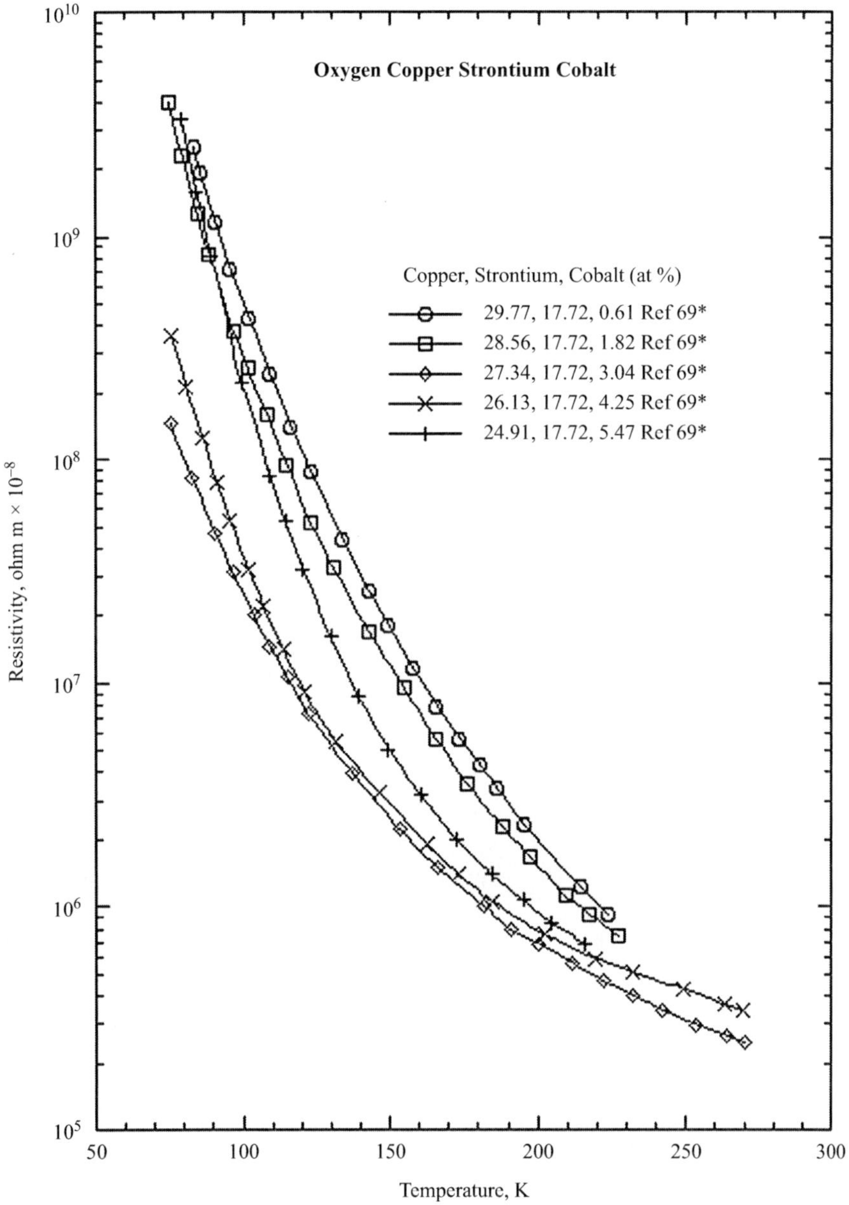
Oxygen Copper Strontium Cobalt
Copper, Strontium, Cobalt (at %)
29.77, 17.72, 0.61 Ref 69*
28.56, 17.72, 1.82 Ref 69*
27.34, 17.72, 3.04 Ref 69*
26.13, 17.72, 4.25 Ref 69*
24.91, 17.72, 5.47 Ref 69*
Resistivity, ohm m × 10⁻⁸
10¹⁰
10⁹
10⁸
10⁷
10⁶
10⁵
50
100
150
200
250
300
Temperature, K

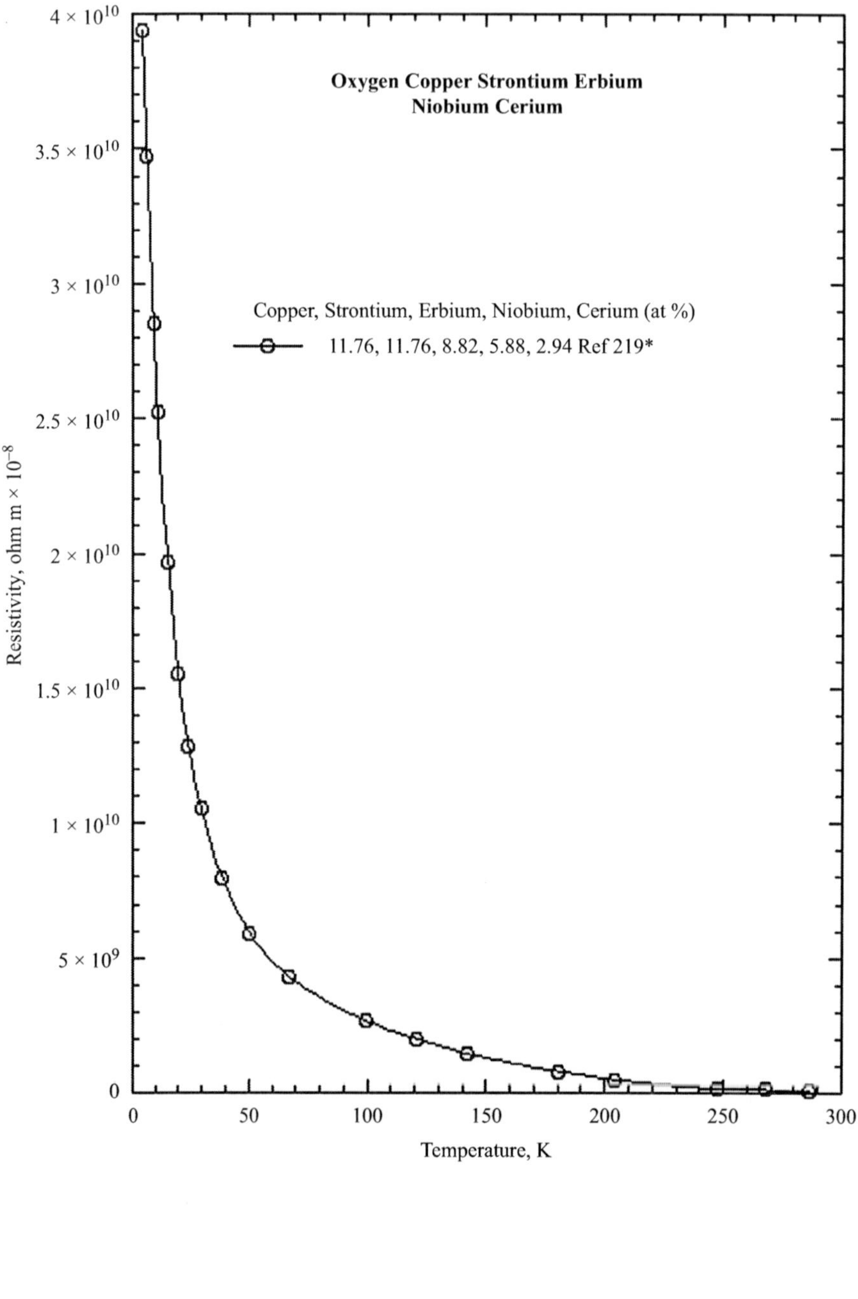

Oxygen Copper Strontium Erbium
Niobium Cerium
Copper, Strontium, Erbium, Niobium, Cerium (at %)
11.76, 11.76, 8.82, 5.88, 2.94 Ref 219*
Resistivity, ohm m × 10^{-8}
Temperature, K
4 × 10^{10}
3.5 × 10^{10}
3 × 10^{10}
2.5 × 10^{10}
2 × 10^{10}
1.5 × 10^{10}
1 × 10^{10}
5 × 10^{9}
0
0
50
100
150
200
250
300

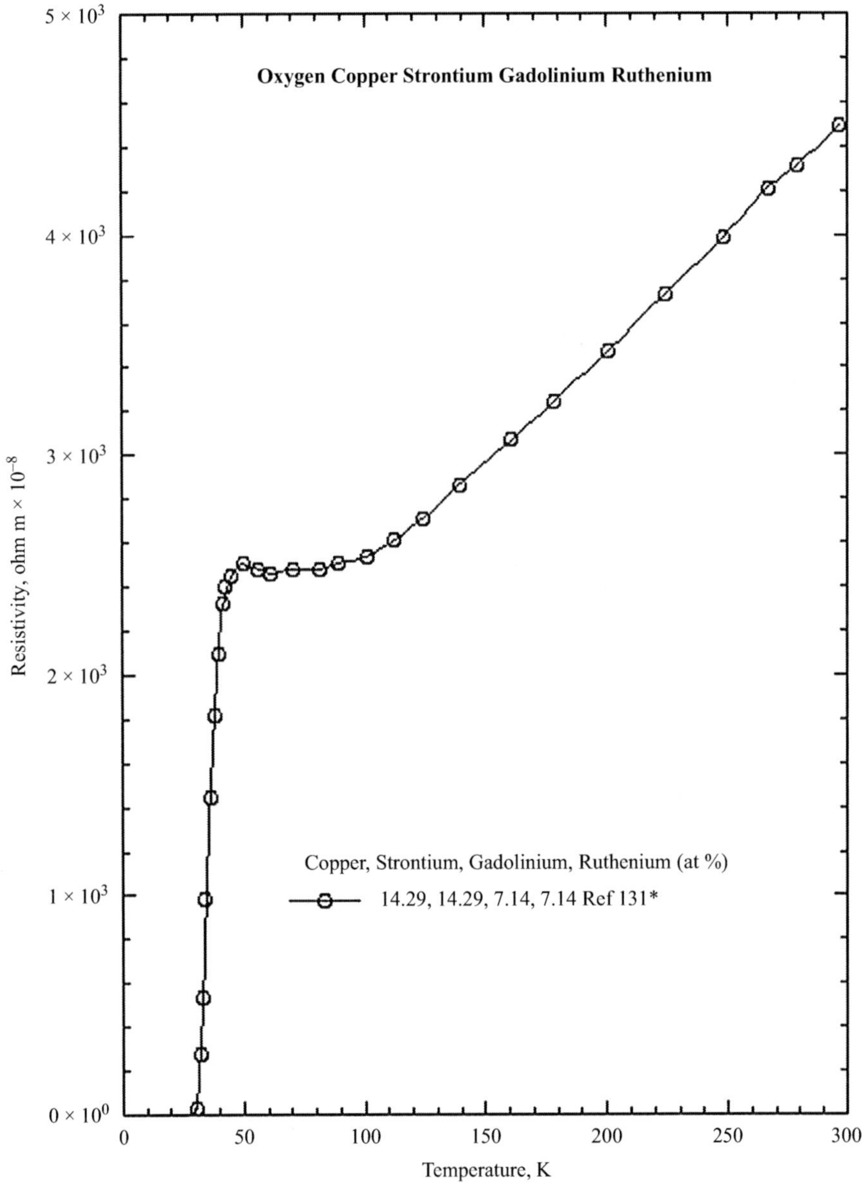
Oxygen Copper Strontium Gadolinium Ruthenium
Copper, Strontium, Gadolinium, Ruthenium (at %)
14.29, 14.29, 7.14, 7.14 Ref 131*
Resistivity, ohm m × 10^{-8}
5×10^3
4×10^3
3×10^3
2×10^3
1×10^3
0×10^0
0
50
100
150
200
250
300
Temperature, K

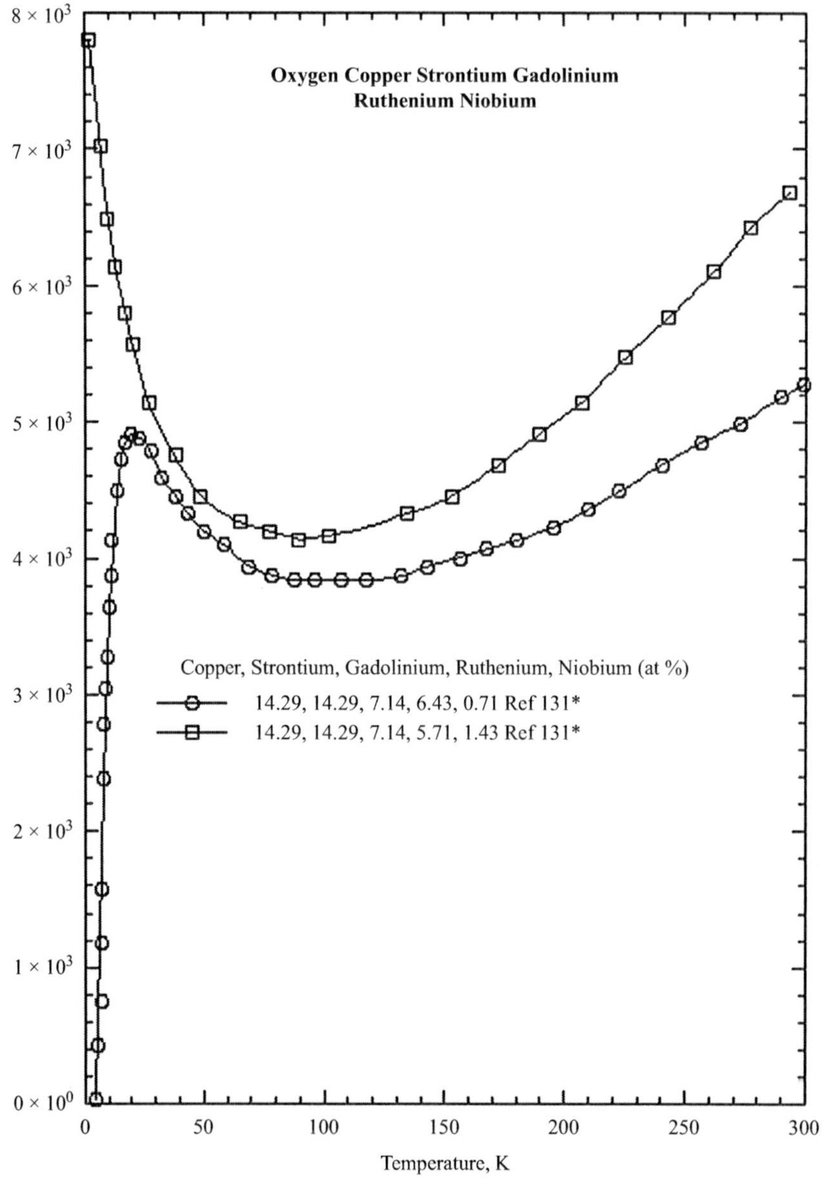

Oxygen Copper Strontium Gadolinium
Ruthenium Niobium
Copper, Strontium, Gadolinium, Ruthenium, Niobium (at %)
14.29, 14.29, 7.14, 6.43, 0.71 Ref 131*
14.29, 14.29, 7.14, 5.71, 1.43 Ref 131*
Resistivity, ohm m × 10⁻⁸
Temperature, K
8 × 10³
7 × 10³
6 × 10³
5 × 10³
4 × 10³
3 × 10³
2 × 10³
1 × 10³
0 × 10⁰
0
50
100
150
200
250
300

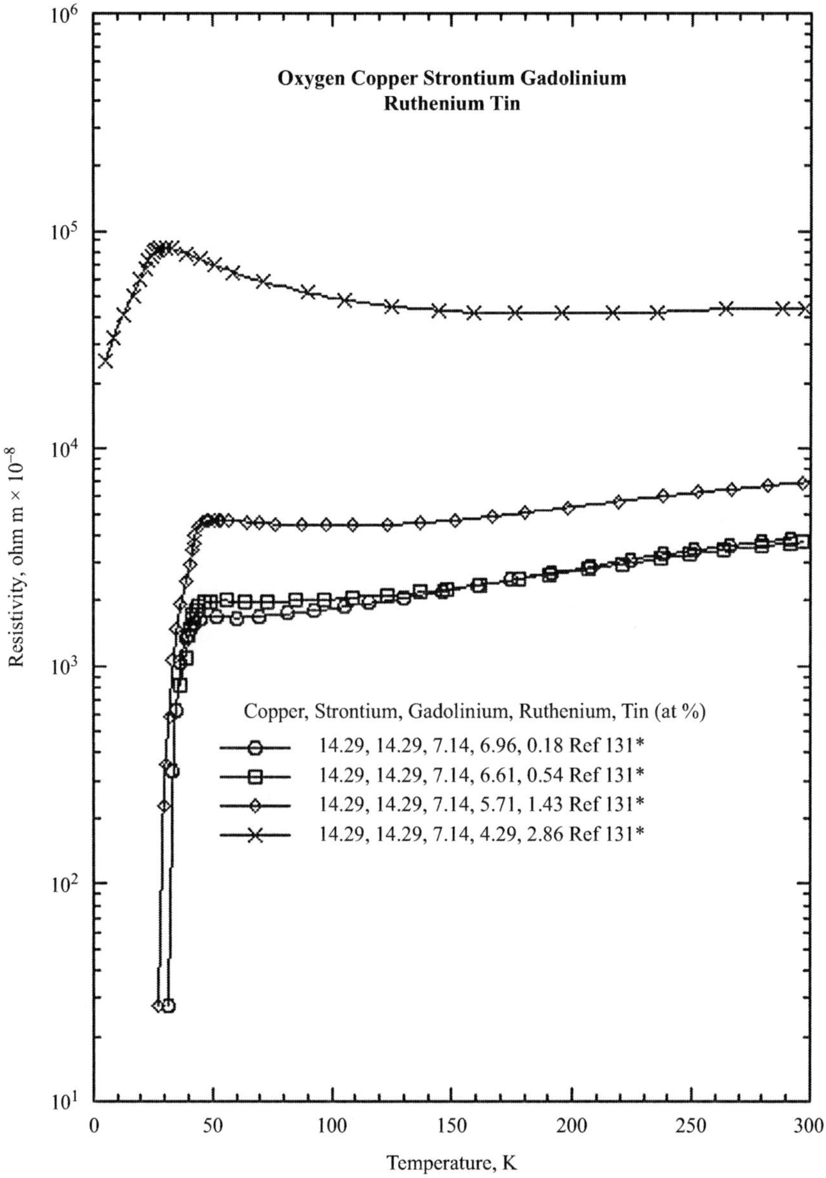
Oxygen Copper Strontium Gadolinium
Ruthenium Tin
Resistivity, ohm m × 10⁻⁸
Temperature, K
Copper, Strontium, Gadolinium, Ruthenium, Tin (at %)
14.29, 14.29, 7.14, 6.96, 0.18 Ref 131*
14.29, 14.29, 7.14, 6.61, 0.54 Ref 131*
14.29, 14.29, 7.14, 5.71, 1.43 Ref 131*
14.29, 14.29, 7.14, 4.29, 2.86 Ref 131*

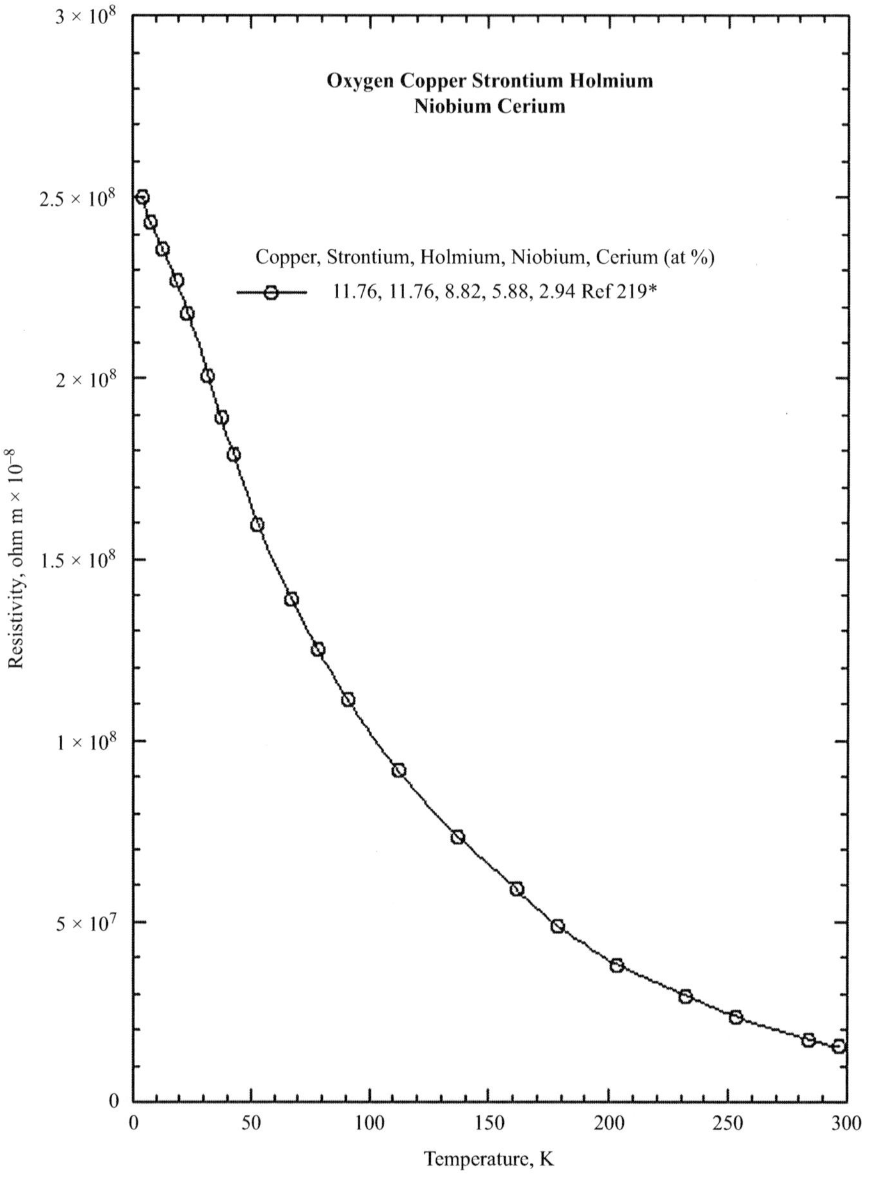
Oxygen Copper Strontium Holmium
Niobium Cerium
Copper, Strontium, Holmium, Niobium, Cerium (at %)
11.76, 11.76, 8.82, 5.88, 2.94 Ref 219*
Resistivity, ohm m × 10⁻⁸
3 × 10⁸
2.5 × 10⁸
2 × 10⁸
1.5 × 10⁸
1 × 10⁸
5 × 10⁷
0
0
50
100
150
200
250
300
Temperature, K

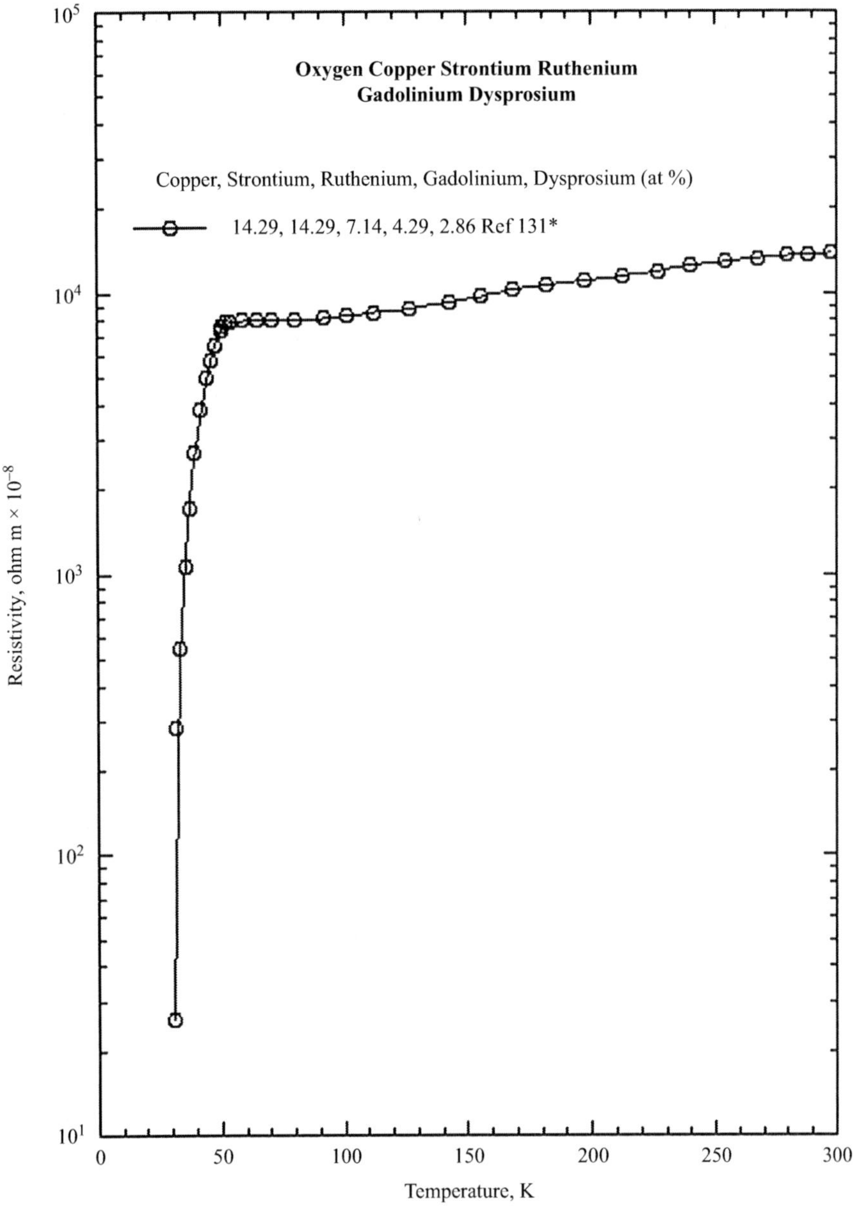
Oxygen Copper Strontium Ruthenium
Gadolinium Dysprosium
Copper, Strontium, Ruthenium, Gadolinium, Dysprosium (at %)
14.29, 14.29, 7.14, 4.29, 2.86 Ref 131*
Resistivity, ohm m × 10−8
Temperature, K
10^5
10^4
10^3
10^2
10^1
0
50
100
150
200
250
300

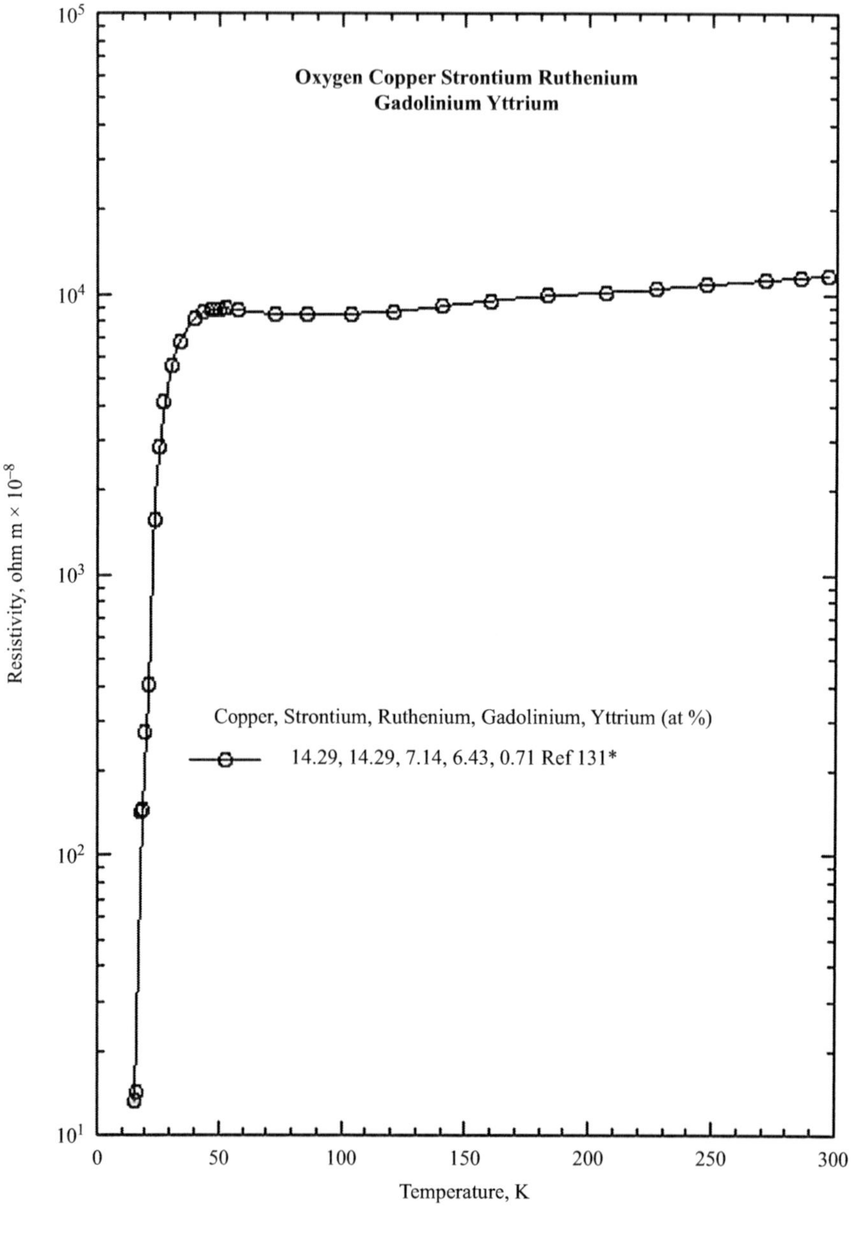
Oxygen Copper Strontium Ruthenium
Gadolinium Yttrium
Copper, Strontium, Ruthenium, Gadolinium, Yttrium (at %)
14.29, 14.29, 7.14, 6.43, 0.71 Ref 131*
Resistivity, ohm m × 10⁻⁸
Temperature, K
10^5
10^4
10^3
10^2
10^1
0
50
100
150
200
250
300

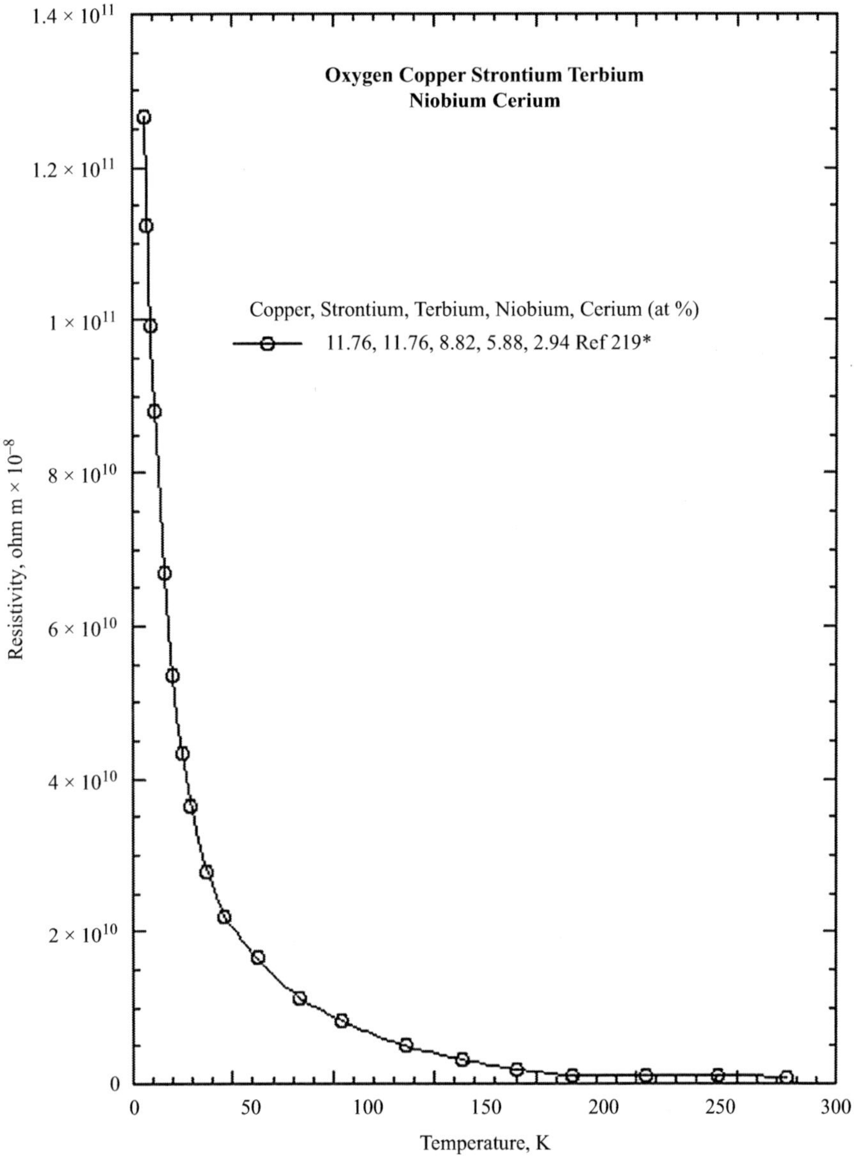
Oxygen Copper Strontium Terbium Niobium Cerium
Copper, Strontium, Terbium, Niobium, Cerium (at %)
11.76, 11.76, 8.82, 5.88, 2.94 Ref 219*
Resistivity, ohm m × 10−8
1.4 × 10^11
1.2 × 10^11
1 × 10^11
8 × 10^10
6 × 10^10
4 × 10^10
2 × 10^10
0
0
50
100
150
200
250
300
Temperature, K

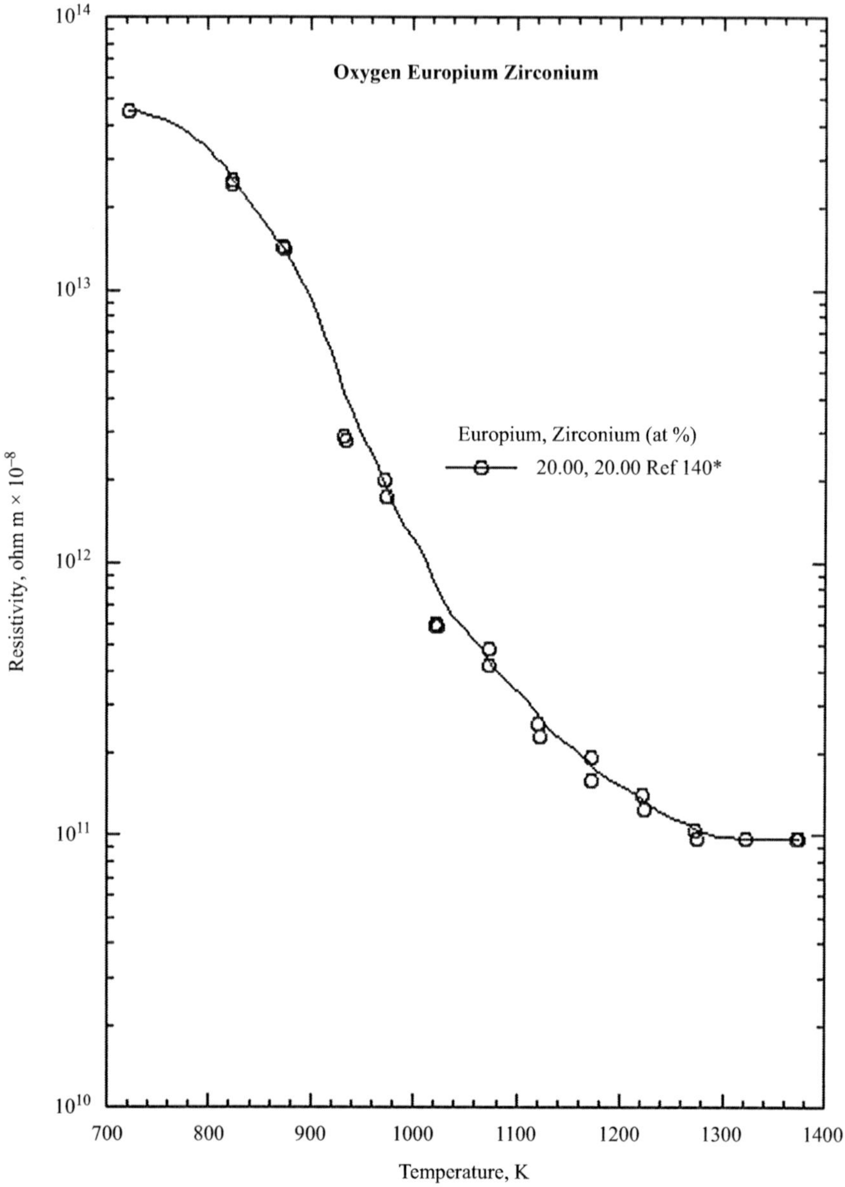
Oxygen Europium Zirconium
Europium, Zirconium (at %)
20.00, 20.00 Ref 140*
Resistivity, ohm m × 10⁻⁸
10^14
10^13
10^12
10^11
10^10
700
800
900
1000
1100
1200
1300
1400
Temperature, K

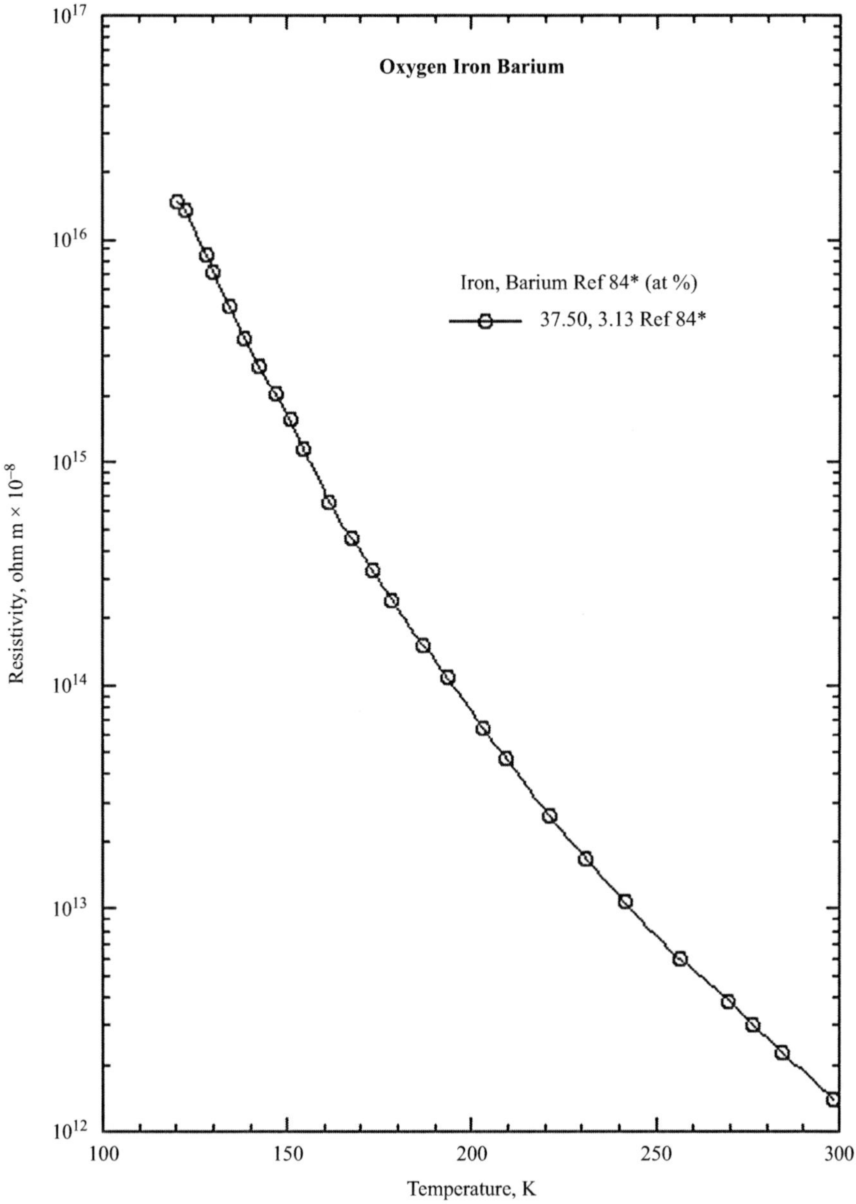
Oxygen Iron Barium
Iron, Barium Ref 84* (at %)
37.50, 3.13 Ref 84*
Resistivity, ohm m × 10⁻⁸
Temperature, K
10^17
10^16
10^15
10^14
10^13
10^12
100
150
200
250
300

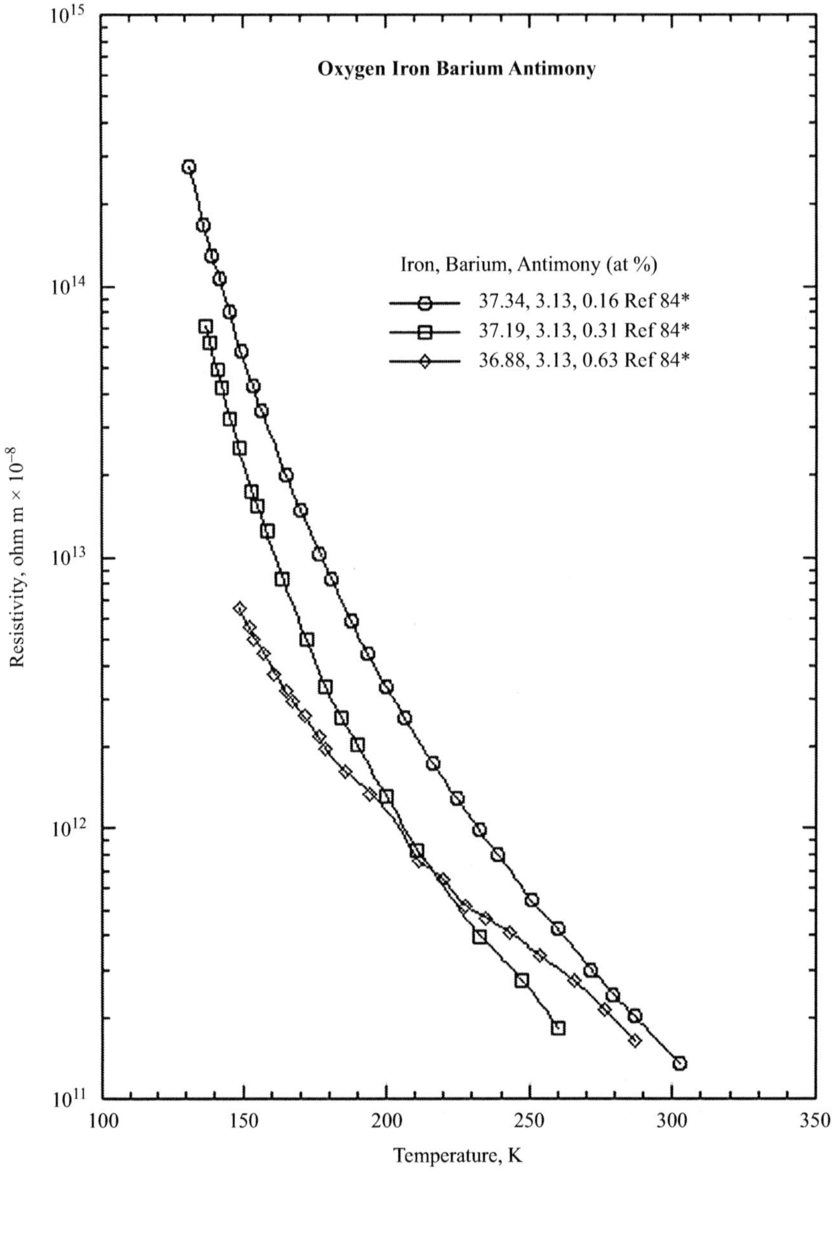
Oxygen Iron Barium Antimony
Iron, Barium, Antimony (at %)
37.34, 3.13, 0.16 Ref 84*
37.19, 3.13, 0.31 Ref 84*
36.88, 3.13, 0.63 Ref 84*
Resistivity, ohm m × 10⁻⁸
Temperature, K
10^15
10^14
10^13
10^12
10^11
100
150
200
250
300
350

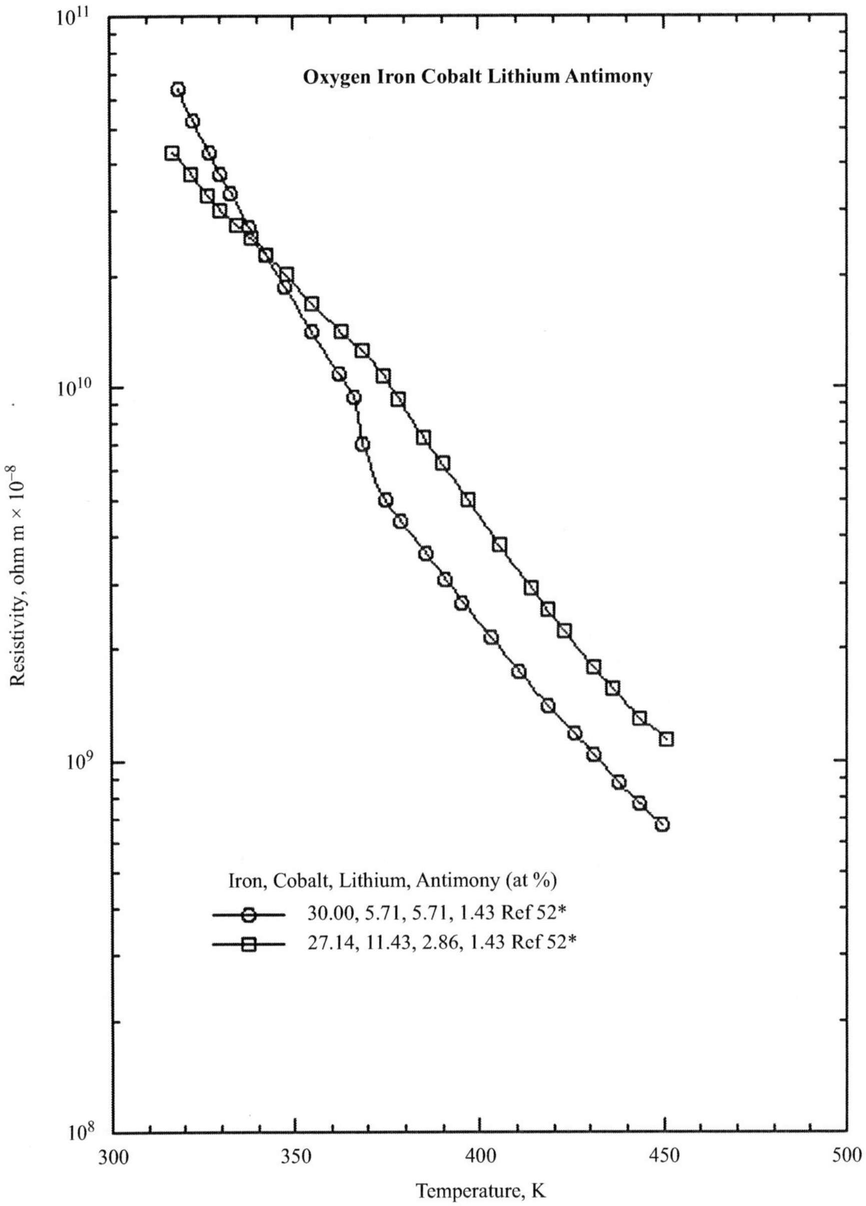

Oxygen Iron Cobalt Lithium Antimony
Iron, Cobalt, Lithium, Antimony (at %)
30.00, 5.71, 5.71, 1.43 Ref 52*
27.14, 11.43, 2.86, 1.43 Ref 52*
Resistivity, ohm m × 10⁻⁸
Temperature, K

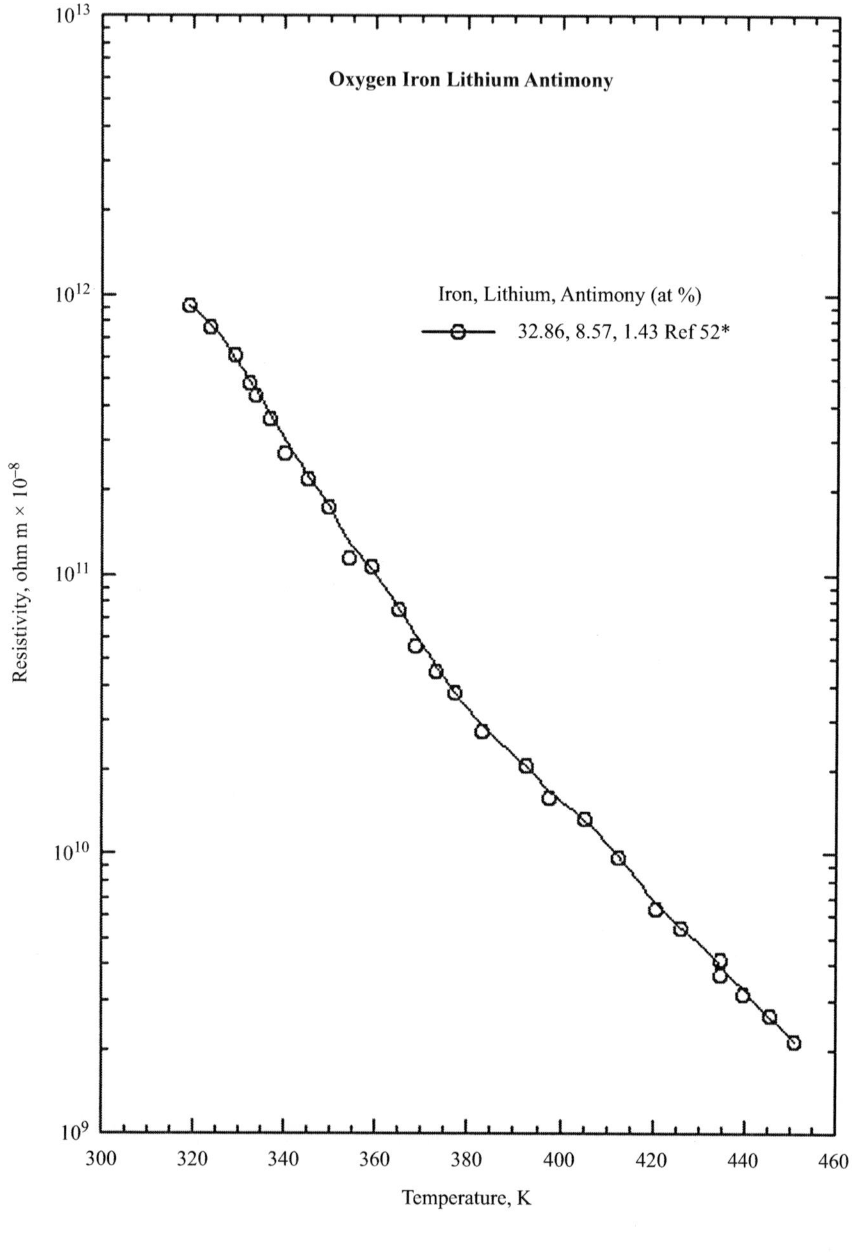
Oxygen Iron Lithium Antimony
Iron, Lithium, Antimony (at %)
32.86, 8.57, 1.43 Ref 52*
Resistivity, ohm m × 10⁻⁸
Temperature, K
10^13
10^12
10^11
10^10
10^9
300
320
340
360
380
400
420
440
460

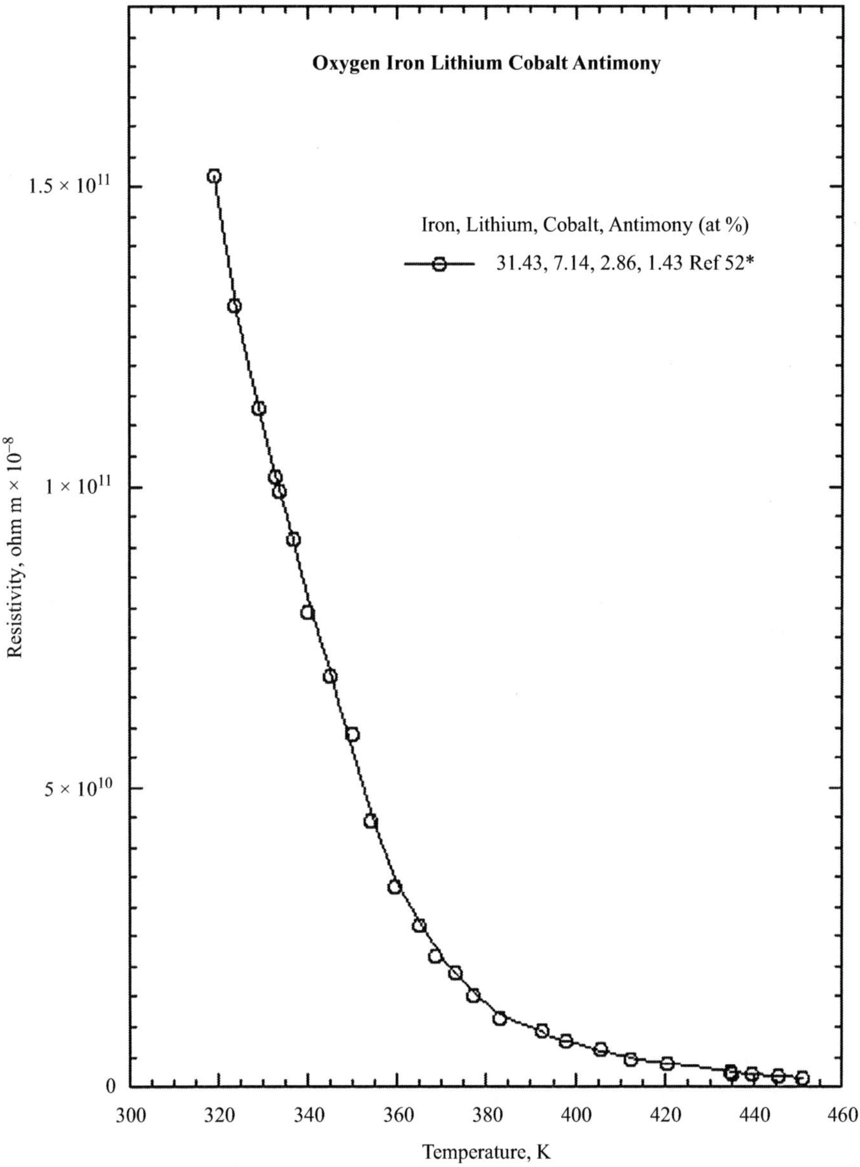

Oxygen Iron Lithium Cobalt Antimony
Iron, Lithium, Cobalt, Antimony (at %)
31.43, 7.14, 2.86, 1.43 Ref 52*
Resistivity, ohm m × 10⁻⁸
1.5 × 10¹¹
1 × 10¹¹
5 × 10¹⁰
0
300
320
340
360
380
400
420
440
460
Temperature, K

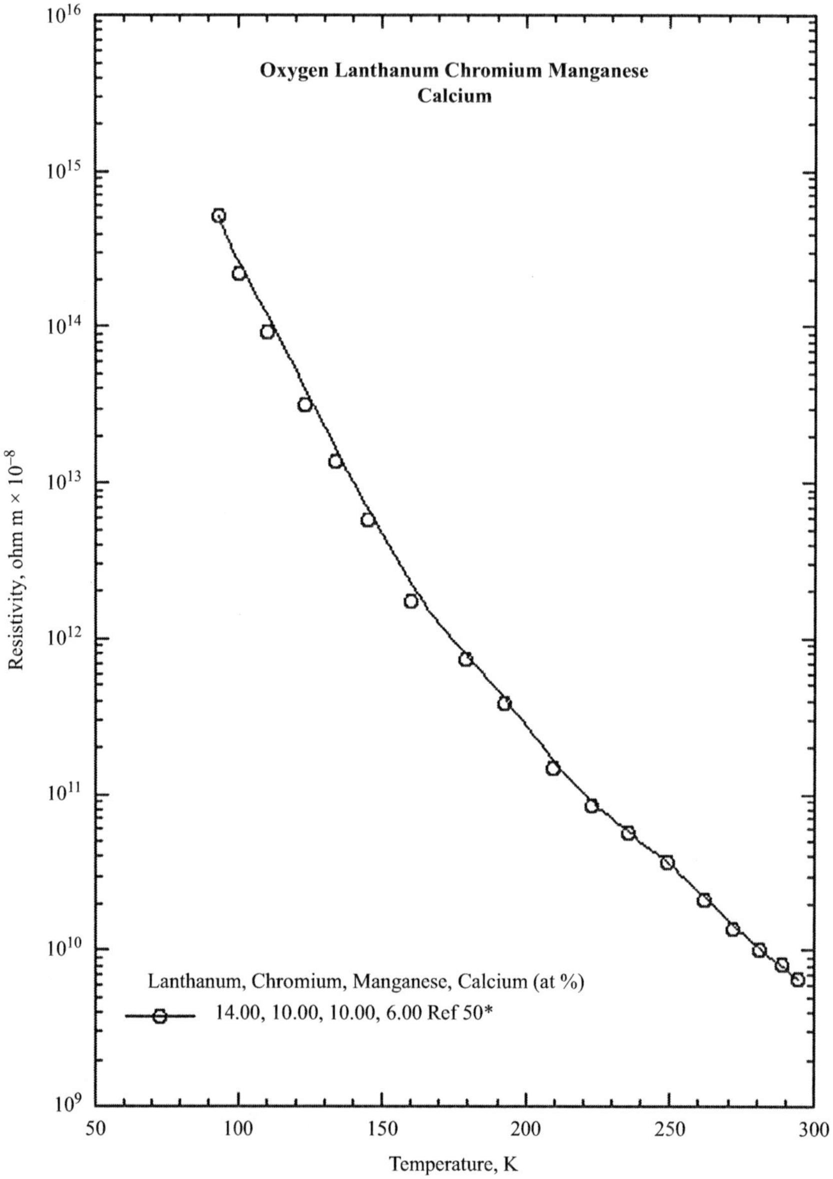

Oxygen Lanthanum Chromium Manganese
Calcium
Resistivity, ohm m × 10⁻⁸
10^16
10^15
10^14
10^13
10^12
10^11
10^10
10^9
50
100
150
200
250
300
Temperature, K
Lanthanum, Chromium, Manganese, Calcium (at %)
14.00, 10.00, 10.00, 6.00 Ref 50*

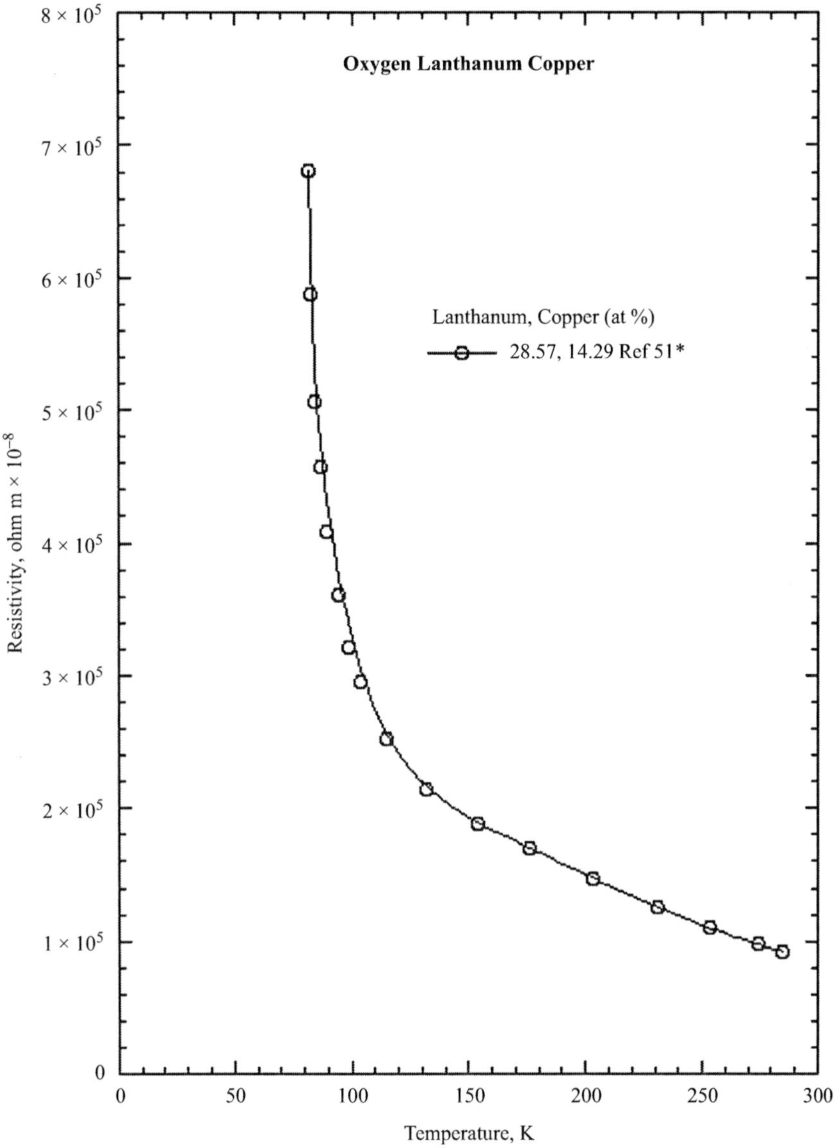

Oxygen Lanthanum Copper
Lanthanum, Copper (at %)
28.57, 14.29 Ref 51*
Resistivity, ohm m × 10^{-8}
Temperature, K
8 × 10^5
7 × 10^5
6 × 10^5
5 × 10^5
4 × 10^5
3 × 10^5
2 × 10^5
1 × 10^5
0
0
50
100
150
200
250
300

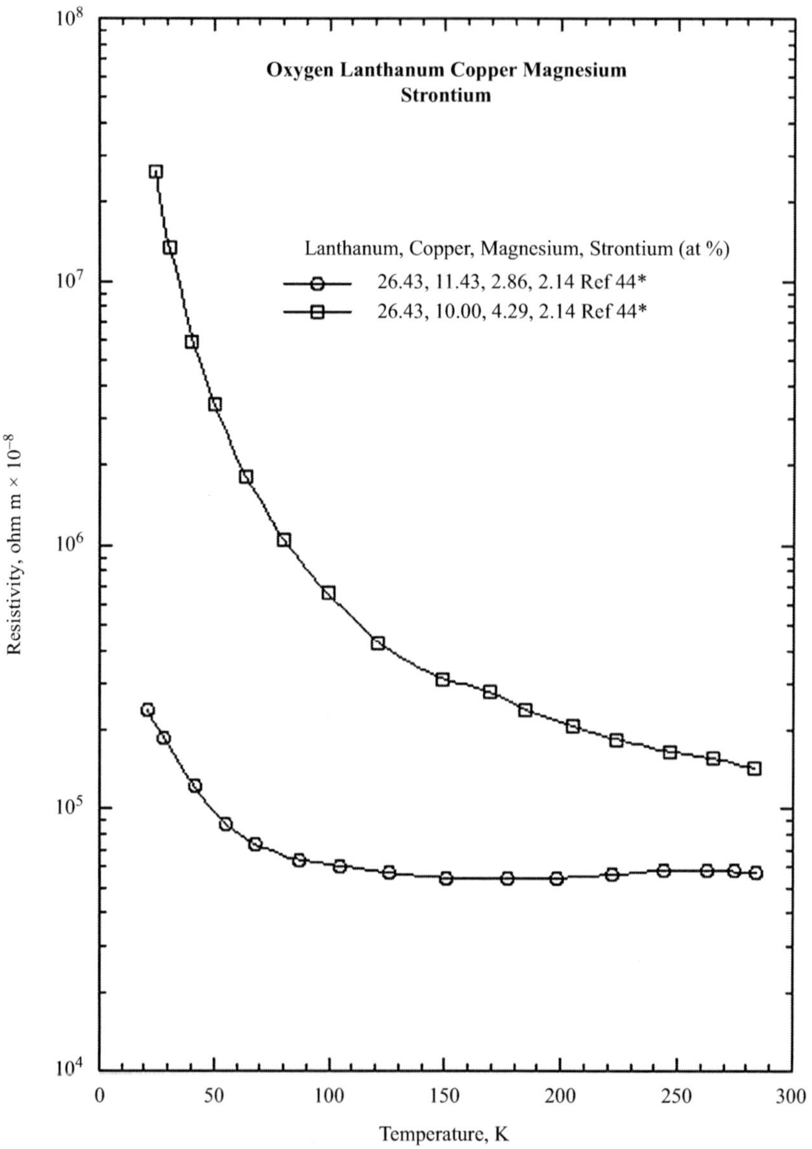
Oxygen Lanthanum Copper Magnesium
Strontium
Lanthanum, Copper, Magnesium, Strontium (at %)
26.43, 11.43, 2.86, 2.14 Ref 44*
26.43, 10.00, 4.29, 2.14 Ref 44*
Resistivity, ohm m × 10⁻⁸
Temperature, K
10⁸
10⁷
10⁶
10⁵
10⁴
0
50
100
150
200
250
300

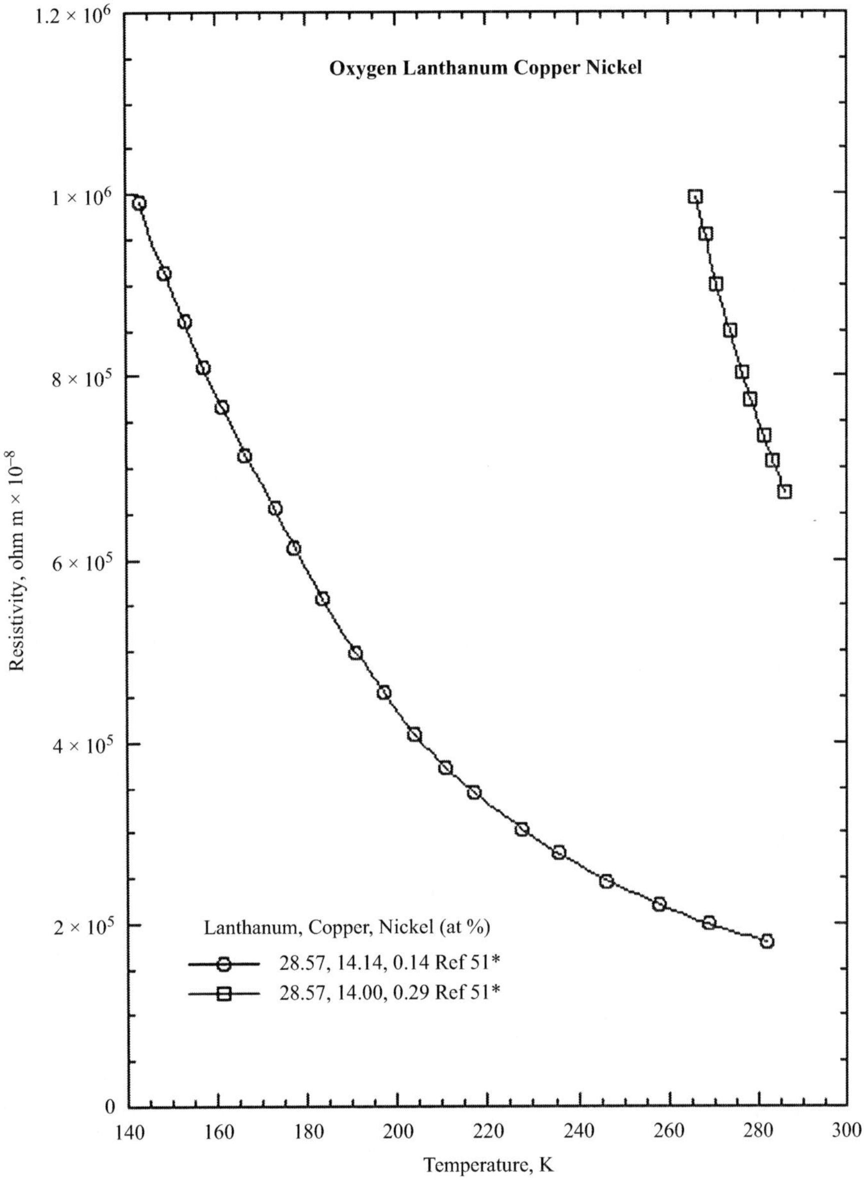

Oxygen Lanthanum Copper Nickel
Resistivity, ohm m × 10⁻⁸
1.2 × 10⁶
1 × 10⁶
8 × 10⁵
6 × 10⁵
4 × 10⁵
2 × 10⁵
0
140
160
180
200
220
240
260
280
300
Temperature, K
Lanthanum, Copper, Nickel (at %)
28.57, 14.14, 0.14 Ref 51*
28.57, 14.00, 0.29 Ref 51*

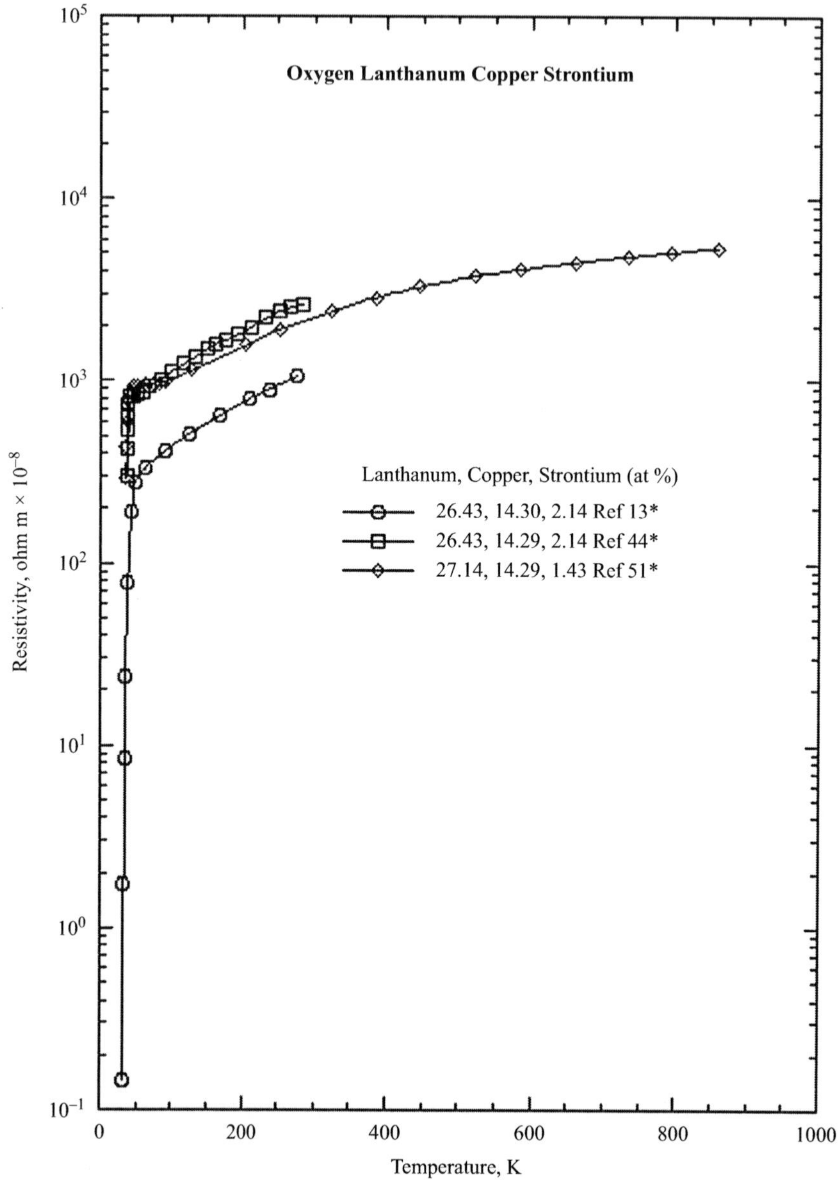

Oxygen Lanthanum Copper Strontium
Lanthanum, Copper, Strontium (at %)
26.43, 14.30, 2.14 Ref 13*
26.43, 14.29, 2.14 Ref 44*
27.14, 14.29, 1.43 Ref 51*
Resistivity, ohm m × 10⁻⁸
Temperature, K

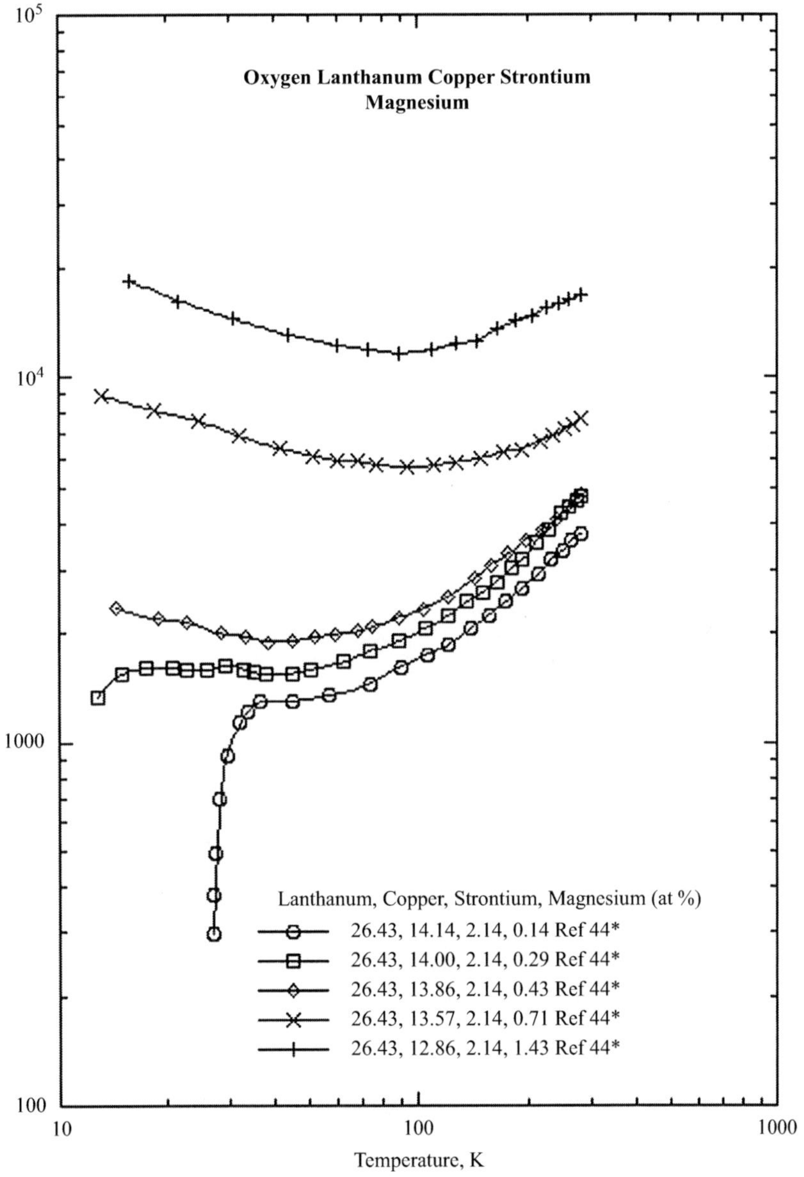
Oxygen Lanthanum Copper Strontium Magnesium
Resistivity, ohm m × 10⁻⁸
Temperature, K
10⁵
10⁴
1000
100
10
100
1000
Lanthanum, Copper, Strontium, Magnesium (at %)
26.43, 14.14, 2.14, 0.14 Ref 44*
26.43, 14.00, 2.14, 0.29 Ref 44*
26.43, 13.86, 2.14, 0.43 Ref 44*
26.43, 13.57, 2.14, 0.71 Ref 44*
26.43, 12.86, 2.14, 1.43 Ref 44*

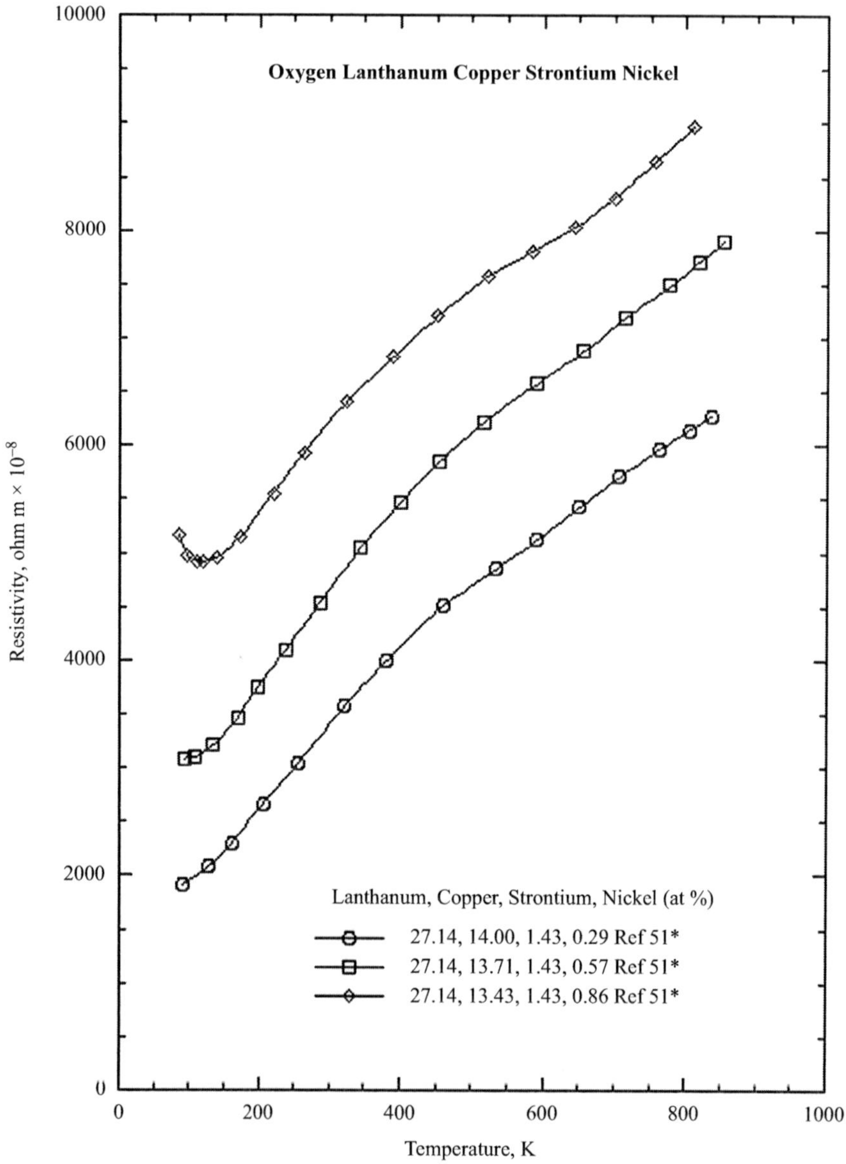
Oxygen Lanthanum Copper Strontium Nickel
Resistivity, ohm m × 10^{-8}
Temperature, K
Lanthanum, Copper, Strontium, Nickel (at %)
27.14, 14.00, 1.43, 0.29 Ref 51*
27.14, 13.71, 1.43, 0.57 Ref 51*
27.14, 13.43, 1.43, 0.86 Ref 51*
0
2000
4000
6000
8000
10000
0
200
400
600
800
1000

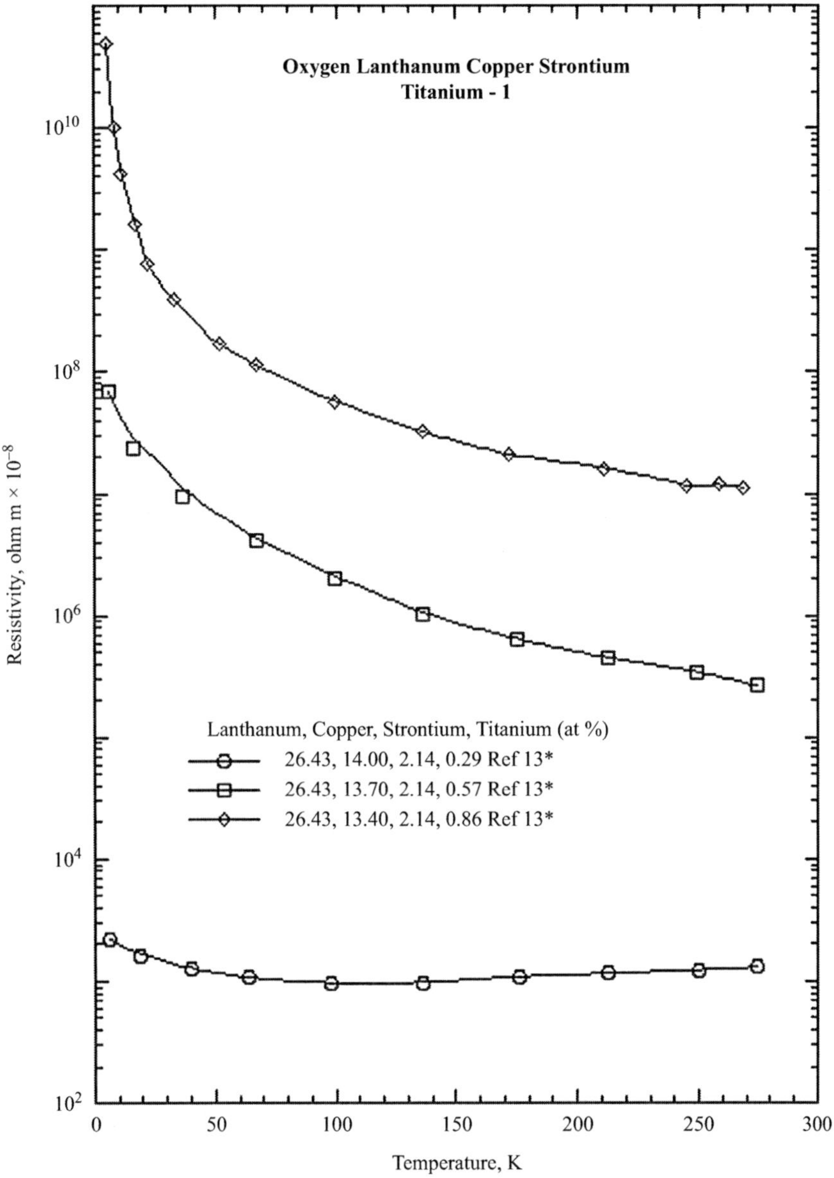

Oxygen Lanthanum Copper Strontium
Titanium - 1
Lanthanum, Copper, Strontium, Titanium (at %)
26.43, 14.00, 2.14, 0.29 Ref 13*
26.43, 13.70, 2.14, 0.57 Ref 13*
26.43, 13.40, 2.14, 0.86 Ref 13*
Resistivity, ohm m × 10^−8
10^10
10^8
10^6
10^4
10^2
0
50
100
150
200
250
300
Temperature, K

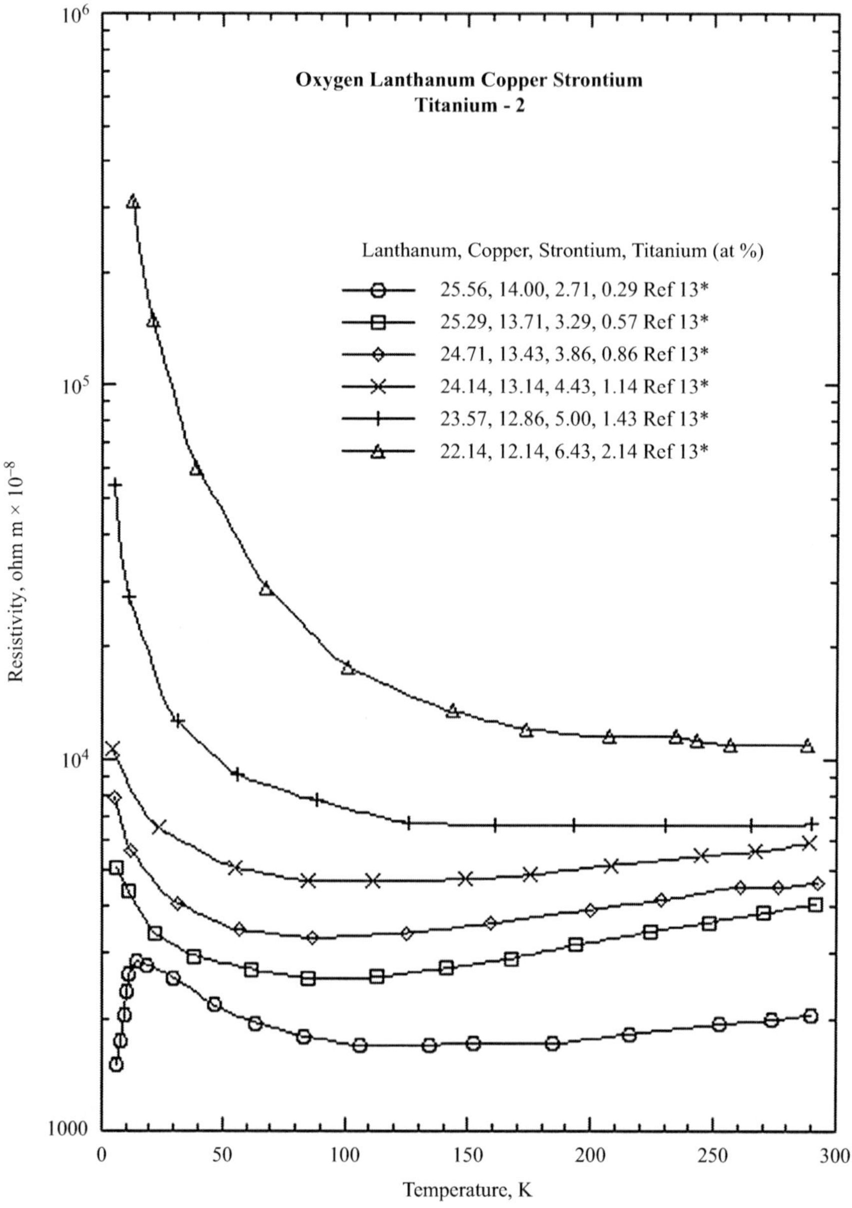
Oxygen Lanthanum Copper Strontium
Titanium - 2
Lanthanum, Copper, Strontium, Titanium (at %)
25.56, 14.00, 2.71, 0.29 Ref 13*
25.29, 13.71, 3.29, 0.57 Ref 13*
24.71, 13.43, 3.86, 0.86 Ref 13*
24.14, 13.14, 4.43, 1.14 Ref 13*
23.57, 12.86, 5.00, 1.43 Ref 13*
22.14, 12.14, 6.43, 2.14 Ref 13*
Resistivity, ohm m × 10⁻⁸
Temperature, K
10⁶
10⁵
10⁴
1000
0
50
100
150
200
250
300

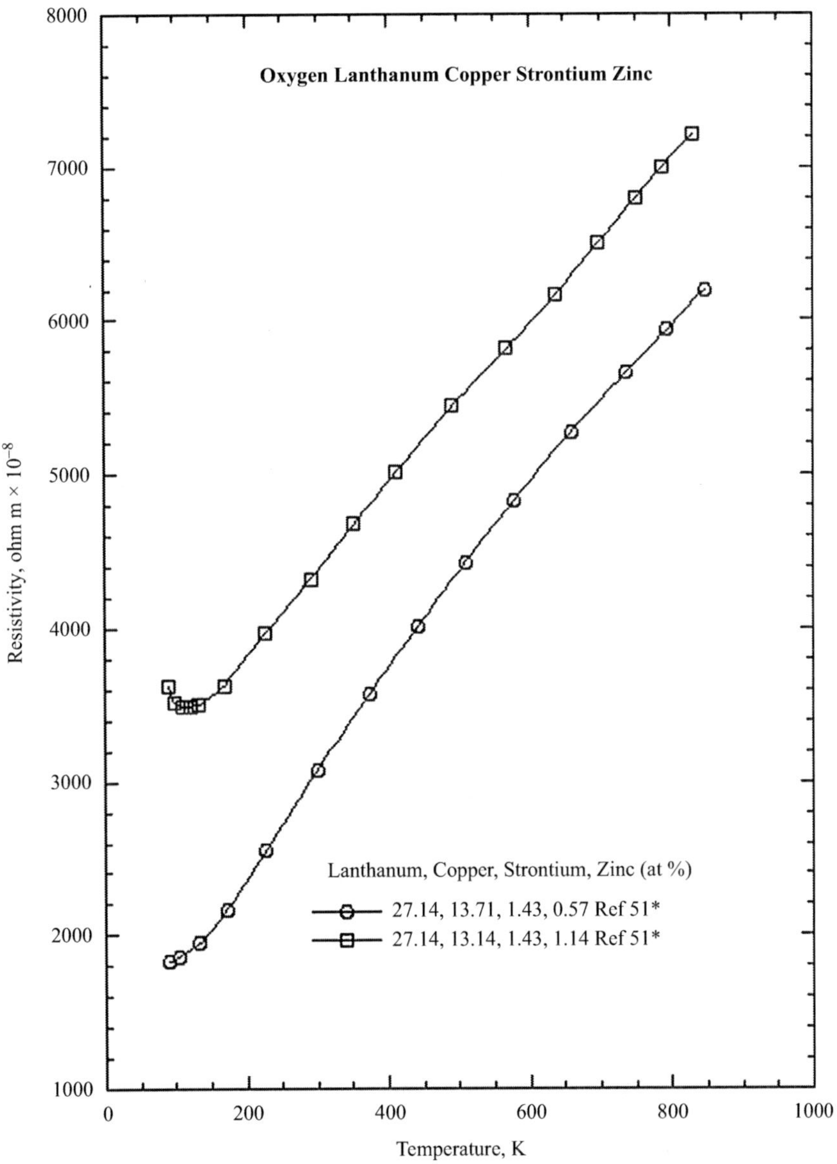
Oxygen Lanthanum Copper Strontium Zinc
Lanthanum, Copper, Strontium, Zinc (at %)
27.14, 13.71, 1.43, 0.57 Ref 51*
27.14, 13.14, 1.43, 1.14 Ref 51*
Resistivity, ohm m × 10⁻⁸
Temperature, K
8000
7000
6000
5000
4000
3000
2000
1000
0
200
400
600
800
1000

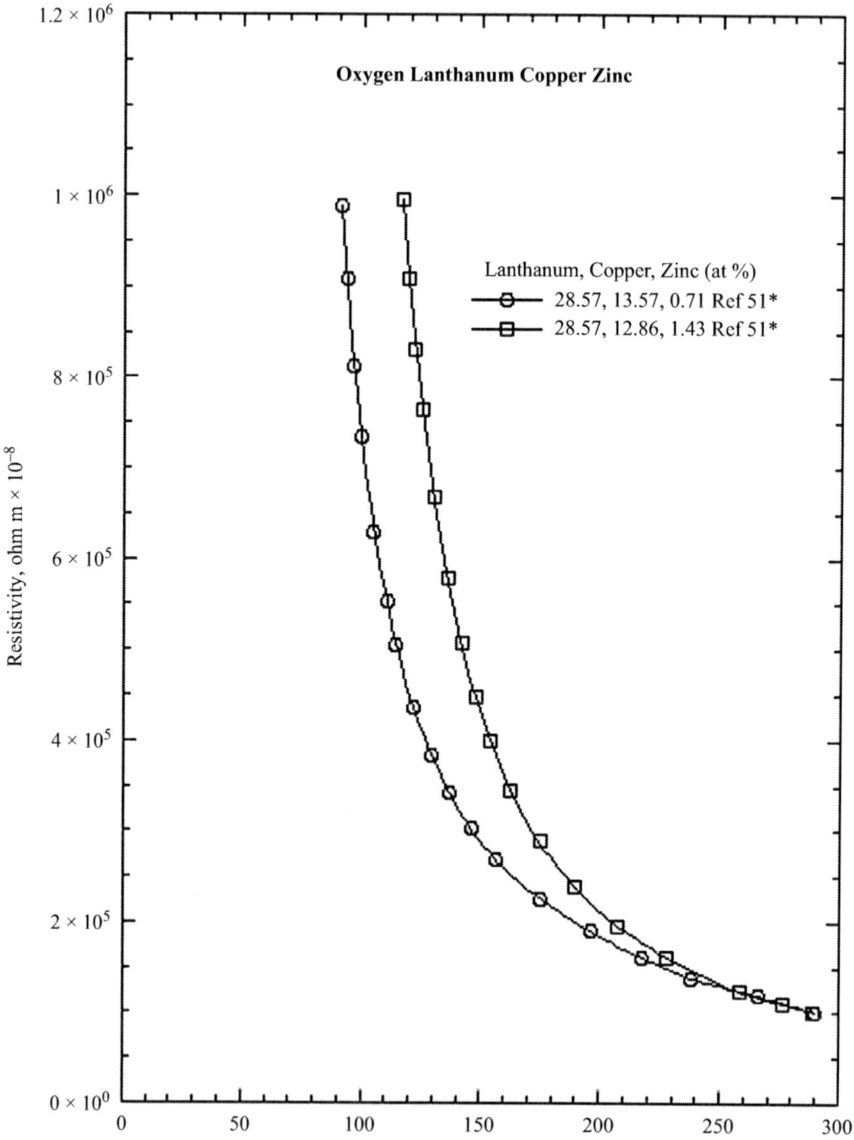
Oxygen Lanthanum Copper Zinc
Lanthanum, Copper, Zinc (at %)
28.57, 13.57, 0.71 Ref 51*
28.57, 12.86, 1.43 Ref 51*
Resistivity, ohm m × 10^-8
1.2 × 10^6
1 × 10^6
8 × 10^5
6 × 10^5
4 × 10^5
2 × 10^5
0 × 10^0
0
50
100
150
200
250
300
Temperature, K

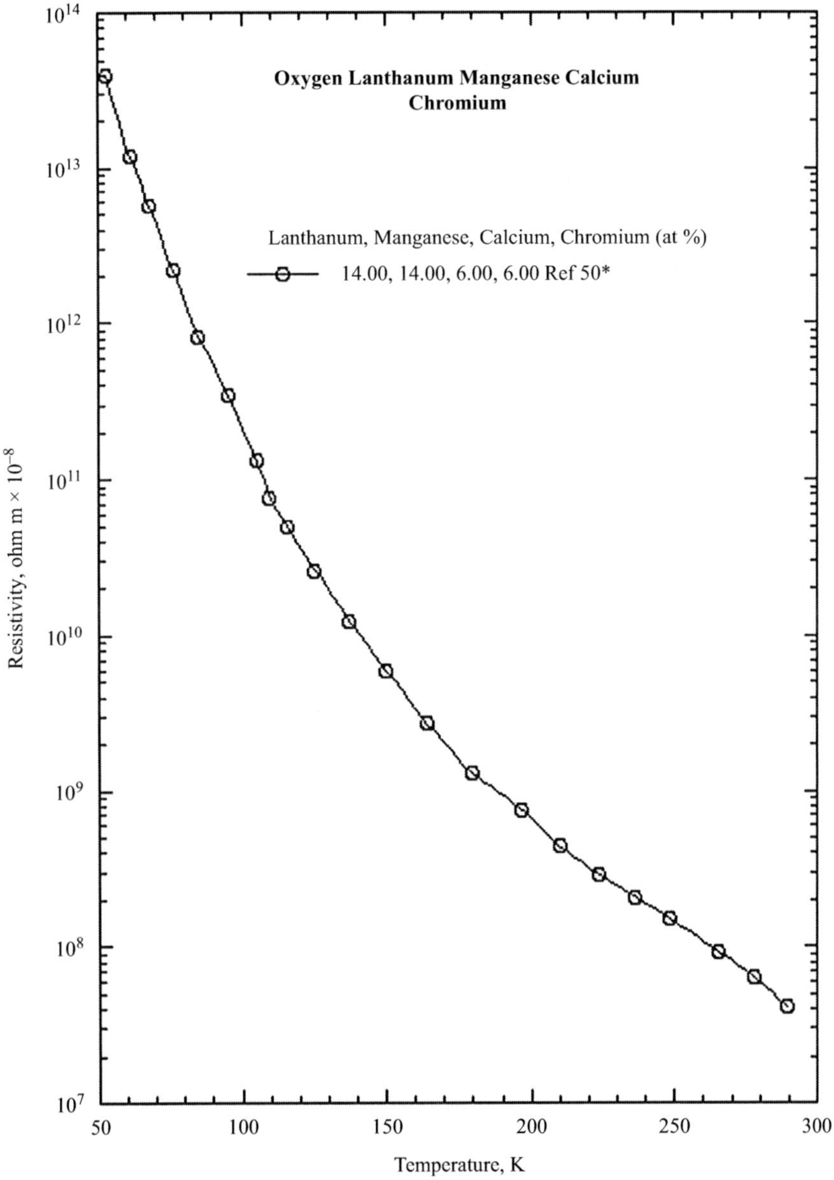
Oxygen Lanthanum Manganese Calcium Chromium
Lanthanum, Manganese, Calcium, Chromium (at %)
14.00, 14.00, 6.00, 6.00 Ref 50*
Resistivity, ohm m × 10−8
Temperature, K
10^14
10^13
10^12
10^11
10^10
10^9
10^8
10^7
50
100
150
200
250
300

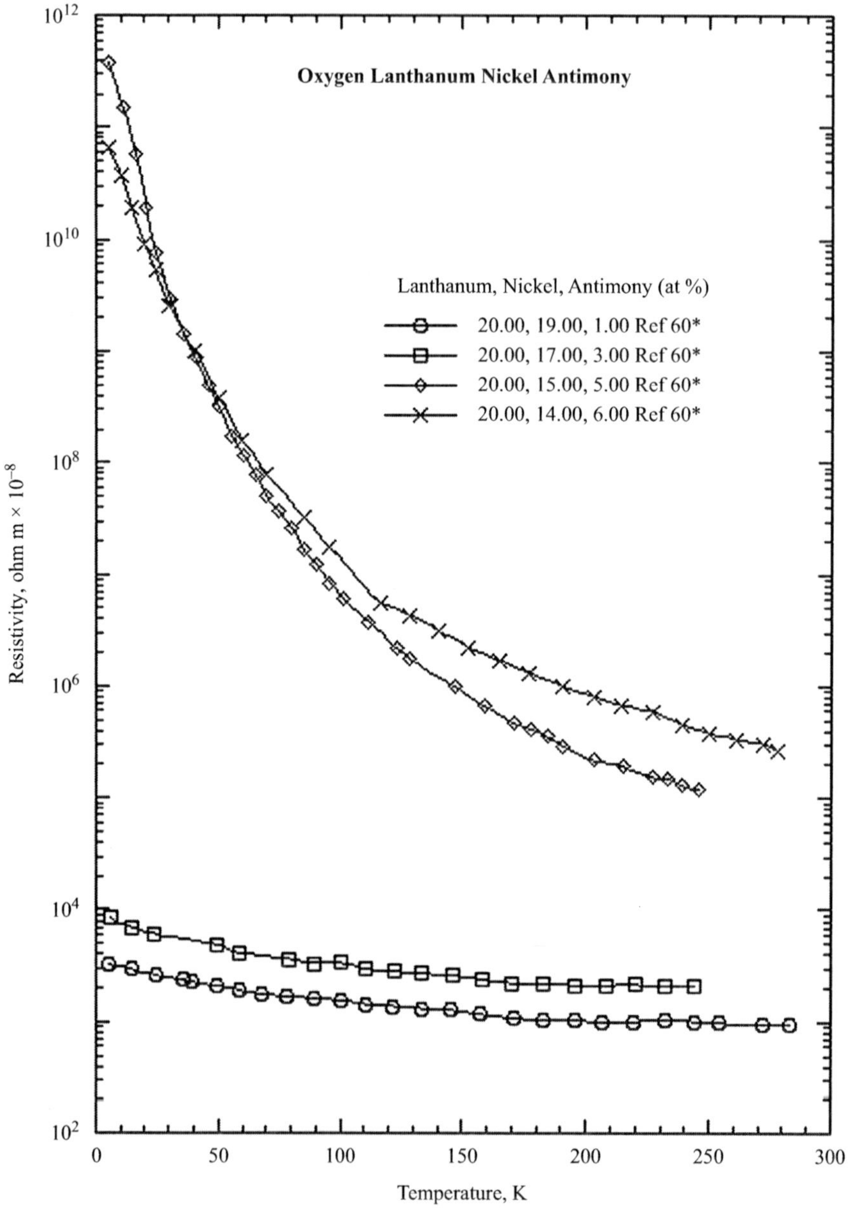
Oxygen Lanthanum Nickel Antimony
Lanthanum, Nickel, Antimony (at %)
20.00, 19.00, 1.00 Ref 60*
20.00, 17.00, 3.00 Ref 60*
20.00, 15.00, 5.00 Ref 60*
20.00, 14.00, 6.00 Ref 60*
Resistivity, ohm m × 10⁻⁸
Temperature, K
0
50
100
150
200
250
300
10²
10⁴
10⁶
10⁸
10¹⁰
10¹²

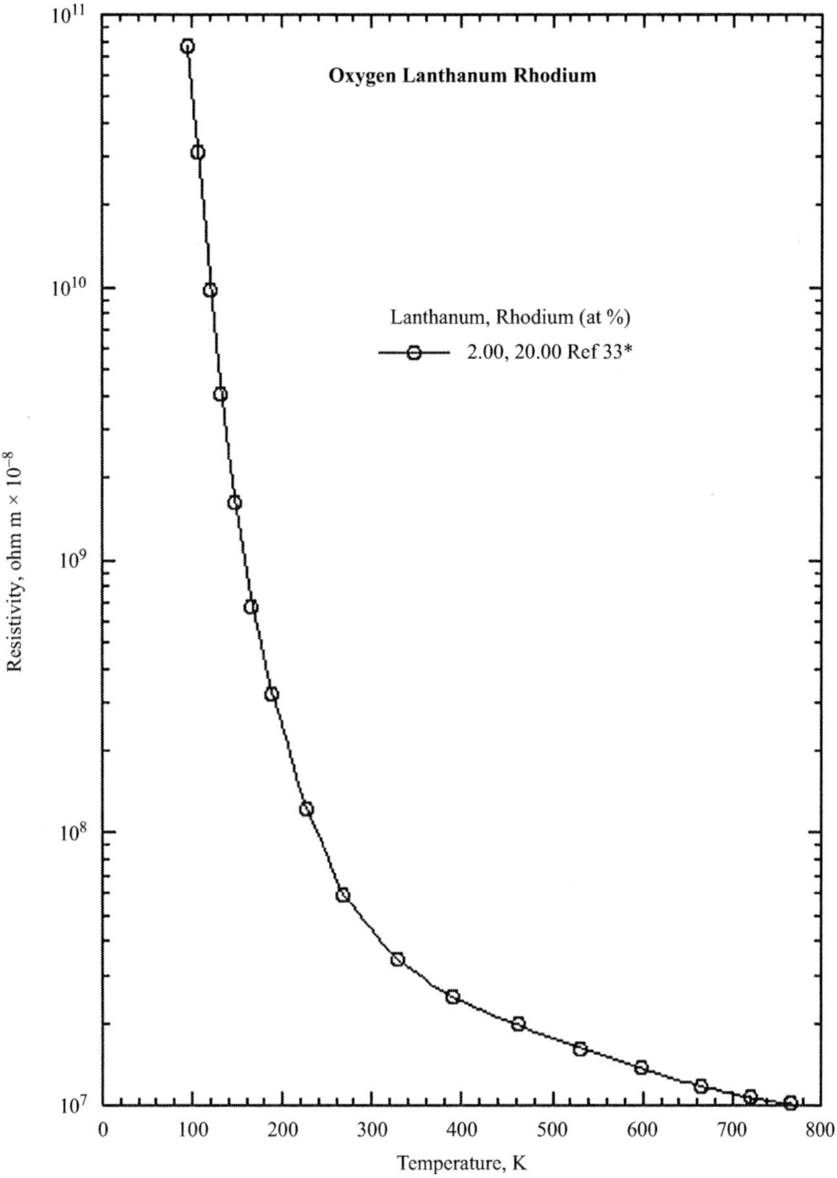

Oxygen Lanthanum Rhodium
Lanthanum, Rhodium (at %)
2.00, 20.00 Ref 33*
Resistivity, ohm m × 10^{-8}
Temperature, K
10^{11}
10^{10}
10^{9}
10^{8}
10^{7}
0
100
200
300
400
500
600
700
800

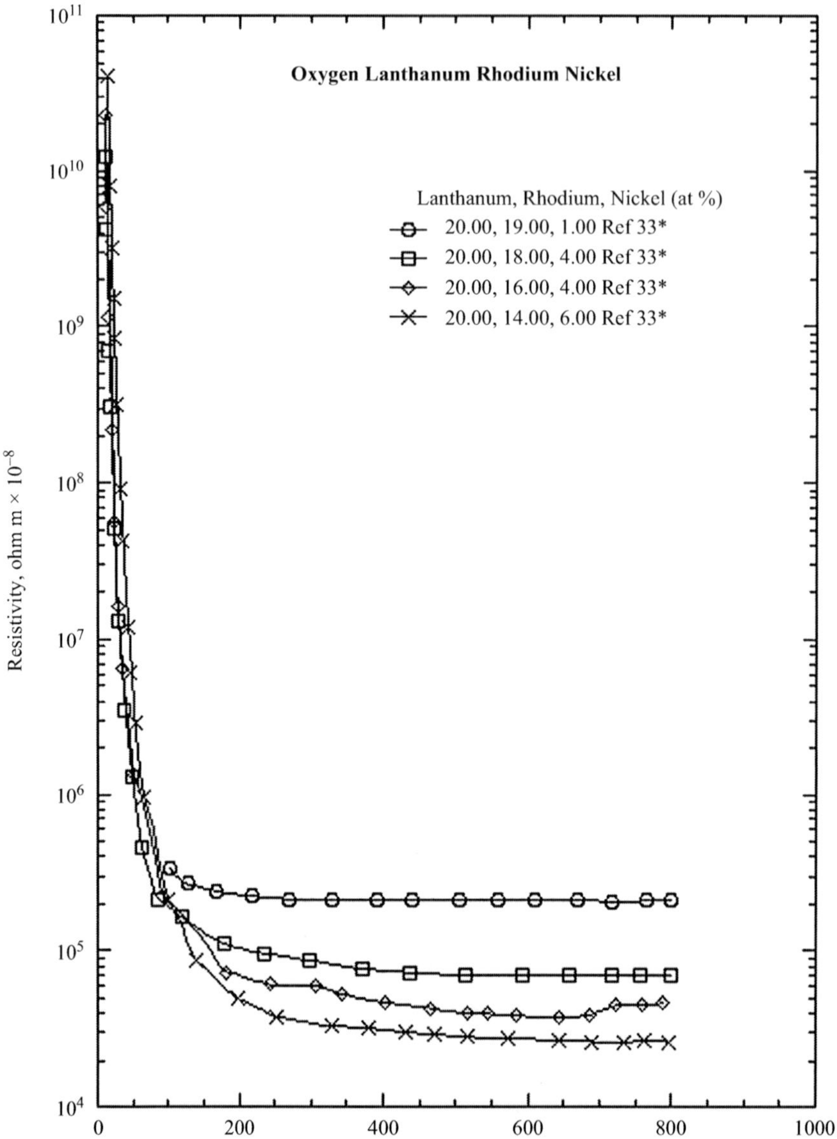
Oxygen Lanthanum Rhodium Nickel
Lanthanum, Rhodium, Nickel (at %)
20.00, 19.00, 1.00 Ref 33*
20.00, 18.00, 4.00 Ref 33*
20.00, 16.00, 4.00 Ref 33*
20.00, 14.00, 6.00 Ref 33*
Resistivity, ohm m × 10⁻⁸
Temperature, K
10^11
10^10
10^9
10^8
10^7
10^6
10^5
10^4
0
200
400
600
800
1000

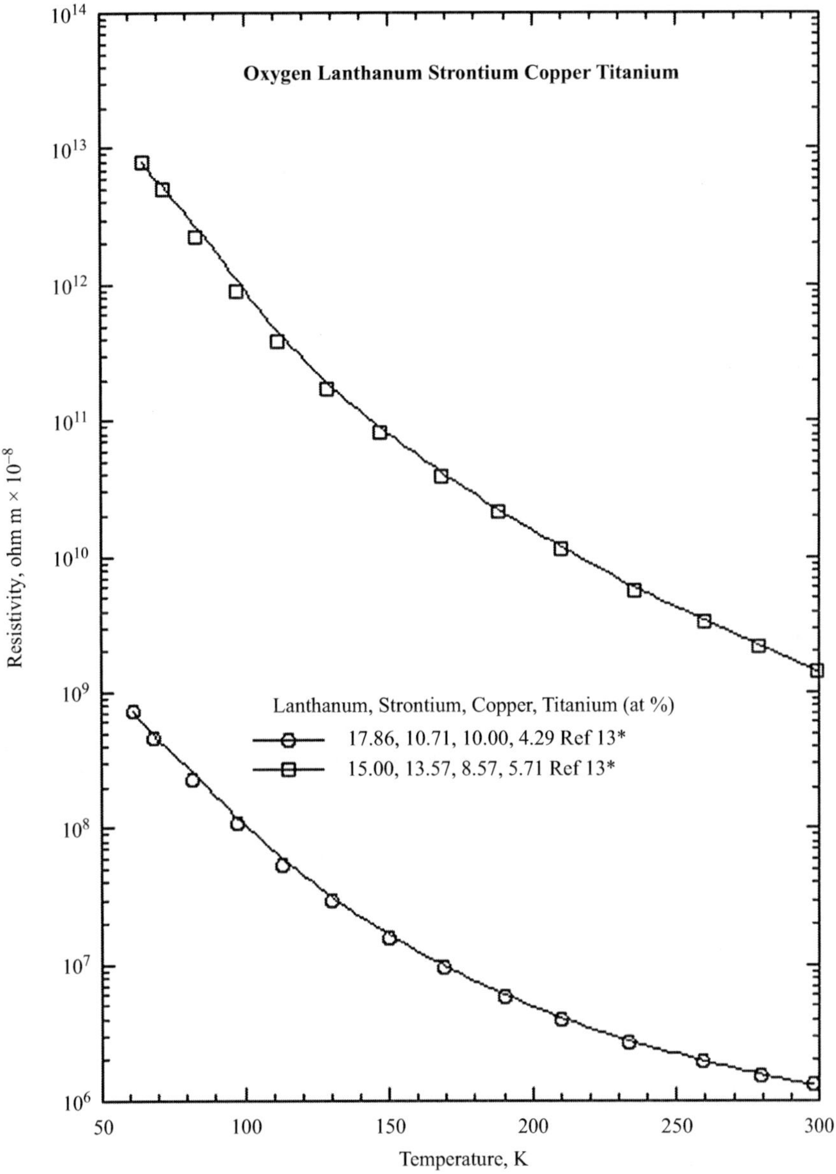

Oxygen Lanthanum Strontium Copper Titanium
Resistivity, ohm m × 10⁻⁸
10¹⁴
10¹³
10¹²
10¹¹
10¹⁰
10⁹
10⁸
10⁷
10⁶
50
100
150
200
250
300
Temperature, K
Lanthanum, Strontium, Copper, Titanium (at %)
17.86, 10.71, 10.00, 4.29 Ref 13*
15.00, 13.57, 8.57, 5.71 Ref 13*

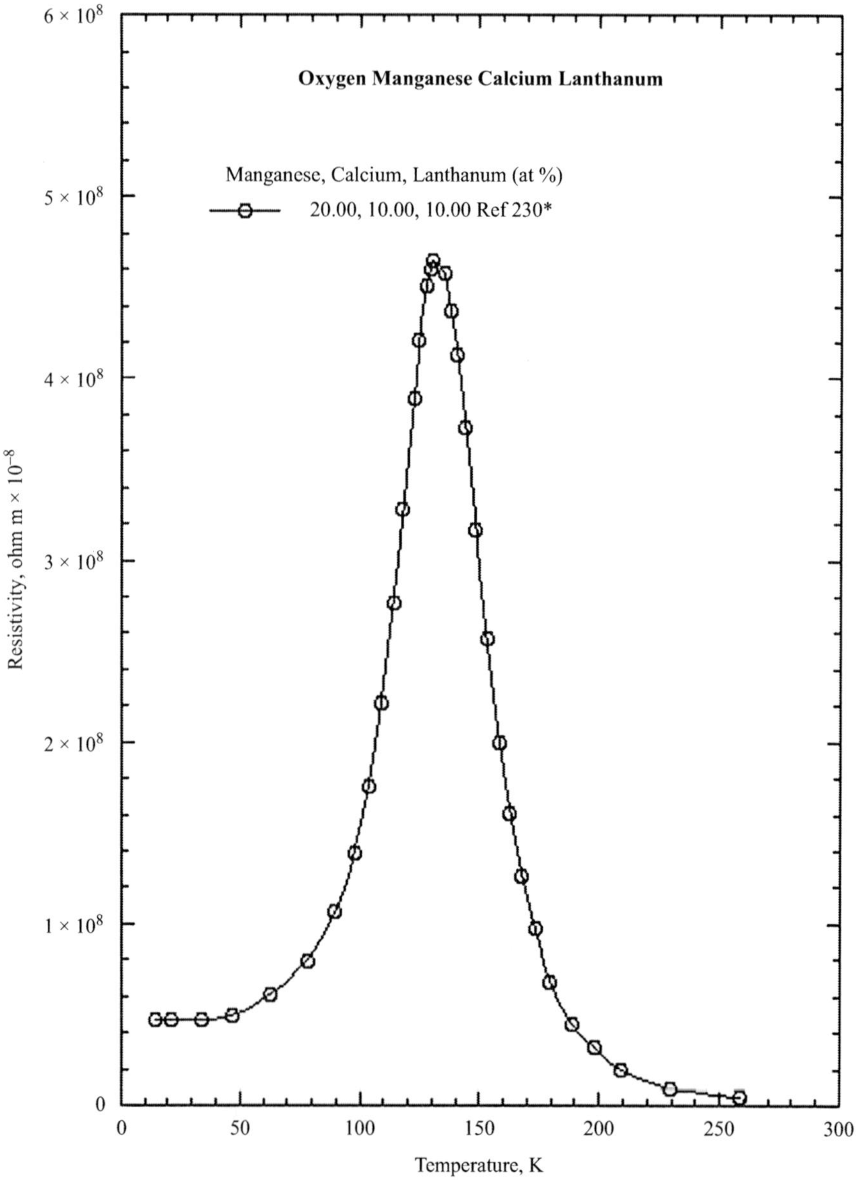
Oxygen Manganese Calcium Lanthanum
Manganese, Calcium, Lanthanum (at %)
20.00, 10.00, 10.00 Ref 230*
Resistivity, ohm m × 10⁻⁸
6 × 10⁸
5 × 10⁸
4 × 10⁸
3 × 10⁸
2 × 10⁸
1 × 10⁸
0
0
50
100
150
200
250
300
Temperature, K

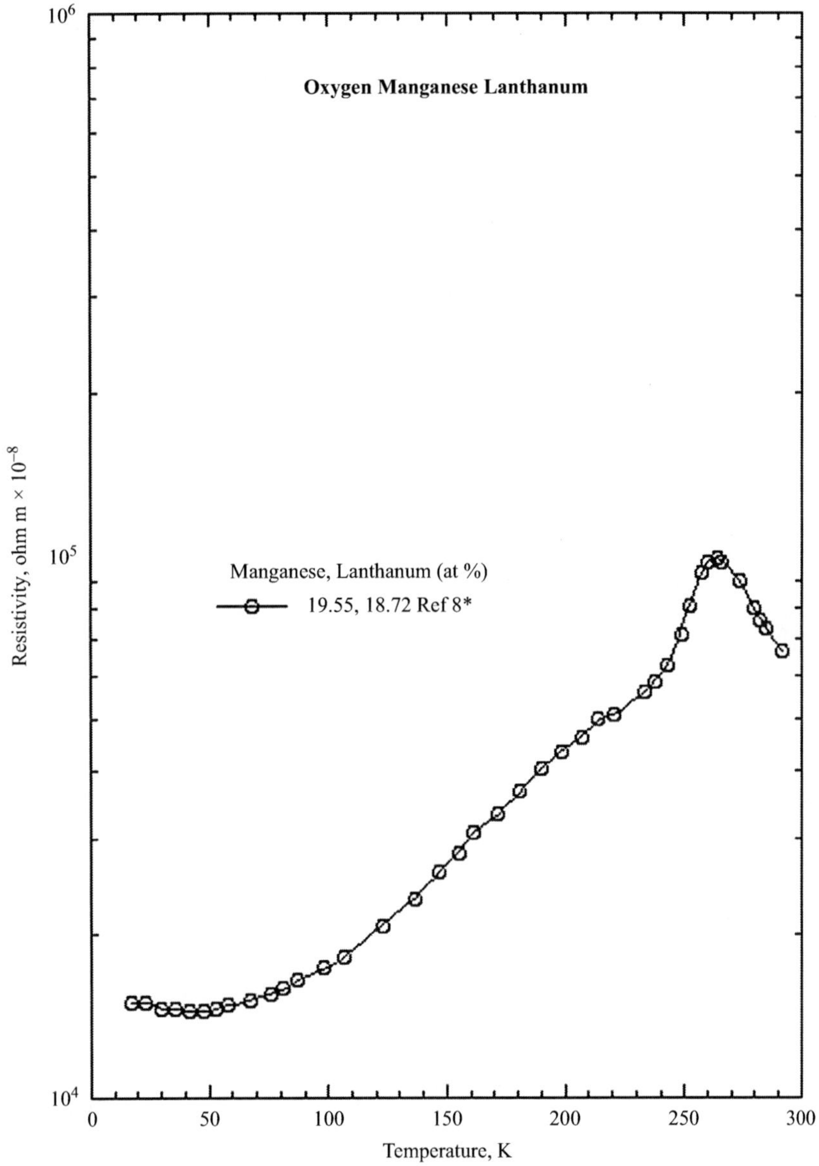
Oxygen Manganese Lanthanum
Manganese, Lanthanum (at %)
19.55, 18.72 Ref 8*
Resistivity, ohm m × 10⁻⁸
Temperature, K
10⁶
10⁵
10⁴
0
50
100
150
200
250
300

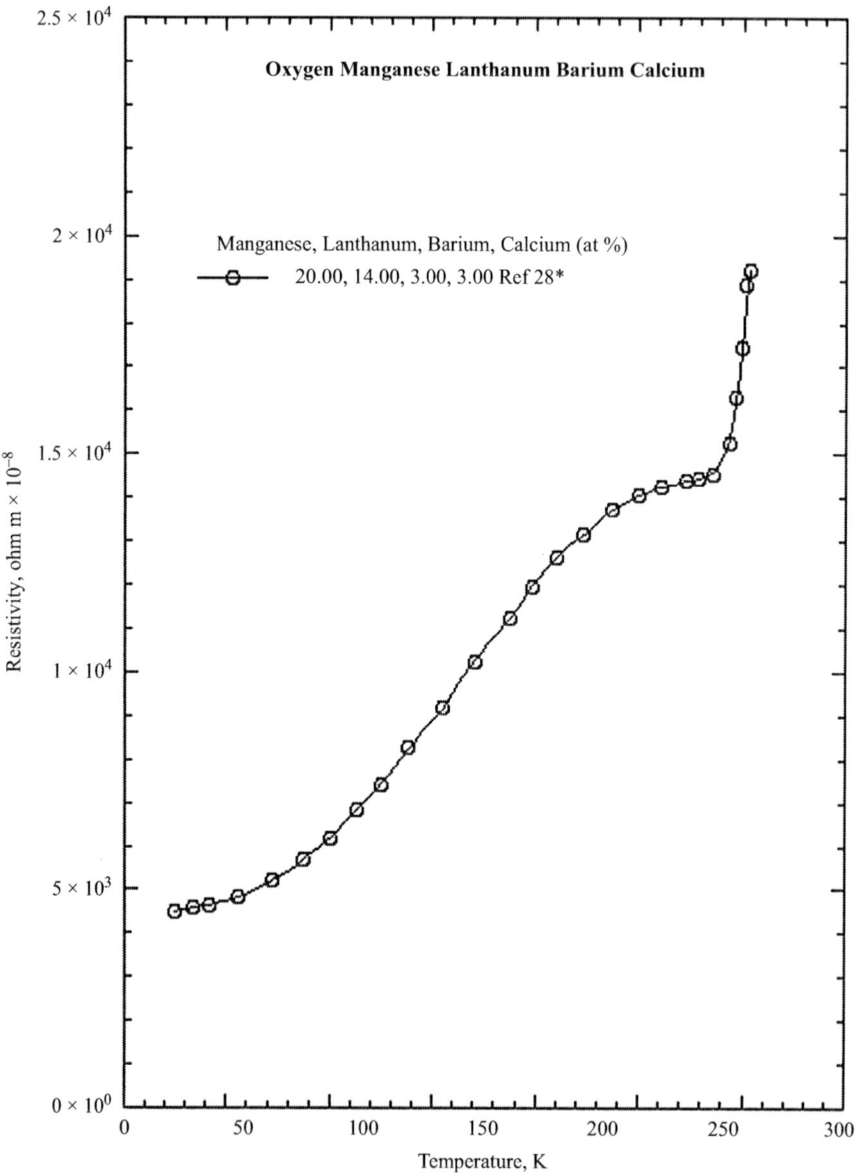

Oxygen Manganese Lanthanum Barium Calcium
Manganese, Lanthanum, Barium, Calcium (at %)
20.00, 14.00, 3.00, 3.00 Ref 28*
Resistivity, ohm m × 10⁻⁸
2.5 × 10⁴
2 × 10⁴
1.5 × 10⁴
1 × 10⁴
5 × 10³
0 × 10⁰
0
50
100
150
200
250
300
Temperature, K

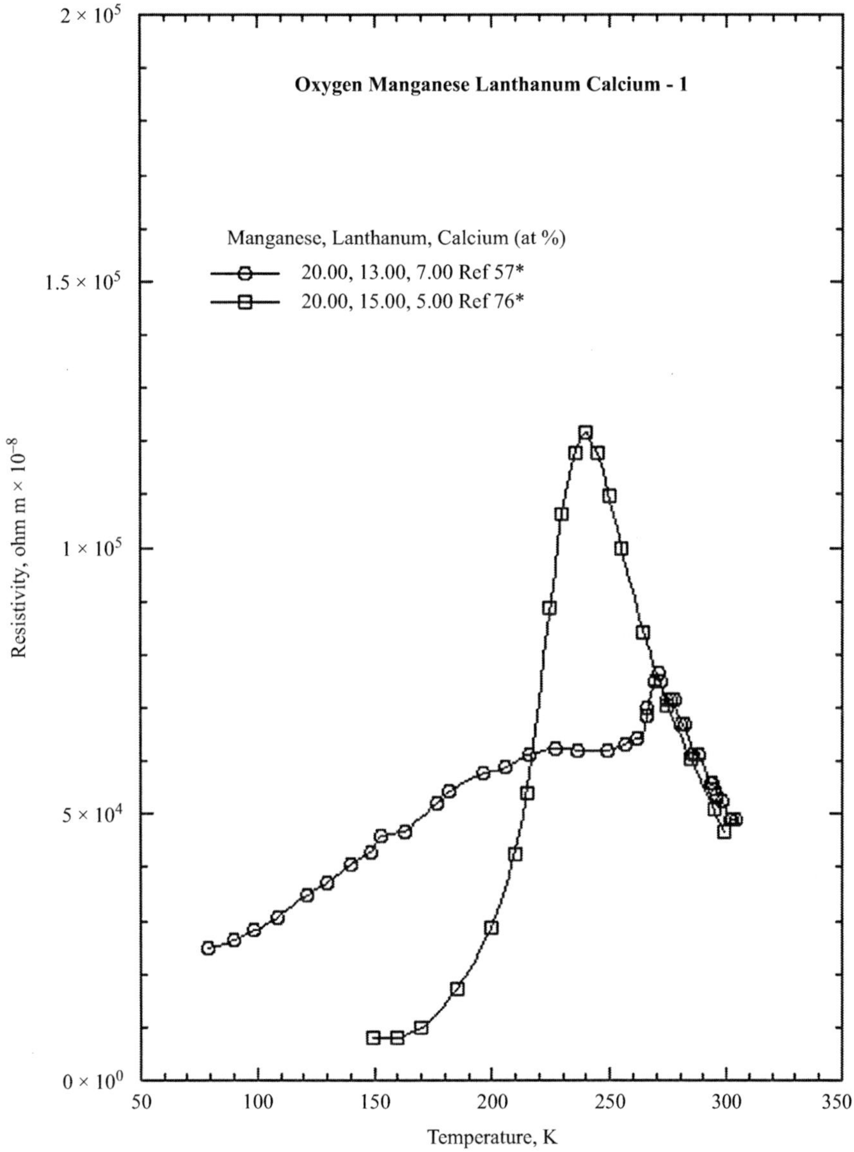
Oxygen Manganese Lanthanum Calcium - 1
Manganese, Lanthanum, Calcium (at %)
20.00, 13.00, 7.00 Ref 57*
20.00, 15.00, 5.00 Ref 76*
Resistivity, ohm m × 10⁻⁸
Temperature, K
2 × 10⁵
1.5 × 10⁵
1 × 10⁵
5 × 10⁴
0 × 10⁰
50
100
150
200
250
300
350

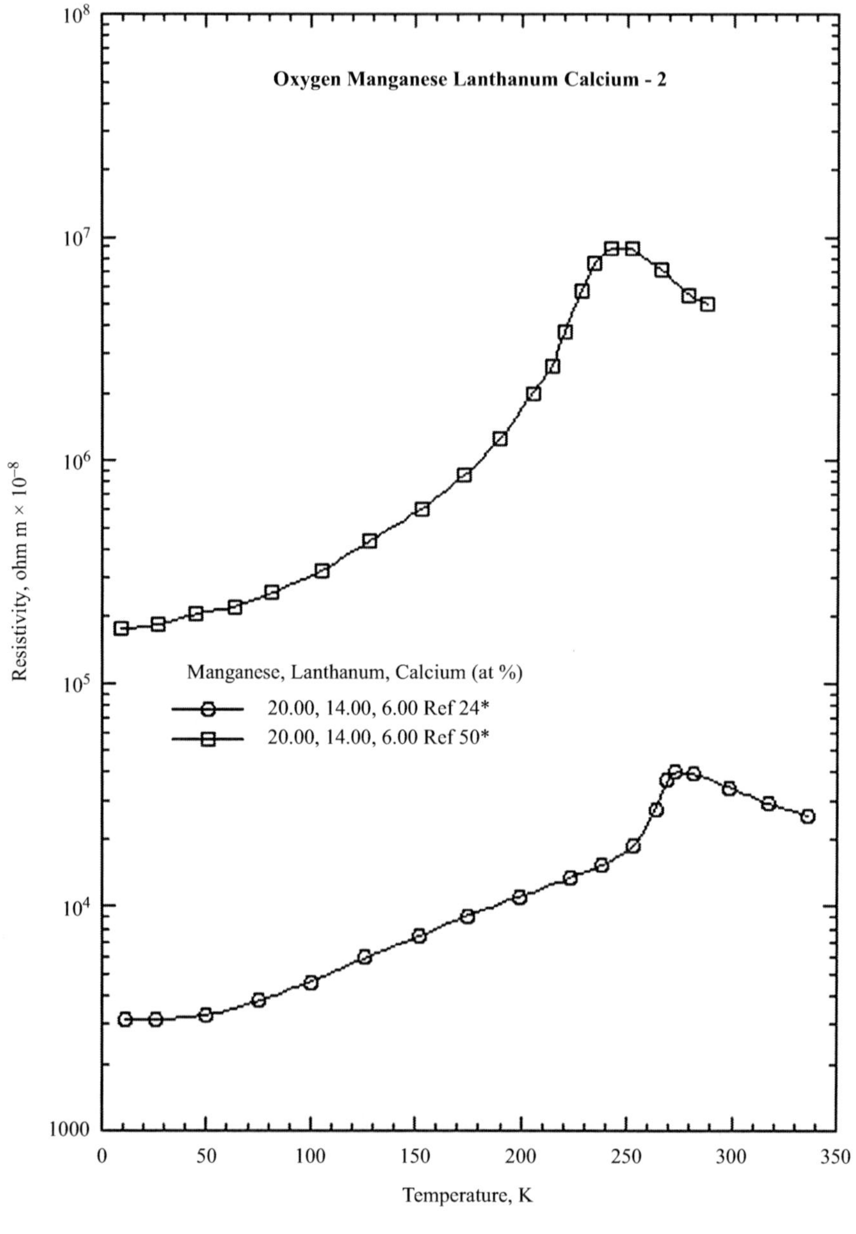

Oxygen Manganese Lanthanum Calcium - 2
Resistivity, ohm m × 10⁻⁸
Temperature, K
Manganese, Lanthanum, Calcium (at %)
20.00, 14.00, 6.00 Ref 24*
20.00, 14.00, 6.00 Ref 50*

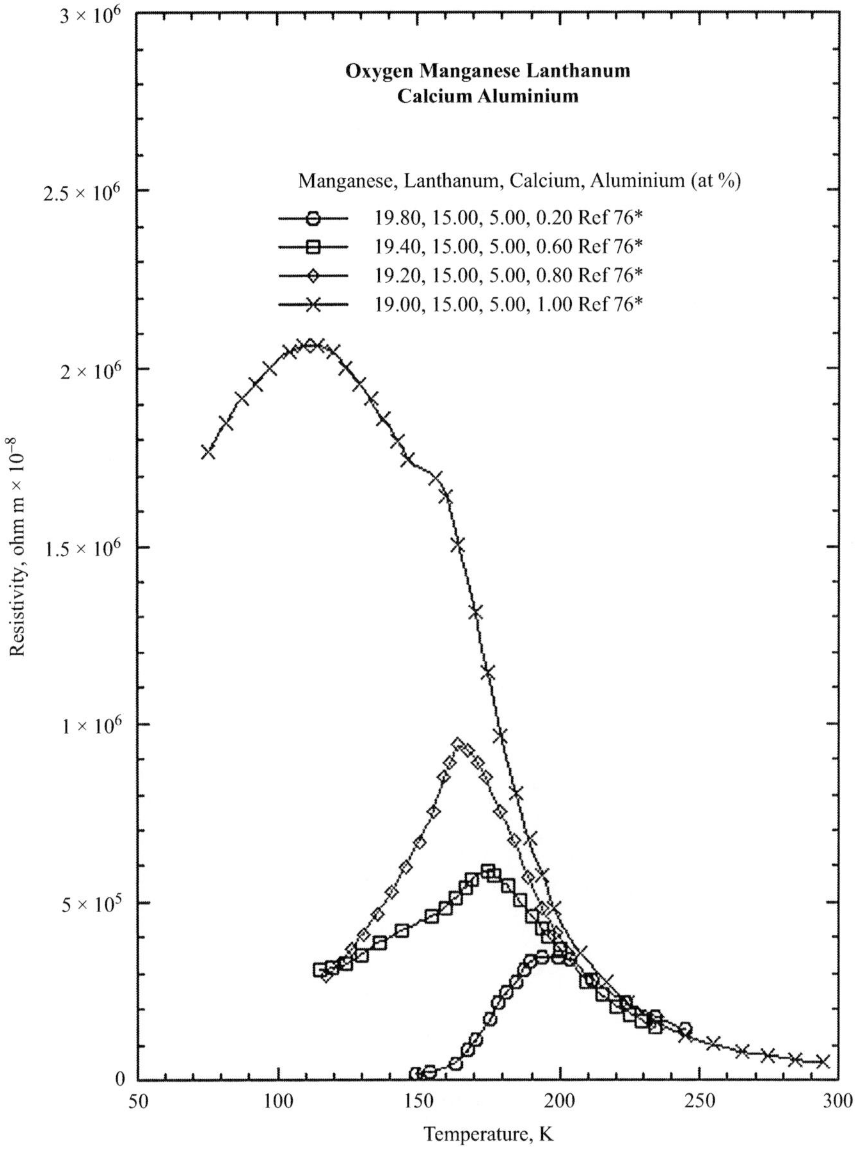
Oxygen Manganese Lanthanum
Calcium Aluminium
Manganese, Lanthanum, Calcium, Aluminium (at %)
19.80, 15.00, 5.00, 0.20 Ref 76*
19.40, 15.00, 5.00, 0.60 Ref 76*
19.20, 15.00, 5.00, 0.80 Ref 76*
19.00, 15.00, 5.00, 1.00 Ref 76*
Resistivity, ohm m × 10⁻⁸
Temperature, K
3 × 10⁶
2.5 × 10⁶
2 × 10⁶
1.5 × 10⁶
1 × 10⁶
5 × 10⁵
0
50
100
150
200
250
300

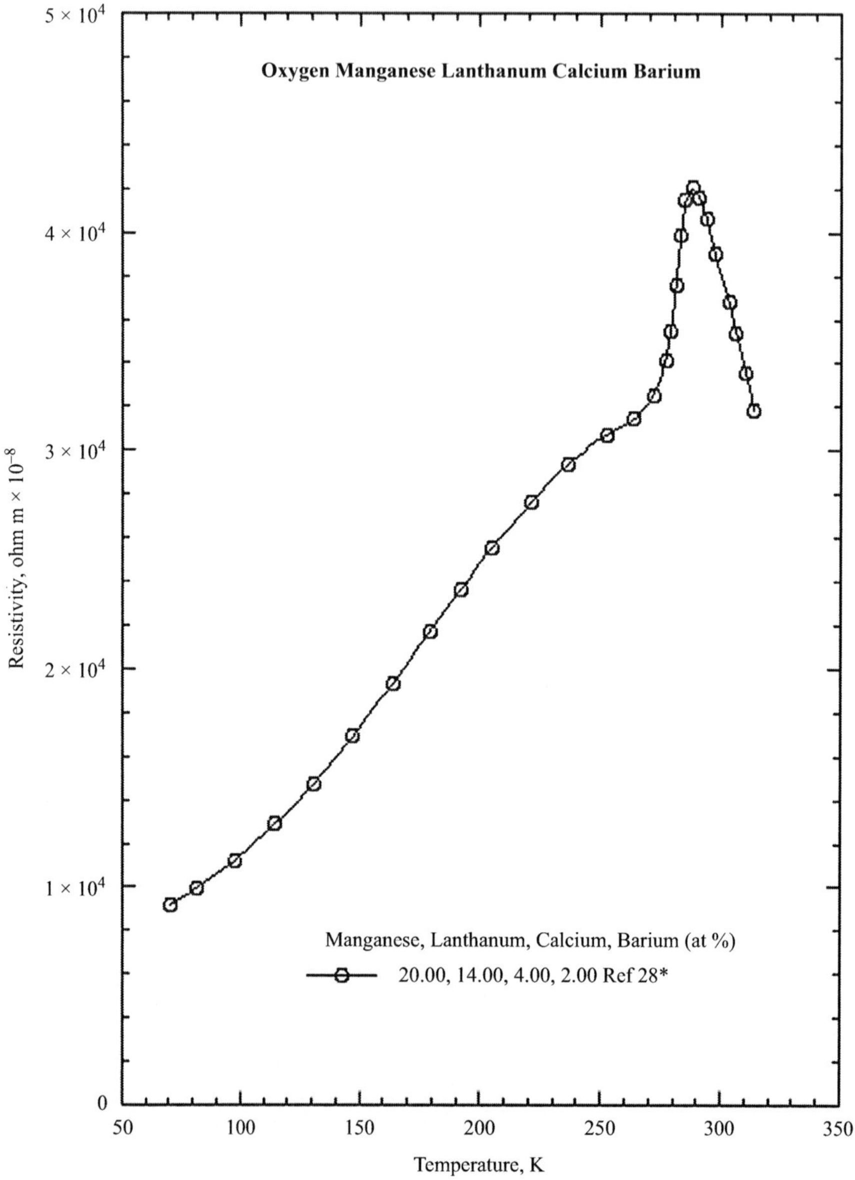
Oxygen Manganese Lanthanum Calcium Barium
Resistivity, ohm m × 10⁻⁸
5 × 10⁴
4 × 10⁴
3 × 10⁴
2 × 10⁴
1 × 10⁴
0
50
100
150
200
250
300
350
Temperature, K
Manganese, Lanthanum, Calcium, Barium (at %)
20.00, 14.00, 4.00, 2.00 Ref 28*

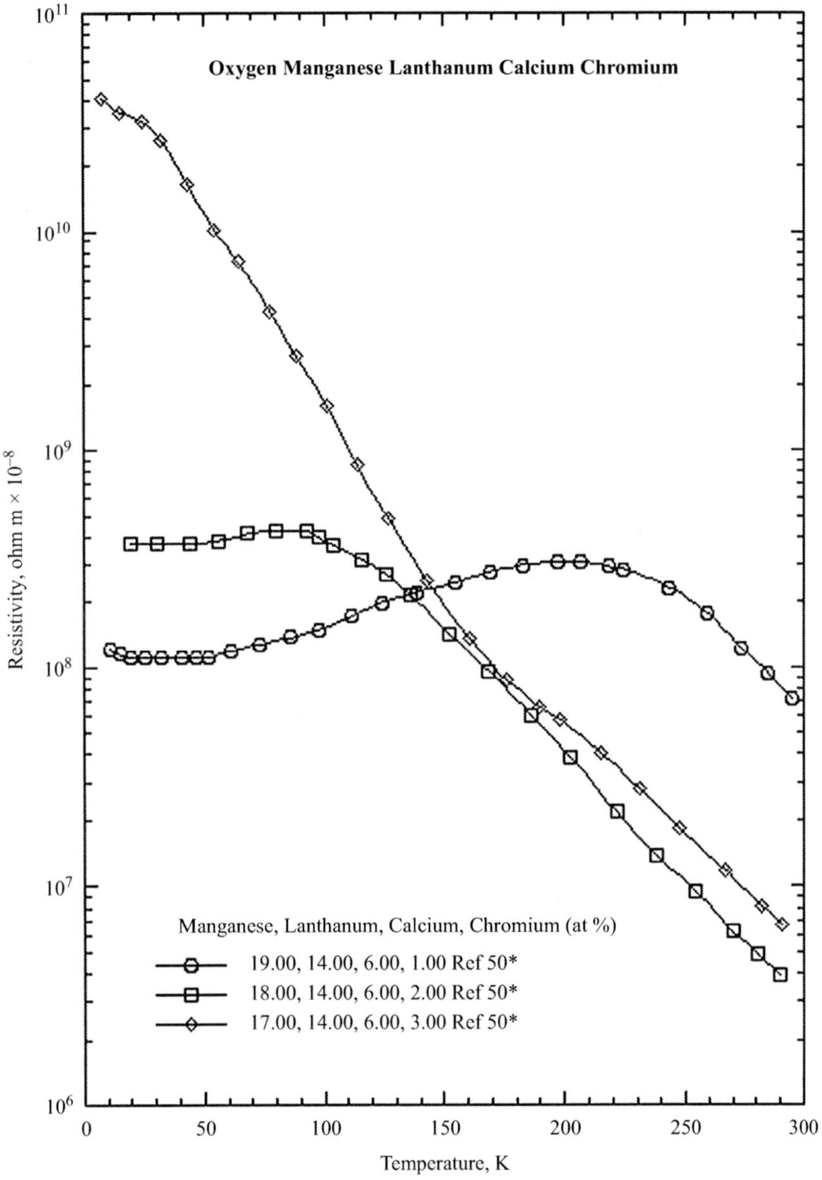
Oxygen Manganese Lanthanum Calcium Chromium
10^{11}
10^{10}
10^{9}
10^{8}
10^{7}
10^{6}
Resistivity, ohm m × 10^{-8}
0
50
100
150
200
250
300
Temperature, K
Manganese, Lanthanum, Calcium, Chromium (at %)
19.00, 14.00, 6.00, 1.00 Ref 50*
18.00, 14.00, 6.00, 2.00 Ref 50*
17.00, 14.00, 6.00, 3.00 Ref 50*

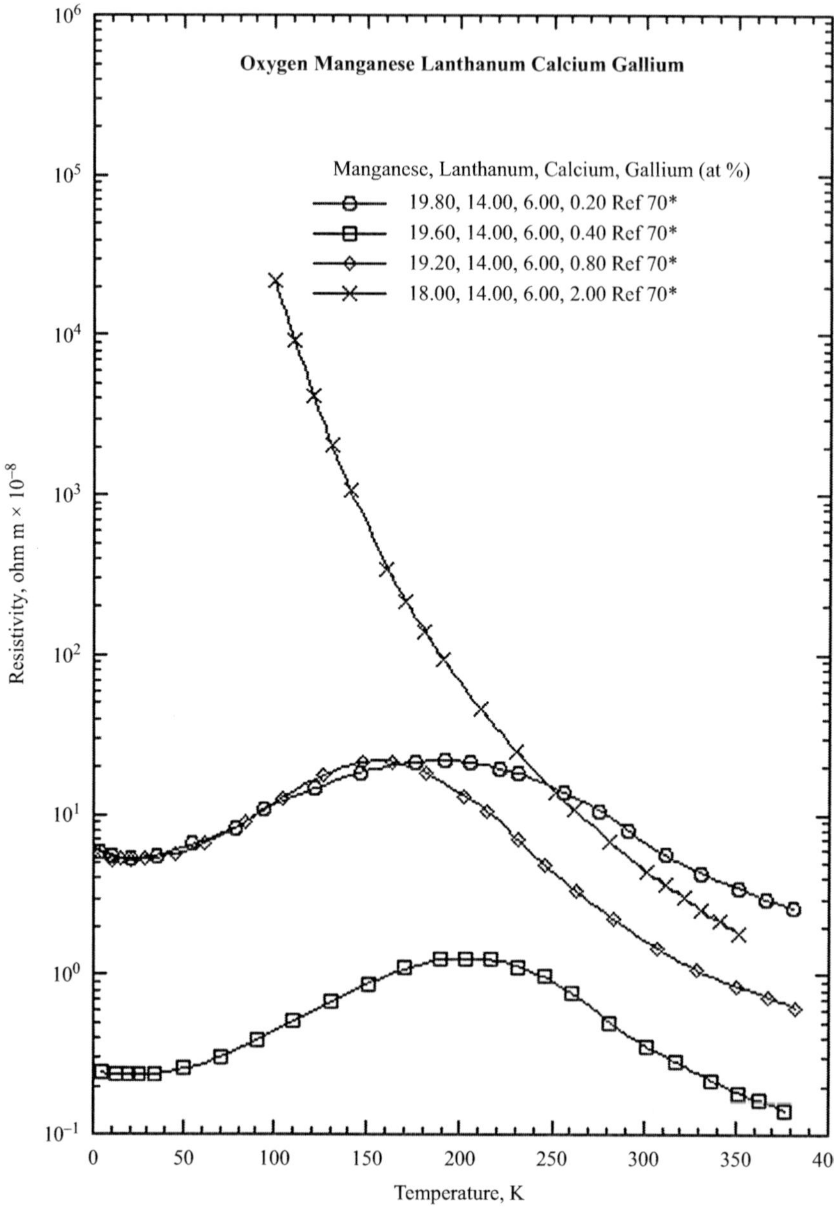
Oxygen Manganese Lanthanum Calcium Gallium
Manganese, Lanthanum, Calcium, Gallium (at %)
19.80, 14.00, 6.00, 0.20 Ref 70*
19.60, 14.00, 6.00, 0.40 Ref 70*
19.20, 14.00, 6.00, 0.80 Ref 70*
18.00, 14.00, 6.00, 2.00 Ref 70*
Resistivity, ohm m × 10⁻⁸
Temperature, K
0
50
100
150
200
250
300
350
400

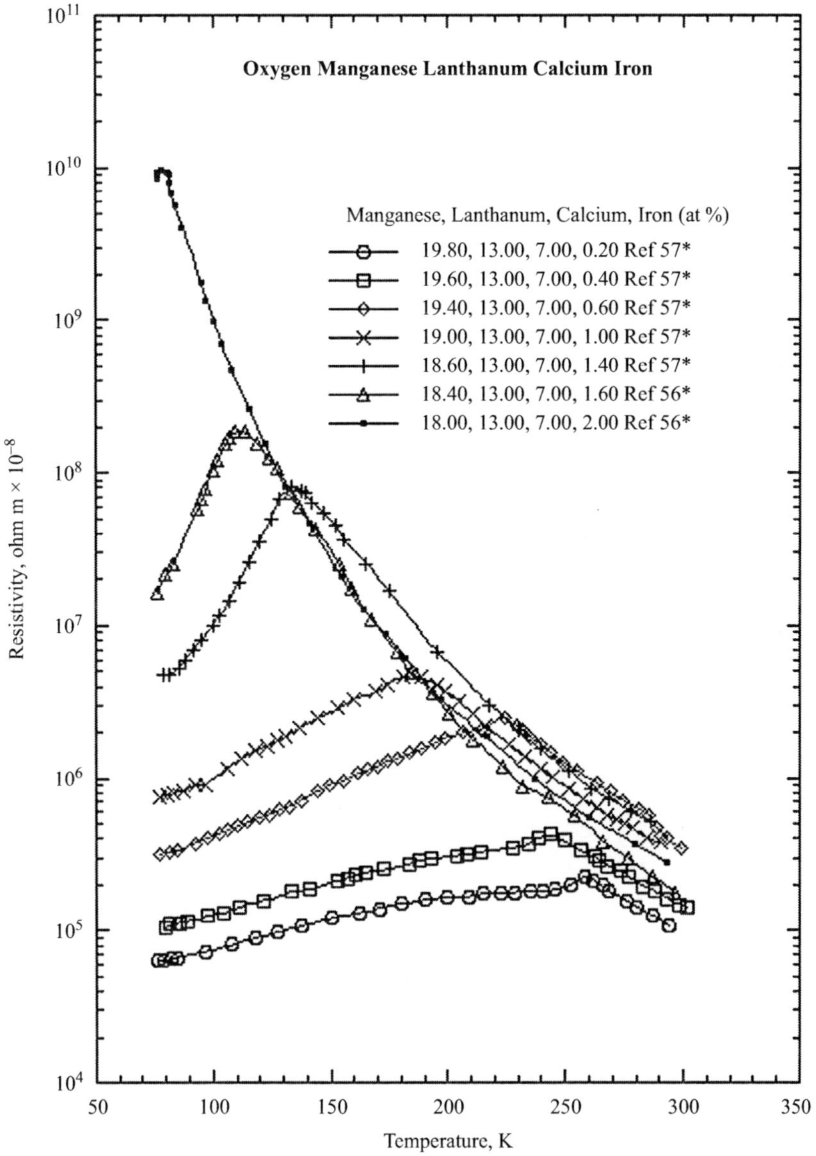

Oxygen Manganese Lanthanum Calcium Iron
Manganese, Lanthanum, Calcium, Iron (at %)
19.80, 13.00, 7.00, 0.20 Ref 57*
19.60, 13.00, 7.00, 0.40 Ref 57*
19.40, 13.00, 7.00, 0.60 Ref 57*
19.00, 13.00, 7.00, 1.00 Ref 57*
18.60, 13.00, 7.00, 1.40 Ref 57*
18.40, 13.00, 7.00, 1.60 Ref 56*
18.00, 13.00, 7.00, 2.00 Ref 56*
Resistivity, ohm m × 10⁻⁸
Temperature, K

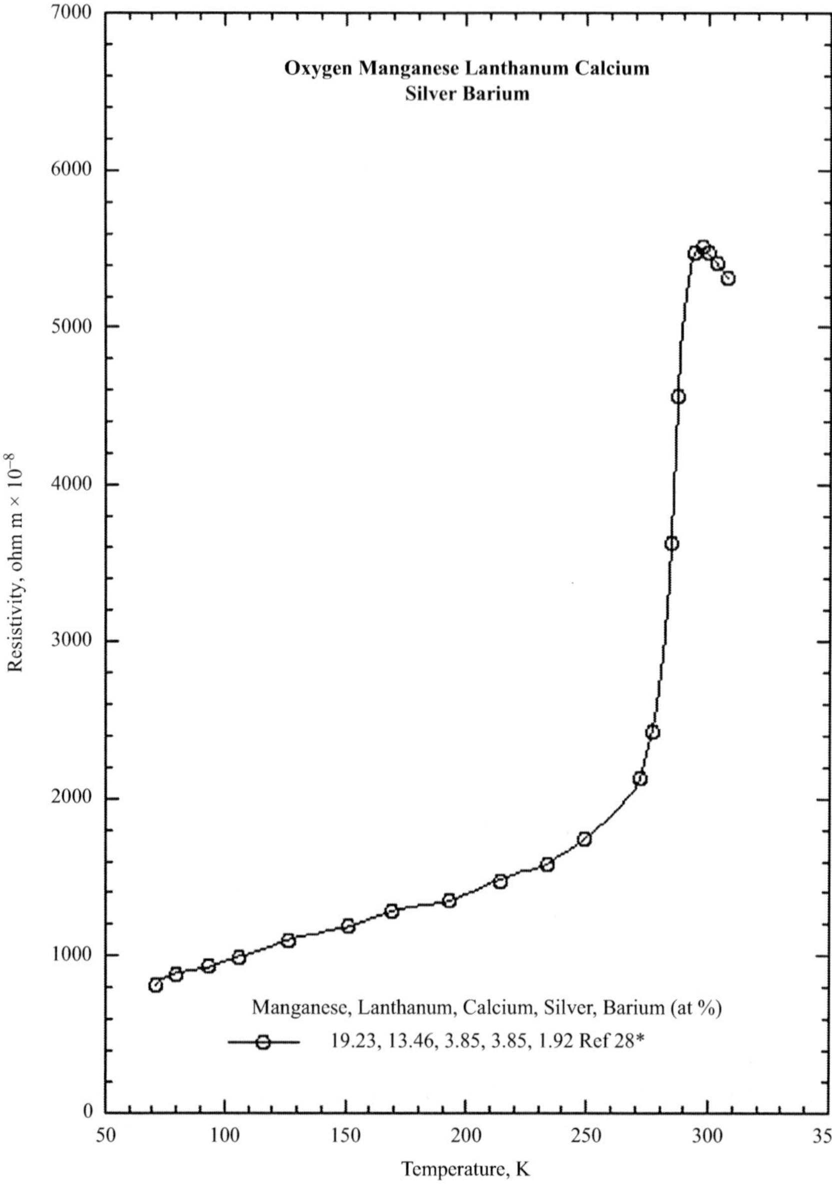
Oxygen Manganese Lanthanum Calcium
Silver Barium
Resistivity, ohm m × 10−8
Temperature, K
Manganese, Lanthanum, Calcium, Silver, Barium (at %)
19.23, 13.46, 3.85, 3.85, 1.92 Ref 28*
0
1000
2000
3000
4000
5000
6000
7000
50
100
150
200
250
300
350

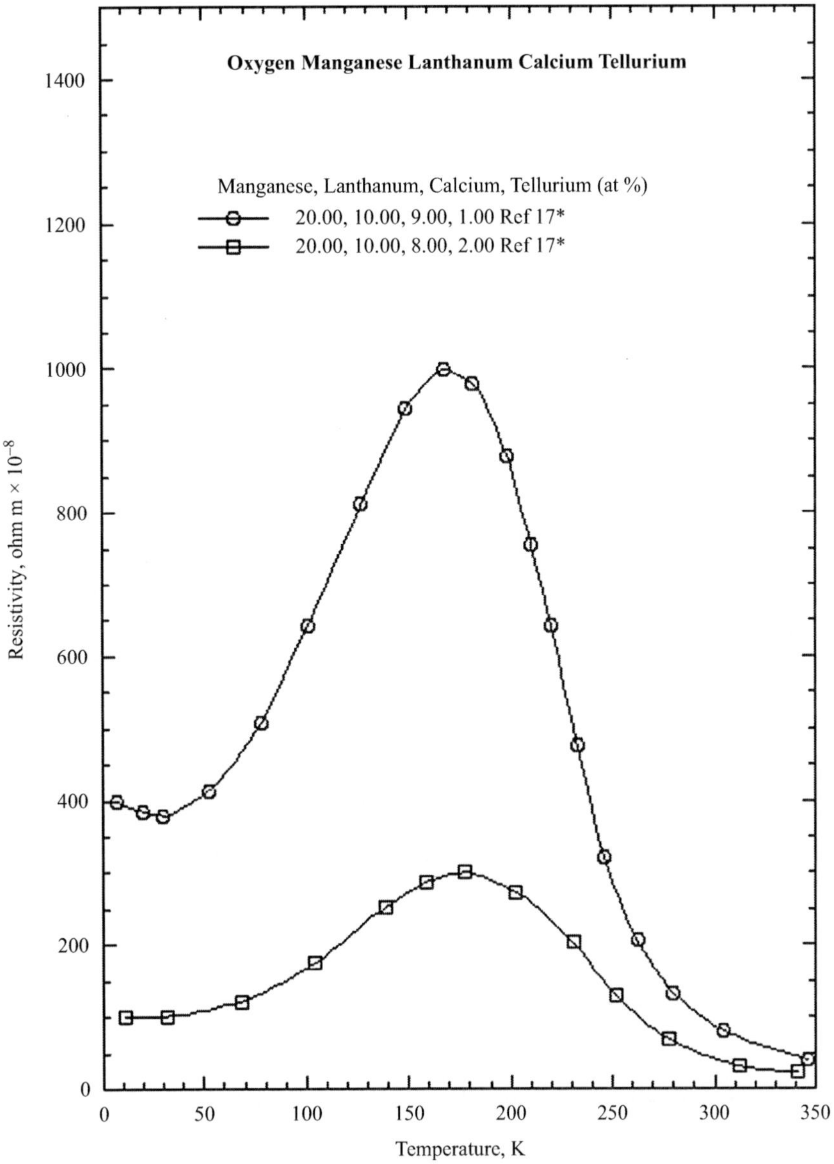
Oxygen Manganese Lanthanum Calcium Tellurium
Manganese, Lanthanum, Calcium, Tellurium (at %)
20.00, 10.00, 9.00, 1.00 Ref 17*
20.00, 10.00, 8.00, 2.00 Ref 17*
Resistivity, ohm m × 10⁻⁸
Temperature, K
0
200
400
600
800
1000
1200
1400
0
50
100
150
200
250
300
350

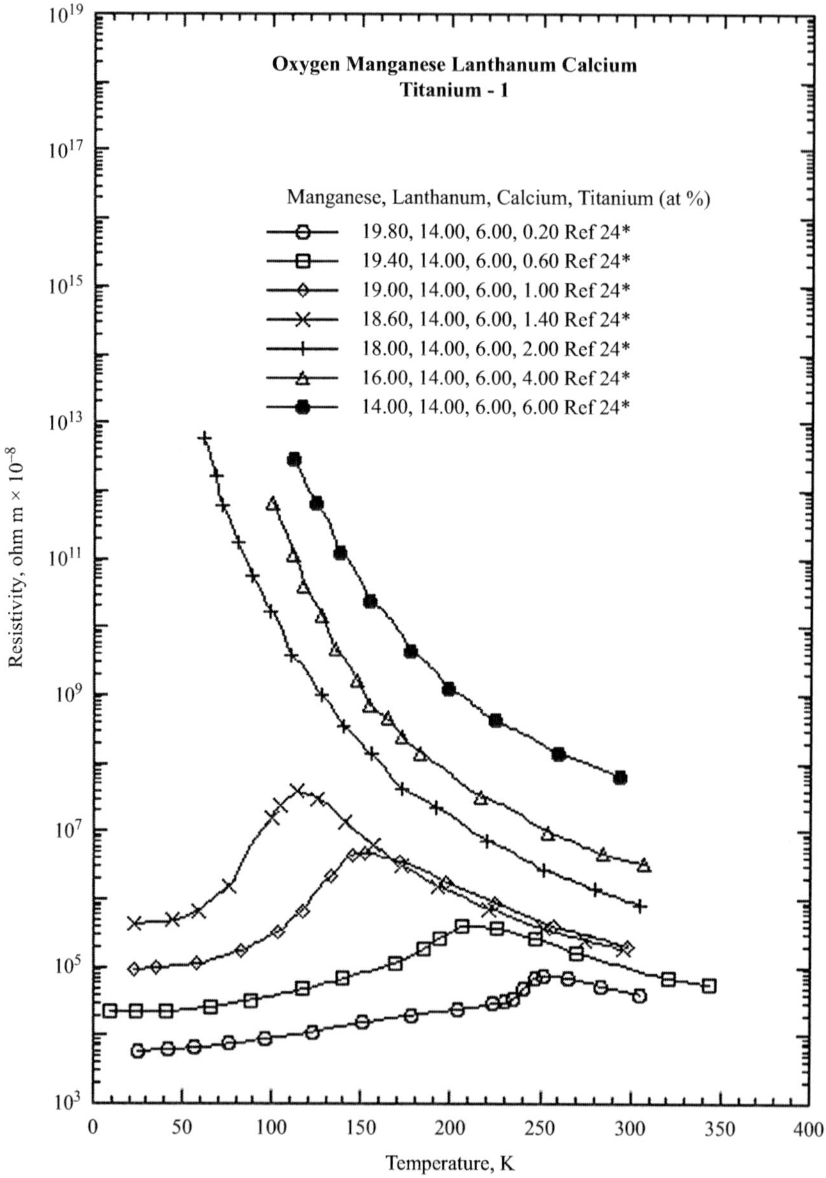
Oxygen Manganese Lanthanum Calcium
Titanium - 1
Manganese, Lanthanum, Calcium, Titanium (at %)
19.80, 14.00, 6.00, 0.20 Ref 24*
19.40, 14.00, 6.00, 0.60 Ref 24*
19.00, 14.00, 6.00, 1.00 Ref 24*
18.60, 14.00, 6.00, 1.40 Ref 24*
18.00, 14.00, 6.00, 2.00 Ref 24*
16.00, 14.00, 6.00, 4.00 Ref 24*
14.00, 14.00, 6.00, 6.00 Ref 24*
Resistivity, ohm m × 10⁻⁸
Temperature, K
10³
10⁵
10⁷
10⁹
10¹¹
10¹³
10¹⁵
10¹⁷
10¹⁹
0
50
100
150
200
250
300
350
400

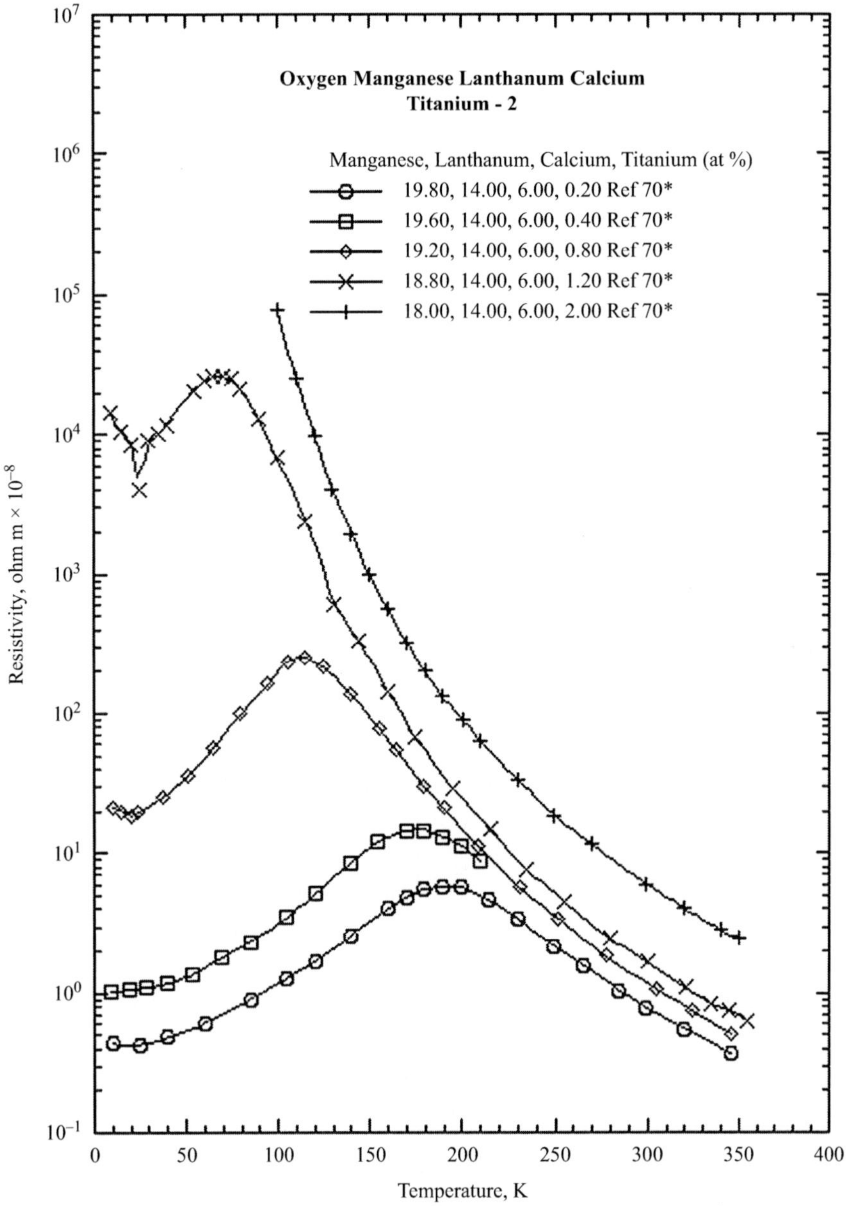
Oxygen Manganese Lanthanum Calcium
Titanium - 2
Manganese, Lanthanum, Calcium, Titanium (at %)
19.80, 14.00, 6.00, 0.20 Ref 70*
19.60, 14.00, 6.00, 0.40 Ref 70*
19.20, 14.00, 6.00, 0.80 Ref 70*
18.80, 14.00, 6.00, 1.20 Ref 70*
18.00, 14.00, 6.00, 2.00 Ref 70*
Resistivity, ohm m × 10⁻⁸
Temperature, K
10⁷
10⁶
10⁵
10⁴
10³
10²
10¹
10⁰
10⁻¹
0
50
100
150
200
250
300
350
400

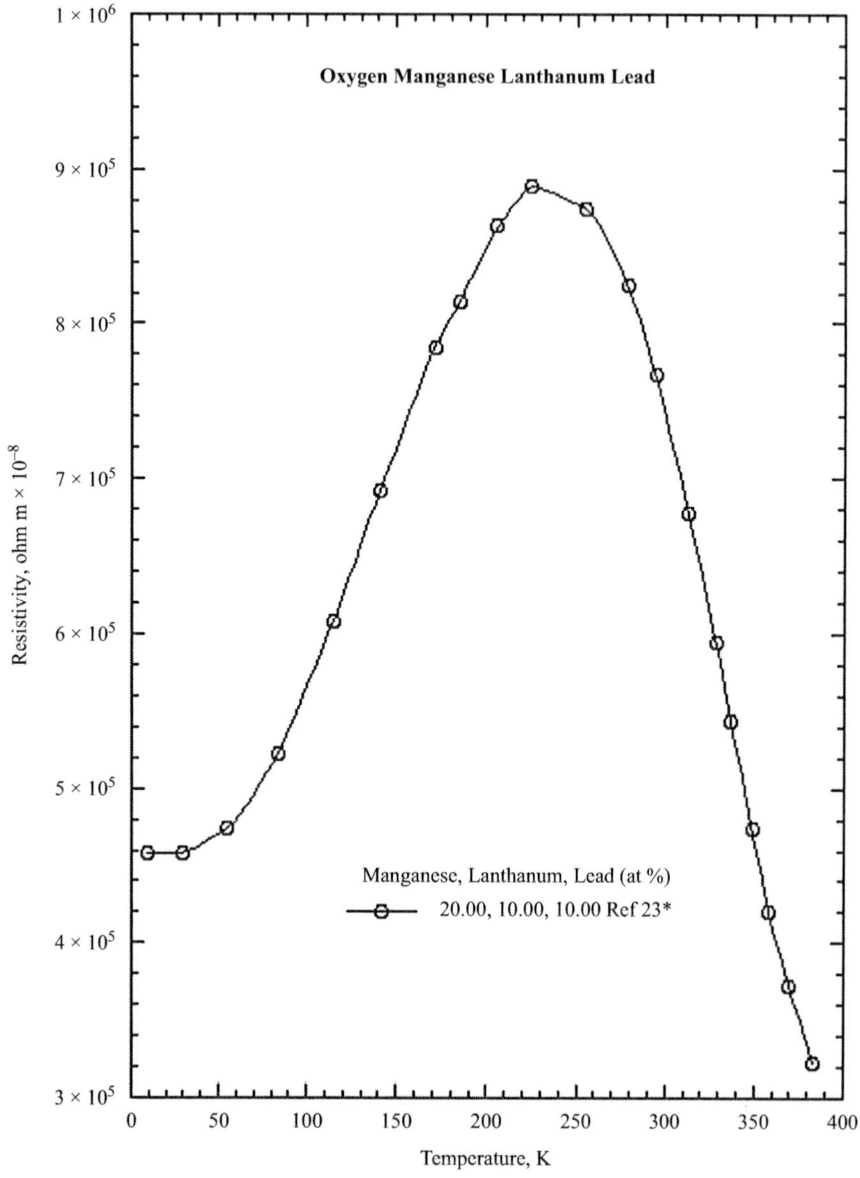
Oxygen Manganese Lanthanum Lead
Manganese, Lanthanum, Lead (at %)
20.00, 10.00, 10.00 Ref 23*
Resistivity, ohm m × 10^-8
Temperature, K
1 × 10^6
9 × 10^5
8 × 10^5
7 × 10^5
6 × 10^5
5 × 10^5
4 × 10^5
3 × 10^5
0
50
100
150
200
250
300
350
400

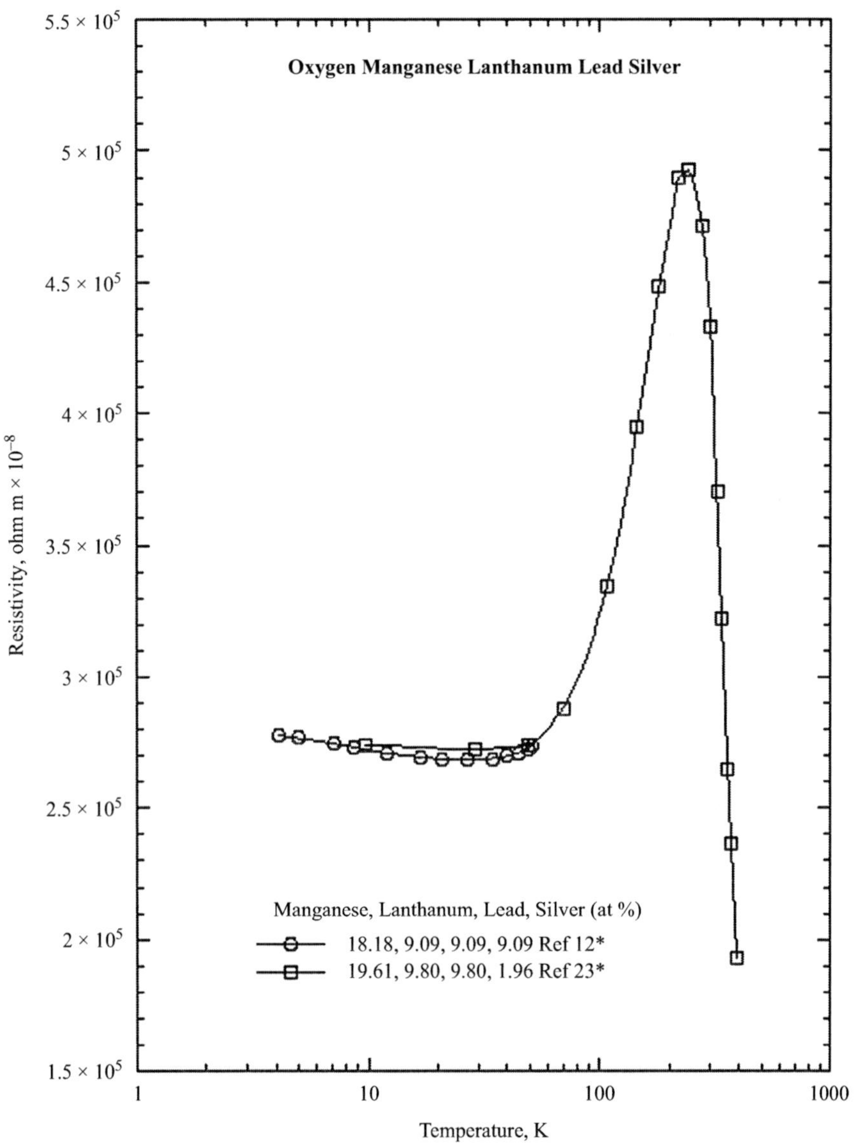
Oxygen Manganese Lanthanum Lead Silver
Resistivity, ohm m × 10⁻⁸
5.5 × 10⁵
5 × 10⁵
4.5 × 10⁵
4 × 10⁵
3.5 × 10⁵
3 × 10⁵
2.5 × 10⁵
2 × 10⁵
1.5 × 10⁵
1
10
100
1000
Temperature, K
Manganese, Lanthanum, Lead, Silver (at %)
18.18, 9.09, 9.09, 9.09 Ref 12*
19.61, 9.80, 9.80, 1.96 Ref 23*

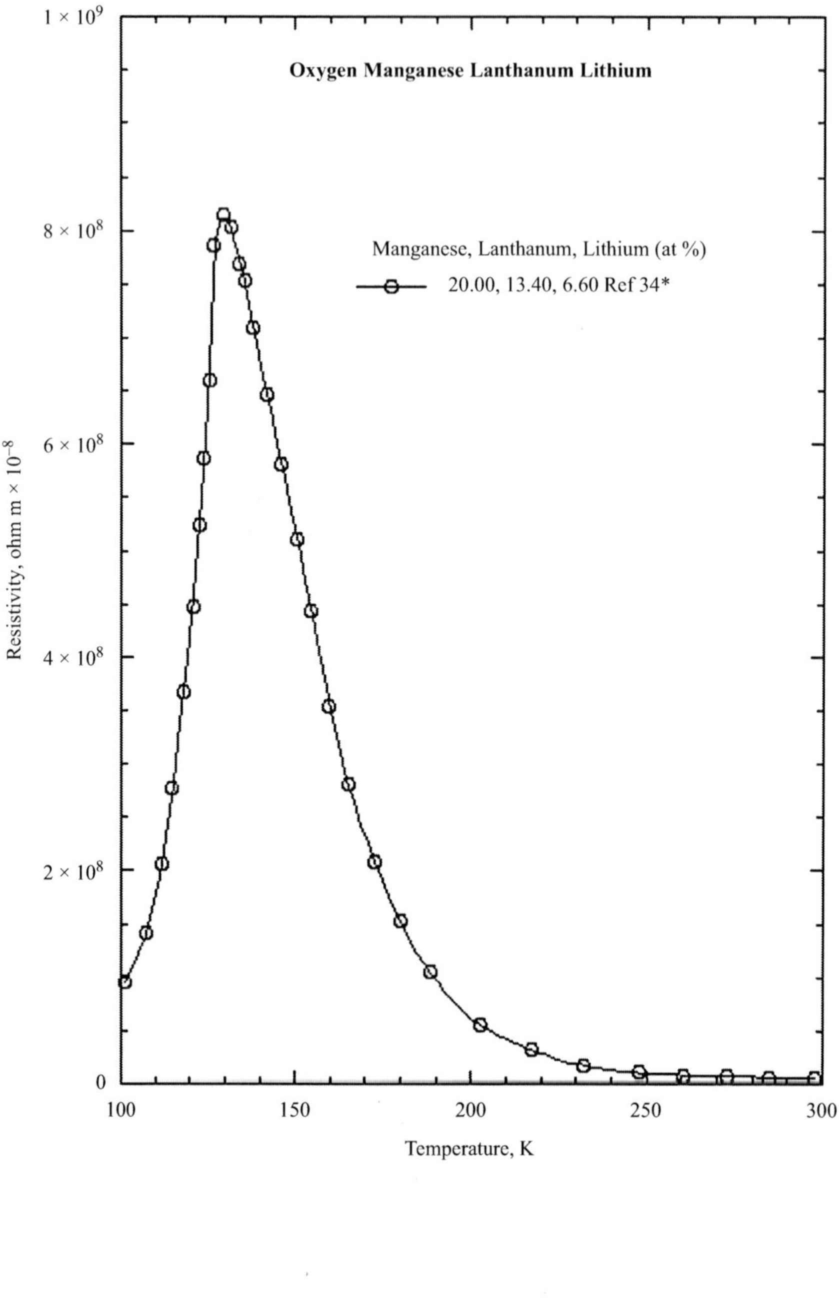

Oxygen Manganese Lanthanum Lithium
Manganese, Lanthanum, Lithium (at %)
20.00, 13.40, 6.60 Ref 34*
Resistivity, ohm m × 10⁻⁸
Temperature, K
1 × 10⁹
8 × 10⁸
6 × 10⁸
4 × 10⁸
2 × 10⁸
0
100
150
200
250
300

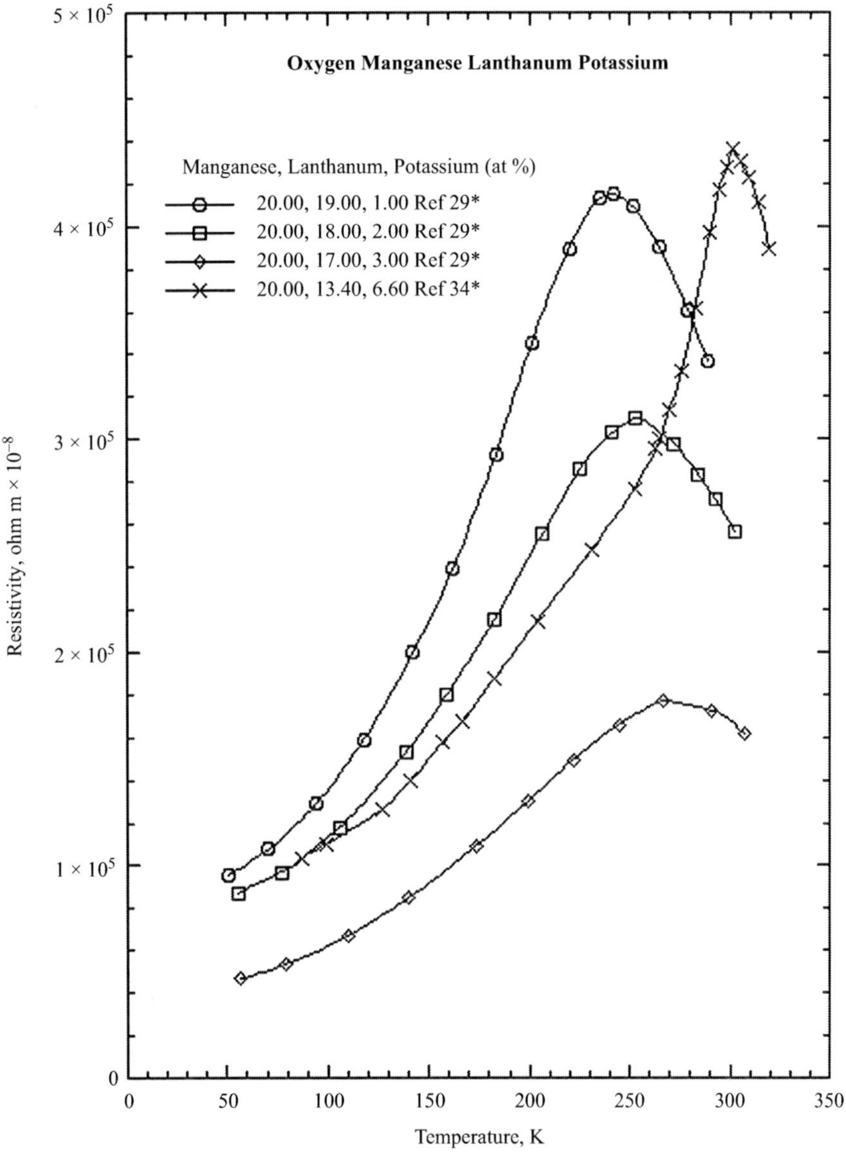
Oxygen Manganese Lanthanum Potassium
Manganese, Lanthanum, Potassium (at %)
20.00, 19.00, 1.00 Ref 29*
20.00, 18.00, 2.00 Ref 29*
20.00, 17.00, 3.00 Ref 29*
20.00, 13.40, 6.60 Ref 34*
Resistivity, ohm m × 10⁻⁸
5 × 10⁵
4 × 10⁵
3 × 10⁵
2 × 10⁵
1 × 10⁵
0
0
50
100
150
200
250
300
350
Temperature, K

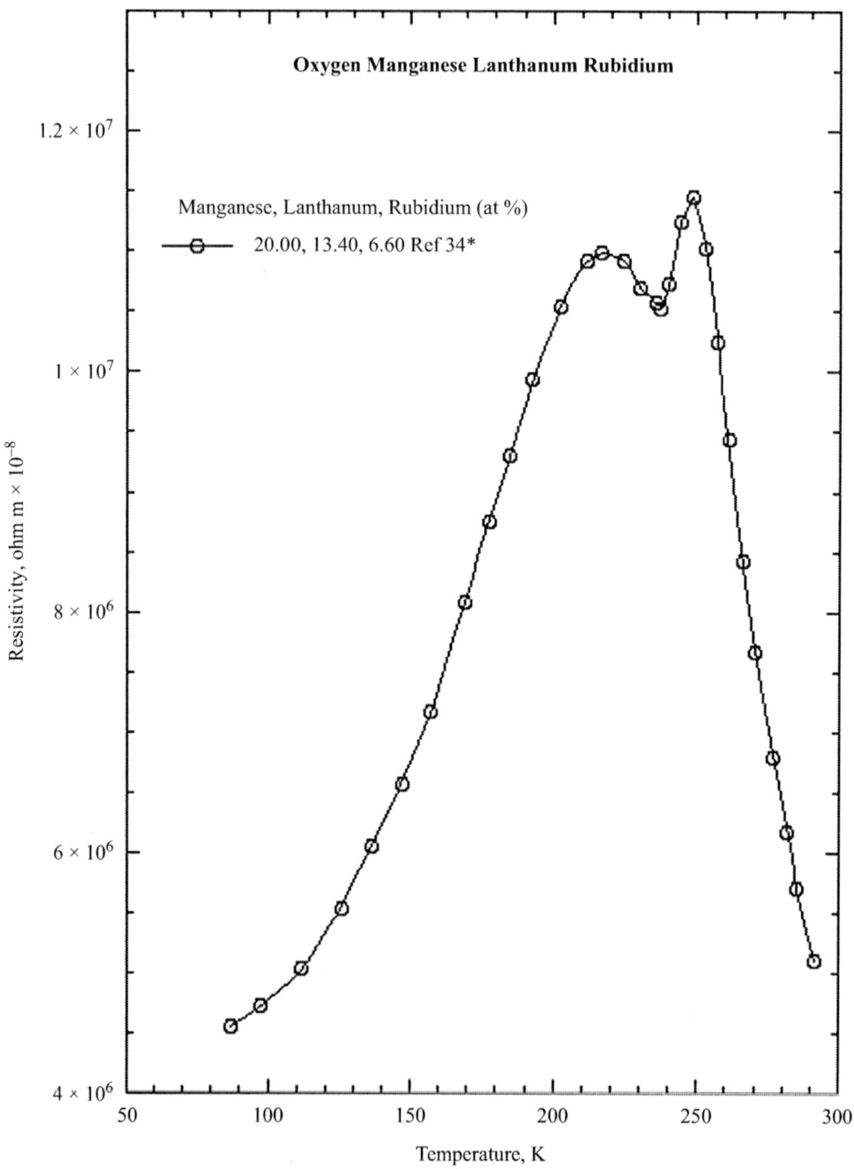
Oxygen Manganese Lanthanum Rubidium
Manganese, Lanthanum, Rubidium (at %)
20.00, 13.40, 6.60 Ref 34*
1.2 × 10^7
1 × 10^7
8 × 10^6
6 × 10^6
4 × 10^6
Resistivity, ohm m × 10^−8
50
100
150
200
250
300
Temperature, K

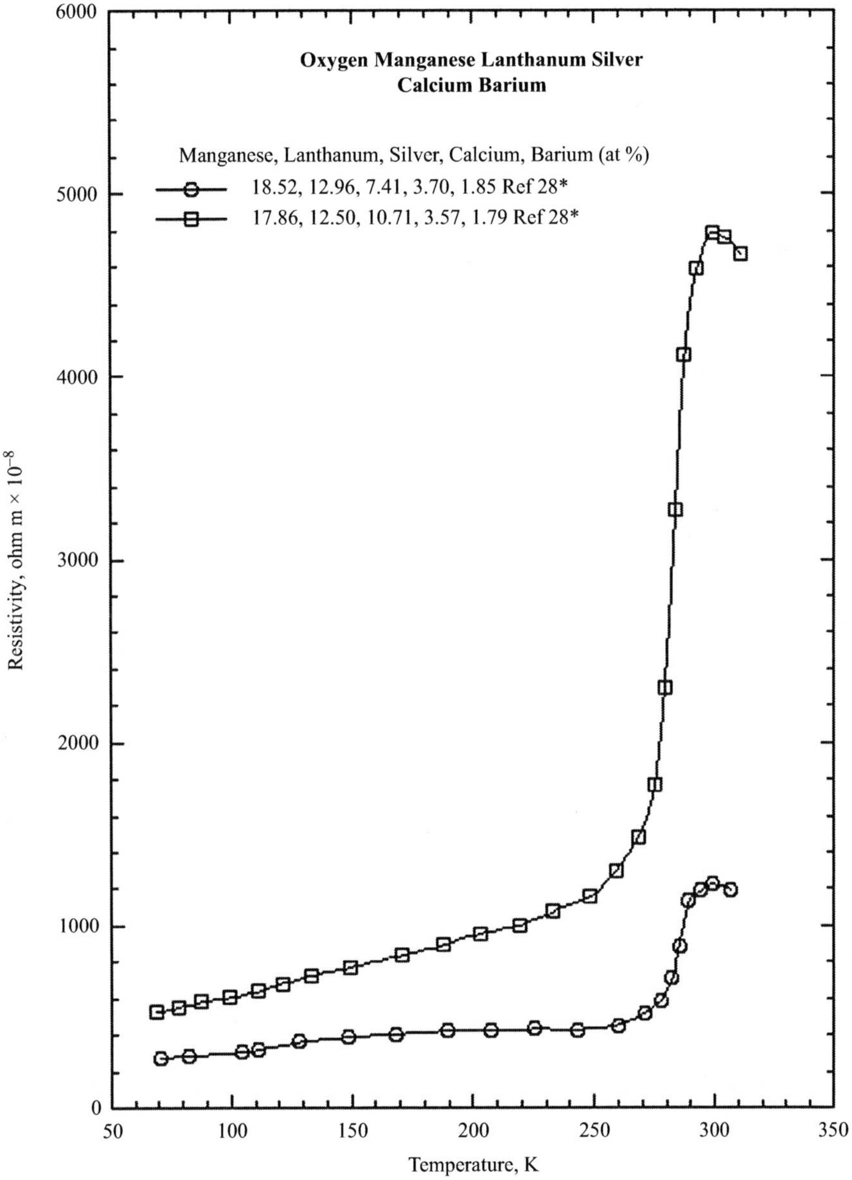
Oxygen Manganese Lanthanum Silver
Calcium Barium
Manganese, Lanthanum, Silver, Calcium, Barium (at %)
18.52, 12.96, 7.41, 3.70, 1.85 Ref 28*
17.86, 12.50, 10.71, 3.57, 1.79 Ref 28*
Resistivity, ohm m × 10⁻⁸
Temperature, K
0
1000
2000
3000
4000
5000
6000
50
100
150
200
250
300
350

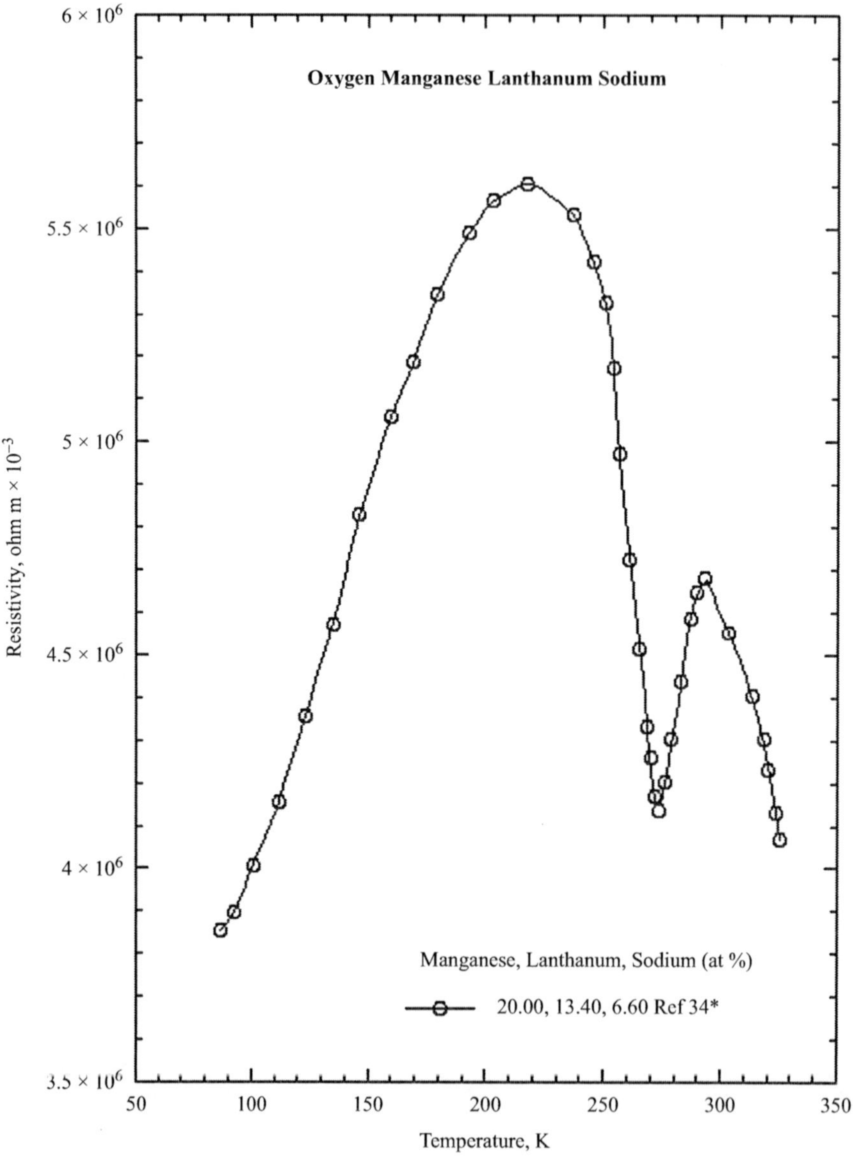
Oxygen Manganese Lanthanum Sodium
Resistivity, ohm m × 10⁻³
Temperature, K
6 × 10^6
5.5 × 10^6
5 × 10^6
4.5 × 10^6
4 × 10^6
3.5 × 10^6
50
100
150
200
250
300
350
Manganese, Lanthanum, Sodium (at %)
20.00, 13.40, 6.60 Ref 34*

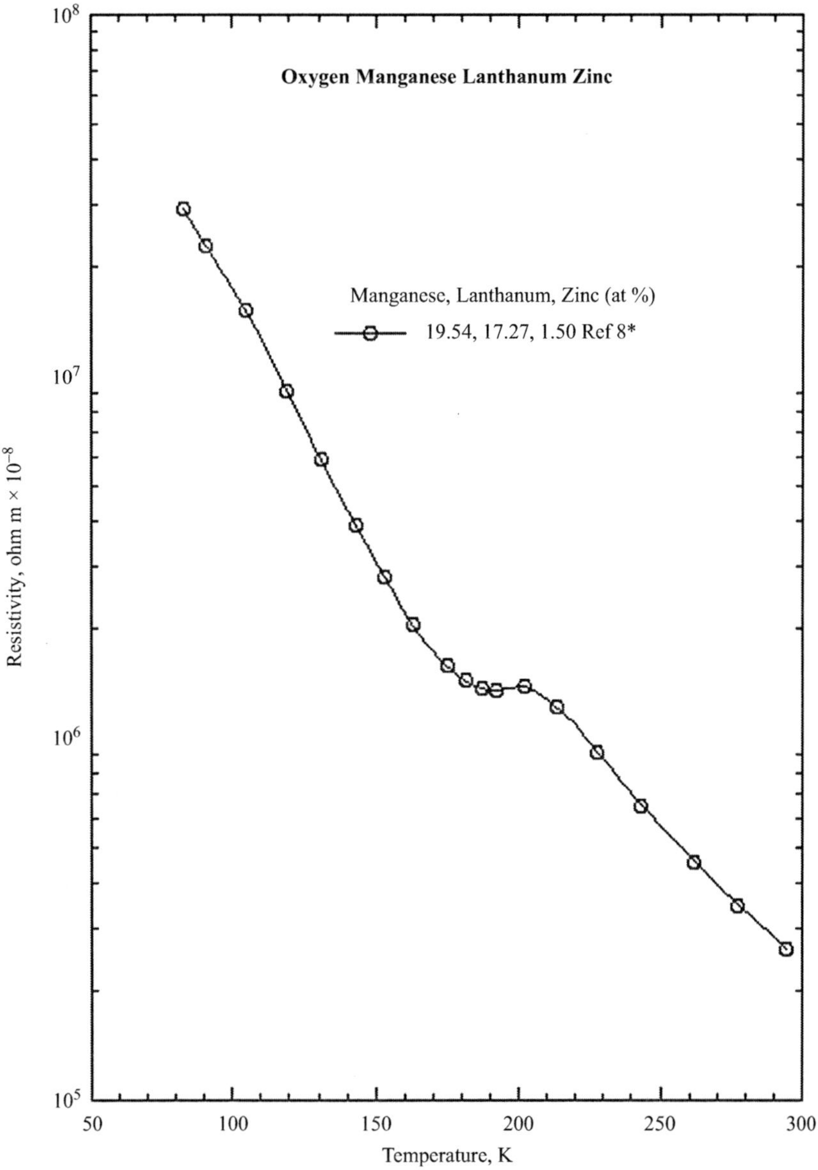
Oxygen Manganese Lanthanum Zinc
Manganese, Lanthanum, Zinc (at %)
19.54, 17.27, 1.50 Ref 8*
Resistivity, ohm m × 10⁻⁸
10⁸
10⁷
10⁶
10⁵
50
100
150
200
250
300
Temperature, K

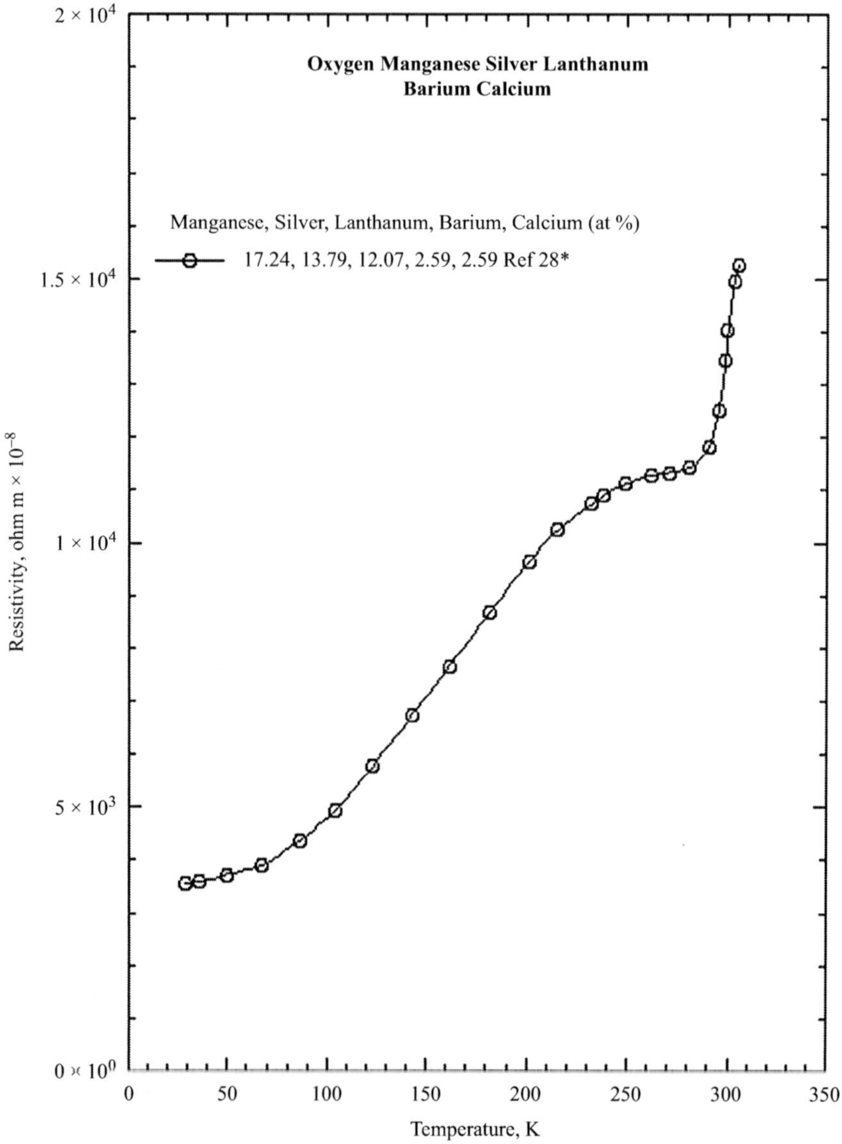
Oxygen Manganese Silver Lanthanum
Barium Calcium
Manganese, Silver, Lanthanum, Barium, Calcium (at %)
17.24, 13.79, 12.07, 2.59, 2.59 Ref 28*
2 × 10^4
1.5 × 10^4
1 × 10^4
5 × 10^3
0 × 10^0
Resistivity, ohm m × 10^-8
0
50
100
150
200
250
300
350
Temperature, K

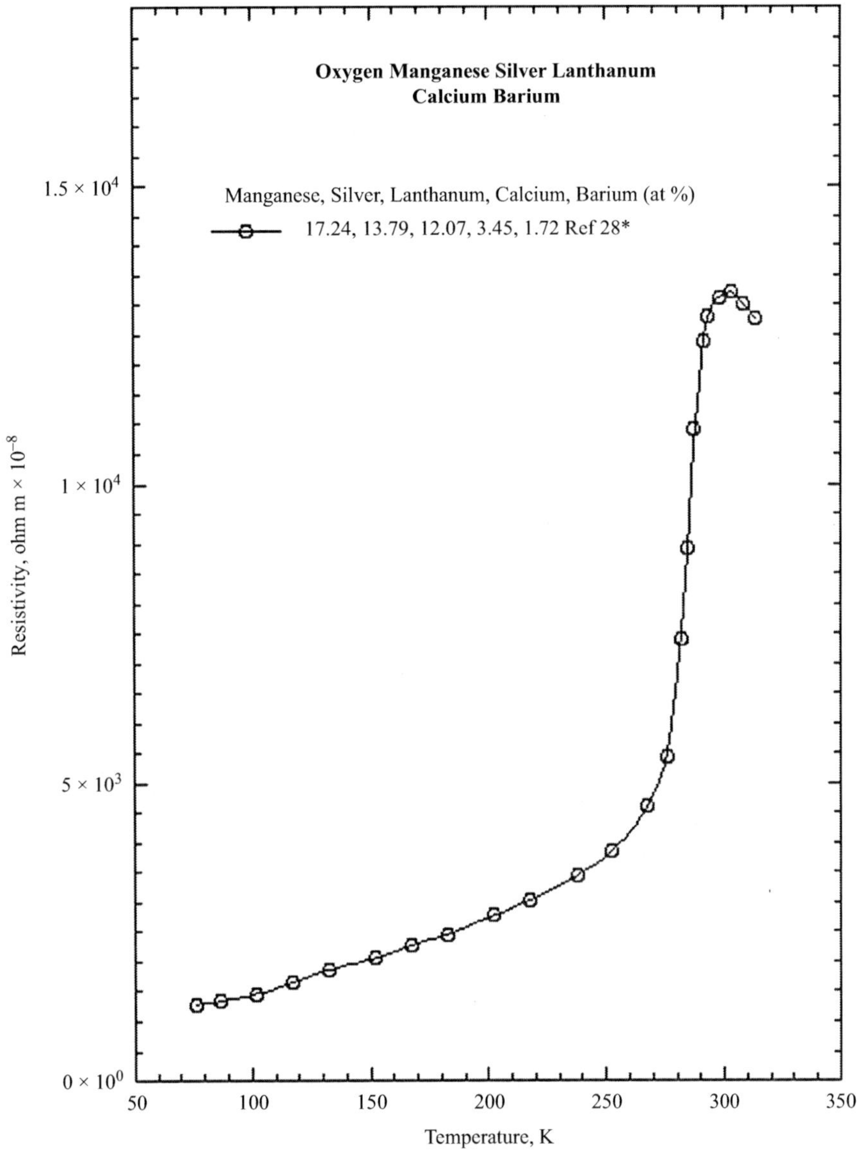
Oxygen Manganese Silver Lanthanum
Calcium Barium
Manganese, Silver, Lanthanum, Calcium, Barium (at %)
17.24, 13.79, 12.07, 3.45, 1.72 Ref 28*
Resistivity, ohm m × 10⁻⁸
1.5 × 10⁴
1 × 10⁴
5 × 10³
0 × 10⁰
50
100
150
200
250
300
350
Temperature, K

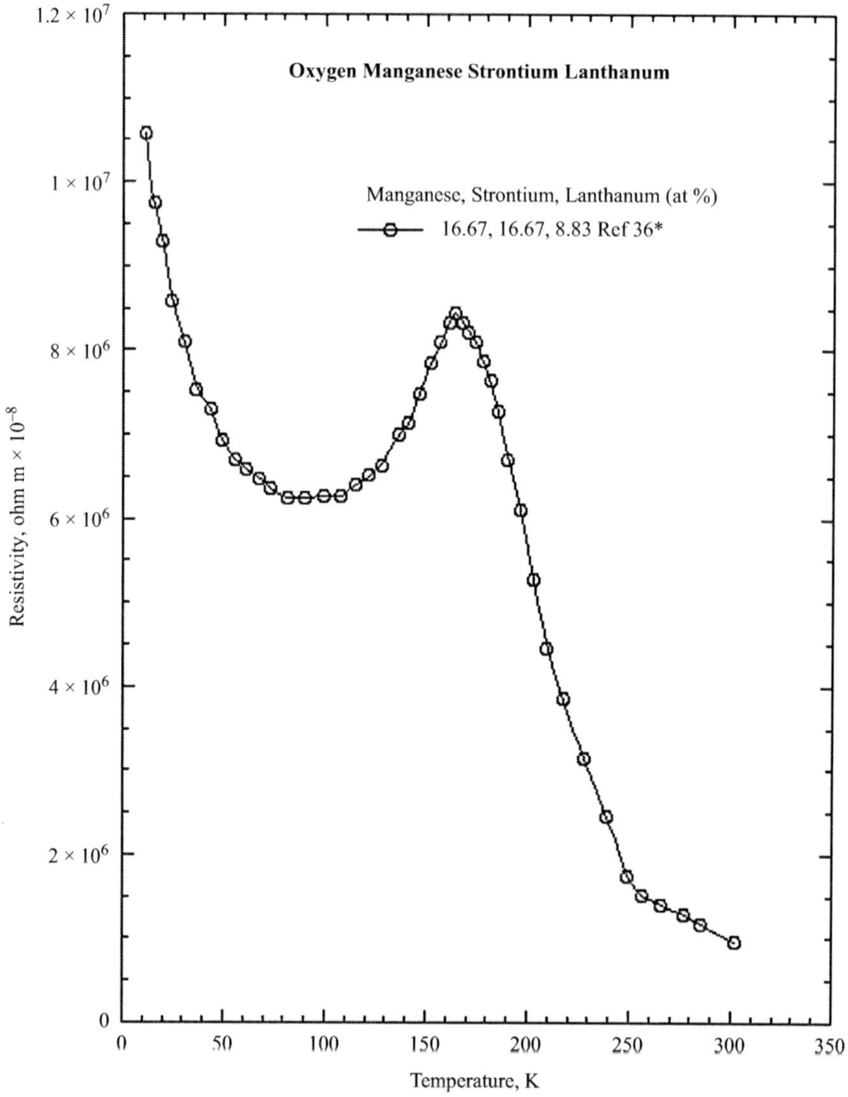
Oxygen Manganese Strontium Lanthanum
Manganese, Strontium, Lanthanum (at %)
16.67, 16.67, 8.83 Ref 36*
Resistivity, ohm m × 10⁻⁸
Temperature, K
1.2 × 10⁷
1 × 10⁷
8 × 10⁶
6 × 10⁶
4 × 10⁶
2 × 10⁶
0
0
50
100
150
200
250
300
350

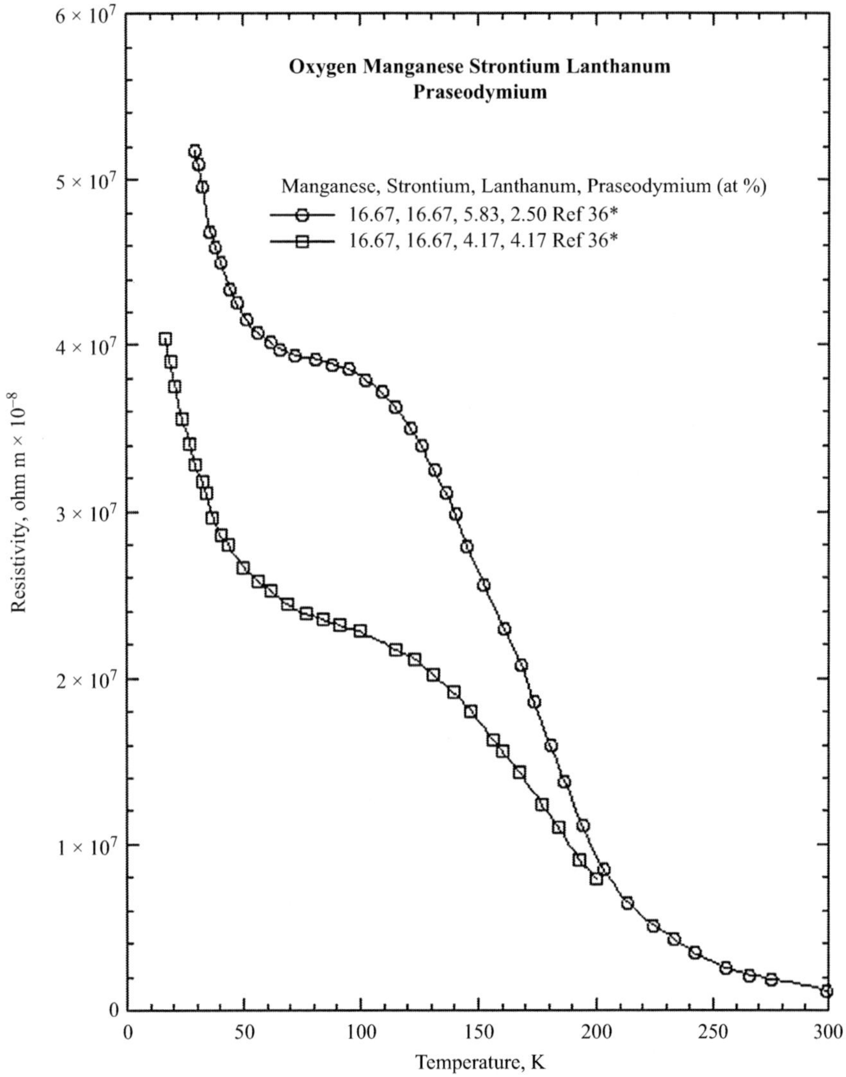
Oxygen Manganese Strontium Lanthanum Praseodymium
Manganese, Strontium, Lanthanum, Praseodymium (at %)
16.67, 16.67, 5.83, 2.50 Ref 36*
16.67, 16.67, 4.17, 4.17 Ref 36*
Resistivity, ohm m × 10⁻⁸
Temperature, K
6 × 10⁷
5 × 10⁷
4 × 10⁷
3 × 10⁷
2 × 10⁷
1 × 10⁷
0
0
50
100
150
200
250
300

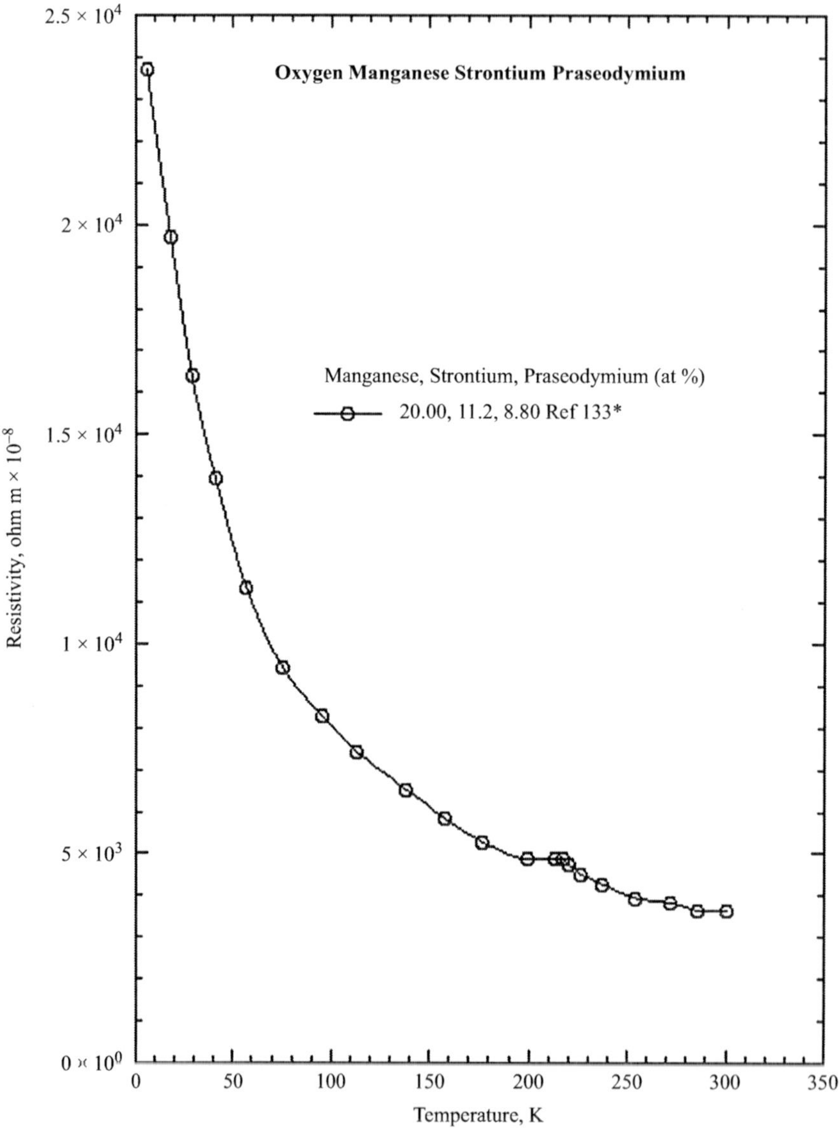
Oxygen Manganese Strontium Praseodymium
Manganese, Strontium, Praseodymium (at %)
20.00, 11.2, 8.80 Ref 133*
Resistivity, ohm m × 10⁻⁸
2.5 × 10⁴
2 × 10⁴
1.5 × 10⁴
1 × 10⁴
5 × 10³
0 × 10⁰
0
50
100
150
200
250
300
350
Temperature, K

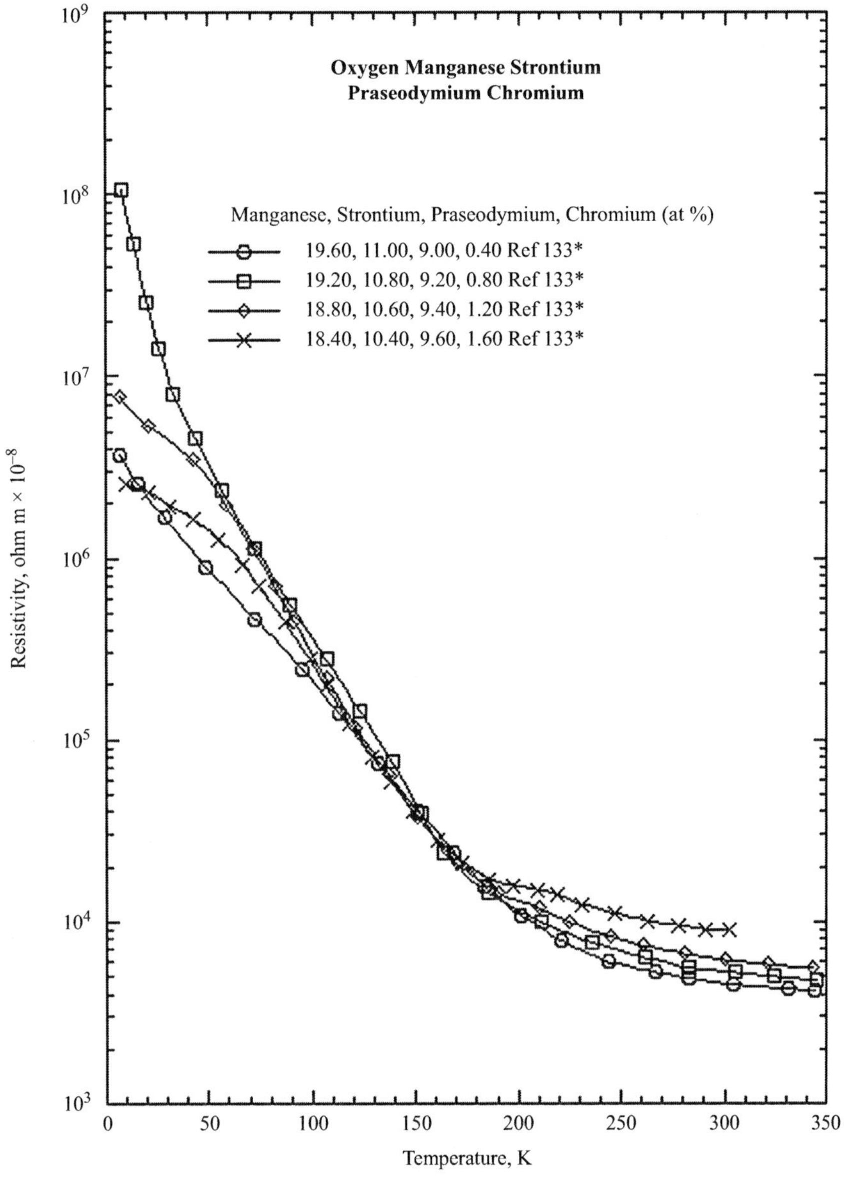

Oxygen Manganese Strontium
Praseodymium Chromium
Manganese, Strontium, Praseodymium, Chromium (at %)
19.60, 11.00, 9.00, 0.40 Ref 133*
19.20, 10.80, 9.20, 0.80 Ref 133*
18.80, 10.60, 9.40, 1.20 Ref 133*
18.40, 10.40, 9.60, 1.60 Ref 133*
Resistivity, ohm m × 10⁻⁸
Temperature, K

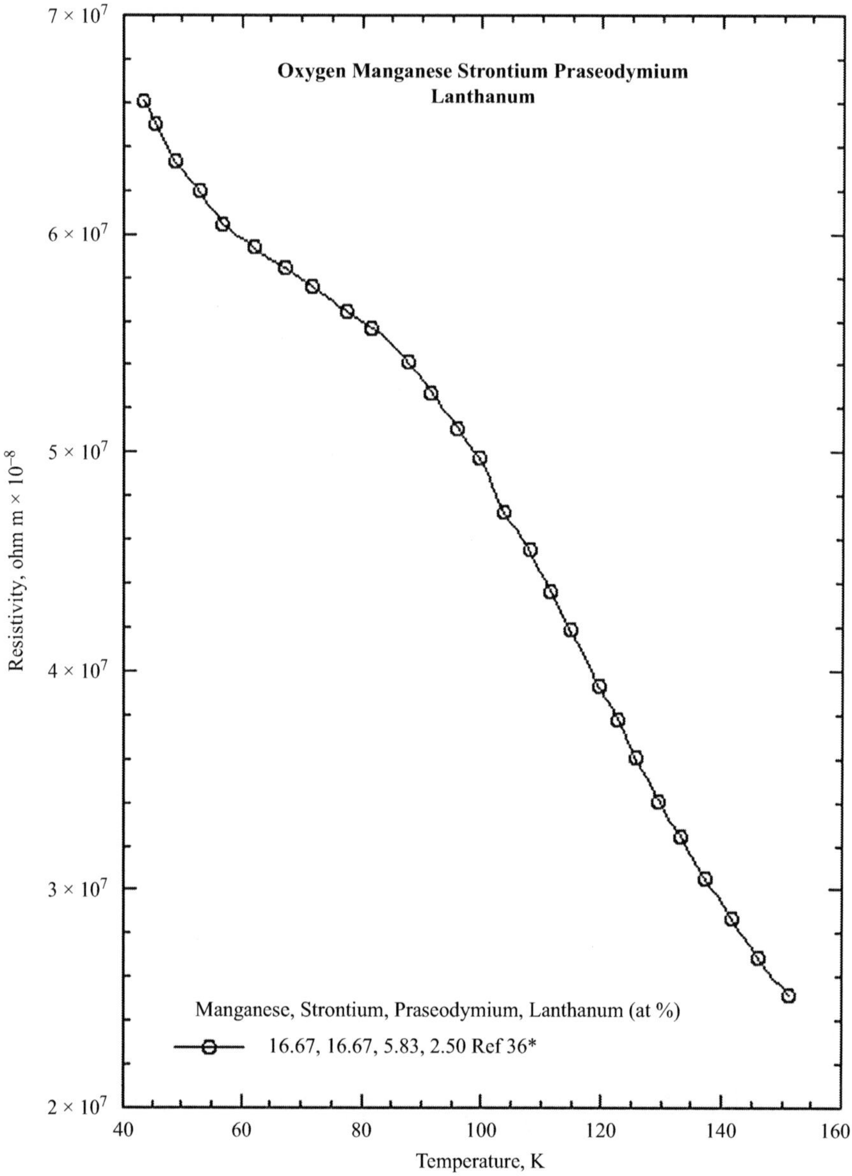
Oxygen Manganese Strontium Praseodymium Lanthanum
Manganese, Strontium, Praseodymium, Lanthanum (at %)
16.67, 16.67, 5.83, 2.50 Ref 36*
Resistivity, ohm m × 10–8
Temperature, K
7 × 10^7
6 × 10^7
5 × 10^7
4 × 10^7
3 × 10^7
2 × 10^7
40
60
80
100
120
140
160

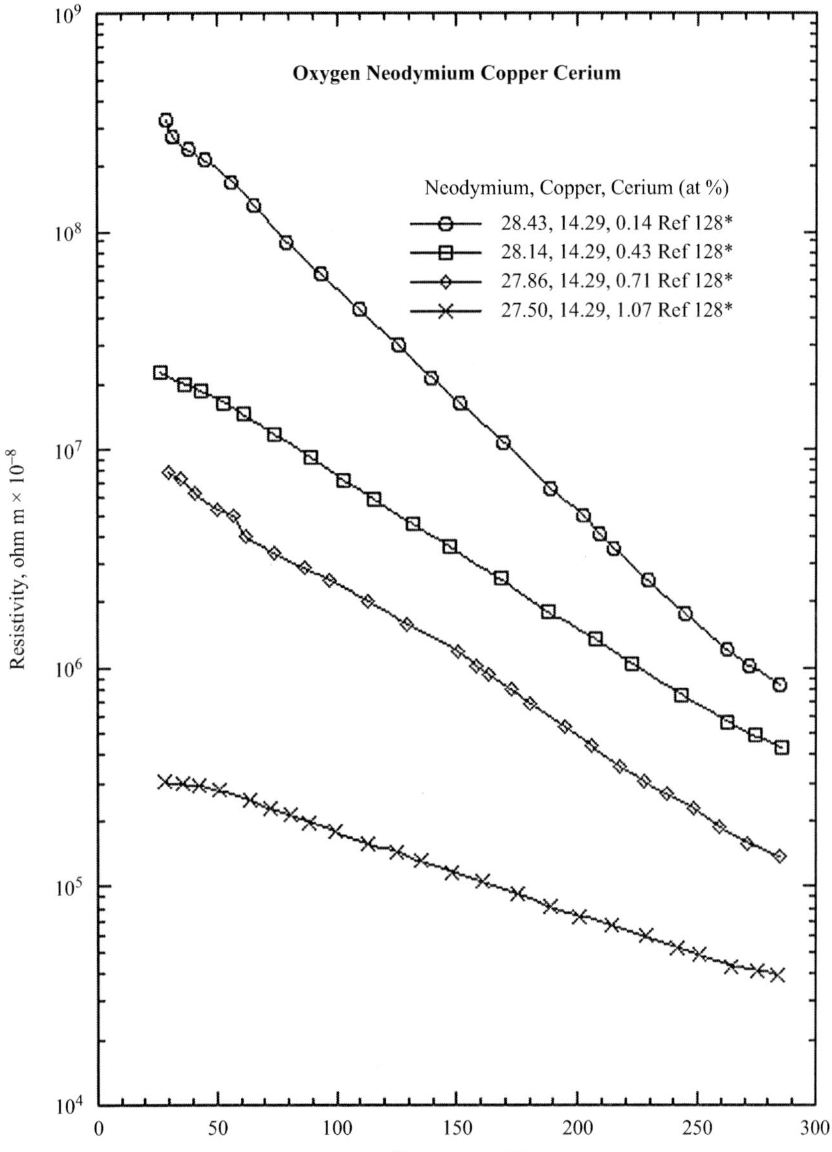

Oxygen Neodymium Copper Cerium
Neodymium, Copper, Cerium (at %)
28.43, 14.29, 0.14 Ref 128*
28.14, 14.29, 0.43 Ref 128*
27.86, 14.29, 0.71 Ref 128*
27.50, 14.29, 1.07 Ref 128*
Resistivity, ohm m × 10⁻⁸
Temperature, K

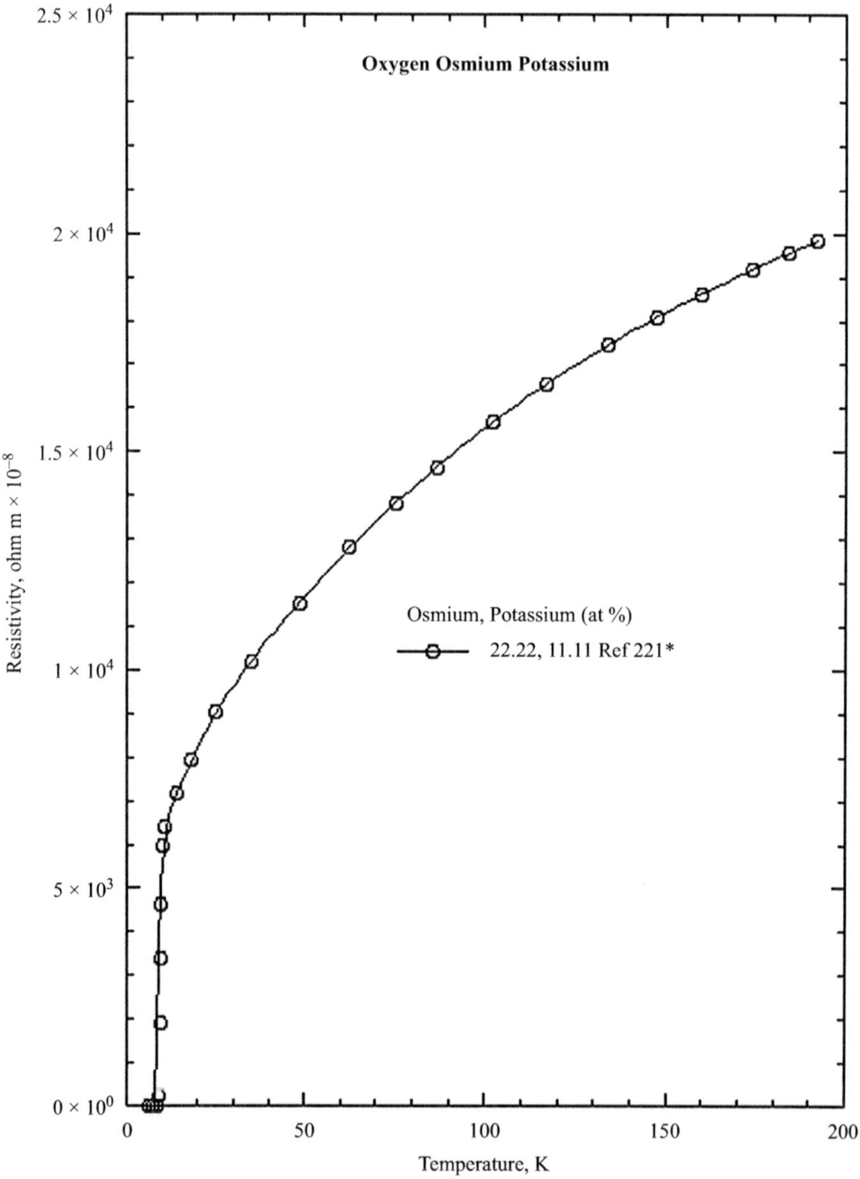
Oxygen Osmium Potassium
Osmium, Potassium (at %)
22.22, 11.11 Ref 221*
Resistivity, ohm m × 10⁻⁸
2.5 × 10⁴
2 × 10⁴
1.5 × 10⁴
1 × 10⁴
5 × 10³
0 × 10⁰
0
50
100
150
200
Temperature, K

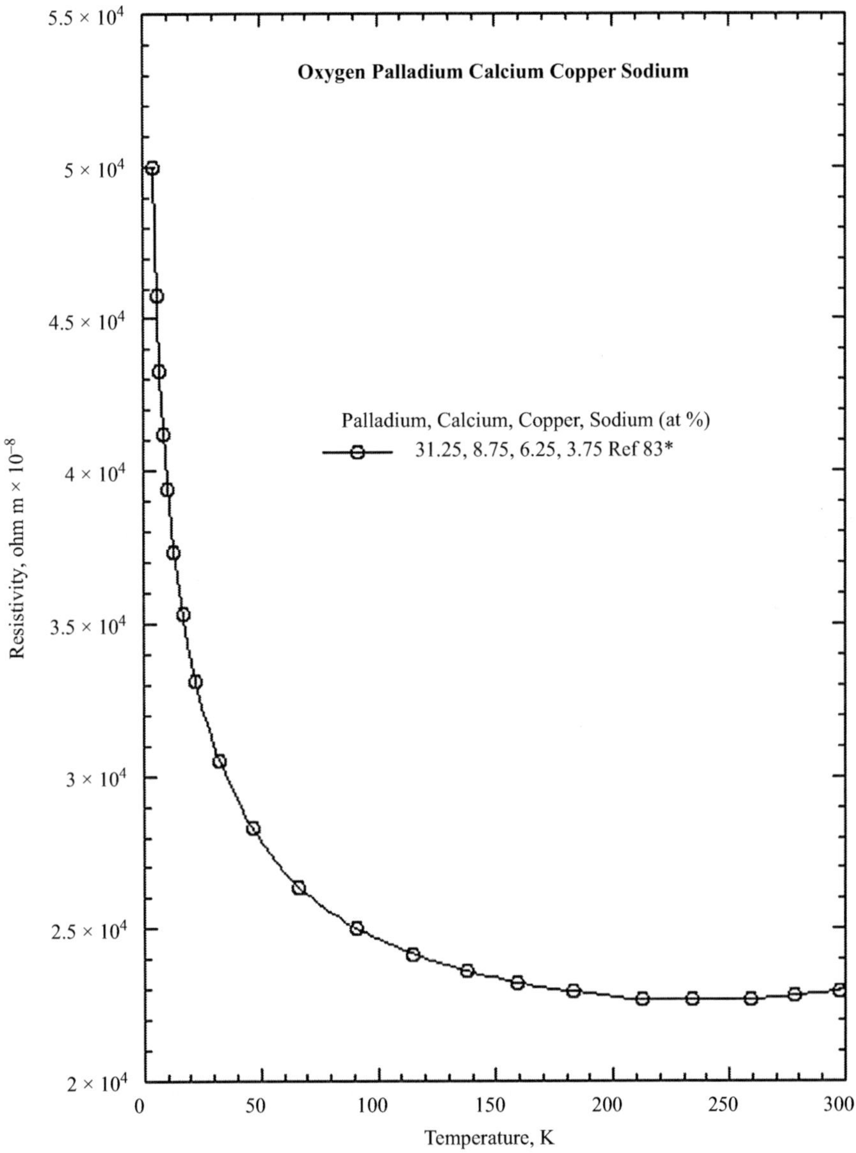
Oxygen Palladium Calcium Copper Sodium
Palladium, Calcium, Copper, Sodium (at %)
31.25, 8.75, 6.25, 3.75 Ref 83*
Resistivity, ohm m × 10⁻⁸
5.5 × 10⁴
5 × 10⁴
4.5 × 10⁴
4 × 10⁴
3.5 × 10⁴
3 × 10⁴
2.5 × 10⁴
2 × 10⁴
0
50
100
150
200
250
300
Temperature, K

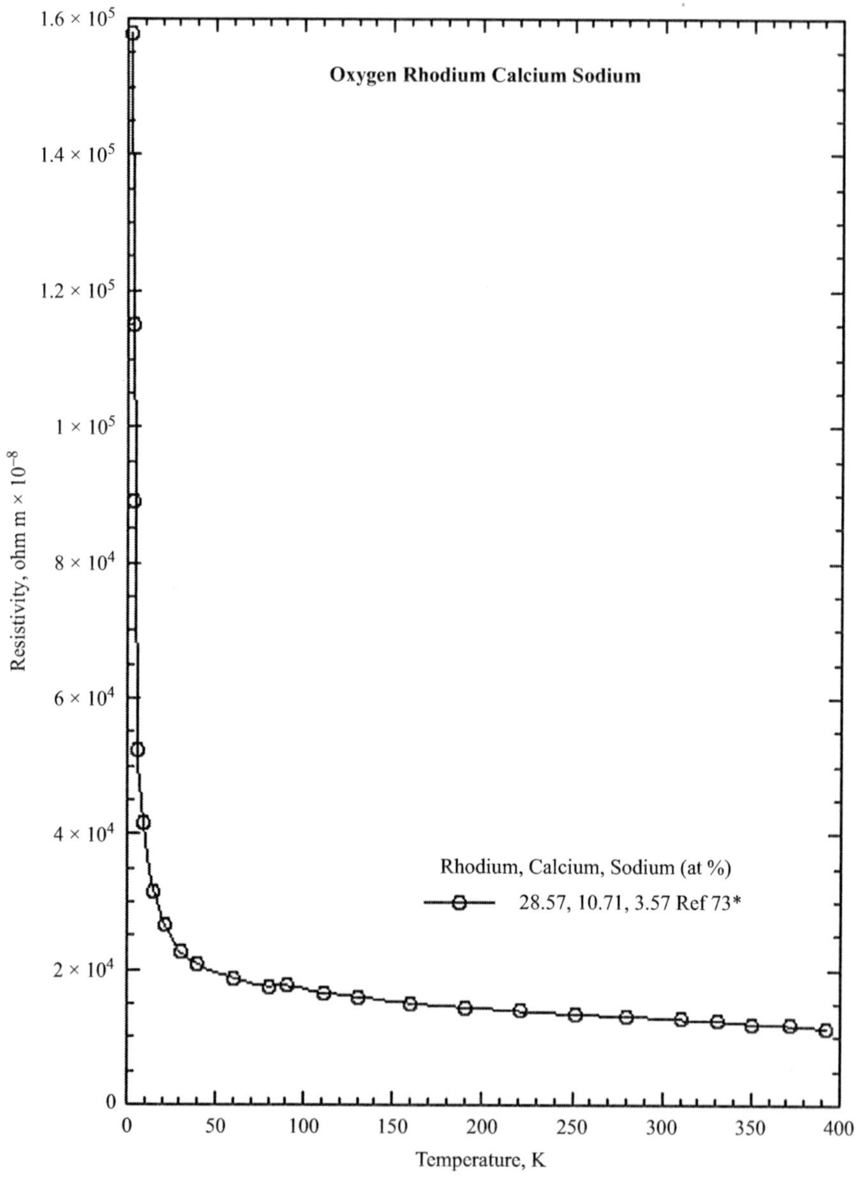
Oxygen Rhodium Calcium Sodium
Rhodium, Calcium, Sodium (at %)
28.57, 10.71, 3.57 Ref 73*
Resistivity, ohm m × 10^−8
1.6 × 10^5
1.4 × 10^5
1.2 × 10^5
1 × 10^5
8 × 10^4
6 × 10^4
4 × 10^4
2 × 10^4
0
0
50
100
150
200
250
300
350
400
Temperature, K

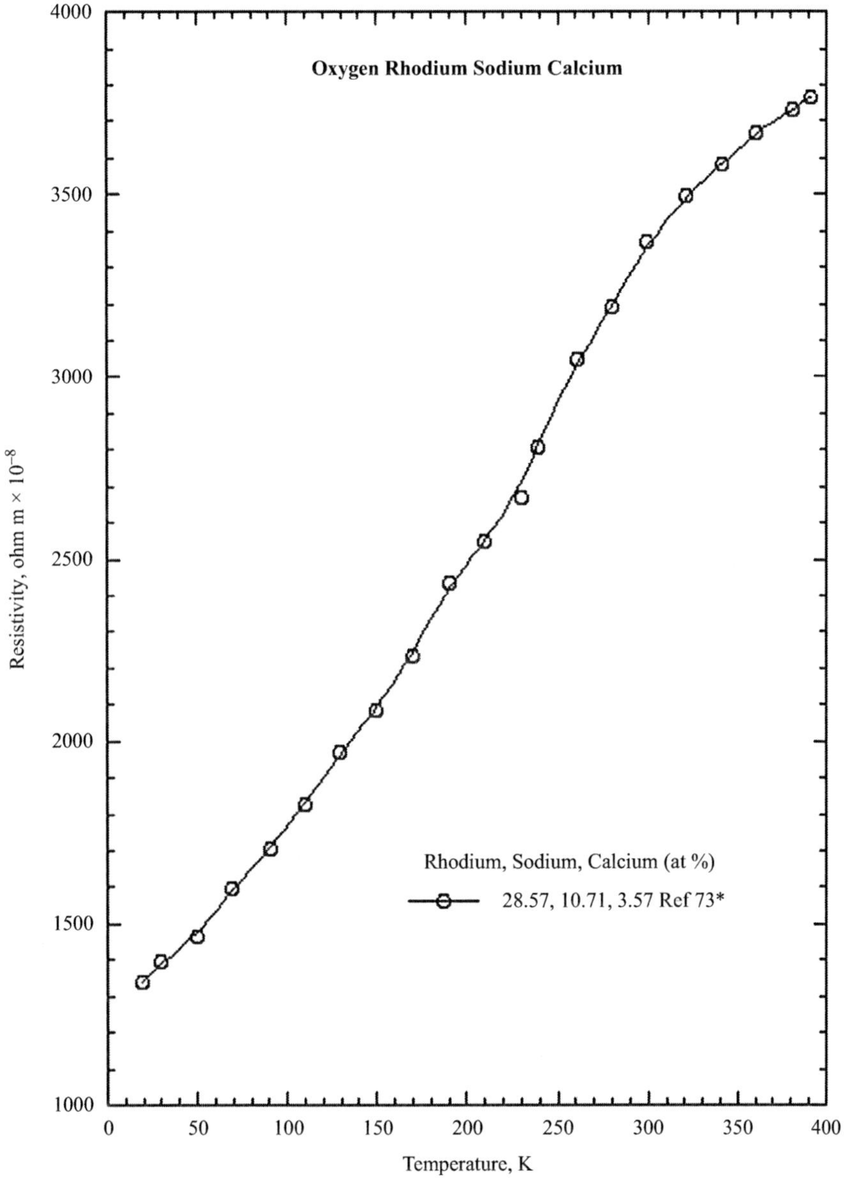

Oxygen Rhodium Sodium Calcium
Rhodium, Sodium, Calcium (at %)
28.57, 10.71, 3.57 Ref 73*
Resistivity, ohm m × 10⁻⁸
Temperature, K
1000
1500
2000
2500
3000
3500
4000
0
50
100
150
200
250
300
350
400

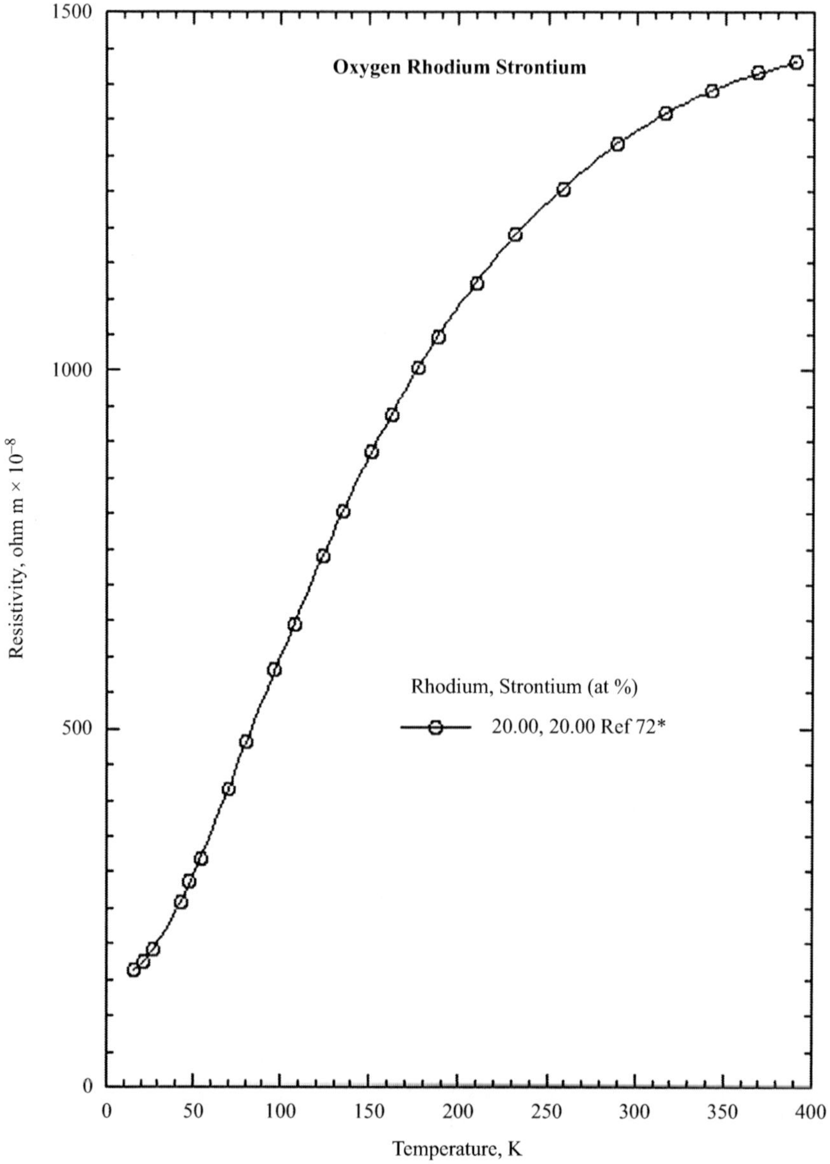
Oxygen Rhodium Strontium
Rhodium, Strontium (at %)
20.00, 20.00 Ref 72*
Resistivity, ohm m × 10−8
Temperature, K
0
500
1000
1500
0
50
100
150
200
250
300
350
400

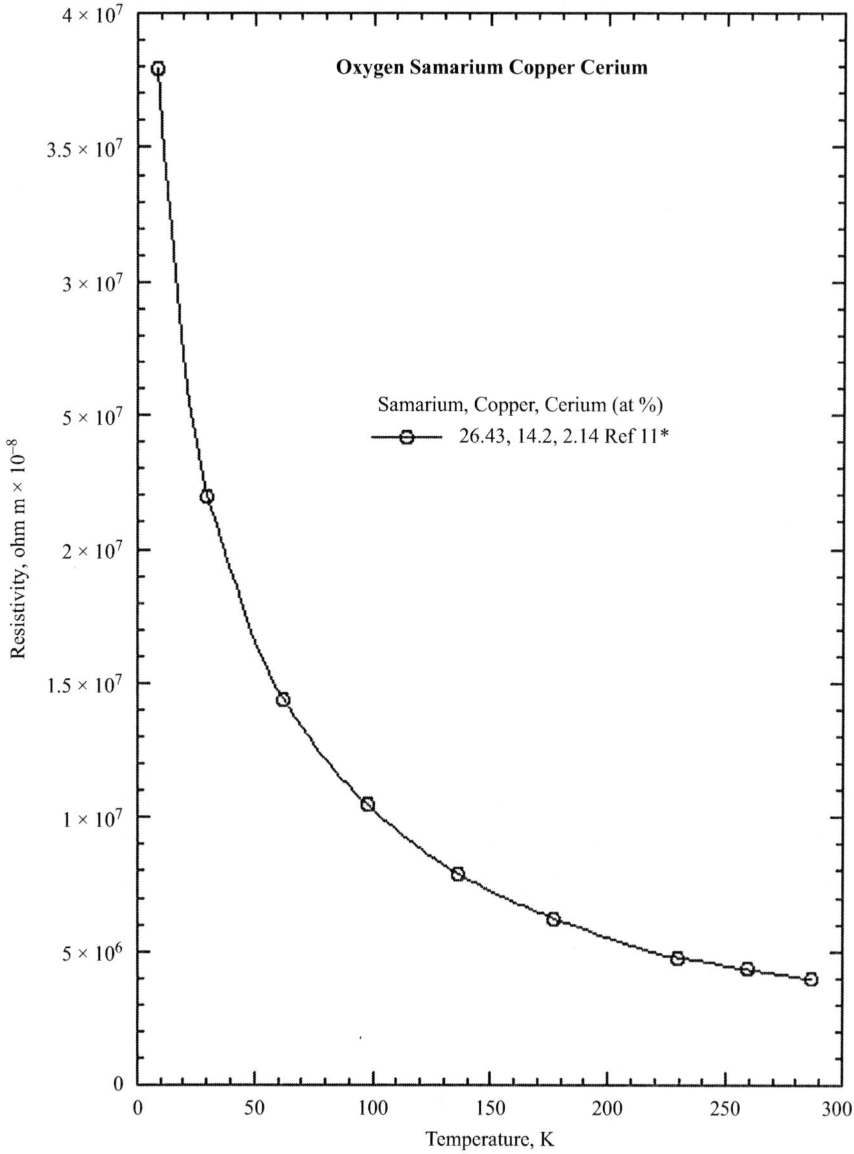
Oxygen Samarium Copper Cerium
Samarium, Copper, Cerium (at %)
26.43, 14.2, 2.14 Ref 11*
Resistivity, ohm m × 10⁻⁸
Temperature, K
4 × 10⁷
3.5 × 10⁷
3 × 10⁷
5 × 10⁷
2 × 10⁷
1.5 × 10⁷
1 × 10⁷
5 × 10⁶
0
0
50
100
150
200
250
300

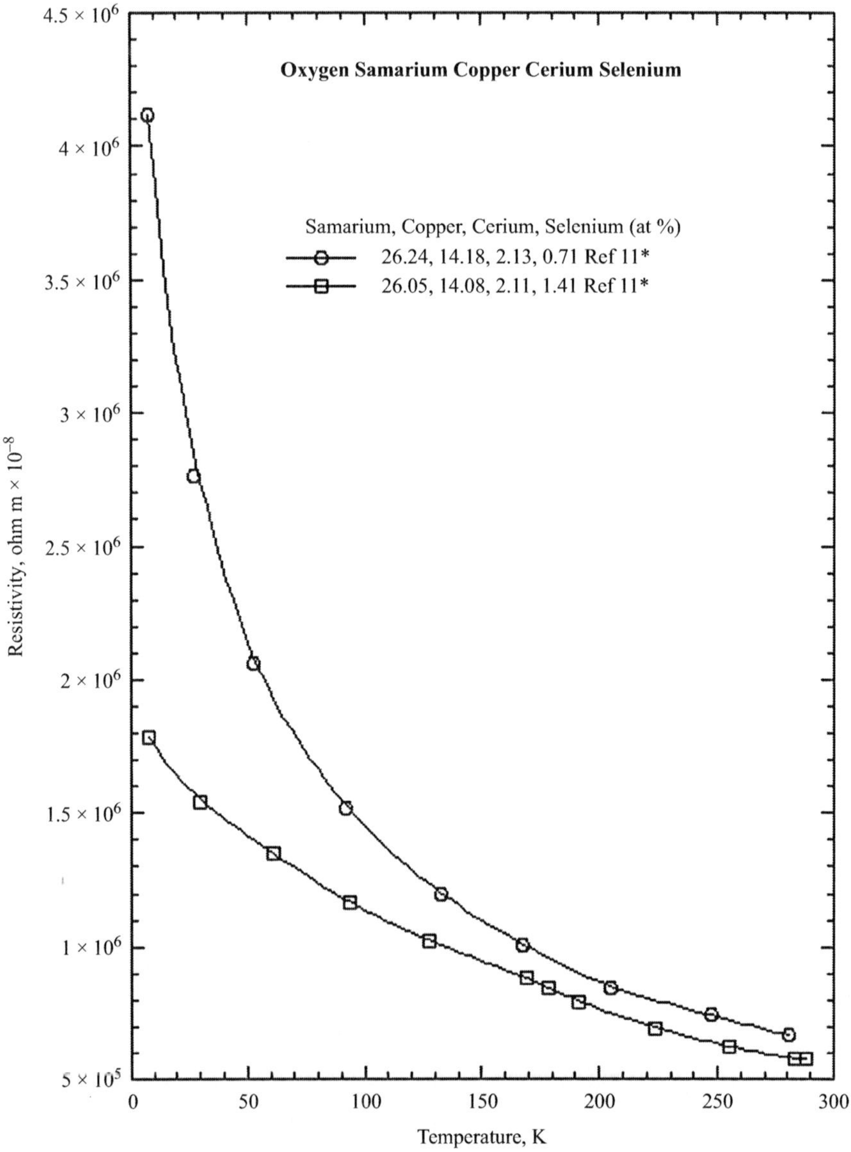
Oxygen Samarium Copper Cerium Selenium
Samarium, Copper, Cerium, Selenium (at %)
26.24, 14.18, 2.13, 0.71 Ref 11*
26.05, 14.08, 2.11, 1.41 Ref 11*
Resistivity, ohm m × 10^-8
Temperature, K
4.5 × 10^6
4 × 10^6
3.5 × 10^6
3 × 10^6
2.5 × 10^6
2 × 10^6
1.5 × 10^6
1 × 10^6
5 × 10^5
0
50
100
150
200
250
300

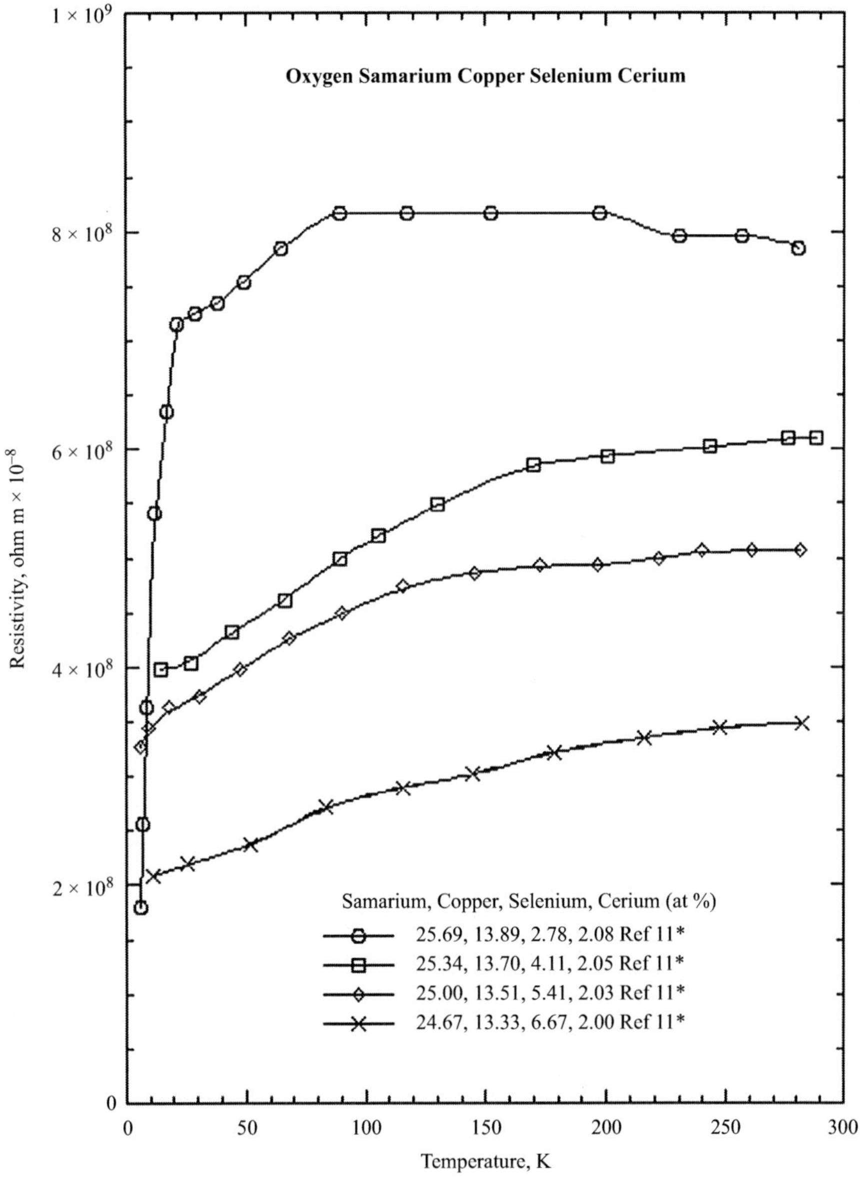
Oxygen Samarium Copper Selenium Cerium
1 × 10^9
8 × 10^8
6 × 10^8
4 × 10^8
2 × 10^8
0
Resistivity, ohm m × 10^−8
0
50
100
150
200
250
300
Temperature, K
Samarium, Copper, Selenium, Cerium (at %)
25.69, 13.89, 2.78, 2.08 Ref 11*
25.34, 13.70, 4.11, 2.05 Ref 11*
25.00, 13.51, 5.41, 2.03 Ref 11*
24.67, 13.33, 6.67, 2.00 Ref 11*

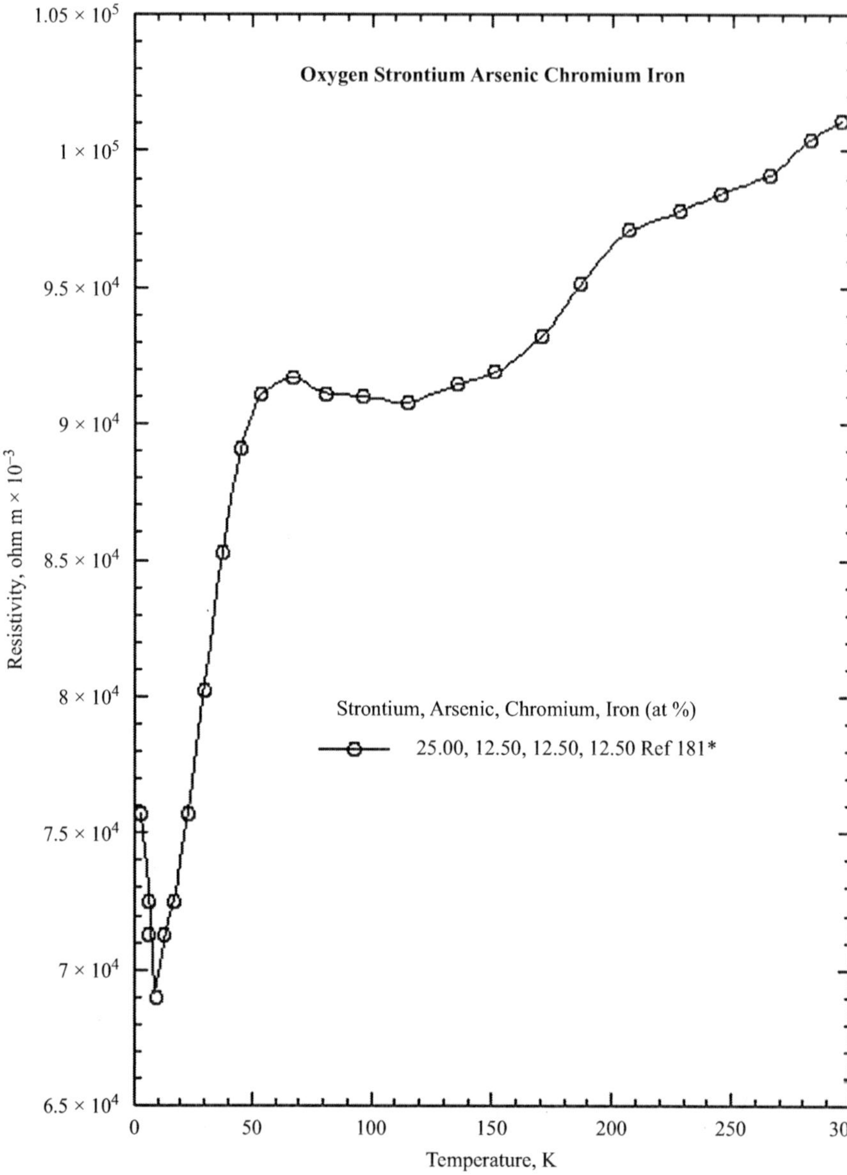
Oxygen Strontium Arsenic Chromium Iron
Strontium, Arsenic, Chromium, Iron (at %)
25.00, 12.50, 12.50, 12.50 Ref 181*
Resistivity, ohm m × 10^-3
1.05 × 10^5
1 × 10^5
9.5 × 10^4
9 × 10^4
8.5 × 10^4
8 × 10^4
7.5 × 10^4
7 × 10^4
6.5 × 10^4
0
50
100
150
200
250
300
Temperature, K

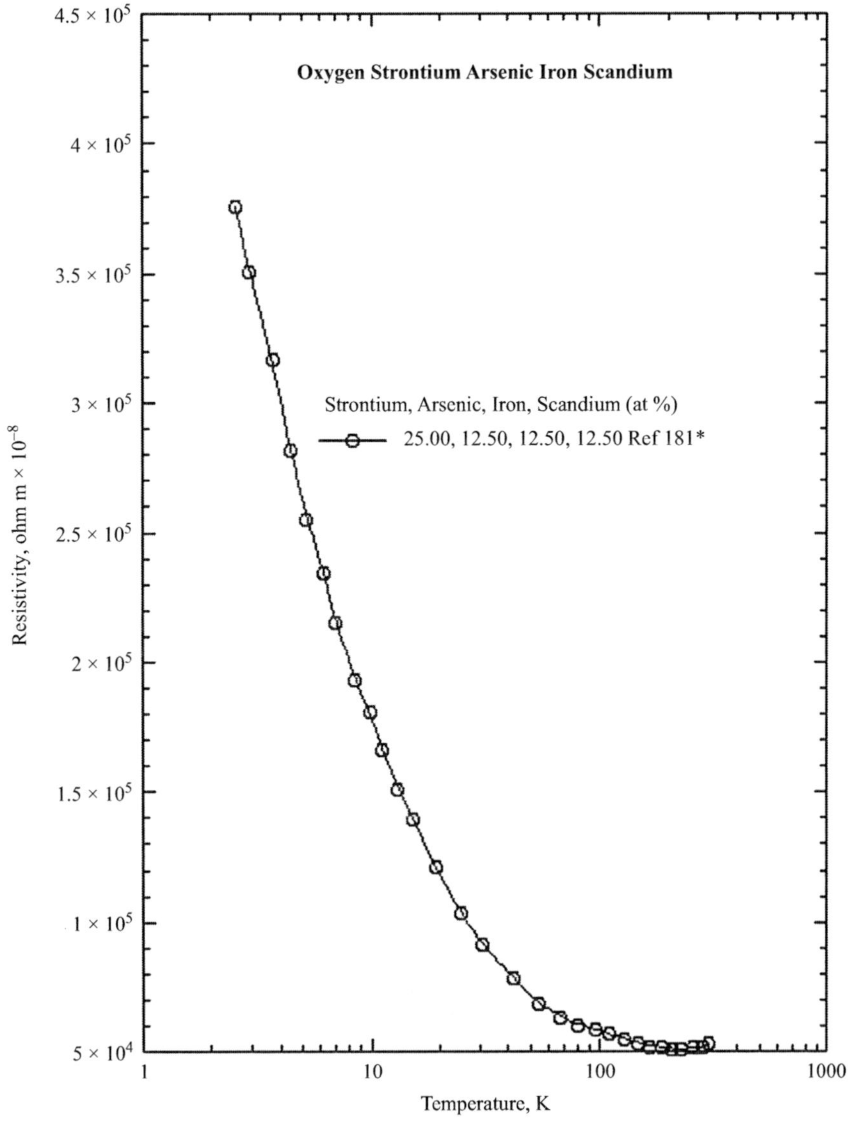
Oxygen Strontium Arsenic Iron Scandium
Strontium, Arsenic, Iron, Scandium (at %)
25.00, 12.50, 12.50, 12.50 Ref 181*
Resistivity, ohm m × 10⁻⁸
4.5 × 10⁵
4 × 10⁵
3.5 × 10⁵
3 × 10⁵
2.5 × 10⁵
2 × 10⁵
1.5 × 10⁵
1 × 10⁵
5 × 10⁴
1
10
100
1000
Temperature, K

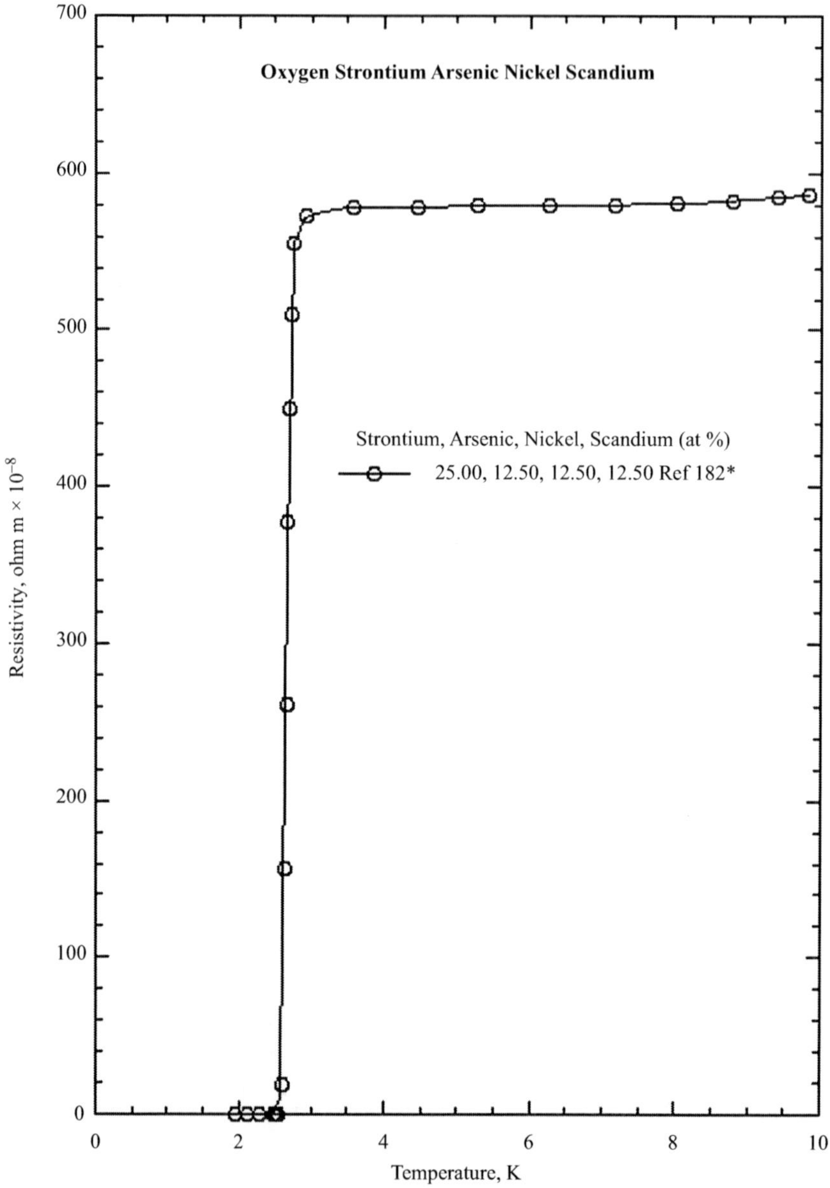
Oxygen Strontium Arsenic Nickel Scandium
Strontium, Arsenic, Nickel, Scandium (at %)
25.00, 12.50, 12.50, 12.50 Ref 182*
Resistivity, ohm m × 10⁻⁸
Temperature, K
0
100
200
300
400
500
600
700
2
4
6
8
10

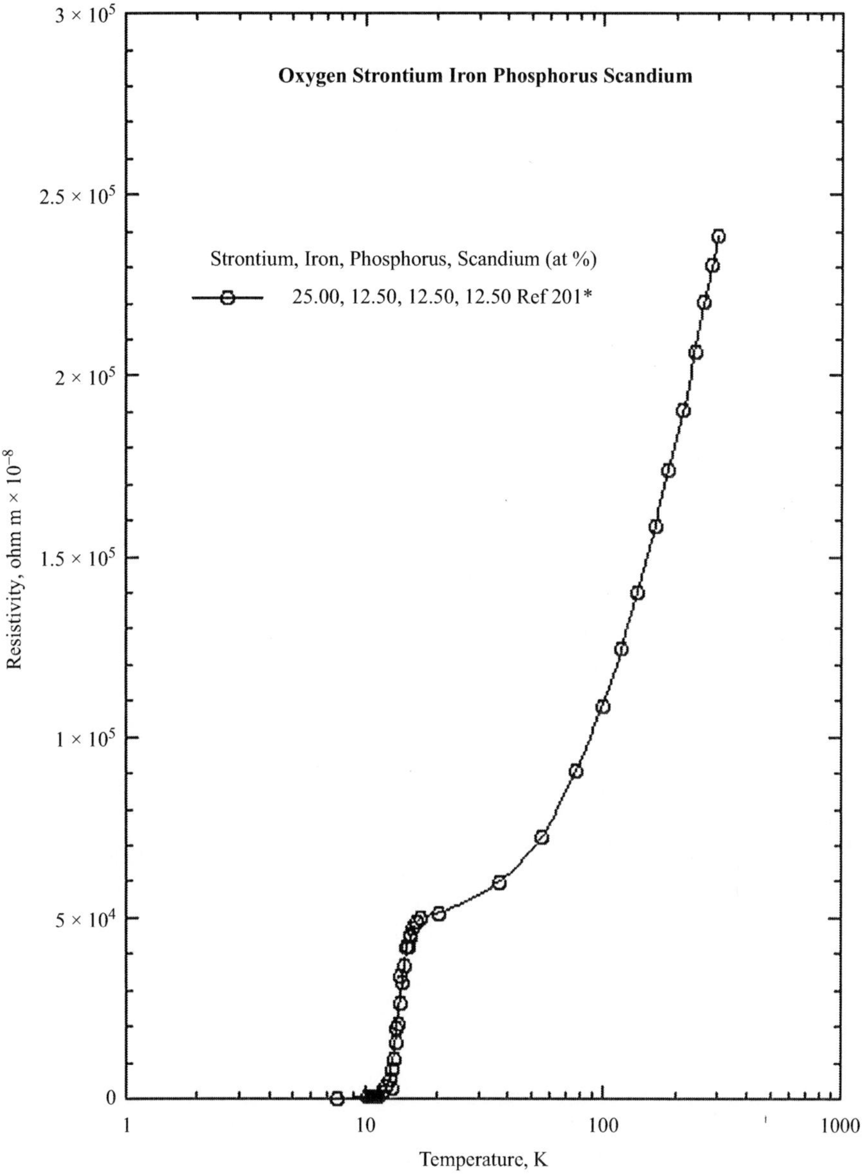
Oxygen Strontium Iron Phosphorus Scandium
Strontium, Iron, Phosphorus, Scandium (at %)
25.00, 12.50, 12.50, 12.50 Ref 201*
Resistivity, ohm m × 10−8
3 × 105
2.5 × 105
2 × 105
1.5 × 105
1 × 105
5 × 104
0
1
10
100
1000
Temperature, K

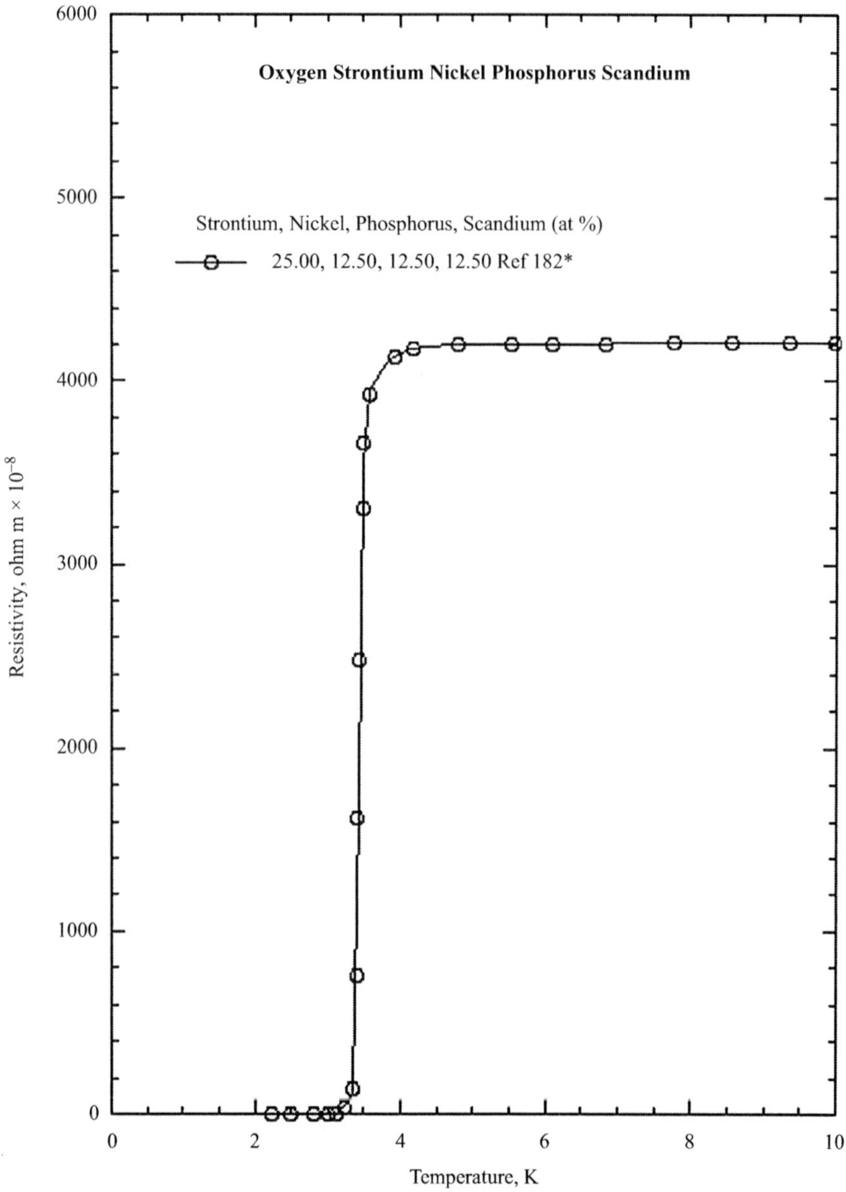
Oxygen Strontium Nickel Phosphorus Scandium
Strontium, Nickel, Phosphorus, Scandium (at %)
25.00, 12.50, 12.50, 12.50 Ref 182*
Resistivity, ohm m × 10⁻⁸
Temperature, K
0
1000
2000
3000
4000
5000
6000
0
2
4
6
8
10

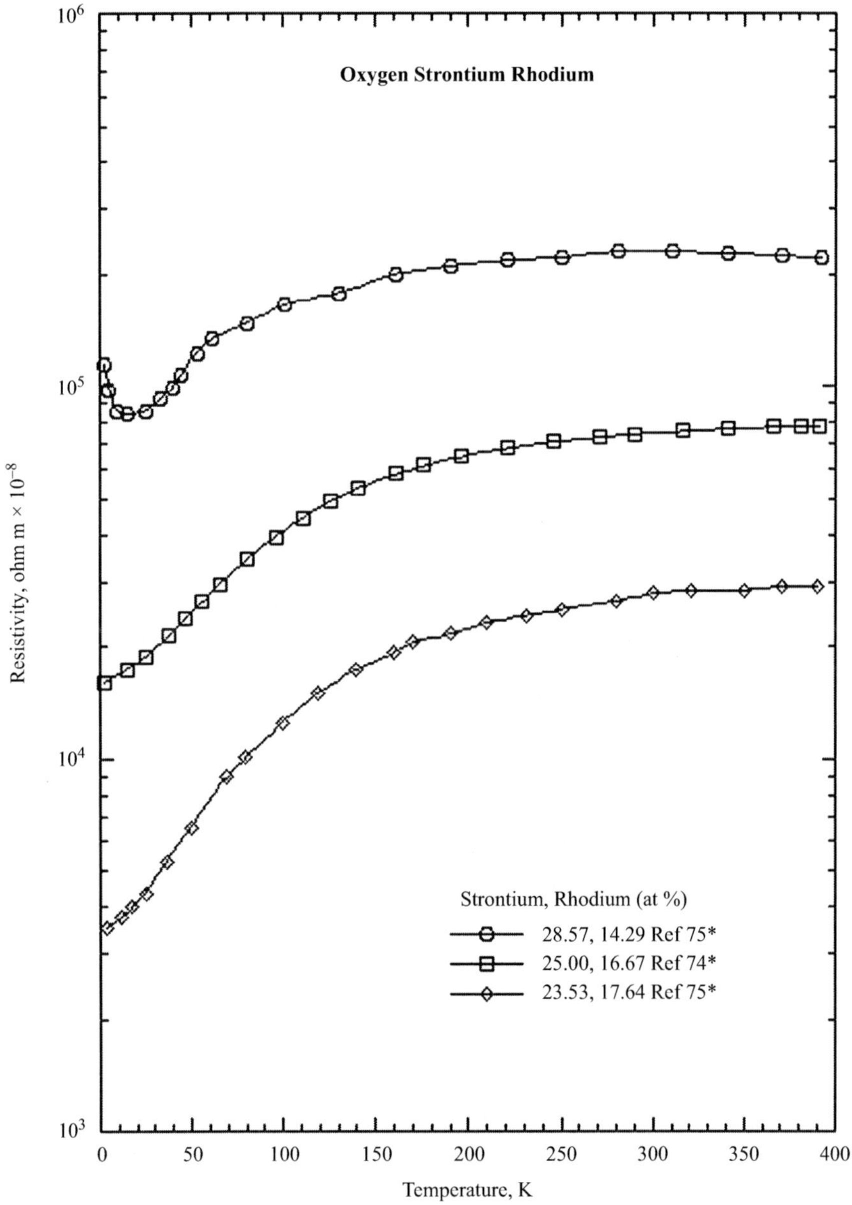

Oxygen Strontium Rhodium
Resistivity, ohm m × 10⁻⁸
10⁶
10⁵
10⁴
10³
0
50
100
150
200
250
300
350
400
Temperature, K
Strontium, Rhodium (at %)
28.57, 14.29 Ref 75*
25.00, 16.67 Ref 74*
23.53, 17.64 Ref 75*

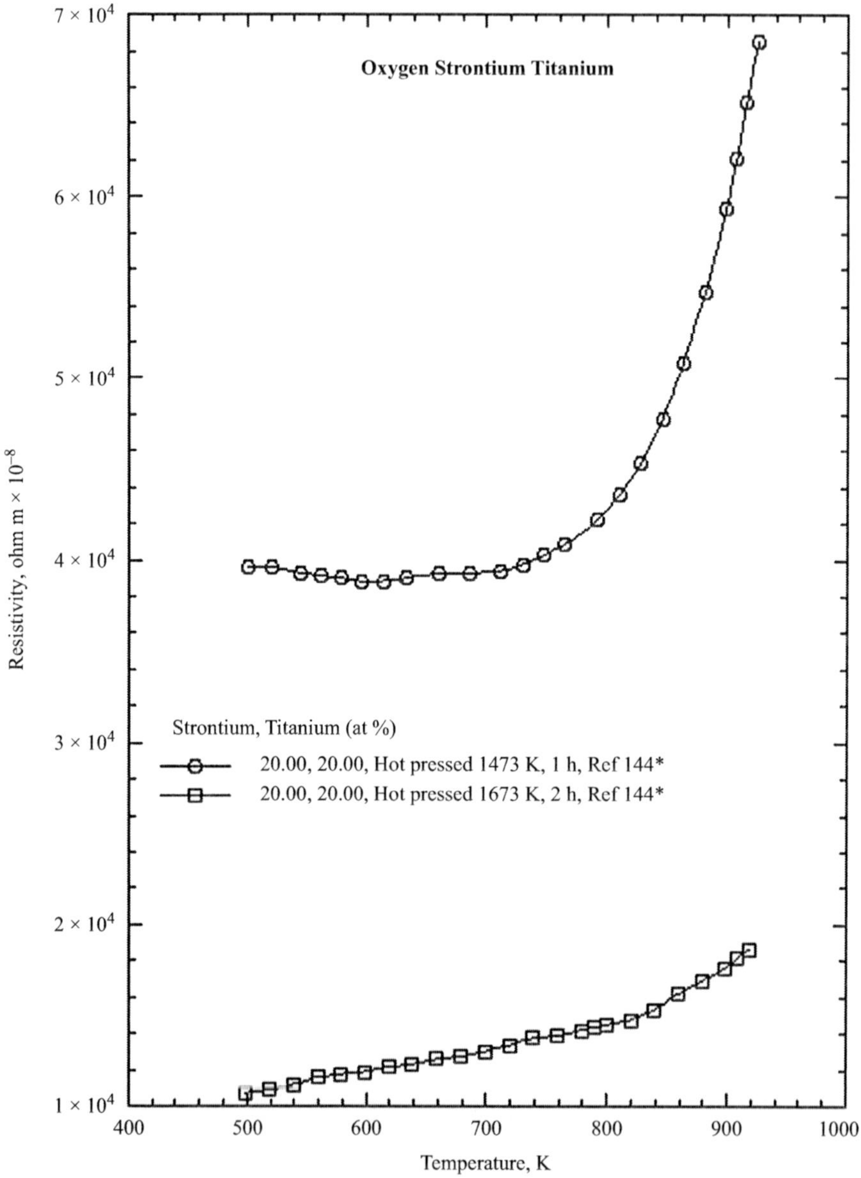
Oxygen Strontium Titanium
7 × 10^4
6 × 10^4
5 × 10^4
4 × 10^4
3 × 10^4
2 × 10^4
1 × 10^4
Resistivity, ohm m × 10^-8
400
500
600
700
800
900
1000
Temperature, K
Strontium, Titanium (at %)
20.00, 20.00, Hot pressed 1473 K, 1 h, Ref 144*
20.00, 20.00, Hot pressed 1673 K, 2 h, Ref 144*

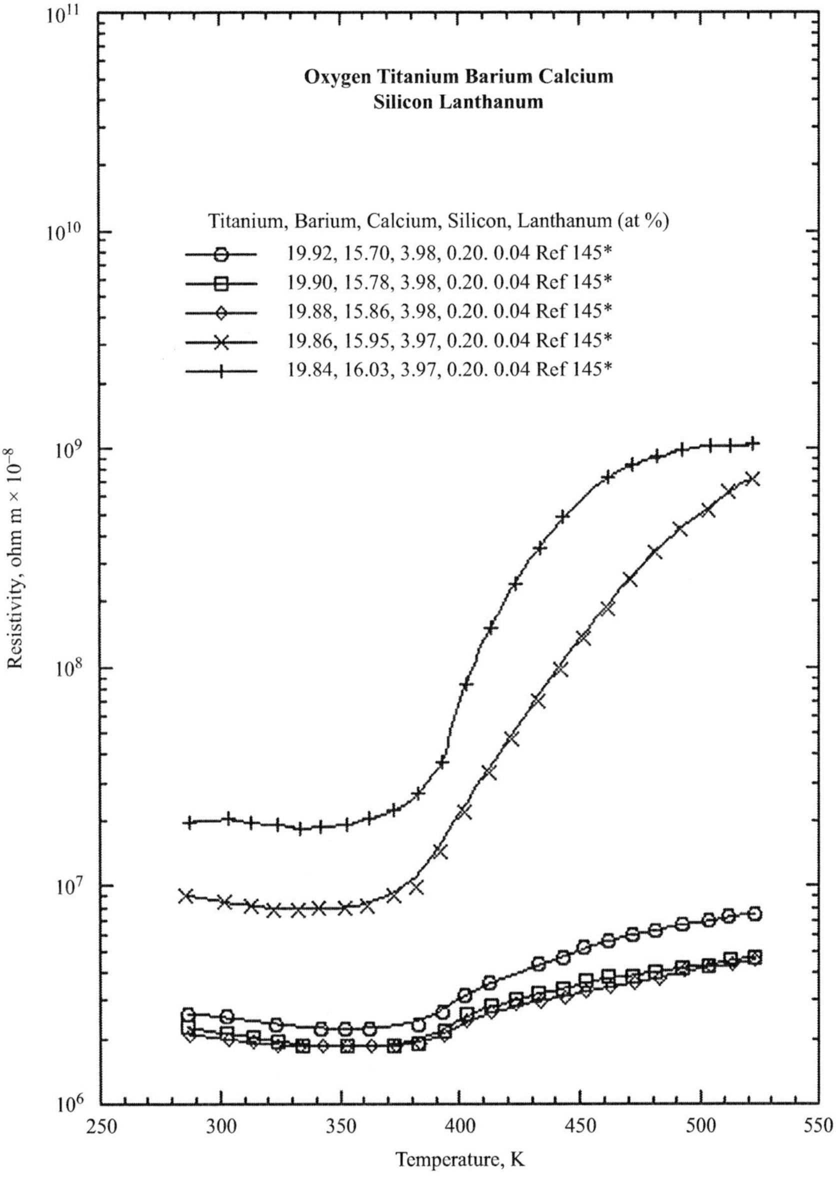
Oxygen Titanium Barium Calcium
Silicon Lanthanum
Titanium, Barium, Calcium, Silicon, Lanthanum (at %)
19.92, 15.70, 3.98, 0.20. 0.04 Ref 145*
19.90, 15.78, 3.98, 0.20. 0.04 Ref 145*
19.88, 15.86, 3.98, 0.20. 0.04 Ref 145*
19.86, 15.95, 3.97, 0.20. 0.04 Ref 145*
19.84, 16.03, 3.97, 0.20. 0.04 Ref 145*
Resistivity, ohm m × 10⁻⁸
Temperature, K
10^11
10^10
10^9
10^8
10^7
10^6
250
300
350
400
450
500
550

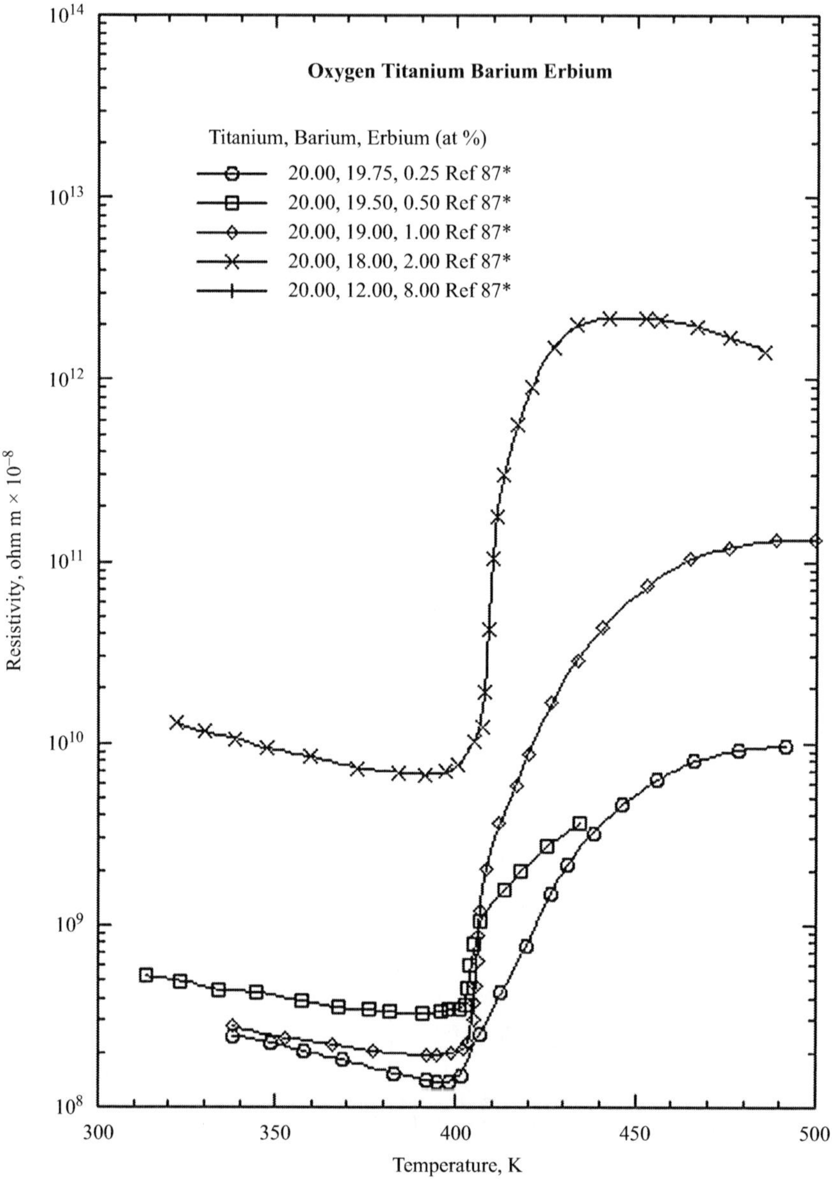
Oxygen Titanium Barium Erbium
Titanium, Barium, Erbium (at %)
20.00, 19.75, 0.25 Ref 87*
20.00, 19.50, 0.50 Ref 87*
20.00, 19.00, 1.00 Ref 87*
20.00, 18.00, 2.00 Ref 87*
20.00, 12.00, 8.00 Ref 87*
Resistivity, ohm m × 10−8
Temperature, K
300
350
400
450
500

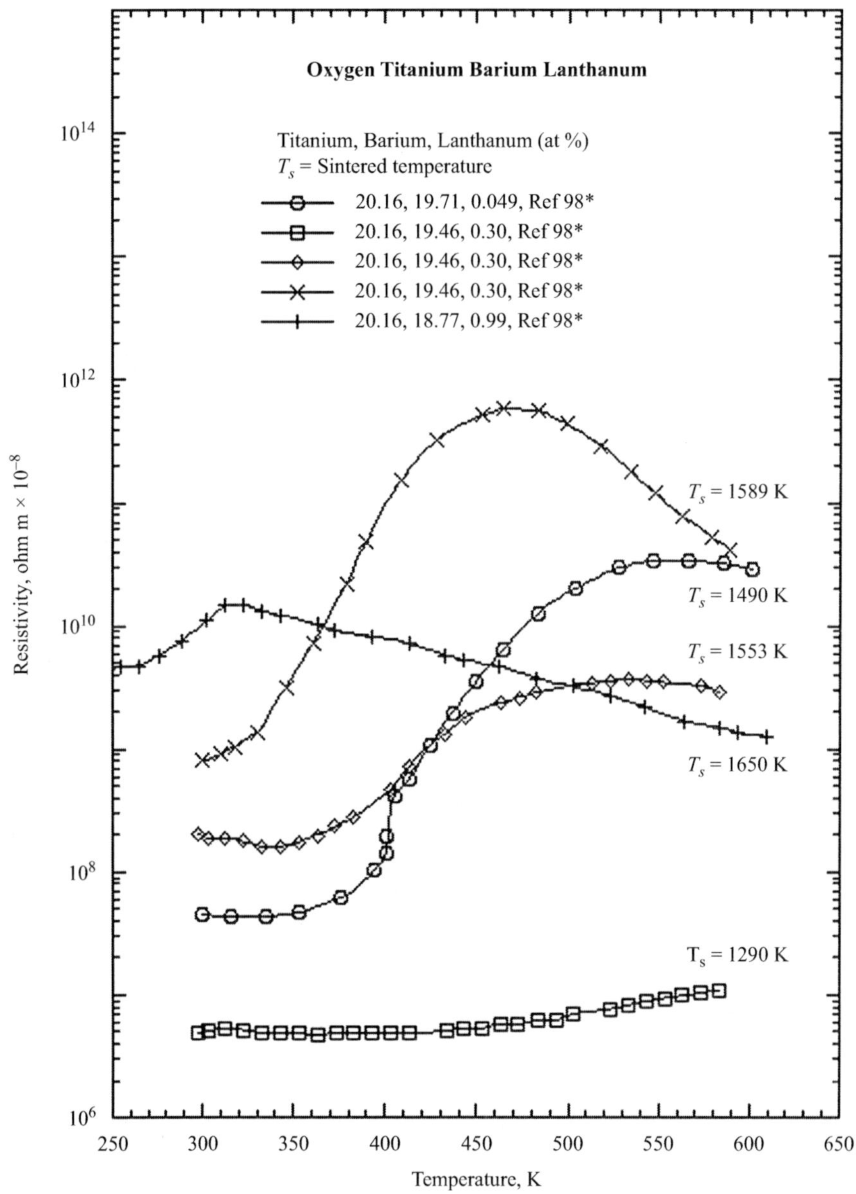
Oxygen Titanium Barium Lanthanum
Titanium, Barium, Lanthanum (at %)
T_s = Sintered temperature
20.16, 19.71, 0.049, Ref 98*
20.16, 19.46, 0.30, Ref 98*
20.16, 19.46, 0.30, Ref 98*
20.16, 19.46, 0.30, Ref 98*
20.16, 18.77, 0.99, Ref 98*
T_s = 1589 K
T_s = 1490 K
T_s = 1553 K
T_s = 1650 K
T_s = 1290 K
Resistivity, ohm m × 10^{-8}
10^{14}
10^{12}
10^{10}
10^{8}
10^{6}
250
300
350
400
450
500
550
600
650
Temperature, K

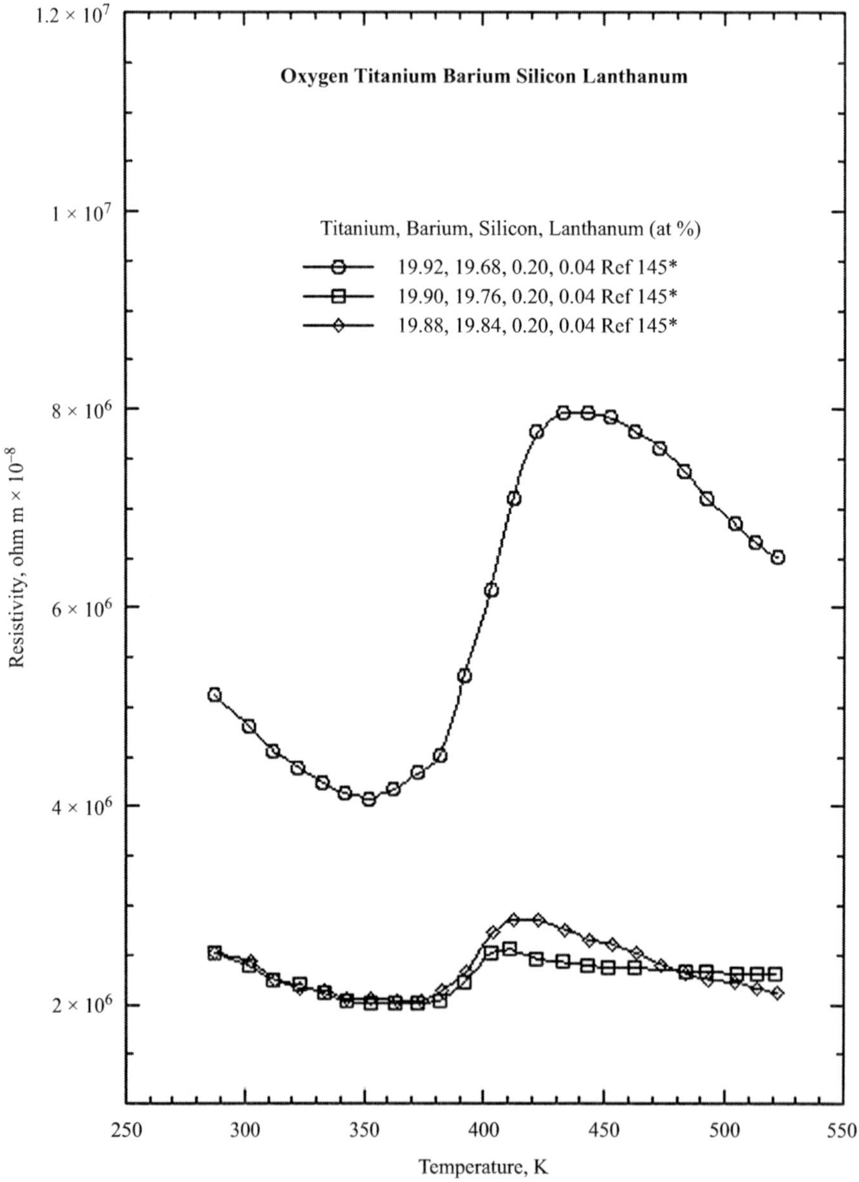
Oxygen Titanium Barium Silicon Lanthanum
Titanium, Barium, Silicon, Lanthanum (at %)
19.92, 19.68, 0.20, 0.04 Ref 145*
19.90, 19.76, 0.20, 0.04 Ref 145*
19.88, 19.84, 0.20, 0.04 Ref 145*
1.2 × 10^7
1 × 10^7
8 × 10^6
6 × 10^6
4 × 10^6
2 × 10^6
Resistivity, ohm m × 10^-8
250
300
350
400
450
500
550
Temperature, K

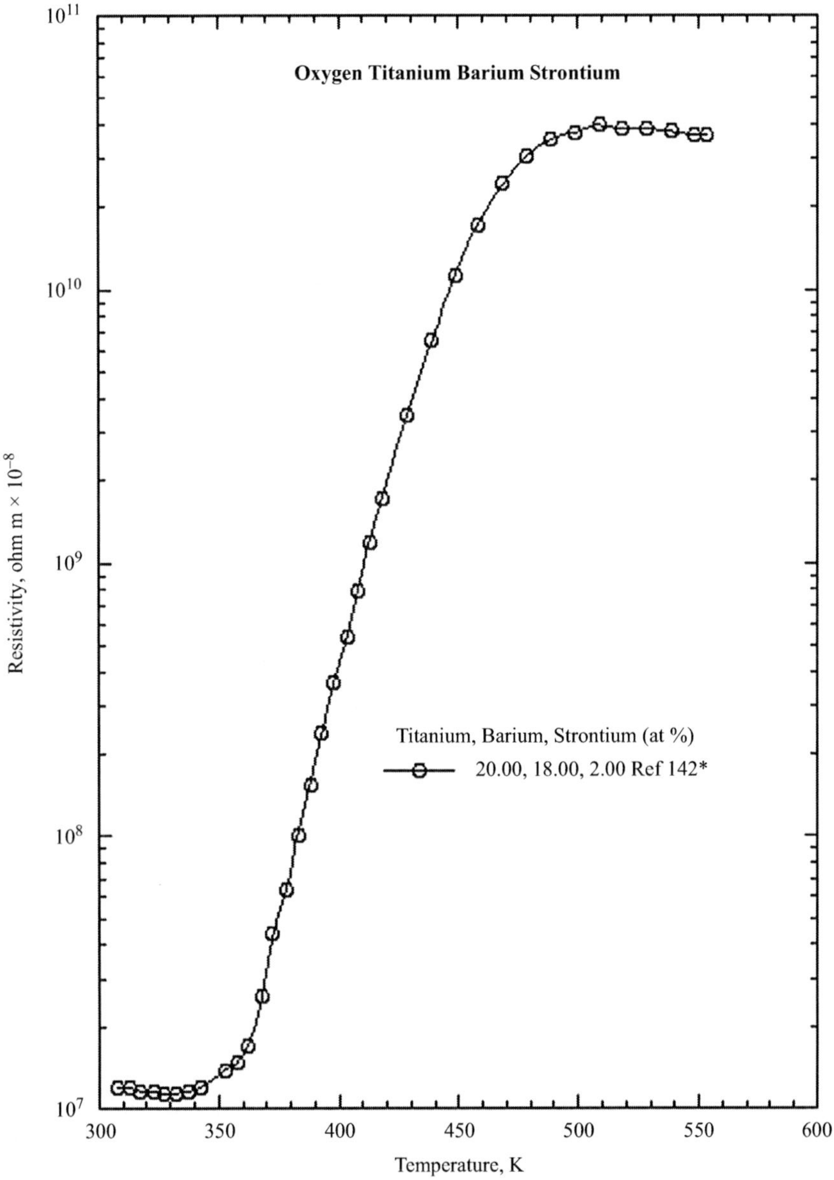

Oxygen Titanium Barium Strontium
Titanium, Barium, Strontium (at %)
20.00, 18.00, 2.00 Ref 142*
Resistivity, ohm m × 10⁻⁸
Temperature, K
10¹¹
10¹⁰
10⁹
10⁸
10⁷
300
350
400
450
500
550
600

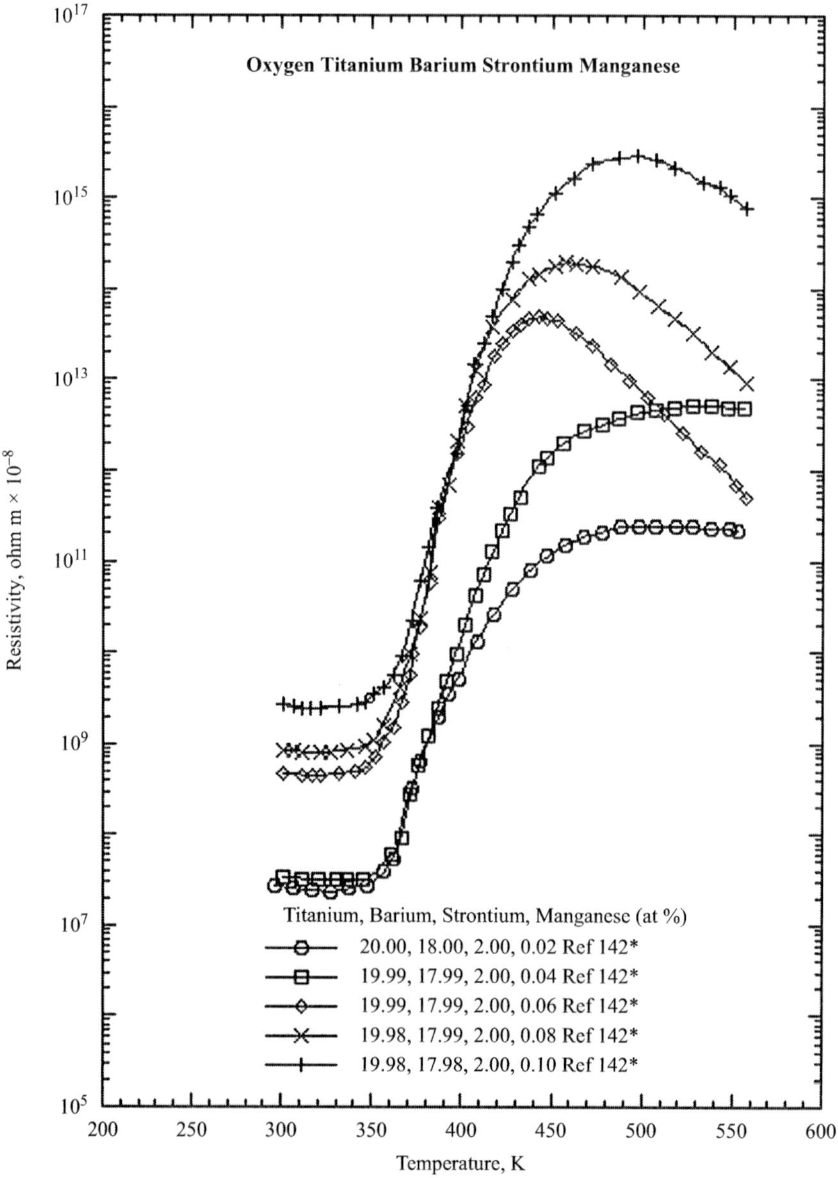
Oxygen Titanium Barium Strontium Manganese
Resistivity, ohm m × 10⁻⁸
10¹⁷
10¹⁵
10¹³
10¹¹
10⁹
10⁷
10⁵
200
250
300
350
400
450
500
550
600
Temperature, K
Titanium, Barium, Strontium, Manganese (at %)
20.00, 18.00, 2.00, 0.02 Ref 142*
19.99, 17.99, 2.00, 0.04 Ref 142*
19.99, 17.99, 2.00, 0.06 Ref 142*
19.98, 17.99, 2.00, 0.08 Ref 142*
19.98, 17.98, 2.00, 0.10 Ref 142*

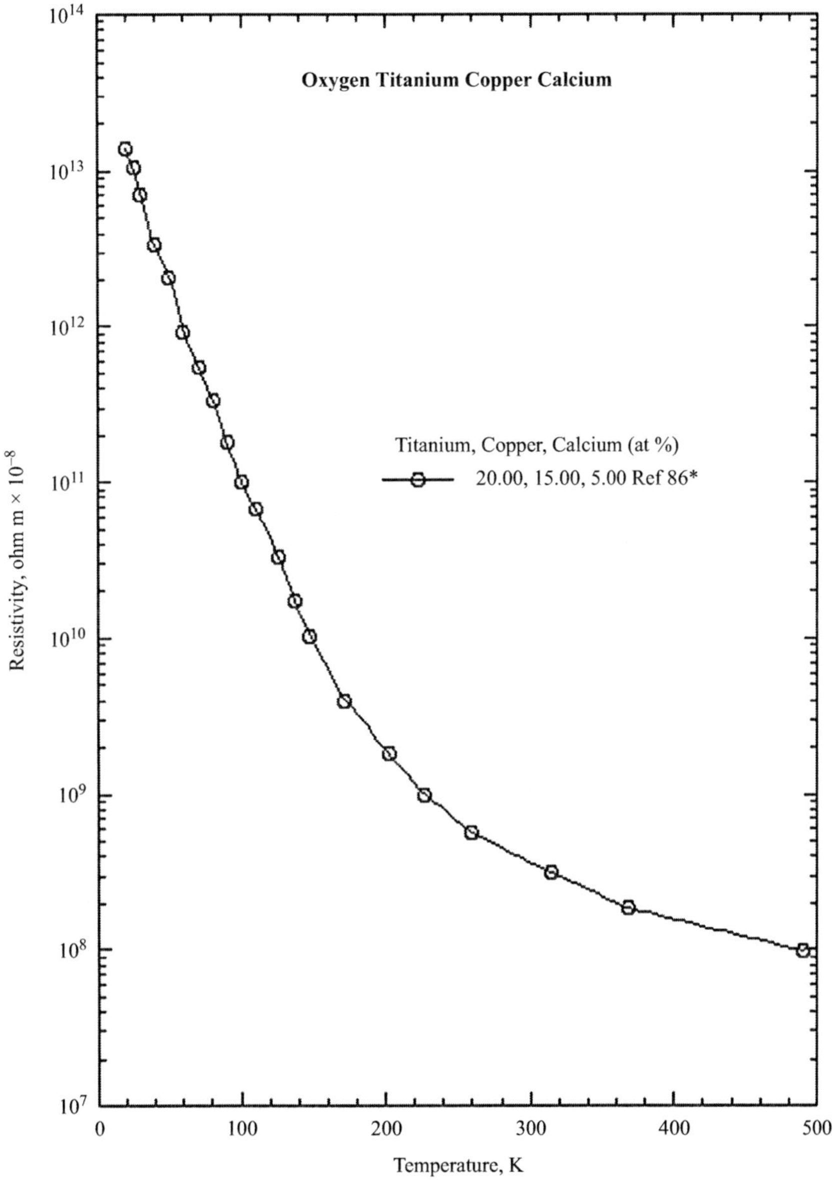

Oxygen Titanium Copper Calcium
Titanium, Copper, Calcium (at %)
20.00, 15.00, 5.00 Ref 86*
Resistivity, ohm m × 10⁻⁸
10^{14}
10^{13}
10^{12}
10^{11}
10^{10}
10^{9}
10^{8}
10^{7}
0
100
200
300
400
500
Temperature, K

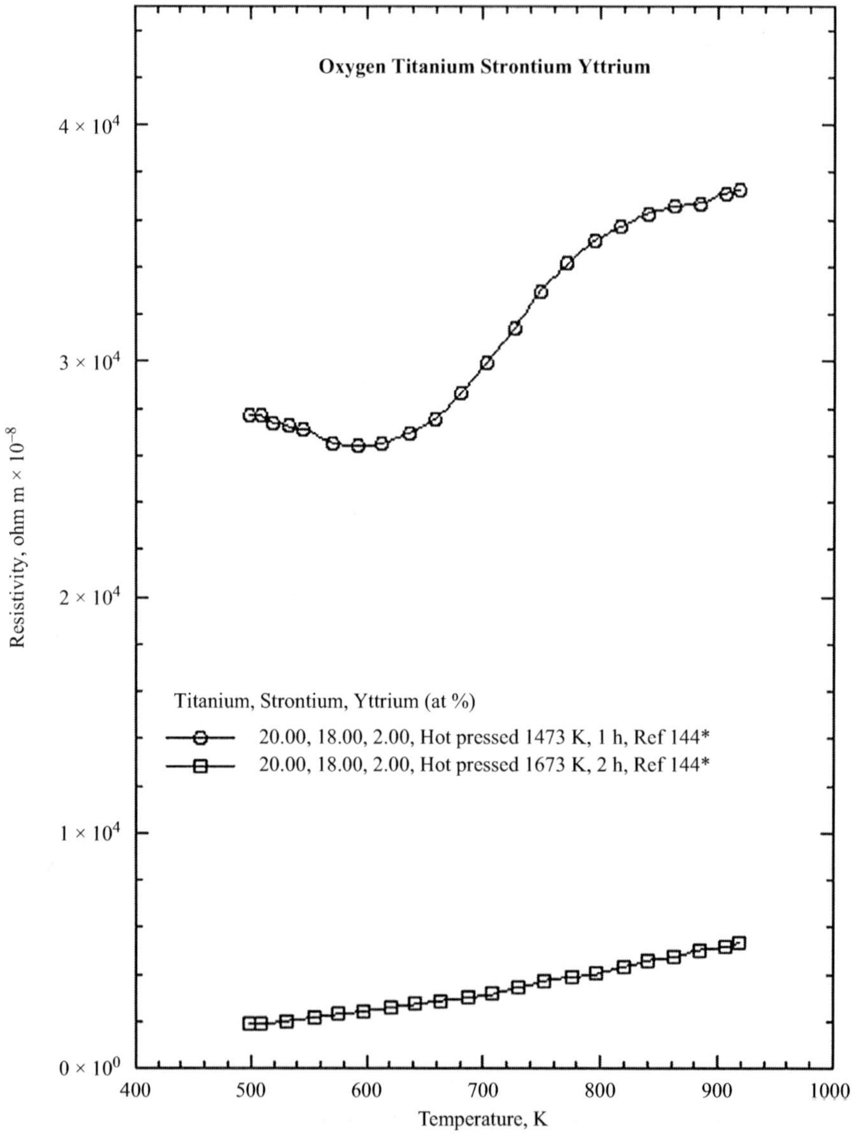
Oxygen Titanium Strontium Yttrium
4 × 10^4
3 × 10^4
2 × 10^4
1 × 10^4
0 × 10^0
Resistivity, ohm m × 10^-8
Titanium, Strontium, Yttrium (at %)
20.00, 18.00, 2.00, Hot pressed 1473 K, 1 h, Ref 144*
20.00, 18.00, 2.00, Hot pressed 1673 K, 2 h, Ref 144*
400
500
600
700
800
900
1000
Temperature, K

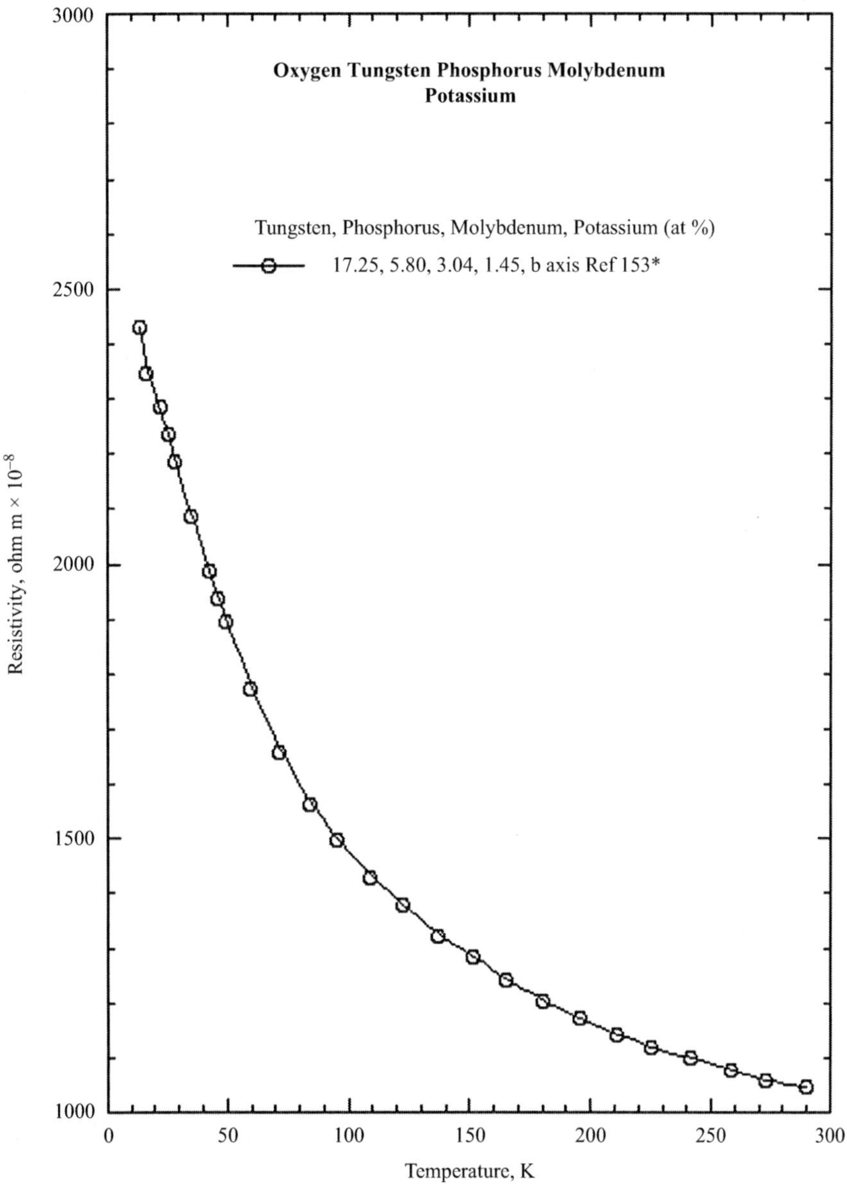

Oxygen Tungsten Phosphorus Molybdenum Potassium
Tungsten, Phosphorus, Molybdenum, Potassium (at %)
17.25, 5.80, 3.04, 1.45, b axis Ref 153*
Resistivity, ohm m × 10⁻⁸
Temperature, K
3000
2500
2000
1500
1000
0
50
100
150
200
250
300

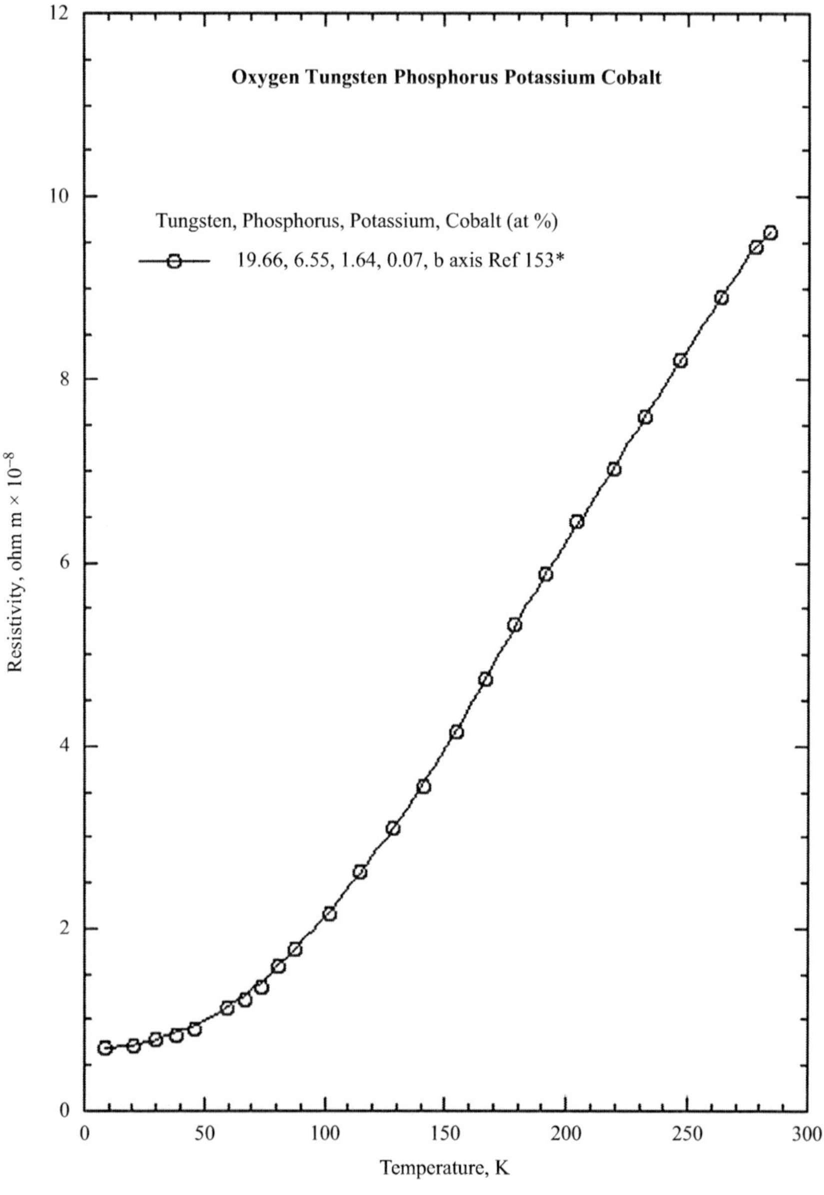
Oxygen Tungsten Phosphorus Potassium Cobalt
Tungsten, Phosphorus, Potassium, Cobalt (at %)
19.66, 6.55, 1.64, 0.07, b axis Ref 153*
Resistivity, ohm m × 10⁻⁸
Temperature, K
0
2
4
6
8
10
12
0
50
100
150
200
250
300

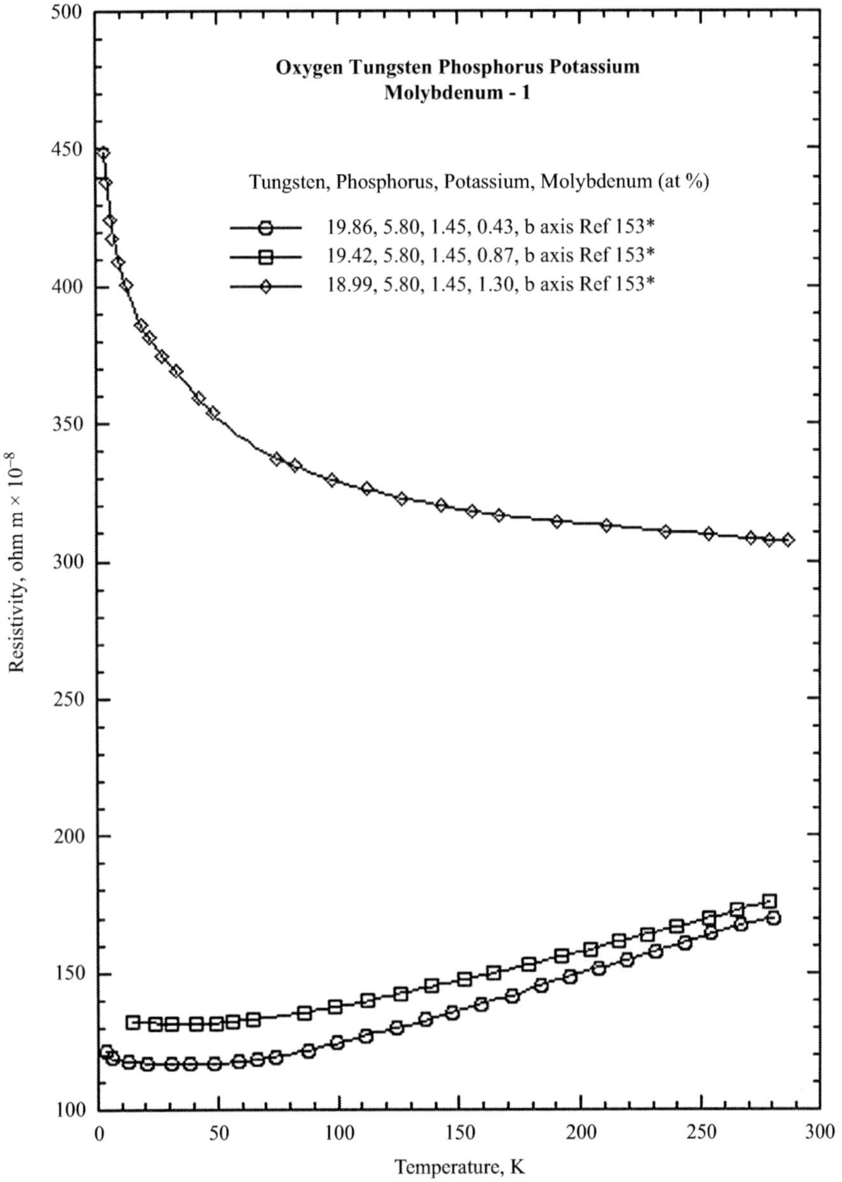
Oxygen Tungsten Phosphorus Potassium
Molybdenum - 1
Tungsten, Phosphorus, Potassium, Molybdenum (at %)
19.86, 5.80, 1.45, 0.43, b axis Ref 153*
19.42, 5.80, 1.45, 0.87, b axis Ref 153*
18.99, 5.80, 1.45, 1.30, b axis Ref 153*
Resistivity, ohm m × 10^{-8}
Temperature, K
500
450
400
350
300
250
200
150
100
0
50
100
150
200
250
300

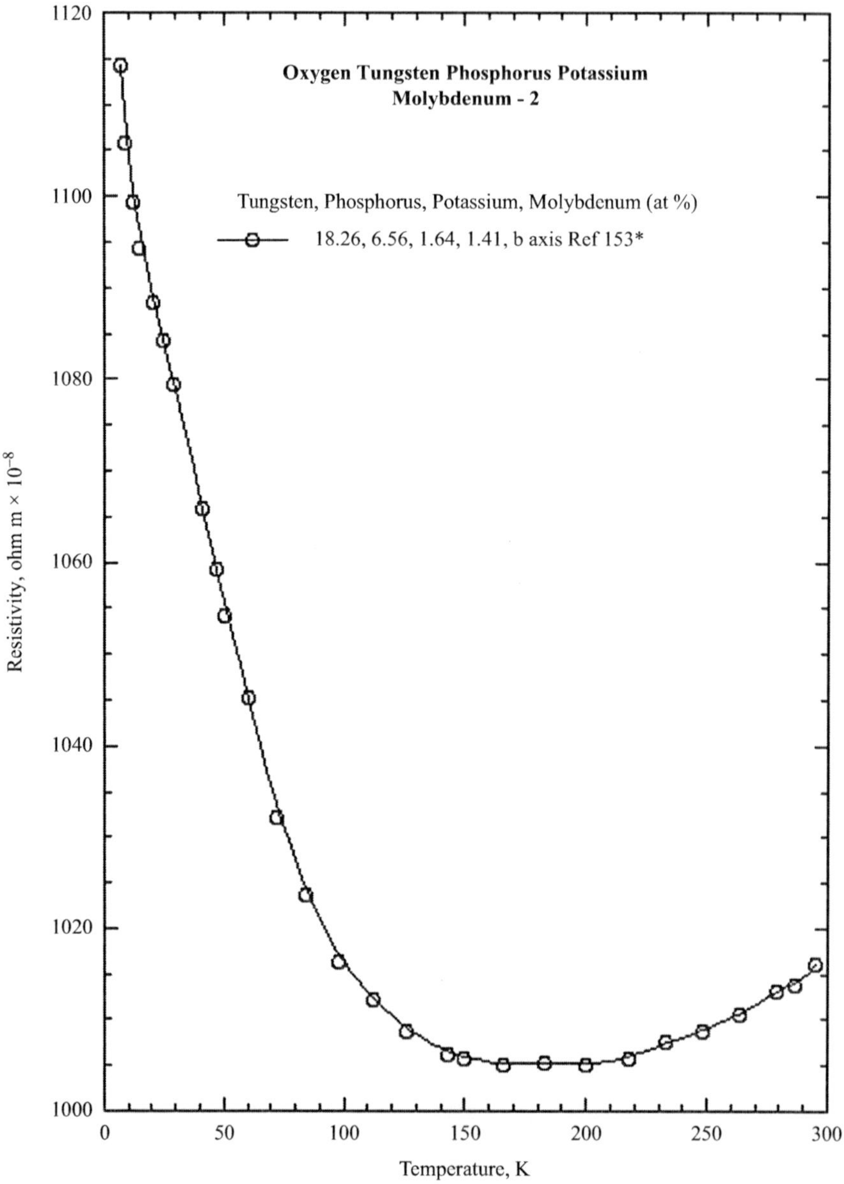
Oxygen Tungsten Phosphorus Potassium
Molybdenum - 2
Tungsten, Phosphorus, Potassium, Molybdenum (at %)
18.26, 6.56, 1.64, 1.41, b axis Ref 153*
Resistivity, ohm m × 10−8
Temperature, K
1000
1020
1040
1060
1080
1100
1120
0
50
100
150
200
250
300

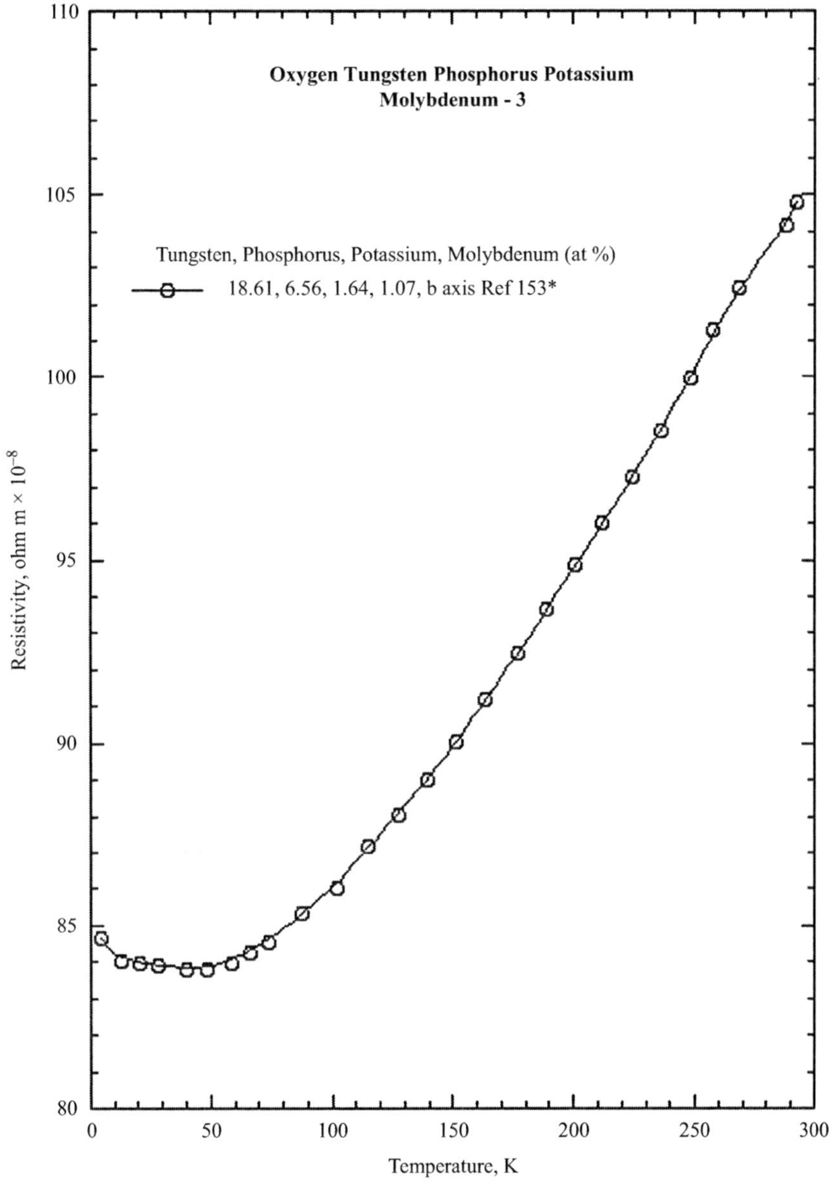

Oxygen Tungsten Phosphorus Potassium
Molybdenum - 3
Tungsten, Phosphorus, Potassium, Molybdenum (at %)
18.61, 6.56, 1.64, 1.07, b axis Ref 153*
Resistivity, ohm m × 10⁻⁸
Temperature, K
110
105
100
95
90
85
80
0
50
100
150
200
250
300

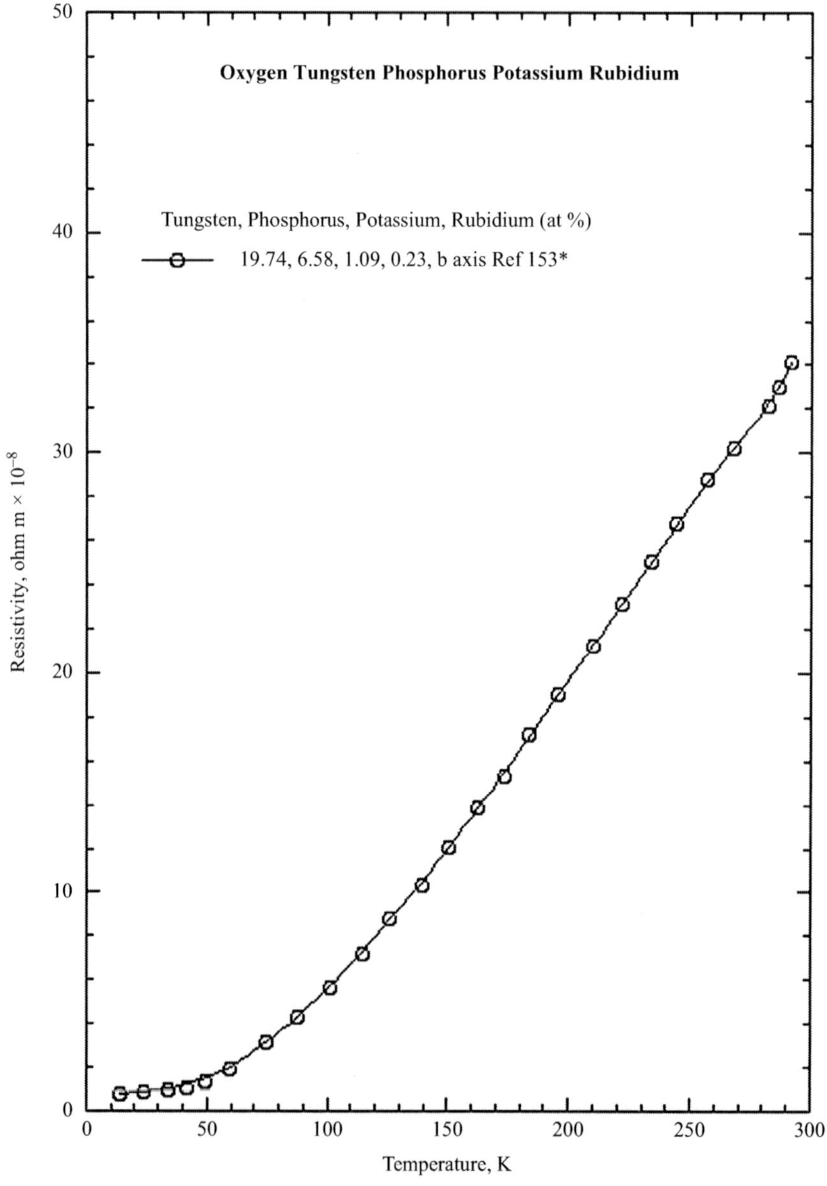
Oxygen Tungsten Phosphorus Potassium Rubidium
Tungsten, Phosphorus, Potassium, Rubidium (at %)
19.74, 6.58, 1.09, 0.23, b axis Ref 153*
Resistivity, ohm m × 10⁻⁸
Temperature, K
0
10
20
30
40
50
0
50
100
150
200
250
300

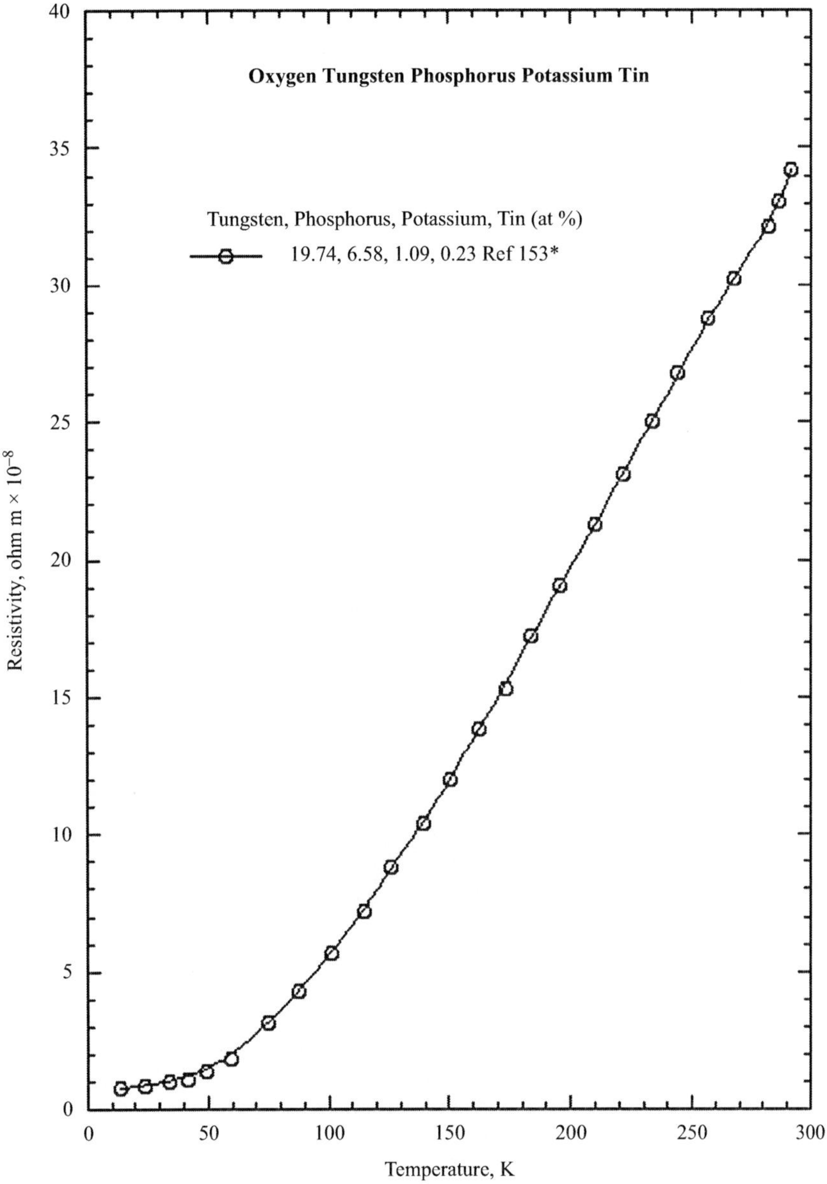
Oxygen Tungsten Phosphorus Potassium Tin
Tungsten, Phosphorus, Potassium, Tin (at %)
19.74, 6.58, 1.09, 0.23 Ref 153*
Resistivity, ohm m × 10⁻⁸
Temperature, K
0
5
10
15
20
25
30
35
40
50
100
150
200
250
300

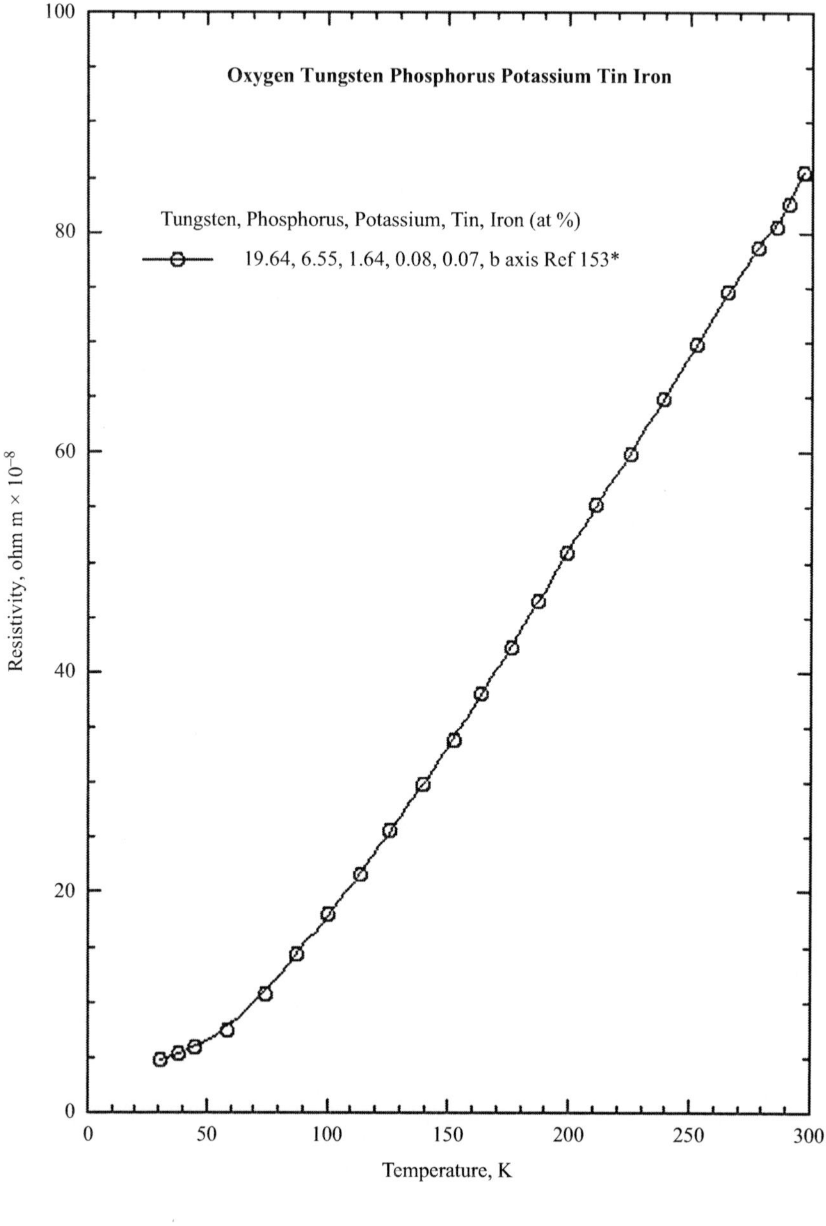
Oxygen Tungsten Phosphorus Potassium Tin Iron
Tungsten, Phosphorus, Potassium, Tin, Iron (at %)
19.64, 6.55, 1.64, 0.08, 0.07, b axis Ref 153*
Resistivity, ohm m × 10⁻⁸
Temperature, K
0
20
40
60
80
100
0
50
100
150
200
250
300

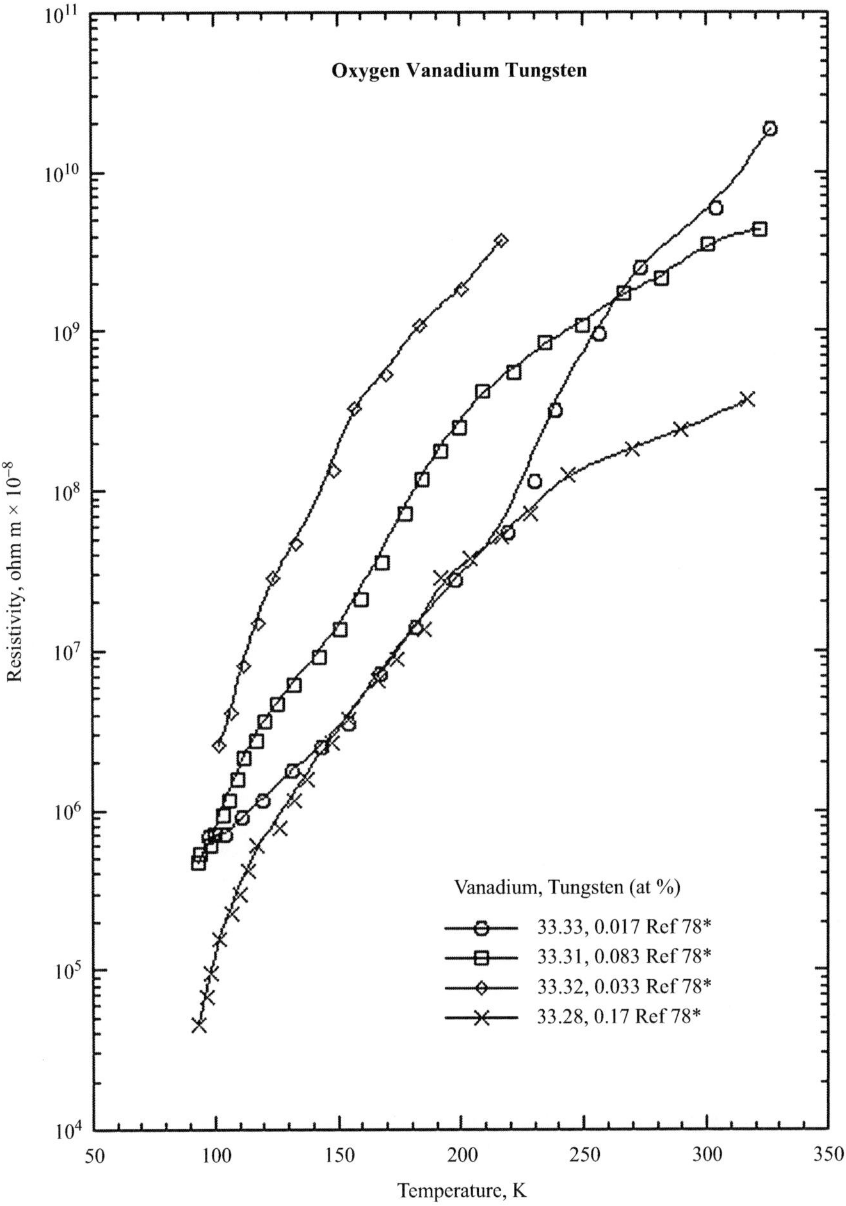
Oxygen Vanadium Tungsten
Resistivity, ohm m × 10^-8
Temperature, K
Vanadium, Tungsten (at %)
33.33, 0.017 Ref 78*
33.31, 0.083 Ref 78*
33.32, 0.033 Ref 78*
33.28, 0.17 Ref 78*

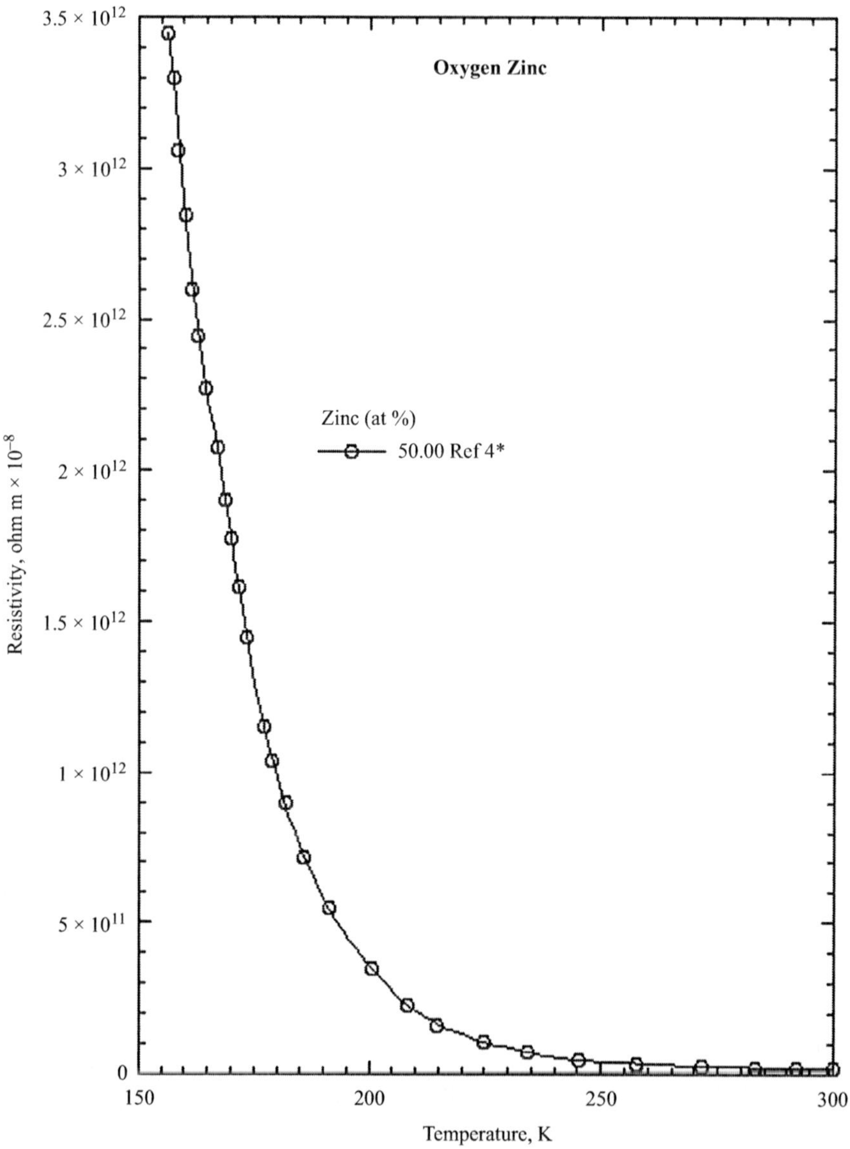
Oxygen Zinc
Zinc (at %)
50.00 Ref 4*
Resistivity, ohm m × 10⁻⁸
3.5 × 10^12
3 × 10^12
2.5 × 10^12
2 × 10^12
1.5 × 10^12
1 × 10^12
5 × 10^11
0
150
200
250
300
Temperature, K

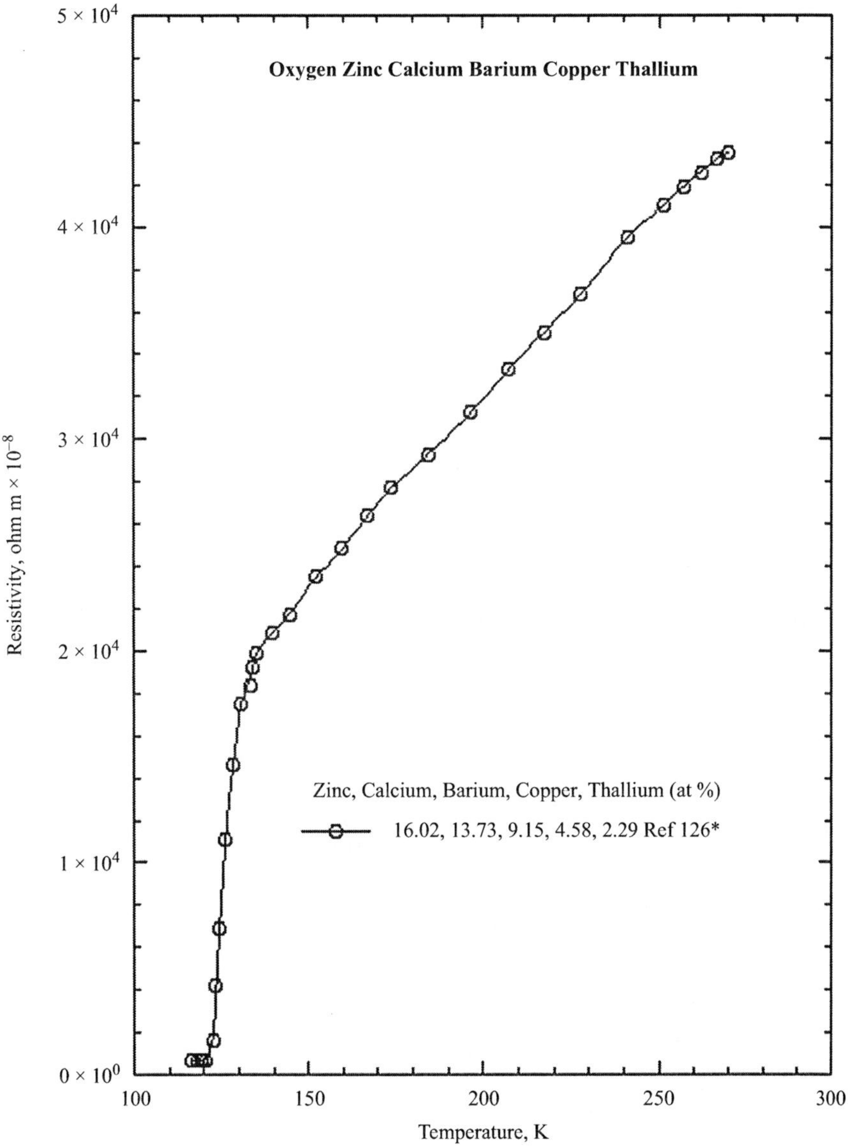
Oxygen Zinc Calcium Barium Copper Thallium
Resistivity, ohm m × 10⁻⁸
5 × 10⁴
4 × 10⁴
3 × 10⁴
2 × 10⁴
1 × 10⁴
0 × 10⁰
100
150
200
250
300
Temperature, K
Zinc, Calcium, Barium, Copper, Thallium (at %)
16.02, 13.73, 9.15, 4.58, 2.29 Ref 126*

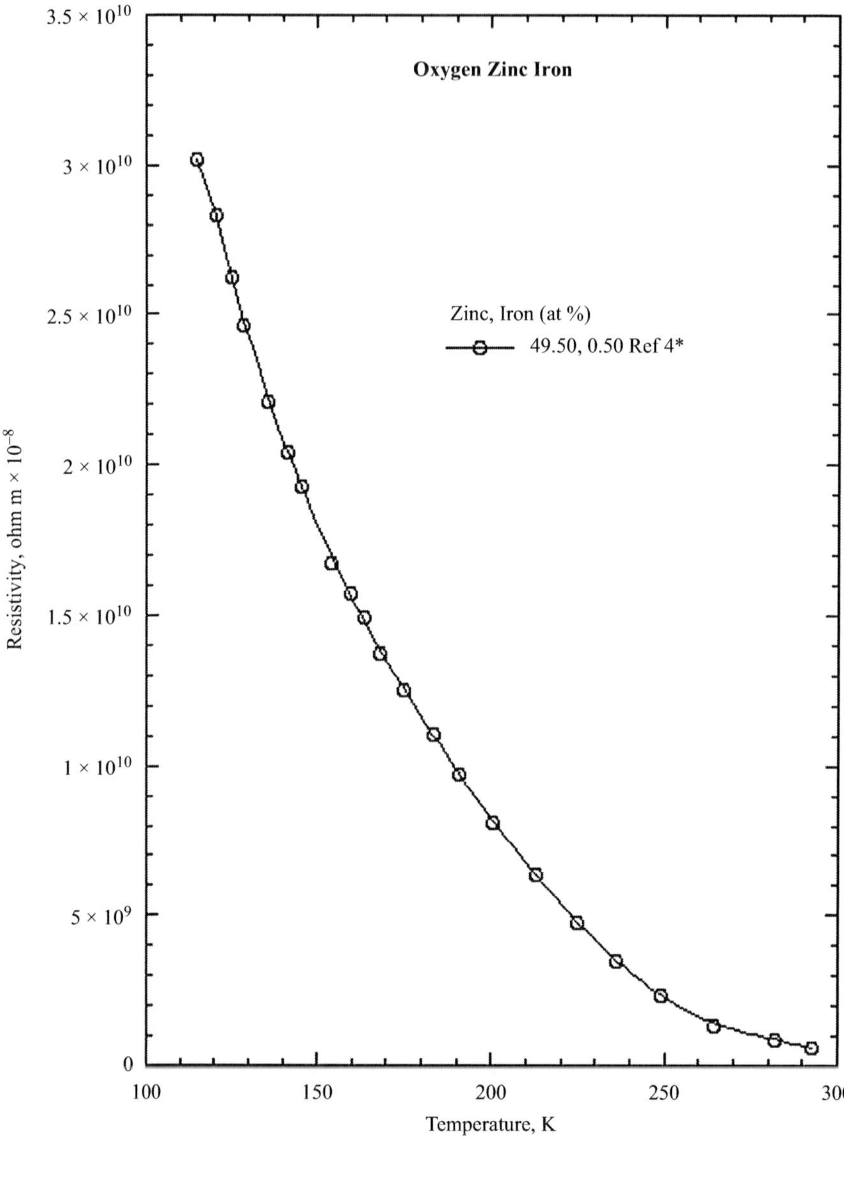
Oxygen Zinc Iron
Zinc, Iron (at %)
49.50, 0.50 Ref 4*
3.5 × 10^10
3 × 10^10
2.5 × 10^10
2 × 10^10
1.5 × 10^10
1 × 10^10
5 × 10^9
0
Resistivity, ohm m × 10^-8
100
150
200
250
300
Temperature, K

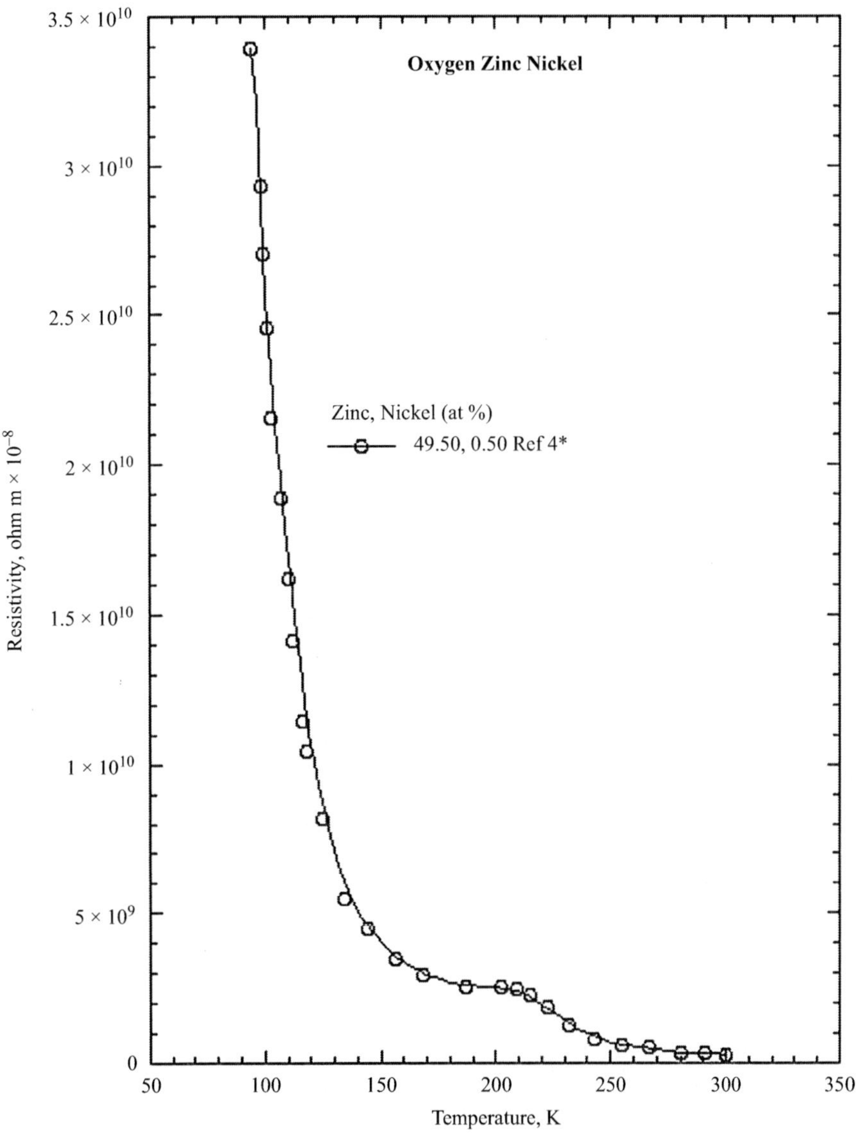
Oxygen Zinc Nickel
Zinc, Nickel (at %)
49.50, 0.50 Ref 4*
Resistivity, ohm m × 10−8
3.5 × 10^10
3 × 10^10
2.5 × 10^10
2 × 10^10
1.5 × 10^10
1 × 10^10
5 × 10^9
0
50
100
150
200
250
300
350
Temperature, K

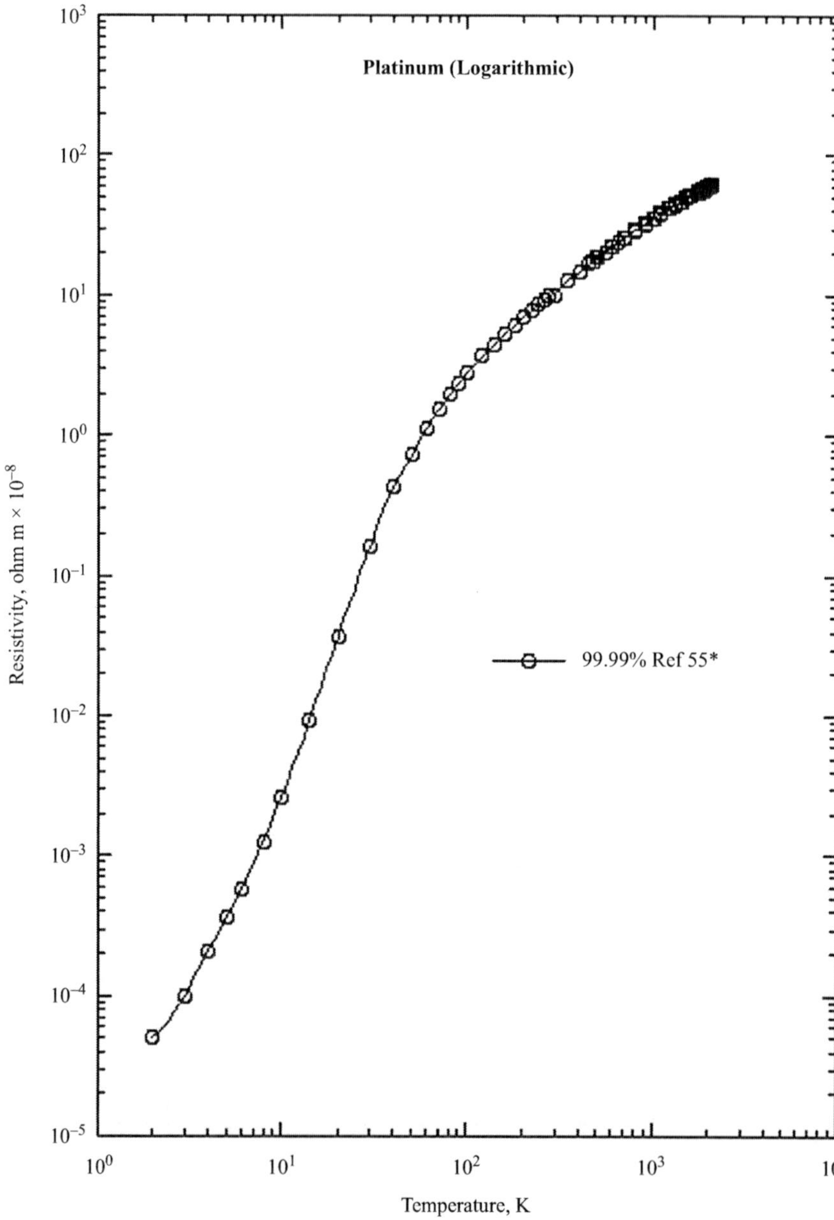
Platinum (Logarithmic)
99.99% Ref 55*
Resistivity, ohm m × 10⁻⁸
Temperature, K
10³
10²
10¹
10⁰
10⁻¹
10⁻²
10⁻³
10⁻⁴
10⁻⁵
10⁰
10¹
10²
10³
10⁴

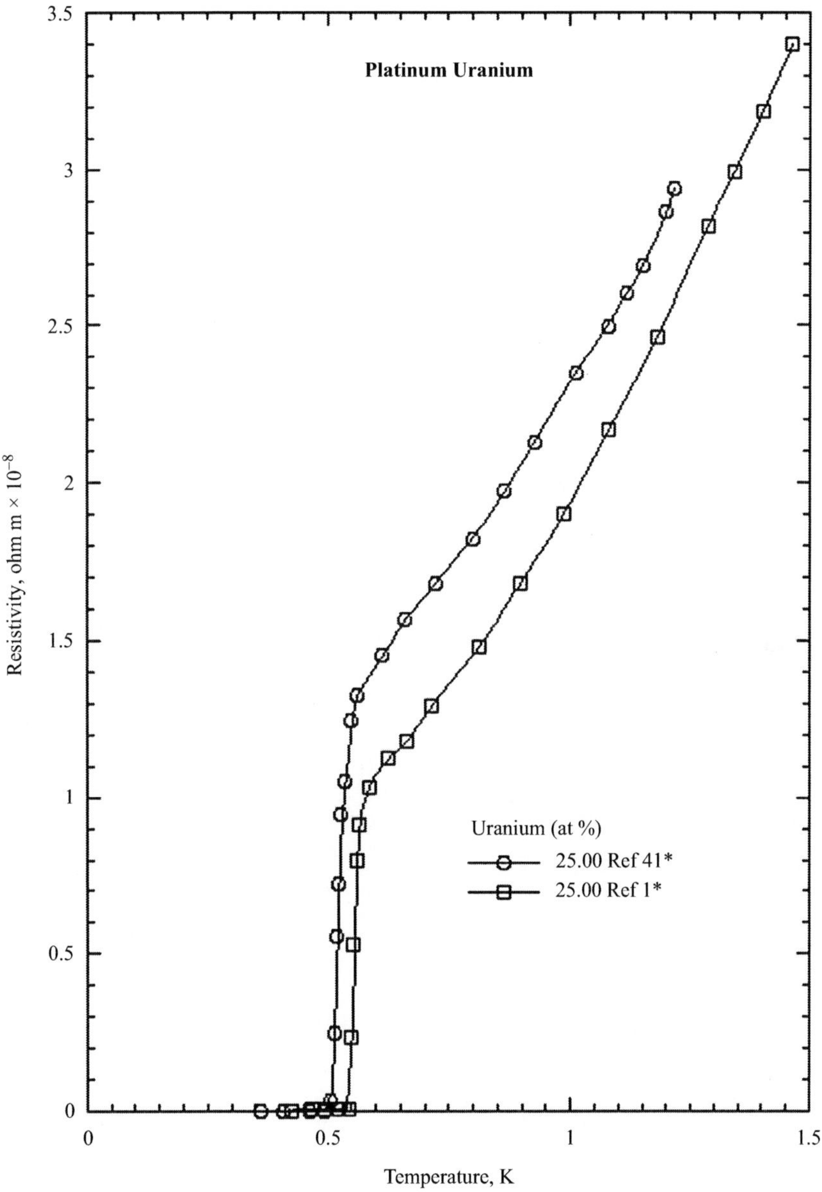
Platinum Uranium
Resistivity, ohm m × 10^−8
Temperature, K
Uranium (at %)
25.00 Ref 41*
25.00 Ref 1*
0
0.5
1
1.5
2
2.5
3
3.5
0
0.5
1
1.5

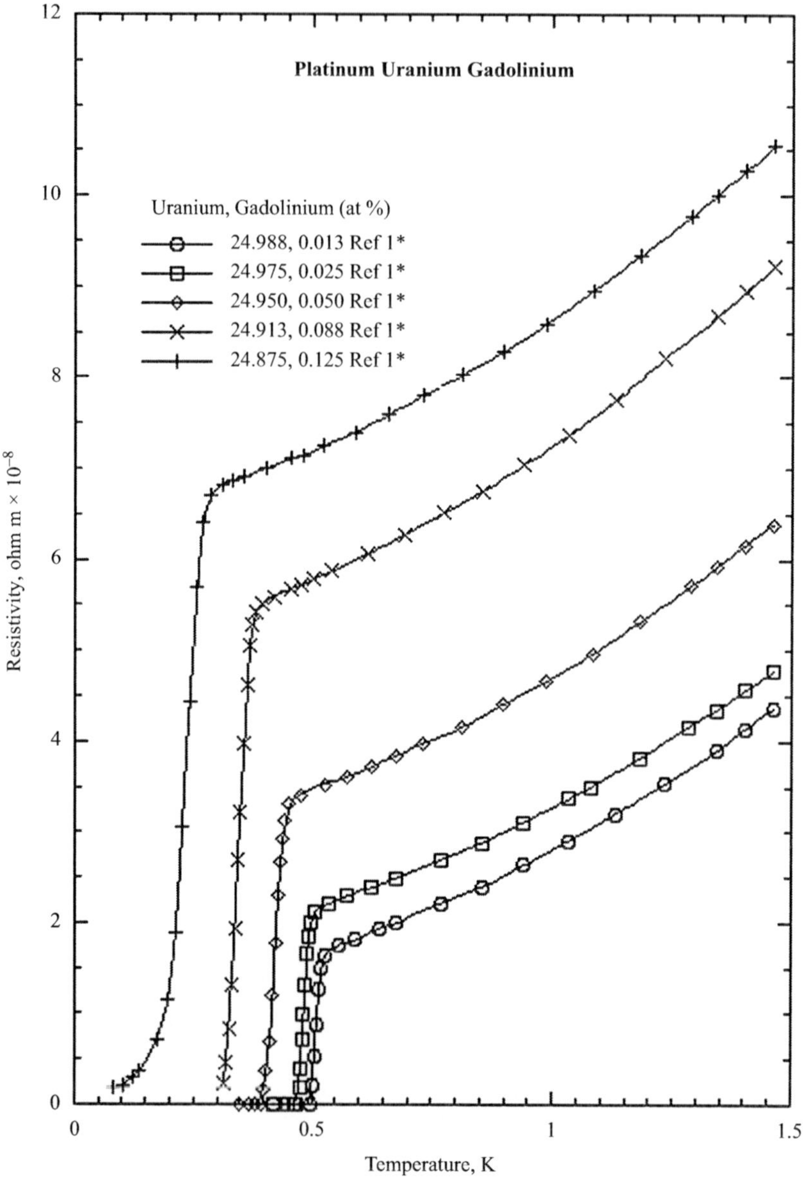
Platinum Uranium Gadolinium
Uranium, Gadolinium (at %)
24.988, 0.013 Ref 1*
24.975, 0.025 Ref 1*
24.950, 0.050 Ref 1*
24.913, 0.088 Ref 1*
24.875, 0.125 Ref 1*
Resistivity, ohm m × 10⁻⁸
Temperature, K
0
0.5
1
1.5
0
2
4
6
8
10
12

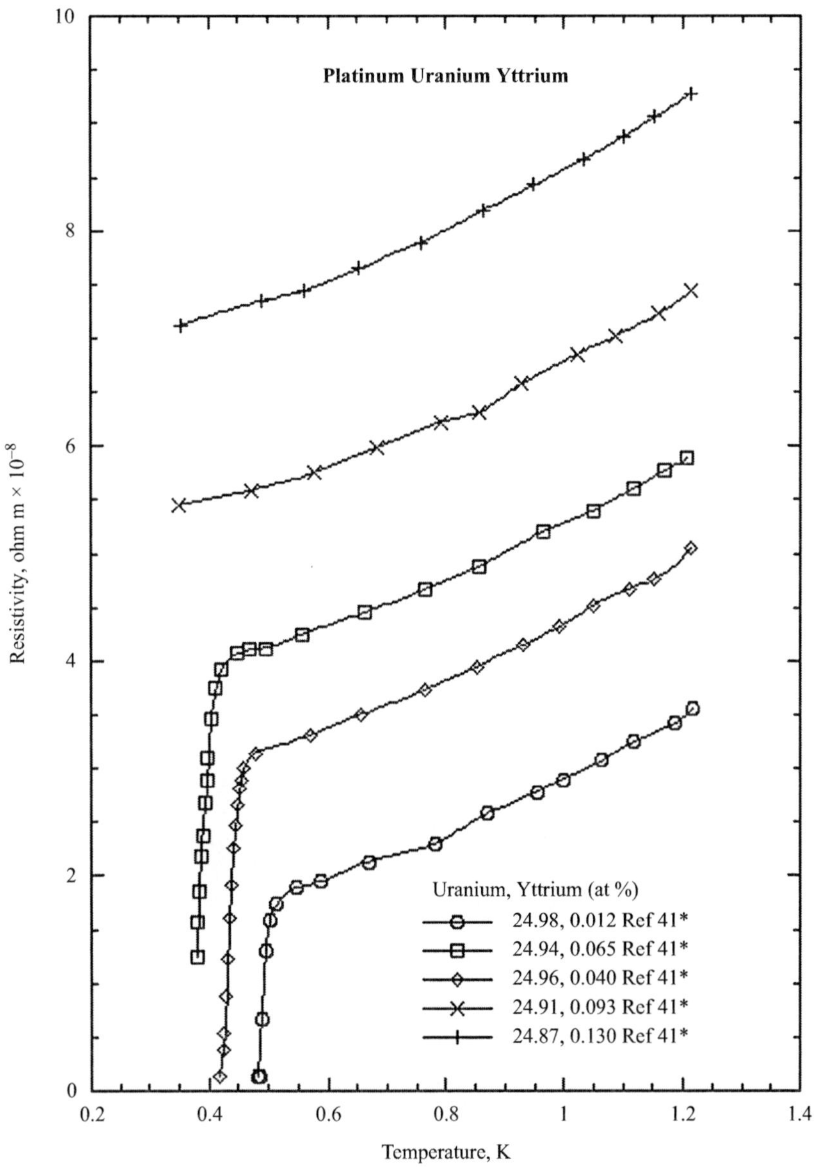
Platinum Uranium Yttrium
Resistivity, ohm m × 10⁻⁸
Temperature, K
Uranium, Yttrium (at %)
24.98, 0.012 Ref 41*
24.94, 0.065 Ref 41*
24.96, 0.040 Ref 41*
24.91, 0.093 Ref 41*
24.87, 0.130 Ref 41*
0
2
4
6
8
10
0.2
0.4
0.6
0.8
1
1.2
1.4

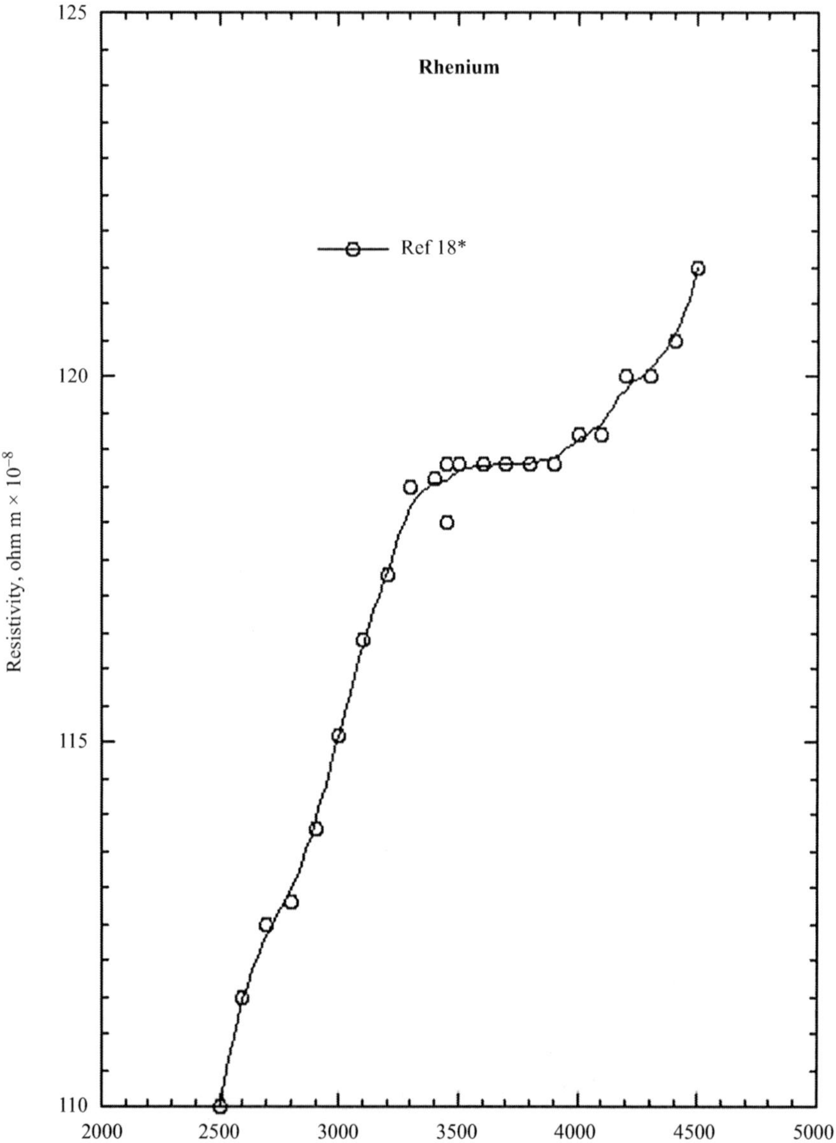
Rhenium
Ref 18*
125
120
115
110
2000
2500
3000
3500
4000
4500
5000
Temperature, K
Resistivity, ohm m × 10⁻⁸

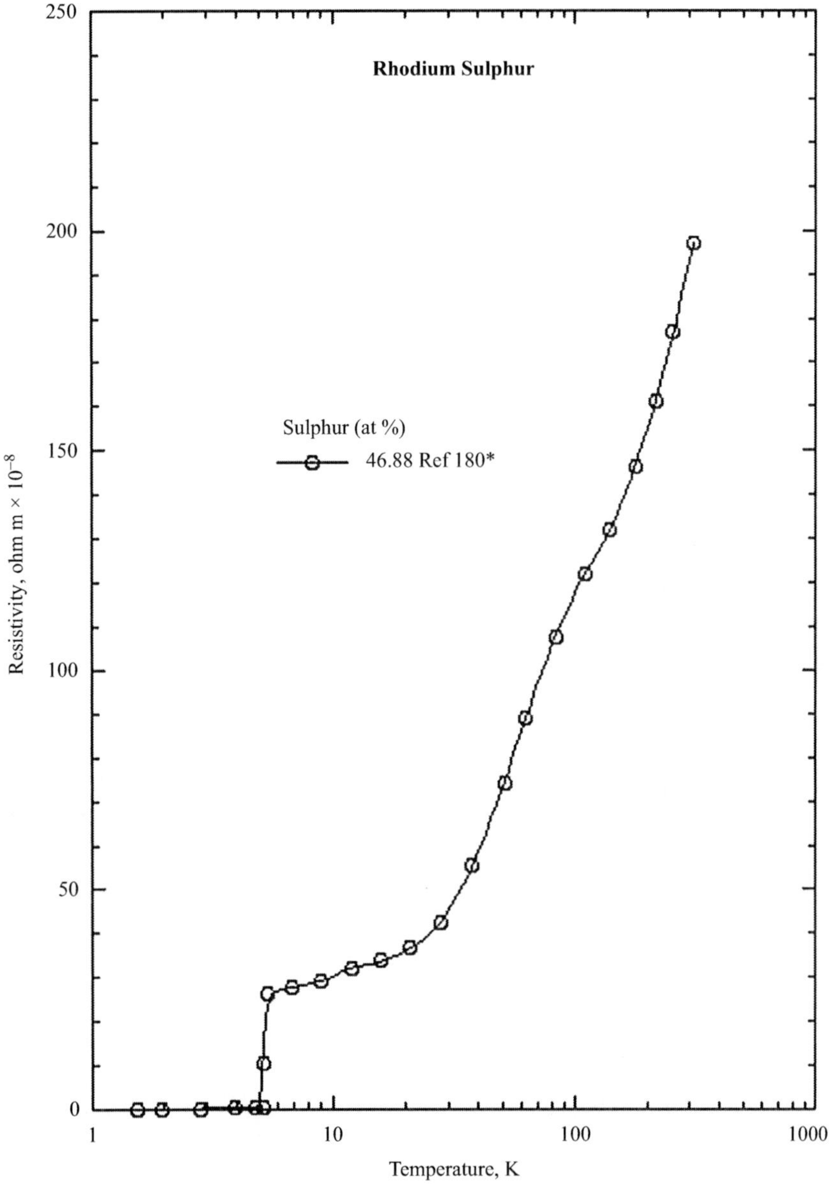
Rhodium Sulphur
Sulphur (at %)
46.88 Ref 180*
Resistivity, ohm m × 10⁻⁸
250
200
150
100
50
0
1
10
100
1000
Temperature, K

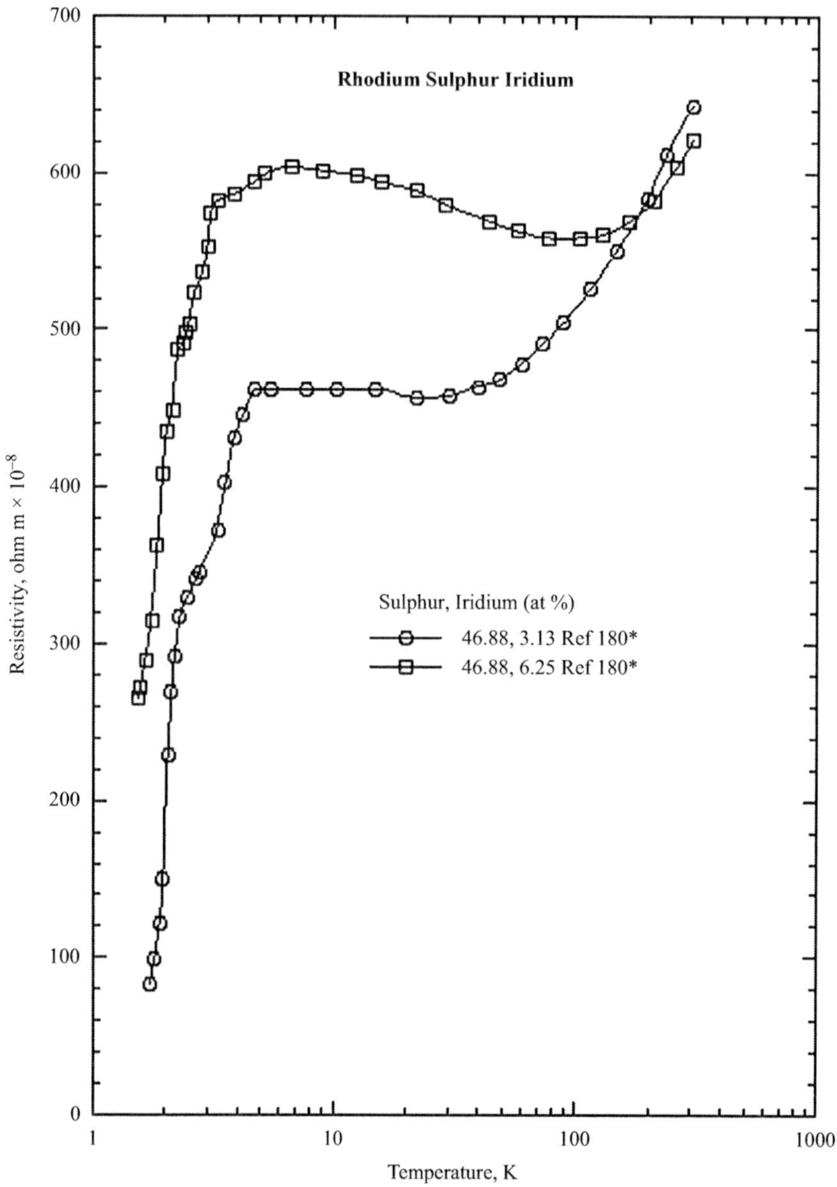

Rhodium Sulphur Iridium
Sulphur, Iridium (at %)
46.88, 3.13 Ref 180*
46.88, 6.25 Ref 180*
Resistivity, ohm m × 10^{-8}
Temperature, K
0
100
200
300
400
500
600
700
1
10
100
1000

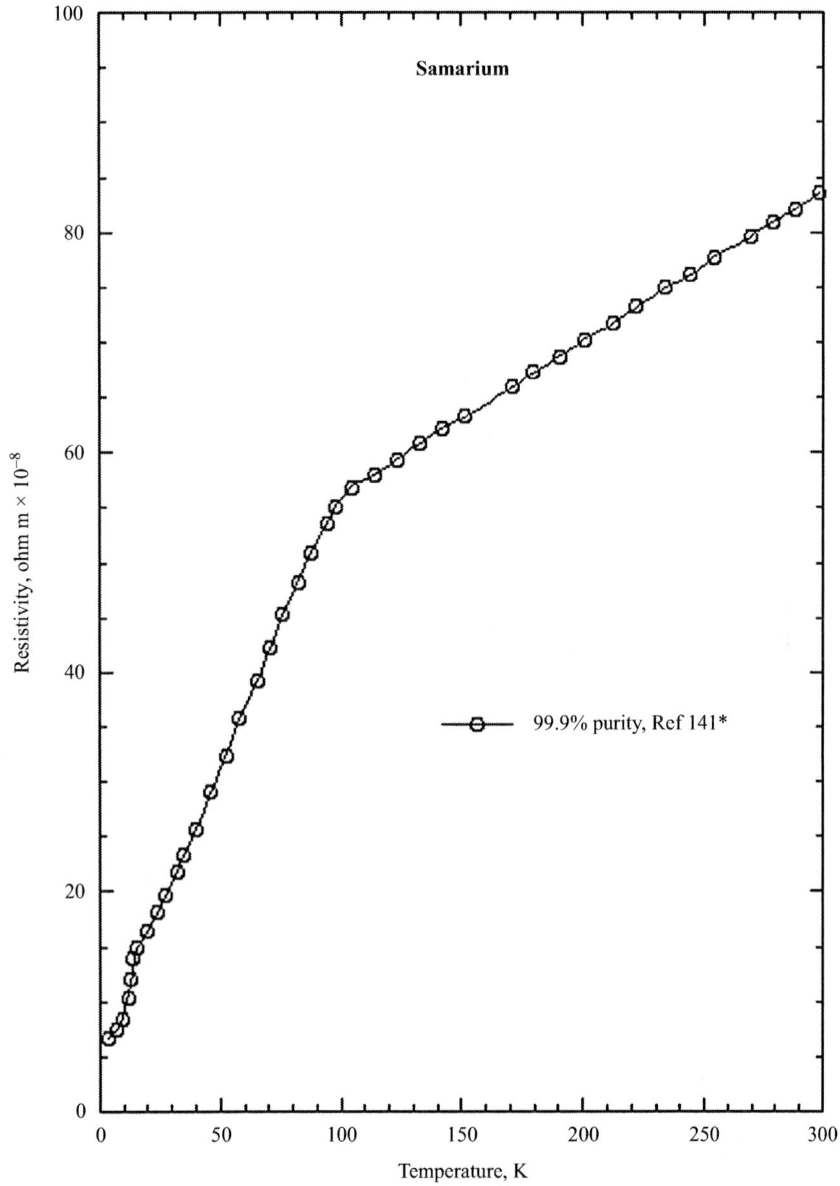
Samarium
99.9% purity, Ref 141*
Resistivity, ohm m × 10⁻⁸
Temperature, K
0
20
40
60
80
100
0
50
100
150
200
250
300

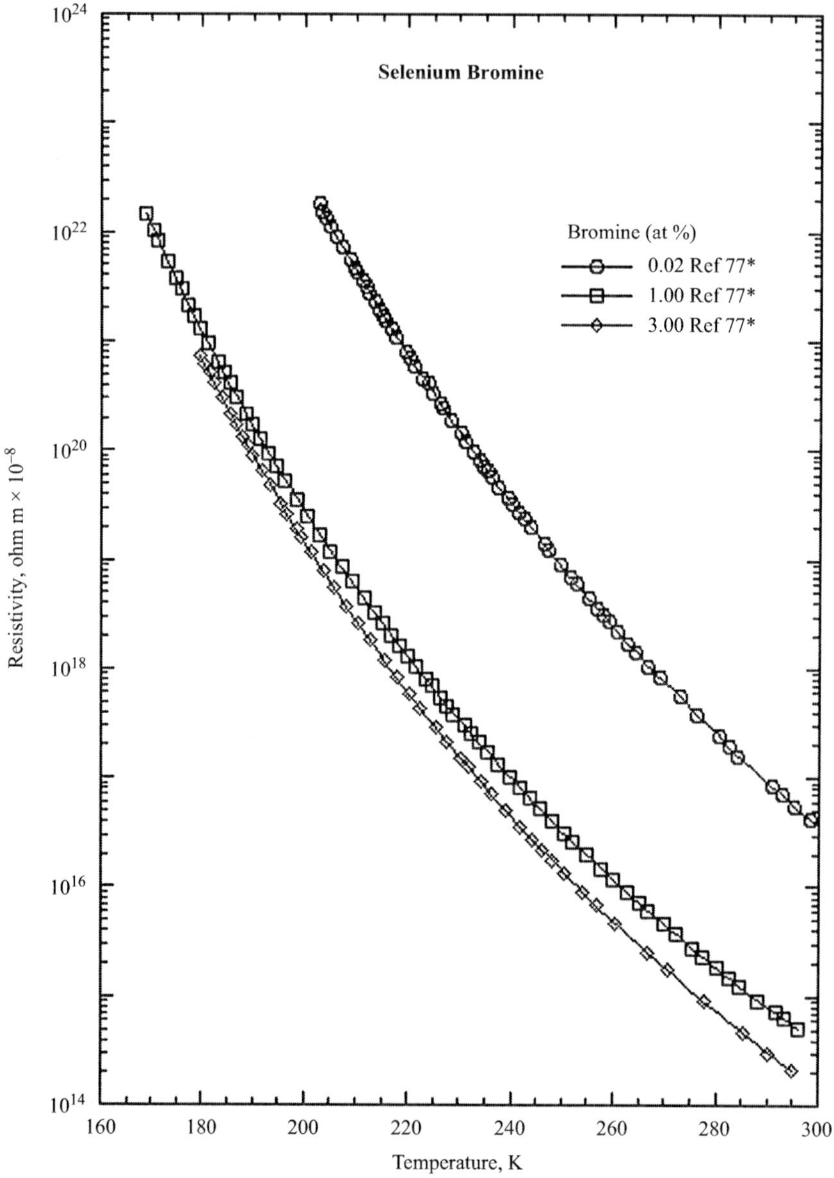
Selenium Bromine
Bromine (at %)
0.02 Ref 77*
1.00 Ref 77*
3.00 Ref 77*
Resistivity, ohm m × 10⁻⁸
Temperature, K
10^24
10^22
10^20
10^18
10^16
10^14
160
180
200
220
240
260
280
300

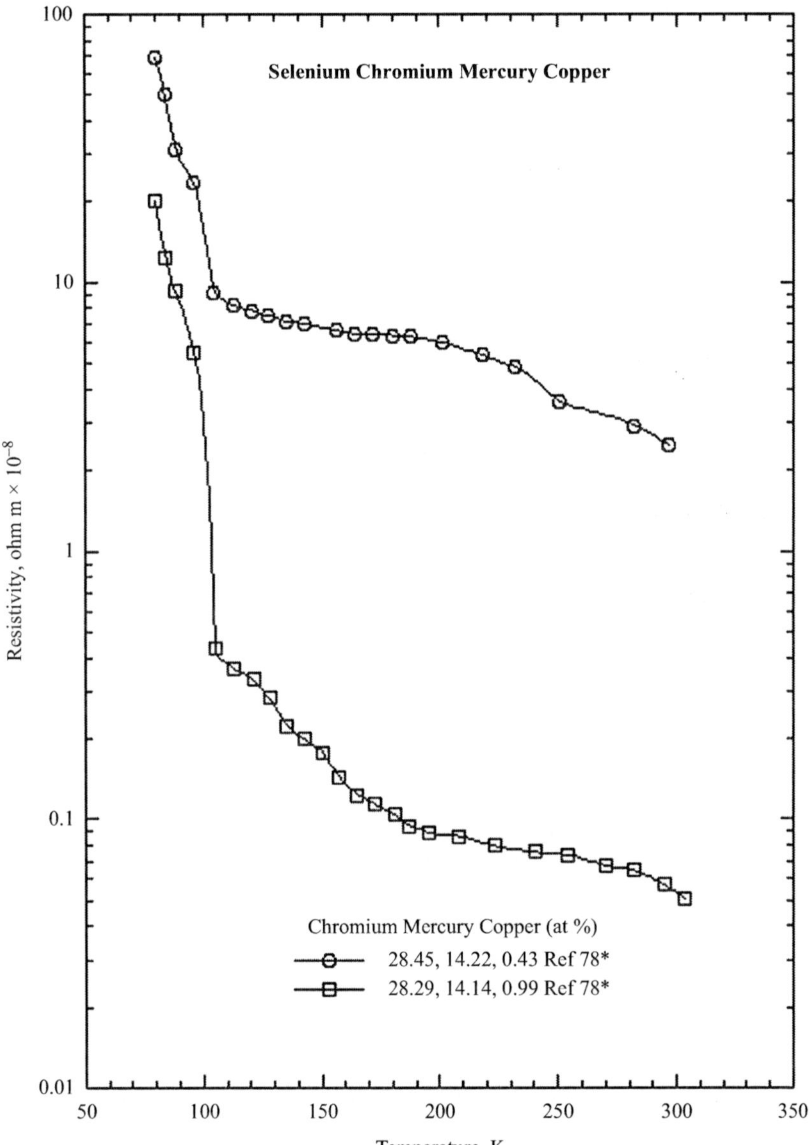

Selenium Chromium Mercury Copper
Resistivity, ohm m × 10⁻⁸
Temperature, K
100
10
1
0.1
0.01
50
100
150
200
250
300
350
Chromium Mercury Copper (at %)
28.45, 14.22, 0.43 Ref 78*
28.29, 14.14, 0.99 Ref 78*

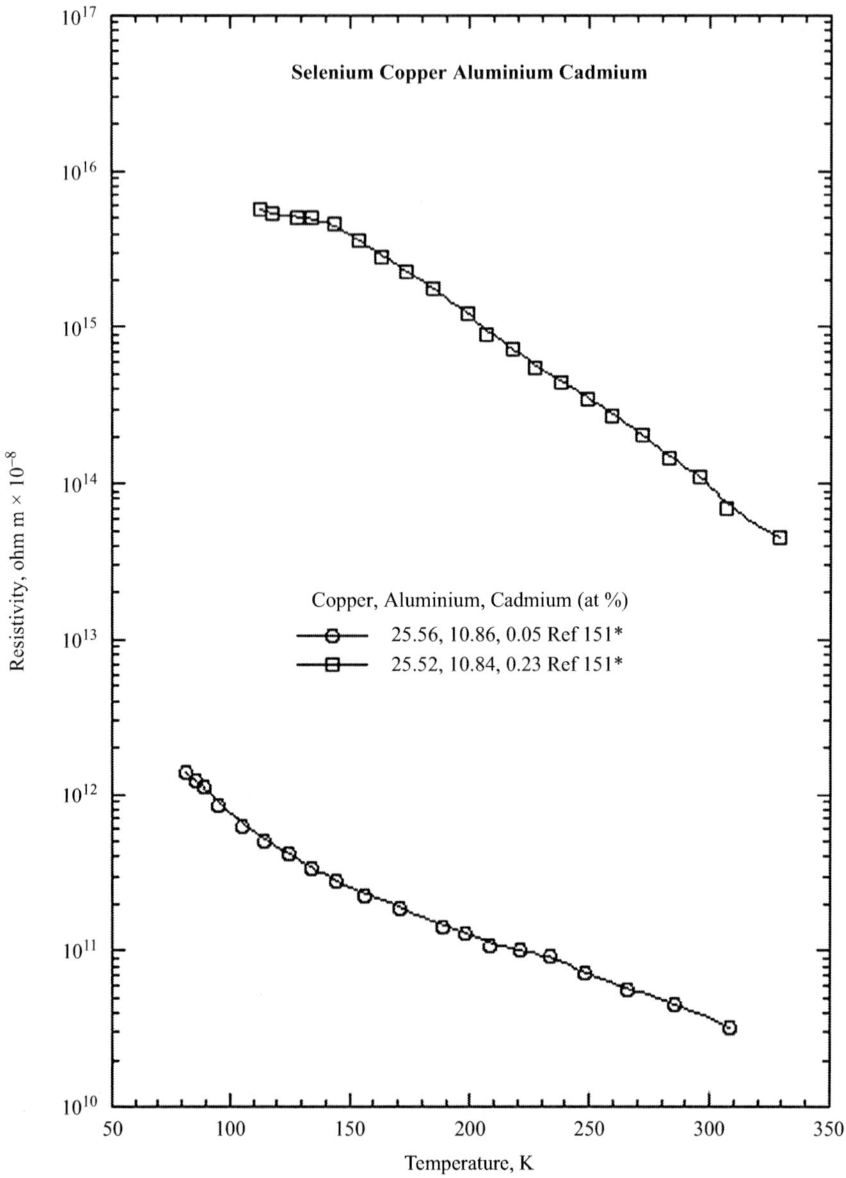
Selenium Copper Aluminium Cadmium
Copper, Aluminium, Cadmium (at %)
25.56, 10.86, 0.05 Ref 151*
25.52, 10.84, 0.23 Ref 151*
Resistivity, ohm m × 10^{-8}
Temperature, K
10^{10}
10^{11}
10^{12}
10^{13}
10^{14}
10^{15}
10^{16}
10^{17}
50
100
150
200
250
300
350

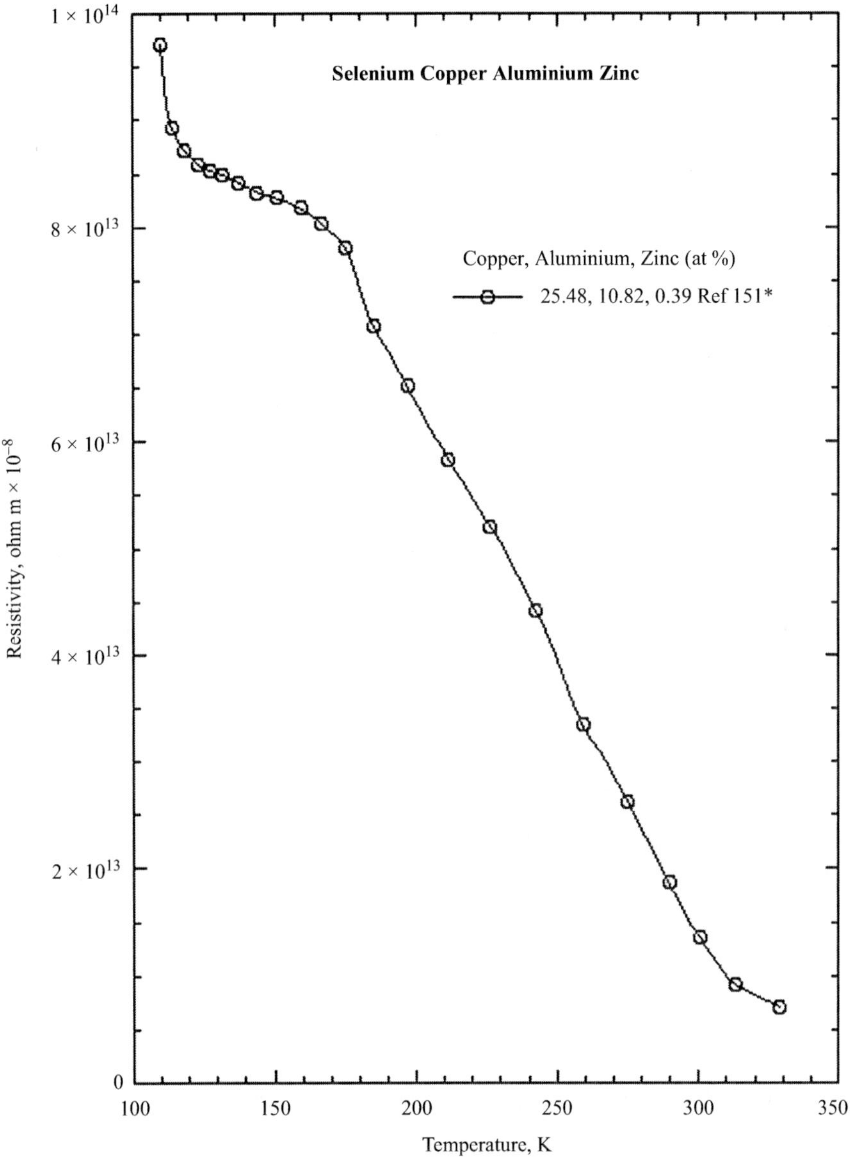
Selenium Copper Aluminium Zinc
Copper, Aluminium, Zinc (at %)
25.48, 10.82, 0.39 Ref 151*
Resistivity, ohm m × 10⁻⁸
Temperature, K
1 × 10¹⁴
8 × 10¹³
6 × 10¹³
4 × 10¹³
2 × 10¹³
0
100
150
200
250
300
350

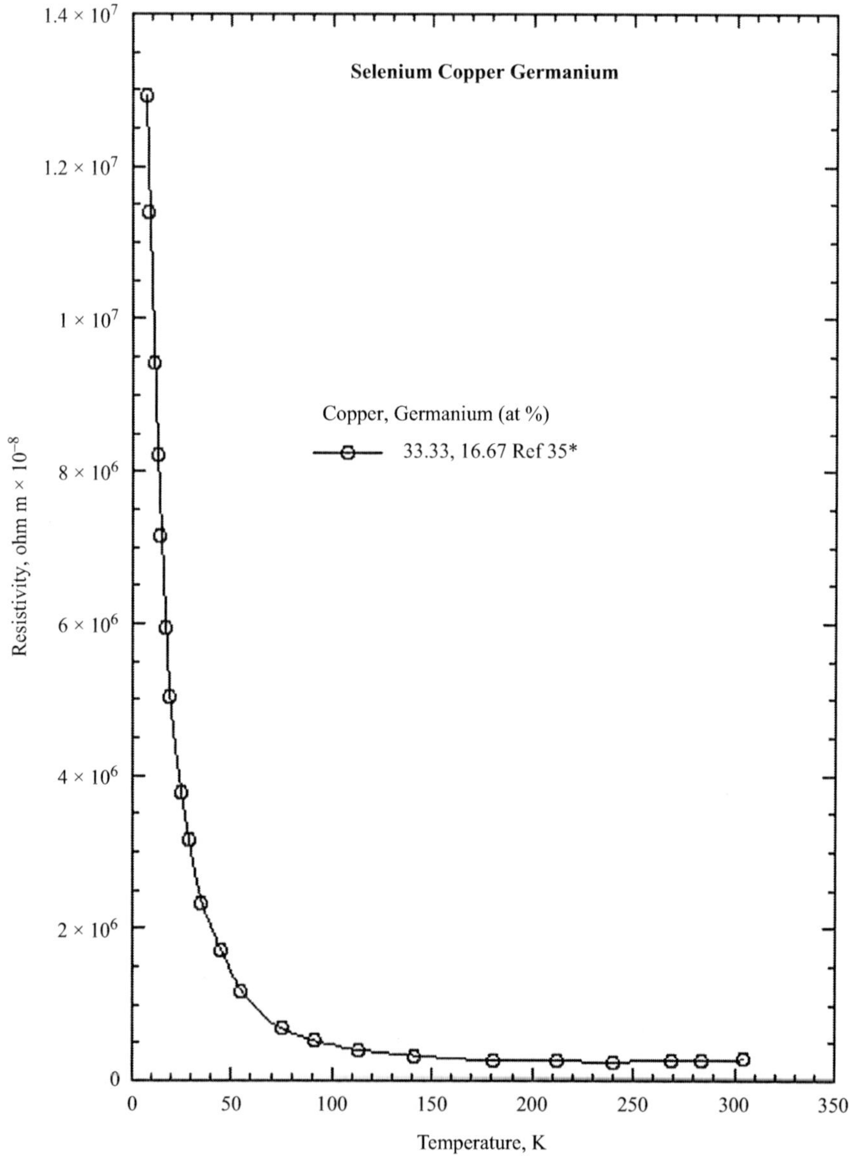
Selenium Copper Germanium
Copper, Germanium (at %)
33.33, 16.67 Ref 35*
Resistivity, ohm m × 10^-8
Temperature, K
1.4 × 10^7
1.2 × 10^7
1 × 10^7
8 × 10^6
6 × 10^6
4 × 10^6
2 × 10^6
0
0
50
100
150
200
250
300
350

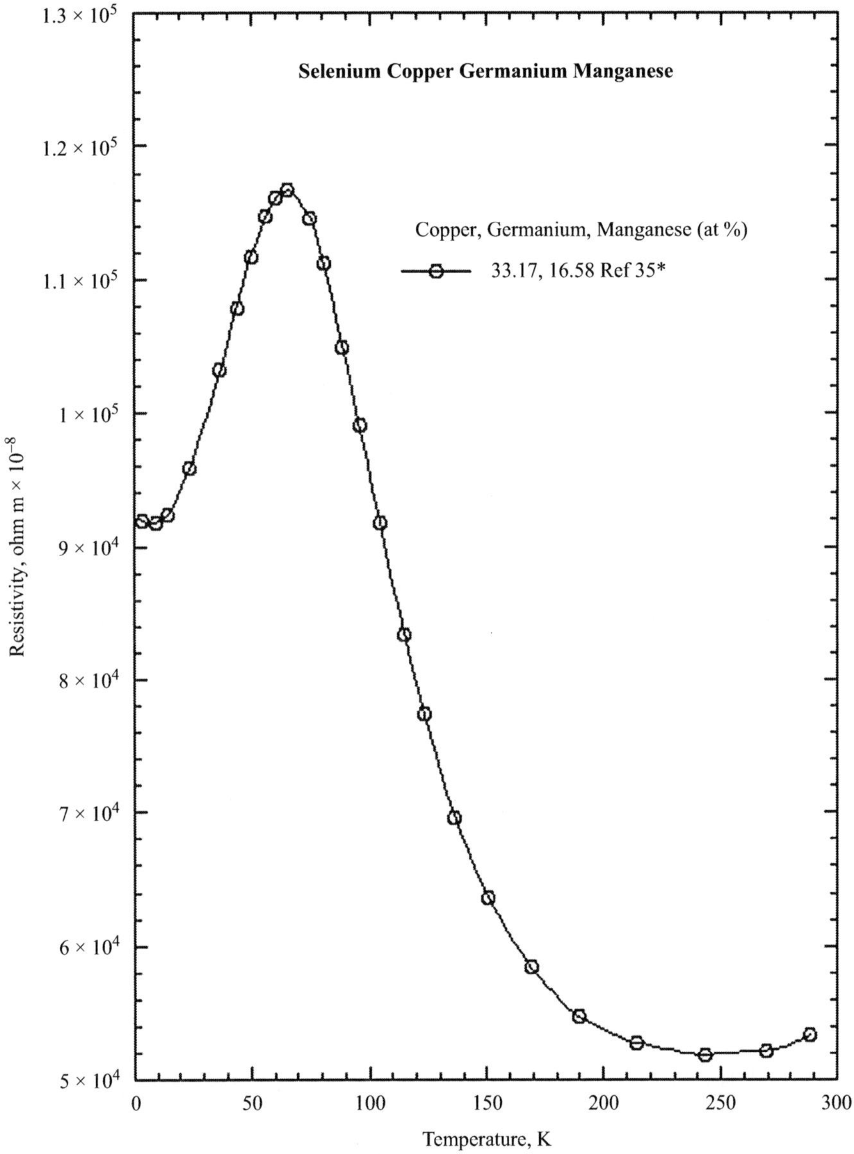
Selenium Copper Germanium Manganese
Copper, Germanium, Manganese (at %)
33.17, 16.58 Ref 35*
Resistivity, ohm m × 10^−8
1.3 × 10^5
1.2 × 10^5
1.1 × 10^5
1 × 10^5
9 × 10^4
8 × 10^4
7 × 10^4
6 × 10^4
5 × 10^4
0
50
100
150
200
250
300
Temperature, K

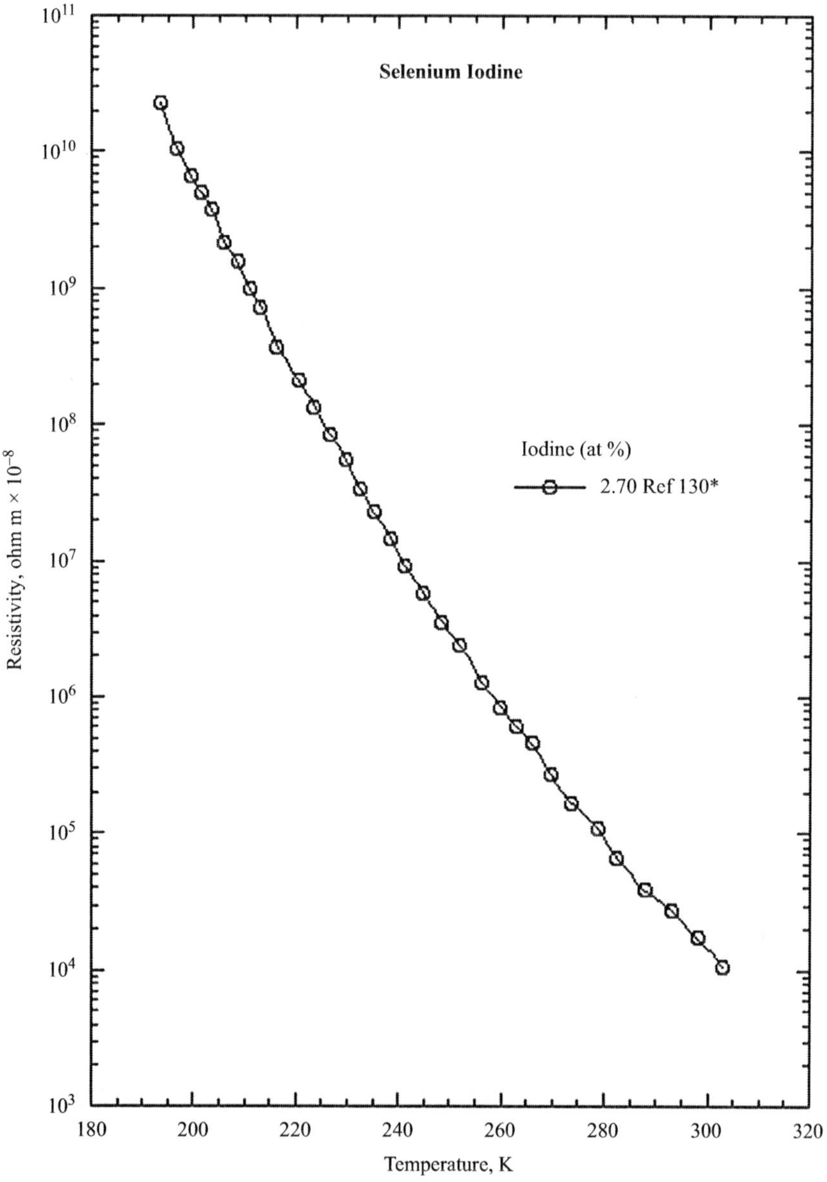

Selenium Iodine
Iodine (at %)
2.70 Ref 130*
Resistivity, ohm m × 10^{-8}
Temperature, K
10^{11}
10^{10}
10^9
10^8
10^7
10^6
10^5
10^4
10^3
180
200
220
240
260
280
300
320

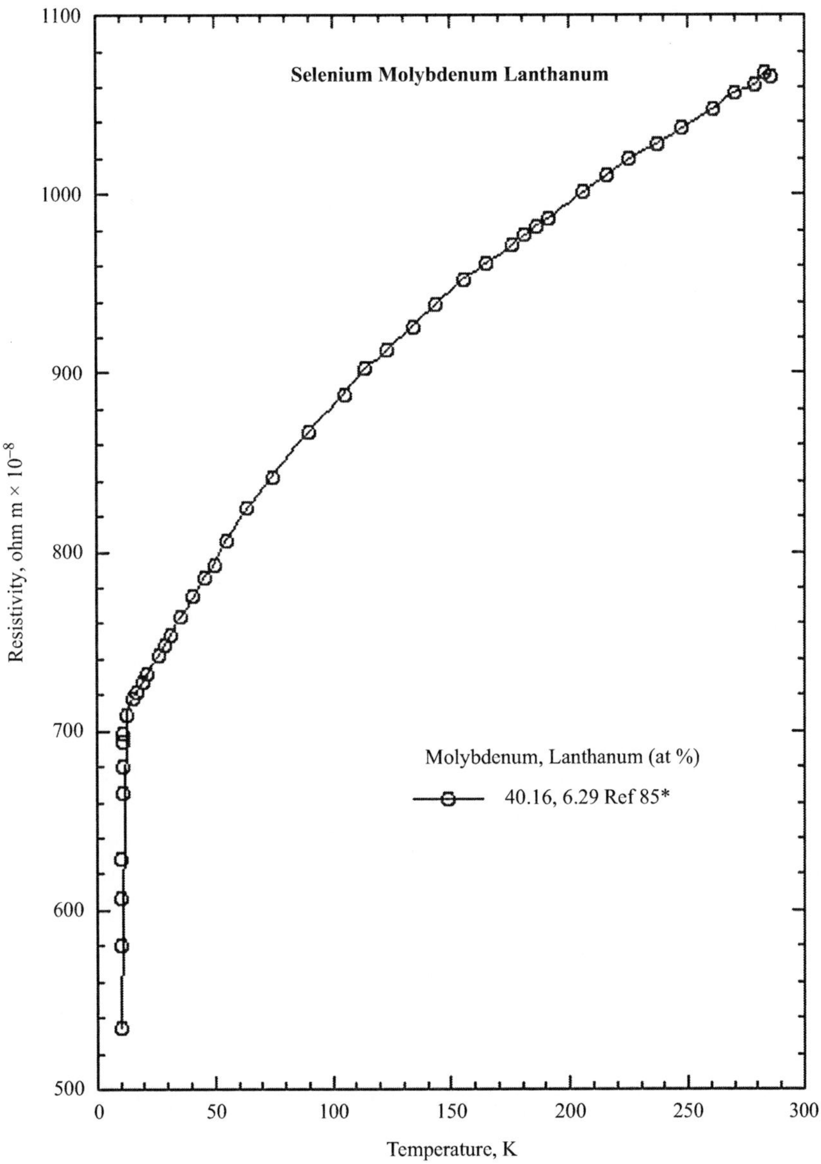
Selenium Molybdenum Lanthanum
Molybdenum, Lanthanum (at %)
40.16, 6.29 Ref 85*
Resistivity, ohm m × 10^{-8}
Temperature, K
1100
1000
900
800
700
600
500
0
50
100
150
200
250
300

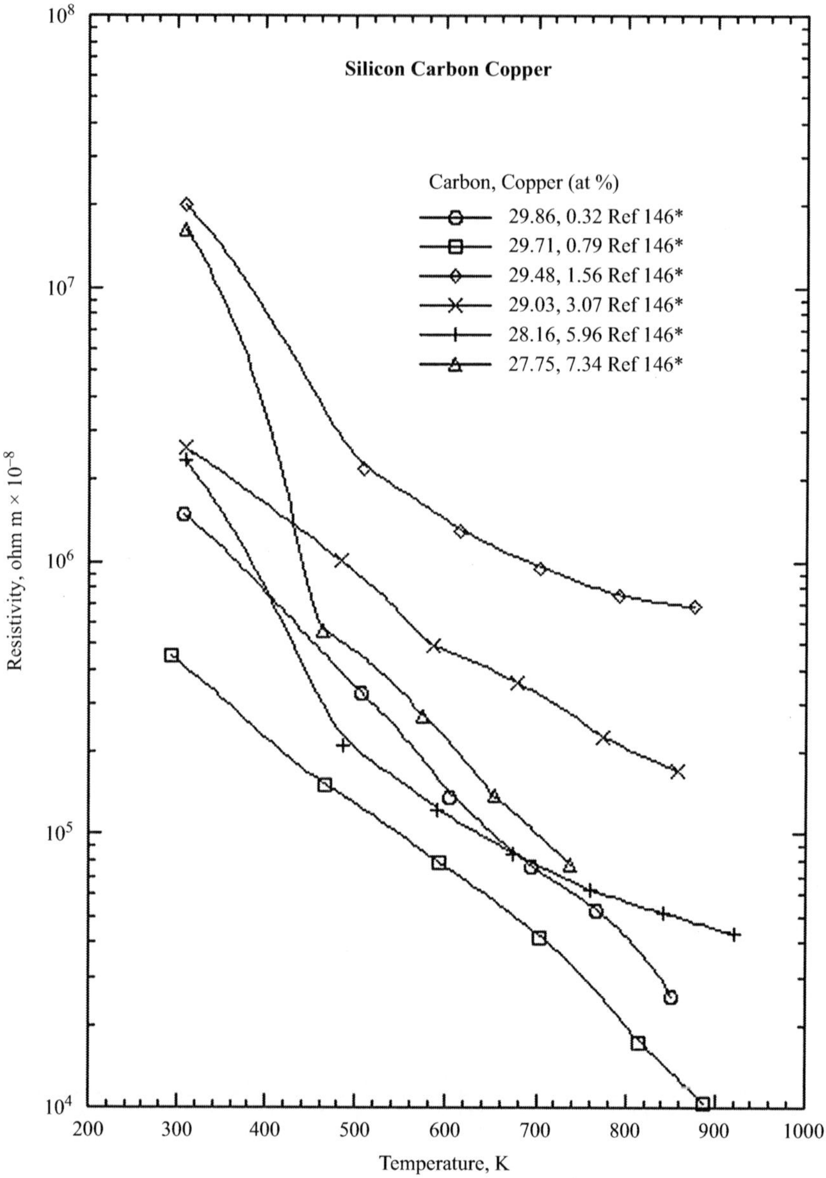
Silicon Carbon Copper
Carbon, Copper (at %)
29.86, 0.32 Ref 146*
29.71, 0.79 Ref 146*
29.48, 1.56 Ref 146*
29.03, 3.07 Ref 146*
28.16, 5.96 Ref 146*
27.75, 7.34 Ref 146*
Resistivity, ohm m × 10⁻⁸
10⁸
10⁷
10⁶
10⁵
10⁴
200
300
400
500
600
700
800
900
1000
Temperature, K

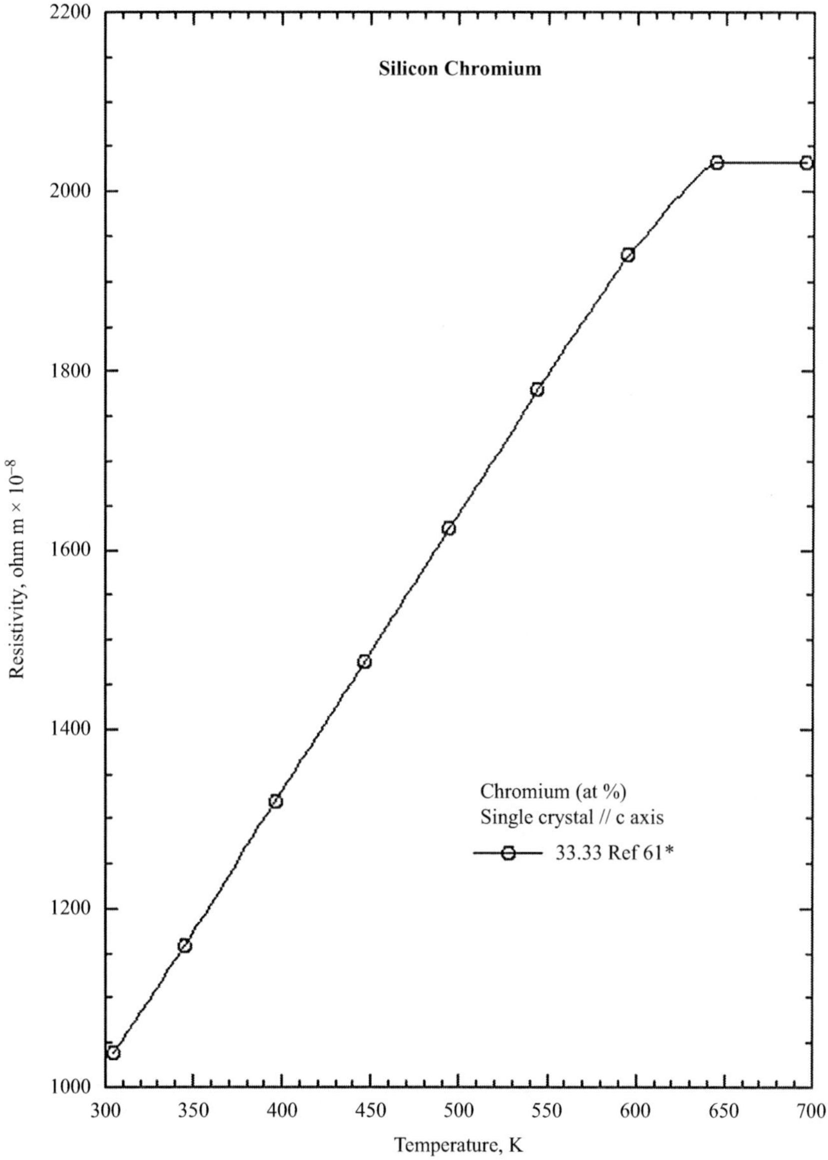
Silicon Chromium
Chromium (at %)
Single crystal // c axis
33.33 Ref 61*
Resistivity, ohm m × 10^{-8}
Temperature, K
2200
2000
1800
1600
1400
1200
1000
300
350
400
450
500
550
600
650
700

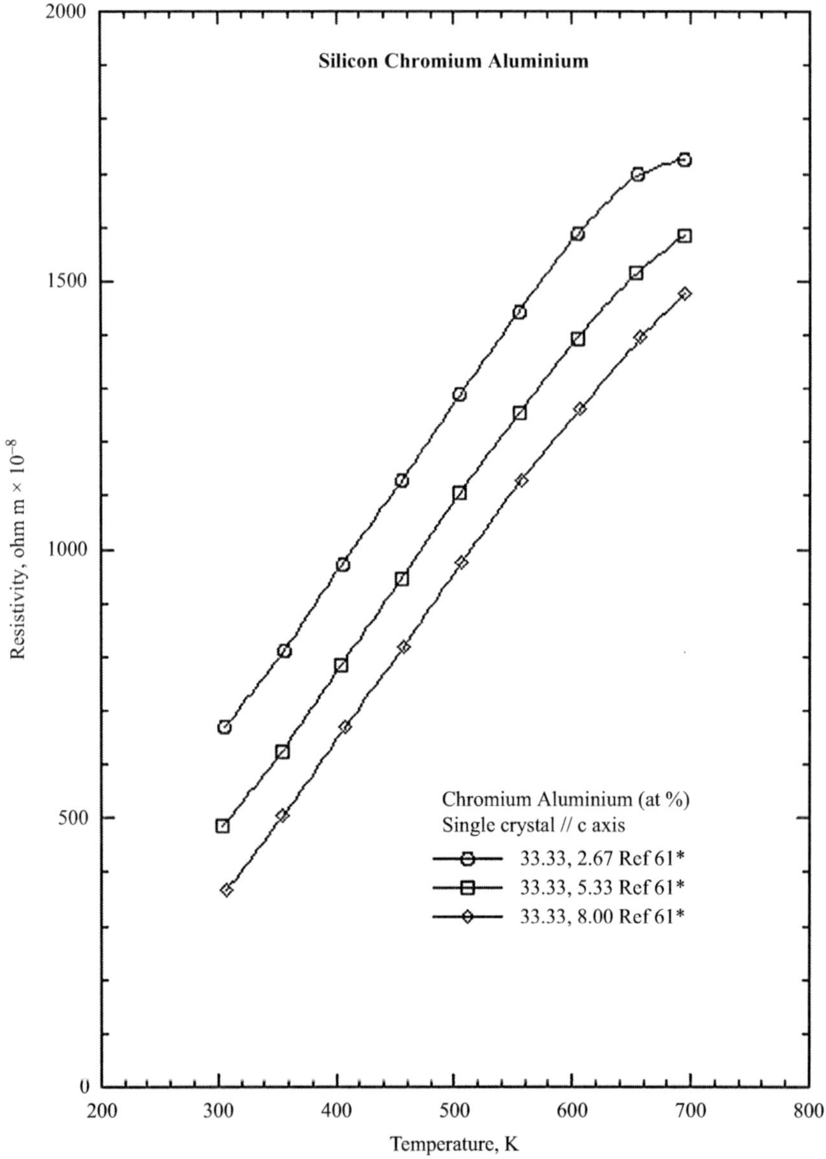
Silicon Chromium Aluminium
Resistivity, ohm m × 10⁻⁸
Temperature, K
Chromium Aluminium (at %)
Single crystal // c axis
33.33, 2.67 Ref 61*
33.33, 5.33 Ref 61*
33.33, 8.00 Ref 61*
0
500
1000
1500
2000
200
300
400
500
600
700
800

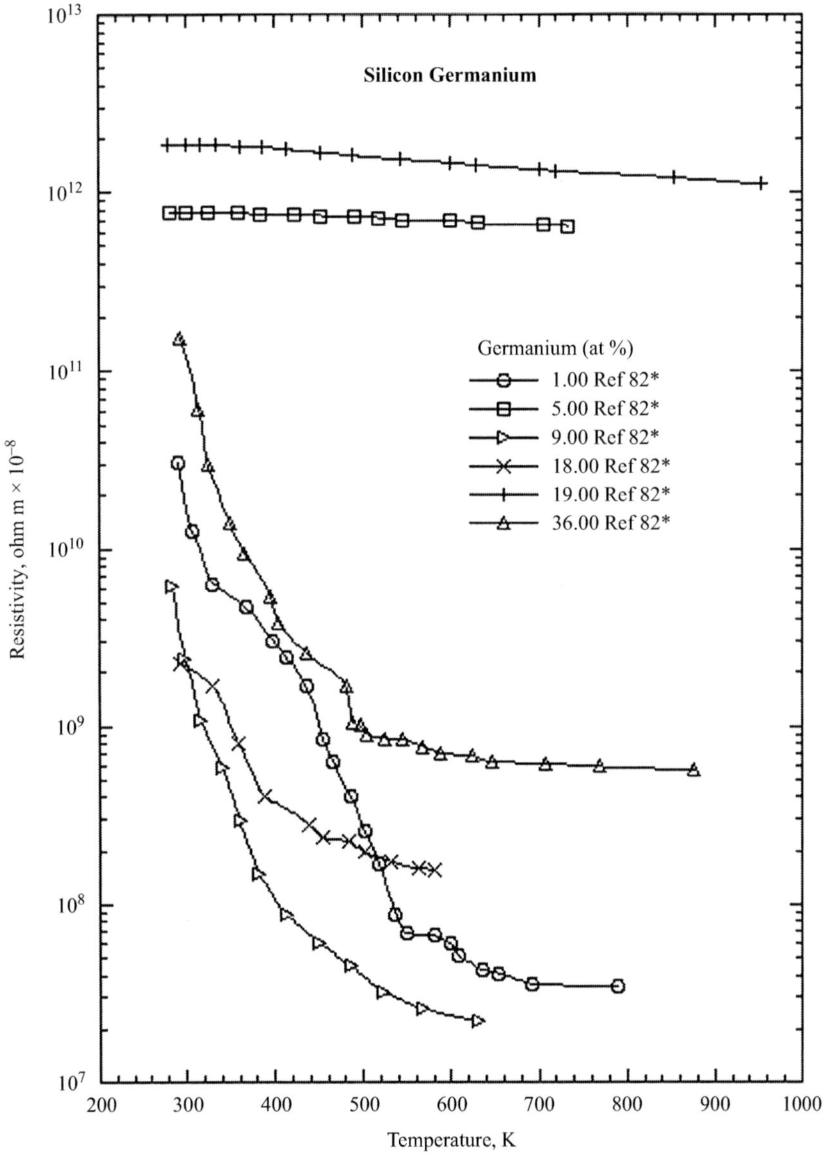
Silicon Germanium
Germanium (at %)
1.00 Ref 82*
5.00 Ref 82*
9.00 Ref 82*
18.00 Ref 82*
19.00 Ref 82*
36.00 Ref 82*
Resistivity, ohm m × 10^{-8}
Temperature, K
10^{13}
10^{12}
10^{11}
10^{10}
10^{9}
10^{8}
10^{7}
200
300
400
500
600
700
800
900
1000

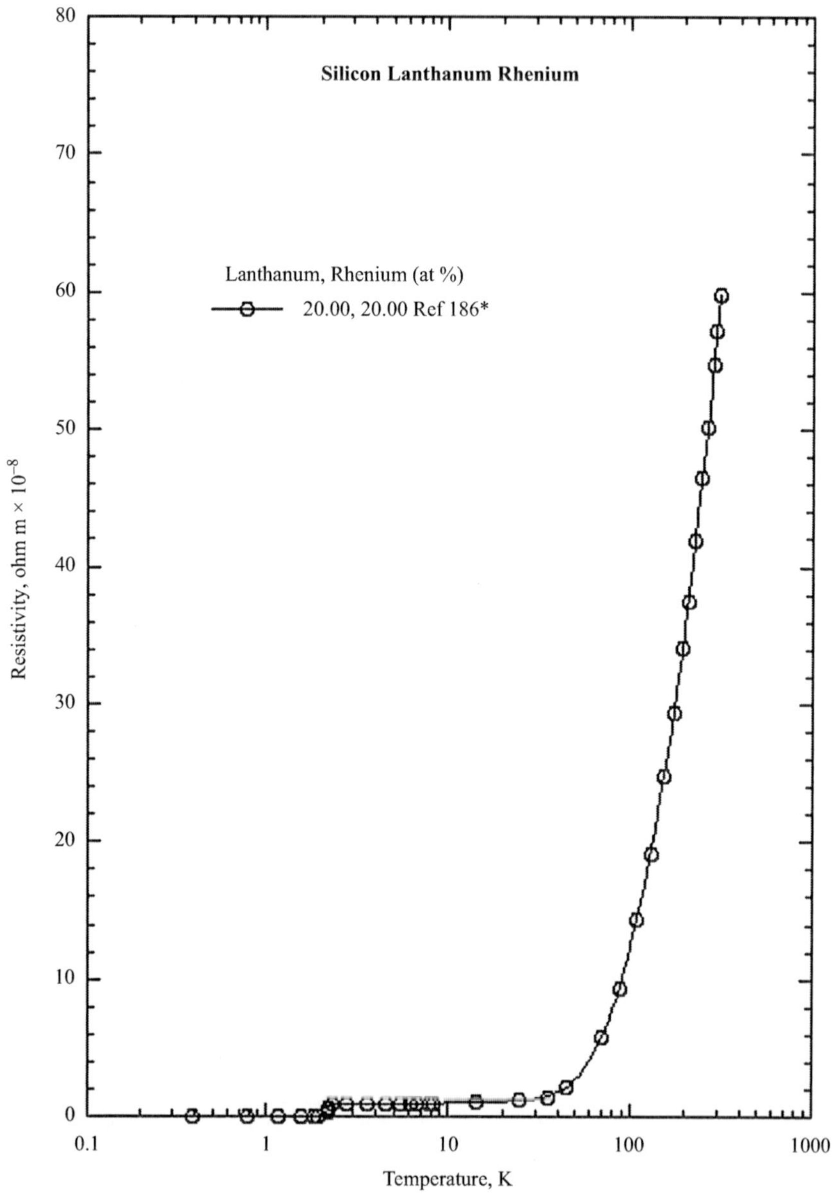
Silicon Lanthanum Rhenium
Lanthanum, Rhenium (at %)
20.00, 20.00 Ref 186*
Resistivity, ohm m × 10^-8
Temperature, K
0
10
20
30
40
50
60
70
80
0.1
1
10
100
1000

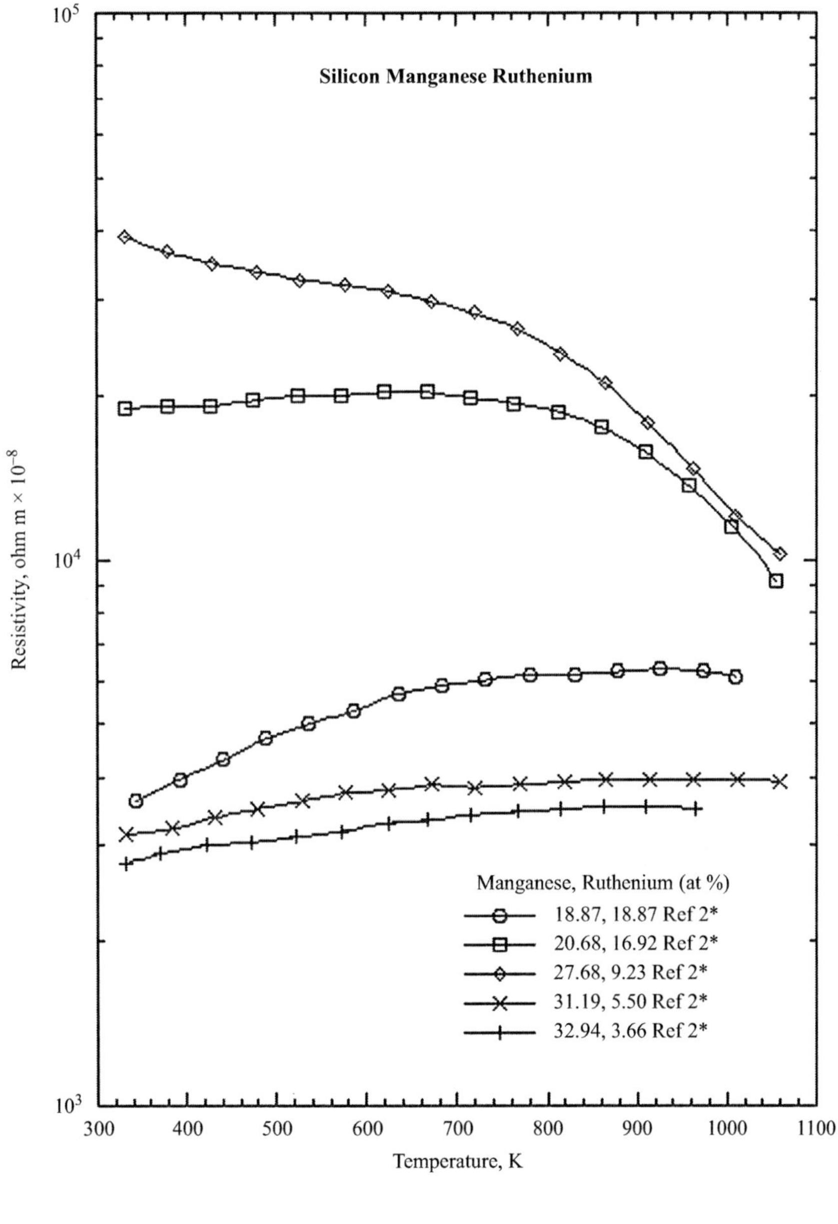
Silicon Manganese Ruthenium
Resistivity, ohm m × 10⁻⁸
10⁵
10⁴
10³
300
400
500
600
700
800
900
1000
1100
Temperature, K
Manganese, Ruthenium (at %)
18.87, 18.87 Ref 2*
20.68, 16.92 Ref 2*
27.68, 9.23 Ref 2*
31.19, 5.50 Ref 2*
32.94, 3.66 Ref 2*

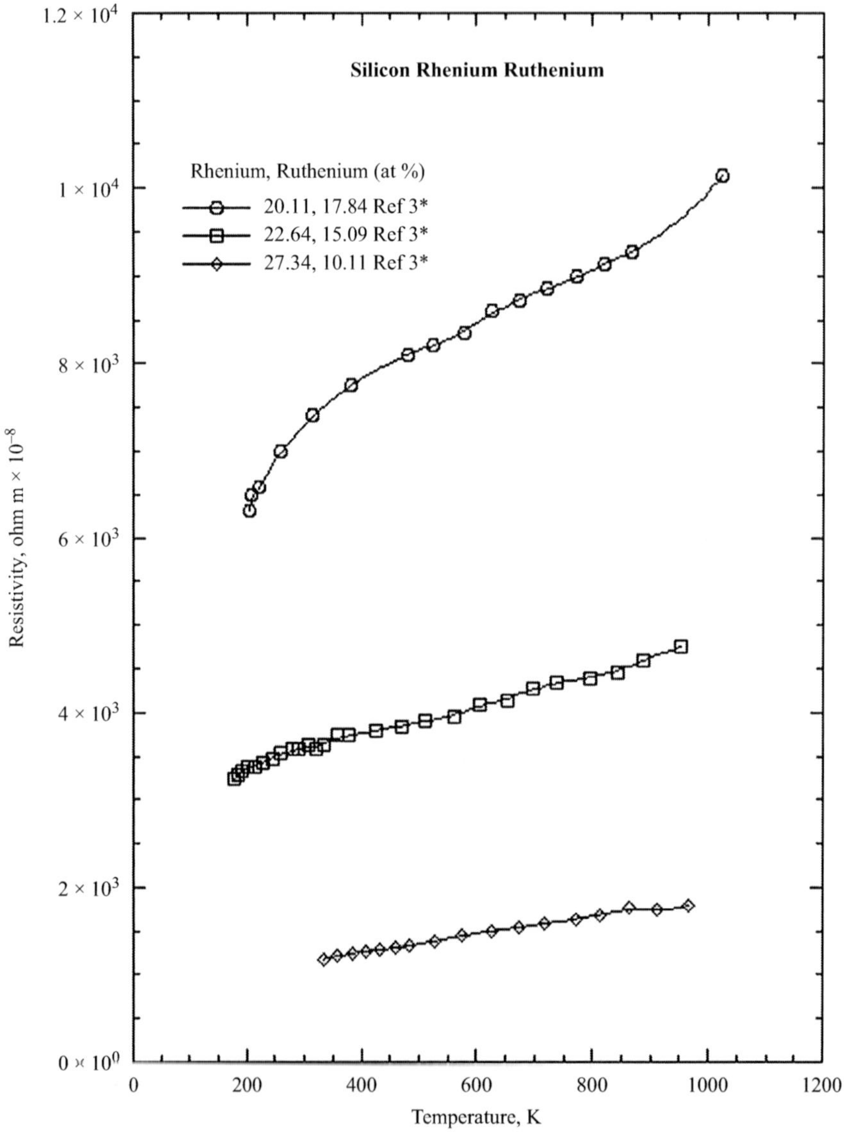
Silicon Rhenium Ruthenium
Rhenium, Ruthenium (at %)
20.11, 17.84 Ref 3*
22.64, 15.09 Ref 3*
27.34, 10.11 Ref 3*
Resistivity, ohm m × 10−8
1.2 × 10^4
1 × 10^4
8 × 10^3
6 × 10^3
4 × 10^3
2 × 10^3
0 × 10^0
0
200
400
600
800
1000
1200
Temperature, K

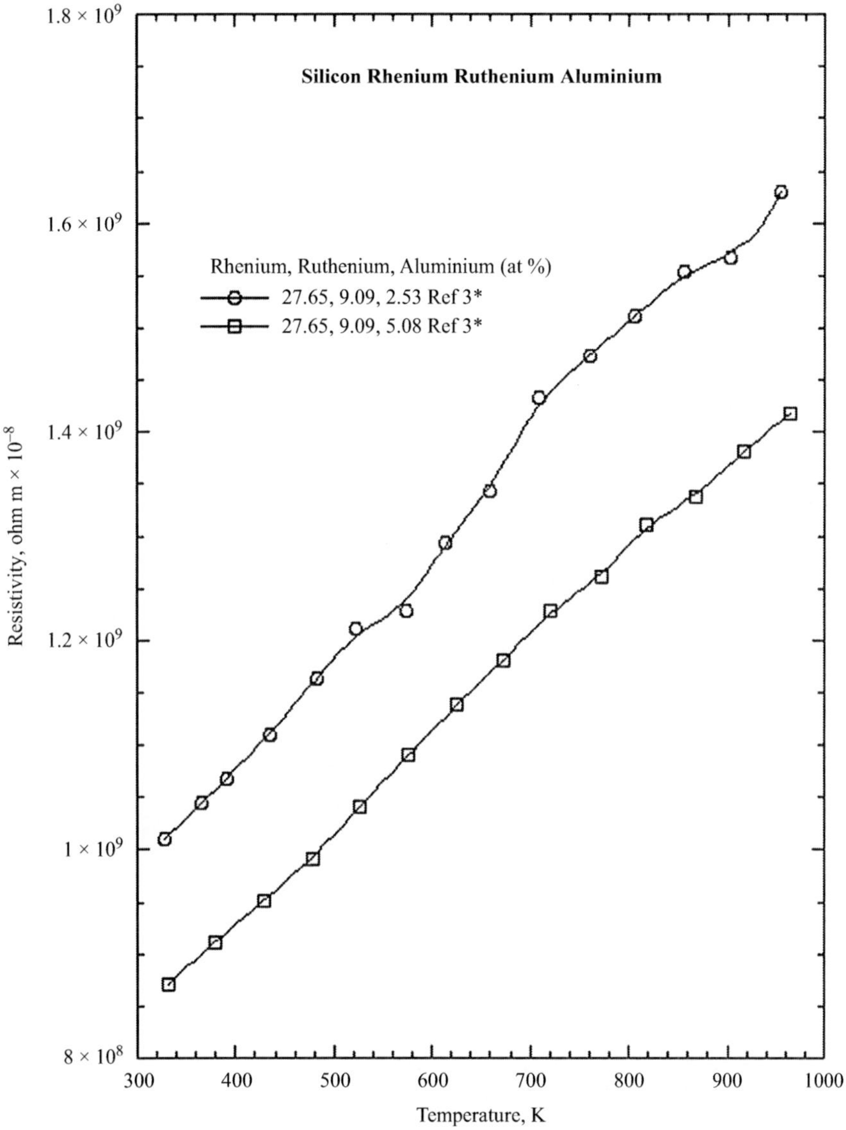
Silicon Rhenium Ruthenium Aluminium
Rhenium, Ruthenium, Aluminium (at %)
27.65, 9.09, 2.53 Ref 3*
27.65, 9.09, 5.08 Ref 3*
Resistivity, ohm m × 10⁻⁸
1.8 × 10⁹
1.6 × 10⁹
1.4 × 10⁹
1.2 × 10⁹
1 × 10⁹
8 × 10⁸
300
400
500
600
700
800
900
1000
Temperature, K

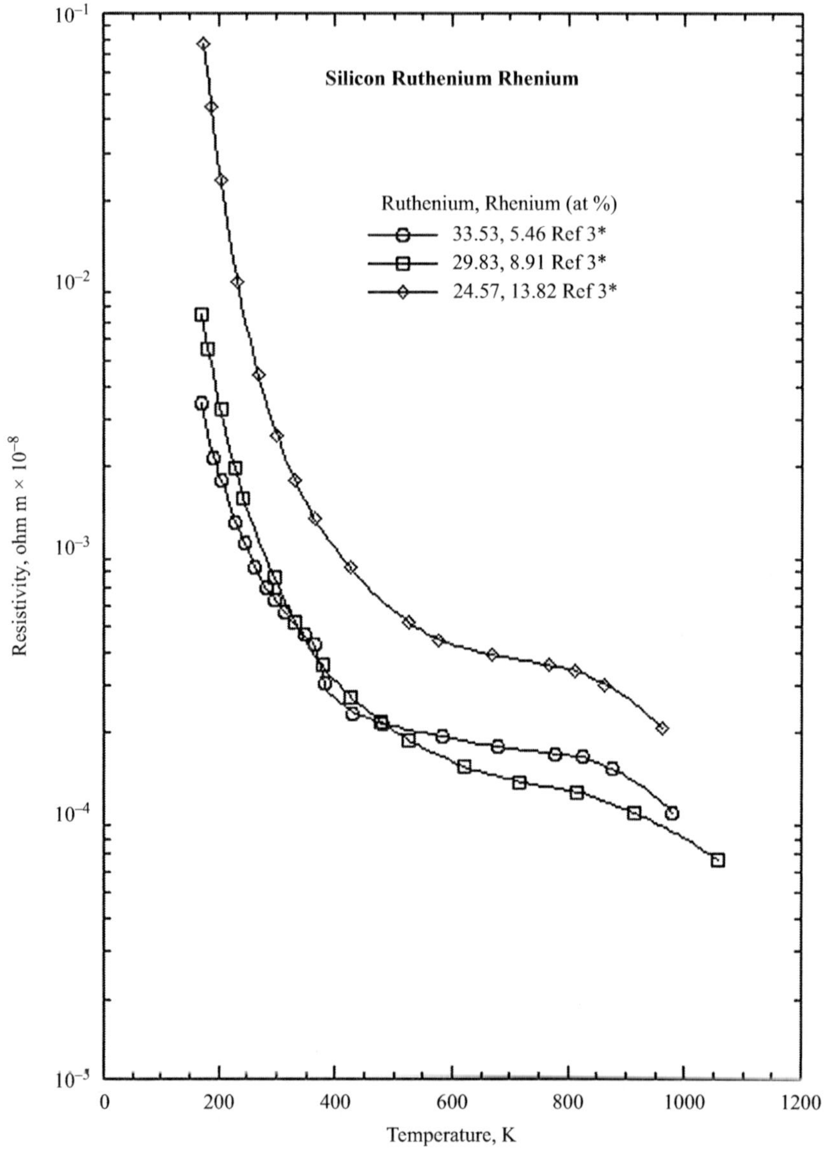
Silicon Ruthenium Rhenium
Ruthenium, Rhenium (at %)
33.53, 5.46 Ref 3*
29.83, 8.91 Ref 3*
24.57, 13.82 Ref 3*
Resistivity, ohm m × 10⁻⁸
Temperature, K
10⁻¹
10⁻²
10⁻³
10⁻⁴
10⁻⁵
0
200
400
600
800
1000
1200

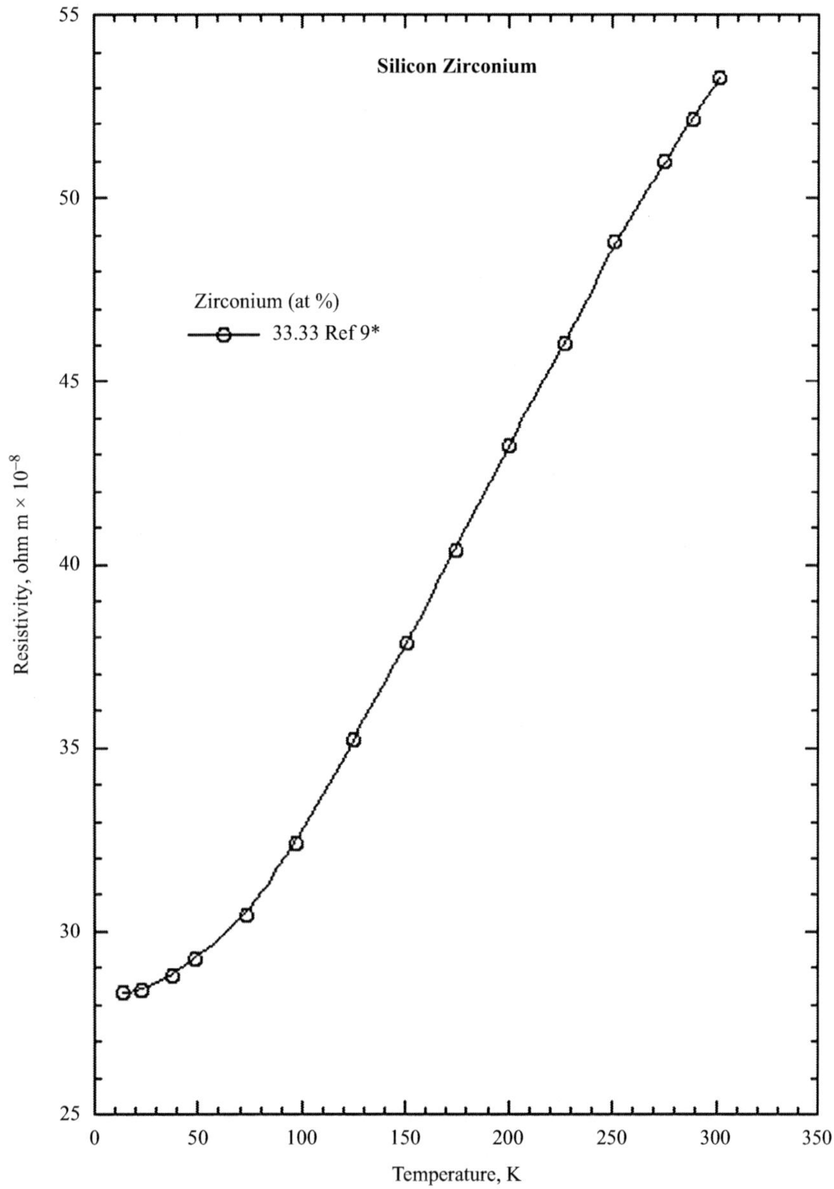
Silicon Zirconium
Zirconium (at %)
33.33 Ref 9*
Resistivity, ohm m × 10−8
Temperature, K
55
50
45
40
35
30
25
0
50
100
150
200
250
300
350

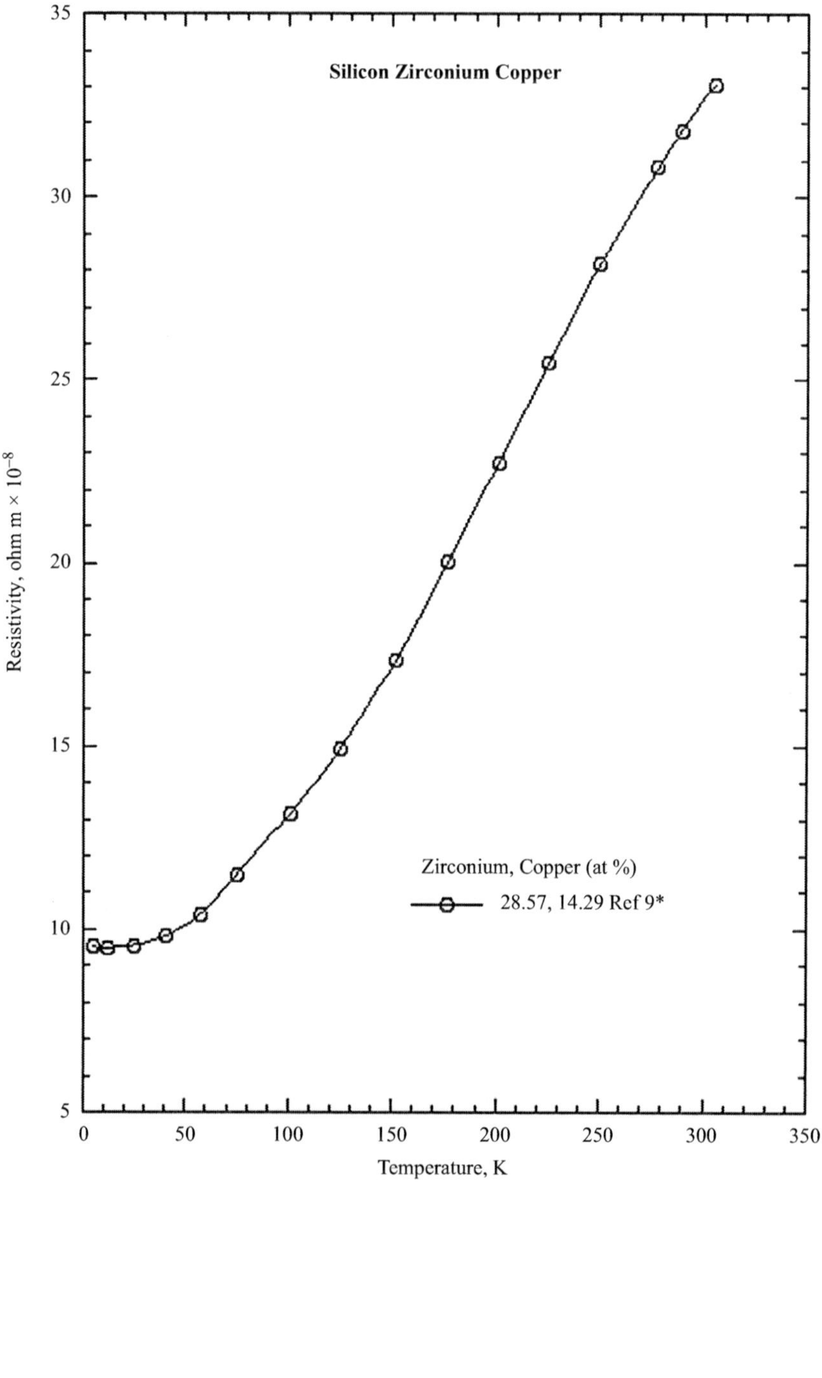
Silicon Zirconium Copper
Resistivity, ohm m × 10⁻⁸
Temperature, K
Zirconium, Copper (at %)
28.57, 14.29 Ref 9*
35
30
25
20
15
10
5
0
50
100
150
200
250
300
350

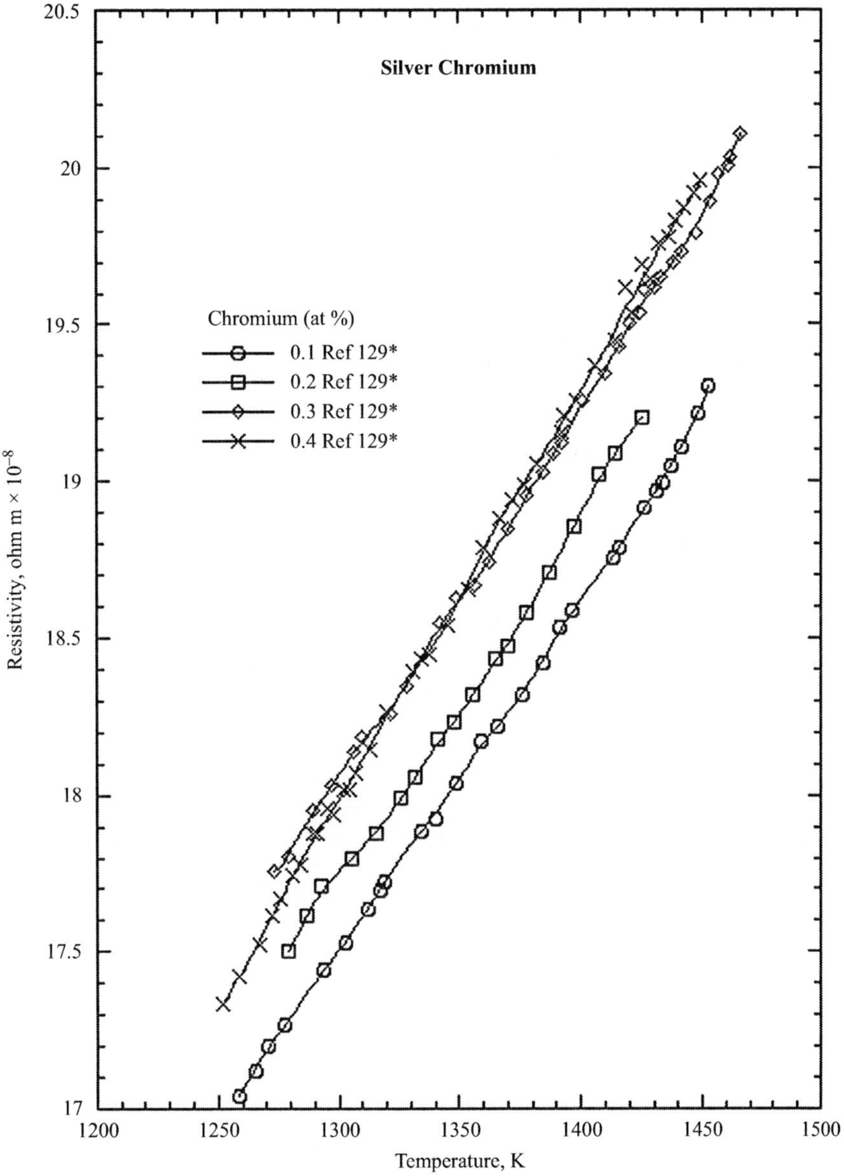
Silver Chromium
Chromium (at %)
0.1 Ref 129*
0.2 Ref 129*
0.3 Ref 129*
0.4 Ref 129*
Resistivity, ohm m × 10⁻⁸
Temperature, K
17
17.5
18
18.5
19
19.5
20
20.5
1200
1250
1300
1350
1400
1450
1500

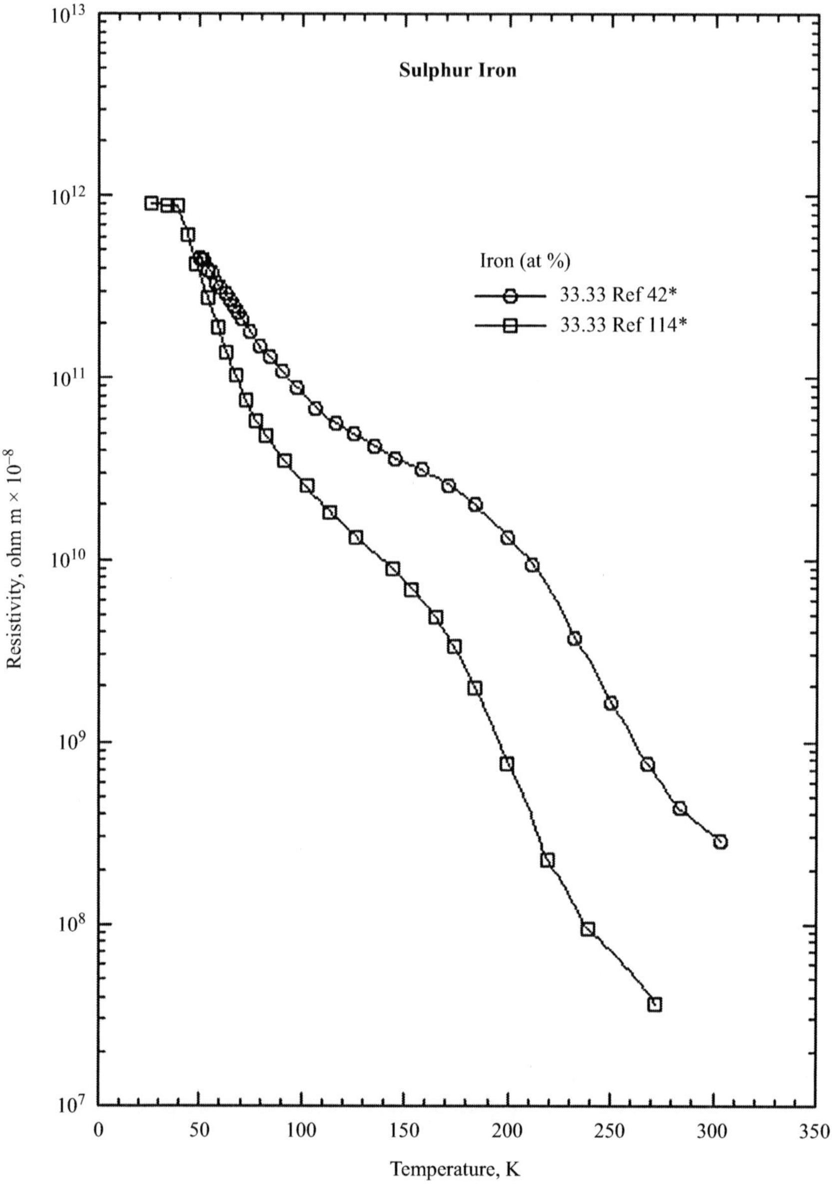
Sulphur Iron
Iron (at %)
33.33 Ref 42*
33.33 Ref 114*
10^{13}
10^{12}
10^{11}
10^{10}
10^{9}
10^{8}
10^{7}
Resistivity, ohm m × 10^{-8}
0
50
100
150
200
250
300
350
Temperature, K

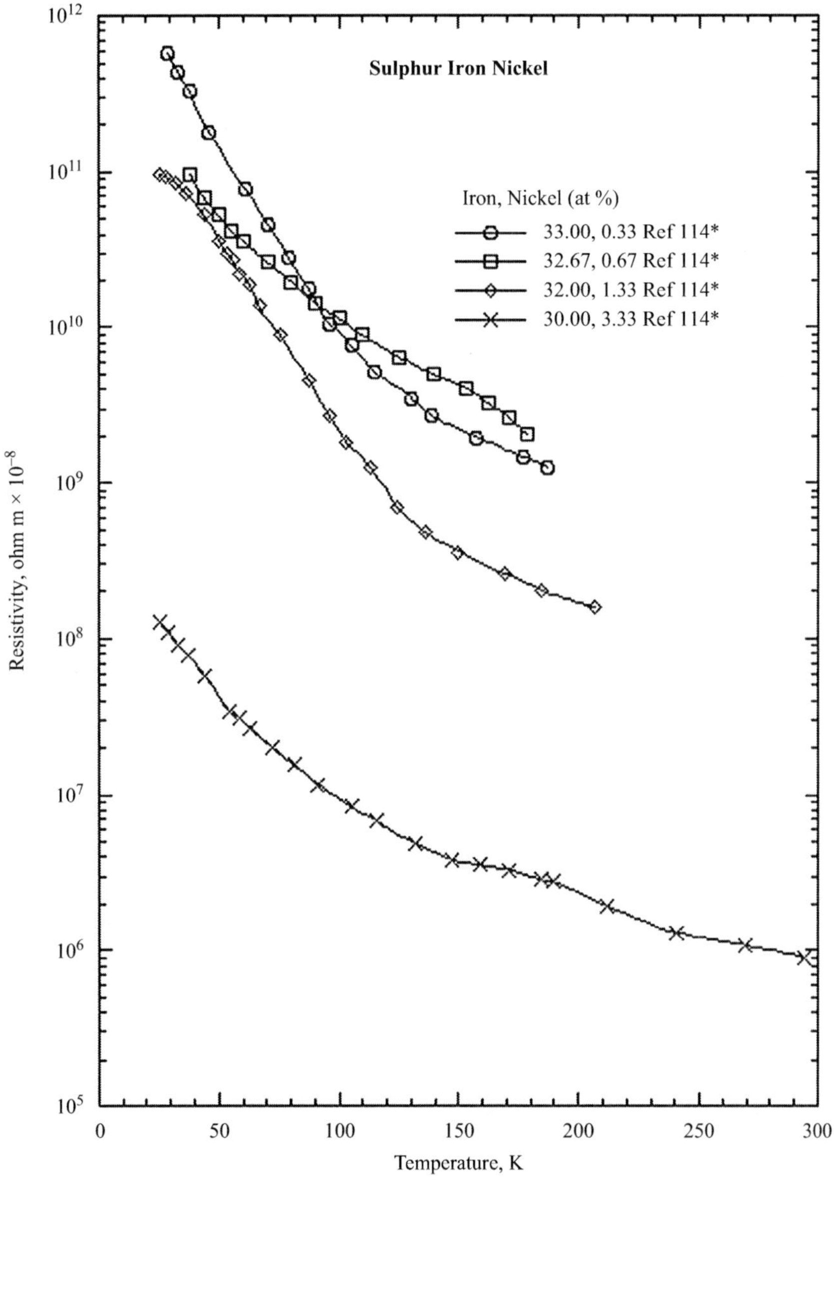
Sulphur Iron Nickel
Iron, Nickel (at %)
33.00, 0.33 Ref 114*
32.67, 0.67 Ref 114*
32.00, 1.33 Ref 114*
30.00, 3.33 Ref 114*
Resistivity, ohm m × 10⁻⁸
Temperature, K
10¹²
10¹¹
10¹⁰
10⁹
10⁸
10⁷
10⁶
10⁵
0
50
100
150
200
250
300

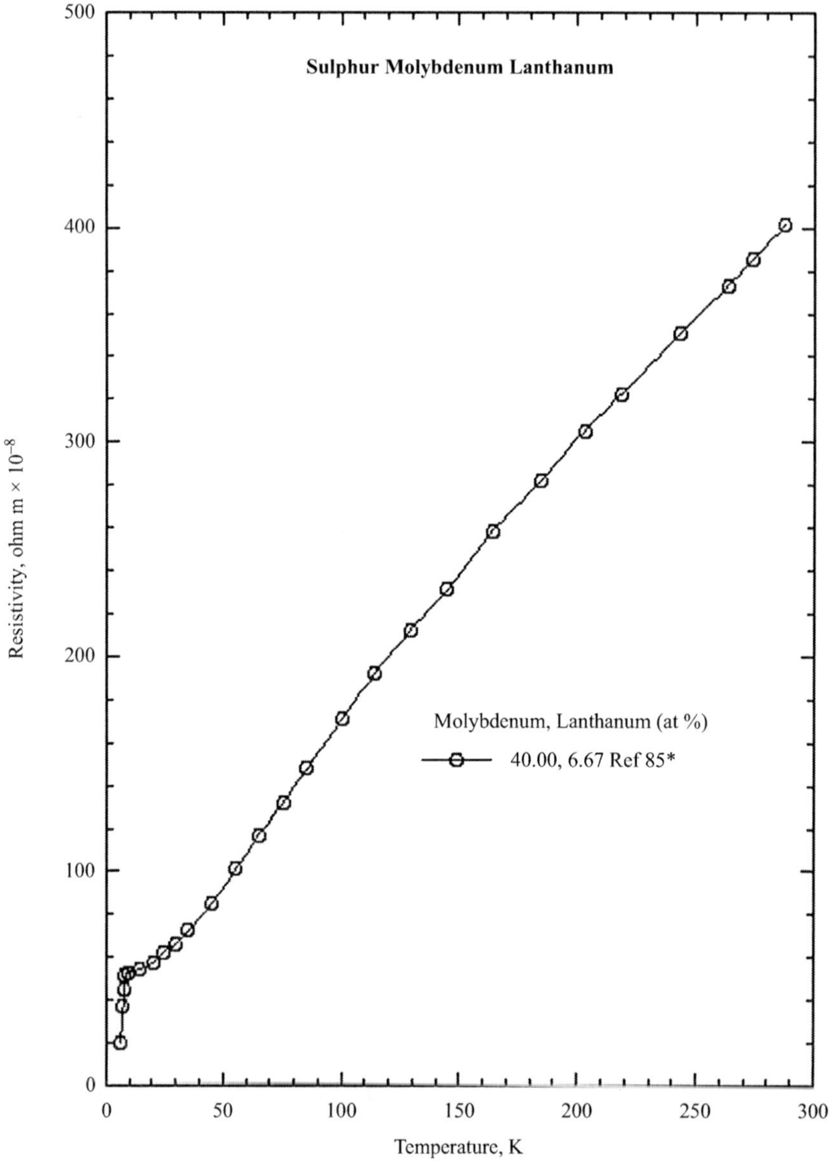
Sulphur Molybdenum Lanthanum
Molybdenum, Lanthanum (at %)
40.00, 6.67 Ref 85*
Resistivity, ohm m × 10⁻⁸
Temperature, K
0
100
200
300
400
500
0
50
100
150
200
250
300

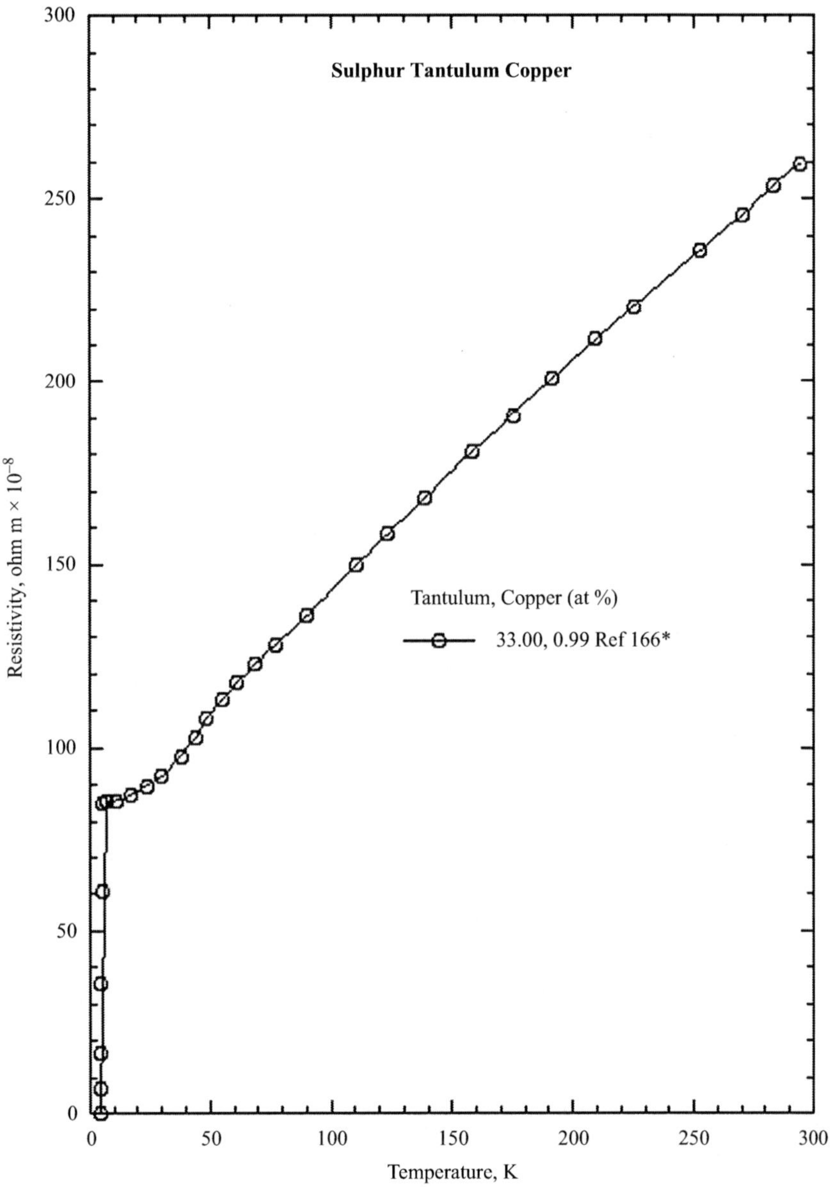
Sulphur Tantulum Copper
Tantulum, Copper (at %)
33.00, 0.99 Ref 166*
Resistivity, ohm m × 10⁻⁸
Temperature, K
0
50
100
150
200
250
300

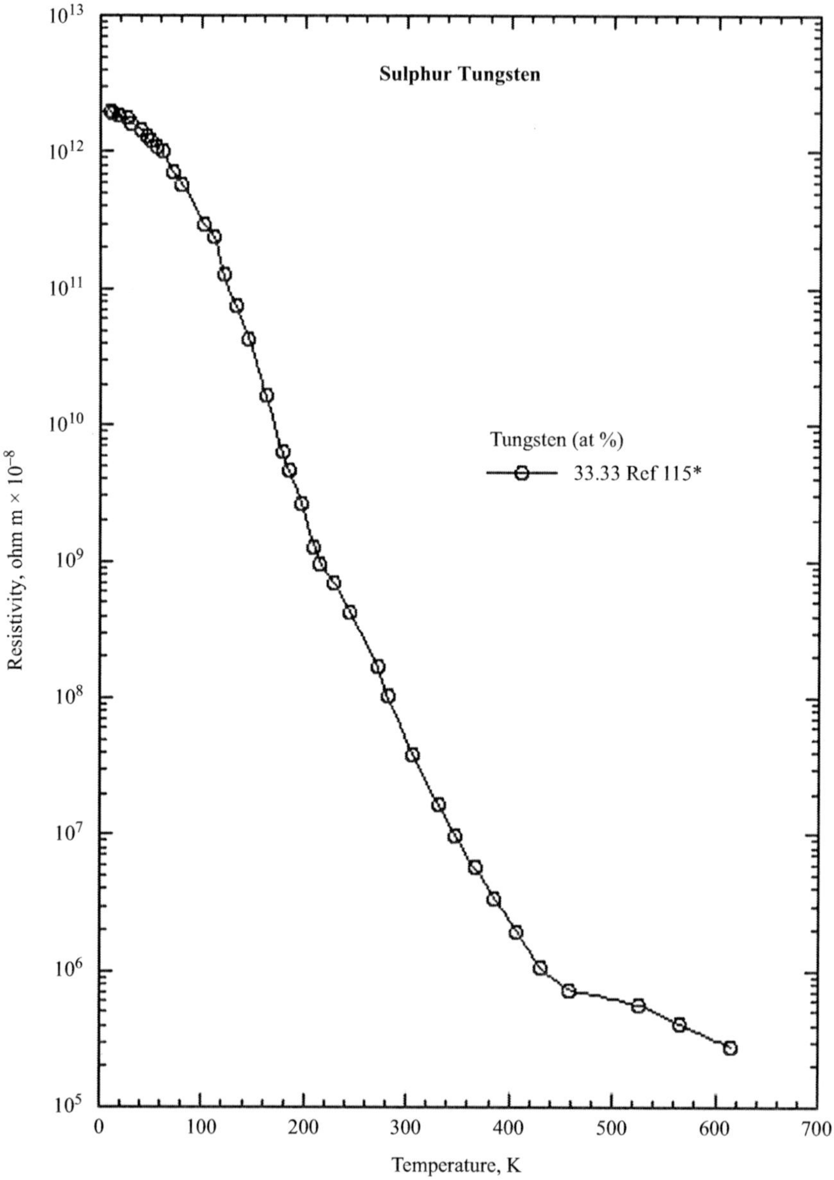

Sulphur Tungsten
Tungsten (at %)
33.33 Ref 115*
Resistivity, ohm m × 10⁻⁸
Temperature, K
10^13
10^12
10^11
10^10
10^9
10^8
10^7
10^6
10^5
0
100
200
300
400
500
600
700

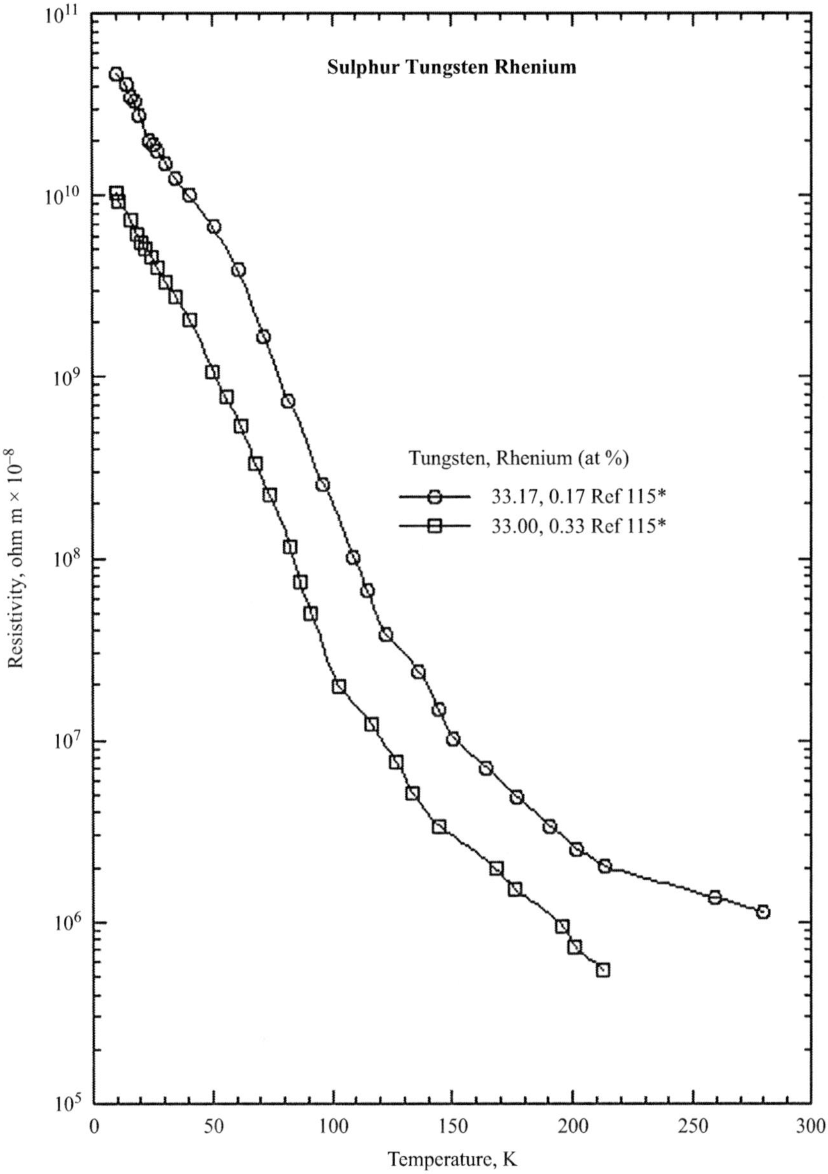
Sulphur Tungsten Rhenium
Tungsten, Rhenium (at %)
33.17, 0.17 Ref 115*
33.00, 0.33 Ref 115*
Resistivity, ohm m × 10⁻⁸
Temperature, K
10¹¹
10¹⁰
10⁹
10⁸
10⁷
10⁶
10⁵
0
50
100
150
200
250
300

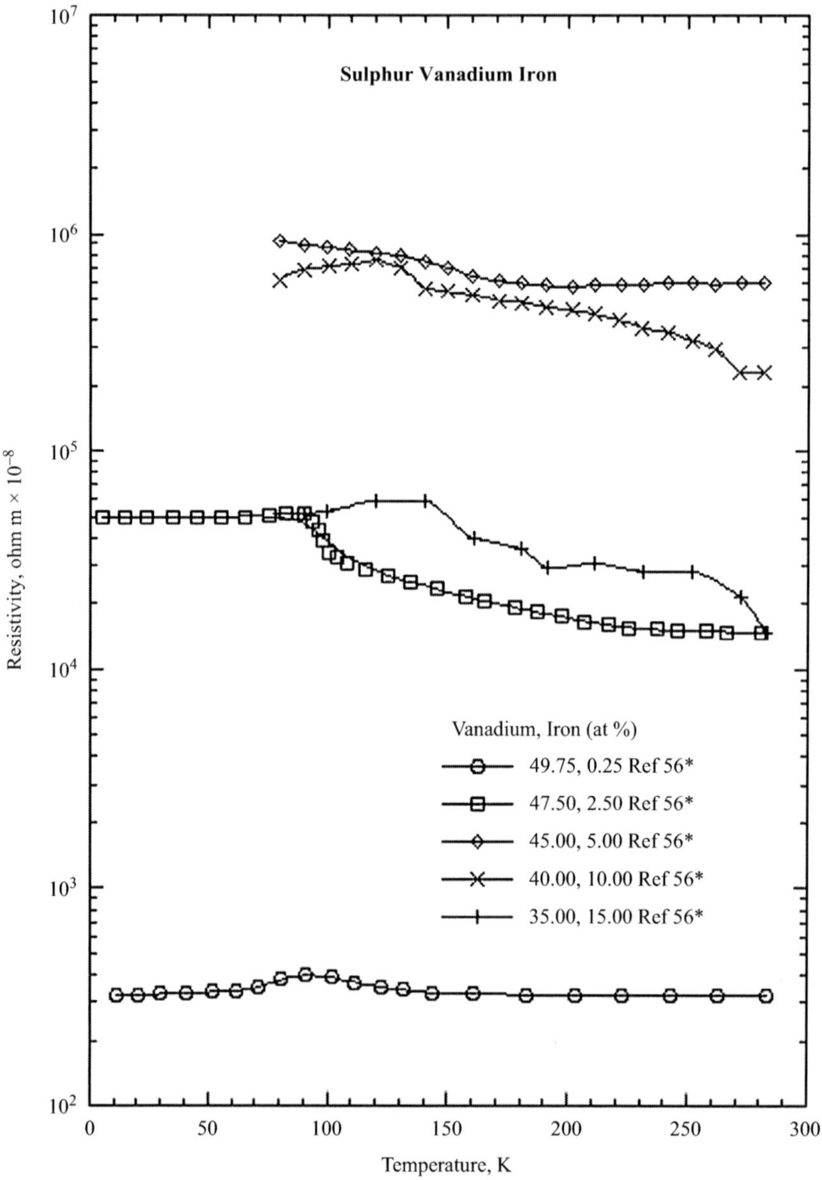
Sulphur Vanadium Iron
Resistivity, ohm m × 10^{-8}
Temperature, K
10^7
10^6
10^5
10^4
10^3
10^2
0
50
100
150
200
250
300
Vanadium, Iron (at %)
49.75, 0.25 Ref 56*
47.50, 2.50 Ref 56*
45.00, 5.00 Ref 56*
40.00, 10.00 Ref 56*
35.00, 15.00 Ref 56*

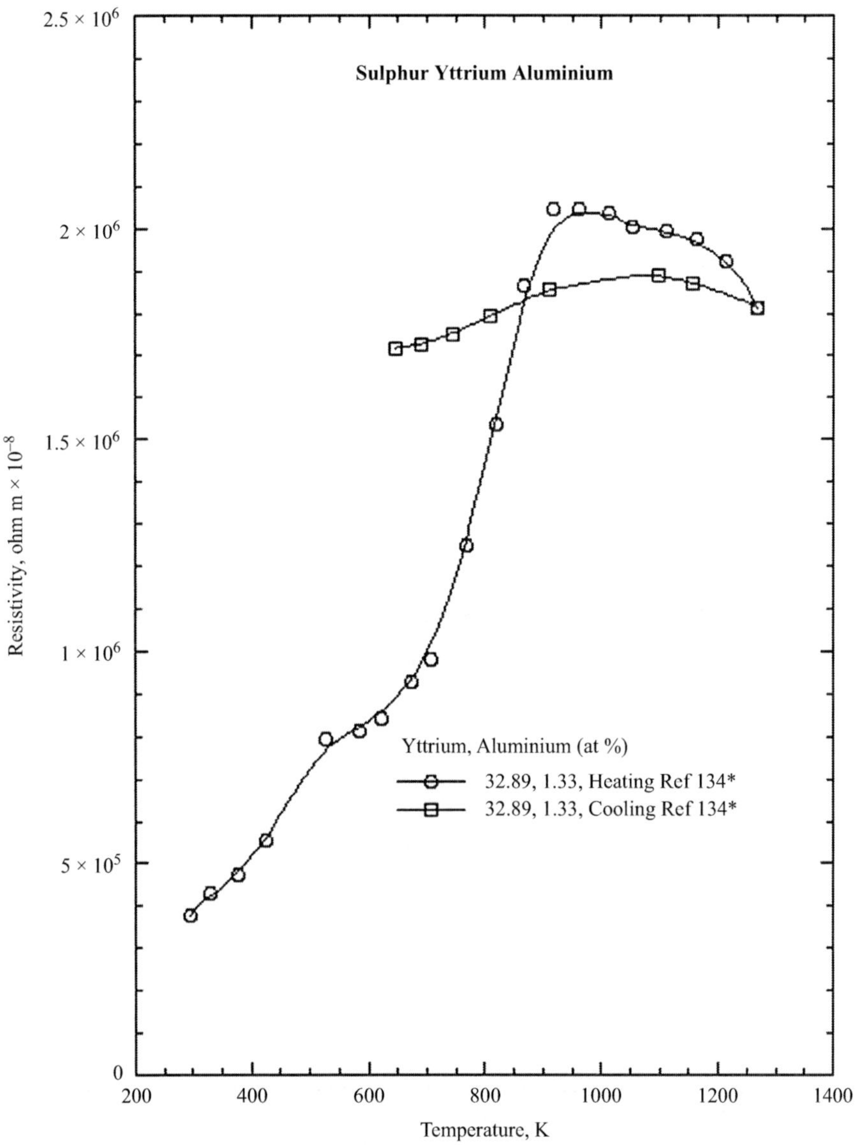

Sulphur Yttrium Aluminium
Resistivity, ohm m × 10⁻⁸
Temperature, K
Yttrium, Aluminium (at %)
32.89, 1.33, Heating Ref 134*
32.89, 1.33, Cooling Ref 134*
2.5 × 10⁶
2 × 10⁶
1.5 × 10⁶
1 × 10⁶
5 × 10⁵
0
200
400
600
800
1000
1200
1400

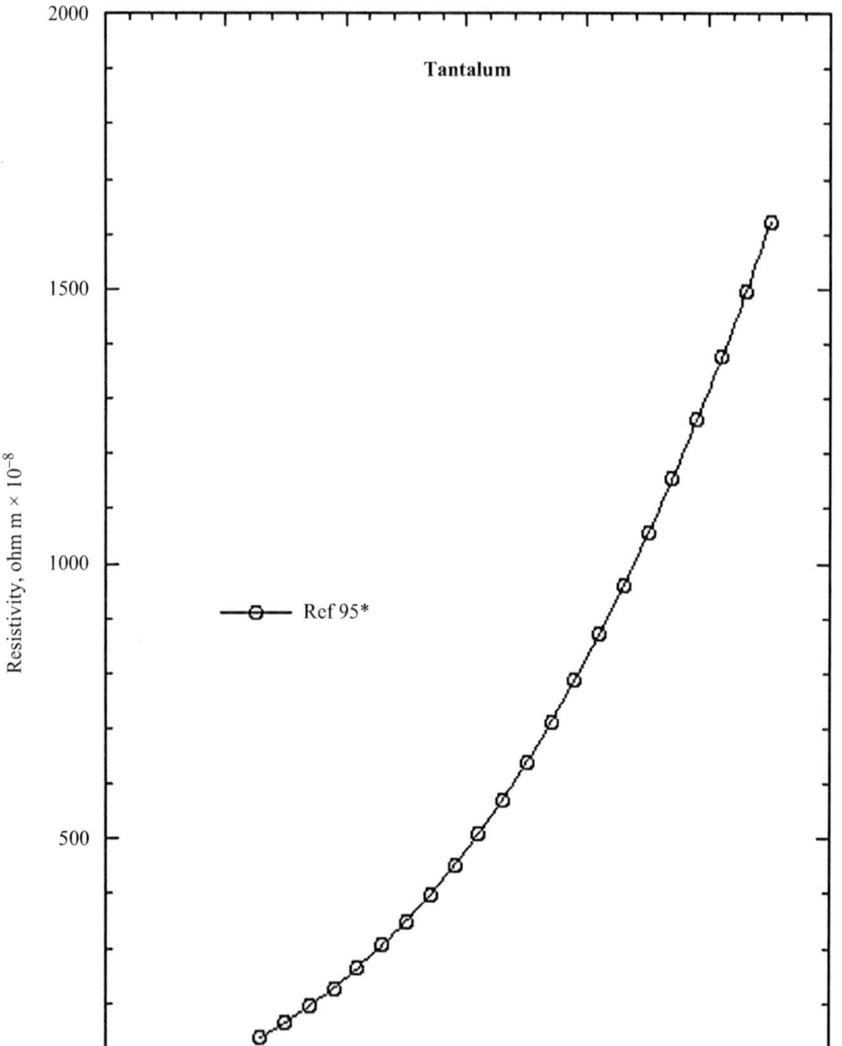
Tantalum
Ref 95*
Resistivity, ohm m × 10⁻⁸
Temperature, K
0
500
1000
1500
2000
2000
3000
4000
5000
6000
7000
8000

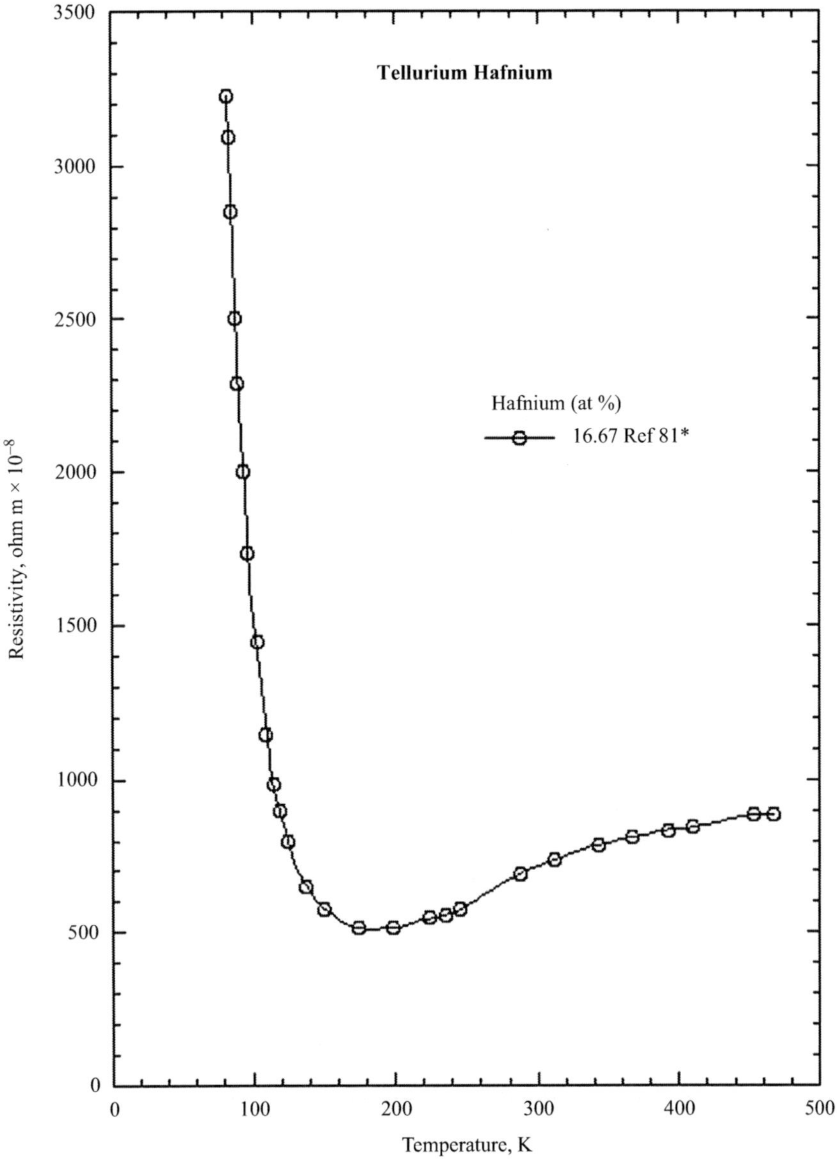
Tellurium Hafnium
Hafnium (at %)
16.67 Ref 81*
Resistivity, ohm m × 10^-8
Temperature, K
0
500
1000
1500
2000
2500
3000
3500
0
100
200
300
400
500

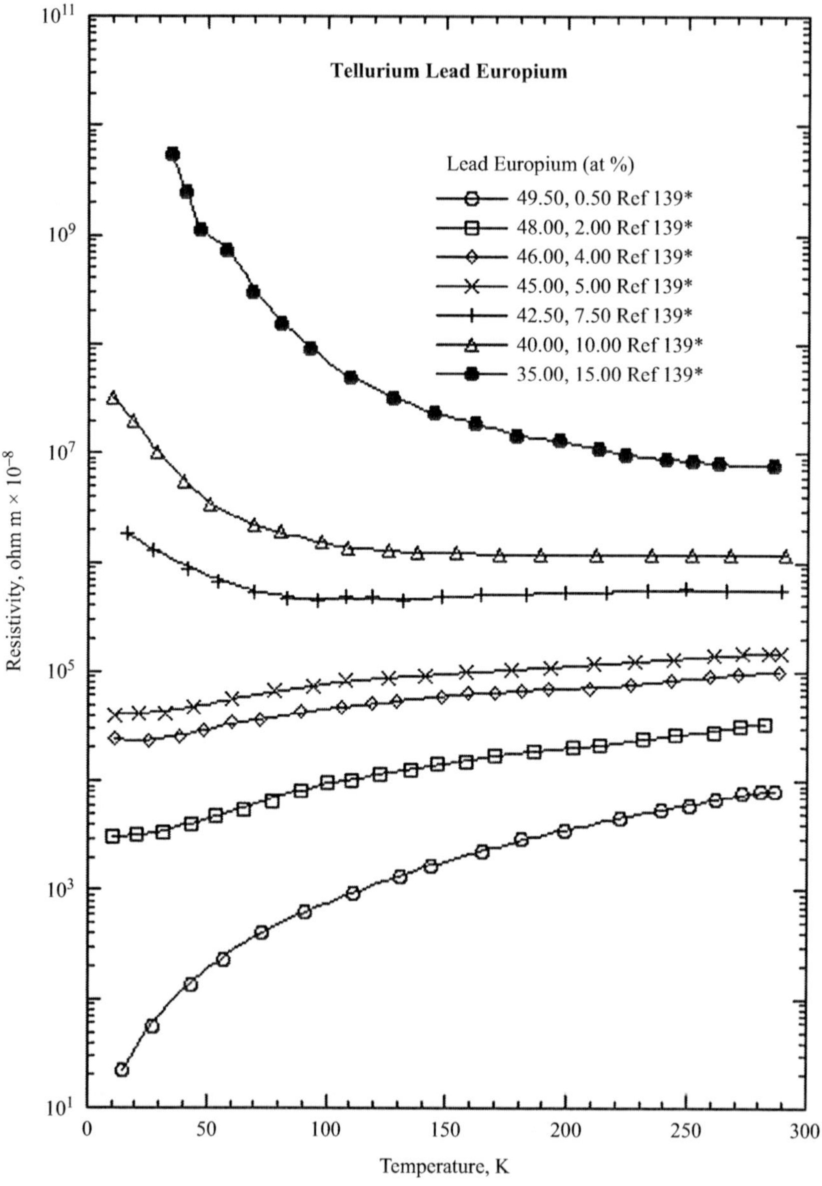
Tellurium Lead Europium
Lead Europium (at %)
49.50, 0.50 Ref 139*
48.00, 2.00 Ref 139*
46.00, 4.00 Ref 139*
45.00, 5.00 Ref 139*
42.50, 7.50 Ref 139*
40.00, 10.00 Ref 139*
35.00, 15.00 Ref 139*
Resistivity, ohm m × 10⁻⁸
Temperature, K
10¹¹
10⁹
10⁷
10⁵
10³
10¹
0
50
100
150
200
250
300

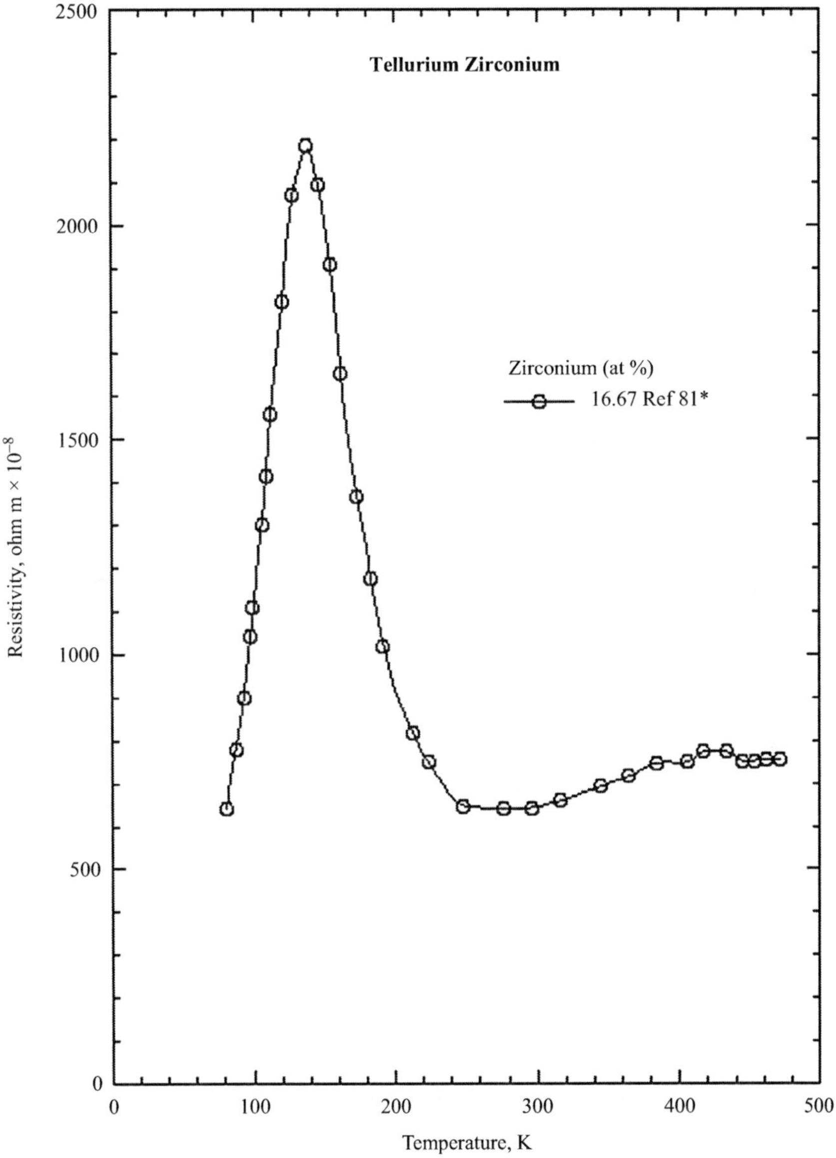
Tellurium Zirconium
Zirconium (at %)
16.67 Ref 81*
Resistivity, ohm m × 10⁻⁸
Temperature, K
0
500
1000
1500
2000
2500
0
100
200
300
400
500

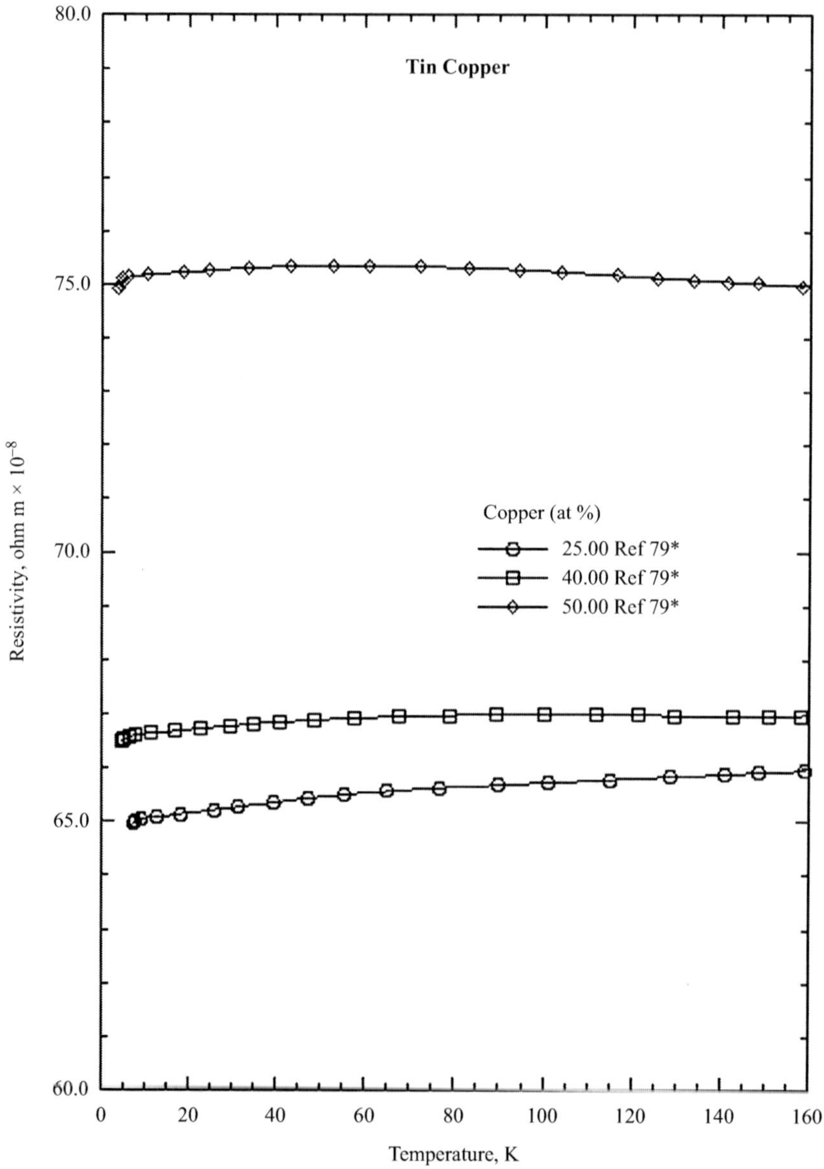

Tin Copper
Copper (at %)
25.00 Ref 79*
40.00 Ref 79*
50.00 Ref 79*
Resistivity, ohm m × 10⁻⁸
Temperature, K
80.0
75.0
70.0
65.0
60.0
0
20
40
60
80
100
120
140
160

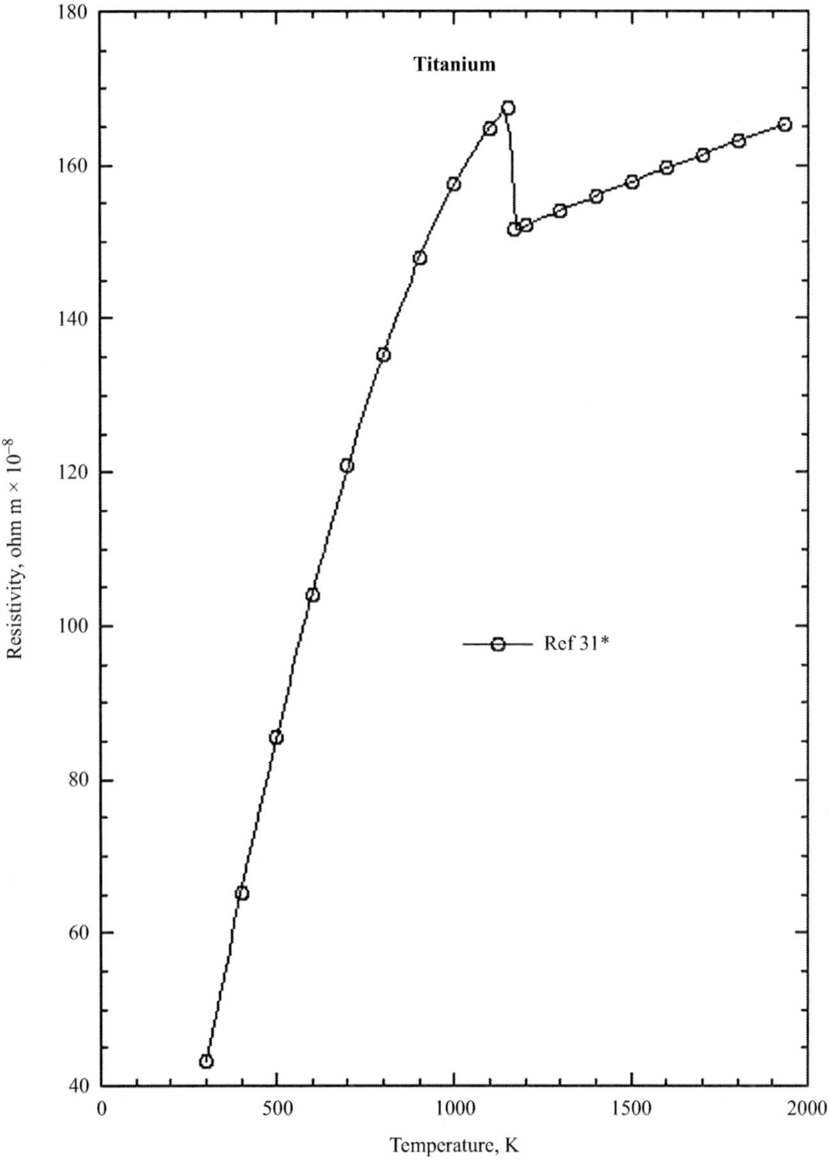

Titanium
Ref 31*
Resistivity, ohm m × 10^−8
Temperature, K
180
160
140
120
100
80
60
40
0
500
1000
1500
2000

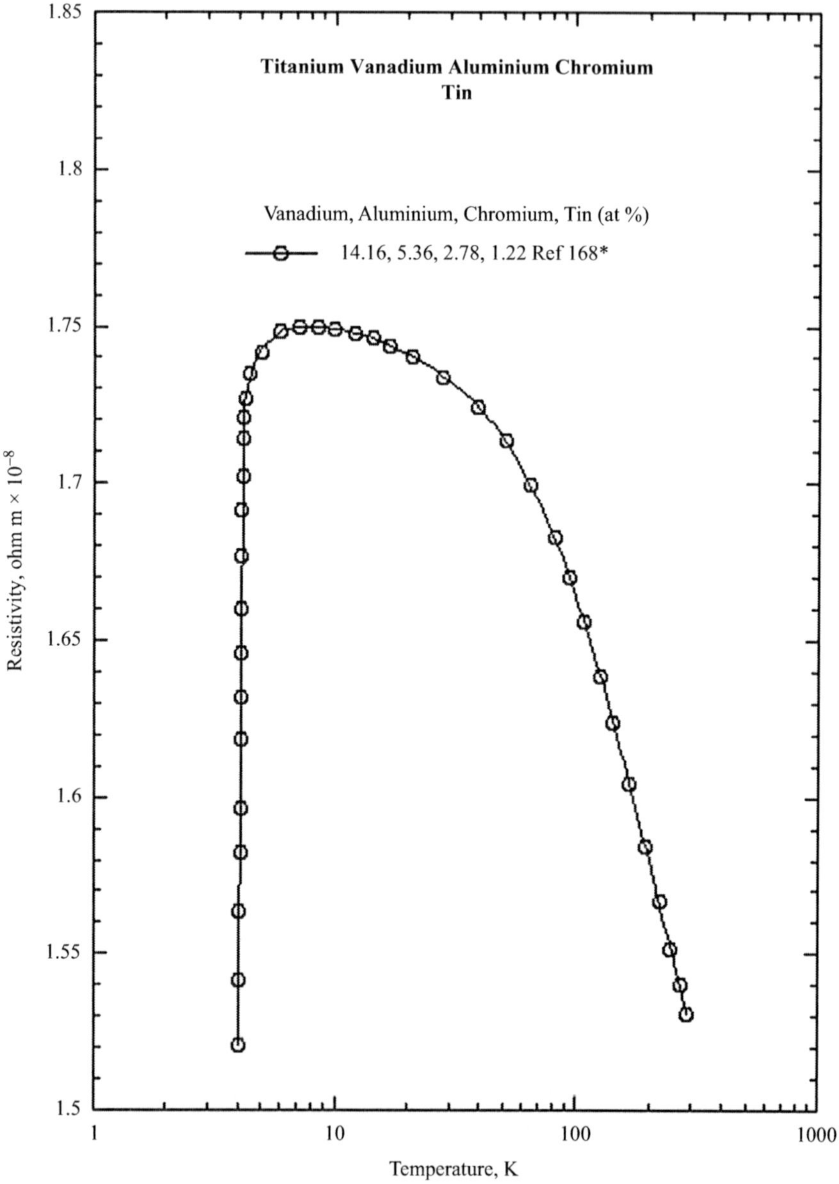
Titanium Vanadium Aluminium Chromium Tin
Vanadium, Aluminium, Chromium, Tin (at %)
14.16, 5.36, 2.78, 1.22 Ref 168*
Resistivity, ohm m × 10⁻⁸
1.85
1.8
1.75
1.7
1.65
1.6
1.55
1.5
1
10
100
1000
Temperature, K

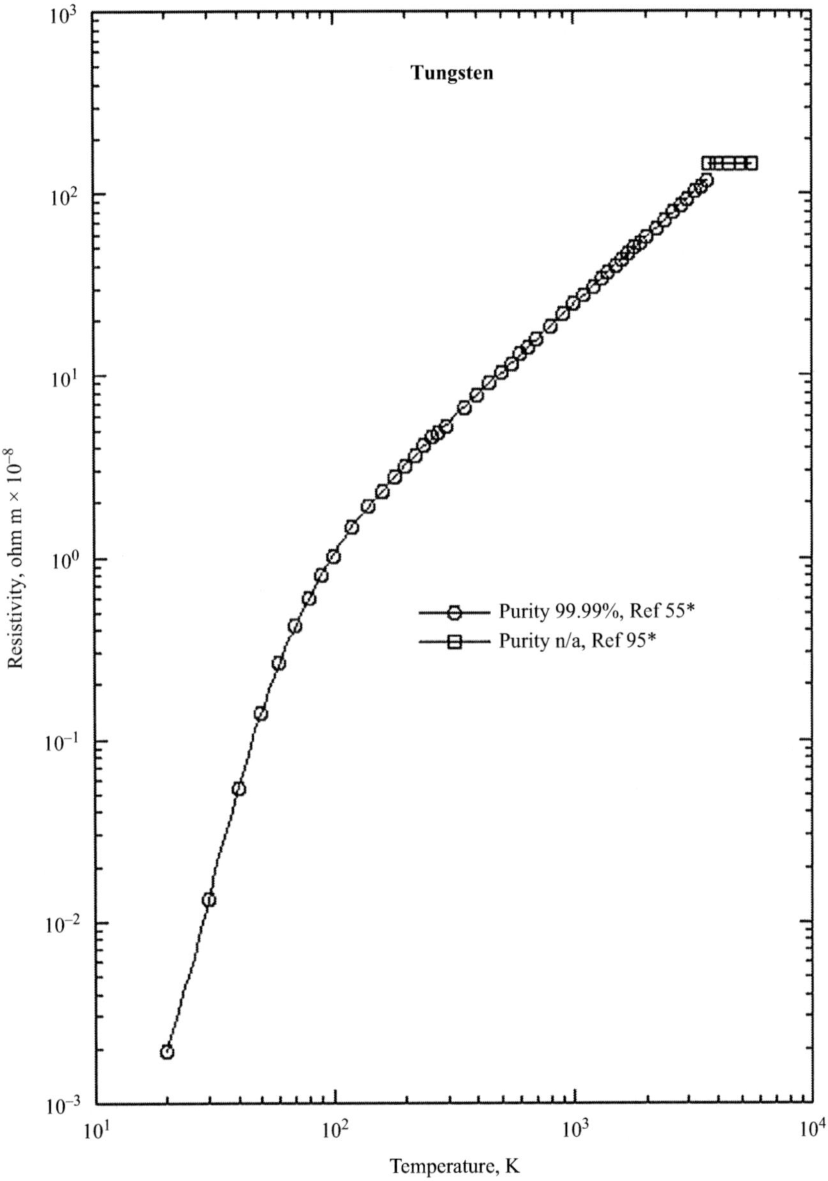
Tungsten
Purity 99.99%, Ref 55*
Purity n/a, Ref 95*
Resistivity, ohm m × 10⁻⁸
Temperature, K
10³
10²
10¹
10⁰
10⁻¹
10⁻²
10⁻³
10¹
10²
10³
10⁴

6.2 Electrical resistivity as a function of temperature and pressure

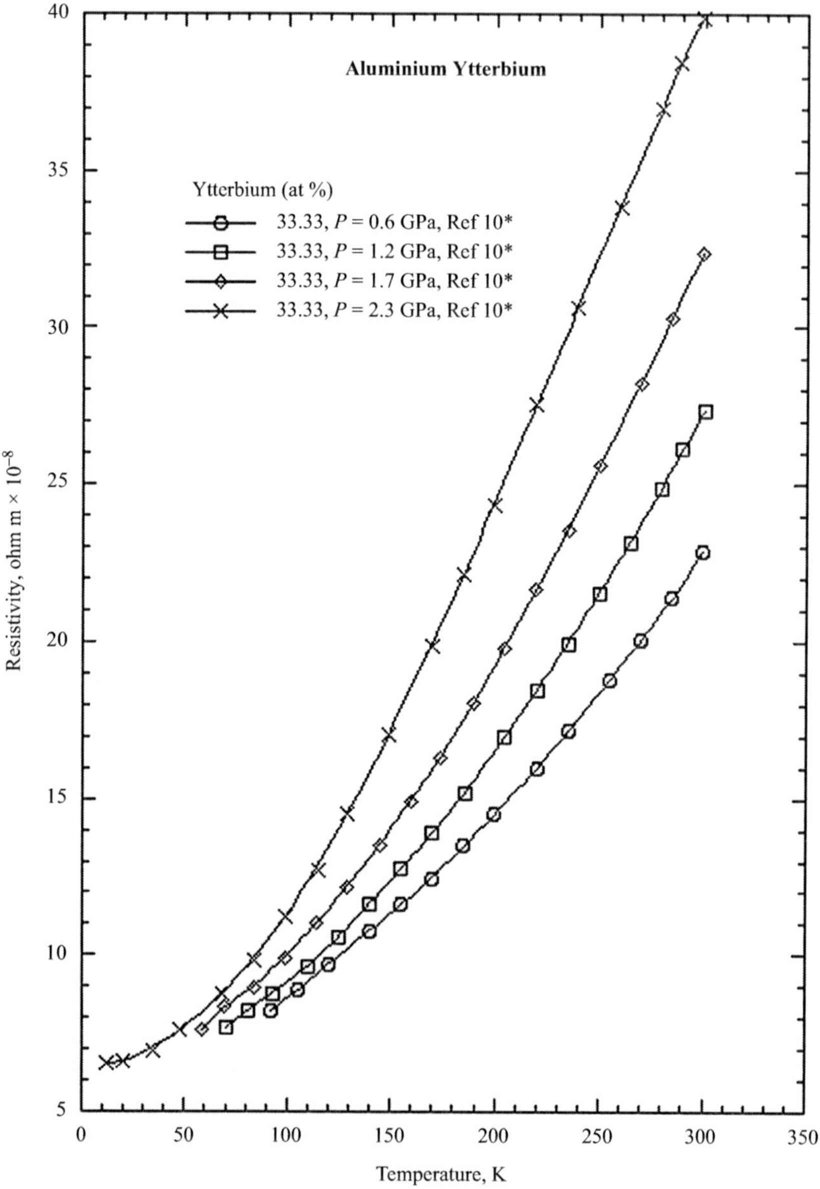

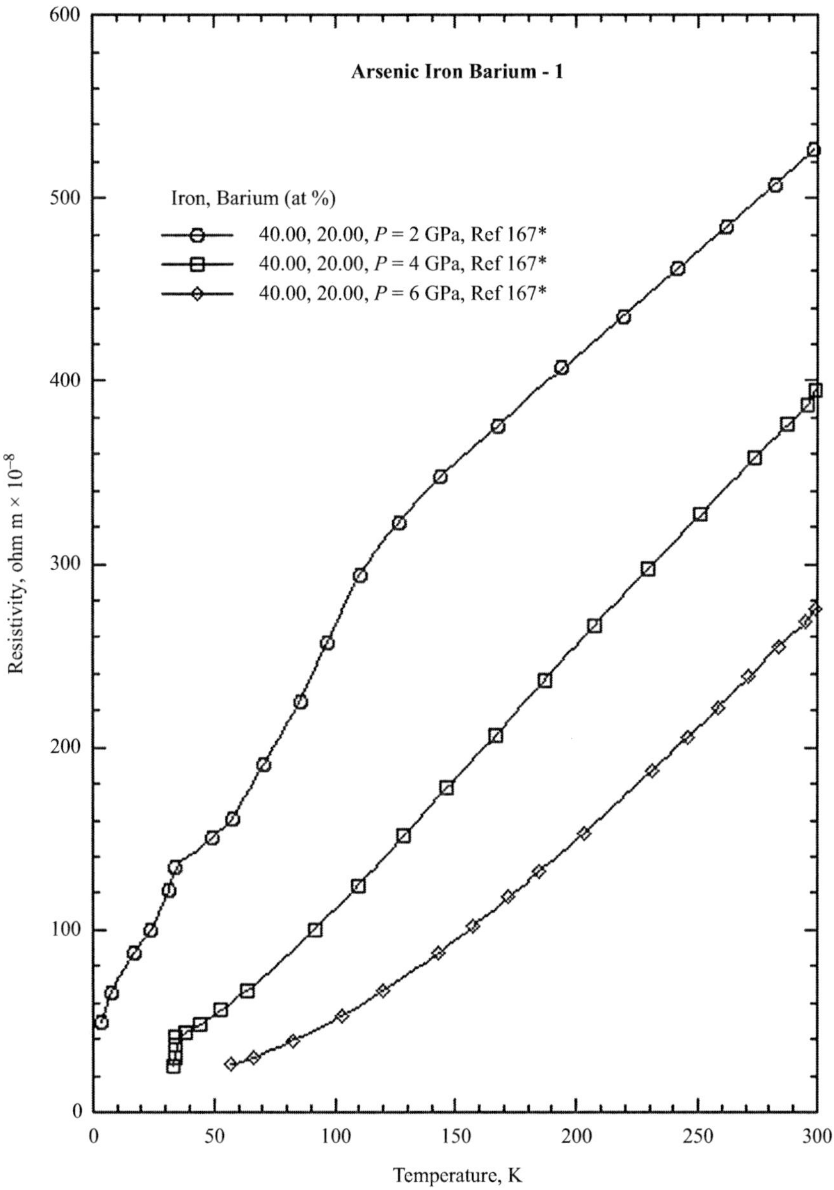
Arsenic Iron Barium - 1
Iron, Barium (at %)
40.00, 20.00, P = 2 GPa, Ref 167*
40.00, 20.00, P = 4 GPa, Ref 167*
40.00, 20.00, P = 6 GPa, Ref 167*
Resistivity, ohm m × 10⁻⁸
Temperature, K
0
100
200
300
400
500
600
0
50
100
150
200
250
300

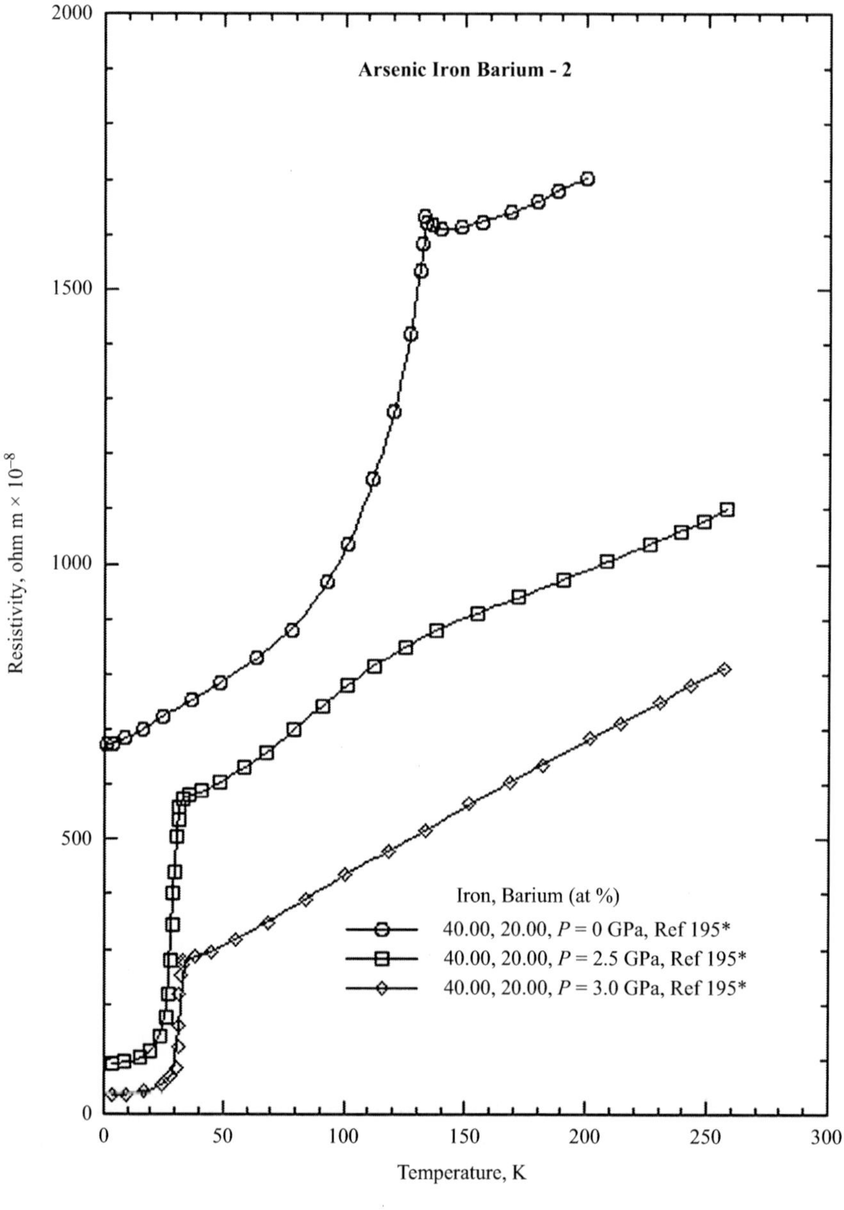

Arsenic Iron Barium - 2
Resistivity, ohm m × 10⁻⁸
2000
1500
1000
500
0
0
50
100
150
200
250
300
Temperature, K
Iron, Barium (at %)
40.00, 20.00, P = 0 GPa, Ref 195*
40.00, 20.00, P = 2.5 GPa, Ref 195*
40.00, 20.00, P = 3.0 GPa, Ref 195*

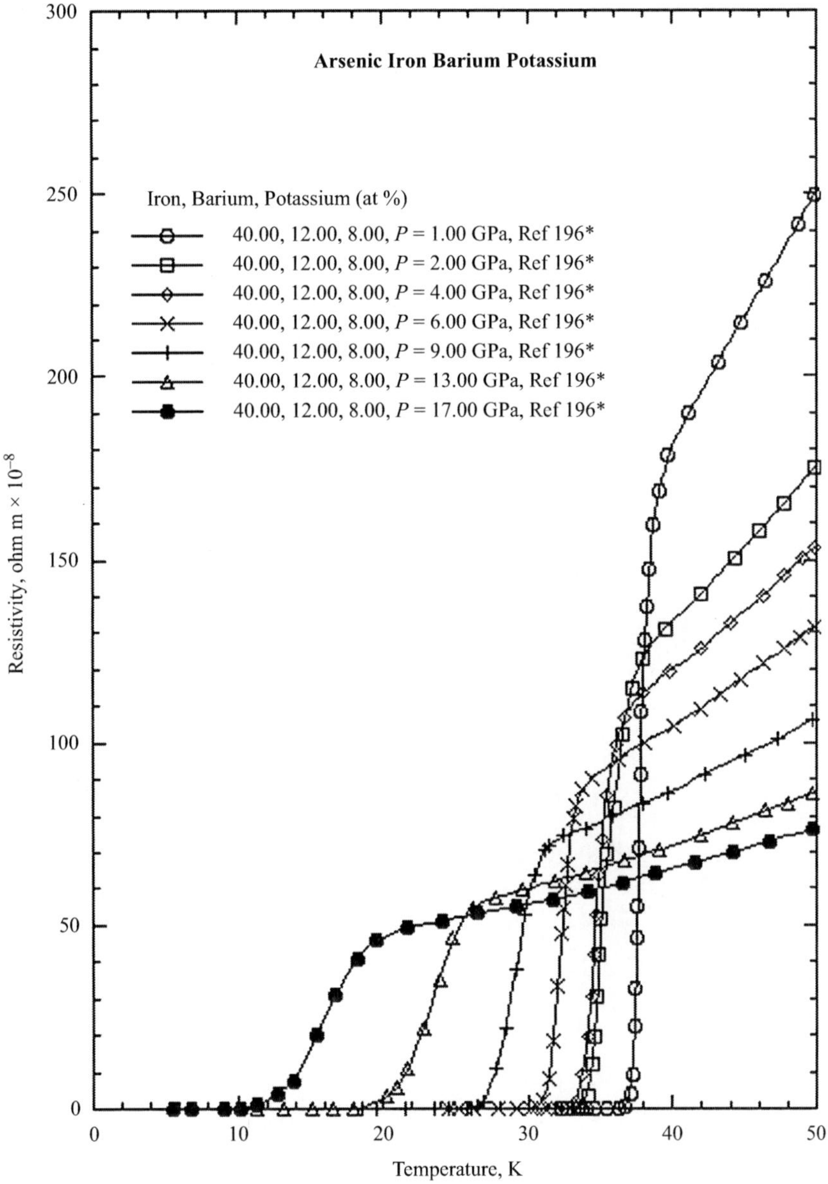

Arsenic Iron Barium Potassium
Iron, Barium, Potassium (at %)
40.00, 12.00, 8.00, P = 1.00 GPa, Ref 196*
40.00, 12.00, 8.00, P = 2.00 GPa, Ref 196*
40.00, 12.00, 8.00, P = 4.00 GPa, Ref 196*
40.00, 12.00, 8.00, P = 6.00 GPa, Ref 196*
40.00, 12.00, 8.00, P = 9.00 GPa, Ref 196*
40.00, 12.00, 8.00, P = 13.00 GPa, Ref 196*
40.00, 12.00, 8.00, P = 17.00 GPa, Ref 196*
Resistivity, ohm m × 10⁻⁸
Temperature, K
0
50
100
150
200
250
300
0
10
20
30
40
50

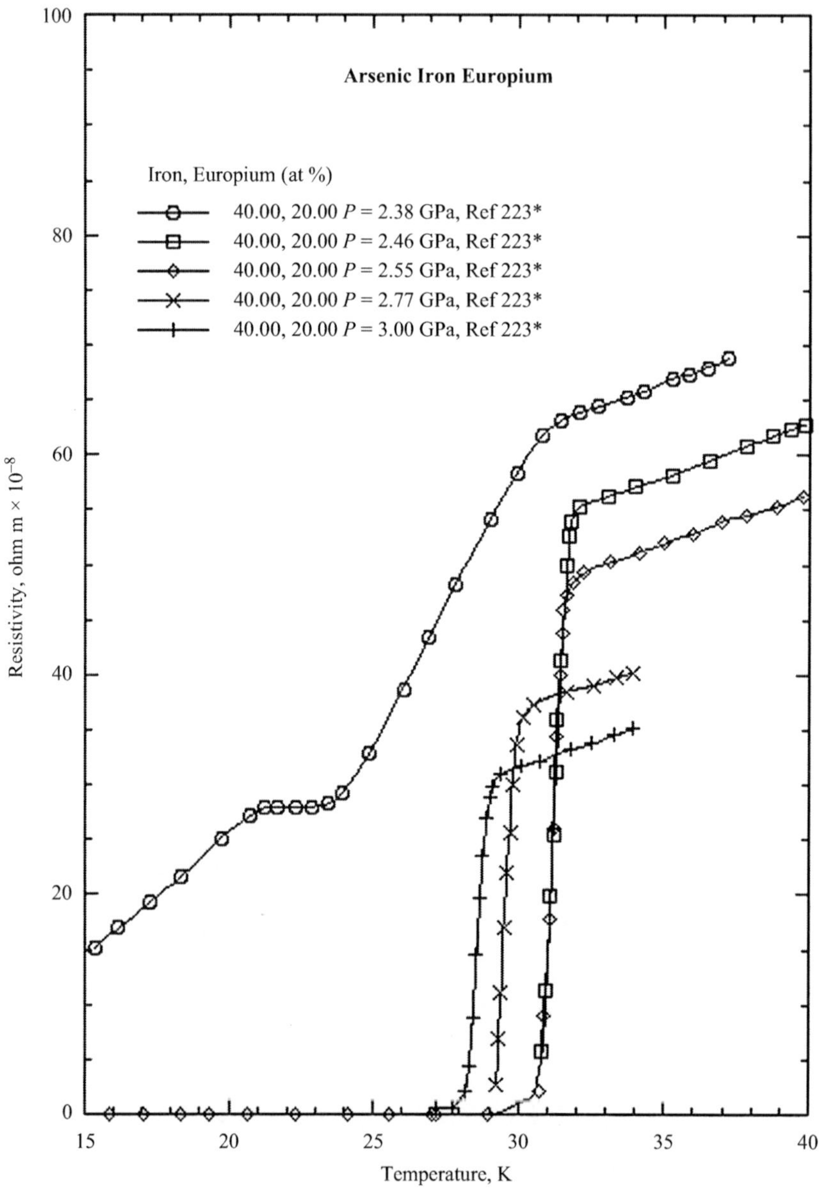
Arsenic Iron Europium
Iron, Europium (at %)
40.00, 20.00 P = 2.38 GPa, Ref 223*
40.00, 20.00 P = 2.46 GPa, Ref 223*
40.00, 20.00 P = 2.55 GPa, Ref 223*
40.00, 20.00 P = 2.77 GPa, Ref 223*
40.00, 20.00 P = 3.00 GPa, Ref 223*
Resistivity, ohm m × 10⁻⁸
Temperature, K
0
20
40
60
80
100
15
20
25
30
35
40

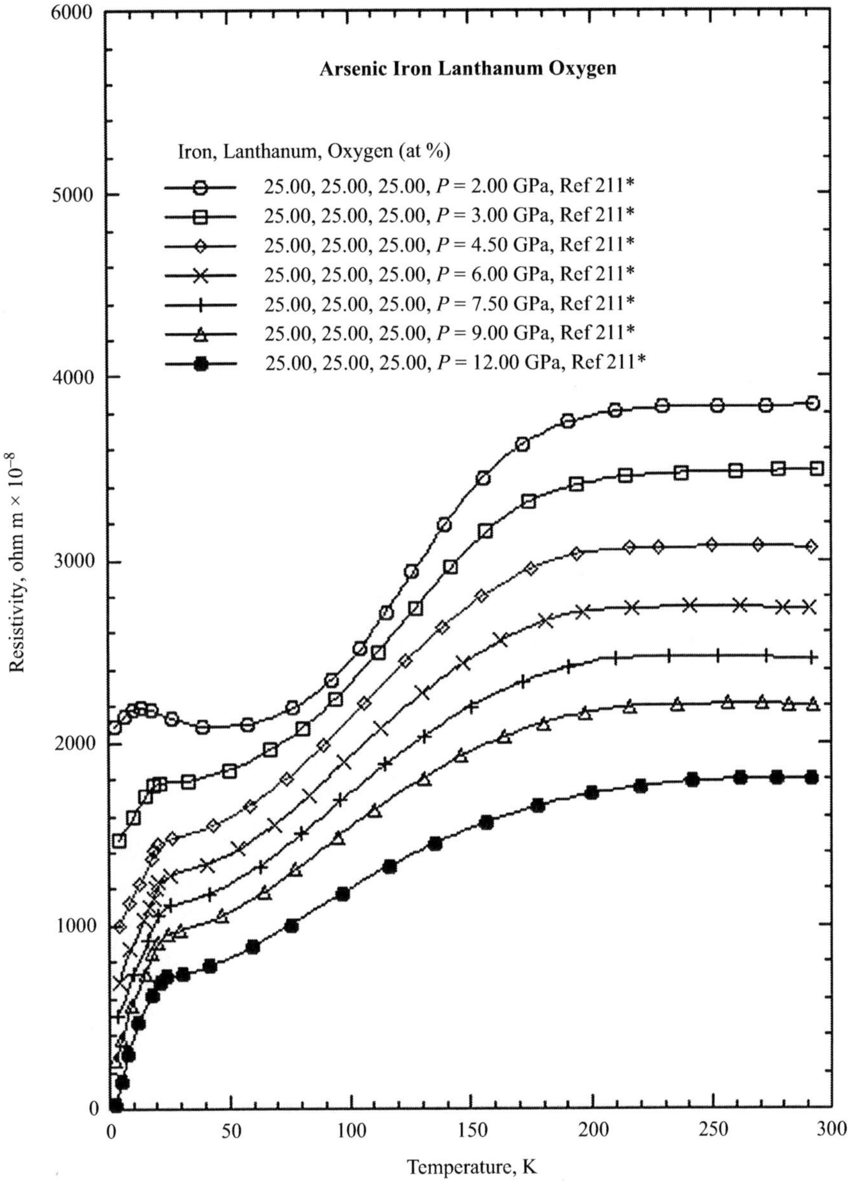
Arsenic Iron Lanthanum Oxygen
Iron, Lanthanum, Oxygen (at %)
25.00, 25.00, 25.00, P = 2.00 GPa, Ref 211*
25.00, 25.00, 25.00, P = 3.00 GPa, Ref 211*
25.00, 25.00, 25.00, P = 4.50 GPa, Ref 211*
25.00, 25.00, 25.00, P = 6.00 GPa, Ref 211*
25.00, 25.00, 25.00, P = 7.50 GPa, Ref 211*
25.00, 25.00, 25.00, P = 9.00 GPa, Ref 211*
25.00, 25.00, 25.00, P = 12.00 GPa, Ref 211*
Resistivity, ohm m × 10⁻⁸
Temperature, K
0
1000
2000
3000
4000
5000
6000
0
50
100
150
200
250
300

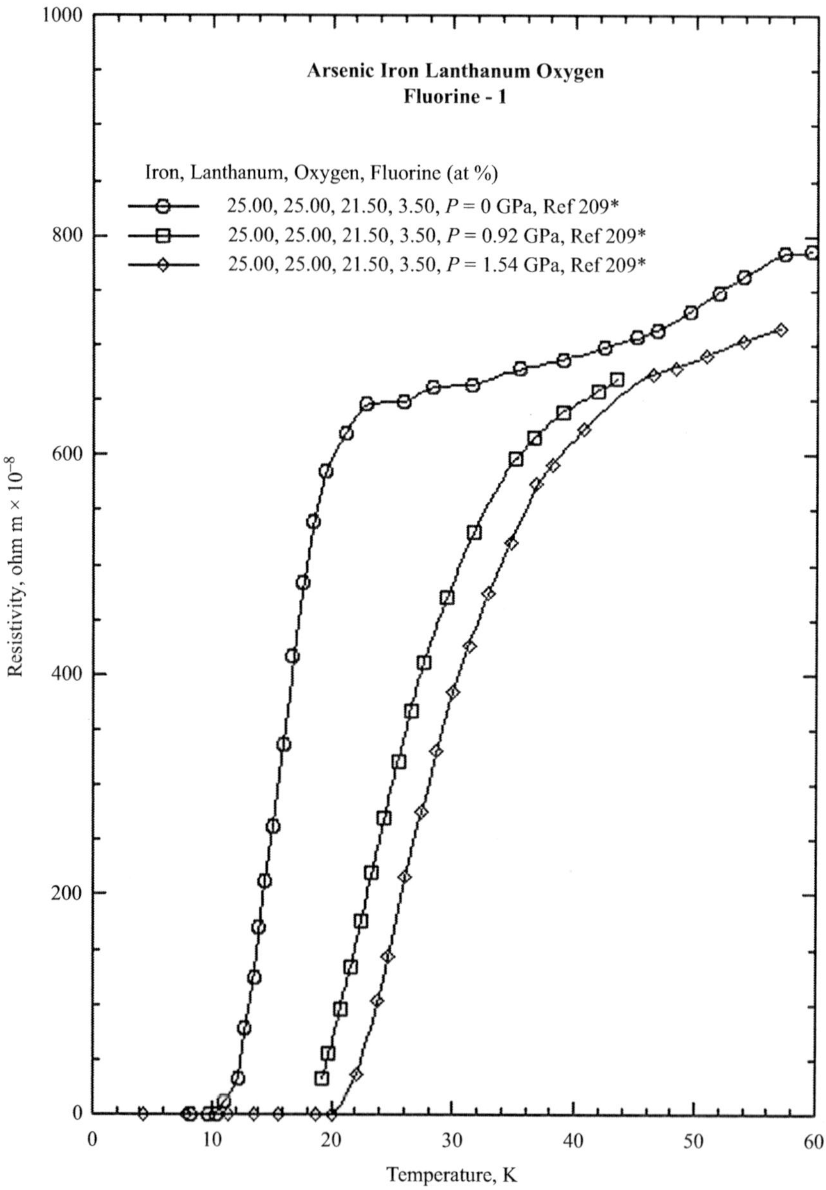
Arsenic Iron Lanthanum Oxygen
Fluorine - 1
Iron, Lanthanum, Oxygen, Fluorine (at %)
25.00, 25.00, 21.50, 3.50, P = 0 GPa, Ref 209*
25.00, 25.00, 21.50, 3.50, P = 0.92 GPa, Ref 209*
25.00, 25.00, 21.50, 3.50, P = 1.54 GPa, Ref 209*
Resistivity, ohm m × 10⁻⁸
Temperature, K
0
10
20
30
40
50
60
200
400
600
800
1000

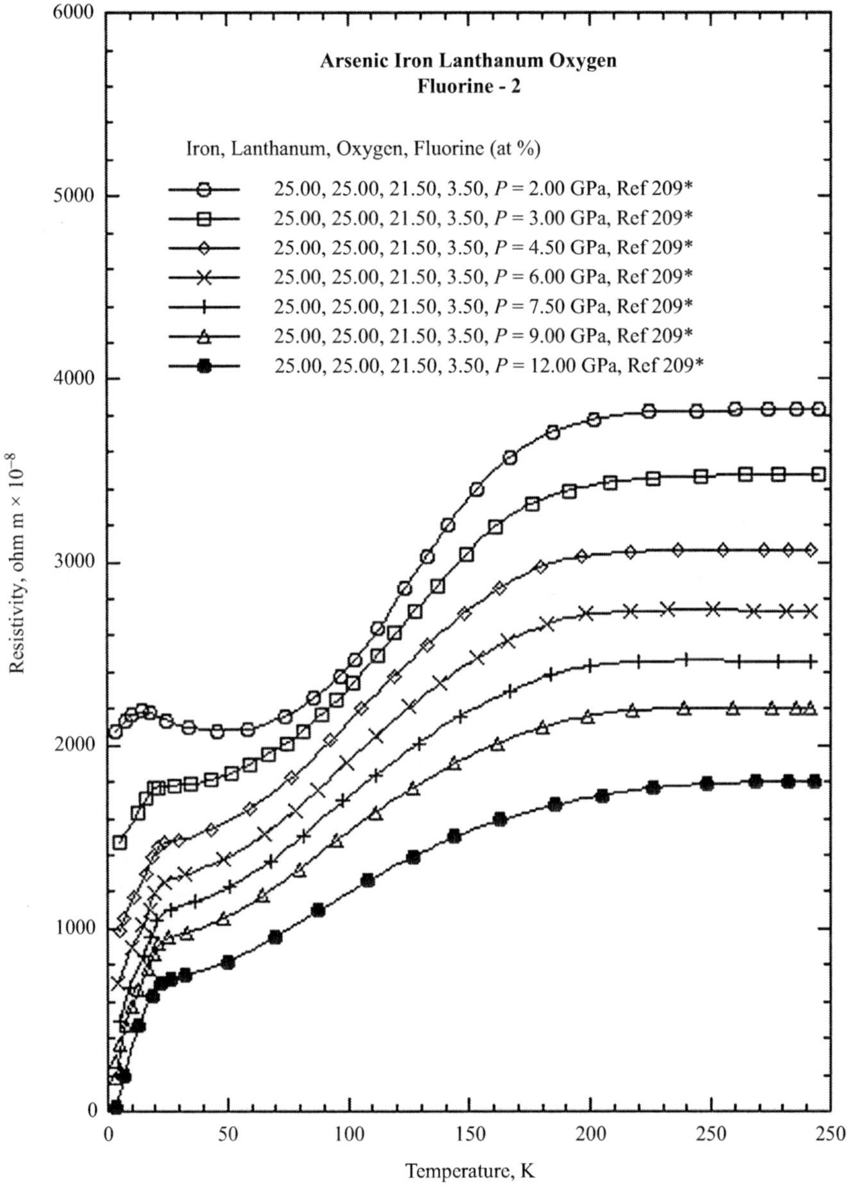
Arsenic Iron Lanthanum Oxygen
Fluorine - 2
Iron, Lanthanum, Oxygen, Fluorine (at %)
25.00, 25.00, 21.50, 3.50, P = 2.00 GPa, Ref 209*
25.00, 25.00, 21.50, 3.50, P = 3.00 GPa, Ref 209*
25.00, 25.00, 21.50, 3.50, P = 4.50 GPa, Ref 209*
25.00, 25.00, 21.50, 3.50, P = 6.00 GPa, Ref 209*
25.00, 25.00, 21.50, 3.50, P = 7.50 GPa, Ref 209*
25.00, 25.00, 21.50, 3.50, P = 9.00 GPa, Ref 209*
25.00, 25.00, 21.50, 3.50, P = 12.00 GPa, Ref 209*
Resistivity, ohm m × 10⁻⁸
Temperature, K
6000
5000
4000
3000
2000
1000
0
0
50
100
150
200
250
250

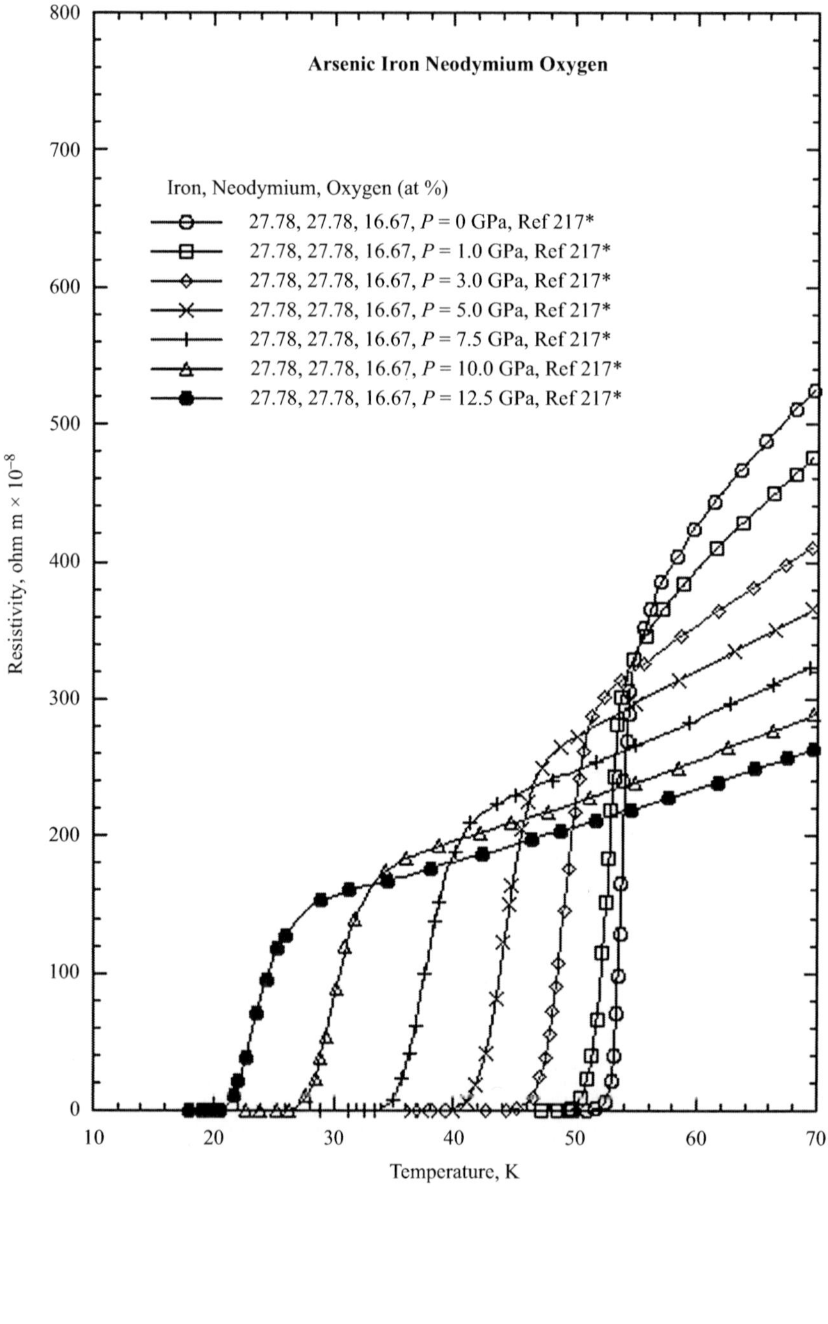
Arsenic Iron Neodymium Oxygen
Iron, Neodymium, Oxygen (at %)
27.78, 27.78, 16.67, P = 0 GPa, Ref 217*
27.78, 27.78, 16.67, P = 1.0 GPa, Ref 217*
27.78, 27.78, 16.67, P = 3.0 GPa, Ref 217*
27.78, 27.78, 16.67, P = 5.0 GPa, Ref 217*
27.78, 27.78, 16.67, P = 7.5 GPa, Ref 217*
27.78, 27.78, 16.67, P = 10.0 GPa, Ref 217*
27.78, 27.78, 16.67, P = 12.5 GPa, Ref 217*
Resistivity, ohm m × 10^−8
Temperature, K
0
100
200
300
400
500
600
700
800
10
20
30
40
50
60
70

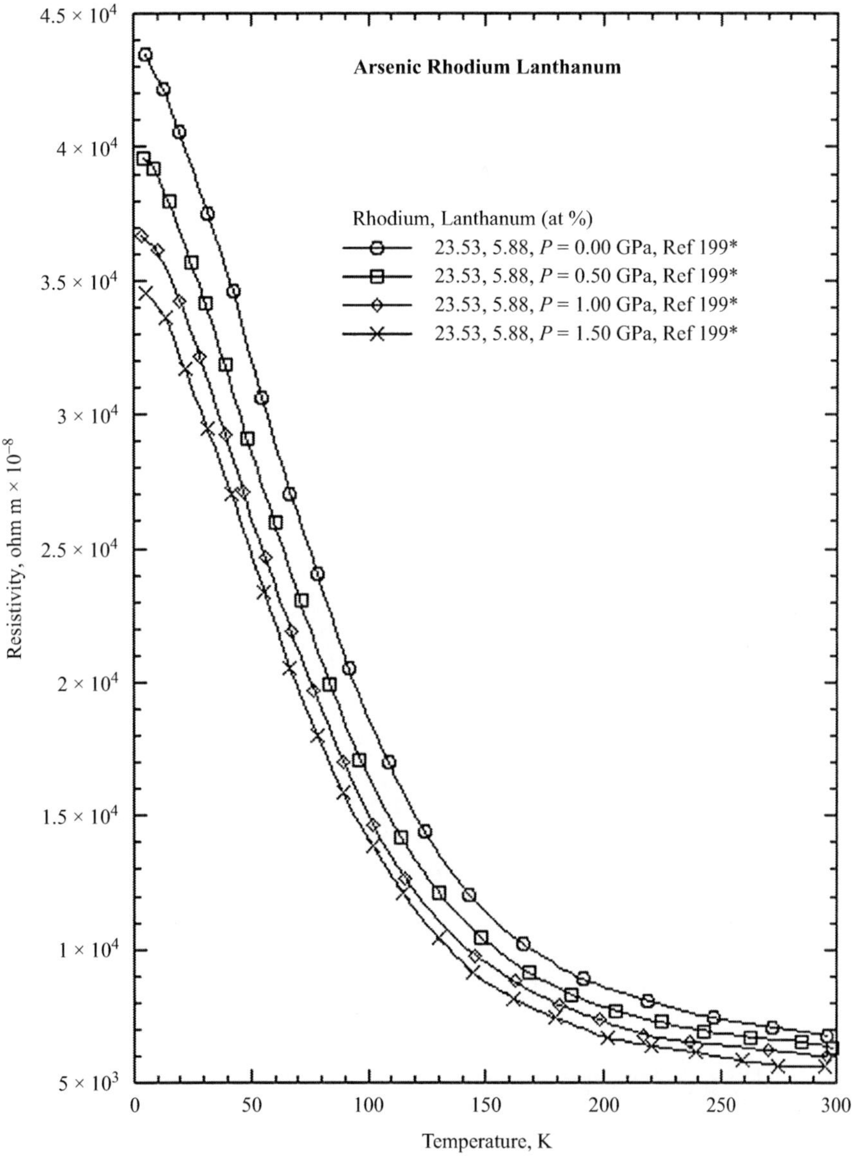
Arsenic Rhodium Lanthanum
Rhodium, Lanthanum (at %)
23.53, 5.88, P = 0.00 GPa, Ref 199*
23.53, 5.88, P = 0.50 GPa, Ref 199*
23.53, 5.88, P = 1.00 GPa, Ref 199*
23.53, 5.88, P = 1.50 GPa, Ref 199*
Resistivity, ohm m × 10^−8
Temperature, K
4.5 × 10^4
4 × 10^4
3.5 × 10^4
3 × 10^4
2.5 × 10^4
2 × 10^4
1.5 × 10^4
1 × 10^4
5 × 10^3
0
50
100
150
200
250
300

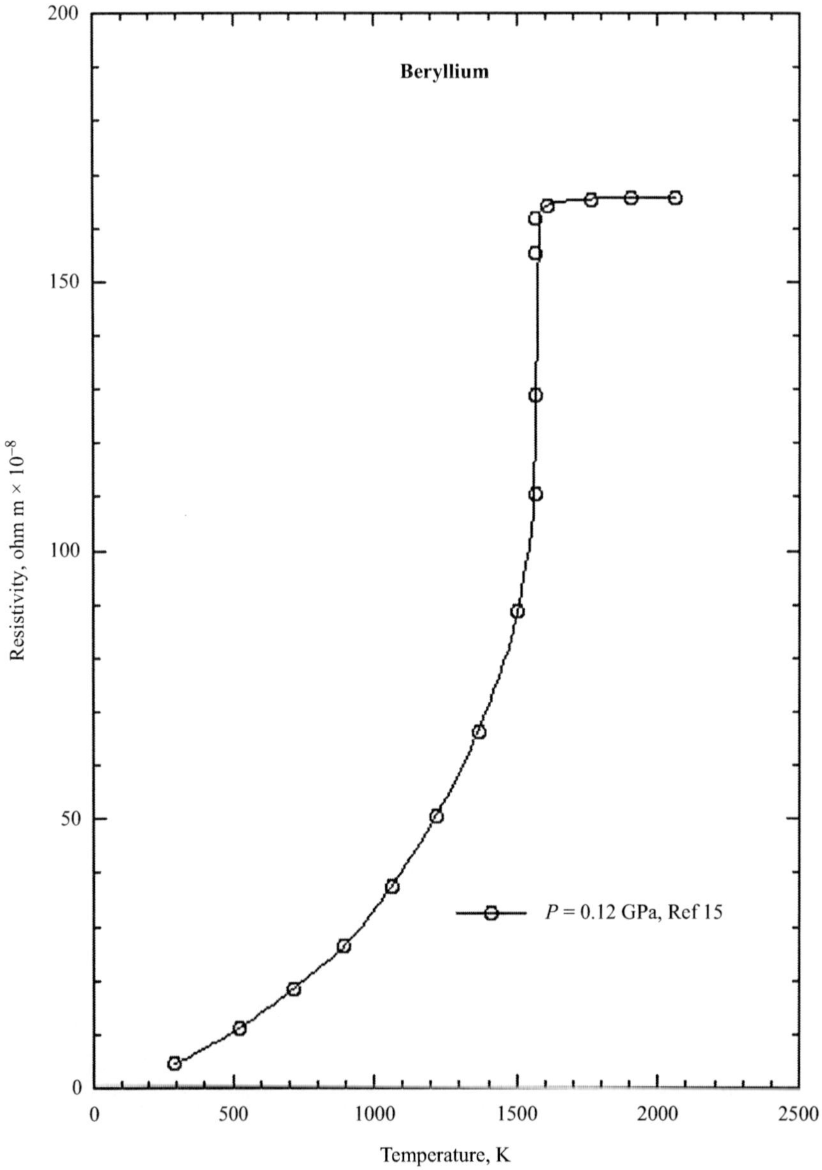
Beryllium
P = 0.12 GPa, Ref 15
Resistivity, ohm m × 10^−8
Temperature, K
0
50
100
150
200
0
500
1000
1500
2000
2500

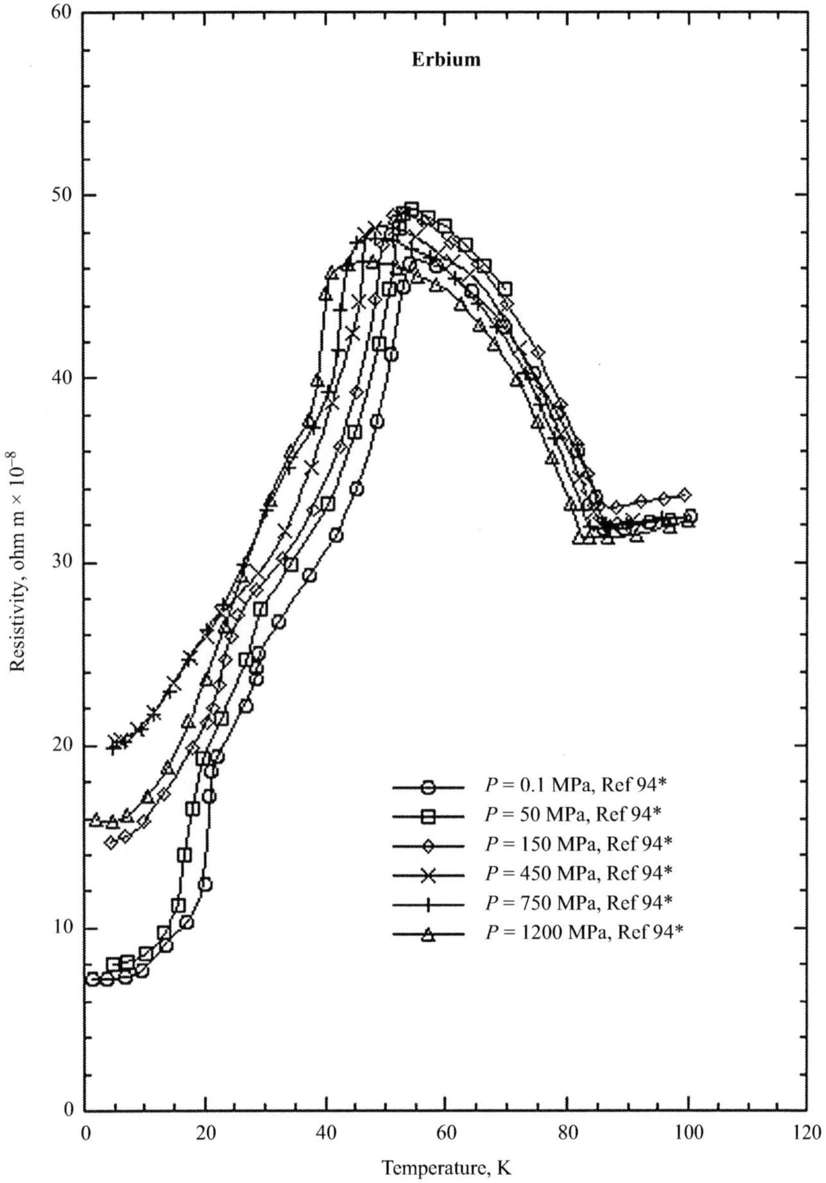
Erbium
Resistivity, ohm m × 10⁻⁸
Temperature, K
P = 0.1 MPa, Ref 94*
P = 50 MPa, Ref 94*
P = 150 MPa, Ref 94*
P = 450 MPa, Ref 94*
P = 750 MPa, Ref 94*
P = 1200 MPa, Ref 94*

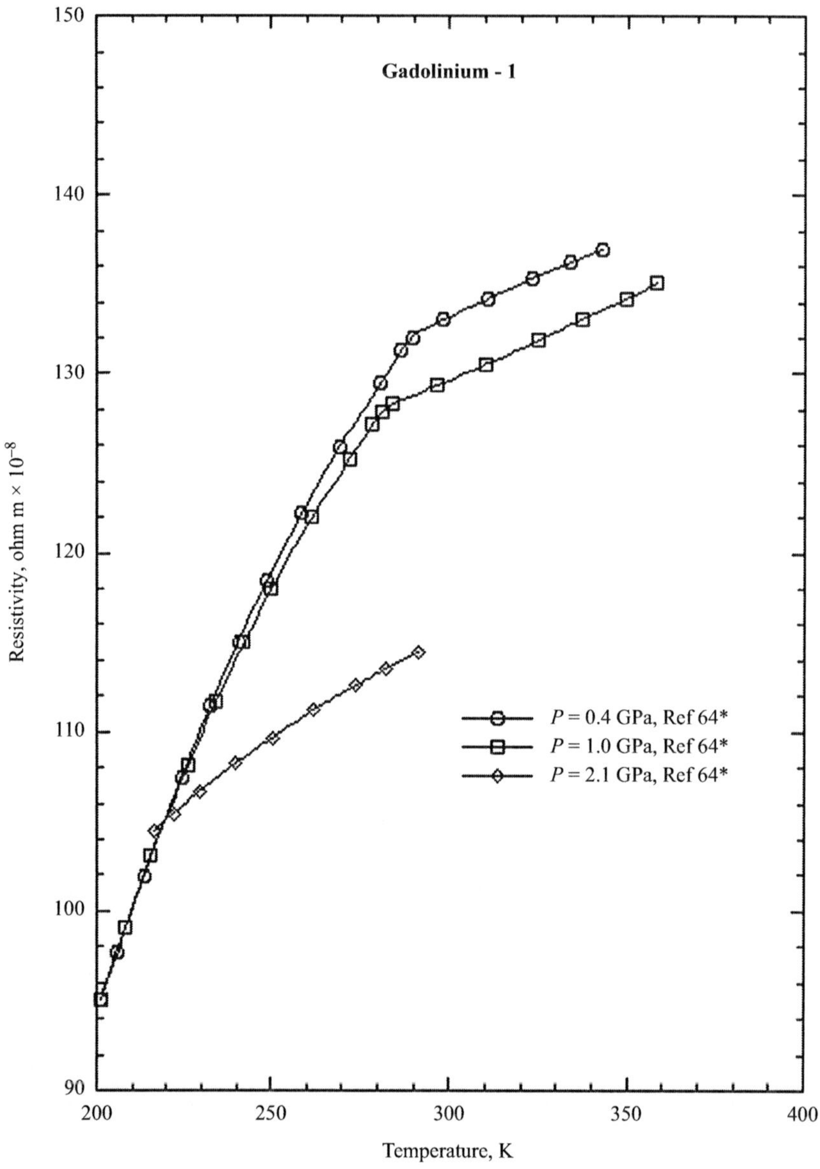

Gadolinium - 1
Resistivity, ohm m × 10^−8
Temperature, K
150
140
130
120
110
100
90
200
250
300
350
400
P = 0.4 GPa, Ref 64*
P = 1.0 GPa, Ref 64*
P = 2.1 GPa, Ref 64*

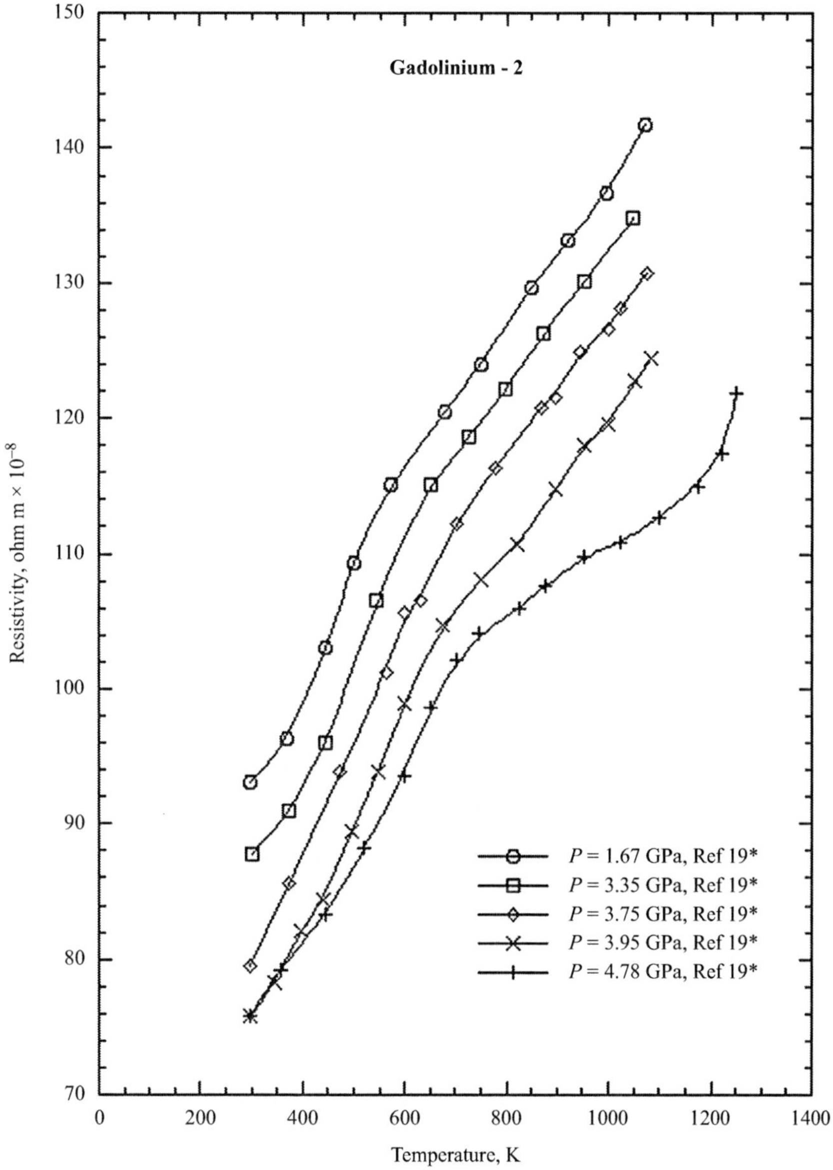
Gadolinium - 2
Resistivity, ohm m × 10⁻⁸
Temperature, K
P = 1.67 GPa, Ref 19*
P = 3.35 GPa, Ref 19*
P = 3.75 GPa, Ref 19*
P = 3.95 GPa, Ref 19*
P = 4.78 GPa, Ref 19*

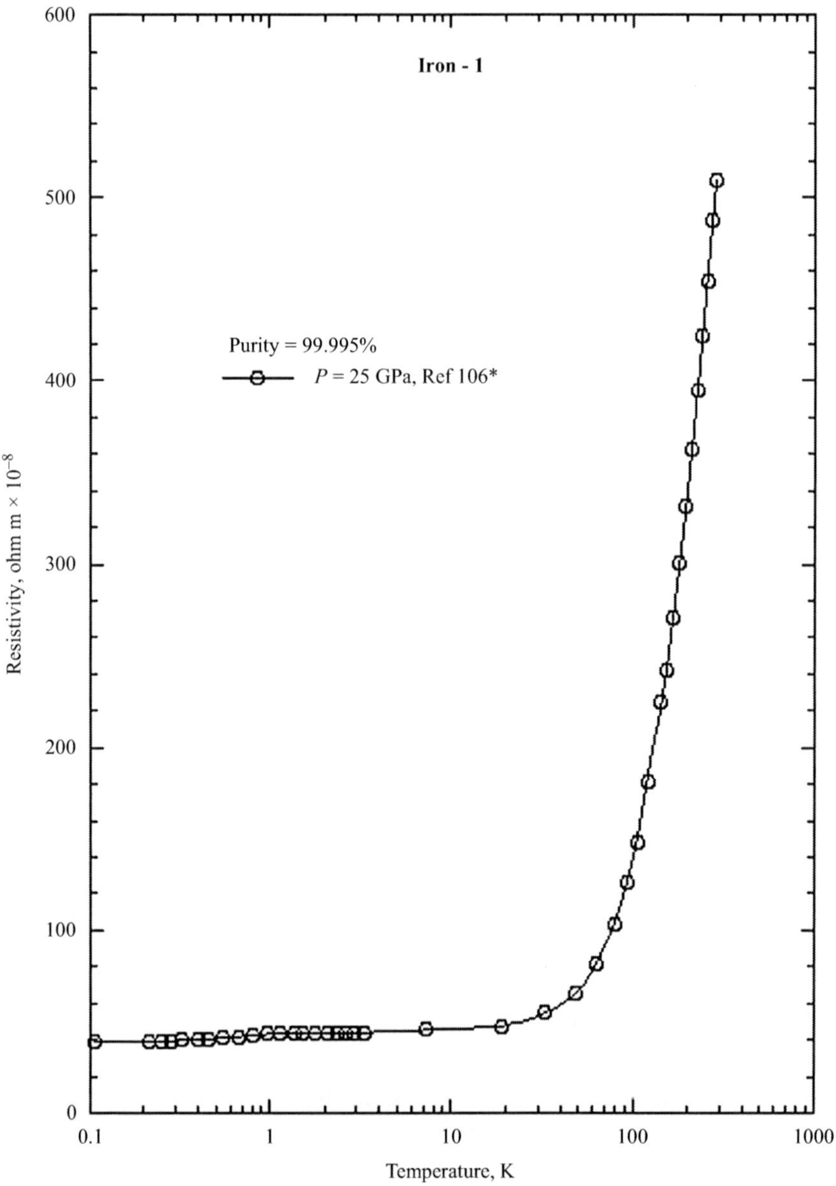
Iron - 1
Purity = 99.995%
P = 25 GPa, Ref 106*
Resistivity, ohm m × 10⁻⁸
Temperature, K
0
100
200
300
400
500
600
0.1
1
10
100
1000

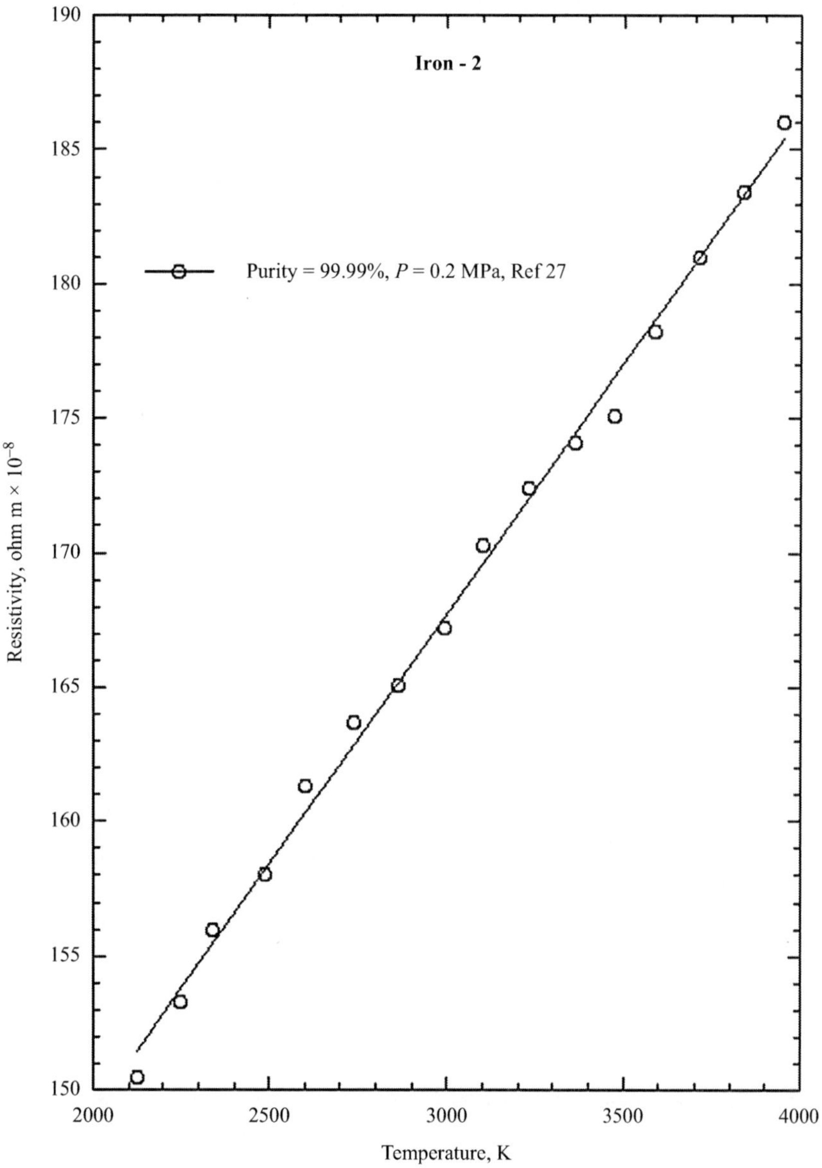

Iron - 2
Purity = 99.99%, P = 0.2 MPa, Ref 27
Resistivity, ohm m × 10⁻⁸
190
185
180
175
170
165
160
155
150
2000
2500
3000
3500
4000
Temperature, K

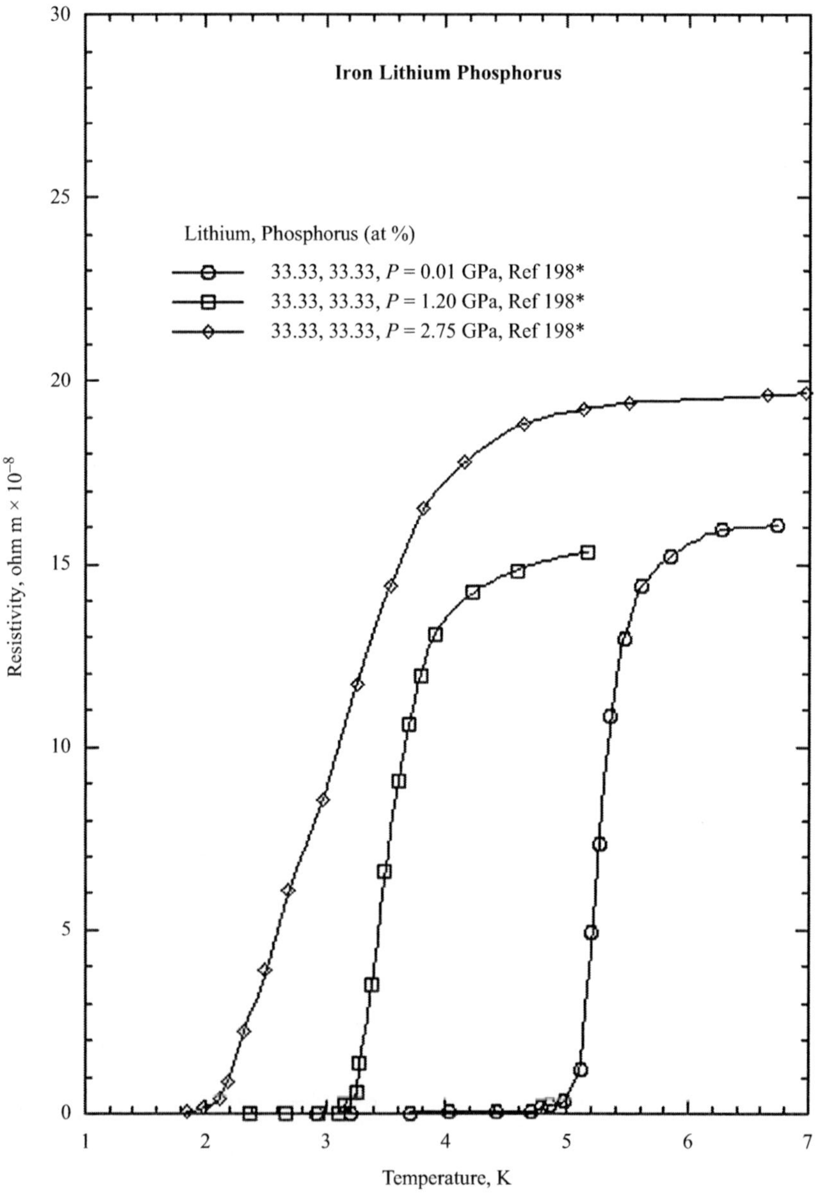

Iron Lithium Phosphorus
Lithium, Phosphorus (at %)
33.33, 33.33, P = 0.01 GPa, Ref 198*
33.33, 33.33, P = 1.20 GPa, Ref 198*
33.33, 33.33, P = 2.75 GPa, Ref 198*
Resistivity, ohm m × 10−8
Temperature, K
0
5
10
15
20
25
30
1
2
3
4
5
6
7

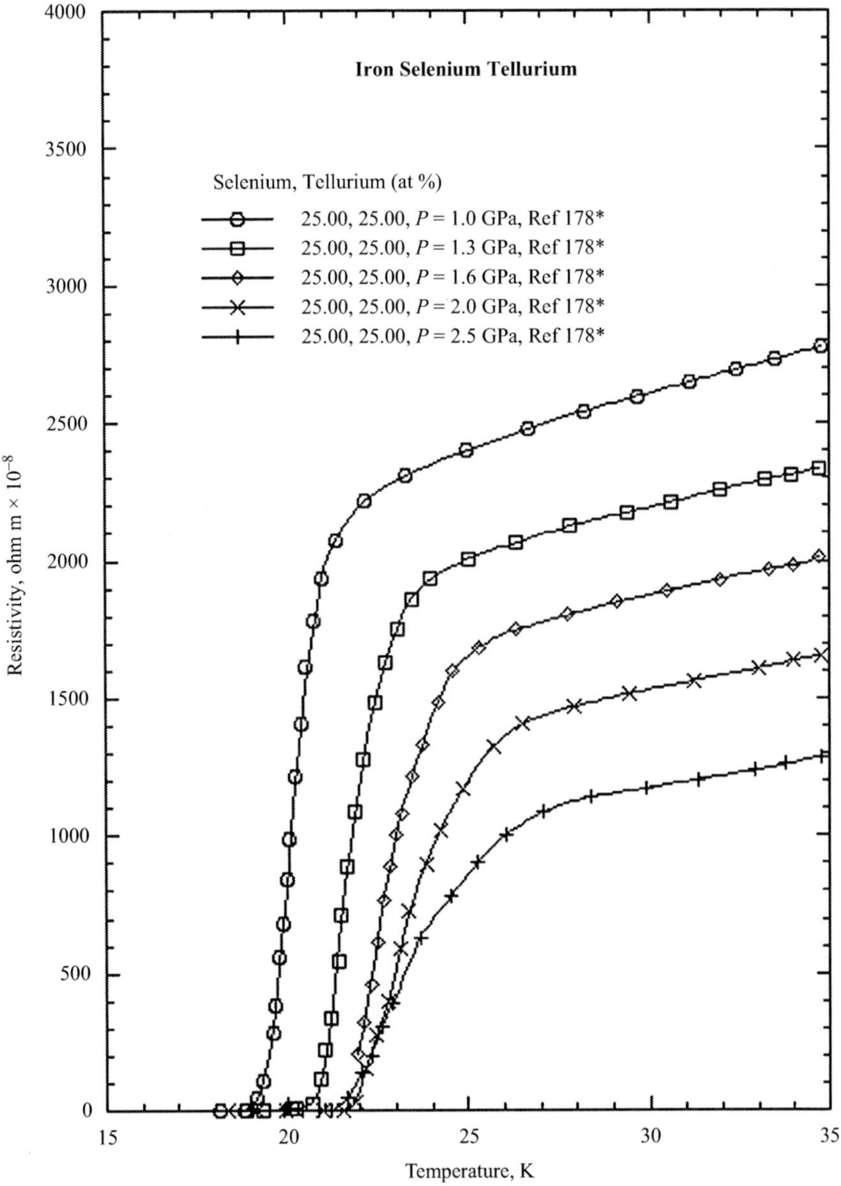
Iron Selenium Tellurium
Selenium, Tellurium (at %)
25.00, 25.00, P = 1.0 GPa, Ref 178*
25.00, 25.00, P = 1.3 GPa, Ref 178*
25.00, 25.00, P = 1.6 GPa, Ref 178*
25.00, 25.00, P = 2.0 GPa, Ref 178*
25.00, 25.00, P = 2.5 GPa, Ref 178*
Resistivity, ohm m × 10⁻⁸
Temperature, K
0
500
1000
1500
2000
2500
3000
3500
4000
15
20
25
30
35

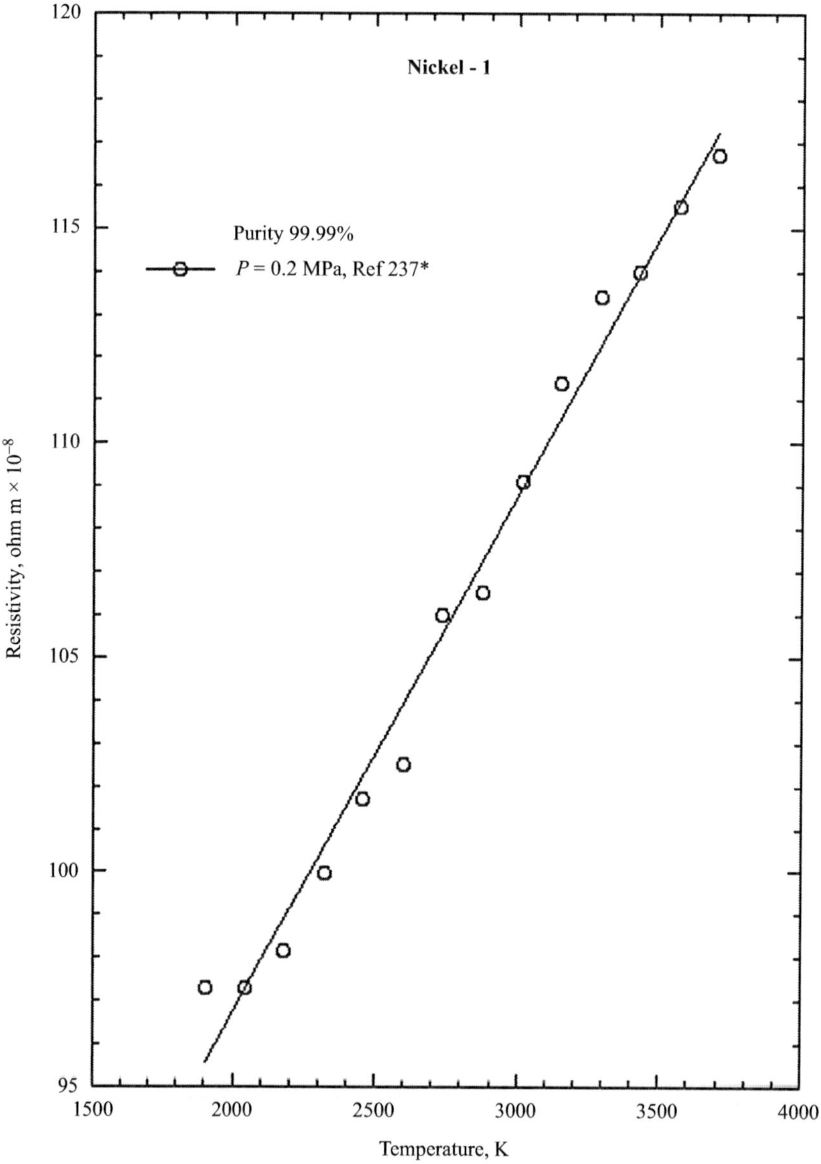
Nickel - 1
Purity 99.99%
P = 0.2 MPa, Ref 237*
Resistivity, ohm m × 10−8
Temperature, K
120
115
110
105
100
95
1500
2000
2500
3000
3500
4000

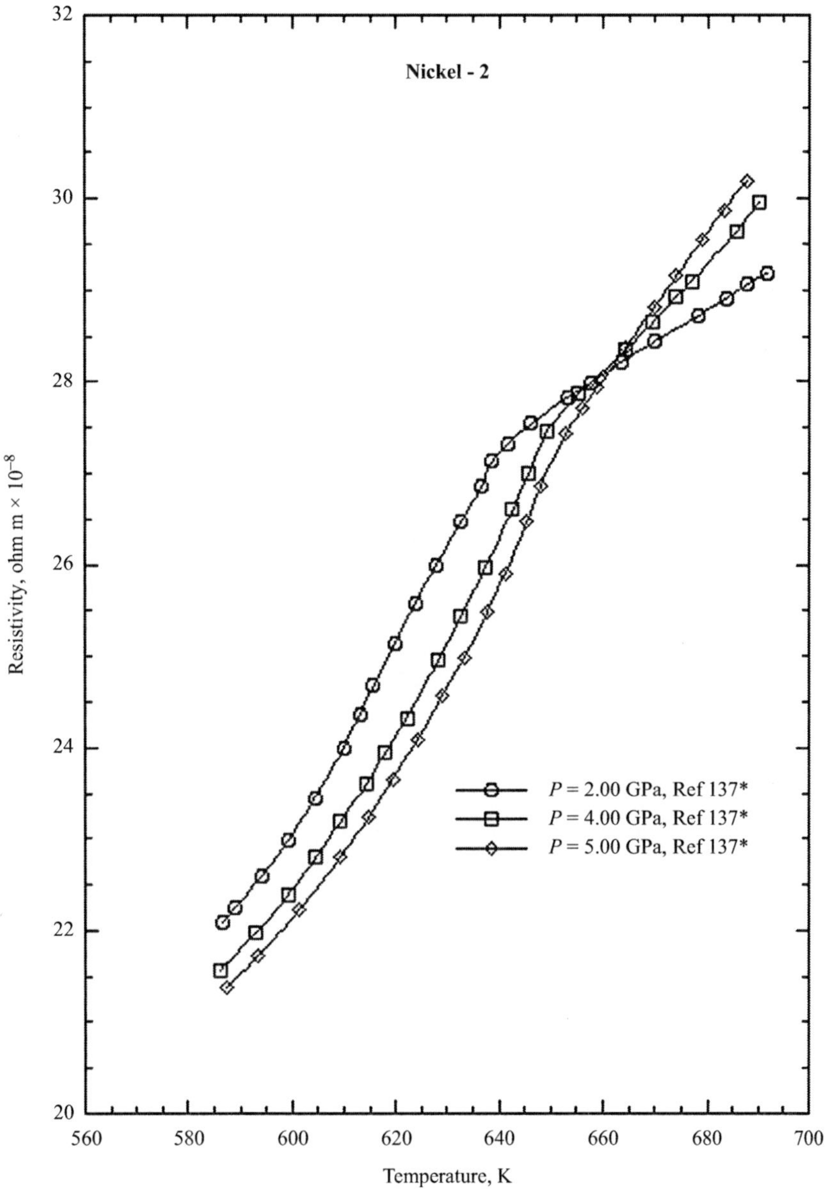
Nickel - 2
Resistivity, ohm m × 10⁻⁸
Temperature, K
32
30
28
26
24
22
20
560
580
600
620
640
660
680
700
P = 2.00 GPa, Ref 137*
P = 4.00 GPa, Ref 137*
P = 5.00 GPa, Ref 137*

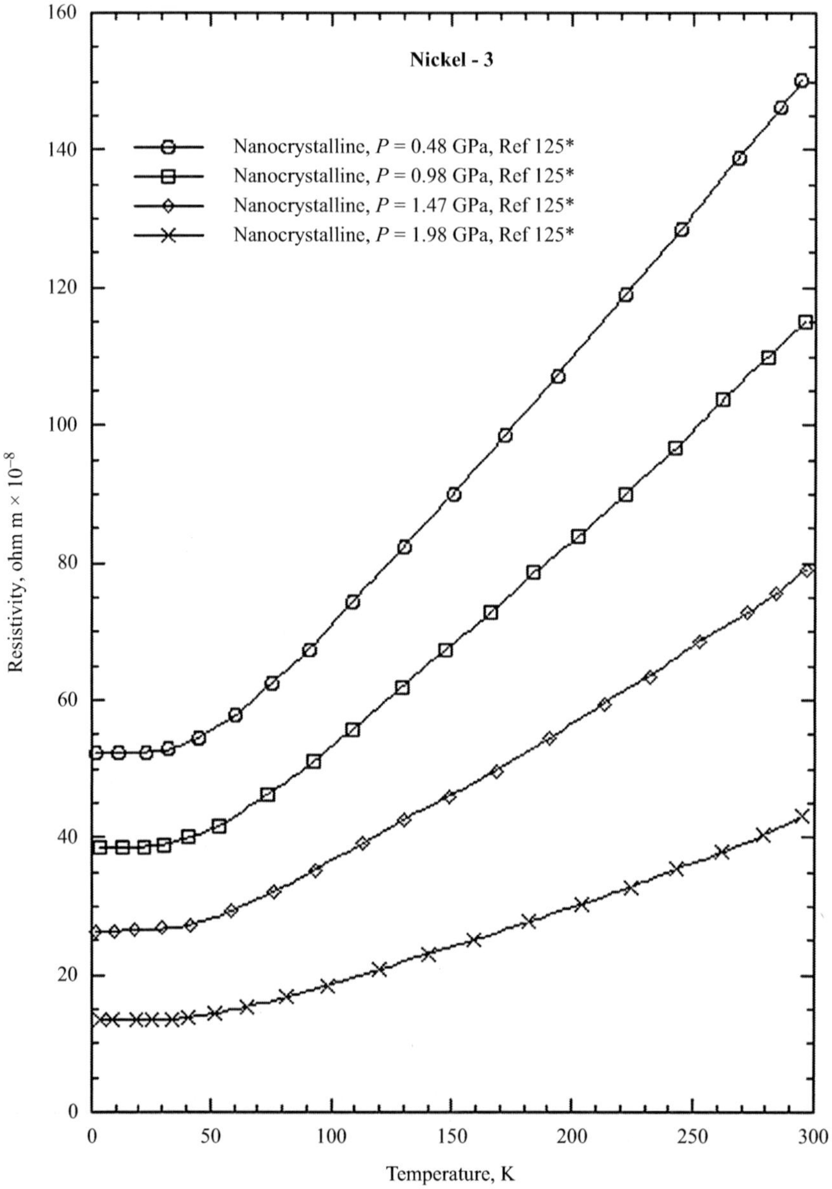
Nickel - 3
Nanocrystalline, P = 0.48 GPa, Ref 125*
Nanocrystalline, P = 0.98 GPa, Ref 125*
Nanocrystalline, P = 1.47 GPa, Ref 125*
Nanocrystalline, P = 1.98 GPa, Ref 125*
Resistivity, ohm m × 10⁻⁸
Temperature, K
0
20
40
60
80
100
120
140
160
0
50
100
150
200
250
300

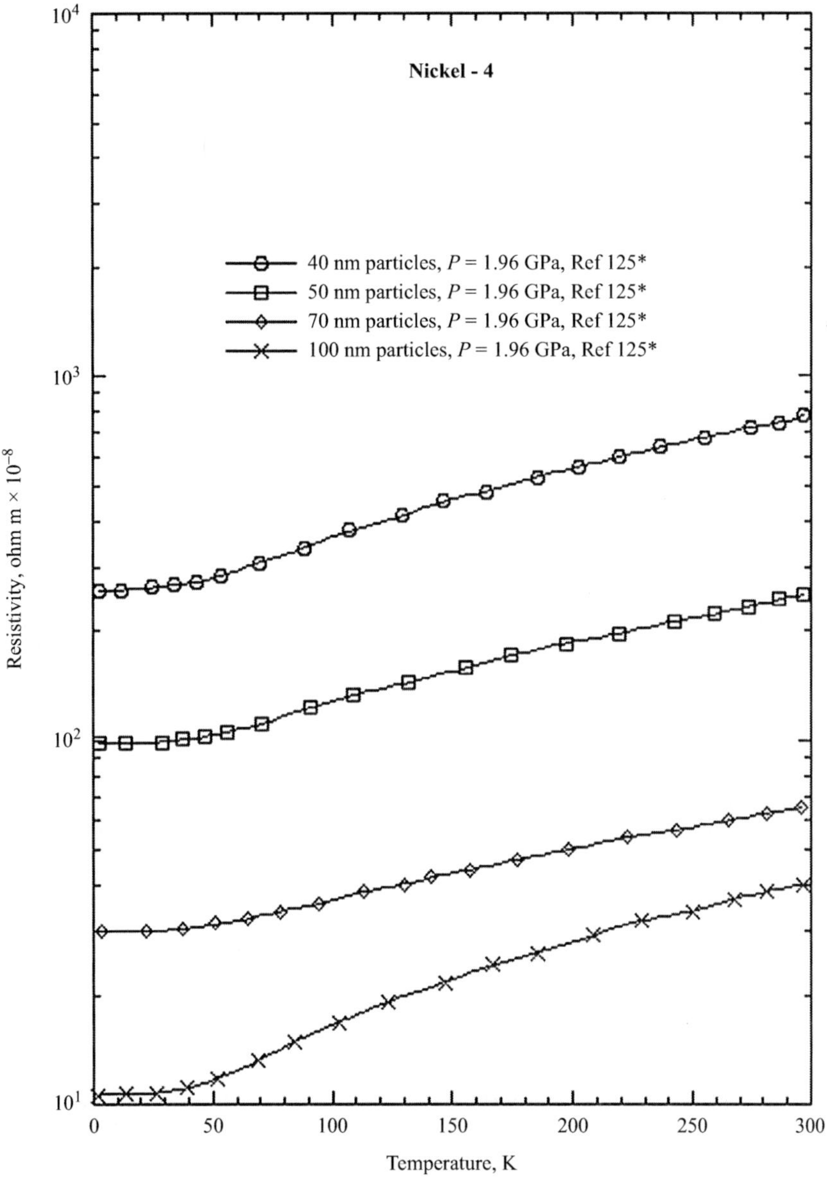
Nickel - 4
40 nm particles, P = 1.96 GPa, Ref 125*
50 nm particles, P = 1.96 GPa, Ref 125*
70 nm particles, P = 1.96 GPa, Ref 125*
100 nm particles, P = 1.96 GPa, Ref 125*
Resistivity, ohm m × 10⁻⁸
10⁴
10³
10²
10¹
0
50
100
150
200
250
300
Temperature, K

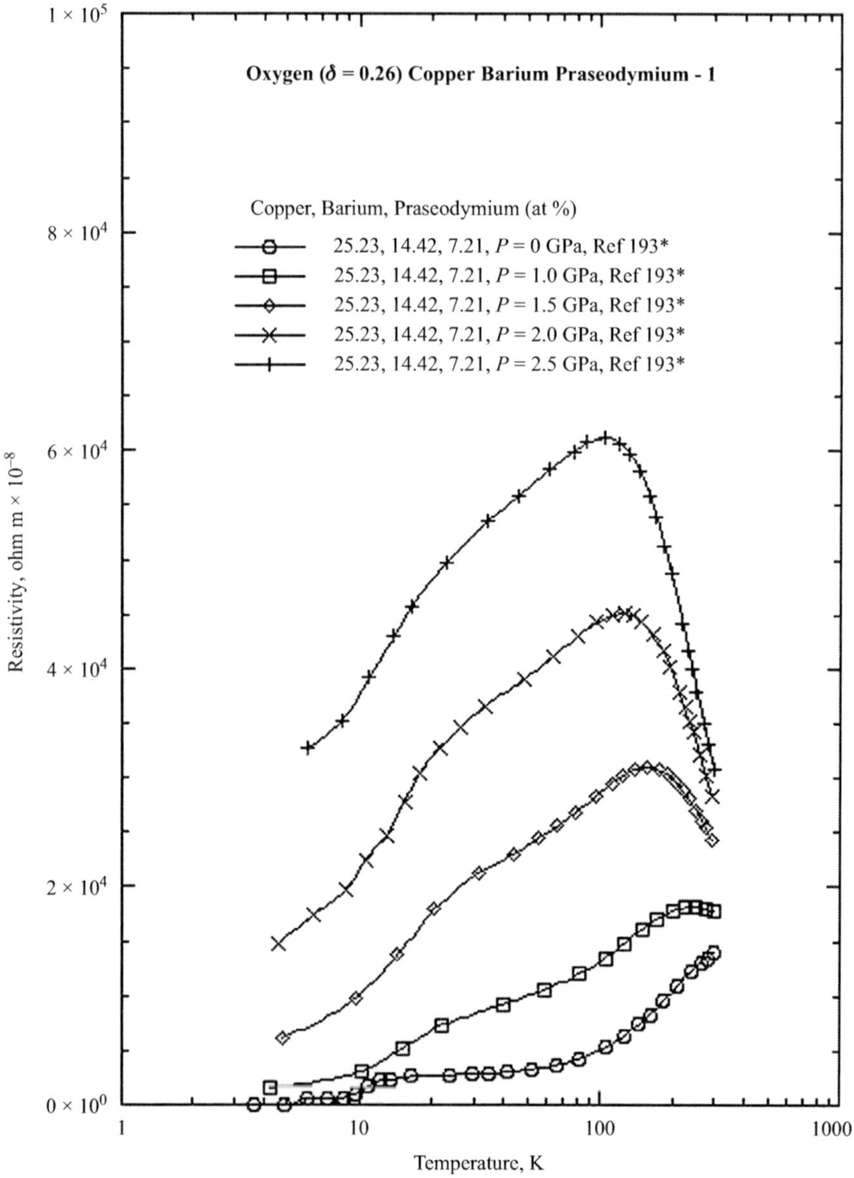

Oxygen (δ = 0.26) Copper Barium Praseodymium - 1
Copper, Barium, Praseodymium (at %)
25.23, 14.42, 7.21, P = 0 GPa, Ref 193*
25.23, 14.42, 7.21, P = 1.0 GPa, Ref 193*
25.23, 14.42, 7.21, P = 1.5 GPa, Ref 193*
25.23, 14.42, 7.21, P = 2.0 GPa, Ref 193*
25.23, 14.42, 7.21, P = 2.5 GPa, Ref 193*
1 × 10^5
8 × 10^4
6 × 10^4
4 × 10^4
2 × 10^4
0 × 10^0
Resistivity, ohm m × 10^-8
1
10
100
1000
Temperature, K

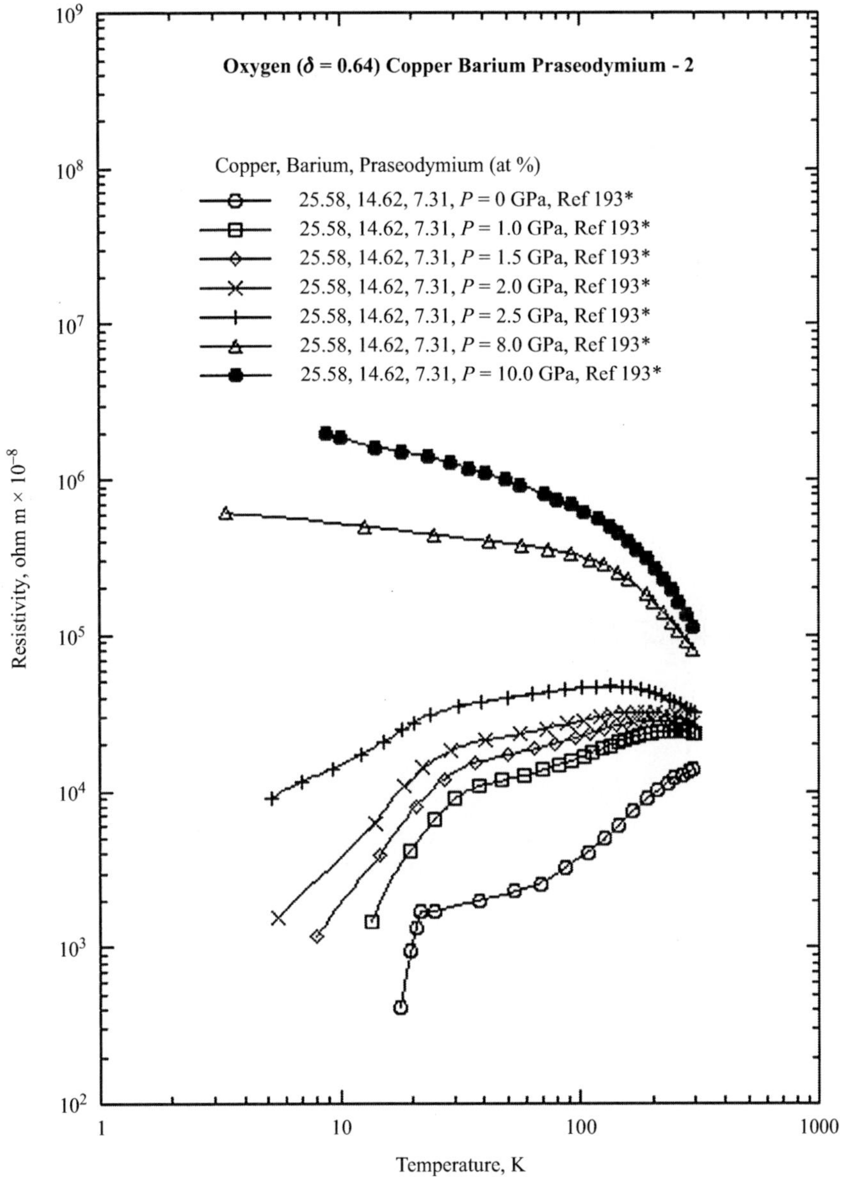
Oxygen (δ = 0.64) Copper Barium Praseodymium - 2
Copper, Barium, Praseodymium (at %)
25.58, 14.62, 7.31, P = 0 GPa, Ref 193*
25.58, 14.62, 7.31, P = 1.0 GPa, Ref 193*
25.58, 14.62, 7.31, P = 1.5 GPa, Ref 193*
25.58, 14.62, 7.31, P = 2.0 GPa, Ref 193*
25.58, 14.62, 7.31, P = 2.5 GPa, Ref 193*
25.58, 14.62, 7.31, P = 8.0 GPa, Ref 193*
25.58, 14.62, 7.31, P = 10.0 GPa, Ref 193*
Resistivity, ohm m × 10⁻⁸
Temperature, K
10⁹
10⁸
10⁷
10⁶
10⁵
10⁴
10³
10²
1
10
100
1000

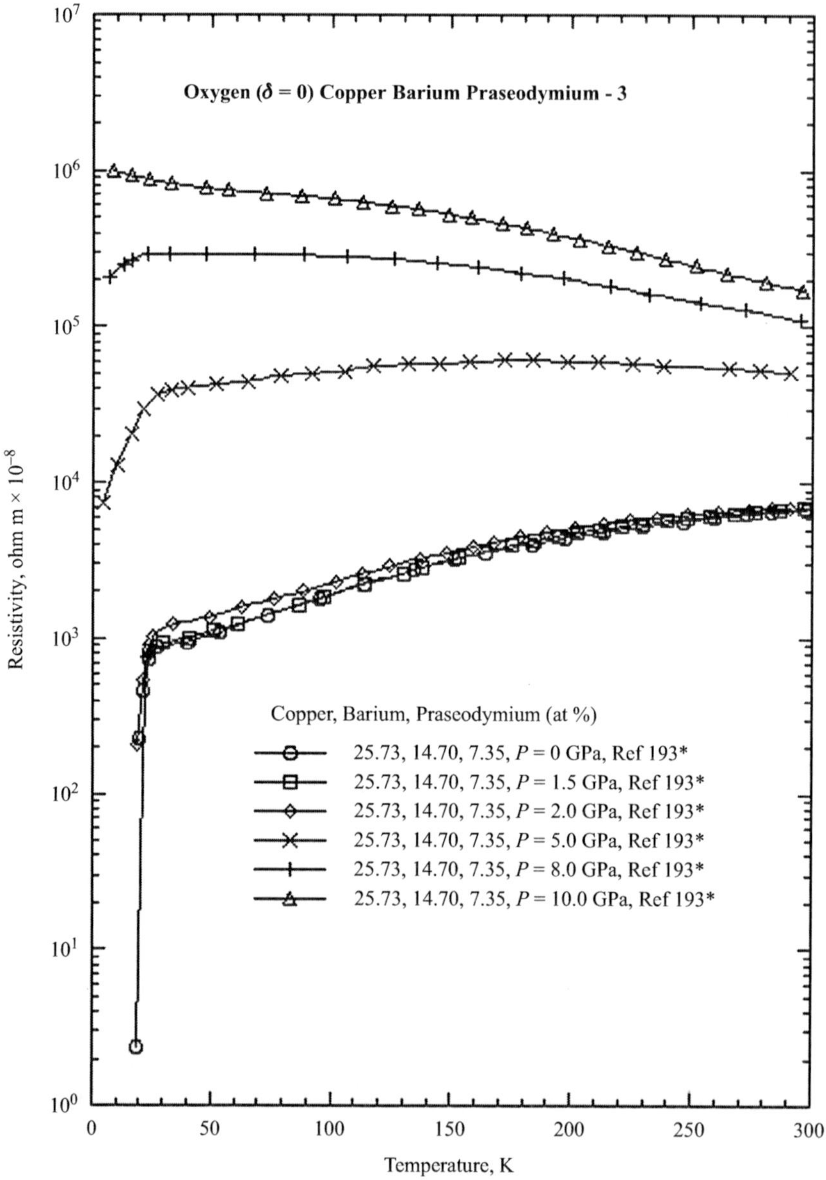
Oxygen (δ = 0) Copper Barium Praseodymium - 3
Resistivity, ohm m × 10⁻⁸
Temperature, K
Copper, Barium, Praseodymium (at %)
25.73, 14.70, 7.35, P = 0 GPa, Ref 193*
25.73, 14.70, 7.35, P = 1.5 GPa, Ref 193*
25.73, 14.70, 7.35, P = 2.0 GPa, Ref 193*
25.73, 14.70, 7.35, P = 5.0 GPa, Ref 193*
25.73, 14.70, 7.35, P = 8.0 GPa, Ref 193*
25.73, 14.70, 7.35, P = 10.0 GPa, Ref 193*
10⁷
10⁶
10⁵
10⁴
10³
10²
10¹
10⁰
0
50
100
150
200
250
300

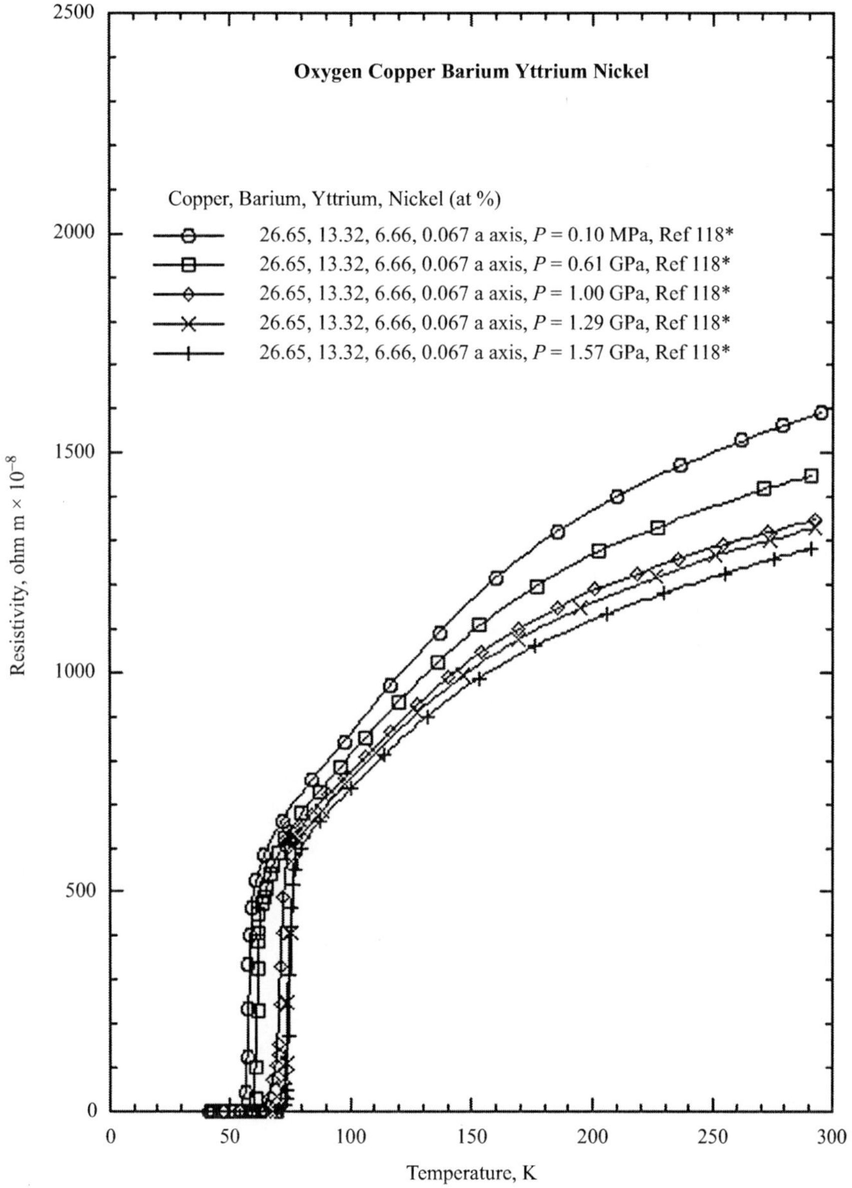
Oxygen Copper Barium Yttrium Nickel
Copper, Barium, Yttrium, Nickel (at %)
26.65, 13.32, 6.66, 0.067 a axis, P = 0.10 MPa, Ref 118*
26.65, 13.32, 6.66, 0.067 a axis, P = 0.61 GPa, Ref 118*
26.65, 13.32, 6.66, 0.067 a axis, P = 1.00 GPa, Ref 118*
26.65, 13.32, 6.66, 0.067 a axis, P = 1.29 GPa, Ref 118*
26.65, 13.32, 6.66, 0.067 a axis, P = 1.57 GPa, Ref 118*
Resistivity, ohm m × 10⁻⁸
2500
2000
1500
1000
500
0
0
50
100
150
200
250
300
Temperature, K

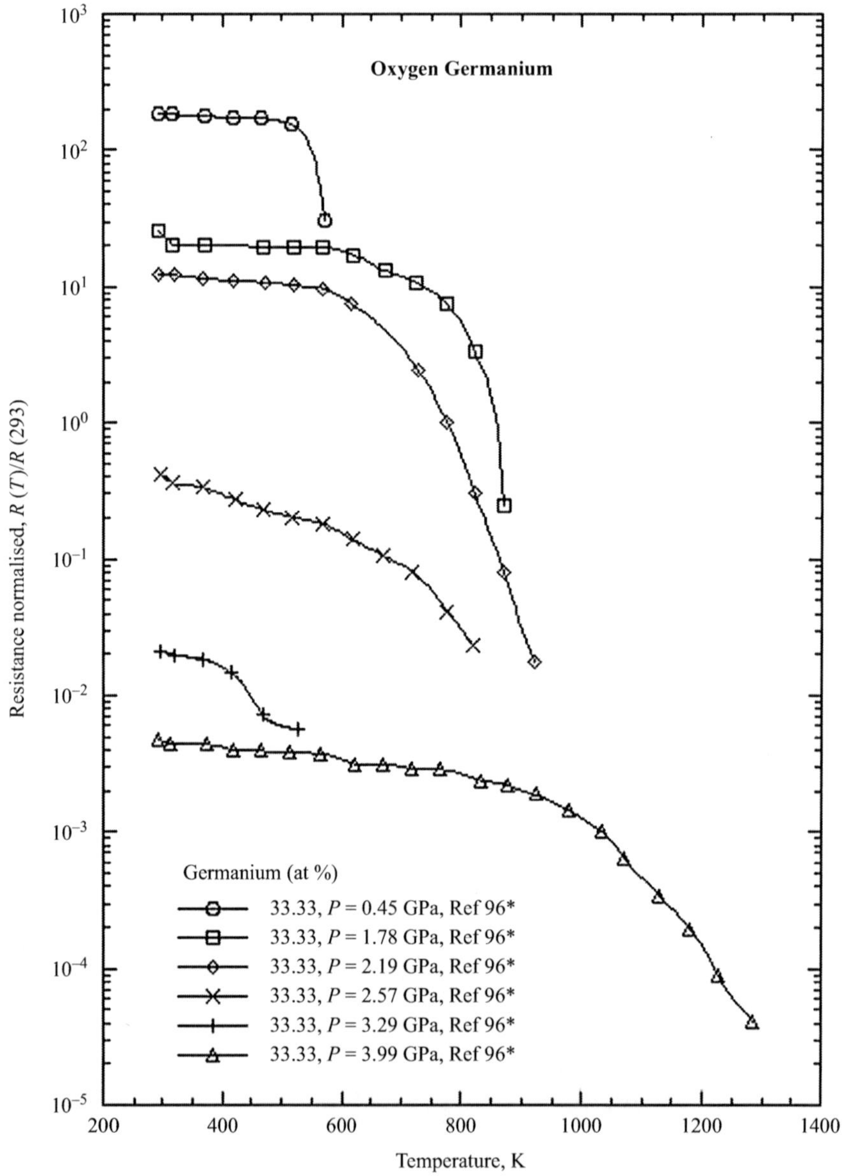
Oxygen Germanium
Resistance normalised, R (T)/R (293)
Temperature, K
Germanium (at %)
33.33, P = 0.45 GPa, Ref 96*
33.33, P = 1.78 GPa, Ref 96*
33.33, P = 2.19 GPa, Ref 96*
33.33, P = 2.57 GPa, Ref 96*
33.33, P = 3.29 GPa, Ref 96*
33.33, P = 3.99 GPa, Ref 96*

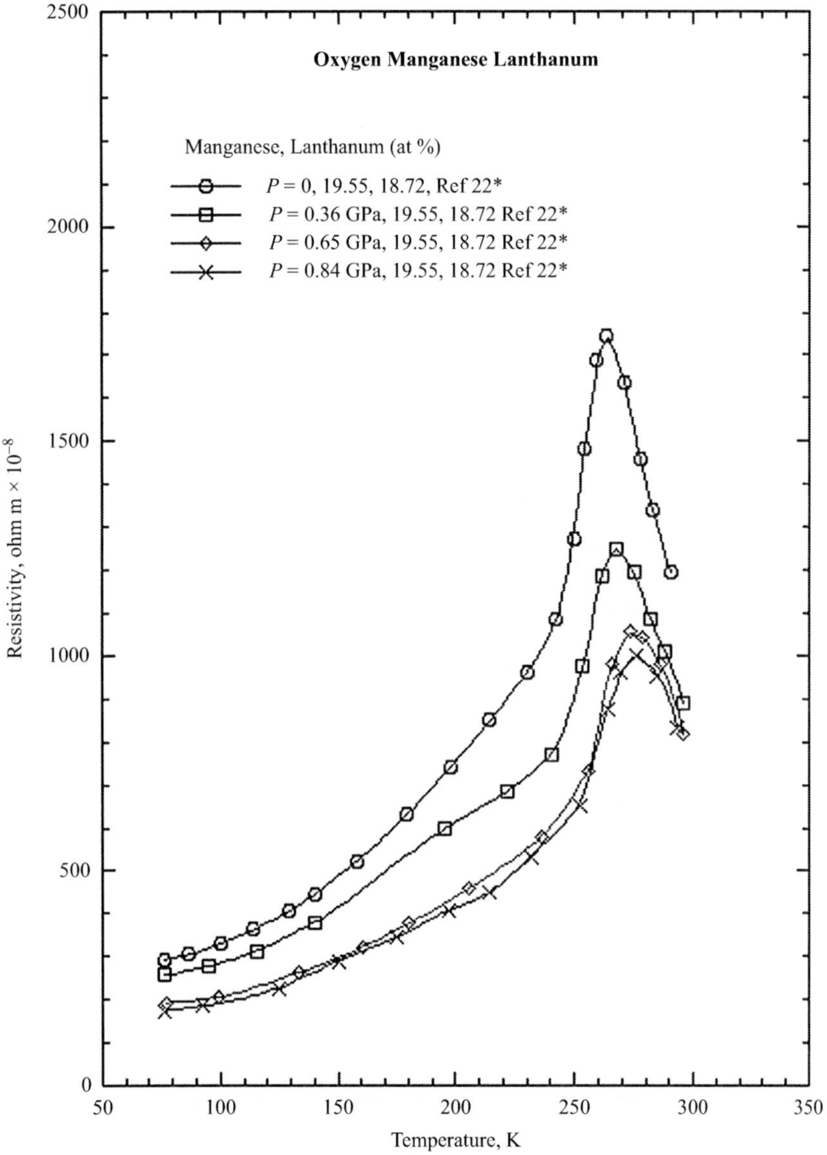
Oxygen Manganese Lanthanum
Manganese, Lanthanum (at %)
P = 0, 19.55, 18.72, Ref 22*
P = 0.36 GPa, 19.55, 18.72 Ref 22*
P = 0.65 GPa, 19.55, 18.72 Ref 22*
P = 0.84 GPa, 19.55, 18.72 Ref 22*
Resistivity, ohm m × 10^−8
Temperature, K
0
500
1000
1500
2000
2500
50
100
150
200
250
300
350

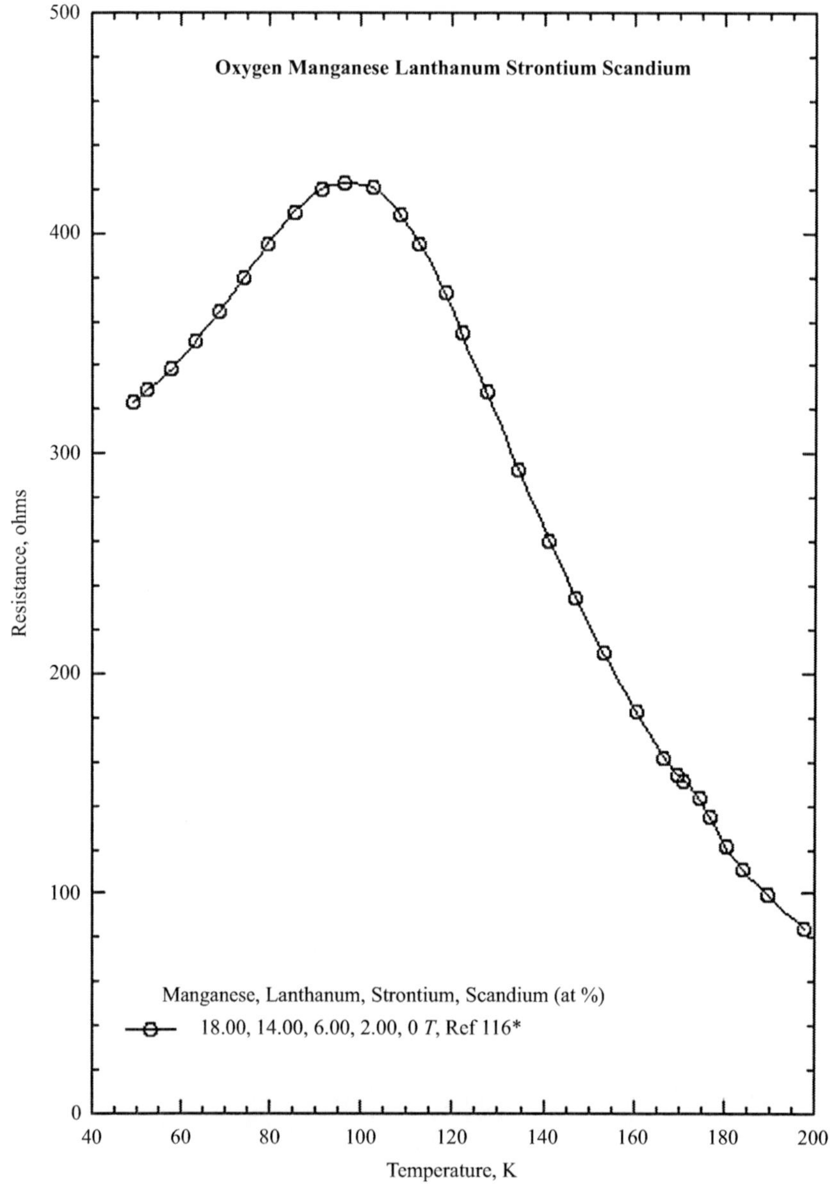
Oxygen Manganese Lanthanum Strontium Scandium
Resistance, ohms
Temperature, K
Manganese, Lanthanum, Strontium, Scandium (at %)
18.00, 14.00, 6.00, 2.00, 0 T, Ref 116*
0
100
200
300
400
500
40
60
80
100
120
140
160
180
200

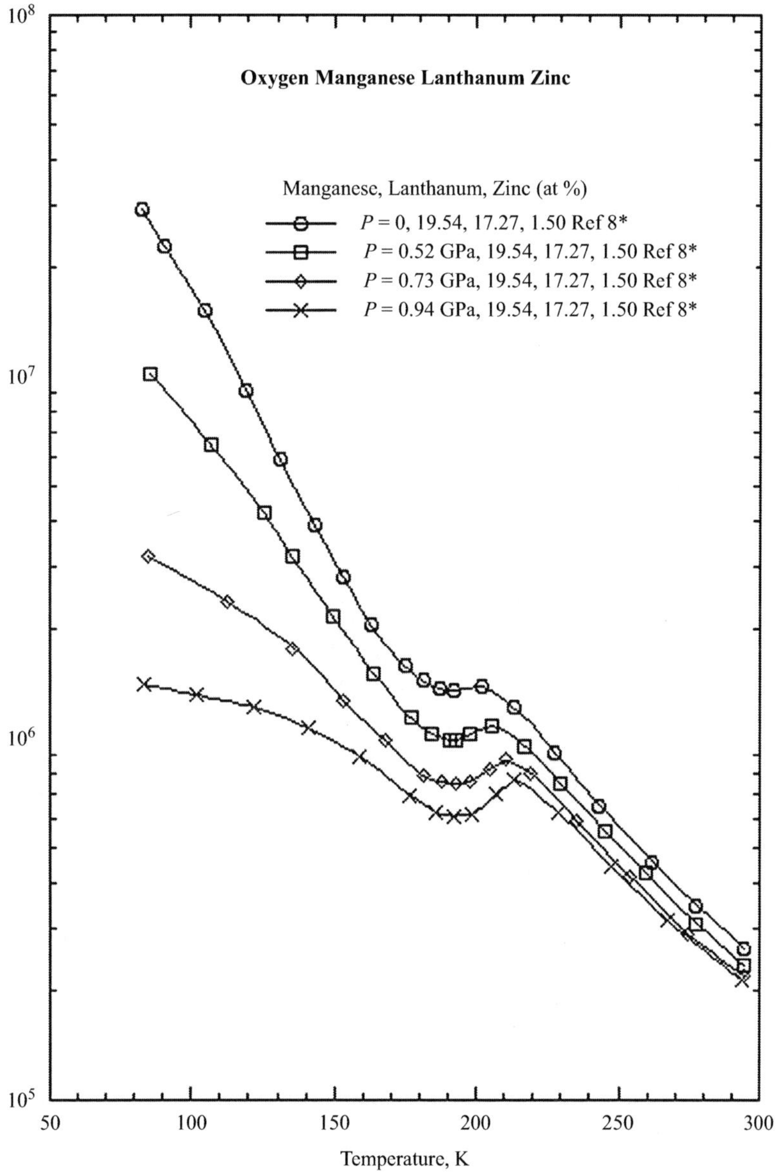

Oxygen Manganese Lanthanum Zinc
Manganese, Lanthanum, Zinc (at %)
P = 0, 19.54, 17.27, 1.50 Ref 8*
P = 0.52 GPa, 19.54, 17.27, 1.50 Ref 8*
P = 0.73 GPa, 19.54, 17.27, 1.50 Ref 8*
P = 0.94 GPa, 19.54, 17.27, 1.50 Ref 8*
Resistivity, ohm m × 10⁻⁸
Temperature, K
10^8
10^7
10^6
10^5
50
100
150
200
250
300

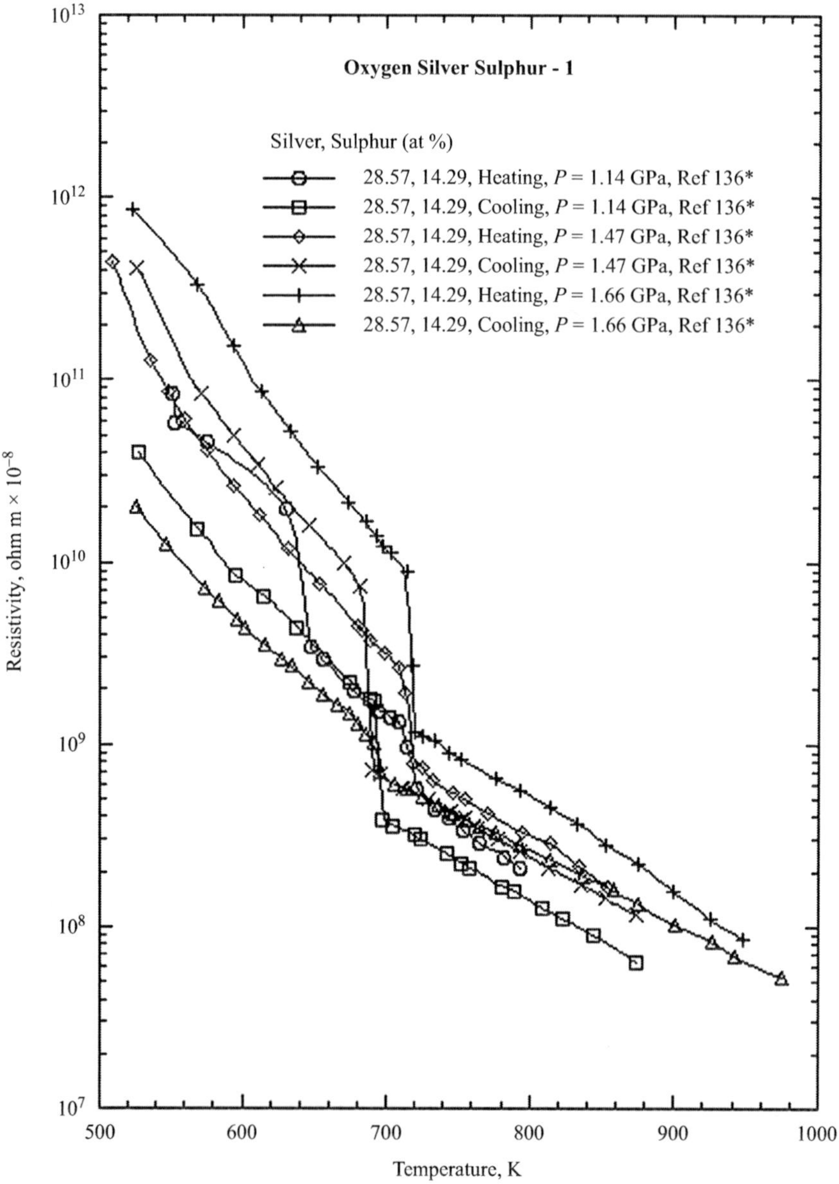
Oxygen Silver Sulphur - 1
Silver, Sulphur (at %)
28.57, 14.29, Heating, P = 1.14 GPa, Ref 136*
28.57, 14.29, Cooling, P = 1.14 GPa, Ref 136*
28.57, 14.29, Heating, P = 1.47 GPa, Ref 136*
28.57, 14.29, Cooling, P = 1.47 GPa, Ref 136*
28.57, 14.29, Heating, P = 1.66 GPa, Ref 136*
28.57, 14.29, Cooling, P = 1.66 GPa, Ref 136*
Resistivity, ohm m × 10⁻⁸
Temperature, K
10¹³
10¹²
10¹¹
10¹⁰
10⁹
10⁸
10⁷
500
600
700
800
900
1000

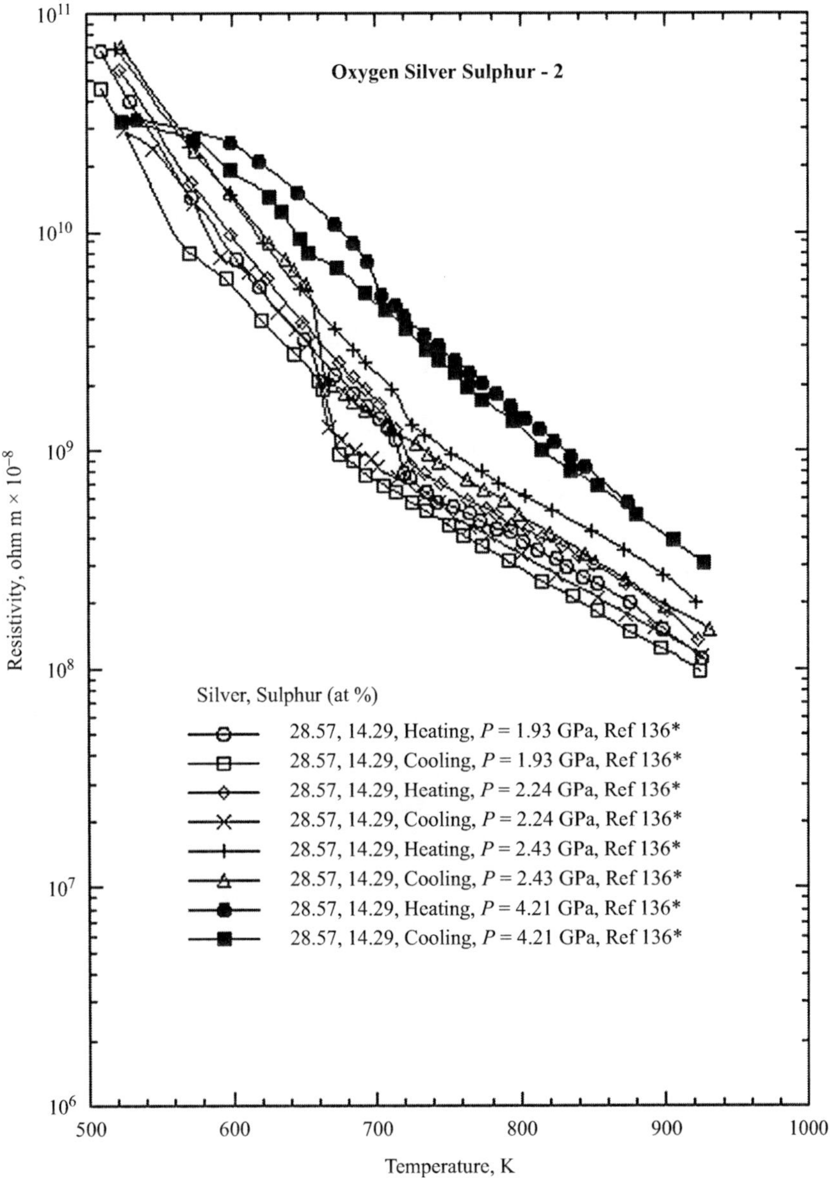
Oxygen Silver Sulphur - 2
Resistivity, ohm m × 10⁻⁸
10¹¹
10¹⁰
10⁹
10⁸
10⁷
10⁶
500
600
700
800
900
1000
Temperature, K
Silver, Sulphur (at %)
28.57, 14.29, Heating, P = 1.93 GPa, Ref 136*
28.57, 14.29, Cooling, P = 1.93 GPa, Ref 136*
28.57, 14.29, Heating, P = 2.24 GPa, Ref 136*
28.57, 14.29, Cooling, P = 2.24 GPa, Ref 136*
28.57, 14.29, Heating, P = 2.43 GPa, Ref 136*
28.57, 14.29, Cooling, P = 2.43 GPa, Ref 136*
28.57, 14.29, Heating, P = 4.21 GPa, Ref 136*
28.57, 14.29, Cooling, P = 4.21 GPa, Ref 136*

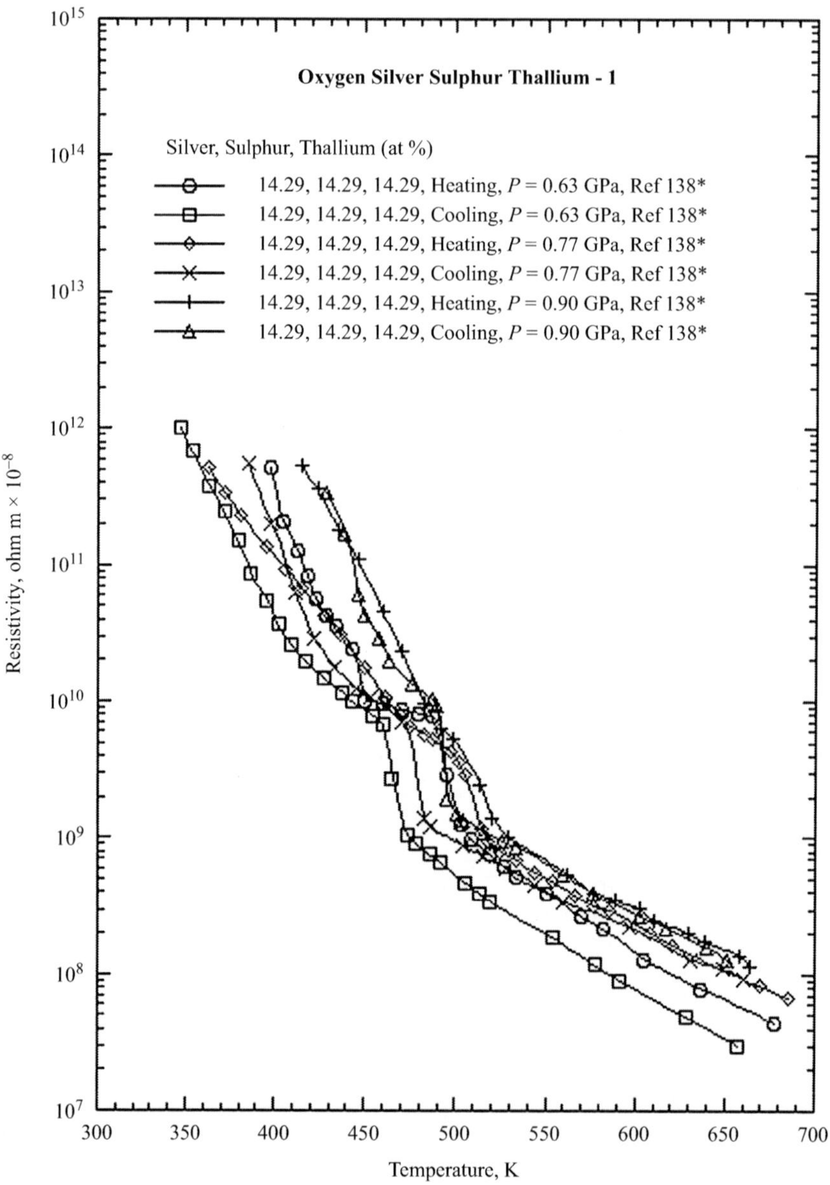

Oxygen Silver Sulphur Thallium - 1
Silver, Sulphur, Thallium (at %)
14.29, 14.29, 14.29, Heating, P = 0.63 GPa, Ref 138*
14.29, 14.29, 14.29, Cooling, P = 0.63 GPa, Ref 138*
14.29, 14.29, 14.29, Heating, P = 0.77 GPa, Ref 138*
14.29, 14.29, 14.29, Cooling, P = 0.77 GPa, Ref 138*
14.29, 14.29, 14.29, Heating, P = 0.90 GPa, Ref 138*
14.29, 14.29, 14.29, Cooling, P = 0.90 GPa, Ref 138*
Resistivity, ohm m × 10⁻⁸
10¹⁵
10¹⁴
10¹³
10¹²
10¹¹
10¹⁰
10⁹
10⁸
10⁷
300
350
400
450
500
550
600
650
700
Temperature, K

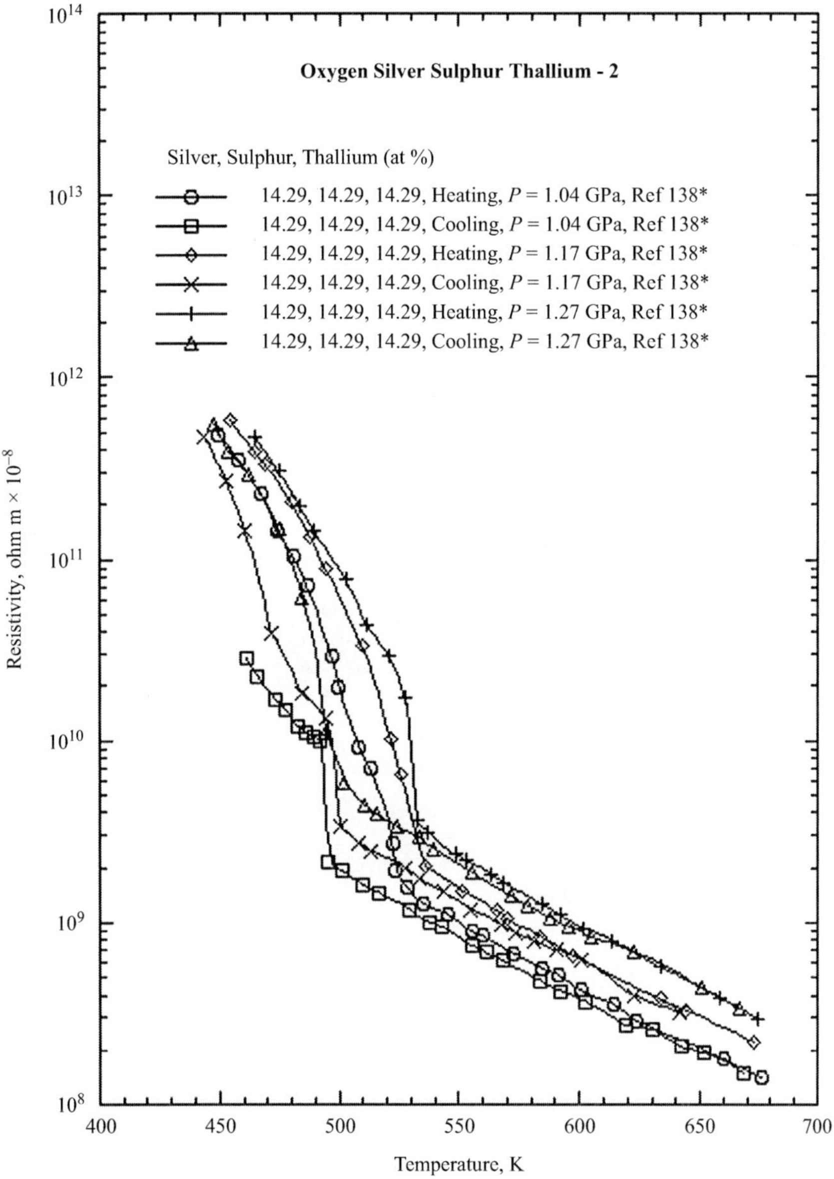

Oxygen Silver Sulphur Thallium - 2
Silver, Sulphur, Thallium (at %)
14.29, 14.29, 14.29, Heating, P = 1.04 GPa, Ref 138*
14.29, 14.29, 14.29, Cooling, P = 1.04 GPa, Ref 138*
14.29, 14.29, 14.29, Heating, P = 1.17 GPa, Ref 138*
14.29, 14.29, 14.29, Cooling, P = 1.17 GPa, Ref 138*
14.29, 14.29, 14.29, Heating, P = 1.27 GPa, Ref 138*
14.29, 14.29, 14.29, Cooling, P = 1.27 GPa, Ref 138*
Resistivity, ohm m × 10⁻⁸
Temperature, K

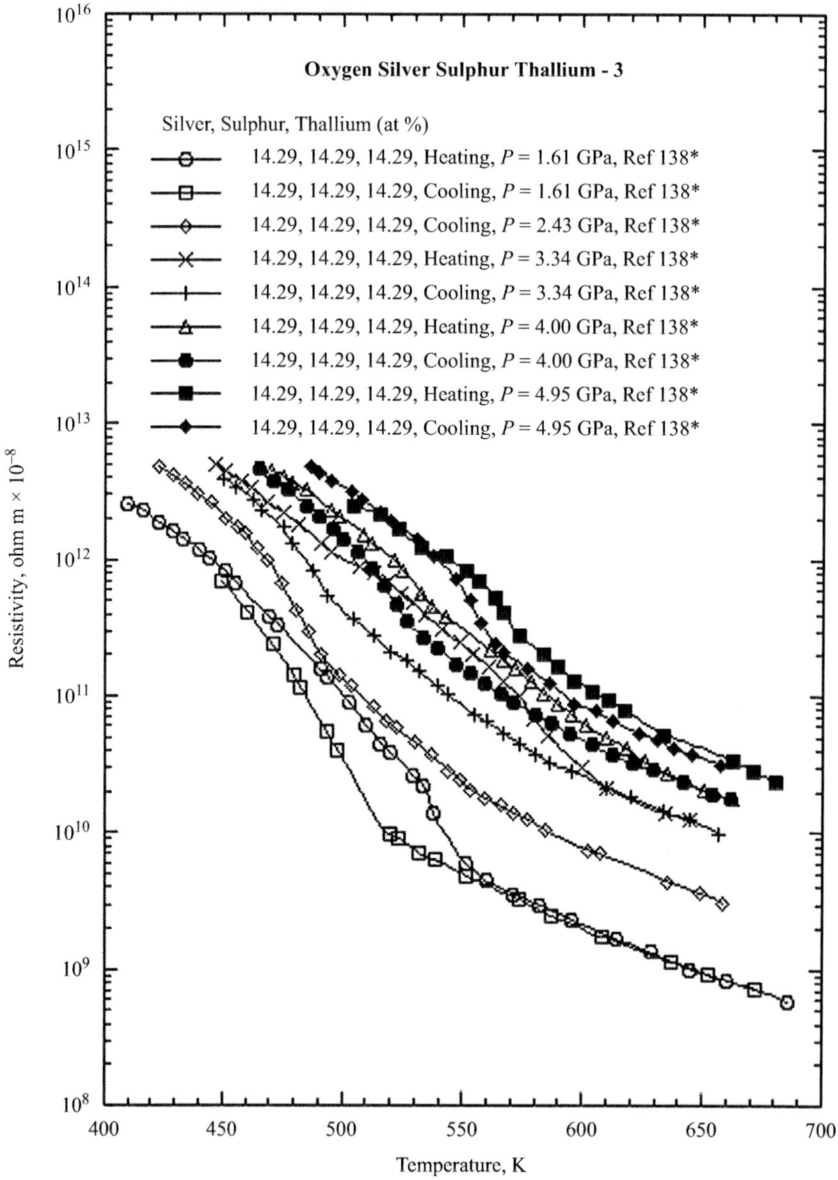
Oxygen Silver Sulphur Thallium - 3
Silver, Sulphur, Thallium (at %)
14.29, 14.29, 14.29, Heating, P = 1.61 GPa, Ref 138*
14.29, 14.29, 14.29, Cooling, P = 1.61 GPa, Ref 138*
14.29, 14.29, 14.29, Cooling, P = 2.43 GPa, Ref 138*
14.29, 14.29, 14.29, Heating, P = 3.34 GPa, Ref 138*
14.29, 14.29, 14.29, Cooling, P = 3.34 GPa, Ref 138*
14.29, 14.29, 14.29, Heating, P = 4.00 GPa, Ref 138*
14.29, 14.29, 14.29, Cooling, P = 4.00 GPa, Ref 138*
14.29, 14.29, 14.29, Heating, P = 4.95 GPa, Ref 138*
14.29, 14.29, 14.29, Cooling, P = 4.95 GPa, Ref 138*
Resistivity, ohm m × 10⁻⁸
Temperature, K
10¹⁶
10¹⁵
10¹⁴
10¹³
10¹²
10¹¹
10¹⁰
10⁹
10⁸
400
450
500
550
600
650
700

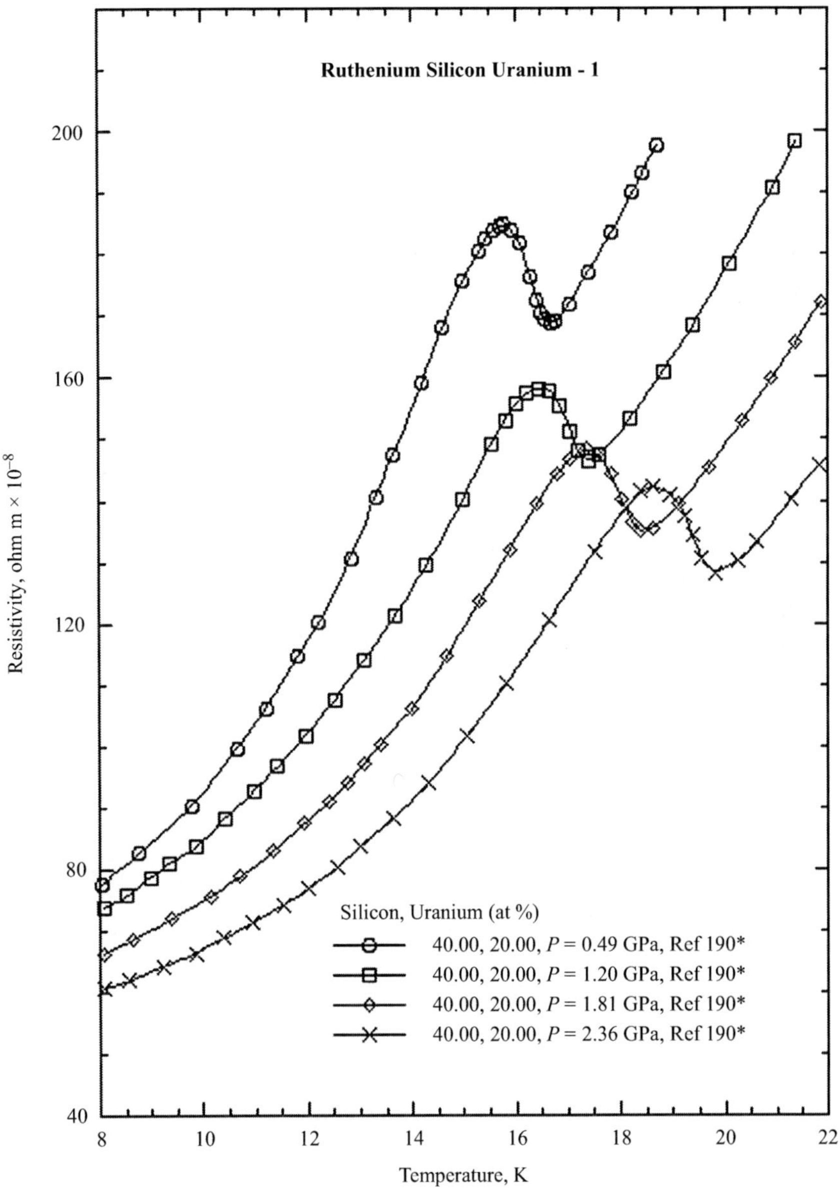

Ruthenium Silicon Uranium - 1
Resistivity, ohm m × 10⁻⁸
40
80
120
160
200
Temperature, K
8
10
12
14
16
18
20
22
Silicon, Uranium (at %)
40.00, 20.00, P = 0.49 GPa, Ref 190*
40.00, 20.00, P = 1.20 GPa, Ref 190*
40.00, 20.00, P = 1.81 GPa, Ref 190*
40.00, 20.00, P = 2.36 GPa, Ref 190*

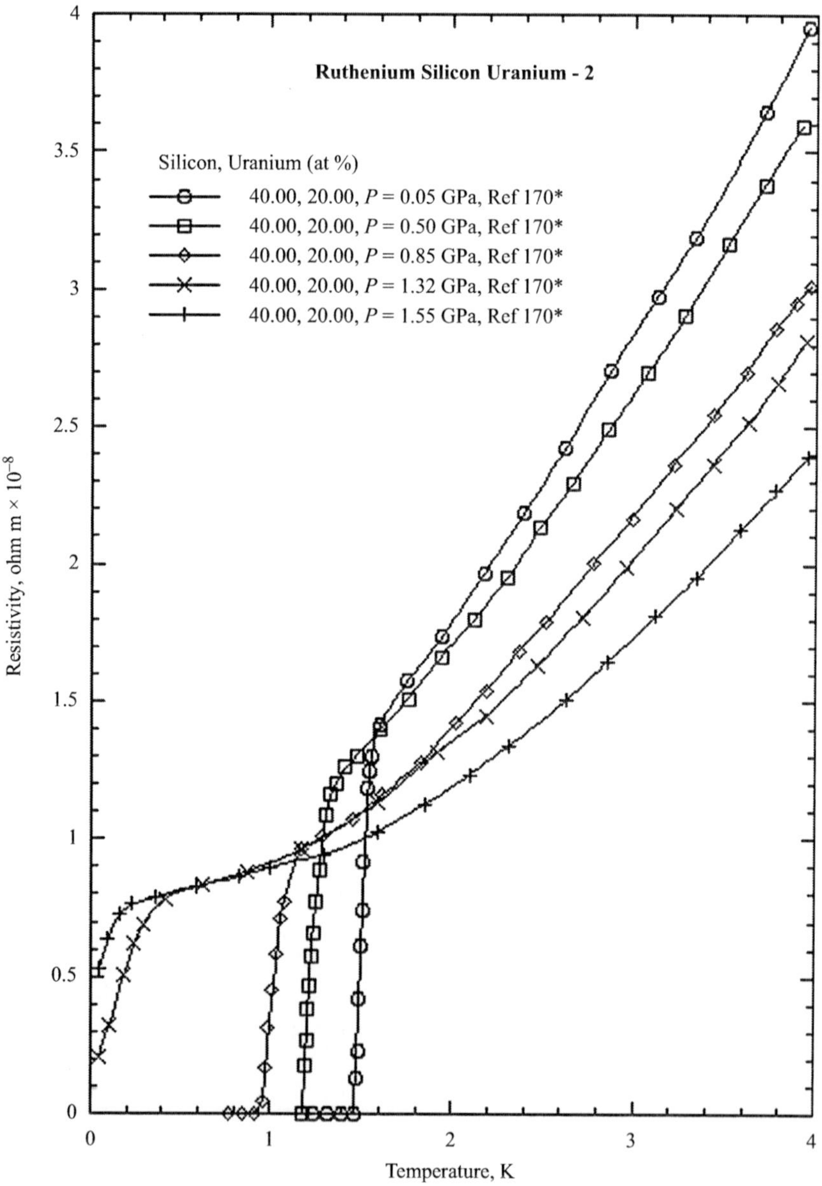
Ruthenium Silicon Uranium - 2
Silicon, Uranium (at %)
40.00, 20.00, P = 0.05 GPa, Ref 170*
40.00, 20.00, P = 0.50 GPa, Ref 170*
40.00, 20.00, P = 0.85 GPa, Ref 170*
40.00, 20.00, P = 1.32 GPa, Ref 170*
40.00, 20.00, P = 1.55 GPa, Ref 170*
Resistivity, ohm m × 10⁻⁸
Temperature, K
0
0.5
1
1.5
2
2.5
3
3.5
4
0
1
2
3
4

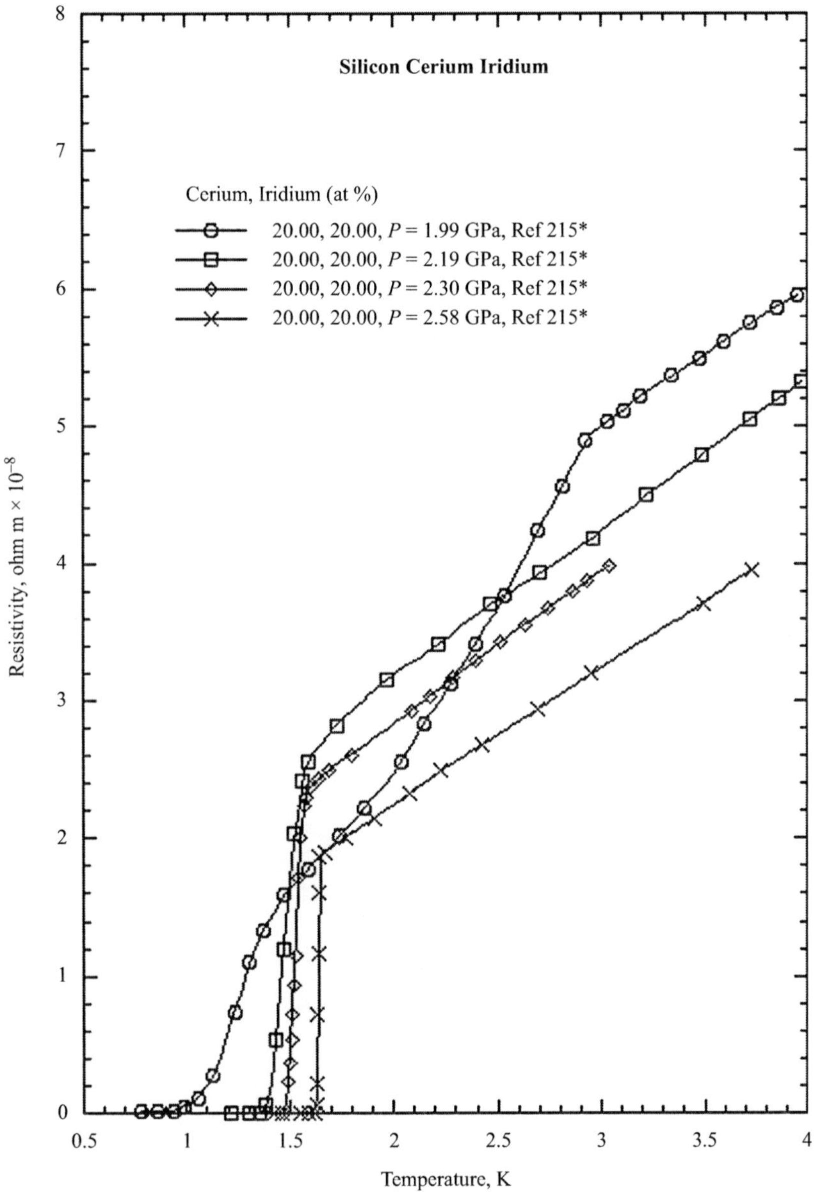
Silicon Cerium Iridium
Cerium, Iridium (at %)
20.00, 20.00, P = 1.99 GPa, Ref 215*
20.00, 20.00, P = 2.19 GPa, Ref 215*
20.00, 20.00, P = 2.30 GPa, Ref 215*
20.00, 20.00, P = 2.58 GPa, Ref 215*
Resistivity, ohm m × 10⁻⁸
Temperature, K
0
1
2
3
4
5
6
7
8
0.5
1
1.5
2
2.5
3
3.5
4

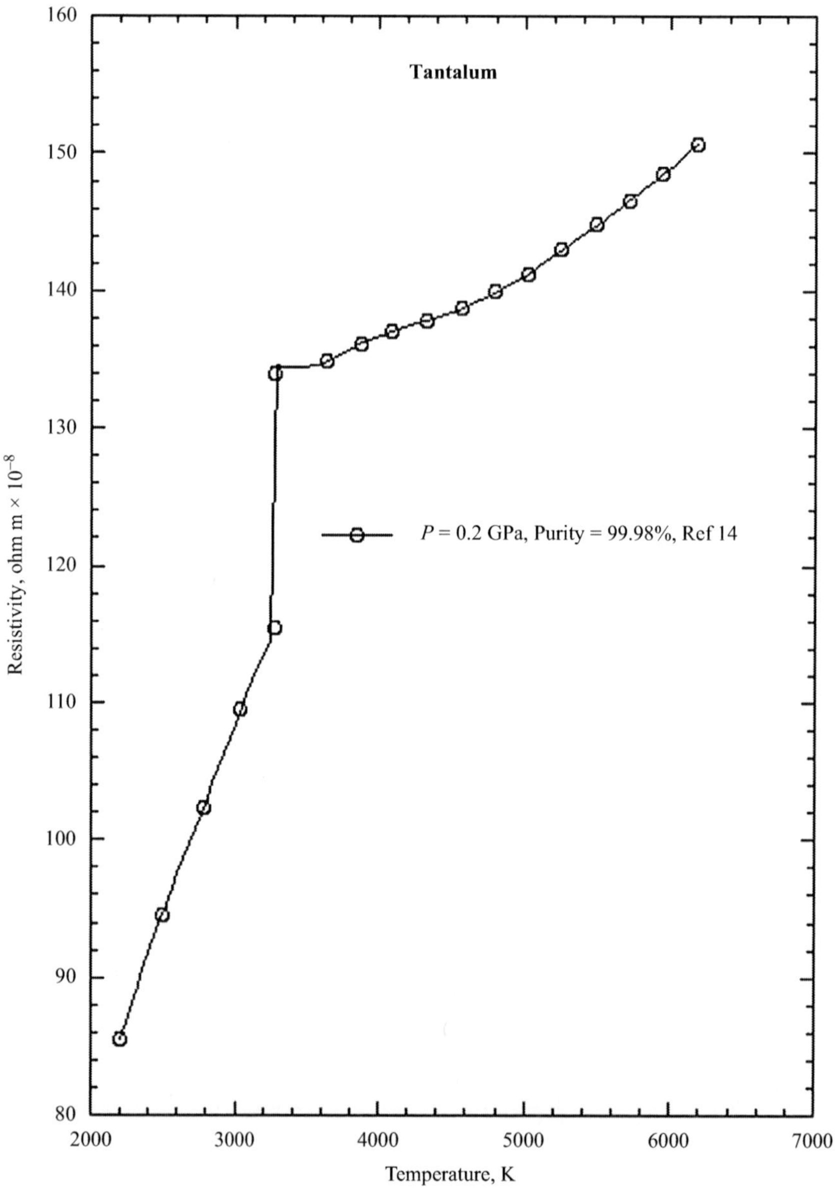
Tantalum
P = 0.2 GPa, Purity = 99.98%, Ref 14
Resistivity, ohm m × 10⁻⁸
Temperature, K
160
150
140
130
120
110
100
90
80
2000
3000
4000
5000
6000
7000

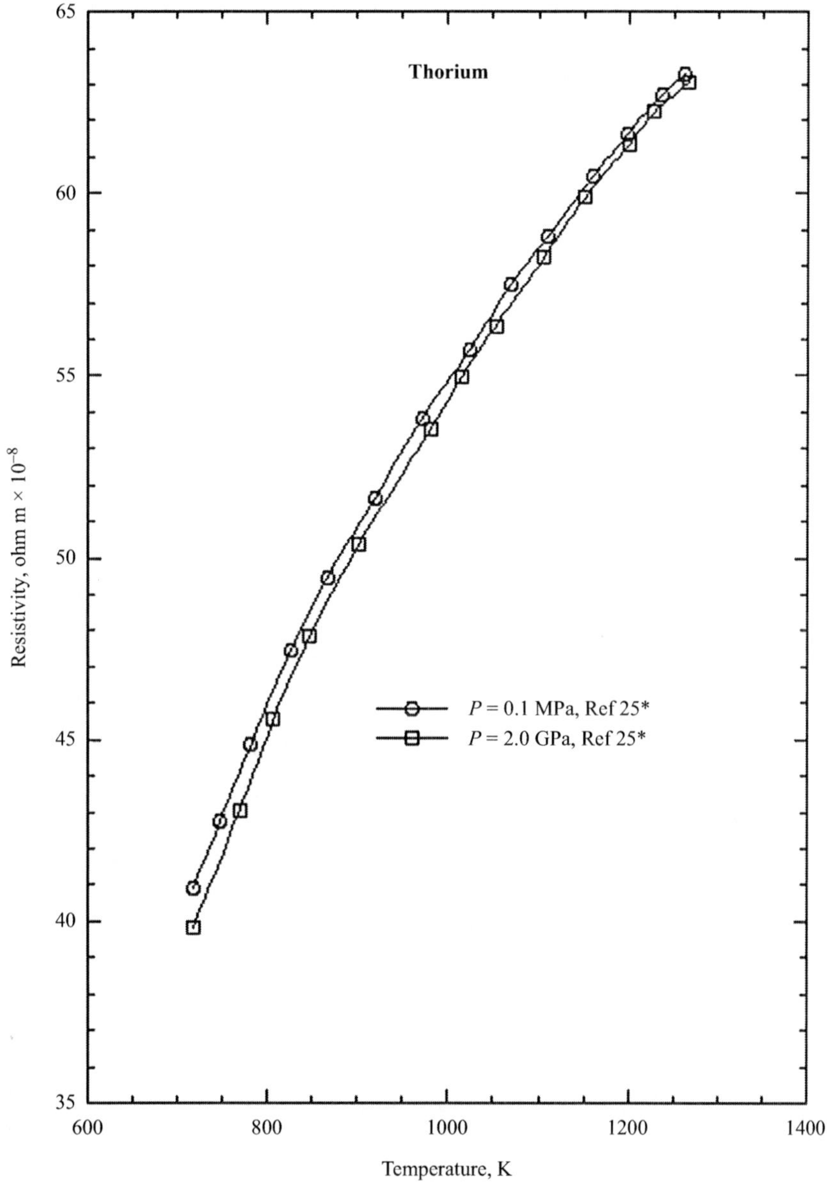
Thorium
P = 0.1 MPa, Ref 25*
P = 2.0 GPa, Ref 25*
Resistivity, ohm m × 10^−8
Temperature, K
35
40
45
50
55
60
65
600
800
1000
1200
1400

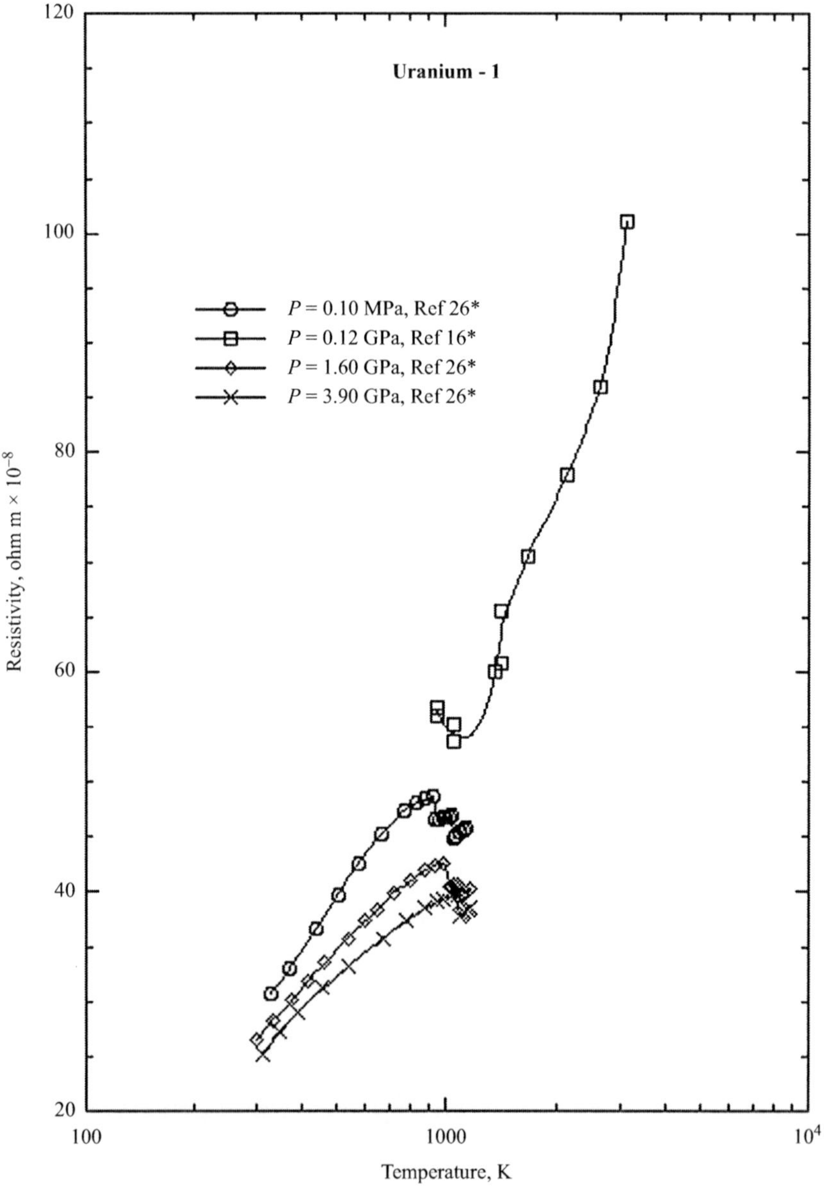

Uranium - 1
P = 0.10 MPa, Ref 26*
P = 0.12 GPa, Ref 16*
P = 1.60 GPa, Ref 26*
P = 3.90 GPa, Ref 26*
Resistivity, ohm m × 10⁻⁸
Temperature, K
120
100
80
60
40
20
100
1000
10⁴

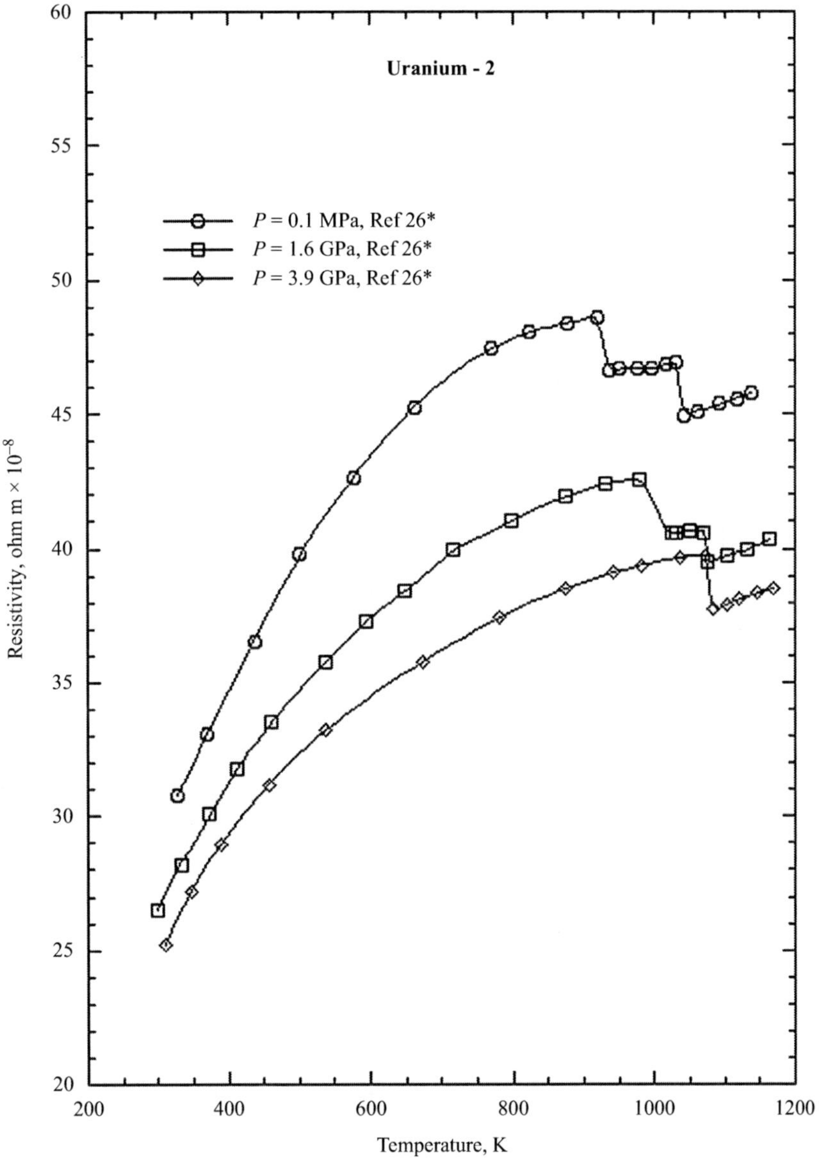
Uranium - 2
P = 0.1 MPa, Ref 26*
P = 1.6 GPa, Ref 26*
P = 3.9 GPa, Ref 26*
Resistivity, ohm m × 10⁻⁸
Temperature, K
60
55
50
45
40
35
30
25
20
200
400
600
800
1000
1200

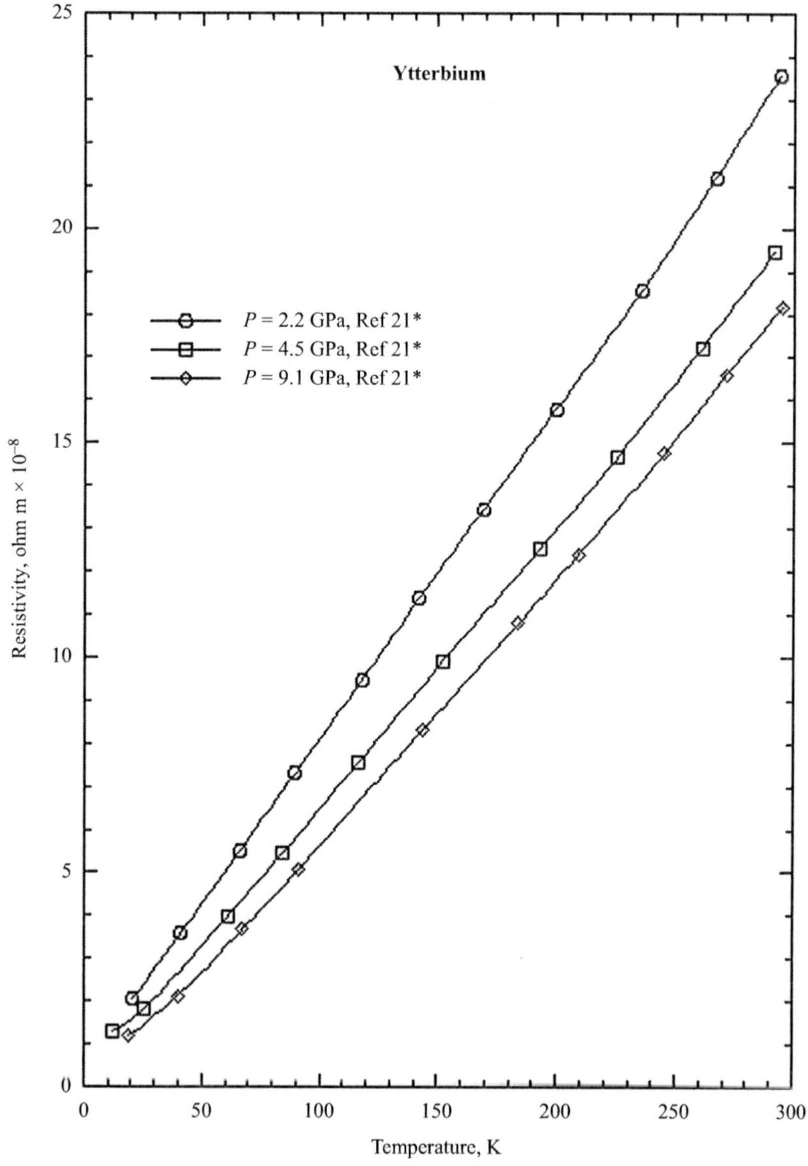
Ytterbium
P = 2.2 GPa, Ref 21*
P = 4.5 GPa, Ref 21*
P = 9.1 GPa, Ref 21*
Resistivity, ohm m × 10⁻⁸
Temperature, K
0
5
10
15
20
25
0
50
100
150
200
250
300

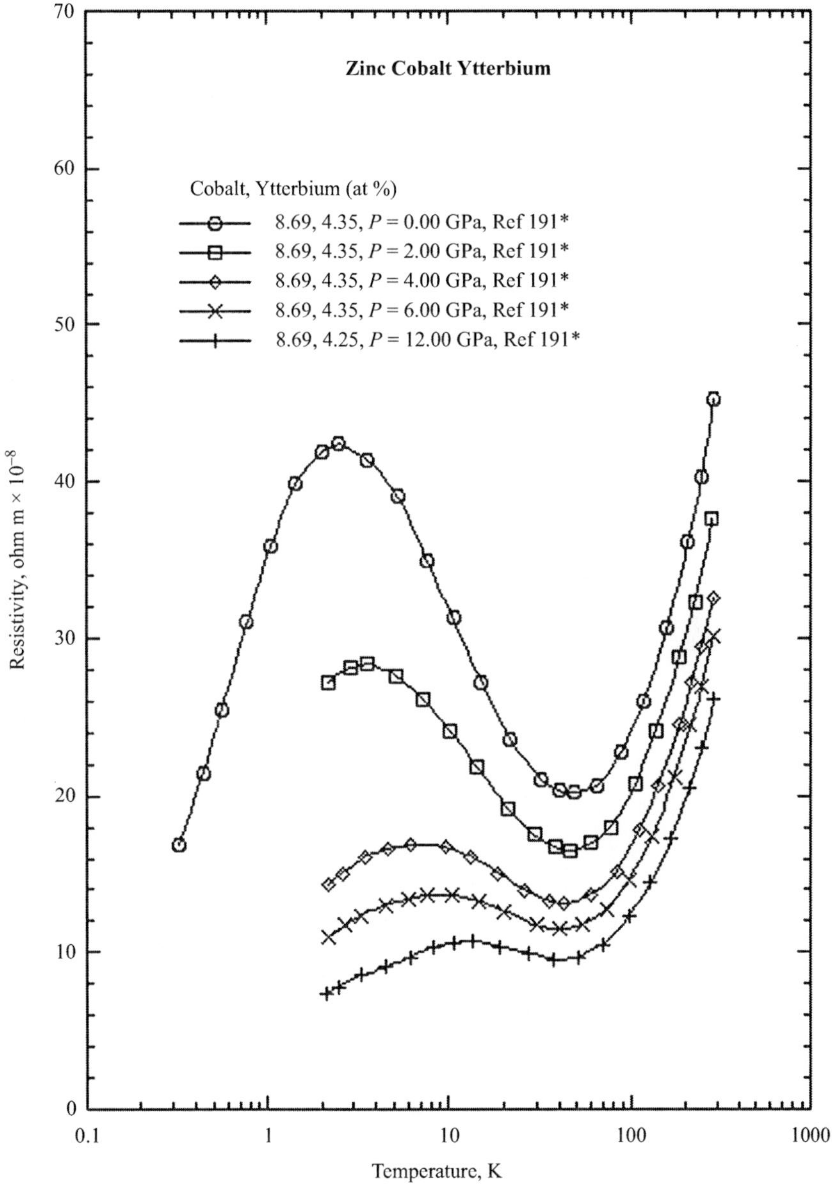
Zinc Cobalt Ytterbium
Cobalt, Ytterbium (at %)
8.69, 4.35, P = 0.00 GPa, Ref 191*
8.69, 4.35, P = 2.00 GPa, Ref 191*
8.69, 4.35, P = 4.00 GPa, Ref 191*
8.69, 4.35, P = 6.00 GPa, Ref 191*
8.69, 4.25, P = 12.00 GPa, Ref 191*
Resistivity, ohm m × 10⁻⁸
Temperature, K
0
10
20
30
40
50
60
70
0.1
1
10
100
1000

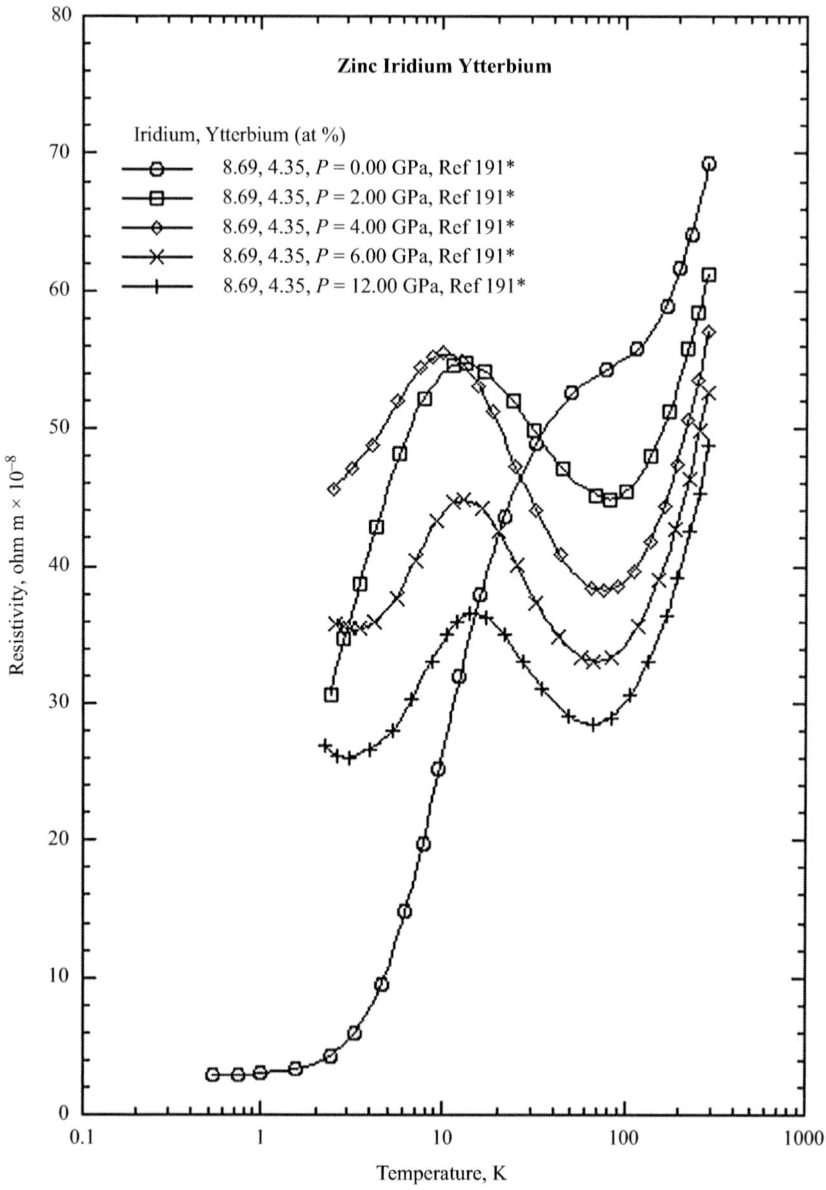
Zinc Iridium Ytterbium
Iridium, Ytterbium (at %)
8.69, 4.35, P = 0.00 GPa, Ref 191*
8.69, 4.35, P = 2.00 GPa, Ref 191*
8.69, 4.35, P = 4.00 GPa, Ref 191*
8.69, 4.35, P = 6.00 GPa, Ref 191*
8.69, 4.35, P = 12.00 GPa, Ref 191*
Resistivity, ohm m × 10⁻⁸
Temperature, K
0
10
20
30
40
50
60
70
80
0.1
1
10
100
1000

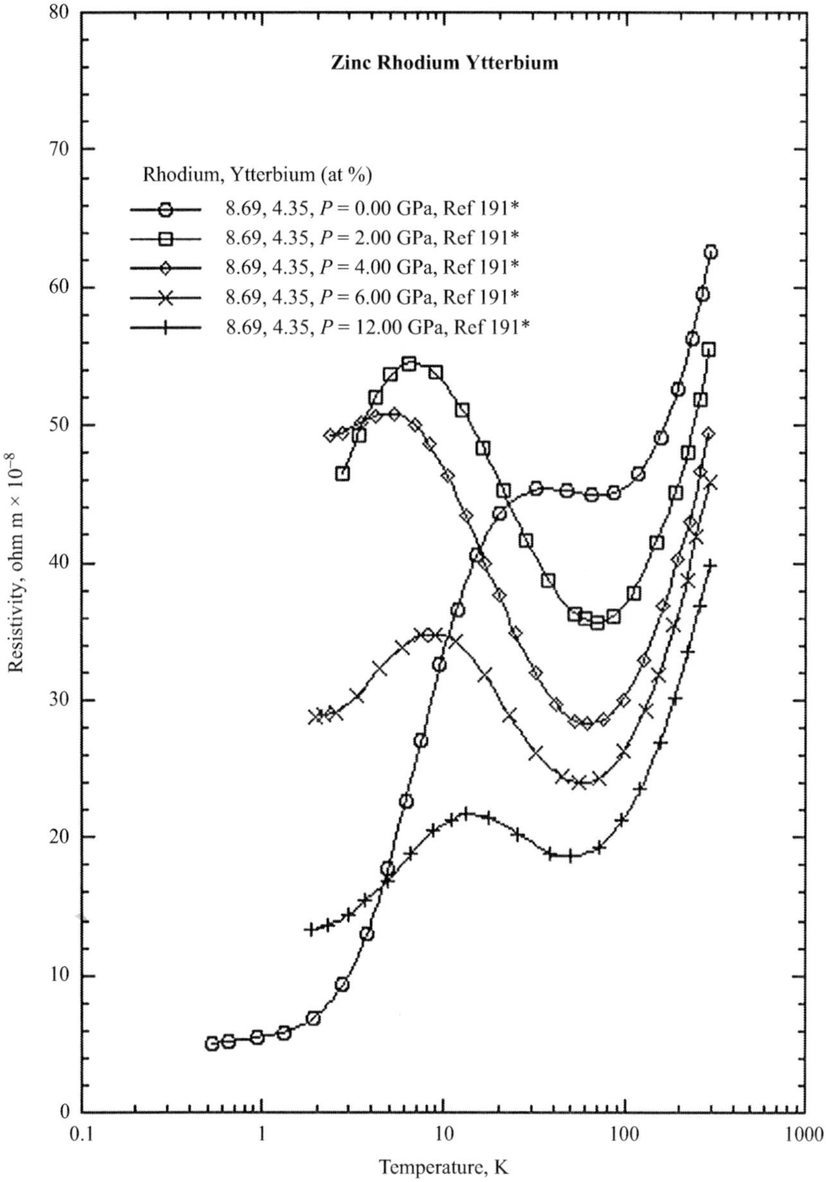
Zinc Rhodium Ytterbium
Rhodium, Ytterbium (at %)
8.69, 4.35, P = 0.00 GPa, Ref 191*
8.69, 4.35, P = 2.00 GPa, Ref 191*
8.69, 4.35, P = 4.00 GPa, Ref 191*
8.69, 4.35, P = 6.00 GPa, Ref 191*
8.69, 4.35, P = 12.00 GPa, Ref 191*
Resistivity, ohm m × 10⁻⁸
Temperature, K
0
10
20
30
40
50
60
70
80
0.1
1
10
100
1000

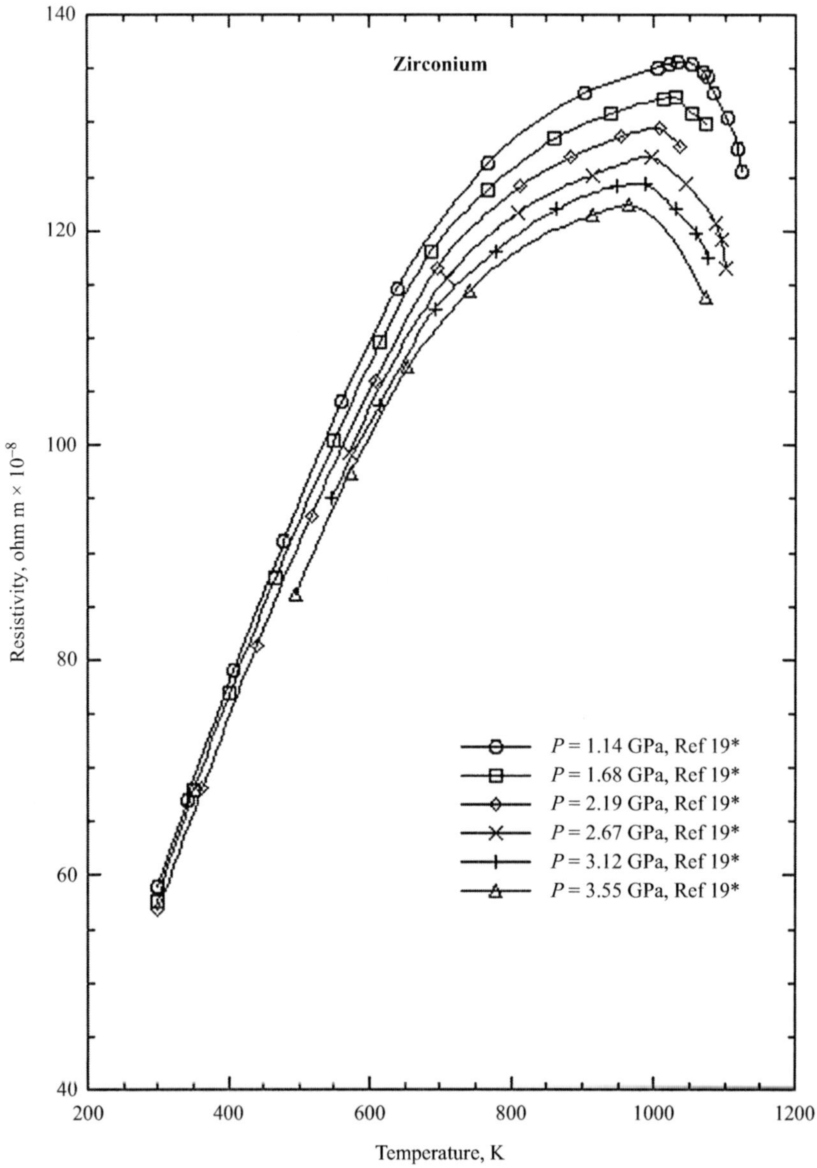
Zirconium
Resistivity, ohm m × 10−8
Temperature, K
140
120
100
80
60
40
200
400
600
800
1000
1200
P = 1.14 GPa, Ref 19*
P = 1.68 GPa, Ref 19*
P = 2.19 GPa, Ref 19*
P = 2.67 GPa, Ref 19*
P = 3.12 GPa, Ref 19*
P = 3.55 GPa, Ref 19*

6.3 Resistance measurements as a function of temperature and pressure

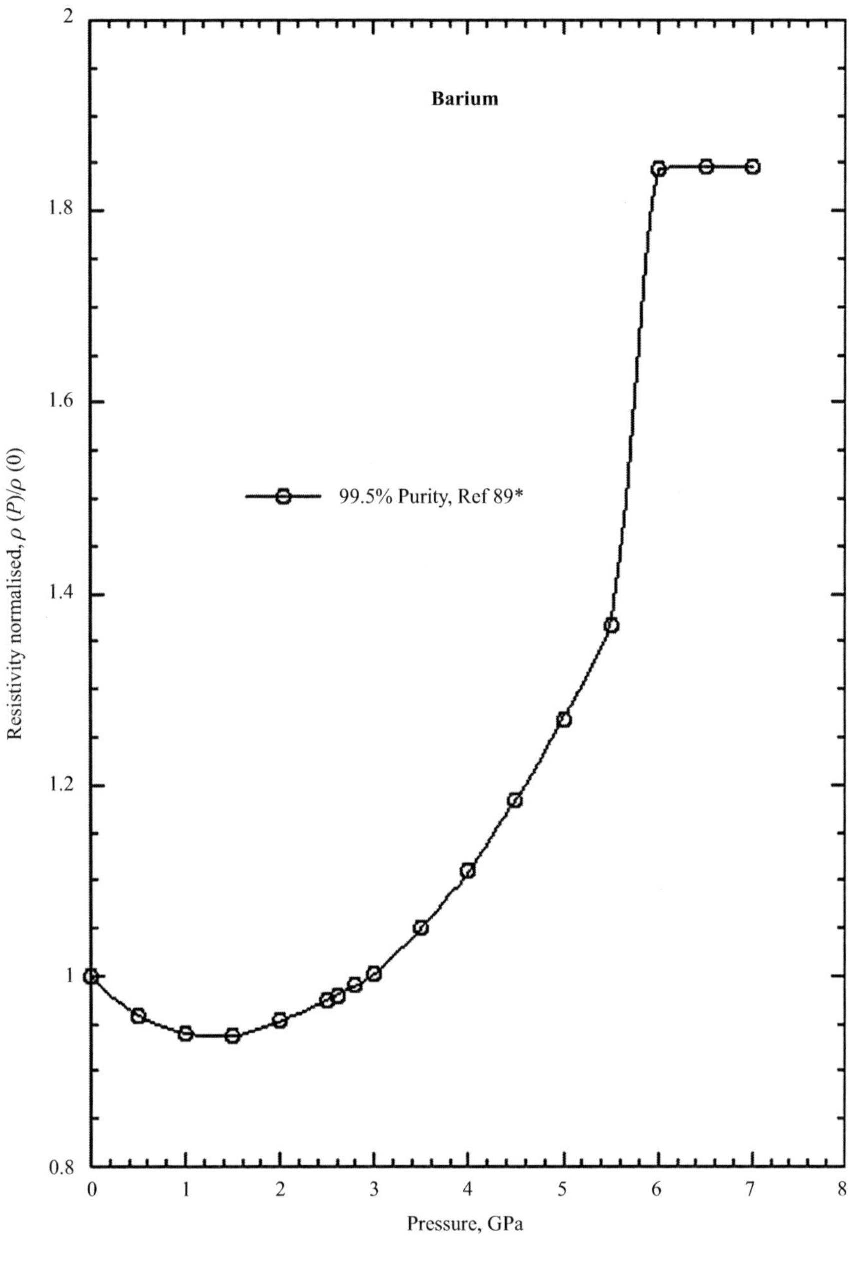

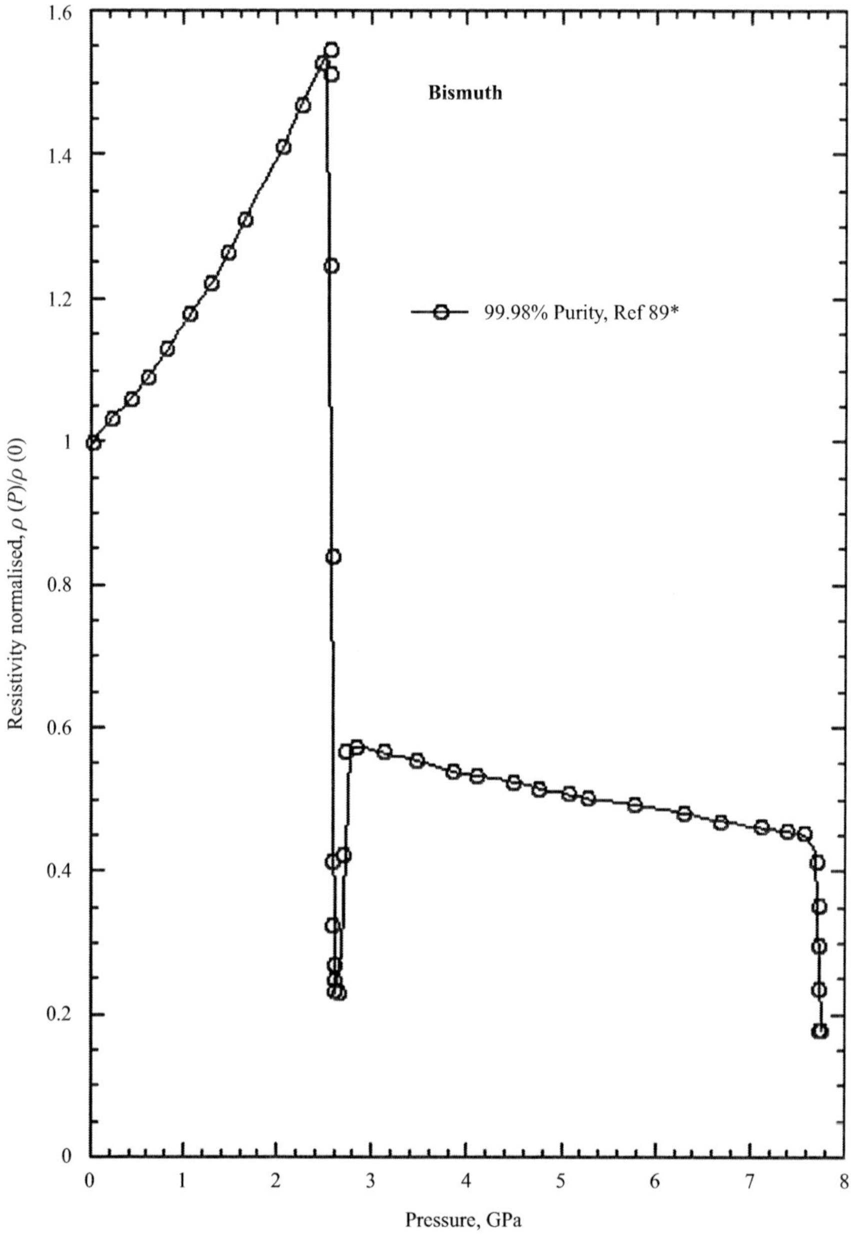
Bismuth
99.98% Purity, Ref 89*
Resistivity normalised, ρ (P)/ρ (0)
Pressure, GPa
0
0.2
0.4
0.6
0.8
1
1.2
1.4
1.6
0
1
2
3
4
5
6
7
8

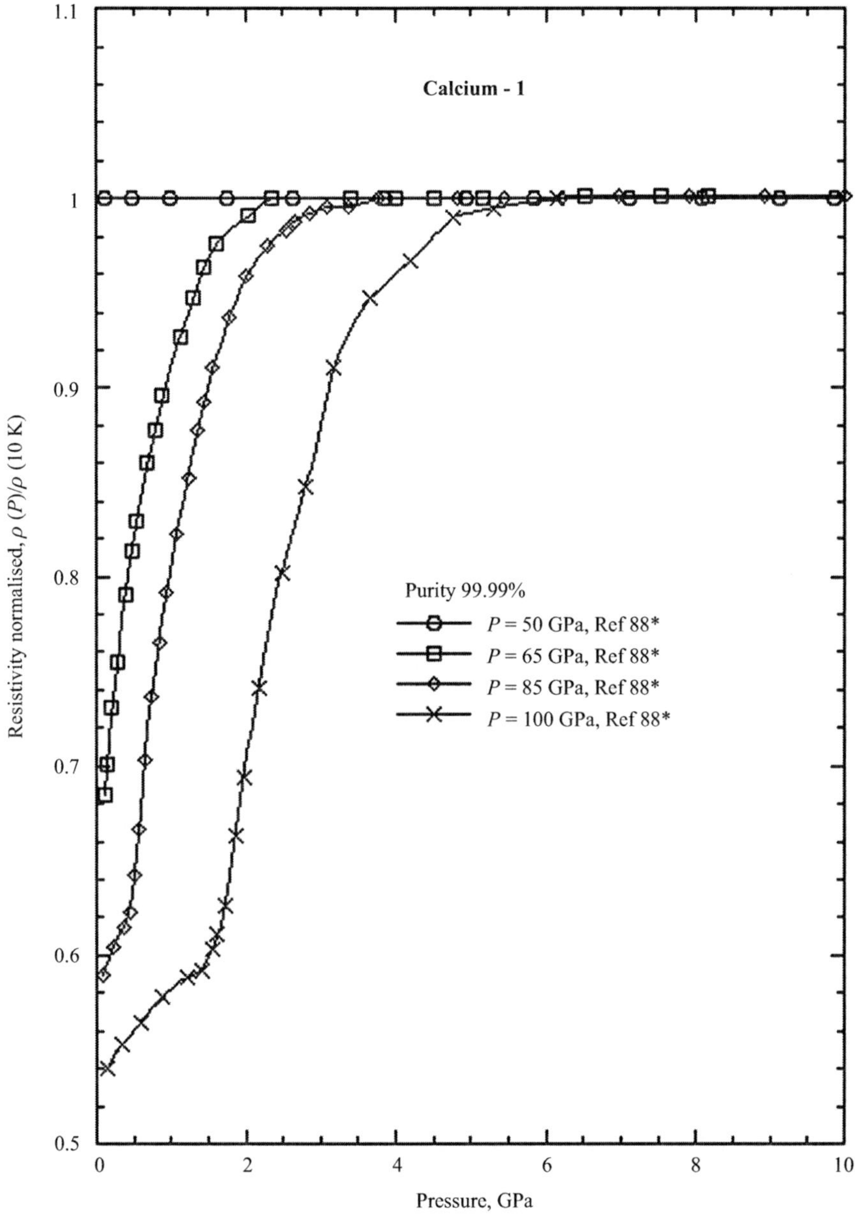
Calcium - 1
Purity 99.99%
P = 50 GPa, Ref 88*
P = 65 GPa, Ref 88*
P = 85 GPa, Ref 88*
P = 100 GPa, Ref 88*
Resistivity normalised, ρ (P)/ρ (10 K)
Pressure, GPa
1.1
1
0.9
0.8
0.7
0.6
0.5
0
2
4
6
8
10

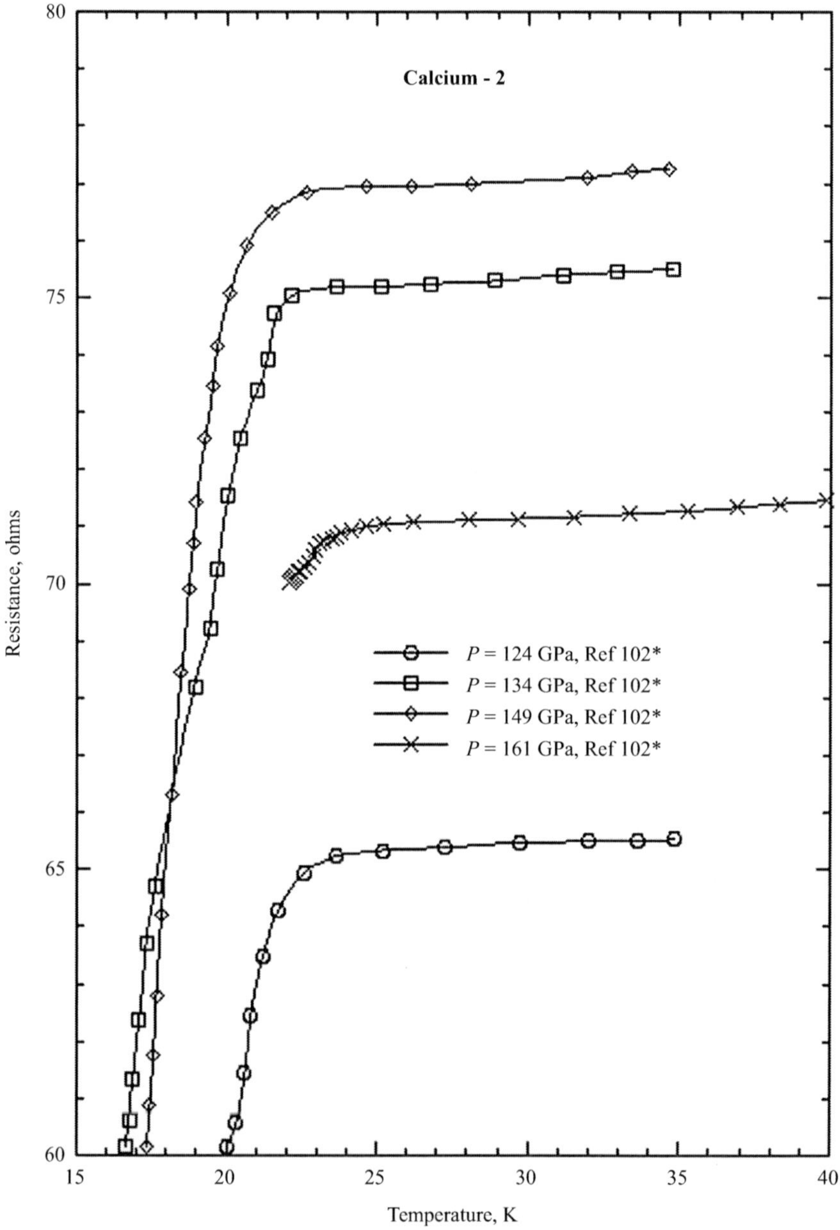
Calcium - 2
Resistance, ohms
Temperature, K
80
75
70
65
60
15
20
25
30
35
40
P = 124 GPa, Ref 102*
P = 134 GPa, Ref 102*
P = 149 GPa, Ref 102*
P = 161 GPa, Ref 102*

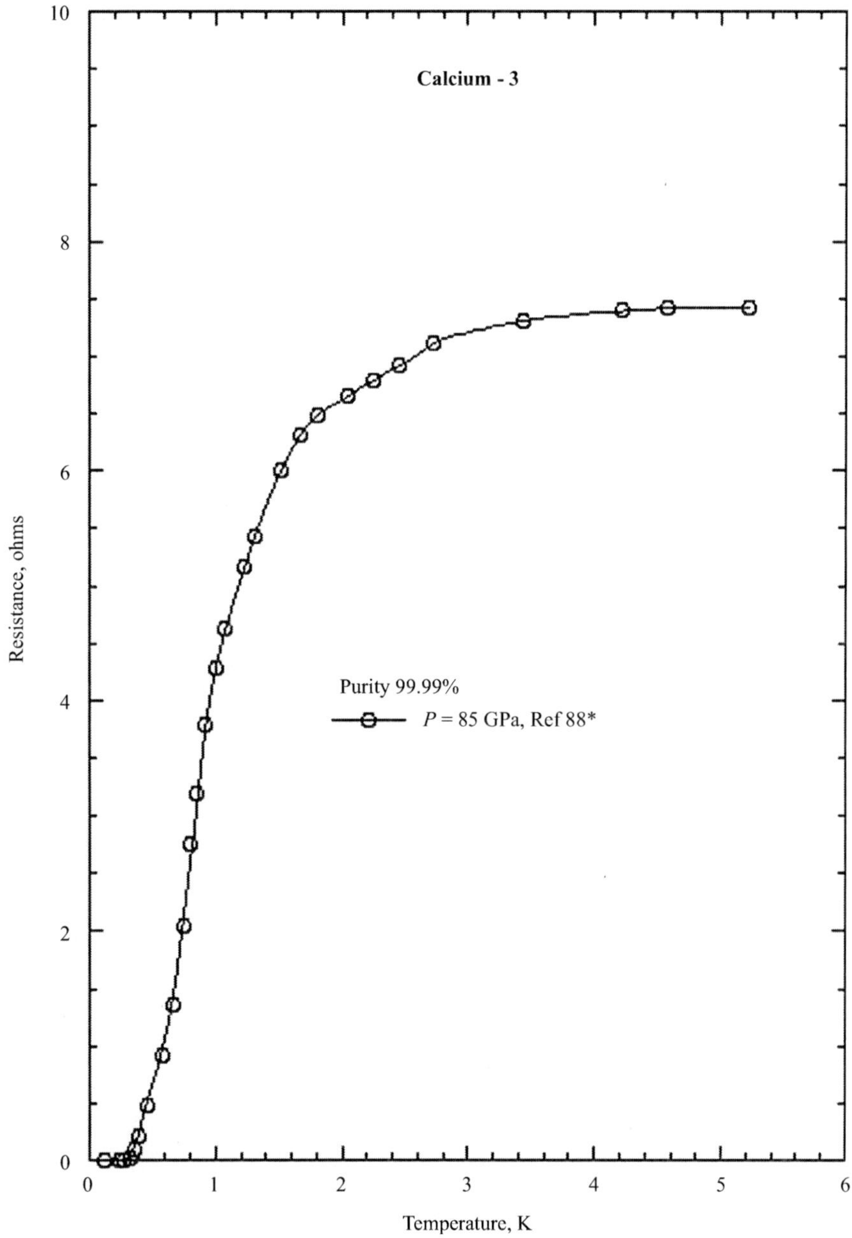
Calcium - 3
Purity 99.99%
P = 85 GPa, Ref 88*
Resistance, ohms
Temperature, K
10
8
6
4
2
0
0
1
2
3
4
5
6

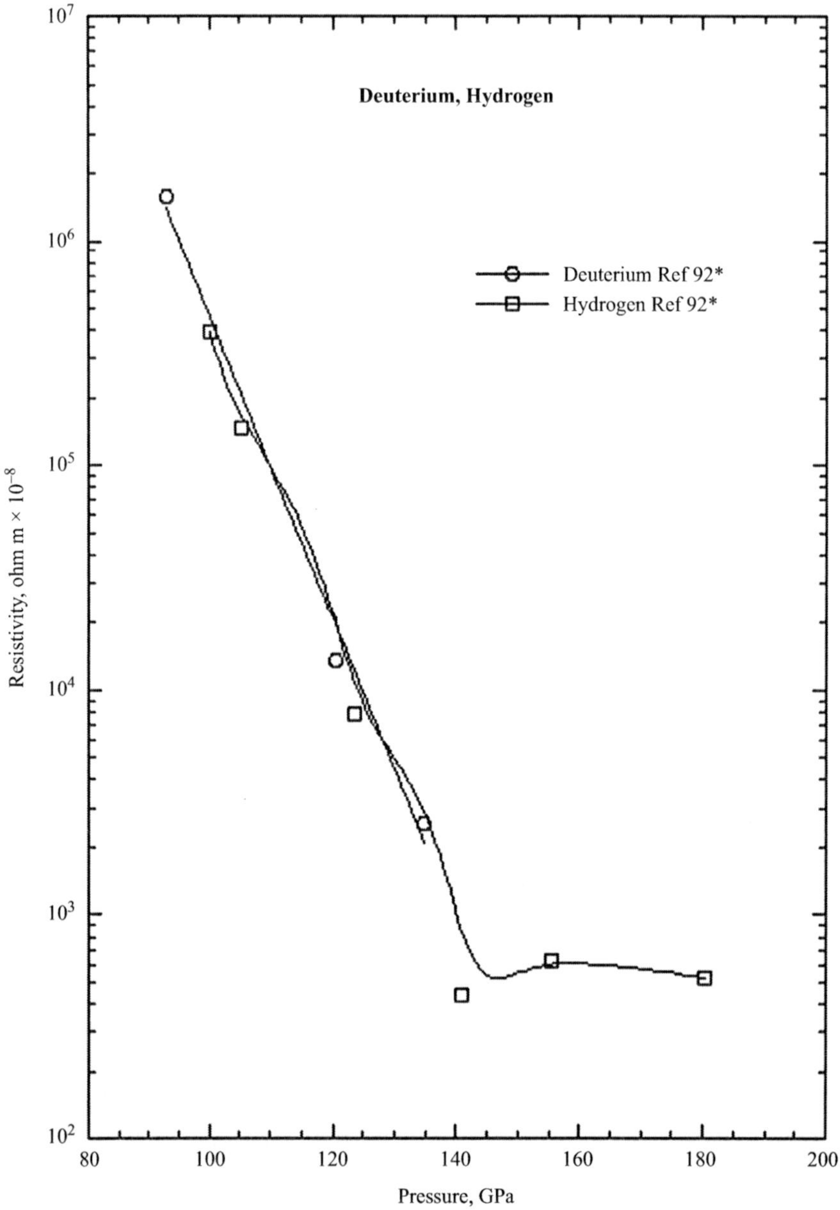

Deuterium, Hydrogen
Deuterium Ref 92*
Hydrogen Ref 92*
Resistivity, ohm m × 10⁻⁸
Pressure, GPa

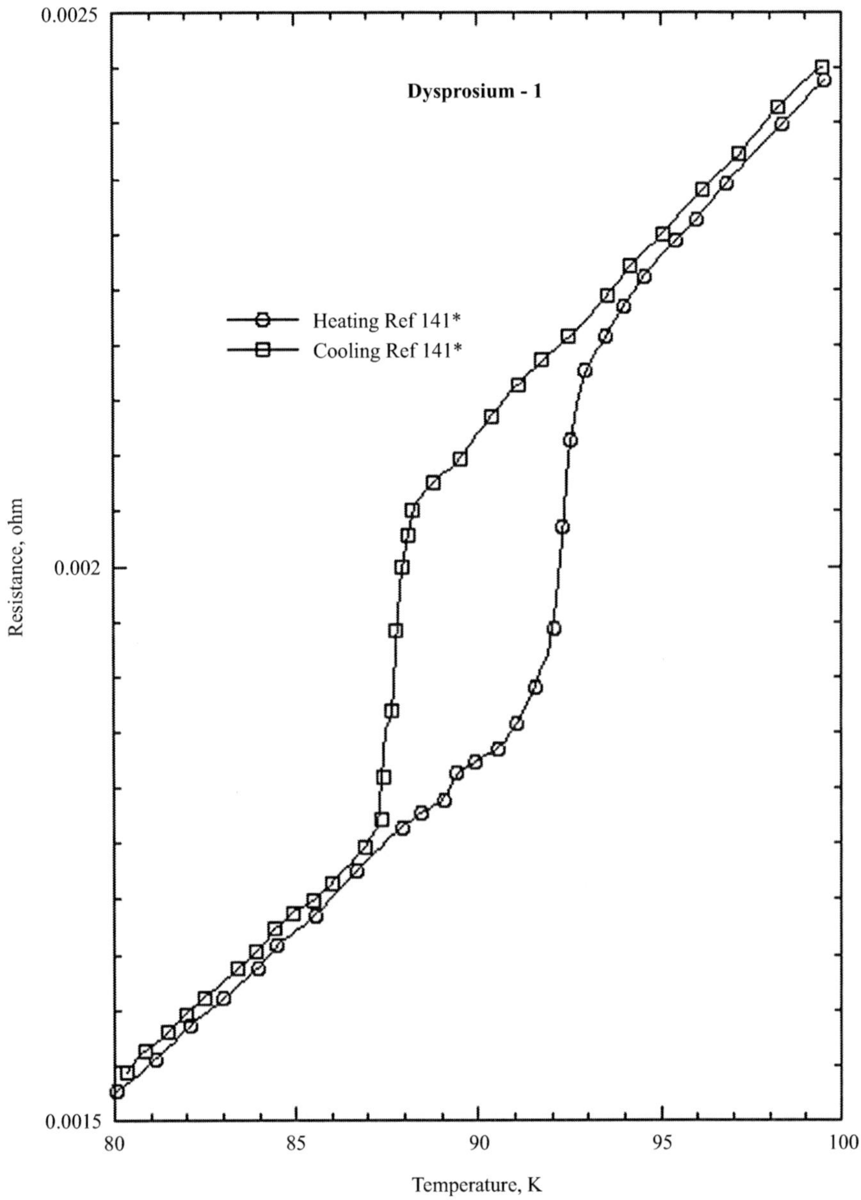
Dysprosium - 1
Heating Ref 141*
Cooling Ref 141*
0.0025
0.002
0.0015
Resistance, ohm
80
85
90
95
100
Temperature, K

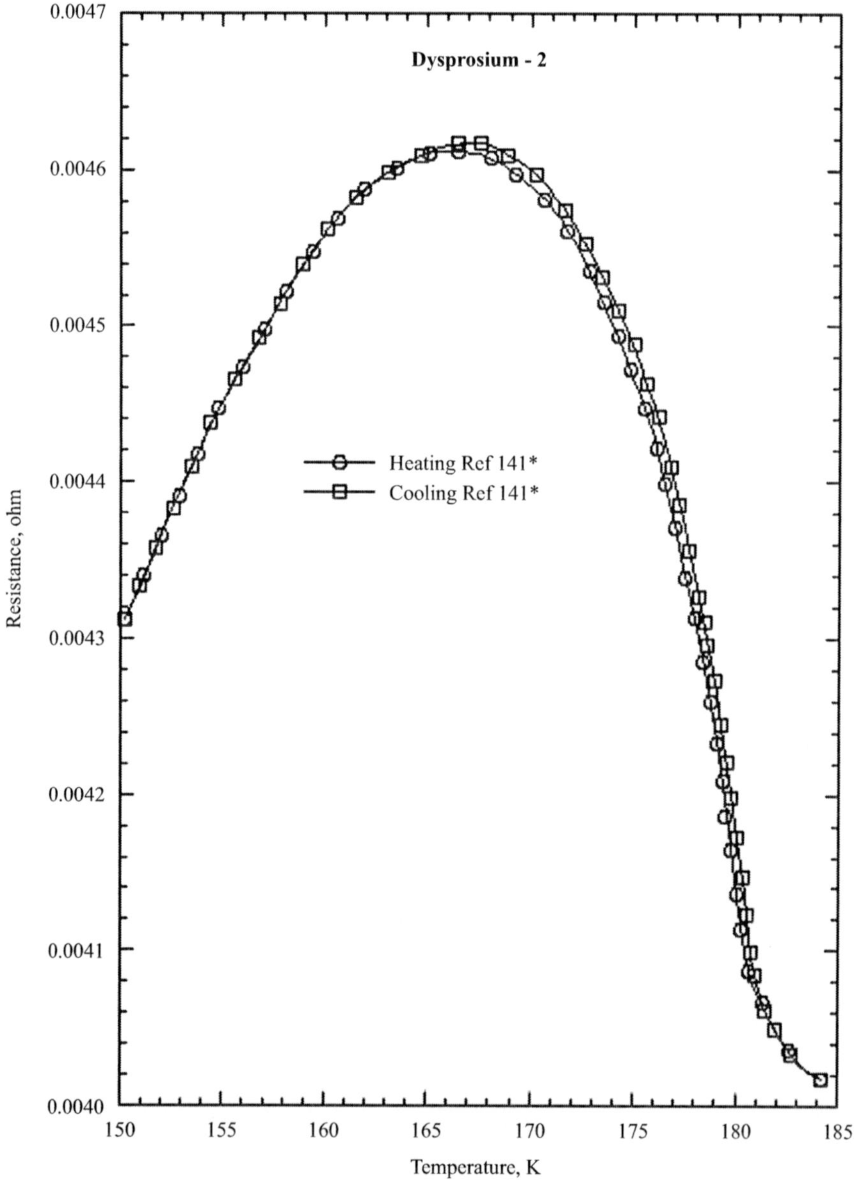
Dysprosium - 2
Heating Ref 141*
Cooling Ref 141*
Resistance, ohm
0.0047
0.0046
0.0045
0.0044
0.0043
0.0042
0.0041
0.0040
150
155
160
165
170
175
180
185
Temperature, K

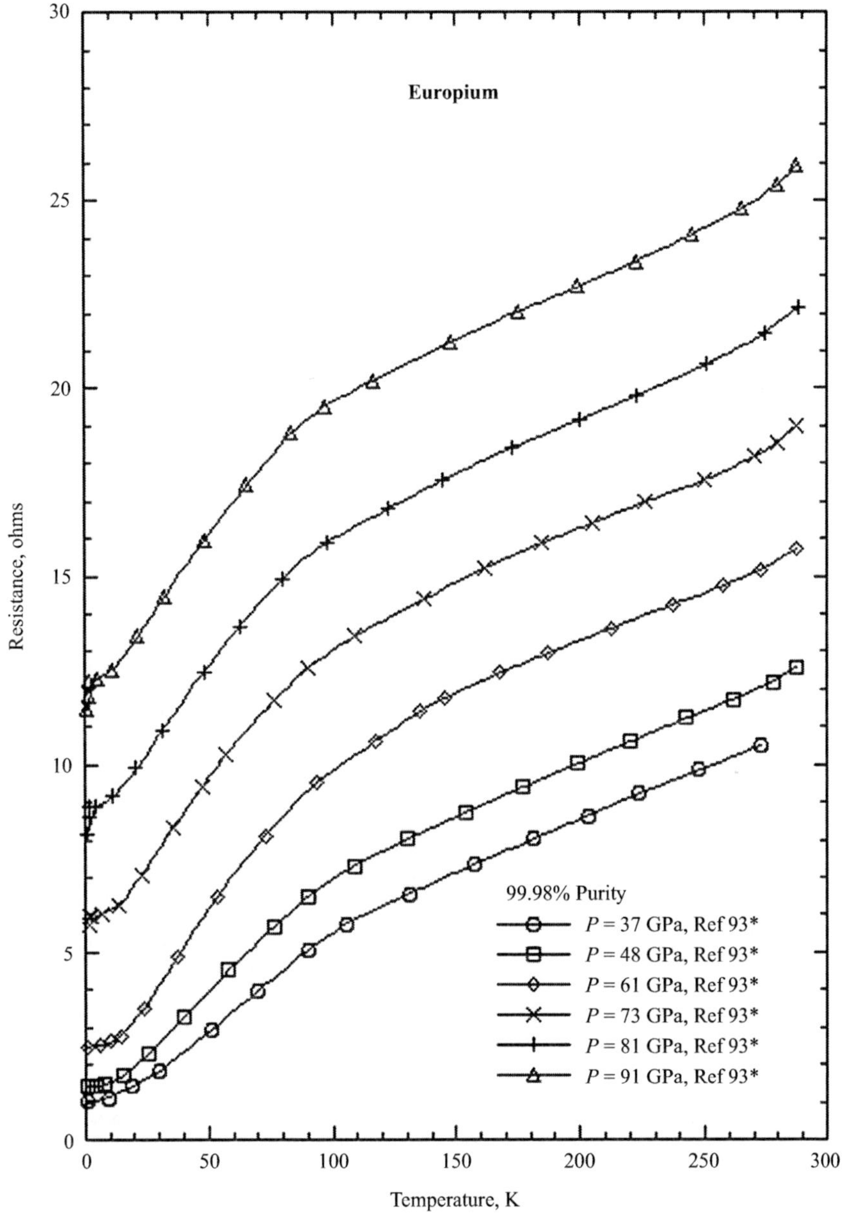
Europium
Resistance, ohms
Temperature, K
0
5
10
15
20
25
30
0
50
100
150
200
250
300
99.98% Purity
P = 37 GPa, Ref 93*
P = 48 GPa, Ref 93*
P = 61 GPa, Ref 93*
P = 73 GPa, Ref 93*
P = 81 GPa, Ref 93*
P = 91 GPa, Ref 93*

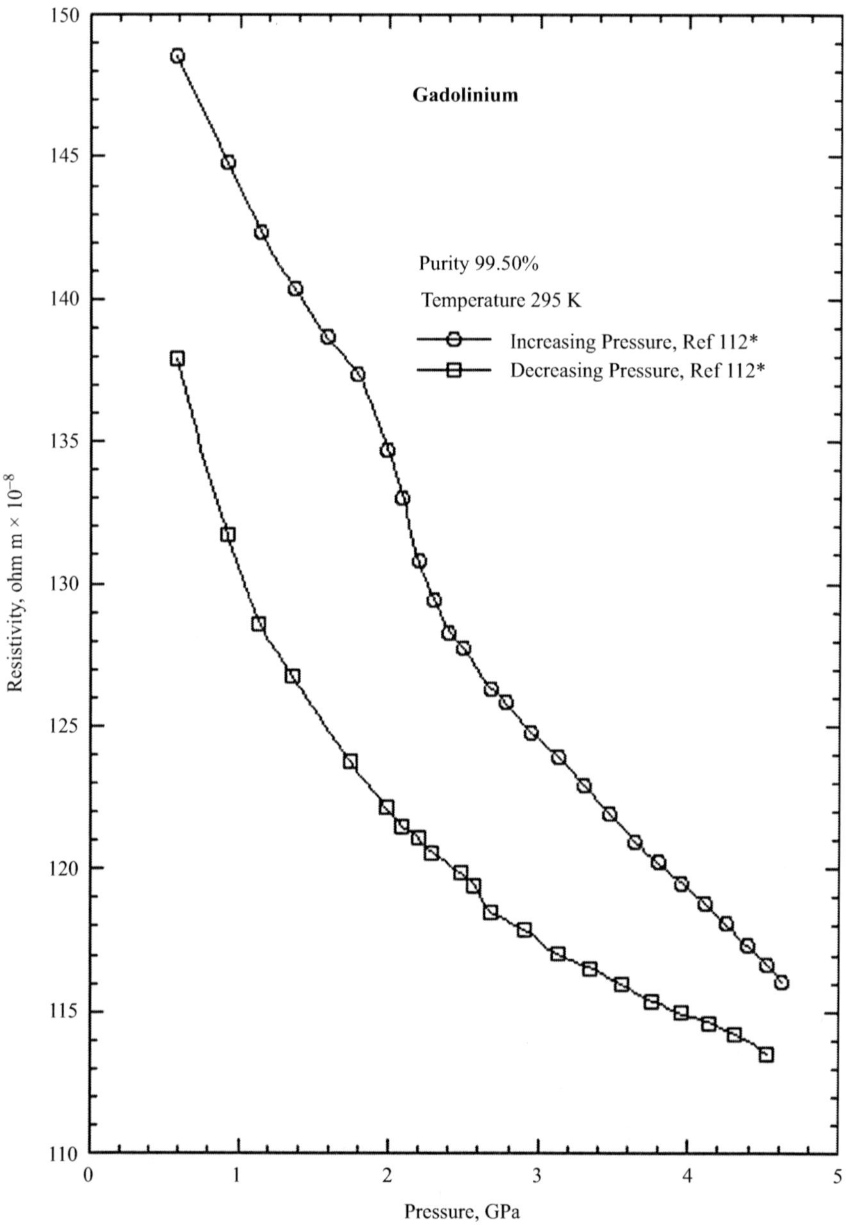
Gadolinium
Purity 99.50%
Temperature 295 K
Increasing Pressure, Ref 112*
Decreasing Pressure, Ref 112*
Resistivity, ohm m × 10⁻⁸
Pressure, GPa
150
145
140
135
130
125
120
115
110
0
1
2
3
4
5

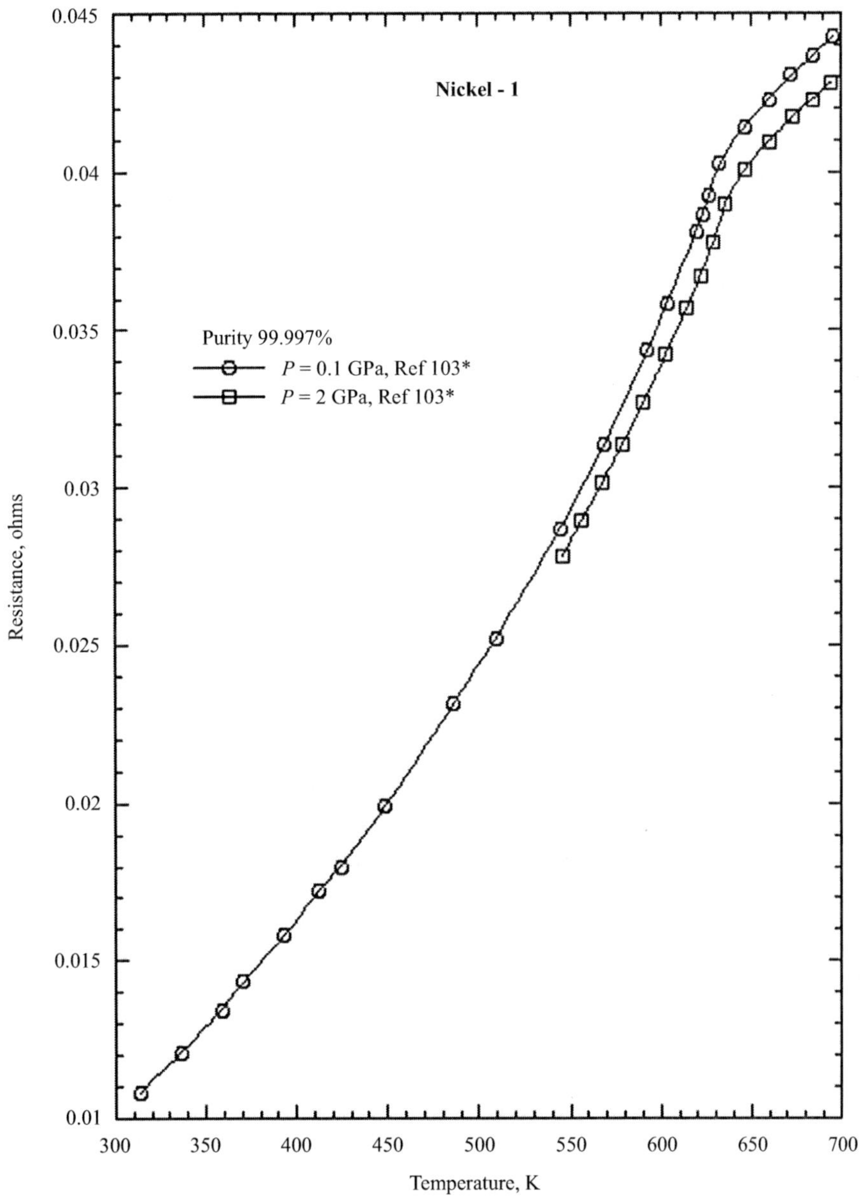
Nickel - 1
Purity 99.997%
P = 0.1 GPa, Ref 103*
P = 2 GPa, Ref 103*
Resistance, ohms
Temperature, K
0.045
0.04
0.035
0.03
0.025
0.02
0.015
0.01
300
350
400
450
500
550
600
650
700

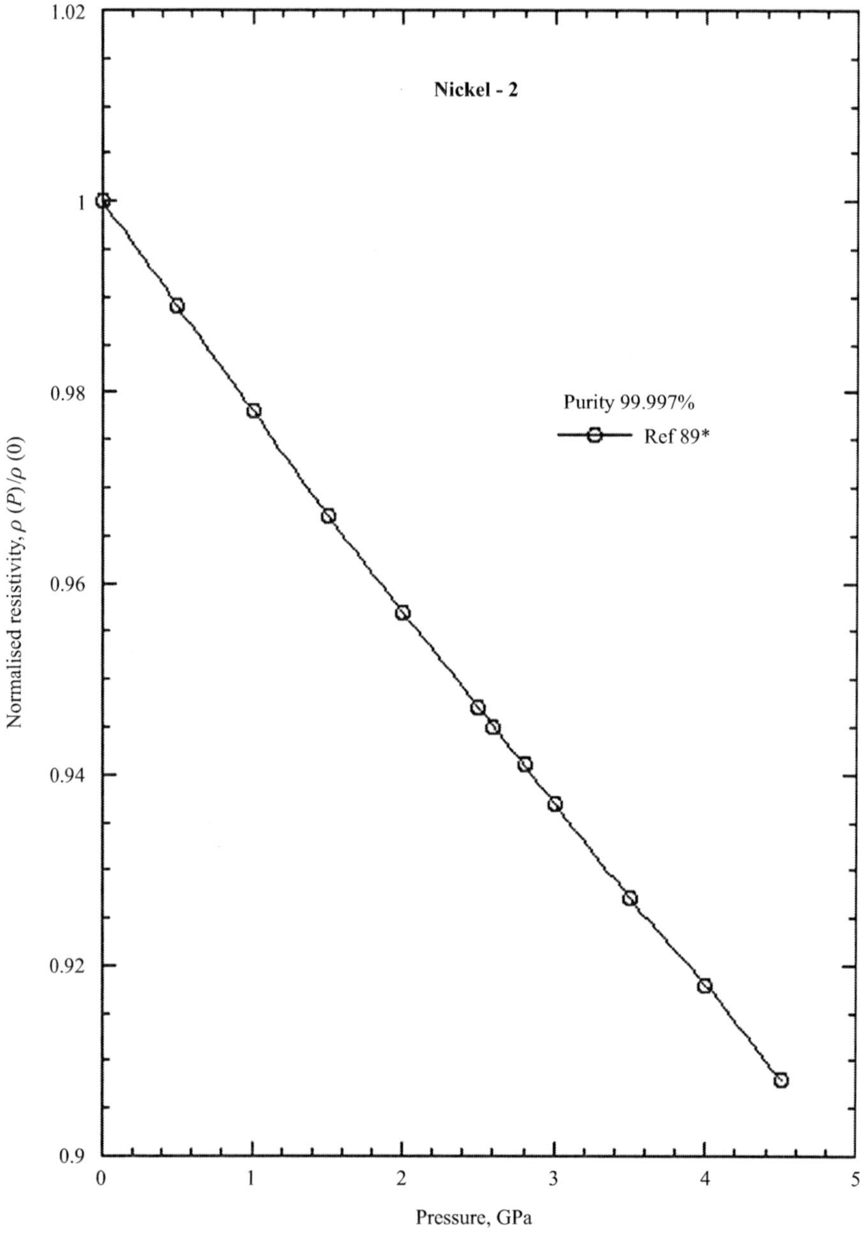
Nickel - 2
Purity 99.997%
Ref 89*
Normalised resistivity, ρ (P)/ρ (0)
1.02
1
0.98
0.96
0.94
0.92
0.9
0
1
2
3
4
5
Pressure, GPa

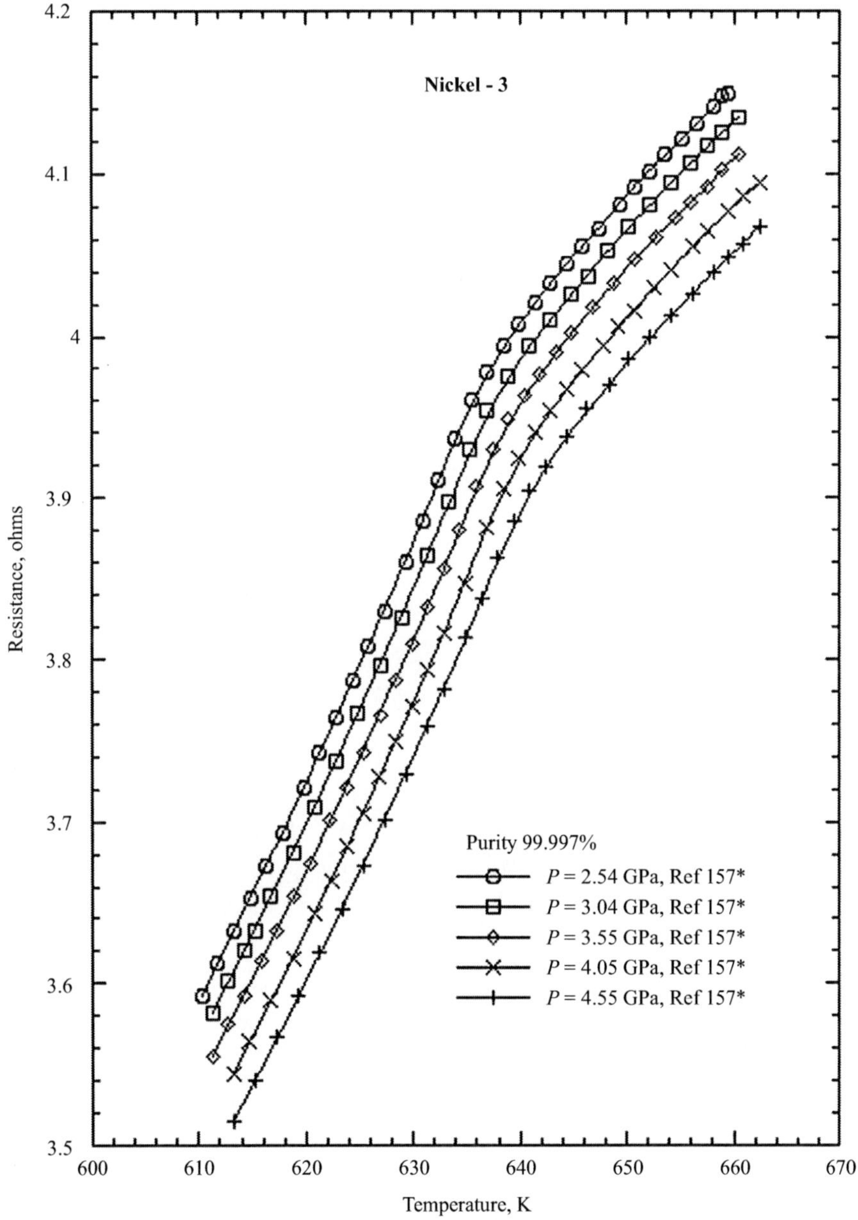
Nickel - 3
Resistance, ohms
Temperature, K
4.2
4.1
4
3.9
3.8
3.7
3.6
3.5
600
610
620
630
640
650
660
670
Purity 99.997%
P = 2.54 GPa, Ref 157*
P = 3.04 GPa, Ref 157*
P = 3.55 GPa, Ref 157*
P = 4.05 GPa, Ref 157*
P = 4.55 GPa, Ref 157*

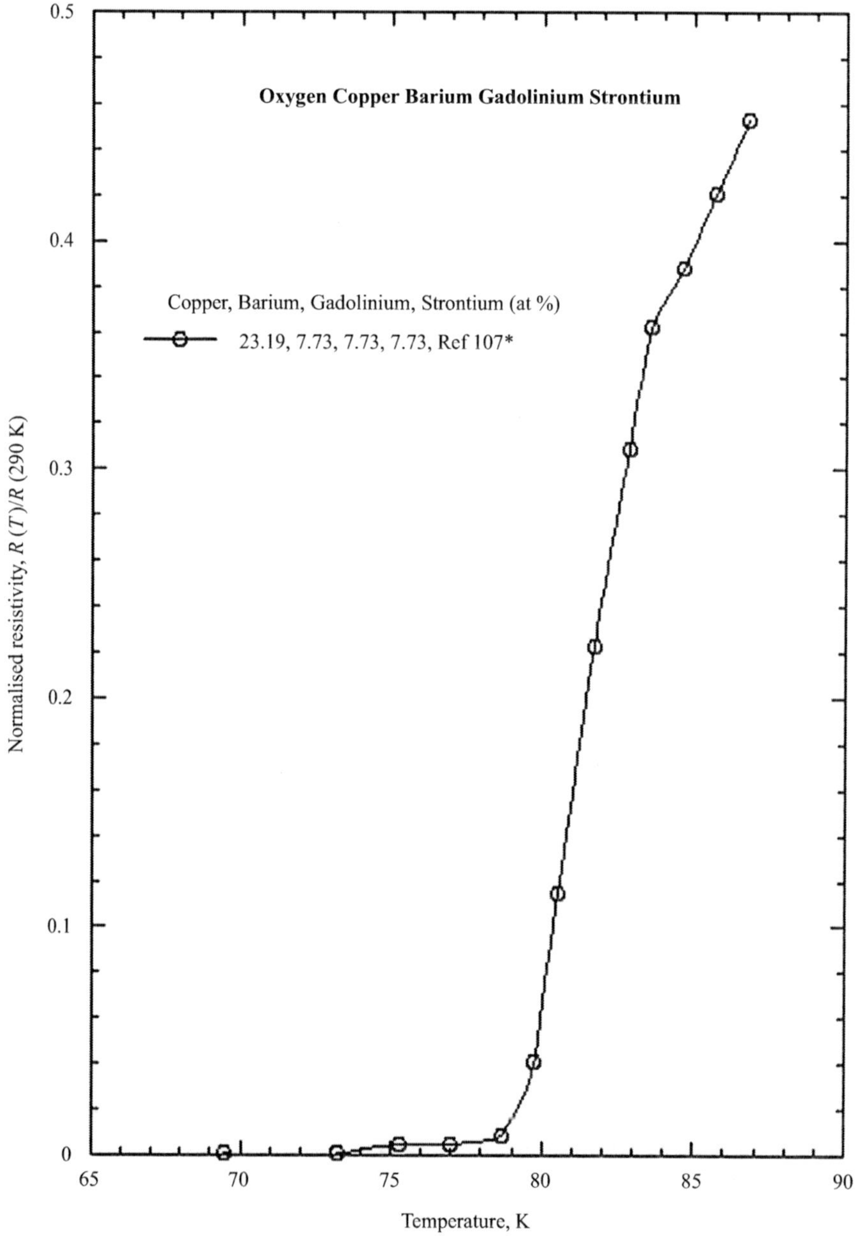
Oxygen Copper Barium Gadolinium Strontium
Copper, Barium, Gadolinium, Strontium (at %)
23.19, 7.73, 7.73, 7.73, Ref 107*
Normalised resistivity, R (T)/R (290 K)
0.5
0.4
0.3
0.2
0.1
0
65
70
75
80
85
90
Temperature, K

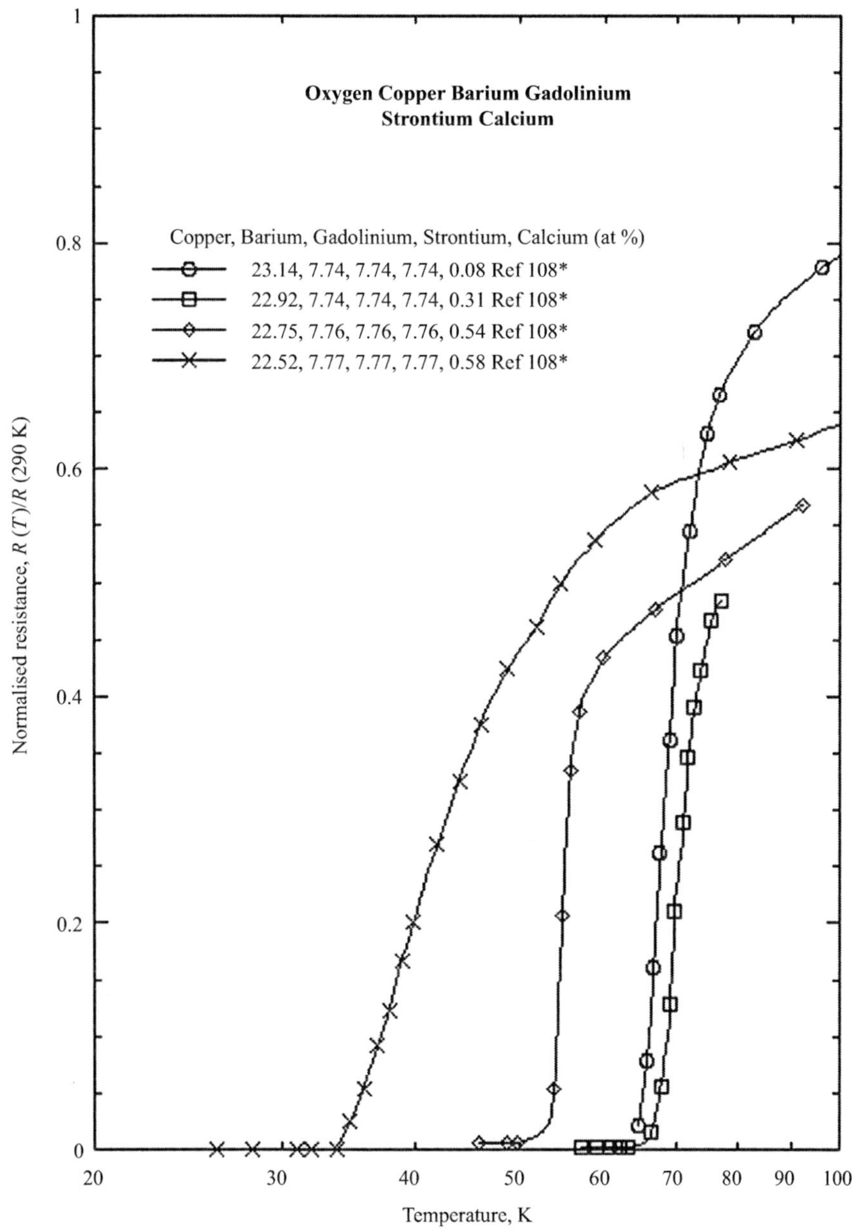
Oxygen Copper Barium Gadolinium
Strontium Calcium
Copper, Barium, Gadolinium, Strontium, Calcium (at %)
23.14, 7.74, 7.74, 7.74, 0.08 Ref 108*
22.92, 7.74, 7.74, 7.74, 0.31 Ref 108*
22.75, 7.76, 7.76, 7.76, 0.54 Ref 108*
22.52, 7.77, 7.77, 7.77, 0.58 Ref 108*
Normalised resistance, R (T)/R (290 K)
Temperature, K
0
0.2
0.4
0.6
0.8
1
20
30
40
50
60
70
80
90
100

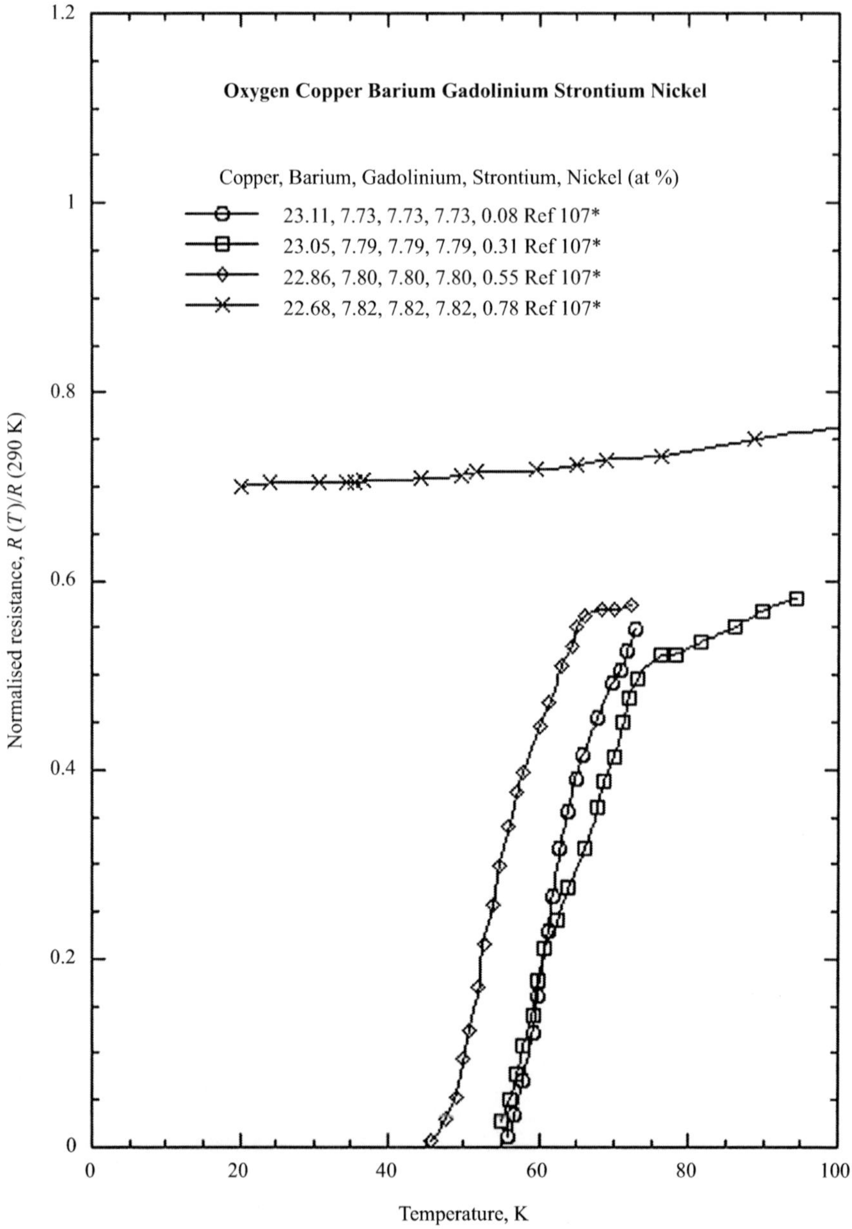
Oxygen Copper Barium Gadolinium Strontium Nickel
Copper, Barium, Gadolinium, Strontium, Nickel (at %)
23.11, 7.73, 7.73, 7.73, 0.08 Ref 107*
23.05, 7.79, 7.79, 7.79, 0.31 Ref 107*
22.86, 7.80, 7.80, 7.80, 0.55 Ref 107*
22.68, 7.82, 7.82, 7.82, 0.78 Ref 107*
Normalised resistance, R (T)/R (290 K)
Temperature, K
1.2
1
0.8
0.6
0.4
0.2
0
0
20
40
60
80
100

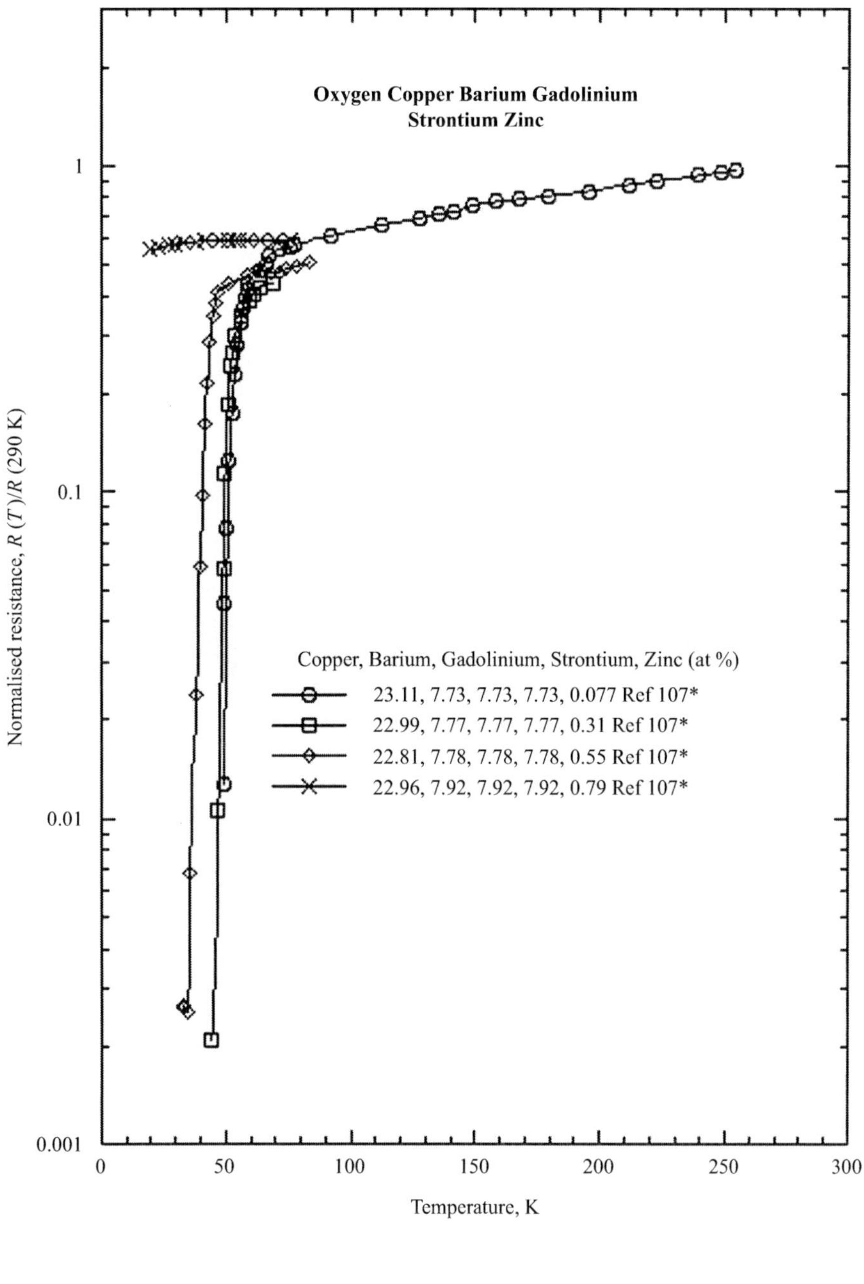
Oxygen Copper Barium Gadolinium
Strontium Zinc
Copper, Barium, Gadolinium, Strontium, Zinc (at %)
23.11, 7.73, 7.73, 7.73, 0.077 Ref 107*
22.99, 7.77, 7.77, 7.77, 0.31 Ref 107*
22.81, 7.78, 7.78, 7.78, 0.55 Ref 107*
22.96, 7.92, 7.92, 7.92, 0.79 Ref 107*
Normalised resistance, R (T)/R (290 K)
Temperature, K
1
0.1
0.01
0.001
0
50
100
150
200
250
300

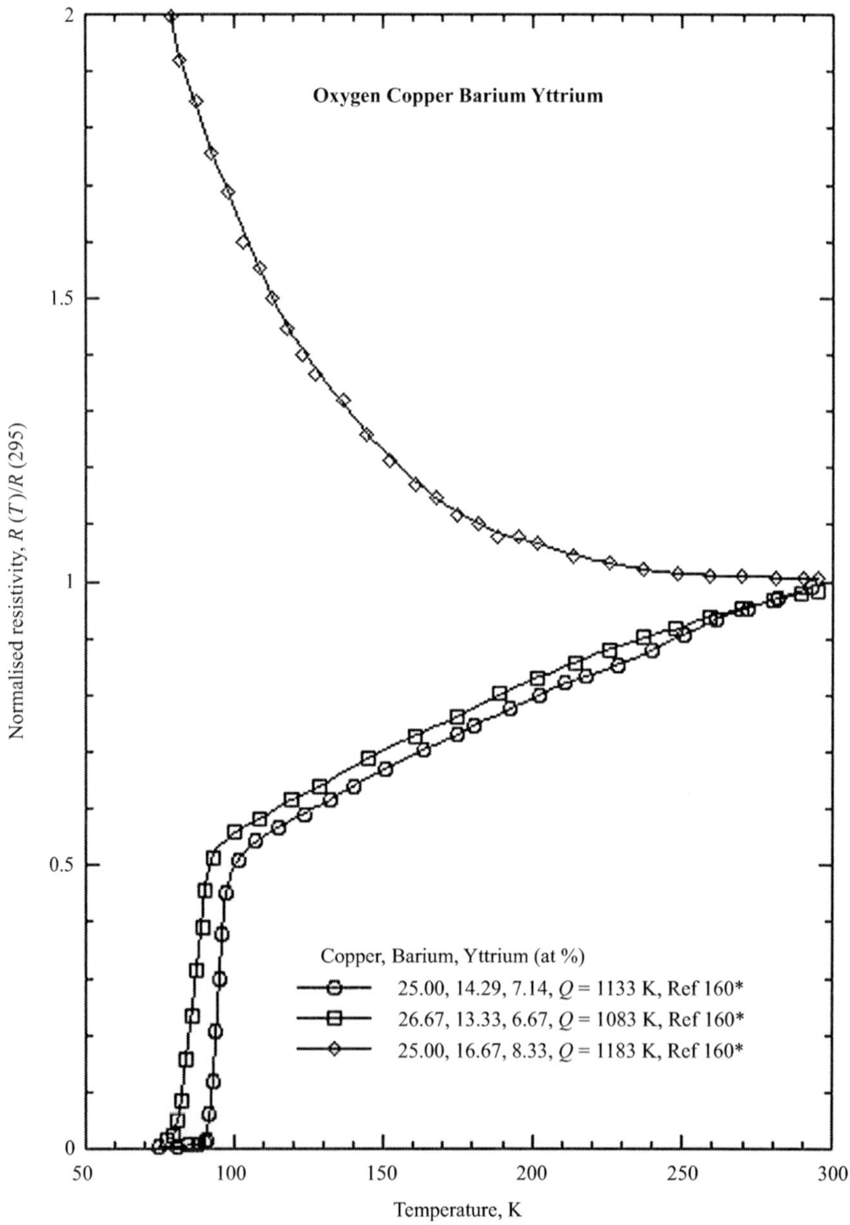

Oxygen Copper Barium Yttrium
Normalised resistivity, R (T)/R (295)
Temperature, K
Copper, Barium, Yttrium (at %)
25.00, 14.29, 7.14, Q = 1133 K, Ref 160*
26.67, 13.33, 6.67, Q = 1083 K, Ref 160*
25.00, 16.67, 8.33, Q = 1183 K, Ref 160*

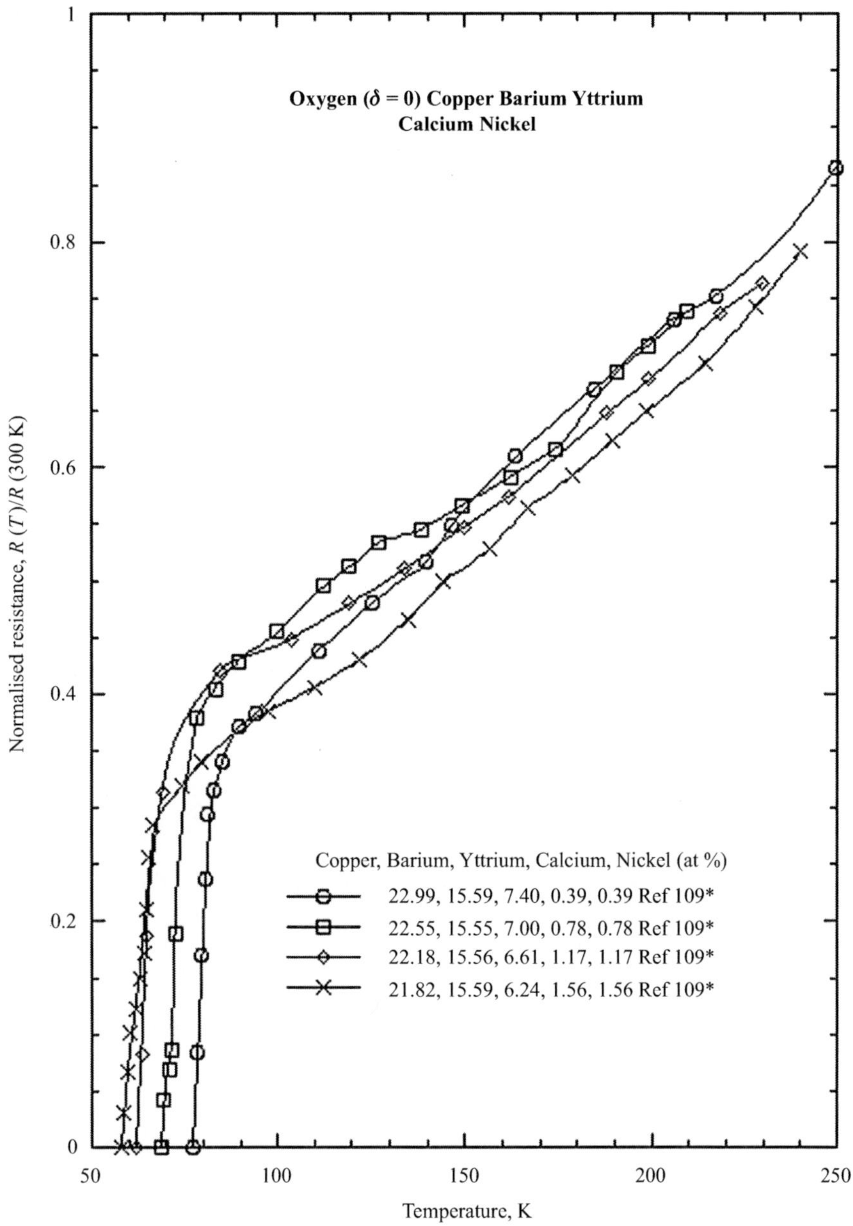
Oxygen (δ = 0) Copper Barium Yttrium
Calcium Nickel
Copper, Barium, Yttrium, Calcium, Nickel (at %)
22.99, 15.59, 7.40, 0.39, 0.39 Ref 109*
22.55, 15.55, 7.00, 0.78, 0.78 Ref 109*
22.18, 15.56, 6.61, 1.17, 1.17 Ref 109*
21.82, 15.59, 6.24, 1.56, 1.56 Ref 109*
Normalised resistance, R (T)/R (300 K)
Temperature, K
1
0.8
0.6
0.4
0.2
0
50
100
150
200
250

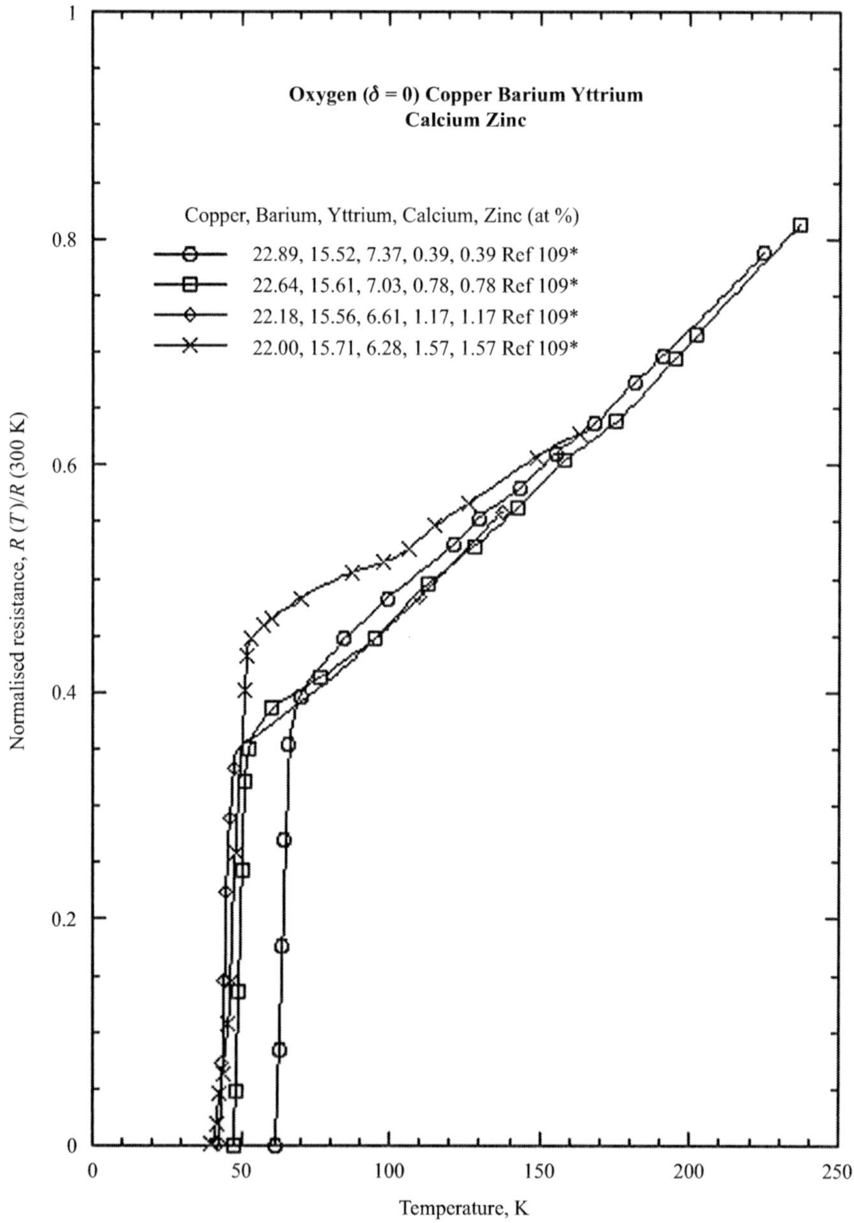
Oxygen (δ = 0) Copper Barium Yttrium
Calcium Zinc
Copper, Barium, Yttrium, Calcium, Zinc (at %)
22.89, 15.52, 7.37, 0.39, 0.39 Ref 109*
22.64, 15.61, 7.03, 0.78, 0.78 Ref 109*
22.18, 15.56, 6.61, 1.17, 1.17 Ref 109*
22.00, 15.71, 6.28, 1.57, 1.57 Ref 109*
Normalised resistance, R (T)/R (300 K)
Temperature, K
0
0.2
0.4
0.6
0.8
1
0
50
100
150
200
250

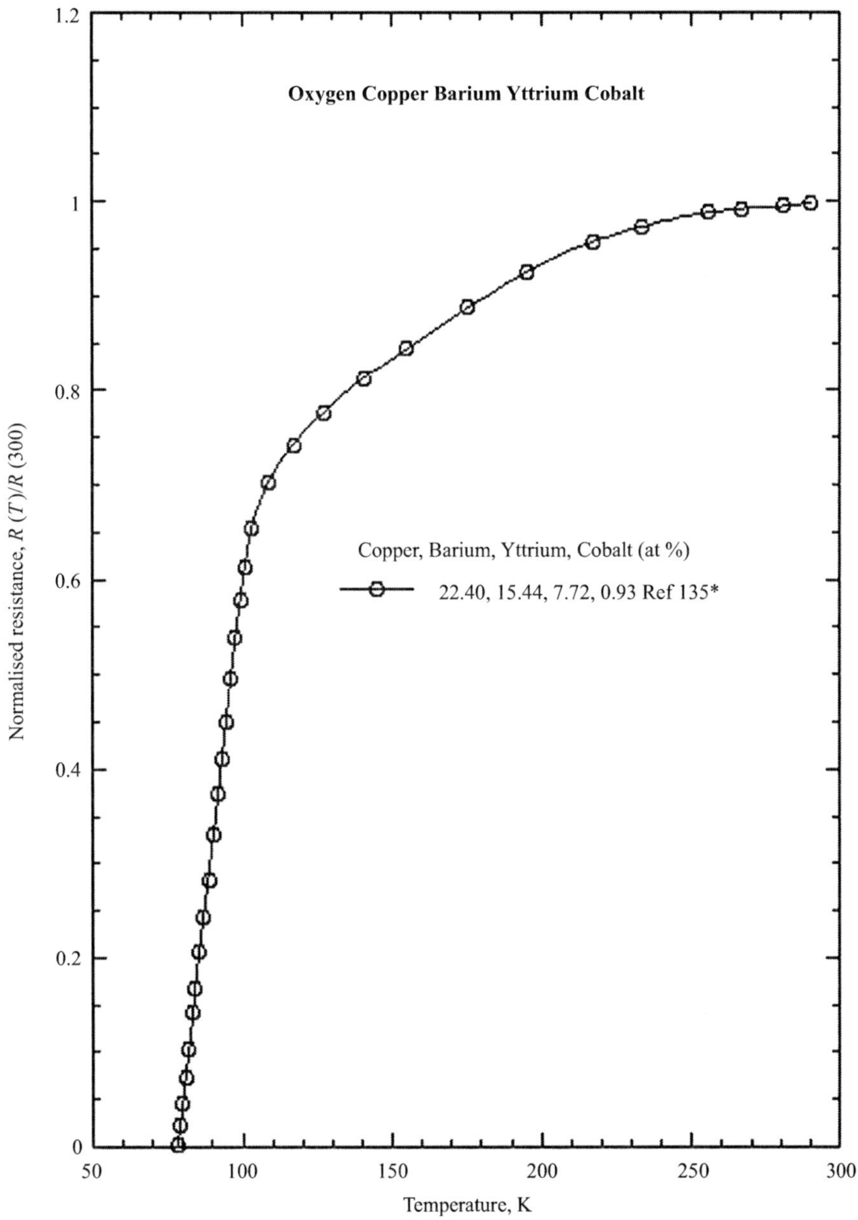

Oxygen Copper Barium Yttrium Cobalt
Copper, Barium, Yttrium, Cobalt (at %)
22.40, 15.44, 7.72, 0.93 Ref 135*
Normalised resistance, R (T)/R (300)
Temperature, K
0
0.2
0.4
0.6
0.8
1
1.2
50
100
150
200
250
300

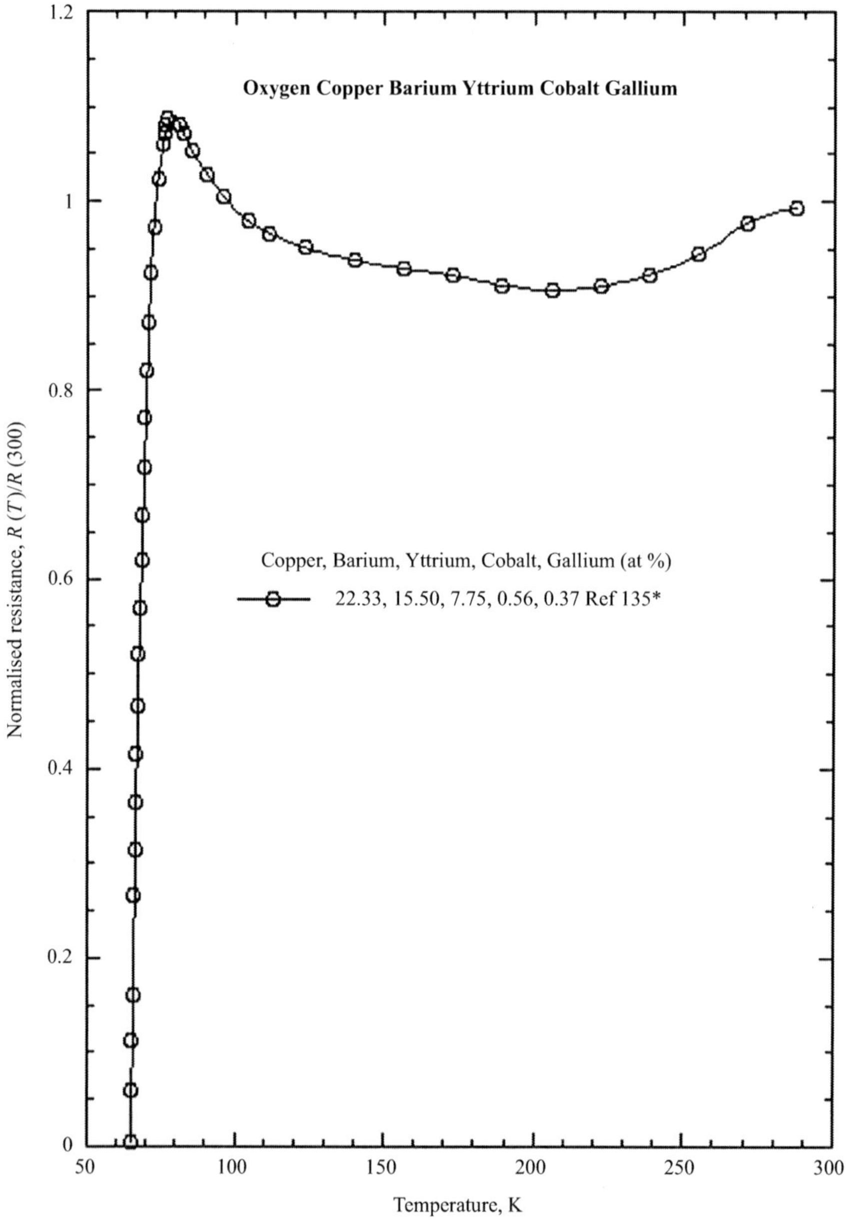
Oxygen Copper Barium Yttrium Cobalt Gallium
Copper, Barium, Yttrium, Cobalt, Gallium (at %)
22.33, 15.50, 7.75, 0.56, 0.37 Ref 135*
Normalised resistance, R (T)/R (300)
Temperature, K
1.2
1
0.8
0.6
0.4
0.2
0
50
100
150
200
250
300

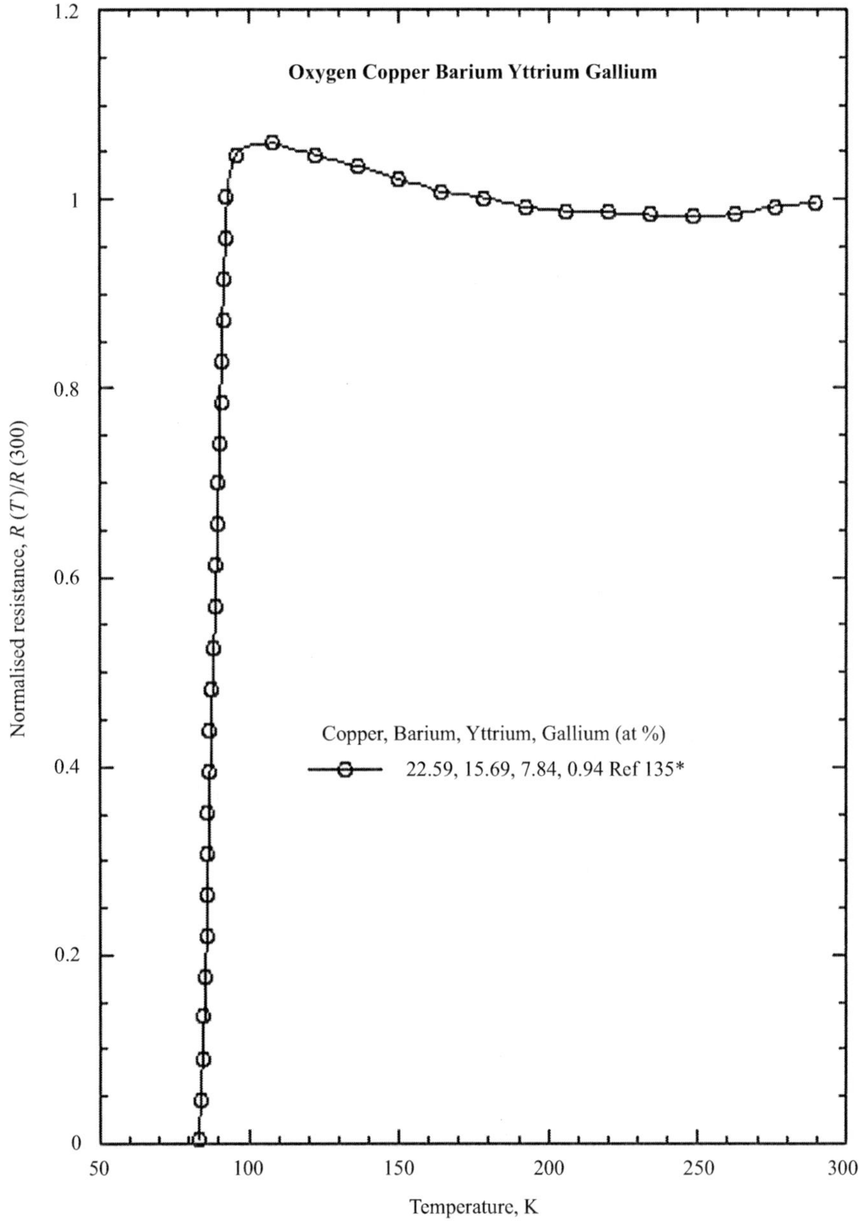
Oxygen Copper Barium Yttrium Gallium
Copper, Barium, Yttrium, Gallium (at %)
22.59, 15.69, 7.84, 0.94 Ref 135*
Normalised resistance, R (T)/R (300)
Temperature, K
0
0.2
0.4
0.6
0.8
1
1.2
50
100
150
200
250
300

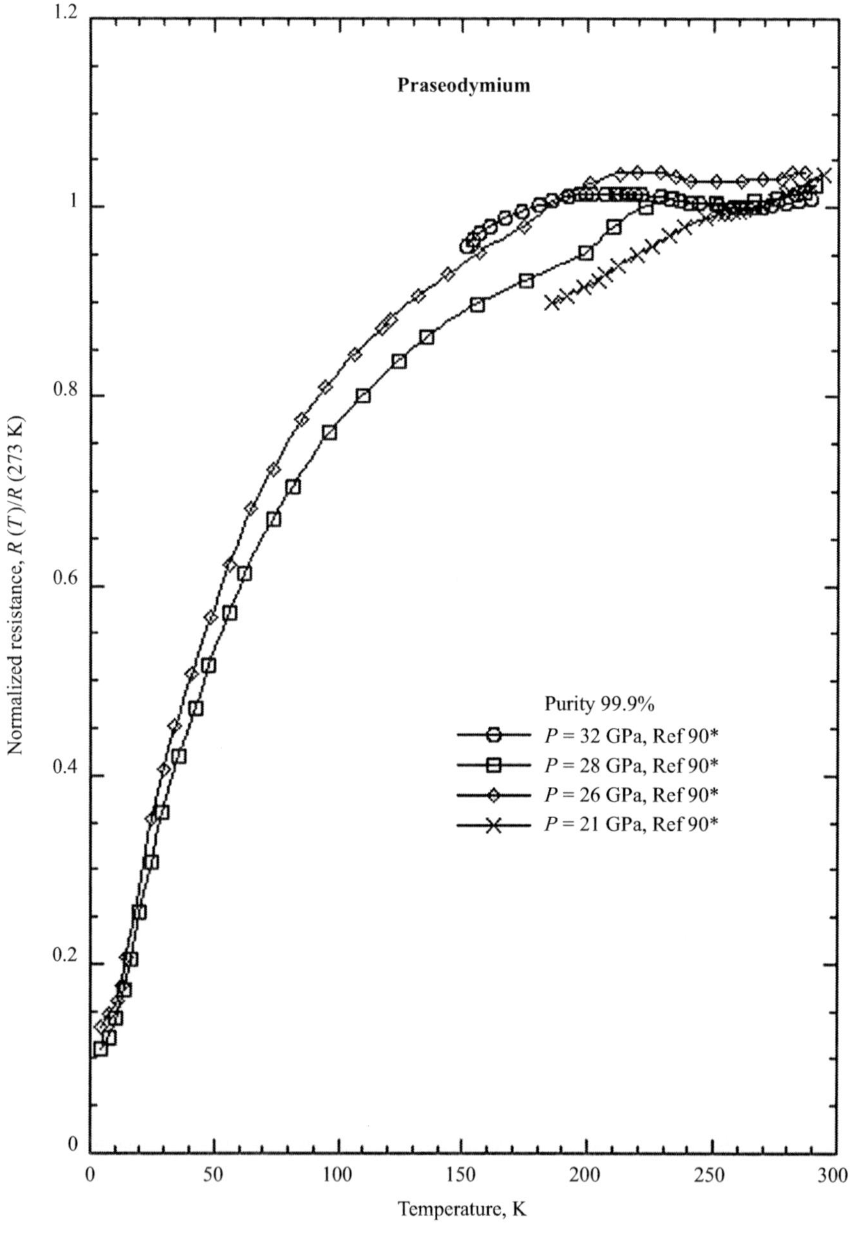
Praseodymium
Purity 99.9%
P = 32 GPa, Ref 90*
P = 28 GPa, Ref 90*
P = 26 GPa, Ref 90*
P = 21 GPa, Ref 90*
Normalized resistance, R (T)/R (273 K)
Temperature, K
0
0.2
0.4
0.6
0.8
1
1.2
0
50
100
150
200
250
300

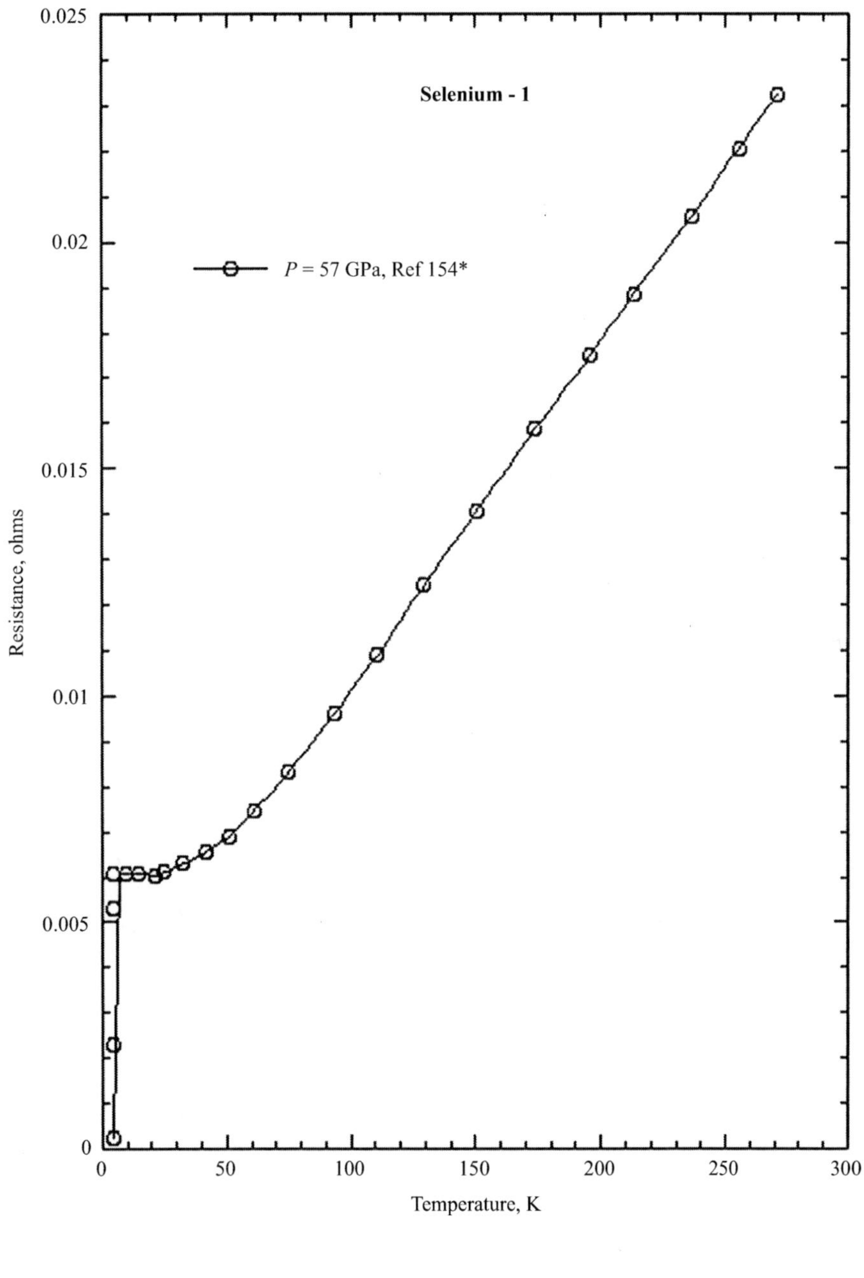
Selenium - 1
P = 57 GPa, Ref 154*
Resistance, ohms
0.025
0.02
0.015
0.01
0.005
0
Temperature, K
0
50
100
150
200
250
300

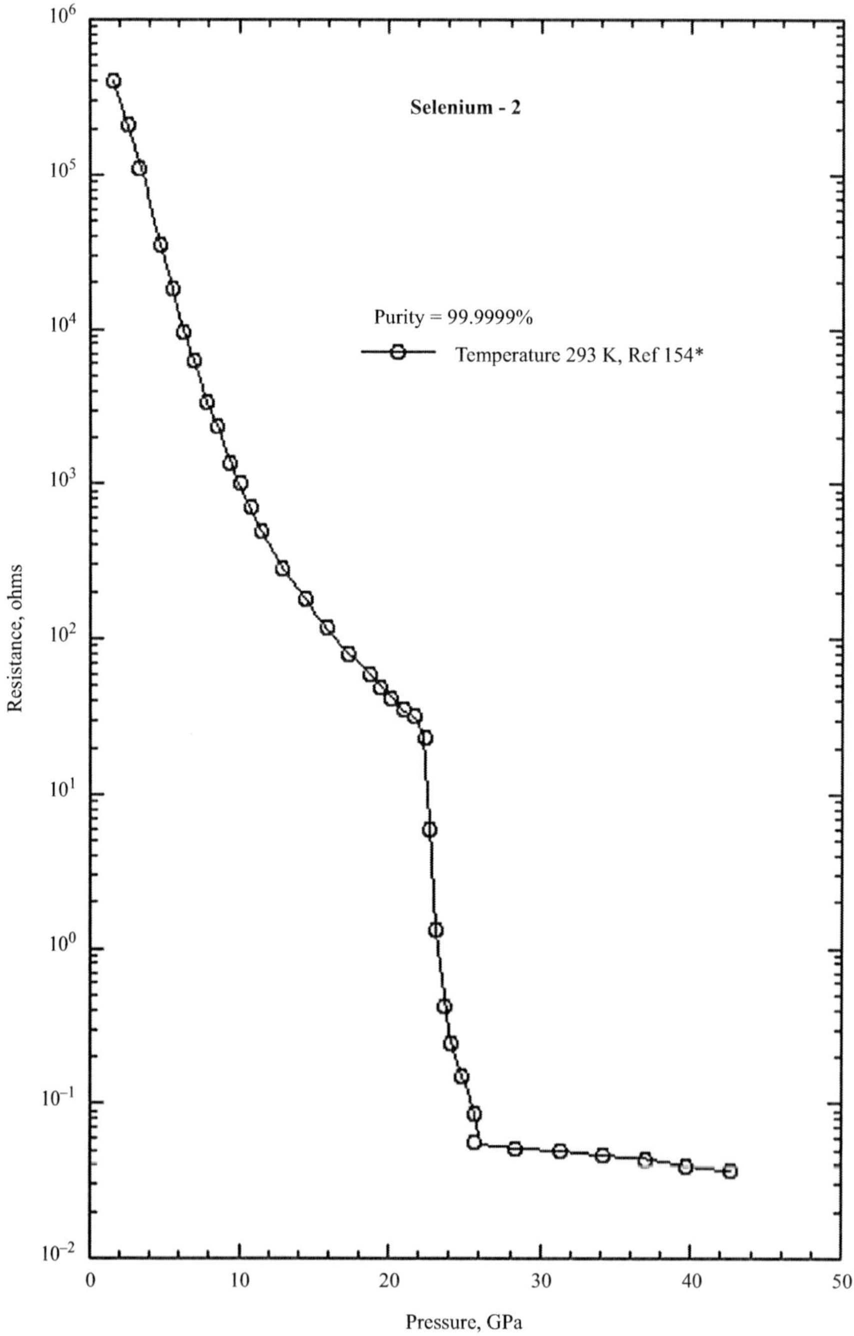
Selenium - 2
Purity = 99.9999%
Temperature 293 K, Ref 154*
Resistance, ohms
Pressure, GPa
10^{6}
10^{5}
10^{4}
10^{3}
10^{2}
10^{1}
10^{0}
10^{-1}
10^{-2}
0
10
20
30
40
50

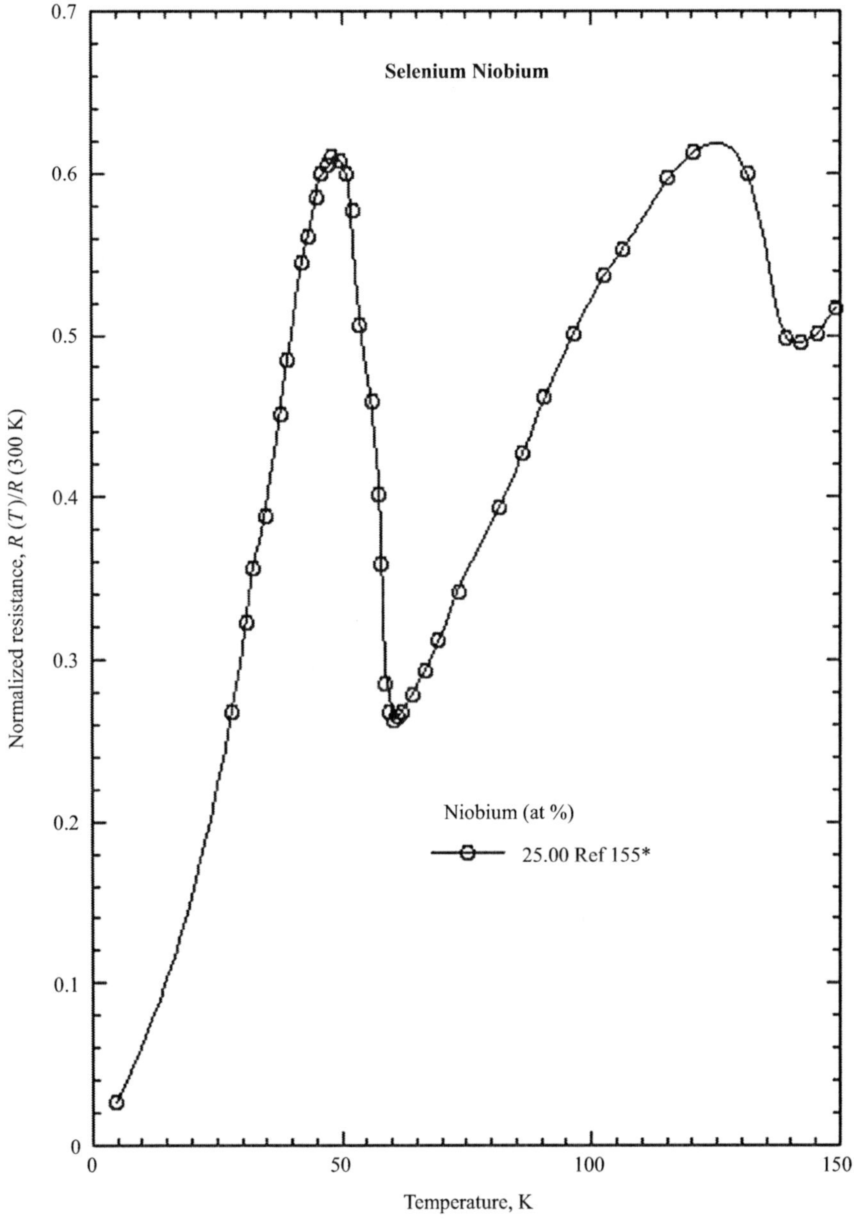
Selenium Niobium
Niobium (at %)
25.00 Ref 155*
Normalized resistance, R (T)/R (300 K)
Temperature, K
0
0.1
0.2
0.3
0.4
0.5
0.6
0.7
0
50
100
150

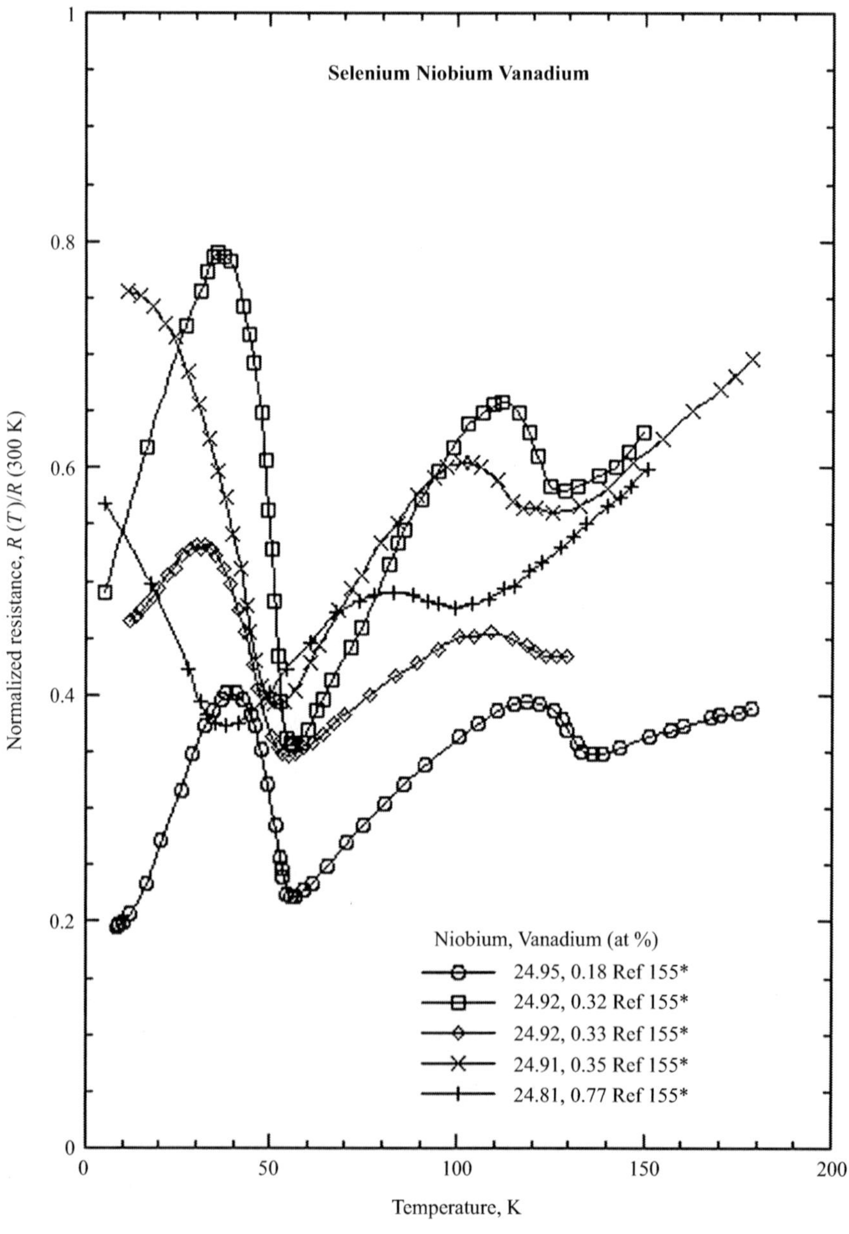
Selenium Niobium Vanadium
Normalized resistance, R (T)/R (300 K)
Temperature, K
Niobium, Vanadium (at %)
24.95, 0.18 Ref 155*
24.92, 0.32 Ref 155*
24.92, 0.33 Ref 155*
24.91, 0.35 Ref 155*
24.81, 0.77 Ref 155*
0
0.2
0.4
0.6
0.8
1
0
50
100
150
200

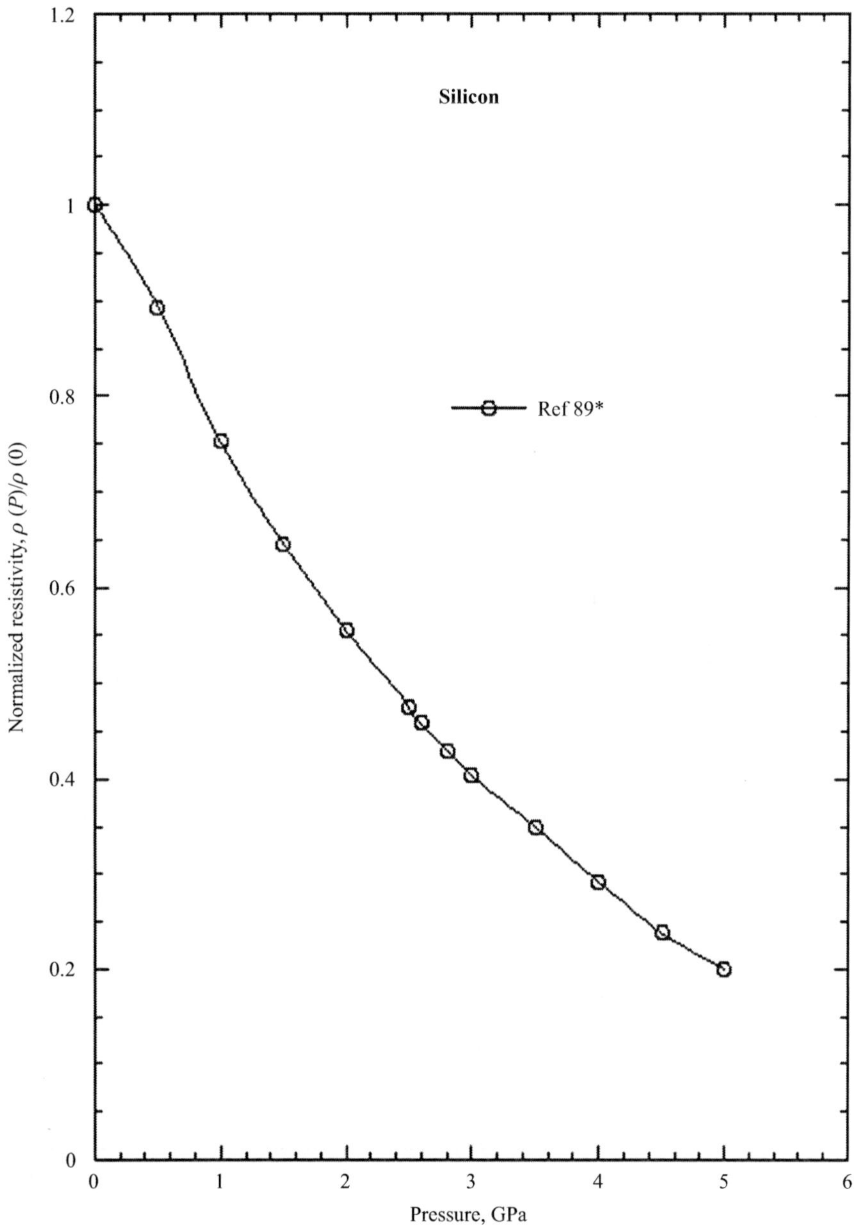
Silicon
Ref 89*
Normalized resistivity, ρ (P)/ρ (0)
Pressure, GPa
1.2
1
0.8
0.6
0.4
0.2
0
0
1
2
3
4
5
6

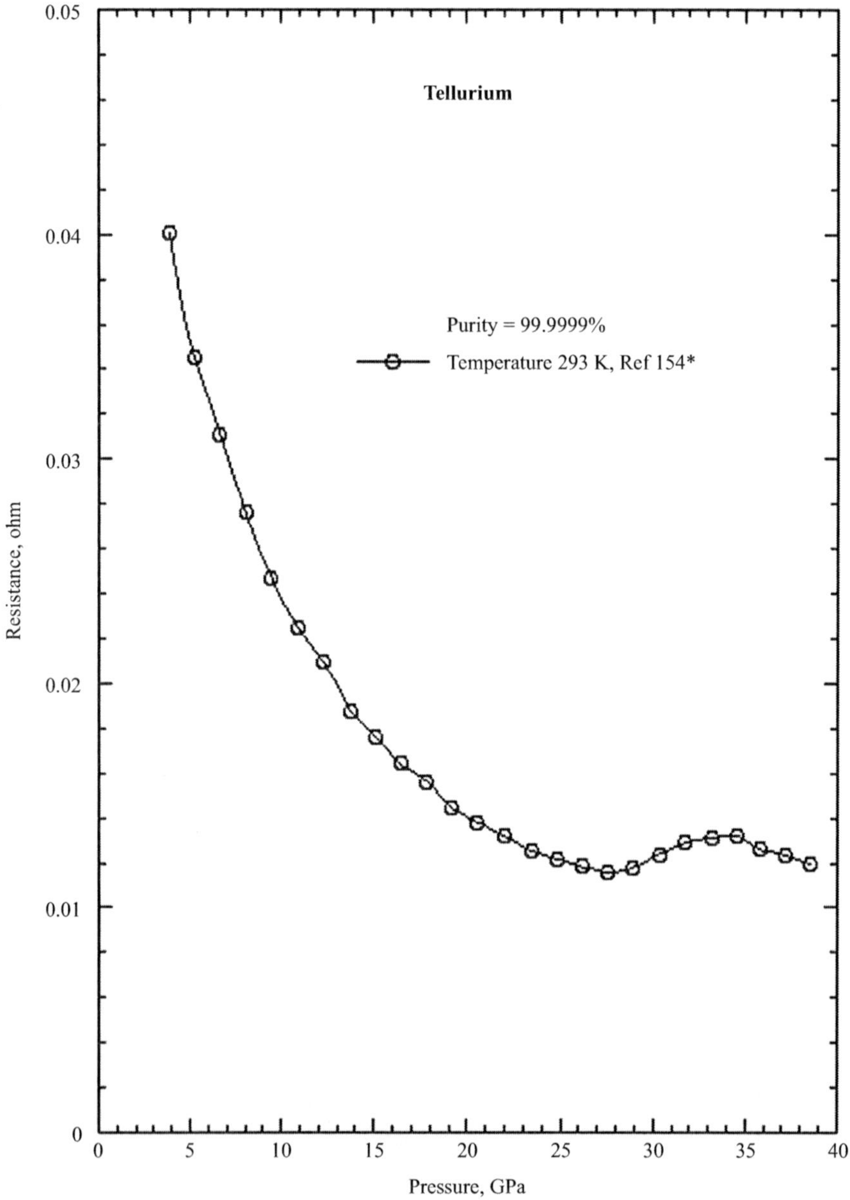
Tellurium
Purity = 99.9999%
Temperature 293 K, Ref 154*
Resistance, ohm
Pressure, GPa
0.05
0.04
0.03
0.02
0.01
0
0
5
10
15
20
25
30
35
40

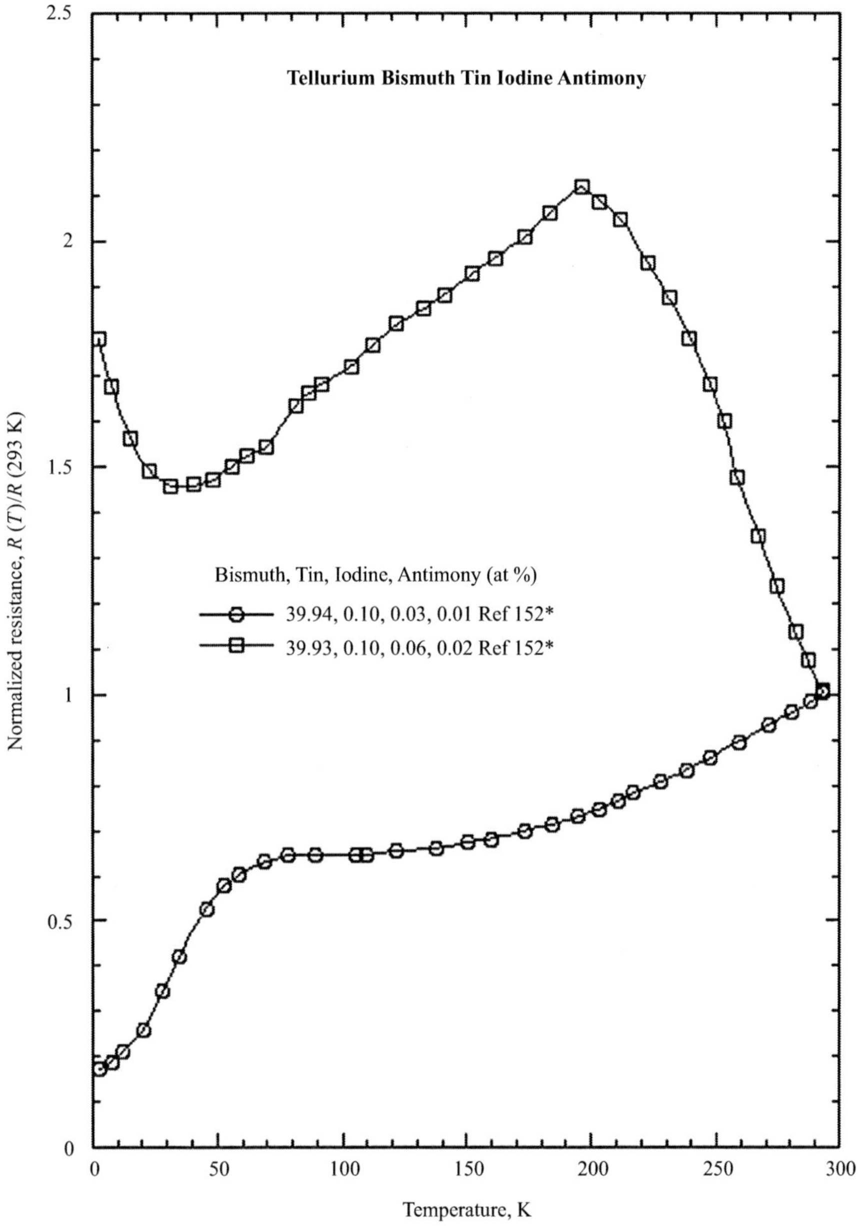
Tellurium Bismuth Tin Iodine Antimony
Bismuth, Tin, Iodine, Antimony (at %)
39.94, 0.10, 0.03, 0.01 Ref 152*
39.93, 0.10, 0.06, 0.02 Ref 152*
Normalized resistance, R (T)/R (293 K)
Temperature, K
0
0.5
1
1.5
2
2.5
0
50
100
150
200
250
300

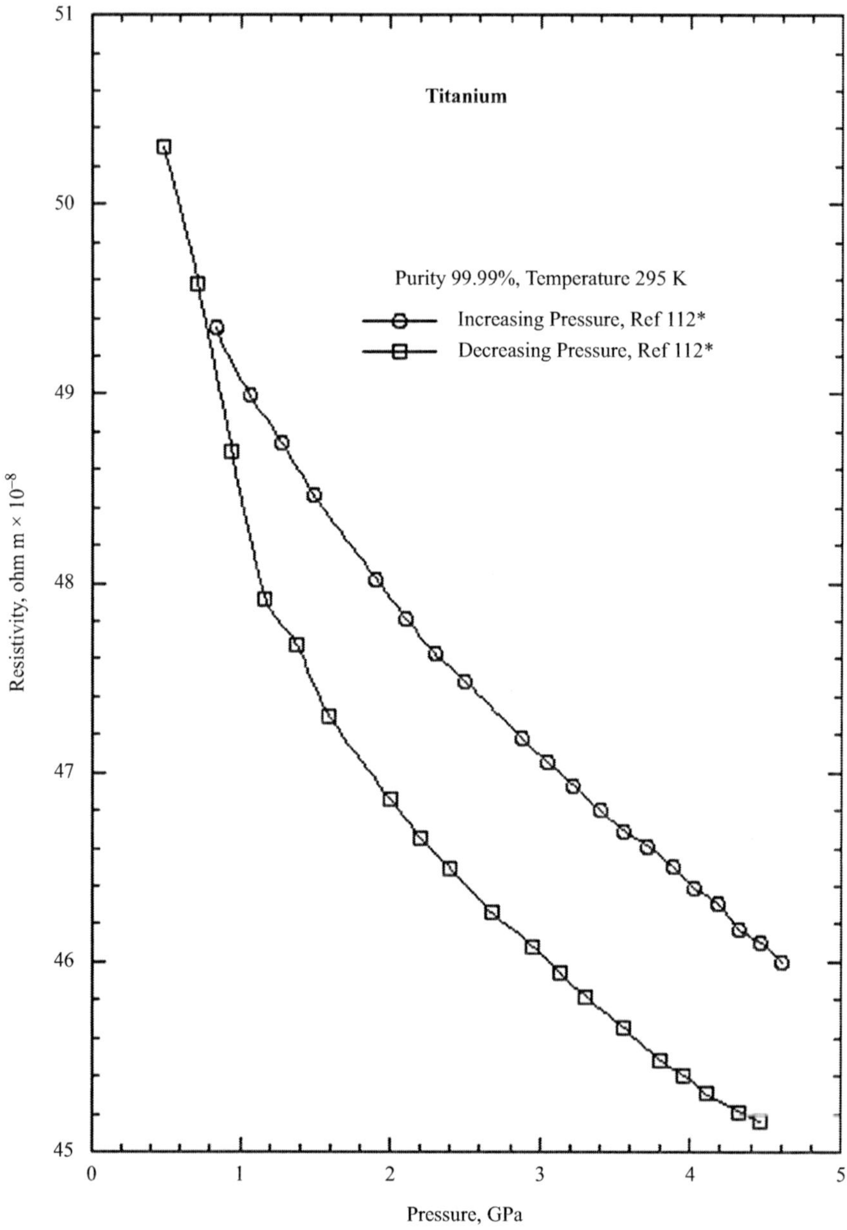
Titanium
Purity 99.99%, Temperature 295 K
Increasing Pressure, Ref 112*
Decreasing Pressure, Ref 112*
Resistivity, ohm m × 10⁻⁸
Pressure, GPa
51
50
49
48
47
46
45
0
1
2
3
4
5

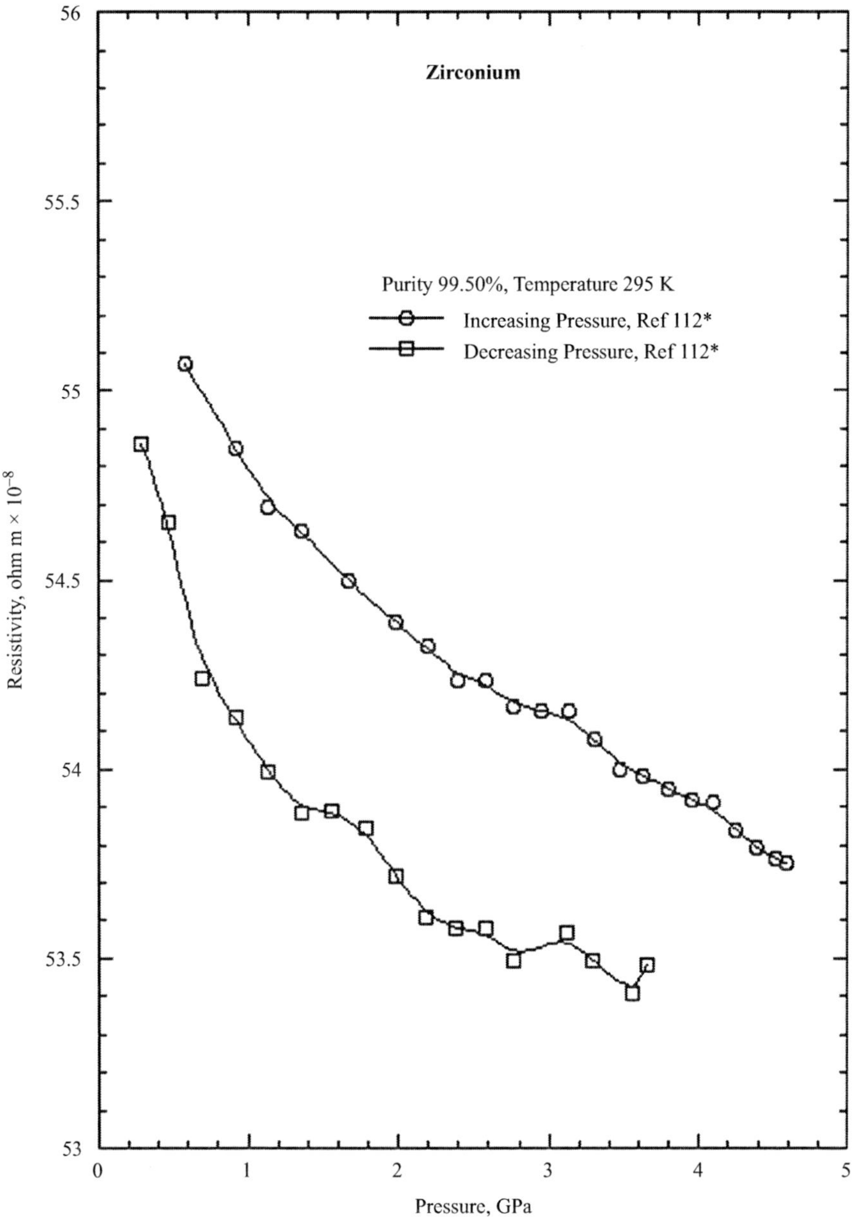
Zirconium
Purity 99.50%, Temperature 295 K
Increasing Pressure, Ref 112*
Decreasing Pressure, Ref 112*
Resistivity, ohm m × 10^-8
Pressure, GPa
56
55.5
55
54.5
54
53.5
53
0
1
2
3
4
5

Chapter 7
Resistivity references

[1] Duijn H.M., van Dijk N.H., de Visser A., Franse J.J.M. 'The influence of Gd and Ni impurities on the superconducting transition temperature of UPt_3'. *Physica B.* 1996, vols 223–4, pp. 44–46

[2] Okamoto N.L., Koyama T., Kishida K., Tanaka K., Inui H. 'Crystal structure and thermoelectric properties of chimney-ladder compounds in the Ru_2Si_3-Mn_4Si_7 pseudobinary system'. *Acta Mater.* 2009, vol. 57, pp. 5036–45

[3] Kishida K., Ishida A., Koyama T., Harada S., Norihiko I.O., Tanaka K., *et al.* 'Thermoelectric properties of ternary and Al-containing quaternary $Ru_{1-x}Re_xSi_y$ chimney-ladder compounds'. *Acta Mater.* 2009, vol. 57, pp. 2010–19

[4] Singh S., Rama N., Sethupathi K., Ramachandra Rao M.S. 'Correlation between electric transport, optical and magnetic properties of transition metal ion doped ZnO'. *J. Appl. Phys.* 2009, vol. 103, 07108-D

[5] Boboridis K. 'Thermophysical property measurements on niobium and titanium by a microsecond-resolution pulse heating technique using high-speed laser polarimetry and radiation thermometry'. *Int. J. Thermophys.* 2002, vol. 23, pp. 279–91

[6] Dobrosavljević A.S., Maglić K.D. 'Heat capacity and electric resistivity of copper research material for calorimetry'. *High Temp. High Press.* 1991, vol. 23, pp. 129–33

[7] Perović N.Lj., Maglić K.D., Stanimirović A.M. 'Specific heat capacity and electrical resistivity of nichrome and constantan thermocouple alloys between 300 K and 1500 K'. *5th International Symposium on Temperature and Thermal Measurement in Industry and Science, TEMPMEKO'93*. Amsterdam: Elsevier, 1993, pp. 367–76

[8] Markovich V., Rozenberg E., Gorodetsky G., Mogilyansky D., Revzin B., Pelleg J. 'Pressure effect on magnetic and transport properties of Zn doped $La_{0.91}Mn_{0.95}O_3$ manganite'. *J. App. Phys.* 2001, vol. 90, no. 5, pp. 2347–51

[9] Setton M., Van der Spiegel J. 'Structural and electrical properties of $ZrSi_2$ and Zr_2CuSi_4 formed by rapid thermal processing'. *J. App. Phys.* 1991, vol. 70, no. 1, pp. 193–7

[10] Notatari H., Saiga Y., Kato Y., Iwata K., Katano S., Fujiwara T., *et al.* 'Effect of pressure on the electrical resistivity of $YbAl_2$'. *J. Phys. Soc. Jpn.* 2007, vol. 76, supp. A, pp. 80–81

[11] Wu Y.L., Xiong Y.F., Li L.F. 'Effect of selenium on superconductivity in high-pressure synthesis $Sm_{1.85}Ce_{0.15}CuO_4$'. *Physica C.* 2006, vol. 434, pp. 157–60

[12] Auslender M.I., Rozenberg Kar'kin A.E., Chaudhuri B.K., Gorodetsky G. 'The nature of the low-temperature minimum of resistivity in ceramic manganites'. *J. Alloy Compd.* 2001, vol. 326, pp. 81–84

[13] Wang C., Sun Y., Qu Z., Zhang Y. 'Suppression of superconductivity by the localization of hole carriers in Ti-doped $La_{1.85}Sr_{0.15}CuO_4$'. *Phys. Rev. B.* 2006, vol. 73, 144518

[14] Berthault A., Arles L., Matricon J. 'High-pressure, high-temperature thermophysical measurements on tantalum and tungsten'. *Int. J. Thermophys.* 1986, vol. 7, pp. 167–79

[15] Boivineau M., Arlès L., Vermeulen J.M., Thévenin Th. 'Thermophysical properties of solid and liquid beryllium'. *Int. J. Thermophys.* 1993, vol. 14, pp. 427–39

[16] Boivineau M., Arlès L., Vermeulen J.M., Thévenin Th. 'High-pressure thermophysical properties of solid and liquid uranium'. *Physica B.* 1993, vol. 190, pp. 31–39

[17] Ang R., Sun Y.P., Zhu X.B., Song W.H. 'Influence of tellurium on the perovskite manganite $La_{0.5}Ca_{0.5}MnO_3$'. *Solid State Commun.* 2006, vol. 138, pp. 505–10

[18] Thévenin Th., Arlès L., Boivineau M., Vermeulen J.M. 'Thermophysical properties of rhenium'. *Int. J. Thermophys.* 1993, vol. 14, pp. 441–8

[19] Balog P.S., Secco R.A. 'High pressure and temperature behaviour of electrical resistivity of hcp metals Ti, Zr and Gd'. *J. Phys. Condens. Matter.* 1999, vol. 11, pp. 1273–87

[20] Roro K.T., Prinsloo H.L., Alberts H.L. 'Influence of V and Mn on the electrical transport properties of a Cr + 1.2 at. % Ga alloy'. *J. Alloy Compd.* 2005, vol. 393, pp. 16–25

[21] Spendeler L., Jaccard D., Sierro J. 'High pressure transport properties of pure ytterbium in the metallic bcc phase'. *Phys. Lett. A.* 1993, vol. 177, pp. 375–8

[22] Markovich V., Rozenberg E., Gorodetsky G., Revzin B., Pelleg J., Felner I. 'Effect of pressure and magnetic field on the resistance of the self-doped manganese perovskite $La_{0.91}Mn_{0.95}O_3$'. *Phys. Rev. B.* 2000, vol. 62, pp. 186–90

[23] Rozenberg E., Auslender M., Felner I., Gorodetsky G. 'Low-temperature resistivity minimum in ceramic manganites'. *J. App. Phys.* 2000, vol. 88, pp. 2578–82

[24] Liu X., Xu X., Zhang Y. 'Effect of Ti dopant on the carrier density collapse in colossal magnetoresistance material $La_{0.7}Ca_{0.3}Mn_{1-y}Ti_yO_3$'. *Phys. Rev. B.* 2000, vol. 62, pp. 112–19

[25] Sahu P.Ch., Yousuf M., Govinda Rajan K. 'Electrical resistivity of thorium up to 10 GPa and 1300 K'. *Physica B.* 1989, vol. 160, pp. 177–82

[26] Sahu P.Ch., Yousuf M., Govinda Rajan K. 'Electrical resistivity and phase-transition behaviour of uranium under pressure and temperature'. *Physica B.* 1992, vol. 183, pp. 145–55

[27] Hixson R.S., Winkler M.A., Hodgdon M.L. 'Sound speed and thermophysical properties of liquid iron and nickel'. *Phys. Rev. B.* 1990, vol. 42, pp. 6485–91

[28] Tripathi R., Awana V.P.S., Panwar N., Bhalla G.L., Habermier H.U., Agarwal S.K., *et al.* 'Enhanced room temperature coefficient of resistance and magneto-resistance of Ag-added $La_{0.7}Ca_{0.3\text{-}x}Ba_xMnO_3$ composites'. *J. Phys. D: J Appl. Phys.* 2009, vol. 42, 175002

[29] Das S., Dey T.K. 'Electrical conductivity and low field magnetoresistance in polycrystalline $La_{1\text{-}x}K_xMnO_3$ pellets prepared by pyrophoric method'. *Solid State Commns.* 2005, vol. 134, pp. 837–42

[30] Maglić K.D., Perović N.Lj., Vuković G.S., Zeković Lj.P. 'Specific heat and electrical resistivity of niobium measured by subsecond calorimetric technique'. *Int. J. Thermophys.* 1994, vol. 15, pp. 963–72

[31] Maglić K.D., Pavičić D.Z. 'Thermal and electrical properties of titanium between 300 and 1900 K'. *Int. J. Thermophys.* 2001, vol. 22, pp. 1833–41

[32] Pavičić D.Z., Maglić K.D. 'Specific heat and electrical resistivity of 53% niobium-47% titanium alloy measured by subsecond calorimetric technique'. *Int. J. Thermophys.* 2002, vol. 23, pp. 1319–25

[33] Shibasaki S., Takahashi Y., Terasaki I. 'Thermoelectric properties of $LaRh_{1\text{-}x}Ni_xO_3$'. *J. Phys. Condens. Matter.* 2009, vol. 21, pp. 1–4

[34] Kalyana Lakshmi Y., Venkataiah M., Vithal M., Venugopal Reddy P. 'Magnetic and electrical behavior of $La_{1\text{-}x}A_xMnO_3$ (A = Li, Na, K and Rb) manganites'. *Physica B.* 2008, vol. 403, pp. 3059–66

[35] Marcano G. 'Effect of Mn-doping on the electrical properties of Cu_2GeSe_3'. *J. Phys. Chem. Solids.* 2005, vol. 66, pp. 2086–9

[36] Bhatia S.N., Mohapatra N., Kundaliya D., Malik S.K. 'Transport, magnetic and thermal properties of Pr doped $LaSr_2Mn_2O_7$: a Kondo-type behaviour of resistivity'. *J. Phys. Condens. Matter.* 2006, vol. 18, pp. 7179–92

[37] Khan N.A., Baber N., Iqbal M.Z., Ul Haq, A. 'Simple method for direct synthesis $YBa_2Cu_4O_8$ at atmospheric oxygen pressure'. *Appl. Phys. Lett.* 1993, vol. 63, pp. 257–9

[38] Righini F., Spišiak J., Brussolino G.C., Gualano M. 'Thermophysical properties by a pulse heating reflectometric technique: niobium 1100 to 2700 K'. *Int. J. Thermophys.* 1999, vol. 20, pp. 1107–16

[39] Takabatake T., Iga F., Yoshino T., Echizen Y., Katoh K., Kobayashi K., *et al.* 'Ce and Yb-based Kondo semiconductors'. *J. Magn. Magn. Mater.* 1998, vols 171–81, pp. 277–82

[40] Ateş A., Yildirim M., Gürbulak B. 'Investigation of the electrical properties of Ho-doped InSe single crystal'. *Physica E.* 2004, vol. 21, pp. 85–90

[41] Vorenkamp T., Koziol Z., de Visser A., Bakker K., Franse J.J.M. 'Impurity effects in superconducting UPt_3'. *J. Appl. Phys.* 1991, vol. 69, pp. 5487–9

[42] Ho C.H., Huang C.E., Wu C.C. 'Preparation and characterization of Ni-incorporated FeS_2 single crystals'. *J. Cryst. Growth*. 2004, vol. 270, pp. 535–41

[43] Soni A., Okram G.S. 'Resistivity and thermopower measurement setups in the temperature range of 5–325 K'. *Rev. Sci. Inst.* 2008, vol. 79, 125103–125103-4

[44] Wu X.S., Lin J., Liu W.J., Chen W.M., Jin X., Jiang S.S., *et al.* 'Structure studies of $La_{1.85}Sr_{0.15}CuO_4$ doped with Mg at high doping level'. *J. Supercond.* 1997, vol. 10, pp. 159–64

[45] Mori T., Tanaka T. 'Effect of transition metal doping and carbon doping on thermoelectric properties of YB_{66} single crystals'. *J. Solid State Chem.* 2006, vol. 179, pp. 2889–94

[46] Mori T. 'High temperature thermoelectric properties of B_{12} icosahedral cluster-containing rare earth boride crystals'. *J. App. Phys.* 2005, vol. 97, pp. 3702–4

[47] Maglić K.D., Perović N., Vuković G. 'Specific heat and electric resistivity of molybdenum between 400 and 2500 K'. *High Temp. High Press.* 1997, vol. 29, pp. 97–102

[48] Okamoto Y., Aruga A., Kasai H., Morimoto J., Miyakawa T., Fujimoto S. 'Temperature dependence of thermoelectric properties of SiC/B_4C'. *AIP Conference Proceedings 316, Thirteenth International Conference on Thermoelectrics*. Maryland, USA: American Institute of Physics, 1994, pp. 92–95

[49] Shibasaki S., Kobayashi W., Terasaki I. 'Mg substitution effects of new thermoelectric Rb oxides'. *2007 International Conference on Thermoelectrics*. Piscataway, USA: Institute of Electrical and Electronics Engineers, 2007, pp. 133–5

[50] Cabeza O., Barca O., Francesconi G., Bari M.A., Severack C., Muirhead C.M., *et al.* 'Magnetic and electrical properties of Cr-doped $La_{0.7}Ca_{0.3}MnO_3$'. *J. Magn. Magn. Mater.* 1999, vols 196–7, pp. 504–5

[51] Sera M., Sato H., Hiroi M., Kobayashi N. 'Cu site substitution effect on the Hall coefficient of $La_{2-x}Sr_xCuO_4$'. *Solid State Commun.* 1994, vol. 90, pp. 803–6

[52] Radhapiyari L., Phanjoubam S., Sarma H.N.K., Prakash C. 'Influence of Co^{2+} on the electrical and magnetic properties of Li-Sb ferrites'. *Mater. Letts.* 2000, vol. 44, pp. 65–69

[53] He T., Huang Q., Ramirez A.P., Wang Y., Regan K.A., Rogado N., *et al.* 'Superconductivity in the non-oxide Perovskite $MgCNi_3$'. *Nature* 2001, vol. 411, pp. 54–56

[54] Koike Y., Okubo T., Fujiwara A., Noji T., Saito Y. 'Trial of intercalation of Br and Li into $Bi_2Sr_2Ca_{n-1}Cu_nO_{2n+4}$ (n = 1, 2, 3)'. *Solid State Commun.* 1991, vol. 79, pp. 501–5

[55] White C.K., Minges M.L. 'Thermophysical properties of some key solids: An update'. *Int. J. Thermophys.* 1997, vol. 18, pp. 1269–327

[56] Loseva G.V., Ovchinnikov S.G., Balaev A.D., Ivanova N.B., Kiselev N.I. 'Transition from the Kondo regime to long-range magnetic order in the $Fe_xV_{1-x}S$ system'. *Phys. Solid State.* 2000, vol. 42, pp. 1322–4

[57] Hasanain S.K., Nadeem M., Shah W.H., Akhtar M.J., Hasan M.M. 'Effects of iron doping on the transport and magnetic behaviour in $La_{0.65}Ca_{0.35}Mn_{1-y}Fe_yO_3$'. *J. Phys. Condens. Matter.* 2000, vol. 12, pp. 9007–17

[58] Young S., Xiaojun X., Yuheng Z. 'Effects of Fe doping in $La_{0.67}Sr_{0.33}CoO_3$'. *Phys. Rev. B.* 2000, vol. 62, pp. 5289–92

[59] Mhiaoui S., Gasser J.G., Ben Abdellah A. 'On the thermopower of liquid antimony'. *13th International Conference on Liquid and Amorphous Metals. J. Phy. Conf. Ser.* 2008, vol. 98, pp. 1–9

[60] Alvarez I., Veiga M.L., Pico C. 'Structural characterization and electrical behaviour of $LaNi_{1-x}Sb_xO_3$ ($0 \leq x \geq 1/3$) perovskites'. *Solid State Ionics.* 1996, vol. 91, pp. 265–71

[61] Pan Z.J., Zhang L.T., Wu J.S. 'Effects of Al doping on the transport performances of $CrSi_2$ single crystals'. *Scripta Mater.* 2007, vol. 56, pp. 245–8

[62] Xu G.J., Grivel J.C., Abrahamsen A.B., Chen X.P., Andersen N.H. 'Structure and superconductivity of double-doped $Mg_{1-x}(Al_{0.5}Li_{0.5})xB_2$'. *Physica C.* 2003, vol. 399, pp. 8–14

[63] Xu G.J., Grivel J.C., Abrahamsen A.B., Chen X.P., Andersen N.H. 'Superconducting properties of Zn and Al double-doped $Mg_{1-x}(Zn_{0.5}Al_{0.5})_xB_2$'. *Physica C.* 2004, vol. 403, pp. 113–18

[64] Jacobsson P., Sundqvist B. 'Thermal conductivity and electrical resistivity of gadolinium as functions of pressure and temperature'. *Phys. Rev. B.* 1989, vol. 40, pp. 9541–51

[65] Larica C., Nunes E., Freitas J.C.C. 'Low field magnetic studies of some $Gd_{1-x}La_x$ alloys'. *J. Phys. Condens. Matter.* 1999, vol. 11, pp. 821–31

[66] Dudáš J., Feher A., Janoš Š. 'The temperature hysteresis of the electrical resistance of dysprosium near the order-disorder transition'. *J. Less Common Met.* 1987, vol. 134, pp. L9–11

[67] Isoda Y., Nagai T., Fujiu H., Imai Y., Shinohara Y. 'Thermoelectric properties of Sb-doped $Mg_2Si_{0.5}Sn_{0.5}$'. *25th International Conference on Thermoelectrics.* Piscataway, USA: Institute of Electrical and Electronics Engineers, 2006, pp. 406–10

[68] Isoda Y., Nagai T., Fujiu H., Imai Y., Shinohara Y. 'The effect of bismuth doping on thermoelectric properties of $Mg_2Si_{0.5}Sn_{0.5}$'. *26th International Conference on Thermoelectrics.* Piscataway, USA: Institute of Electrical and Electronics Engineers, 2007, pp. 251–5

[69] Wang J., Lin Y., Zou H., Pu S., Shi J. 'Structural transition, electrical and magnetic properties of the B-site Co doped $Sr_{14}Cu_{24}O_{41}$'. *J. Phys. Condens. Matter.* 2009, vol. 21, pp. 1–10

[70] Cao D., Bridges F., Anderson M., Ramirez A.P., Olapinski M., Subramanian M.A., *et al.* 'Local distortions in $La_{0.7}Ca_{0.3}Mn_{1-b}A_bO_3$ (A = Ti and Ga) colossal magnetoresistance samples: Correlations with magnetization and evidence for cluster formation'. *Phys. Rev. B.* 2001, vol. 64, pp. 1–14

[71] Joshi A.G., Kuberkar D.G., Baldha G.J., Kulkarni R.G. 'Effect of Mo-Ca substitution on the superconductivity of $GdBa_2Cu_3O_{7-\delta}$'. *Physica C.* 1997, vol. 291, pp. 25–33

[72] Yamaura Y., Takayama-Muromachi E. 'Enhanced paramagnetism of the 4d itinerant electrons in the rhodium oxide perovskite $SrRhO_3$'. *Phys. Rev. B.* 2001, vol. 64, pp. 1–5

[73] Yamaura K., Huang Q., Molodovan M., Young D.P., Sato A., Baba Y., *et al.* 'High-pressure synthesis, crystal structure determination, and a Ca substitution study of the metallic rhodium oxide $NaRh_2O_4$'. *Chem. Mater.* 2005, vol. 17, pp. 359–65

[74] Yamaura K., Huang Q., Young D.P., Noguchi Y., Takayama-Muromachi E. 'Crystal structure and electronic and magnetic properties of the bilayered rhodium oxide $Sr_3Rh_2O_7$'. *Phys. Rev. B.* 2002, vol. 66, 134431

[75] Yamaura K., Huang Q., Young D.P., Takayama-Muromachi E. 'Crystal structure and magnetic properties of the trilayered perovskite $Sr_4Rh_3O_{10}$: A new member of the strontium rhodate family'. *Chem. Mater.* 2004, vol. 16, pp. 3424–30

[76] Bathe R., Shinde S.R., Gapechup K.M., Adhi K.P., Patil S.I. 'Effect of aluminium doping on the magneto-transport properties of $La_{0.75}Ca_{0.25}MnO_3$'. *J. Magn. Magn. Mater.* 2003, vol. 256, pp. 425–9

[77] Assunção M.C. 'Effect of halogen impurities on the transport properties of selenium glasses'. *J. Non-Cryst. Solids.* 1991, vol. 136, pp. 81–90

[78] Groń T., Duda A., Krajewski A., Kusz J., Warczewski J., Nikiforov K.G. 'The electrical conductivity of the strongly defective $HgCr_2Se_4$ single crystals'. *Radiat. Eff. Defects S.* 2002, vol. 157, pp. 1111–16

[79] Korn D., Mürer W., Zibold G. 'Electrical resistivity of amorphous Sn-Cu alloys'. *Phys. Lett.* 1974, vol. 47A, pp. 117–18

[80] Kleinschmidt P. 'Electrical conductivity measurements on doped VO_2 single crystals'. *Phys. Lett.* 1974, vol. 47A, pp. 205–6

[81] Littleton R.T., Jeffries J., Kaeser M.A., Long M., Tritt T.M. 'High temperature transport probe for thermopower and resistivity measurements'. *1998 Fall Materials Research Society, Symposium Z: Thermoelectric Materials.* Materials Science and Engineering Department, Clemson University, Clemson, SC 29634 USA, 1998

[82] Yonenaga I., Goto T., Tang X.F., Yamaguchi S. 'Thermal and electrical properties of Czochralski grown germanium-silicon alloys'. *18th International Conference on Thermoelectrics.* Piscataway, USA: Institute of Electrical and Electronics Engineers, 1999, pp. 436–9

[83] Kanazawa M., Tsuda N. 'Logarithmic temperature dependence of resistivity for $Ca_{0.7}Na_{0.3}Pd_3O_4$ doped with Cu'. *J. Phys. Soc. Jpn.* 2000, vol. 69, pp. 4112–13

[84] Brahma P., Banerjee S., Chakraborty S., Chakravorty D. 'Small polaron and bipolaron transport in antimony oxide doped barium hexaferrites'. *J. App. Phys.* 2000, vol. 88, pp. 6526–8

[85] Peña O., Le Berre F., Padiou J., Marchand T. 'Single crystal studies of the Chevrel-phase superconductor $La_xMo_6Se_8$'. *J. Solid State Chem.* 1998, vol. 136, pp. 160–6
[86] Chiodelli G., Massarotti V., Capsomi D., Bini M., Azzoni, C.B., Mozzati M.C., *et al.* 'Electric and dielectric properties of pure and doped $CaCu_3Ti_4O_{12}$ perovskite materials'. *Solid State Commun.* 2004, vol. 132, pp. 241–6
[87] Viviani M., Buscaglia V., Buscaglia L., Mitoseriu L., Testino A., Nanni P., *et al.* 'Analysis of conductivity and PTCR effect in Er-doped $BaTiO_3$ ceramics'. *J. Eur. Ceram. Soc.* 2004, vol. 34, pp. 1221–5
[88] Okada S., Shimizu K., Kobayashi T.C., Amaya K., Endo S. 'Superconductivity of calcium under high pressures'. *J. Phys. Soc. Jpn.* 1996, vol. 65, pp. 1924–6
[89] Andersson G., Sundqvist B., Bäckström G. 'A high-pressure cell for electrical resistance measurements at hydrostatic pressures up to 8 GPa: Results for Bi, Ba, Ni and Si'. *J. Appl. Phys.* 1989, vol. 65, pp. 3943–50
[90] Tateiwa N., Nakagawa A., Fujio K., Kawae T., Takeda K. 'Pressure-induced valence change in the rare earth metals: The case of praseodymium'. *J. Alloy Compd.* 2006, vols 408–12, pp. 244–7
[91] Shrivastava V., Jha A.K., Mendiratta R.G. 'Structural and electrical studies in La-substituted $SrBi_2Nb_2O_9$ ferroelectric ceramics'. *Physica B.* 2006, vol. 371, pp. 337–42
[92] Weir S.T., Mitchell A.C., Nellis W.J. 'Metallization of fluid molecular hydrogen at 140 GPa'. *Phys. Rev. Lett.* 1996, vol. 76, pp. 1860–3
[93] Debessai M., Matsuoka T., Hamlin J.J., Schilling J.S. 'Pressure-induced superconducting state of europium metal at low temperatures'. *Phys. Rev. Lett.* 2009, vol. 102, 197002
[94] Ellerby M., McEwen K.A., Bauer E., Hauser R., Jensen J. 'Pressure-dependent resistivity and magnetoresistivity of erbium'. *Phys. Rev. B.* 2000, vol. 61, pp. 6790–7
[95] Pottlacher G., Kaschnitz E., Jäger H. 'Investigations of thermophysical properties of liquid metals with a rapid resistive heating technique'. *J. Non-Cryst Solids.* 1993, vols 156–8, pp. 374–8
[96] Ault K.M., Secco R.A. 'High pressure conductivity study of the α-quartz $\rightarrow$ rutile transformation in GeO_2'. *Solid State Commun.* 1996, vol. 98, pp. 449–52
[97] Makovec D., Ule N., Drofenik M. 'Positive temperature coefficient of resistivity effect in highly donor-doped barium titanate'. *J. Am. Ceram. Soc.* 2001, vol. 84, pp. 1273–80
[98] Langhammer H.T., Makovec D., Pu Y., Abicht H.P., Drofenik M. 'Grain boundary reoxidation of donor-doped barium titanate ceramics'. *J. Eur. Ceram. Soc.* 2006, vol. 26, pp. 2899–2907
[99] Sasmal K., Lv B., Lorenz B., Guloy A., Chen F., Xue Y.Y., *et al.* 'Superconducting Fe-based compounds $(A_{1-x}Sr_x)$ Fe_2As_2 with A = K and Cs with transition temperatures up to 37K'. *Phys. Rev. Lett.* 2008, vol. 101, 107007

[100] Secco E.A., Secco R.A. 'Cation conductivity in mixed thallium halides'. *Solid State Ionics.* 1999, vol. 118, pp. 37–42

[101] Nomura T., Kim S.W., Kamihara Y., Hirano M., Sushko P.V., Kato K., *et al.* 'Crystallographic phase transition and high-T_c superconductivity in LaFeAsO:F'. *Supercond. Sci. Technol.* 2008, vol. 21, 125028

[102] Yabuuchi T., Matsuoka T., Nakamoto Y., Shimizu K. 'Superconductivity of Ca exceeding 25 K at megabar pressure'. *J. Phys. Soc. Jpn.* 2006, vol. 75, 083703

[103] Sundqvist B. 'Electrical resistance of nickel in the range 300–725 K and 0–2 GPa'. *Phys. Rev. B.* 1988, vol. 38, pp. 283–9

[104] Okada H., Takahashi H., Takahashi H., Matuishi S., Hirano M., Hosono H. 'Pressure effect on iron-based superconductors $Ca(Fe_{1-x}Co_x)AsF$'. *J. Phys. Conf. Ser.* 2010, vol. 200, 012151

[105] Wang S.F., Dai S.Y., Zhou Y.L., Zhu Y.B., Chen Z.H., Lü H.B., *et al.* 'Effect of Cd doping on structure and superconductivity in $Mg_{0.5}Cd_{0.5}B_2$'. *J. Supercond.* 2002, vol. 17, pp. 397–400

[106] Shimizu K., Kimura T., Furomoto S., Takeda K., Kontani K., Onuki Y., *et al.* 'Superconductivity in the non-magnetic state of iron under pressure'. *Nature.* 2001, vol. 412, pp. 316–18

[107] Chong T.V., Ishii O., Kambe S. 'The study on Zn and Ni substituted GdBaSr $Cu_3O_{7-\delta}$ superconductor'. *Physica C.* 2008, vol. 468, pp. 1214–16

[108] Chong T.V., Kambe S., Ishii O. 'Effect of Ca doping on the superconducting properties of $GdBaSrCu_3O_{7-\delta}$ bulk samples'. *Physica C.* 2009, vol. 469, pp. 985–7

[109] Gunasekaran R.A., Ganguly J.V., Yakhmi, J.V. 'Superconducting behaviour of co-doped $Y_{1-x}Ca_xBa_2Cu_{3-x}M_xO_{7-\delta}$ ($M = Ni$ or Zn and $0.0 \leq x \leq 0.30$)'. *Physica C.* 1995, vol. 243, pp. 160–6

[110] Mori T., Zhang F. 'Low-temperature magnetism of the compound $GdB_{18}Si_5$'. *J. Phys. Condens. Matter.* 2002, vol. 14, pp. 11831–6

[111] Awana V.P.S., Vajpayee A., Mudgel M., Ganesan V., Awasthi A.M., Bhalla G.L., *et al.* 'Physical property characterization of bulk MgB_2 superconductor'. *Eur. Phys. J. B.* 2008, vol. 62, pp. 281–94

[112] Balog P.S., Secco R.A. 'High pressure and temperature behaviour of electrical resistivity of HCP metals Ti, Zr and Gd'. *J. Phys. Condens. Matter.* 1999, vol. 11, pp. 1273–87

[113] Mugdel M., Sharath Chandra L.S., Ganesan V., Bhalla G.L., Kishan H., Awana V.P.S. 'Enhanced critical parameters of nanocarbon doped MgB_2 superconductor'. *J. App. Phys.* 2009, vol. 106, 033904

[114] Ho C.H., Hsich M.H., Huang Y.S. 'Compensation and carrier conduction in synthetic $Fe_{1-x}Ni_xS_2$ ($0 \leq x \leq 0.1$) single crystals'. *J. Electrochem. Soc.* 2008, vol. 155, pp. H254–8

[115] Yen P.C., Huang Y.S., Tiong K.K. 'The growth and characterization of rhenium-doped WS_2 single crystals'. *J. Phys. Condens. Matter.* 2004, vol. 16, pp. 2171–80

[116] Yan C., Huang Y., Wang Z., Zhang L., Gao S., Liao C., *et al.* 'Structural, magnetic and transport properties of Sc-doped $La_{0.7}Sr_{0.3}MnO_3$'. *Chinese Sci. Bull.* 2000, vol. 45, pp. 810–14

[117] Sefat A.S., McGuire M.A., Sales B.C., Jin R., Howe J.Y., Mandrus D. 'Electronic correlations in the superconductor $LaFeAsO_{0.89}F_{0.11}$ with low carrier-density'. *Phys. Rev. B.* 2008, vol. 77, 174503

[118] Zhou J.S., Goodenough J.B., Dabrowski B. 'Anomalous transport properties in a $YBa_2Cu_4O_8$ crystal'. *Phys. Rev. B.* 1998, vol. 58, pp. R2956–9

[119] Xia Y., Ponnambalam V., Bhattacharta S., Pope A.I., Poon S.J., Tritt, T.M. 'Electrical transport properties of TiCoSb half-Heusler phases that exhibit high resistivity'. *J. Phys. Condens. Matter.* 2001, vol. 13, pp. 1–13

[120] Ren Z.A., Che G.C., Dong X.L., Yang J., Lu W., Yi W., *et al.* 'Superconductivity and phase diagram in iron-based arsenic-oxides $ReFeAsO_{1-\delta}$ (Re = rare earth metal) without fluorine doping'. *Eur. Phys. Lett.* 2008, vol. 83, 17002

[121] Pękala K., Pękala M. 'Low temperature transport properties of nanocrystalline Cu, Fe and Ni'. *Nanostruc. Mater.* 1995, vol. 6, pp. 819–22

[122] White G.K., Woods S.B. 'Electrical and thermal resistivity of the transition elements at low temperatures'. *Phil. Trans. R Soc. Lond. A.* 1959, vol. 12, pp. 273–302

[123] Laubitz M.J., Matsumura T., Kelly P.J. 'Transport properties of the ferromagnetic metals. II. Nickel'. *Can. J. Phys.* 1976, vol. 54, pp. 92–102

[124] White G.K. *Experimental Techniques in Low Temperature Physics* (Oxford, Oxford University Press, 1959)

[125] Okram G.S., Soni A., Rawat R. 'Anomalous electrical transport behaviour in nanocrystalline nickel'. *Nanotechnology.* 2008, vol. 19, 185711

[126] Khan N.A., Mumtaz M. 'Absence of pair-breaking mechanism in $Cu_{0.5}Tl_{0.5}Ba_2Ca_3Cu_{4-y}Zn_yO12\text{-}\delta$'. *Phys. Rev. B.* 2008, vol. 77, 054507

[127] Mori N., Okano H., Furuya A. 'A comparative study of thermoelectric properties in (Pr, Y)$Ba_2Cu_3O_7$ and $PrBa_2Cu_3O_7$-Ag percolative systems'. *Phys. Stat. Sol. (a).* 2006, vol. 203, pp. 2828–31

[128] Mori N., Kameyama T., Enomoto H., Ozaki H., Takano Y., Sekizawa K. 'Thermoelectric power and resistivity in $Nd_{2-x}Ce_xCuO_4$ system'. *J. Alloy Compd.* 2006, vols 408–12, pp. 1222–5

[129] Terzieft P., Luck R., Auchet J. 'Electronic properties of dilute Cr in liquid silver'. *J. Alloy Compd.* 1998, vol. 274, pp. 148–52

[130] Wenzhong L., Buyin L., Ming F., Zhengwei H. 'Novel thermistor of Bi-doped Ba(Sn,Sb)O_3 with linear negative temperature coefficient'. *Sensor. Actuat.* 2000, vol. 80, pp. 38–41

[131] McCrone J.E., Tallon J.L., Cooper J.R., MacLaughlin A.C., Attfield J.P., Bernhard C. 'Magneto-transport properties of doped $RuSr_2GdCu_2O_8$'. *Phys. Rev. B.* 2003, vol. 68, 064514

[132] Chen G.F., Li Z., Wu D., Li G., Hu W.Z., Dong J., *et al.* 'Superconductivity at 41 K and its competition with spin-density-wave instability in layered $CeO_{1-x}F_xFeAs$'. *Phys. Rev. Lett.* 2008, vol. 100, 247002

[133] Jirák Z., Hejtmánek J., Kní ek K., Maryško M. 'Magnetoresistive behaviour of Cr-doped manganites $Pr_{0.44}Sr_{0.56}MnO_3$'. *J. App. Phys.* 2003, vol. 93, pp. 8083–5
[134] Michiels J., Gschneidner K.A. 'Electrical properties of yttrium sequisulfide (Y_2S_3) mechanically alloyed with copper, boron and aluminum'. *J. Alloy Compd.* 1997, vol. 347, pp. 9–14
[135] Dogra A., Rayaprol S., Shah N.A., Kuberkar D.J. 'Effect of Co – Ga paired substitution on superconductivity in $YBa_2Cu_3O_{7-\delta}$'. *Mod. Phys. Lett. B.* 2004, vol. 18, pp. 485–92
[136] Secco R.A., Secco E.A. 'Structural and non-structural factors in fast ion conduction in Ag_2SO_4 at high pressure'. *Phys. Rev. B.* 1997, vol. 56, pp. 3099–104
[137] Yousuf M., Sahu P.C., Rajan G. 'High-pressure and high-temperature electrical resistivity of ferromagnetic transition metals: Nickel and iron'. *Phys. Rev. B.* 1986, vol. 34, pp. 8086–100
[138] Secco R.A., Secco E.A. 'Effect sof pressure on the fast ion conductor $AgTlSO_4$'. *J. Phys. Chem. Solids.* 1995, vol. 56, pp. 1045–51
[139] Prinz A., Brunthaler G., Ueta Y., Springholz G., Bauer G., Grabecki G., *et al.* 'Electron localization in n-$Pb_{1-x}Eu_xTe$'. *Phys. Rev. B.* 1999, vol. 59, pp. 983–90
[140] Viallet V., Marucco J.F., Saint J., Herbst-Ghysel M., Dragoe N. 'Structural, magnetic and electrical properties of a perovskite containing divalent europium $EuZrO_3$'. *J. Alloy Compd.* 2008, vol. 461, pp. 346–50
[141] Dudáš J. 'The influence of the magnetic spin system on electrical resistivity in some rare earth metals'. *Elektrotechn. Čas.* 1994, vol. 45, pp. 310–14
[142] Jianquan Q., Zhilun G., Yajing W., Longt L. 'Influence of manganese on PTCR effect in $BaTiO_3$-based ceramics doped with Bi_2O_3 vapor'. *Mater. Chem. Phys.* 2002, vol. 73, pp. 97–100
[143] Fong C.H., Luo J.Y., Yeh K.W., Chen T.K., Huang T.W., Wu P.M., *et al.* 'Superconductivity in the PbO-type structure α-FeSe'. *PNAS.* 2008, vol. 105, no. 38, pp. 14262–4
[144] Ito M., Matsuda T. 'Thermoelectric properties of non-doped and y-doped $SrTiO_3$ polycrystals synthesized by polymerized complex process and hot pressing'. *J. Alloy Compd.* 2009, vol. 477, pp. 473–7
[145] Niimi H., Mihara K., Sakabe Y., Kuwabara M. 'Influence of Ba/Ti ratio on the positive temperature coefficient of resistivity characteristics of Ca-doped semiconducting $BaTiO_3$ fired in reducing atmosphere and reoxidized in air'. *J. Am. Ceram. Soc.* 2007, vol. 90, pp. 1817–21
[146] Okamoto Y., Kato K., Asai K., Morimoto J., Miyakawa T., Aruga A. 'Thermoelectric characteristics of sintered SiC/Cu semiconductors'. *J. Jpn. Soc. Powder Metall.* 2009, vol. 56, pp. 477–83
[147] Pulham R.J., Hubberstay P., Hamptenmacher P. 'Solutions of aluminium in liquid lithium: Electrical resistivity of liquid alloys'. *J. Chem. Soc. Faraday Trans.* 1994, vol. 90, pp. 2753–5

[148] Inai H., Okamoto Y., Morimoto J. 'Thermoelectric properties of sintered SiC doped with Ge'. *Mater. Sci. Forum.* 1999, vols 308–11, pp. 659–64

[149] Okamoto Y., Inai H., Morimoto J. 'Temperature dependence on the thermoelectric properties of Si doped SiC'. *J. Jpn. Soc. Powder Metall.* 1998, vol. 45, pp. 905–8

[150] Okamoto Y., Kato K., Morimoto J., Miyakawa T. 'A study for thermoelectric properties of Ni doped SiC sintered thermoelectric semiconductor'. *16th International Conference on Thermoelectrics*. Piscataway, USA: Institute of Electrical and Electronics Engineers,1997, pp. 236–9

[151] Chihibu S., Shishikura M., Ino J., Matsumoto S. 'Electrical and optical properties of $CuAlSe_2$ grown by iodine chemical vapour transport'. *J. App. Phys.* 1991, vol. 70, pp. 1648–55

[152] Tahar M.Z., Nemov S.A., Popov D.I., Svechnikova T.E. 'Transport properties of Sn and SbI_3 doped single crystal p-Bi_2Te_3'. *25th International Conference on Low Temperature Physics. J. Phys. Conf. Ser.* UK: Institute of Physics Publishing Ltd, 2009, vol. 150, 022082

[153] Wang E., Greenblatt M., El-Idrissi Rachidi I., Canadell E., Whangbo M.H. 'Anisotropic electronic properties of the diphosphate tungsten bronzes $K_2P_8W_{24}O_{88}$, $K_2P_8W_{28}O_{100}$ and their substituted compounds'. *J. Solid State Chem.* 1989, vol. 80, pp. 266–75

[154] Akahama Y., Kobayashi M., Kawamura H. 'Pressure-induced superconductivity and phase transition in selenium and tellurium'. *Solid State Commun.* 1992, vol. 84, pp. 803–6

[155] Aronson M.C., Salamon M.B., Christenson K.K., Ghiron K. 'Effects of doping on the electronic properties of $NbSe_3$'. *Phys. Rev. B*. 1988, vol. 38, pp. 468–75

[156] Weller T.E., Ellerby M., Saxena S.S., Smith R.P., Skipper N.T. 'Superconductivity in the intercalated graphite compounds C_6Yb and C_6Ca'. *Nature Phys*. 2005, vol. 1, pp. 39–41

[157] Decker D.L., Chen W. 'High-precision measurement of electrical resistivity of nickel near the ferromagnetic phase transition at high pressure'. *Phys. Rev. B.* 1992, vol. 46, pp. 8237–43

[158] Leitch A.A., Yu X., Winter S.M., Secco R.A., Dube P.A., Oakley R.T. 'Structure and property correlations in heavy atom radical conductors'. *J. Am. Chem. Soc.* 2009, vol. 131, pp. 7112–25

[159] Khan N.A., Irfan M., Nawaz S. 'Normal pressure synthesis of Mg-doped $Cu_{0.5}Tl_{0.5}Ba_2Ca_2Mg_2Cu_5O_{14-\delta}$ superconductor'. *Physica C.* 2007, vol. 455, pp. 63–66

[160] Khan N.A., Mazhar M., Maqsood A. 'A novel method for the direct synthesis of the $Y_2Ba_4Cu_7O_{15-\delta}$ superconductor'. *Supercond. Sci. Technol.* 2002, vol. 15, pp. 577–80

[161] Irfan M., Hassan N., Khan N.A. 'Fluctuation-induced conductivity of five planar $Cu_{0.5}Tl_{0.5}Ba_2Ca_{4-x}Mg_xCu_5O_{14-\delta}$ (x = 1, 2) superconductors'. *Physica C.* 2009, vol. 469, pp. 86–90

[162] Khurram A.A., Khan N.A., Ahadian M.M., Iraji-Zad A. 'Enhanced inter-plane coupling of Mg doped $Cu_{0.5}Tl_{0.5}Ba_2Ca_{2-x}Cu_3O_{10-\delta}$ superconductors: XPS and FTIR studies'. *Physica C.* 2008, vol. 468, pp. 405–10

[163] Liao C.Z., Dong C., Zeng L.M., He B., Cao W.H., Yang L.H. 'Crystal structure and physical properties of the new ternary compound $MgNi_7B_3$'. *J. Alloy Compd.* 2010, vol. 493, pp. 31–34

[164] Yang J., Shen X.-L., Lu W., Yi W., Li Z.-C., Ren Z.-A., *et al.* 'Superconductivity in some heavy rare-earth iron arsenide $REFFeAsO_{1-\delta}$ (RE = Ho, Y, Dy and Tb) compounds'. *New J. Phys.* 2008, vol. 11, 025005

[165] Nasir N., Melnychenko-Koblyuk N., Grytsiv A., Rogl P., Bauer E., Royanian E., *et al.* 'Crystal structure and physical properties of $EPCo_{4.7}Ge_9$ (EP = Sr, Ba, Eu)'. *Intermetallics.* 2009, vol. 17, pp. 471–6

[166] Zhu Z.D., Sun Y.P., Zhu X.B., Luo X., Wang B.S., Li G., *et al.* 'Single crystal growth and characterizations of $Cu_{0.03}TaS_2$ superconductors'. *J. Cryst. Growth.* 2008, vol. 311, pp. 218–21

[167] Ishikawa F., Eguchi N., Kodama M., Fujimaki K., Einaja M., Ohmura A., *et al.* 'Zero-resistance superconducting phase in $BaFe_2As_2$ under high pressure'. *Phys. Rev. B.* 2011, vol. 79, 172506

[168] Ren Z.A., Wei L., Jie Y., Yi W., Shen X.L., Li Z.C., *et al.* 'Superconductivity at 55 K in iron-based F-doped layered quaternary compound $Sm[O_{1-x}F_x]FeAs$'. *Chin. Phys. Lett.* 2008, vol. 25, 2215

[169] Prakash J., Singh S.J., Patnaik S., Ganguli A.K. 'Superconductivity at 31.3 K in Yb-doped La(O/F)FeAs superconductors'. *J. Chem. Soc.* 2010, vol. 122, pp. 43–46

[170] Hassinger E., Matsuda T.D., Knobel G., Taufour V., Aoki D., Flouquet J. 'Inelastic contribution of the resistivity in the hidden order in URu_2Si_2'. *J. Phys. Conf. Ser.* 2011, vol. 273, 012031

[171] Baumbach R.E., Hamlin J.J., Janoschek M., Lum I.K., Maple M.B. 'Magnetic, thermal and transport properties of the actinide based noncentrosymmetric compounds $Th_2Fe_{12}P_7$ and $U_2Fe_{12}P_7$'. *J. Phys. Condens. Matter.* 2011, vol. 23, 094222

[172] Sakai H., Yoshimura K., Ohno H., Kato H., Kambe S., Walstedt R.E., *et al.* 'Superconductivity in a pyrochlore oxide $Cd_2Re_2O_7$'. *J. Phys. Condens. Matter.* 2001, vol. 13, pp. L785–90

[173] Ogino H., Shimizu Y., Ushiyama K., Kawaguchi N., Kishio K., Shimoyama J. 'Superconductivity above 40 K observed in a new iron arsenide oxide $(Fe_2As_2)(Ca_4(Mg,Ti)_3O_y)$'. *App. Phys. Exp.* 2010, vol. 3, 063103

[174] Wikus P., Hertel S.A., Leman S.W., McCarthy K.A., Ojeda S.M., Figueroa-Feliciano E. 'The electrical resistance and thermal conductivity of Ti 15V-3Cr-3Sn-3Al at cryogenic temperatures'. *Cryogenics.* 2011, vol. 51, pp. 41–44

[175] Ronning F., Kurita N., Bauer E.D., Scott B.L., Park T., Klimczuk T., *et al.* 'The first order phase transition and superconductivity in $BaNi_2As_2$ single crystals'. *J. Phys. Condens. Matter.* 2008, vol. 20, 342203

[176] Tapp J.H., Tang Z., Lv B., Sasmal K., Lorenz B., Chu P.C.W., *et al.* 'LiFeAs: An intrinsic FeAs-based superconductor with $T_c = 18$ K'. *Phys. Rev. B.* 2008, vol. 78, 060505 (R)

[177] Tropeano M., Pallecchi I., Cimberle M.R., Ferdeghini C., Lamura G., Vignolo M., *et al.* 'Transport and superconducting properties of Fe-based superconductors: a comparison between $SmFeAsO_{1-x}F_x$ and $Fe_{1+y}Te_{1-x}Se_x$'. *Supercond. Sci. Technol.* 2010, vol. 23, 054001

[178] Horigane K., Takeshita N., Lee C.H., Hiraka H., Yamada K. 'First investigation of pressure effects on transition from superconductive to metallic phase in $FeSe_{0.5}Te_{0.5}$'. *J. Phys. Soc. Jpn.* 2009, vol. 78, 063705

[179] Pandya S., Sherif S., Sharath Chandra L.S., Ganesan V. 'Resistive broadening in sulphur doped FeTe'. *Supercond. Sci. Technol.* 2010, vol. 23, 075015

[180] Naren H.R., Tamizhavel A., Ramakrishan S. 'Effect of iridium substitution in $Rh_{17}S_{15}$'. *J. Phys. Conf. Ser.* 2010, vol. 200, 012052

[181] Ogino H., Katsura Y., Horii S., Kishio, K., Shimoyama J. 'New iron based arsenide oxides $(Fe_2As_2)(Sr_4M_2O_6)$ (M = Sc, Cr)'. *Supercond. Sci. Technol.* 2009, vol. 22, 085001

[182] Matsumura Y., Ogino H., Honi S., Katsura Y., Kishio K., Shimotama H. 'New series of nickel-based pnictide oxide superconductors $(Ni_2Pn_2)(Sr_4Sc_2O_6)$ (Pn = P, As)'. *Appl. Phys. Express.* 2009, vol. 2, 063007

[183] Wu G., Xie Y.L., Chen H., Zhong M., Liu R.H., Shi B.C., *et al.* 'Superconductivity at 56 K in samarium-doped SrFeAsF'. *J. Phys. Condens. Matter.* 2009, vol. 21, 142203

[184] Zhu X., Han F., Cheng P., Mu G., Shen B., Fang L., *et al.* 'Superconductivity in fluoride-arsenide $Sr_{1-x}La_xFeAsF$ compounds'. *EPL.* 2009, vol. 85, 17011

[185] Luo H., Wang Z., Yang H., Cheng P., Zhu X., Wen H.H. 'Growth and characterization of $A_{1-x}K_xFe_2As_2$ (A = Ba, Sr) single crystals with $x = 0 - 0.4$'. *Supercond. Sci. Technol.* 2008, vol. 21, 125014

[186] Anand V.K., Hillier A.D., Adroja D.T., Strydom A.M., Michor H., McEwen K.A., *et al.* 'Specific heat and μSR study on the noncentrosymmetric superconductor $LaRhSi_3$'. *Phys. Rev. B.* 2011, vol. 83, 064522

[187] Qi Y., Gao Z., Wang L., Wang D., Zhang X., Ma Y. 'Superconductivity at 34.7 K in the iron arsenide $Eu_{0.7}Na_{0.3}Fe_2As_2$'. *New J. Phys.* 2008, vol. 10, 123003

[188] Wu G., Chen H., Wu T., Xie Y.L., Yan Y.J., Lin R.H., *et al.* 'Different resistivity response to spin-density wave and superconductivity at 20 K in $Ca_{1-x}Na_xFe_2As_2$'. *J. Phys. Condens. Matter.* 2008, vol. 20, 422201

[189] Martinella A., Ferretti M., Manfrinetti P., Palenzona A., Tropeano M., Cimberle M.R., *et al.* 'Synthesis, crystal structure, microstructure, transport and magnetic properties of SmFeAsO and SmFeAs $(O_{0.93}F_{0.07})$'. *Supercond. Sci. Technol.* 2008, vol. 21, 095017

[190] Jeffries J.R., Butch N.P., Yukich B.T., Maple M.B. 'The evolution of the ordered states of single-crystal URu_2Si_2 under pressure'. *J. Phys. Condens. Matter.* 2008, vol. 20, 095225

[191] Mabsubayashi K., Saiga Y., Matsumoto T., Uwatoko Y. 'Pressure-induced variation of Kondo behaviour on the heavy fermion compounds YbT_2Zn_{20} (T = Co, Rh, Ir)'. *J. Phys. Conf. Ser.* 2009, vol. 200, 012112

[192] Uribe Laverde M.A., Téllez L., Roa-Rojas J. 'Magnetic, electric and equilibrium properties of $YBa_2Cu_{3-x}(PO_4)_xO_{7-\delta}$ high temperature superconducting system'. *Mod. Phys. Lett. B.* 2009, vol. 23, pp. 807–13

[193] Ishikawa F., Fukuda K., Hiura Y., Onda Y., Nakayama A., Yamada Y., *et al.* 'Pressure effects on the electrical resistivity of $Pr_2Ba_4Cu_7O_{15-\delta}$ oxide superconductor'. *J. Phys. Conf. Ser.* 2009, vol. 150, 052079

[194] Ren Z.A., Yang J., Lu W., Yi W., Che G.C., Dong X.L., *et al.* 'Superconductivity at 52 K in iron based F doped layered quarternary compound $Pr[O_{1-x}F_x]FeAs$'. *Mater. Res. Innov.* 2008, vol. 12, pp. 105–6

[195] Yamazaki T., Takeshita N., Kobayashi R., Saito T., Yamada Y., Kondo K., *et al.* 'Pressure and K doping induced superconductivity in $BaFe_2As_2$'. *J. Phys. Conf. Ser.* 2011, vol. 273, 012096

[196] Yamazaki T., Takeshita K., Kondo K., Kobayashi R., Yamada Y., Fukazawa H., *et al.* '^{75}As-NMR study of the iron pnictide $Ba_{1-x}K_xFe_2As_2$ under high pressure'. *J. Phys. Conf. Ser.* 2010, vol. 215, 012041

[197] Rotter M., Tegel M., Scheiienberg I., Schappacher F.M., Pöttgen R., Deisenhofer J., *et al.* 'Competition of magnetism and superconductivity in underdoped $(Ba_{1-x}K_x)Fe_2As_2$'. *New J. Phys.* 2009, vol. 11, 025014

[198] Mydeen K., Lengyel E., Deng Z., Wang X.C., Jin C.Q., Nicklas M. 'Temperature-pressure phase diagram of the superconducting iron pnictide LiFeP'. *Phys. Rev. B.* 2010, vol. 82, 014514

[199] Arii K., Igawa K., Okada H., Takahashi H., Imai M., Akaishi M., *et al.* 'Pressure studies in filled skutterudite $La_{0.8}Rh_4P_{12}$ and $LaRh_4As_{12}$'. *J. Phys. Conf. Ser.* 2009, vol. 150, 052009

[200] Miyasaka S., Suzuki S., Saijo S., Mikasa Y., Masui T., Tajima S. 'Resistivity, magnetic susceptibility and specific heat studies in superconductor $LaFePO_{1-x}F_x$'. *J. Phys. Conf. Ser.* 2009, vol. 150, 052164

[201] Ogino H., Matsumura Y., Katsura Y., Ushiyama K., Horii S., Kishio K., *et al.* 'Superconductivity at 17 K in $(Fe_2P_2)(Sr_4Sc_2O_6)$: a new superconducting layered pnictide oxide with a thick perovskite oxide layer'. *Supercond. Sci. Technol.* 2009, vol. 23, 075008

[202] Wang W.C., Liu Q.Q., Lv Y.X., Gao W.B., Yang L.X., Yu R.C., *et al.* 'The superconductivity at 18 K in LiFeAs system'. *Solid State Commun.* 2008, vol. 148, pp. 538–40

[203] Chen G.F., Li Z., Wu D., Dong J., Li G., Hu W.Z., *et al.* 'Element substitution effect in transition metal oxypnictide $Re(O_{1-x}F_x)TAs$ (Re = rare earth, T = transition metal)'. *Chin. Phys. Lett.* 2008, vol. 25, pp. 2235–8

[204] Ali T., Rupprecht C., Khan R.T., Bauer E., Hilscher G., Michor H. 'The effect of nitrogen vacancies in $La_3Ni_2B_2N_{3-\delta}$'. *J. Phys. Conf. Ser.* 2010, vol. 200, 012004

[205] Koyano M., Yamanouchi M. 'Electronic properties of inhomogeneous Bi-Sb-Ni composite alloys'. *J. Phys. Conf. Ser.* 2009, vol. 150, 052128

[206] Cao G., Jiang S., Lin X., Wang Cli Y., Ren Z., Tao Q., *et al.* 'Narrow superconducting window in $LaFe_{1-x}Ni_xAsO$'. *Phys. Rev. B.* 2009, vol. 79, 174505

[207] Prakash J., Singh S.J., Ahmed J., Patnaik S., Ganguli A.K. 'Compositionally controlled semimetal to superconducting transition in NaF doped LaOFeAs: Enhancement in T_c due to Na-doping'. *Physica C.* 2009, vol. 469, pp. 300–304

[208] Matsuishi S., Inoue Y., Nomura T., Kamihara Y., Hirano M., Hosono H. 'Effect of 3d transition metal doping on the superconductivity in quaternary fluoroarsenide CaFeAsF'. *New J. Phys.* 2009, vol. 11, 025012

[209] Takahashi H., Okada H., Igawa K., Kamihara Y., Hirano M., Hosono H. 'Pressure studies of (La, Sm)$FeAsO_{1-x}F_x$ and LaFePO'. *Physica C.* 2009, vol. 469, pp. 413–17

[210] Prakash J., Singh S.J., Das D., Patnaik S., Ganguli A.K. 'New oxypnictide superconductors: $PrOFe_{1-x}Co_xAs$'. *J. Solid State Chem.* 2010, vol. 183, pp. 338–43

[211] Okada H., Igawa K., Takahashi H., Kamihara Y., Hirano M., Hosono H., *et al.* 'Superconductivity under high pressure in LaFeAsO'. *J. Phys. Soc. Jpn.* 2008, vol. 72, 113712

[212] Mihalik M., Sechovský V., Diviš M., Gabáni S., Mihalik M. 'Superconductivity and physical properties of a LaRhSn single crystal'. *J. Alloy Compd.* 2007, vol. 452, pp. 241–4

[213] Wu G., Liu R.H., Chen H., Yan Y.J., Wu T., Xie Y.L., *et al.* 'Transport properties and superconductivity in $Ba_{1-x}M_xFe_2As_2$ (M = La and K) with double FeAs layers'. *EPL.* 2008, vol. 84, 27010

[214] Gumeniuk R., Schnelle W., Rosner H., Nicklas M., Leithe-Jasper A., Grin, Y.U. 'Superconductivity in the platinum germanides MPt_4Ge_{12} (M = Rare earth or alkaline-earth metal) with filled skutterudite structure'. *Phys. Rev. Lett.* 2008, vol. 100, 017002

[215] Tateiwa N., Haga Y., Matsuda T.D., Ikeda S., Yamamoto E., Okuda Y., *et al.* 'Large heat capacity jump at the superconducting transition temperature in the non-centrosymmetric superconductor $CeIrSi_3$ under high pressure'. *J. Phys. Conf. Ser.* 2008, vol. 121, 052001

[216] Chen G.F., Li Z., Li G., Zhou J., Wu D., Dong J., *et al.* 'Superconducting properties of Fe-based layered superconductor $LaO_{0.9}F_{0.1-\delta}FeAs$'. *Phys. Rev. Lett.* 2008, vol. 101, 057007

[217] Takeshita N., Iyo A., Eisaki H., Kito H., Ito T. 'Remarkable suppression of T_c by pressure in $NdFeAsO_{1-y}$'. *J. Phys. Soc. Jpn.* 2008, vol. 77, 075003

[218] Nobori M., Nakano T., Oomi G., Canfield P.C., Cho B.K. 'Effect of pressure on the superconductivity and magnetism in $HoNi_2B_2C$'. *J. Phys. Conf. Ser.* 2010, vol. 200, 012146

[219] Balamurugan S. 'Syntheses, magnetic and transport properties of $NbSr_2(RE_{1.5}Ce_{0.5})Cu_2O_{10}$; RE = Tb, Ho, Er series'. *Mod. Phys. Lett.* 2010, vol. 24, pp. 1043–57

[220] Wang L., Gao Z., Qi Y., Zhang X., Wang D., Ma Y. 'Structural and critical current properties in polycrystalline $SmFeAsO_{1-x}F_x$'. *Supercond. Sci. Technol.* 2009, vol. 22, 015019

[221] Yonezawa S., Muraoka Y., Matsushita Y., Hiroi Z. 'Superconductivity in a pyrochlore-related oxide KOs_2O_6'. *J. Phys. Condens. Matter.* 2004, vol. 16, pp. L9–12

[222] Ghorbani S.R., Rostamabadi E. 'The normal state transport properties of $NdBa_{2-x}La_xCu_3O_{7-\delta}$: Evidence of localization hole by La'. *Physica C.* 2008, vol. 468, pp. 60–65

[223] Kurita N., Kimata M., Kodama K., Harada A., Tomita M., Suzuki H.S., *et al.* 'High-pressure electrical resistivity measurements of $EuFe_2As_2$ single crystals'. *J. Phys. Conf. Ser.* 2011, vol. 273, 012098

[224] Bortolozo A.D., Fisk Z., Sant'anna O.H., Dos Santos C.A.M., Machado A.J.S. 'Superconductivity in Nb_2InC'. *Physica C.* 2009, vol. 469, pp. 256–8

[225] Klemkiene T., Raudonis R., Beganskiene A., Zalga A., Grigoraviciute I., Kareiva A. 'Scandium and gallium substitution effects in the $(Y_{1-x}Sc_x)Ba_2Cu_4O_8$ and $(Y_{1-x}Sc_x)Ba_2Cu_4O_8$ superconducting oxides'. *Mater. Chem. Phys.* 2010, vol. 119, pp. 208–13

[226] Khurram A.A., Ullah A., Khan N. 'Superconductivity in Co doped $Cu_{0.5}Tl_{0.5}Ba_2(CaM)Cu_{2.95}Co_{0.05}O_{10-\delta}$ (M = Mg,Be) samples'. *J. Alloy Compd.* 2009, vol. 481, pp. 65–69

[227] Roumié M., Awad R., Ibrahim I.H., Zein A., Zahraman K., Nsouli B. 'PIXIE and RBS analysis of Tl-1223 superconducting phase substituted by scandium'. *Nucl. Instrum. Meth. B.* 2008, vol. 266, pp. 133–9

[228] Rao A., Das A., Chakraborty T., Gahtori B., Agarwal S.K., Sarkar C.K., *et al.* 'Electrical and thermal transport properties of $EuBa_2(Cu_{1-x}Mn_x)_3O_{7-\delta}$'. *J. Phys. Condens. Matter.* 2008, vol. 20, 485212

[229] Chong S.V., Hashimoto S., Yamaguchi H., Kadowaki K. 'Neodymium-doping induced superconductivity in 1111-SrFeAsF iron-pnictide system'. *J. Supercond. Nov. Magn.* 2010, vol. 23, pp. 1479–84

[230] Awana V.P.S., Tripathi R., Balamurugan S., Kumar A., Dogra A., Kishan H. 'Thermal hysteresis in electrical transport of charge ordered $La_{0.5}Ca_{0.5}MnO_3$ manganites'. *J. Alloy Compds.* 2009, vol. 475, pp. L13–L16

[231] Mudgel M., Awana V.P.S., Kishan H., Felner I., Alvarez G.A., Bhalla G.L. 'Superconductivity of various borides: The role of stretched c-parameter'. *J. App. Phys.* 2009, vol. 105, 07E313

[232] Mudgel M., Awana V.P.S., Lal R., Kishan H., Sharth Chandra L.S., Ganesan V., *et al.* 'Anomalous thermoelectric power of the $Mg_{1-x}Al_xB_2$ system with $x = 0.0 - 1.0$'. *J. Phys. Condens. Matter.* 2008, vol. 20, 095205

[233] Layek S., Anand V.K., Hossain Z. 'Valence fluctuation in $Ce_2Co_3Ge_5$ and crystal field effect in $Pr_2Co_3Ge_5$'. *J. Magn. Magn. Mater.* 2009, vol. 321, pp. 3447–52

[234] Wang P., Stadnik M., Wang C., Cao G.H. 'Transport, magnetic and ^{57}Fe and ^{155}Gd Mössbauer spectroscopic properties of GdFeAsO and slightly

overdoped superconductor $Gd_{0.84}Th_{0.16}FeAsO$'. *J. Phys. Condens. Matter.* 2010, vol. 22, 145701

[235] Yang J.L., Ren W.J., Li D., Hu W.J., Li B., Zhang Z.D. 'Superconductivity modulated by internal pressure in $Ce_{1-x}Gd_xFeAsO_{0.84}F_{0.16}$ compounds'. *Supercond. Sci. Technol.* 2010, vol. 23, 025003

[236] Aghabagheri S., Mohammadizadeh M.R. 'The bipolaron model in the normal state of Pr-doped $GdBa_2Cu_3O_7$ superconductors'. *Supercond. Sci. Technol.* 2010, vol. 23, 045003